Fundamentals of Signal Enhancement and Array Signal Processing

Fundamentals of Signal Enhancement and Array Signal Processing

Jacob Benesty
INRS, University of Quebec
Montreal, Canada

Israel Cohen
Technion, Israel Institute of Technology
Haifa, Israel

Jingdong Chen
Northwestern Polytechnical University
Xi'an, China

WILEY

IEEE PRESS

This edition first published 2018

Registered Offices
John Wiley & Sons, Inc., 111 River Street, Hoboken, NJ 07030, USA
John Wiley & Sons Singapore Pte. Ltd, 1 Fusionopolis Walk, #07-01 Solaris South Tower, Singapore 138628

Editorial Office
The Atrium, Southern Gate, Chichester, West Sussex, PO19 8SQ, UK

For details of our global editorial offices, customer services, and more information about Wiley products visit us at www.wiley.com.

Wiley also publishes its books in a variety of electronic formats and by print-on-demand. Some content that appears in standard print versions of this book may not be available in other formats.

Library of Congress Cataloging-in-Publication data applied for

Hardback ISBN: 9781119293125

Cover Design by Wiley
Cover Image: © naqiewei/Gettyimages

Set in 10/12pt Warnock by SPi Global, Pondicherry, India

Printed in Singapore by C.OS. Printers Pte Ltd

10 9 8 7 6 5 4 3 2 1

Contents

Preface

Signal enhancement and array signal processing concern the problems of signal estimation, restoration, parameter estimation, and decision-making. These topics lie at the heart of many fundamental applications, such as hands-free voice communications, sonar, radar, ultrasound, seismology, autonomous cars, robotics, and so on. This book is designed as a textbook and its principal goal is to provide a unified introduction to the theory and methods of signal enhancement and array signal processing. The targeted readers are advanced undergraduate and graduate students who are taking – or instructors who are teaching – courses in signal enhancement, array signal processing, and their applications. Of course, practitioners and engineers can also use this book as a reference in designing signal-enhancement and/or array systems.

Since the primary users of this book may come from many different fields, with different background knowledge, we choose to focus on the key principles, theory and methods of signal enhancement and array signal processing from a signal processing perspective without discussing in detail the introductory material related to specific applications. Students are encouraged to read background material from the specialized academic books and research papers in their own field while studying this book.

In most, if not all, application systems, signals are acquired in the time domain. Therefore, comprehensive coverage of the formulation, methods, and algorithms of signal enhancement and array beamforming is provided for this domain. Likewise, thorough coverage of the material in the frequency domain is also presented as the formulation, derivation, analysis, and implementation of signal-enhancement and beamforming algorithms are often carried out in this domain. Readers are assumed to be familiar with Fourier transforms and the short-time Fourier transform (STFT) by which a time-domain signal is mapped to an equivalent sequence in the frequency domain. Readers are also assumed to have some prior knowledge on discrete-time linear systems, linear algebra, statistical signal processing, and stochastic processes.

A solid theoretical understanding always goes hand-in-hand with practical implementations. Therefore, this textbook includes a large number of examples to illustrate important concepts and show how the major algorithms work. MATLAB functions for all the examples can be found on the authors' websites. Besides examples, exercises and problems are provided at the end of every chapter to challenge readers and facilitate their comprehension of the material.

With the long experience of the authors (especially the first one) in both the industry and academia, we hope that this book has been written in such a way that it is easy and pleasant to read without compromising on the rigor of the mathematical developments.

Jacob Benesty
Israel Cohen
Jingdong Chen

About the Companion Website

Don't forget to visit the companion website for this book:

www.wiley.com/go/benesty/arraysignalprocessing

There you will find valuable material designed to enhance your learning, including:

1) Matlab codes used in the book
2) Slides for lectures

Scan this QR code to visit the companion website

1

Introduction

Signal enhancement is a process to either restore a signal of interest or boost the relevant information embedded in the signal of interest and suppress less relevant information from the observation signals. Today, there is almost no field of technical endeavor that is not impacted in some way by this process.

Array signal processing manipulates the signals picked up by the sensors that form an array in order to estimate some specific parameters, enhance a signal of interest, or make a particular decision. The main purpose of this chapter is to:

- define the scope of the field that we call signal enhancement
- present a brief historic overview of this topic
- give some examples of fields where signal enhancement is needed and used
- discuss briefly the principal approaches to signal enhancement
- explain how array signal processing works.

1.1 Signal Enhancement

We human beings rely on our senses to sense the environment around us. Based on this information we build and expand intelligence in our brain to help make decisions and take actions. Similarly, we strive to build systems to help us "see" or "hear" distant events that cannot be reached by our senses. For example, nowadays, sonar systems can hear ships across hundreds of miles of ocean, radar devices can see airplanes from a thousand miles over the horizon, telecommunication systems can connect two or more users from different corners of the world, and high-definition cameras can see events happening on our planet from space. These systems use sensors to measure the physical environment of interest. Signal processing is then applied to extract as much relevant information as possible from the sensors' outputs. Generally, sensors' outputs consist of the signal of interest, which carries very important information, and also a composition of unwanted signals, which is generally termed "noise". This does not contain useful information but interferes with the desired signal. To extract the useful information in the presence of noise, signal enhancement is needed, the objective of which is to:

- enhance the signal-to-noise ratio (SNR)
- restore the signal of interest

Fundamentals of Signal Enhancement and Array Signal Processing, First Edition.
Jacob Benesty, Israel Cohen, and Jingdong Chen.

Companion website: www.wiley.com/go/benesty/arraysignalprocessing

- boost the relevant information while suppressing less relevant information
- improve the performance of signal detection and parameter estimation.

Signal enhancement is a specialized branch of signal processing that has been around for many decades and has profound impact on many fields. In the following subsections, we describe a few areas that routinely use signal enhancement techniques, particularly those developed in the following chapters of this text. Note that we can only cover a few applications, but this should leave the reader with no doubt as to the importance and breadth of application of signal enhancement techniques.

1.1.1 Speech Enhancement and Noise Reduction

In applications related to speech acquisition, processing, recognition, and communications, the speech signal of interest (generally called the "desired speech") can never be recorded in a pure form; it is always immersed in noise. The noise can come from very different sources. For example, microphones that we use to convert acoustic pressure into electronic signals have self-noise, even though the noise floor of popularly used capacitor microphones has been dropping significantly over the years. The associated digital signal processing boards, including preamplifiers, analog-to-digital (A/D) converters, and processors for processing the signals, may also generate noise. Most importantly, noise comes from ambient sources; the environment where we live is full of different kinds of sounds. While the sensors' self and circuit noise is generally white in spectrum, the noise from sound sources in the surrounding environment can vary significantly from one application scenario to another.

Commonly, noise from acoustic environments can be divided into the following four basic categories depending on how the noise is generated:

- *Additive noise* can come from various sources, such as cooling fans, air conditioners, slamming doors, and passing traffic.
- *Echoes* occur due to the coupling between loudspeakers and microphones.
- *Reverberation* is the result of multipath propagation and is introduced by reflections from enclosure surfaces.
- *Interference* comes from concurrent sound sources. In some communication applications, such as teleconferencing, it is possible that each communication site has multiple participants and loudspeakers, so there can be multiple competing sound sources.

Combating these four categories of noise has led to the development of diverse acoustic signal processing techniques. They include noise reduction (or speech enhancement), echo cancellation and suppression, speech dereverberation, and source separation, each of which is a rich subject of research [1–3]. This text presents many methods, algorithms, and techniques that are useful in dealing with additive noise, reverberation, and interference while its major focus, particularly the signal enhancement part from Chapter 2 to Chapter 6, is on reducing additive noise.

Additive noise and the desired speech signal are in general statistically independent. While the noise does not modify the speech characteristics directly, the characteristics of the observation signal are very different from those of the desired speech since it is a mixture of the desired speech signal and noise. Figure 1.1 plots a speech signal recorded in an anechoic (quiet and non-reflective) environment and the same speech

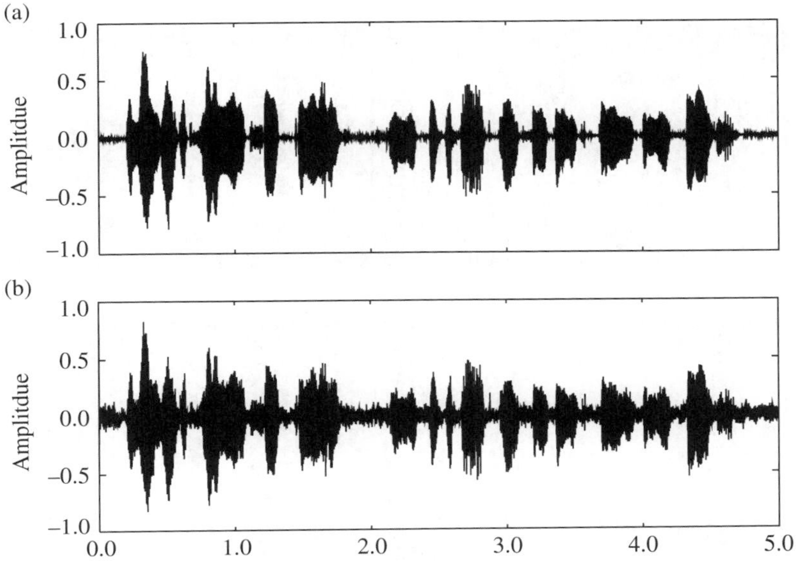

Figure 1.1 (a) A speech signal recorded by a microphone in an anechoic environment and (b) the same speech signal recorded by the same microphone but in a conference room.

signal but recorded in a conference room. The spectrograms of these two signals are shown in Figure 1.2. As can be seen, both the waveform and the spectrogram of the noisy signal are dramatically different from those of the clean speech. The effect of noise may dramatically affect the listener's perception and also machine processing of the observed speech. It is therefore generally required to "clean" the observation signal before it is stored, transmitted, or played (through a loudspeaker, for example). This problem is generally referred to as either noise reduction or speech enhancement.

1.1.2 Underwater Acoustic Signal Enhancement

Over the last few decades, ocean exploration activity for both military and civilian interests has been steadily increasing. As a result, there has been growing demand for underwater communication and signal detection and estimation technologies. Electromagnetic and light waves do not propagate over long distances under water (particularly sea water). In contrast, acoustic waves may propagate across tens or even hundreds of miles under the sea. Therefore, acoustic waves have played an important role in underwater communication and signal detection and estimation. For example, passive sonar systems can detect a submarine from tens of miles away by listening to the sound produced by the submarine, such as from the propellers, engine, and pumps; active sonars transmit sound pulses into the water and listen to the echoes, thereby detecting underwater features such as the location of fish, sunken objects, vessels, and submarines. Underwater wireless communication systems modulate useful information on acoustic carriers with frequencies between a few kilohertz and a few tens of kilohertz and transmit the modulated signal from one end to another through underwater acoustic channels.

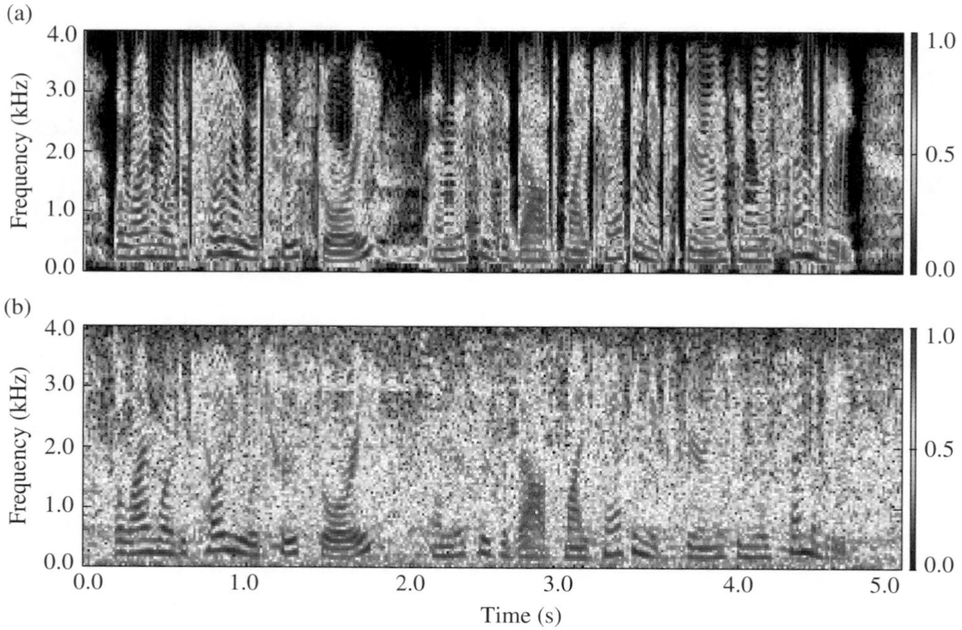

Figure 1.2 (a) The spectrogram of the speech signal in Figure 1.1a; (b) the spectrogram of the speech signal in Figure 1.1b.

However, processing underwater acoustic signals is by no means an easy task. First of all, underwater acoustic channels are generally known as one of the most difficult communication media in use today. Underwater acoustic propagation suffers from the time-varying multipath effect (due to sound reflection at the surface, bottom, and any objects in the vicinity, and also sound refraction in the water), frequency-dependent attenuation (due to absorption and signal spreading loss), and a severe Doppler effect (due to the low speed of sound and motion of the transmitter or receiver or the objects to be detected). Secondly, the ocean is filled with sounds, which interfere with the acoustic signal we are interested in. Underwater sounds are generated by both natural sources, such as marine animals, breaking waves, rain, cracking sea ice, and undersea earthquakes, as well as man-made sources, such as ships, submarines, and military sonars.

Marine animals use sound to obtain detailed information about their surroundings. They rely on sound to communicate, navigate, and feed. For example, dolphins can detect individual prey and navigate around objects underwater by emitting short pulses of sound and listening to the echo. Marine mammal calls can increase ambient noise levels by 20–25 dB in some locations at certain times of year. Blue and fin whales produce low-frequency moans at frequencies of 10–25 Hz, with estimated source levels of up to 190 dB at 1 m. Sounds generated by human activities are also an important part of the total ocean noise. Undersea sound is used for many valuable purposes, including communication, navigation, defense, research and exploration, and fishing.

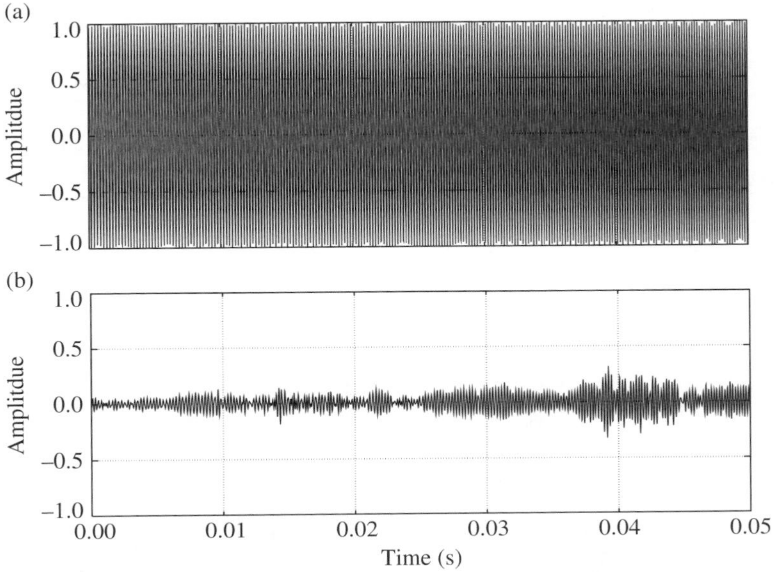

Figure 1.3 A linear frequency modulated signal: (a) emitted by a transmitter of an underwater acoustic communication system and (b) received by a hydrophone six miles away from the transmitter in an underwater environment.

Sounds generated by human activities cover a wide range of frequencies, from a few hertz up to several hundred kilohertz, and a wide range of source levels.

The underwater channel condition and noise sets the ultimate limit on the minimum detectable signal in detection and communication systems. To illustrate how challenging it is to process underwater signals for extracting the useful information, Figure 1.3 plots a linear frequency modulation chirp signal transmitted by an acoustic antenna and the signal received by a hydrophone placed six miles away from the transmitter. The magnitude spectra of these two signals are plotted in Figure 1.4. As seen, the transmitted signal is dramatically distorted by the acoustic channel and noise. Sophisticated signal enhancement techniques, such as those developed in this text, are needed to extract the important parameters or information embedded in the transmitted signal from the received signal.

1.1.3 Signal Enhancement in Radar Systems

A radar system has a transmitter that emits electromagnetic waves (called radar signals) in look directions. When these waves come into contact with an object, they are usually reflected or scattered in many directions. Receivers (usually, but not always, in the same location as the transmitter) are then used to receive the echoes. Through processing the echoes, the radar can determine the range, angle, or velocity of the objects of interest.

The invention of the radar dates back to the late 19th century. Such systems are now used in a broad range of applications, including air defense, traffic control, aircraft anticollision, ocean surveillance, geological observations, meteorological precipitation monitoring, and autonomous cars. In order to estimate the range, angle, or velocity of

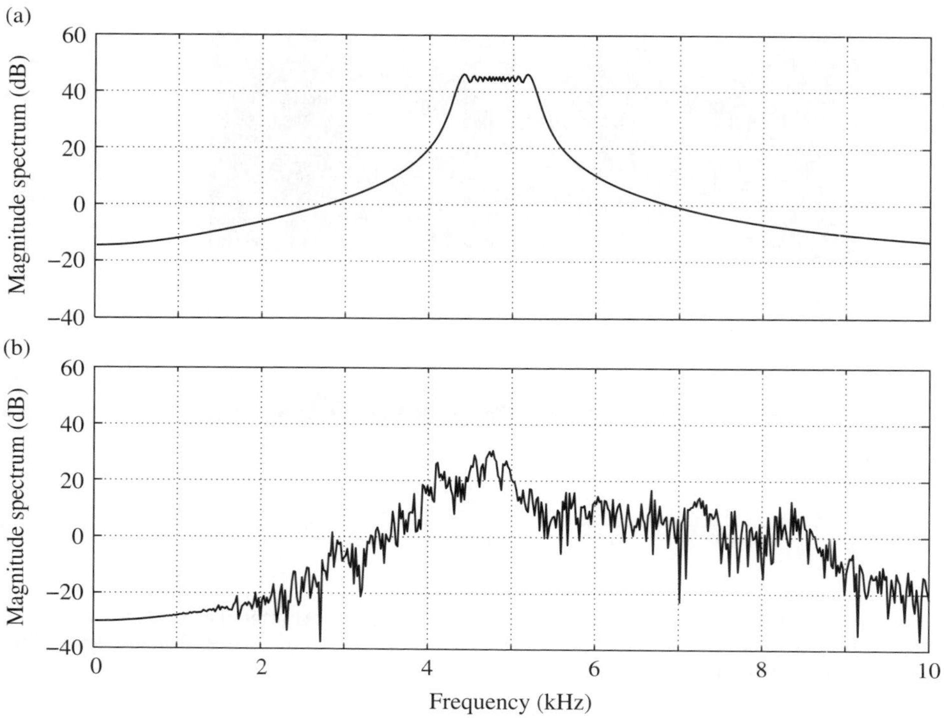

Figure 1.4 The power spectrum of the signal in Figure 1.3a; (b) the power spectrum of the signal in Figure 1.3b.

the objects of interest, radar systems must overcome unwanted signals, which can be divided into the following three categories.

- *Additive noise* is generated by both internal sources (electronics) and external sources (the natural thermal radiation of the background surrounding the target of interest). In modern radar systems, the internal noise is generally lower than the external noise.
- *Clutter* is a term used for echoes returned from targets that are not useful to the radar system user. Clutters can be generated by irrelevant targets, natural objects such as the ground, sea, atmospheric turbulence, ionospheric reflections, and man-made objects such as buildings, as illustrated in Figure 1.5.
- *Jamming* refers to signals received by the radar on its own frequency band but emitted from outside sources. Jamming may be intentional, as with an electronic warfare tactic, or unintentional, as with friendly forces' using equipment that transmits using the same frequency range. It is problematic to radar since the jamming signal only needs to travel one way – from the jammer to the radar receiver – whereas the radar echoes travel two ways – from radar to target and to radar – and are therefore significantly reduced in power by the time they return to the radar receiver. Therefore, jammers can effectively mask targets along the line of sight from the jammer to the radar, even when they are much less powerful than the jammed radars.

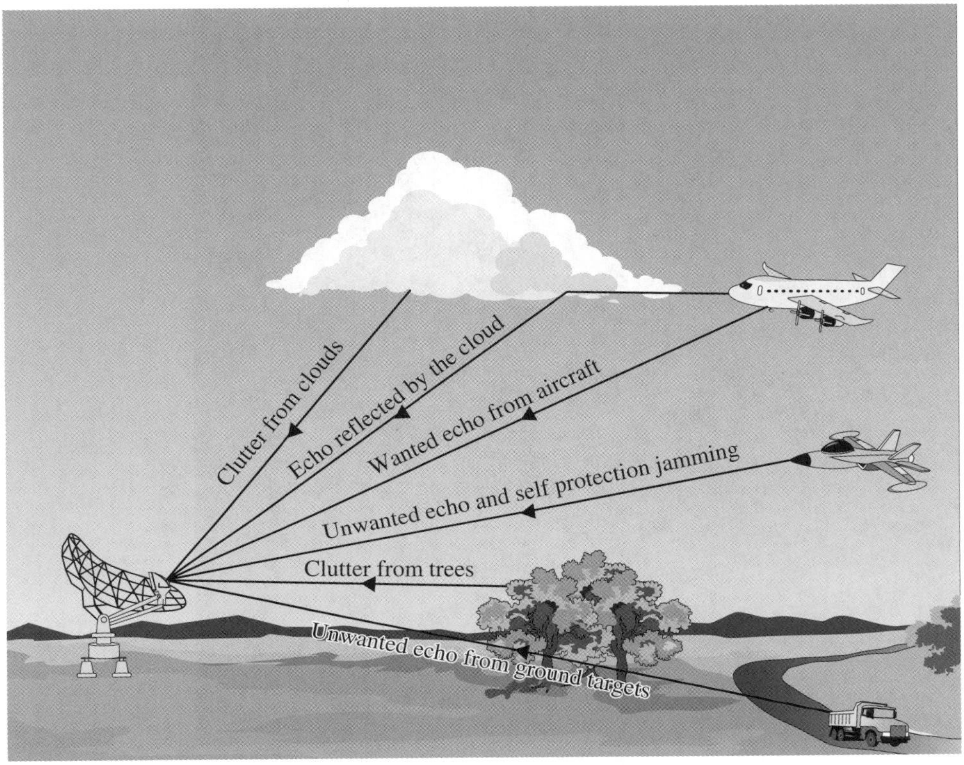

Figure 1.5 An illustration of a radar system and its environments.

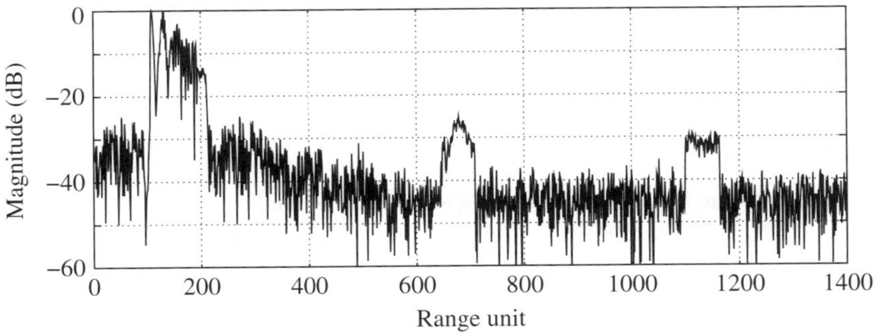

Figure 1.6 Normalized magnitude of an echo received by a pulse radar.

Figure 1.6 plots the magnitude of a signal received by a pulse radar where the transmitted signal is a short pulse. The received signal is composed of an echo returned from a target of interest, two unwanted echoes, and some noise. Figure 1.7 shows a radar image directly mapped from the received signals without using any signal enhancement techniques. Without signal enhancement, it is difficult to determine the position of one target, let alone track multiple targets with high resolution. Therefore, signal

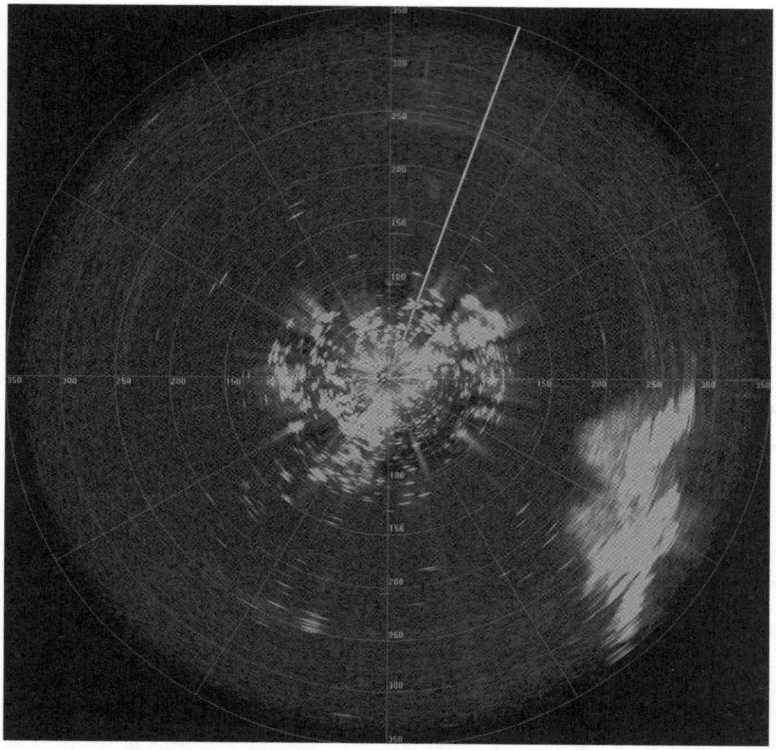

Figure 1.7 Illustration of an image displayed in a radar screen without using signal enhancement.

enhancement techniques, particularly those developed in Chapters 7–11, are needed to deal with additive noise, clutter, and jamming in radar systems. This is done by estimating the important parameters embedded in the radar echo signals.

1.1.4 Signal Enhancement in Ultrasound Systems

Ultrasound refers to sound waves with frequencies greater than 20 kHz, a level which is commonly accepted to be the upper limit of human hearing. This type of high-frequency sound wave is used in many fields for a wide range of applications, such as non-intrusive testing of products and structures, invisible flaw detection, distance measurement, and medical diagnosis, to name just a few. One of the best known ultrasound systems is the sonography instrument that is used in medicine to examine many of the body's internal organs, including – but not limited to – the heart and blood vessels, liver, gallbladder, pancreas, kidneys, and uterus, as well as unborn children (fetus) in pregnant patients.

Typically, a sonography device consists of an array of transmitters, which send short, high-frequency (generally between 2 and 20 MHz) sound pulses into the body. Beamforming is applied to the transmitted pulses so that the ultrasound waves are focused towards a particular point. As the beamformed waves travel toward the desired focal point, they propagate through materials with different densities. With each change in density, reflected waves are produced, some of which propagate back. They are then collected by an array of receivers (the transmitters typically become sensors to receive signals once they have finished generating their respective sound waves). The signal received by each receiver is composed of the wanted echoes and noise. Commonly, noise in sonography is one of two major types:

- *Additive noise* is generated from the sensors, amplifiers, A/D converters, and other electronic system components. It can also come from sources such as background tissues, other organs and anatomical influences, and breathing motion. Generally, this type of noise is independent (or weakly dependent) on the echo signals and is often modeled mathematically as a white Gaussian noise process.
- *Speckle* is the result of three sound scattering effects: specular, diffusive, and diffractive. Specular scattering occurs when the scattering object is large compared to the sound wavelength; diffusive scattering happens when the scattering object is small relative to the wavelength; diffractive scattering occurs mostly for medium-size scattering objects. Unlike additive noise, speckles are generally correlated with the wanted echo signals.

To deal with additive noise and speckles in sonography, beamforming, noise reduction, speckle reduction, and many other enhancement processes are applied to the received signals before high-resolution two-dimensional images are formed to display the distances and intensities of the echoes on the screen.

1.2 Approaches to Signal Enhancement

Signal enhancement is one of the most interesting and appealing yet challenging areas of signal processing. Its objective is generally problem oriented, ranging from simply improving the SNR, boosting relevant information, restoring the signal of interest, to improving measures of which only human subjects can judge the quality. As a result, there is no general rule as what method is optimal and it is quite common that a method that produces the best enhancement result for one application may not be very useful for another. In general, signal enhancement techniques can be classified into one of four broad categories, depending on how the information embedded in the signal and noise are used:

- *time-domain methods,* which directly use temporal information
- *frequency-domain approaches,* which operate on spectra (obtained using the Fourier transform or other time-to-frequency-domain transformations) of a signal
- *spatial-domain techniques,* which acquire and process a signal of interest using an array of sensors
- *combinational methods,* which use temporal, spectral, and spatial information.

Time-domain methods typically achieve signal enhancement by applying a finite-impulse-response filter to the noisy signal that is observed at a sensor. So, the core problem of enhancement is converted to one of designing an optimal filter that can attenuate noise as much as possible while keeping the signal of interest relatively unchanged. The history of this class of methods dates back to the seminal work by Wiener [4], in which the optimal filter from a second-order-statistics viewpoint is achieved through the optimization of the classical mean-squared error (MSE) criterion. The Wiener filter is well known, and has been intensively investigated for signal enhancement [5–8]. However, while it is optimal from the MSE point of view, it introduces signal distortion if the signal to be enhanced is broadband. The amount of signal distortion may not be acceptable for some applications. If this is the case, one may consider using some suboptimal filters that minimize the MSE criterion under certain constraints [5, 6].

The Wiener technique, by its assumption, can only deal with stationary signals. One popularly used approach to extending the Wiener filter to deal with nonstationary signals is to relax the stationarity assumption to one of short-time stationarity. Then, the Wiener filter is computed using signals within only a short-time, sliding window. In this case, the length of the short-time window plays an important role on the tradeoff between the nonstationarity and performance within the short-time window. There are, of course, other ways to deal with signal enhancement of nonstationary signals, for example, combining the Kalman filter and the linear-prediction-coding method [9]. Comprehensive coverage of signal enhancement using temporal information will be given in Chapter 2.

Frequency-domain methods, as the name indicates, explicitly operate on the spectra of the signal to be processed. The root of this class of methods can be traced back to the 1920s, when low-pass, high-pass, and band-pass filters were invented to filter out noise that occupies different frequency bands to the signal of interest. Today, the basic principle of band-pass filtering is still widely used in signal enhancement, but often in more complicated forms, such as comb filters [10] and binary masking [11]. Band-pass filtering, comb filtering, and binary masking are hard-decision methods in the sense that, given a narrow frequency band, they either completely remove the signal component or keep it unchanged. In comparison, a soft decision can be achieved through the spectral enhancement method, which was first developed in the 1960s using analog circuits [12]. A digital-domain version of this method, which is called "spectral subtraction", was then developed in the late 1970s [13]. While it is very useful and is often used as a benchmark against which other techniques are compared, the spectral substraction method has no optimality properties associated with it. An optimal spectral enhancement framework was developed in the early 1980s [14]. This unified a broad class of enhancement algorithms, including spectral substraction, frequency-domain Wiener filtering, and maximum likelihood envelope estimator. Following this work, an optimal spectral amplitude estimator using statistical estimation theory was developed in the early 1980s. Following this work, many statistical spectral estimators were developed, including the minimum mean-squared error (MMSE) estimator [15], the MMSE log-spectral amplitude estimator, the maximum-likelihood (ML) spectral amplitude estimator, the ML spectral power estimator, and the maximum a posteriori (MAP) spectral amplitude estimator. Today, there are still tremendous efforts to find better spectral amplitude estimators, inspired by the work of McAulay and Malpass [14] and Ephraim and Malah [15]. A broad coverage of frequency-domain methods will be given in Chapter 3.

When multiple sensors are used, the spatial information embedded in the sensors' outputs can be exploited to enhance the signal of interest and reduce unwanted noise. This can be done in a straightforward way by extending the single-channel methods of Chapters 2 and 3 to the multichannel cases (see, respectively, Chapters 4 and 5). It can also be done in a different way through array beamforming, which will be discussed in the next section.

1.3 Array Signal Processing

An array consists of a set of sensors positioned at known locations with reference to a common point. The sensors collect signals from sources in their own field of view and the output of each sensor is composed of these source components as well as noise. By processing the sensors' outputs, two groups of functionalities can be achieved: estimation of important parameters of sources and enhancement of some signals of interest.

The history of array signal processing dates back to World War II. Early efforts in this field were mainly focused on parameter estimation: estimating the range, angle, and velocity of the sources of interest. A wide range of processing methods were developed to this end, including fixed beamforming (or spatial filtering), matched filtering, Capon's adaptive beamforming, the MUSIC (Multiple SIgnal Classification) method and its varieties, and the ESPRIT (Estimation of Signal Parameters by Rotational Invariance Techniques) algorithm, to name but a few. The reader is referred to the literature for an in-depth consideration of the problem of array parameter estimation and the associated methods [16–23]. In this book, we choose to focus on the signal enhancement problem with the use of sensor arrays.

The basic principle of signal enhancement using an array of sensors can be illustrated in Figure 1.8. Consider a simple example with a uniformly spaced linear array of M sensors and assume that there is a single source in the farfield such that its spherical wavefront appears planar at the array. If we neglect the propagation attenuation, the signals received at the M sensors can be written as

$$\begin{aligned} Y_m(f) &= X_m(f) + V_m(f) \\ &= X(f)e^{-j2\pi f(t+\tau_{m-1})} + V_m(f) \\ &= X_1(f)e^{-j2\pi f\tau_{m-1}} + V_m(f),\ m = 1, 2, \ldots, M, \end{aligned} \tag{1.1}$$

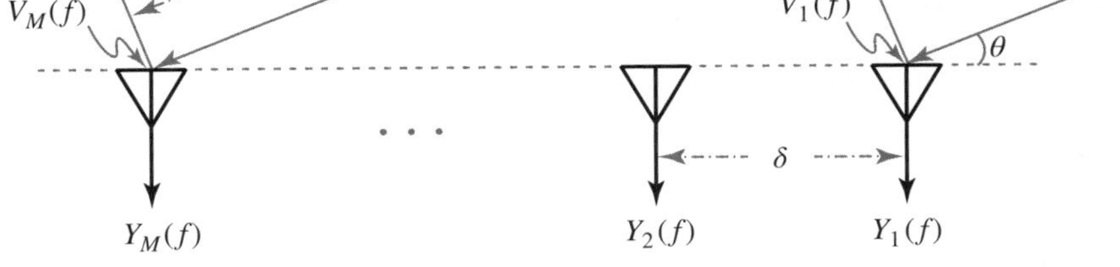

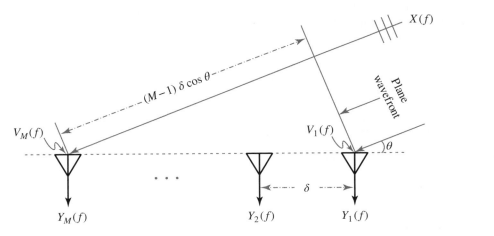

Figure 1.8 Illustration of an array system for signal enhancement.

where f is frequency, t is the propagation time from the source $X(f)$ to sensor 1 (the reference sensor), $X_1(f) = X(f)e^{-\jmath 2\pi ft}$ is the signal component at sensor 1, τ_{m-1} is the relative time delay between the mth sensor and the reference one, and $\jmath = \sqrt{-1}$ is the imaginary unit. It is assumed that $X(f)$ is uncorrelated with $V_m(f)$, $m = 1, 2, \ldots, M$. With a uniform linear array (ULA) and a farfield source, the delay τ_{m-1} can be expressed in the following form according to the geometry shown in Figure 1.8:

$$\tau_{m-1} = (m-1)\delta \cos\theta / c,\ m = 1, 2, \ldots, M, \tag{1.2}$$

where δ is the spacing between two neighboring sensors, c represents velocity of wave propagation, and θ is the signal incidence angle.

Now let us consider processing the M signals $Y_m(f)$, $m = 1, 2, \ldots, M$, to extract the source signal $X(f)$ (up to a delay) and reduce the effect of $V_m(f)$. A straightforward way of doing this is to multiply $Y_m(f)$ by $e^{\jmath 2\pi f \tau_{m-1}}$, and then average the results. This gives an output:

$$\begin{aligned} Z(f) &= \frac{1}{M}\sum_{m=1}^{M} Y_m(f)e^{\jmath 2\pi f \tau_{m-1}} \qquad (1.3) \\ &= X_1(f) + \frac{1}{M}\sum_{m=1}^{M} V_m(f)e^{\jmath 2\pi f \tau_{m-1}}. \end{aligned}$$

To check whether the output $Z(f)$ is less noisy than the input, let us compare the input and output SNRs. The input SNR, according to the signal model given in (1.1), is defined as the SNR at the reference sensor:

$$\mathrm{iSNR}(f) = \frac{\phi_{X_1}(f)}{\phi_{V_1}(f)}, \tag{1.4}$$

where $\phi_{X_1}(f) = E\left[|X_1(f)|^2\right]$ and $\phi_{V_1}(f) = E\left[|V_1(f)|^2\right]$ are the variances of $X_1(f)$ and $V_1(f)$ respectively, and $E[\cdot]$ denotes mathematical expectation.

The output SNR – the SNR of the $Z(f)$ signal – is written as

$$\mathrm{oSNR}(f) = \frac{\phi_{X_1}(f)}{\frac{1}{M^2}E\left[\left|\sum_{m=1}^{M} V_m(f)e^{\jmath 2\pi f \tau_{m-1}}\right|^2\right]}. \tag{1.5}$$

Now, if all the noise signals $V_m(f)$, $m = 1, 2, \ldots, M$, are uncorrelated with each other and have the same variance, it is easy to check that

$$E\left[\left|\sum_{m=1}^{M} V_m(f)e^{\jmath 2\pi f \tau_{m-1}}\right|^2\right] = M \times E\left[|V_1(f)|^2\right]. \tag{1.6}$$

It follows immediately that $\mathrm{oSNR}(f) = M \times \mathrm{iSNR}(f)$. Thus, a simple phase shifting and averaging operation of the sensors' outputs results in an SNR improvement by a factor of M, the number of sensors. The underlying physical principle behind the

SNR improvement can be explained as follows. Through appropriate phase shifting, the signal components from the source of interest have been coherently combined while the noise signals from different sensors add only incoherently as they are uncorrelated with each other, yielding a gain for the overall output signal compared to the noise.

Of course, the above example is only a particular case. More generally, an estimate of the source signal $X(f)$ can be obtained through weighted linear combination of the sensors' outputs:

$$Z(f) = \sum_{m=1}^{M} H_m^*(f) Y_m(f), \tag{1.7}$$

where $H_m(f)$, $m = 1, 2, \ldots, M$, are complex weighting coefficients. The process of finding the appropriate values of $H_m(f)$ so that $Z(f)$ is a good estimate of $X(f)$ is called beamforming. Therefore, the coefficients $H_m(f)$ are also called beamforming coefficients and the vector that consists of all the coefficients is called the beamforming filter or beamformer.

Beamforming has been the central problem of array signal processing ever since sensor arrays were invented, and a large number of algorithms has been developed and described in the literature. By and large, the developed algorithms fall into two major categories – fixed or adaptive beamforming – depending on whether the noise or signal statistics are considered in forming the beamforming filters.

In fixed beamforming, the beamforming filters are designed explicitly using the array geometry information as well as the assumed knowledge of the look direction and noise statistics. Once computed, the coefficients of the beamforming filters will be fixed regardless of the particular application environment. It is for this reason that this design process is called fixed beamforming. The representative algorithms include the delay-and-sum beamformer, the maximum directivity factor beamformer, and the superdirective beamformer. The reader may find a discussion of these algorithms in different contexts and applications in the literature [3, 16–18]. The basic theory and methods for designing fixed beamformers from a narrowband perspective will be covered in Chapter 7. While the principles presented in this chapter are rather general, in designing optimal fixed beamformers, the resulting beamformers may be insufficient to deal with broadband signals as their beampatterns may vary with frequency. Chapters 9 and 10 are also concerned with fixed beamforming, but with focus on processing broadband signals where beampatterns are expected to be the same across a band of frequencies.

In comparison with fixed beamforming, adaptive beamforming algorithms consider using either the noise statistics or the statistics of the array observation data to optimize the beamforming filters. The performance of adaptive beamforming can be more optimal than its fixed counterpart as long as the signal statistics are correctly estimated. The representative algorithms in this category include the minimum variance distortionless response (MVDR) beamformer, which is also known as the Capon's beamformer [27], the linearly constrained minimum variance (LCMV) beamformer, which is also called the Frost's beamformer [28], and the generalized sidelobe canceller [29, 30]. Many applications of adaptive beamforming can be found in the literature [1–3, 31, 32]. The fundamental theory and methods for adaptive beamforming from a frequency-domain perspective will be covered in Chapter 8.

While one may see that most discussion on beamforming in both the literature and this text concerns the frequency domain, it is also possible to formulate this problem in the time domain. Chapter 10 is devoted to a time-domain framework for array beamforming, which can be used to design both fixed and adaptive beamformers as well as narrowband and broadband beamformers.

1.4 Organization of the Book

This book attempts to cover the most basic concepts, fundamental principles, and practical methods of signal enhancement and array beamforming. The material discussed occupies ten chapters, which are divided into two parts.

The first part, Signal Enhancement, consists of five chapters: from Chapter 2 to Chapter 6. We start to discuss the signal enhancement problem in the time domain with a single sensor in Chapter 2. With a single sensor, we show how to exploit the temporal information so as to reduce the level of the additive noise from the sensor's output, thereby enhancing the signal of interest. This chapter presents two fundamental approaches: one deals with the problem from the Wiener filtering perspective and the other from a spectral mode perspective based on the joint diagonalization of the correlation matrices of the signal of interest and the noise. In both approaches, we present the problem formulation and performance measures that can be used to evaluate the signal enhancement performance. Different cost functions are also presented, and based on these we discuss how to derive useful optimal enhancement filters.

Chapter 3 continues the investigation of the single-channel signal enhancement problem, but in the frequency domain. The frequency-domain approach is equivalent to the spectral method discussed in Chapter 2 in the sense that the observation signal at each frequency band can be processed independently from the others. The advantage of the algorithms in this chapter is that they can all be implemented efficiently thanks to the use of the fast Fourier transform. Again, we start from problem formulation and then discuss how to perform signal enhancement with just simple gains at each frequency band. Relevant performance measures are defined and we show how to derive several kinds of enhancement gain, some of which can achieve a compromise between distortion of the desired signal and reduction of the additive noise.

Chapter 4 is basically an extension of Chapter 2. The fundamental difference is that in this chapter we consider the signal enhancement problem with the use of multiple sensors, which are located at distinct positions in the space. In this case, every sensor picks up the signal of interest and noise from its own viewpoint. Now, in addition to the temporal information, the spatial information from the multiple sensors can also be exploited to enhance the signal of interest. As a result, either a better enhancement performance or more flexibility to compromise between noise reduction and desired signal distortion can be achieved, as compared to the single-channel scenario. Similar to Chapter 2, we also discuss two approaches: the Wiener filtering one and the one based on the joint diagonalization of the correlation matrices of the signal of interest and noise.

Chapter 5 deals with the problem of signal enhancement with multiple sensors in the frequency domain. As in Chapter 4, the spatial information embedded in the multiple sensors is exploited to enhance the signal of interest, but in a way that is easier to comprehend. Just like the material in the previous chapters, we present the signal model, problem formulation, and performance measures, and show how to derive different optimal filters. We also discuss the problem in a subspace framework, which is an alternative way to approach the enhancement problem.

Chapter 6 is a unification of the material presented from Chapter 2 to Chapter 5. A general framework is presented here so that the signal enhancement problem in either the time or the frequency domain, with either one sensor or multiple sensors, is treated in a unified framework. Within this framework, we derive a class of optimal linear filters, some of which are well known and some can achieve output signal-to-interference-plus-noise ratios (SINRs) that are between those of the conventional maximum SINR filter and the Wiener filter. This chapter also serves as a bridge between the problem of noise reduction in the previous chapters and the following chapters, on beamforming.

The second part, Array Signal Processing, is about beamforming, and also consists of five chapters, from Chapter 7 to Chapter 11. We start in Chapter 7 by discussing the theory and methods of beamforming from fixed beamformers, which are spatial filters that have the ability to form a main beam pointing in the direction of the signal of interest, while placing sidelobes and nulls in directions other than the look direction. By "fixed", we mean that the beamforming filters are designed before the deployment of the array system and the filters' coefficients do not depend on the array output. Generally, the design of fixed beamformers requires the array geometry information, such as the number of sensors or the location of every sensor relative to a reference point, and the look direction. It is helpful if the directions of interference sources are known as well. To simplify the presentation of the main results, we consider in Chapter 7 only ULAs, and study a number of popularly used fixed beamformers. Note that the generalization of the algorithms in this chapter from ULAs to other geometries is not difficult, in general.

Fixed beamformers are generally robust and easy to implement; but they are at best suboptimal in terms of noise and interference rejection, as neither the statistics of the signal of interest nor those of noise and interference are considered in the beamformer design process. One way to improve the performance of fixed beamformers is through using the statistics of the array outputs or *a priori* information about the source or noise signals, leading to the so-called "adaptive" beamformers, which will be studied in Chapter 8. This chapter discusses several interesting adaptive beamformers, including their derivation, underlying principles, as well as their equivalent forms from a theoretical viewpoint.

In processing broadband signals such as audio and speech, it is desirable, if not a must, to use beamformers that have frequency-invariant beampatterns. One way to achieve this is through differential beamforming, which will be studied in Chapter 9. Differential beamforming is a particular kind of fixed beamforming. It differs from those beamformers in Chapter 7 in that it attempts to measure the differential pressure field of different orders, instead of designing a special beampattern. Besides the property of frequency-invariant beampatterns, differential beamforming can achieve the maximum directivity factor, leading to the highest gains in diffuse noise. As a matter of fact, the so-called "superdirective" beamformer is a particular case of the differential beamformer. In this chapter, we present the fundamental principles underlying differential beamforming, as well as approaches to designing differential beamformers of different orders. One main drawback of differential beamforming as compared to those beamformers in Chapter 7 is white noise amplification. We will present a method that can deal with this problem.

Beampattern design is the most fundamental and important problem in array beamforming. Chapter 10 is dedicated to this issue. Again, for simplicity of presentation, a ULA is assumed. Since we are interested in frequency-invariant beampatterns, the spacing between neighboring sensors must be small, which is assumed here, as it is in Chapter 9. In this chapter, we revisit the definitions of the beampatterns and show some relationships. We then present different techniques for beampattern design. Note that the beampatterns designed in this chapter are similar to the ones obtained with differential sensor arrays in Chapter 9. This makes sense, as most assumptions used in this chapter are the same as those in Chapter 9.

Finally, in Chapter 11, we address the beamforming problem in the time domain. The approach depicted here is broadband in nature. We first describe the time-domain signal model that we adopt, and explain how broadband beamforming works. Then we define several performance measures, some relevant for fixed beamforming while others are more relevant for adaptive beamforming. Finally we show how to derive in great detail three classes of beamformers: fixed, adaptive, and differential. As the reader can see, the algorithms presented in this chapter are more intuitive than the frequency-domain beamformers discussed in previous chapters.

1.5 How to Use the Book

Signal enhancement and array signal processing is a broad subject that finds applications in many different fields. The background description or even the formulation of the problem in the literature is generally field-oriented, typically starting from the physics of wave propagation. Thus many students have been deterred from approaching the subject as it requires confronting, often for the first time, both the physics of wave propagation and the theory of signal processing and optimization. This book is designed as a textbook and it is written with students and instructors from different backgrounds in mind. To help students from the very different backgrounds to quickly get insight into the problem, we choose to focus on the theory, principles, and methods of signal enhancement and array signal processing from a purely signal processing perspective. Readers can then enjoy studying the fundamentals of the problem instead of enduring the introductory material on wave propagation in different types of media.

The material is designed for both advanced undergraduate students and graduate students. A one-semester advanced graduate course can cover virtually all of the text. However, there are also a few other ways to break the material into short courses for teaching advanced undergraduate or junior graduate students. First, a straightforward way is to break the material into two courses: from Chapter 2 to Chapter 6 for a course on signal enhancement and from Chapter 7 to Chapter 11 for a course on array signal processing. Chapter 2 serves as the basis for comprehending the material in Chapters 4, 6, and 11. A short course on signal enhancement and array beamforming in the time domain can be designed using the material presented in Chapters 2, 4, 6, and 11, and the rest of the material can be considered optional. Alternatively, a short course on signal enhancement and array beamforming can be taught based on all the material related to the frequency-domain theory and methods, in Chapters 3, 5, and 7–10.

References

1 Y. Huang, J. Benesty, and J. Chen, *Acoustic MIMO Signal Processing*. Berlin, Germany: Springer-Verlag, 2006.

2 J. Benesty, M. M. Sondhi, and Y. Huang, Eds., *Springer Handbook of Speech Processing*. Berlin, Germany: Springer-Verlag, 2007.

3 J. Benesty, J. Chen, and Y. Huang, *Microphone Array Signal Processing*. Berlin, Germany: Springer-Verlag, 2008.

4 N. Wiener, *Extrapolation, Interpolation, and Smoothing of Stationary Time Series*. New York: Wiley, 1949.

5 J. Benesty, J. Chen, Y. Huang, and S. Doclo, "Study of the Wiener filter for noise reduction," in *Speech Enhancement*, J. Benesty, S. Makino, and J. Chen (eds). Berlin, Germany: Springer-Verlag, 2005, Chapter 2, pp. 9–41.

6 J. Chen, J. Benesty, Y. Huang, and S. Doclo, "New insights into the noise reduction Wiener filter," *IEEE Trans. Audio, Speech, Language Process.*, vol. 14, pp. 1218–1234, Jul. 2006.

7 J. Benesty and J. Chen, *Optimal Time-domain Noise Reduction Filters – A Theoretical Study*. Springer Briefs in Electrical and Computer Engineering, 2011.

8 J. Benesty, J. Chen, Y. Huang, and I. Cohen, *Noise Reduction in Speech Processing*. Berlin, Germany: Springer-Verlag, 2009.

9 K. K. Paliwal and A. Basu, "A speech enhancement method based on Kalman filtering," in *Proc. IEEE ICASSP*, 1987, pp. 177–180.

10 J. S. Lim (ed.), *Speech Enhancement*. Englewood Cliffs, NJ: Prentice-Hall, 1983.

11 D. L. Wang, "On ideal binary mask as the computational goal of auditory scene analysis," in *Speech Separation by Humans and Machines*, P. Divenyi (ed.). Norwell, MA: Kluwer, pp. 181–197.

12 M. R. Schroeder, "Apparatus for suppressing noise and distortion in communication signals," U.S. Patent No. 3,180,936, filed 1 Dec. 1960, issued 27 Apr. 1965.

13 S. F. Boll, "Suppression of acoustic noise in speech using spectral subtraction," *IEEE Trans. Acoust., Speech, Signal Process.*, vol. ASSP-27, pp. 113–120, Apr. 1979.

14 R. J. McAulay and M. L. Malpass, "Speech enhancement using a soft-decision noise suppression filter," *IEEE Trans. Acoust., Speech, Signal Process.*, vol. ASSP-28, pp. 137–145, Apr. 1980.

15 Y. Ephraim and D. Malah, "Speech enhancement using a minimum mean-square error short-time spectral amplitude estimator," *IEEE Trans. Acoust., Speech, Signal Process.*, vol. ASSP-32, pp. 1109–1121, Dec. 1984.

16 H. L. van Trees, *Detection, Estimation, and Modulation Theory, Optimum Array Processing* (Part IV). Hoboken, NJ: Wiley-Interscience, 2002.

17 D. H. Johnson and D. E. Dudgeon, *Array Signal Processing: Concepts and Techniques*. Upper Saddle River, NJ: Prentice Hall, 1993.

18 S. Haykin, *Array Signal Processing*. Upper Saddle River, NJ: Prentice-Hall, 1984.

19 M. Sullivan, *Practical Array Processing*. New York: McGraw-Hill Education, 2008.

20 M. A. Richards, *Fundamentals of Radar Signal Processing*, 2nd edn. New York: McGraw-Hill, 2014.

21 P. S. Naidu, *Sensor Array Signal Processing*, 2nd edn. Boca Raton, FL: CRC Press, 2009.

22 S. Haykin and K. J. R. Liu, *Handbook on Array Processing and Sensor Networks*. Hoboken, NJ: Wiley & Sons, 2010.

23 H. Krim and M. Viberg, "Two decades of array signal processing research: the parametric approach," *IEEE Sig. Process. Mag.*, vol. 13, pp. 67–94, Jul. 1996.

24 J. Benesty and J. Chen, *Study and Design of Differential Microphone Arrays.* Berlin, Germany: Springer-Verlag, 2013.

25 J. Benesty, J. Chen, and I.Cohen, *Design of Circular Differential Microphone Arrays.* Switzerland: Springer, 2015.

26 M. Brandstein and D. Ward (eds), *Microphone Arrays: Signal Processing Techniques and Applications.* Berlin, Germany: Springer-Verlag, 2001.

27 J. Capon, "High resolution frequency-wavenumber spectrum analysis," *Proc. IEEE*, vol. 57, pp. 1408–1418, Aug. 1969.

28 O. L. Frost, III, "An algorithm for linearly constrained adaptive array processing," *Proc. IEEE*, vol. 60, pp. 926–935, Aug. 1972.

29 L. J. Griffiths and C. W. Jim, "An alternative approach to linearly constrained adaptive beamforming," *IEEE Trans. Antennas Propag.*, vol.AP-30, pp. 27–34, Jan. 1982.

30 H. Cox, R. M. Zeskind, and M. M. Owen, "Robust adaptive beamforming," *IEEE Trans. Acoust., Speech, Signal Process.*, vol. ASSP-35, pp. 1365–1376, Oct. 1987.

31 D. G. Manolakis, D. Manolakis, V. K. Ingle, and S. M. Kogon, *Statistical and Adaptive Signal Processing: Spectral Estimation, Signal Modeling, Adaptive Filtering and Array Processing.* Norwood, MA: Artech House, 2005.

32 W. Herbordt, *Sound Capture for Human/Machine Interfaces: Practical Aspects of Microphone Array Signal Processing.* Berlin, Germany: Springer, 2008.

Part I

Signal Enhancement

2

Single-channel Signal Enhancement in the Time Domain

This chapter is dedicated to the study of the signal enhancement problem in the time domain with a single sensor. We show how to fully exploit the temporal information in order to reduce the level of the additive noise from the observations. It is divided into two parts. In the first half, we study this fundamental problem from the classical Wiener filtering perspective. In the second half, we develop a spectral approach, which is based on the joint diagonalization of the desired and noise signal correlation matrices. For both methods, the problem is clearly formulated, performance measures are defined, and useful optimal filters are derived. Examples are also given to show the benefits of both approaches.

2.1 Signal Model and Problem Formulation

In this chapter, we are concerned with the signal enhancement (or noise reduction) problem, in which the desired time-domain signal, $x(t)$, with t being the discrete-time index, needs to be recovered from the noisy observation (sensor signal) [1–3]:

$$y(t) = x(t) + v(t), \tag{2.1}$$

where $v(t)$ is the unwanted additive noise signal, which is assumed to be uncorrelated with $x(t)$. All signals are considered to be real, zero mean, stationary, and broadband.

The signal model given in (2.1) can be put into a vector form by considering the L most recent successive time samples:

$$\mathbf{y}(t) = \mathbf{x}(t) + \mathbf{v}(t), \tag{2.2}$$

where

$$\mathbf{y}(t) = \begin{bmatrix} y(t) & y(t-1) & \cdots & y(t-L+1) \end{bmatrix}^T \tag{2.3}$$

is a vector of length L, superscript^T denotes transpose of a vector or a matrix, and $\mathbf{x}(t)$ and $\mathbf{v}(t)$ are defined in a similar way to $\mathbf{y}(t)$ from (2.3). Since $x(t)$ and $v(t)$ are

Fundamentals of Signal Enhancement and Array Signal Processing, First Edition.
Jacob Benesty, Israel Cohen, and Jingdong Chen.

Companion website: www.wiley.com/go/benesty/arraysignalprocessing

uncorrelated by assumption, the correlation matrix (of size $L \times L$) of the noisy signal can be written as

$$\begin{aligned} \mathbf{R}_\mathbf{y} &= E\left[\mathbf{y}(t)\mathbf{y}^T(t)\right] \\ &= \mathbf{R}_\mathbf{x} + \mathbf{R}_\mathbf{v}, \end{aligned} \tag{2.4}$$

where $E[\cdot]$ denotes mathematical expectation, and $\mathbf{R}_\mathbf{x}=E\left[\mathbf{x}(t)\mathbf{x}^T(t)\right]$ and $\mathbf{R}_\mathbf{v}=E\left[\mathbf{v}(t)\mathbf{v}^T(t)\right]$ are the correlation matrices of $\mathbf{x}(t)$ and $\mathbf{v}(t)$, respectively. We always assume in this chapter that the noise correlation matrix is full rank; in other words, $\text{rank}\left(\mathbf{R}_\mathbf{v}\right) = L$.

The objective of single-channel noise reduction in the time domain is to find a "good" estimate of the sample $x(t)$ from the vector $\mathbf{y}(t)$ [4]. By good, we mean that the additive noise, $v(t)$, is significantly reduced while the desired signal, $x(t)$, is not much distorted.

In the following, we develop two important approaches: the conventional one, which is based on the fundamental Wiener filtering, and the spectral approach, which is based on the spectrum of the desired and noise signals.

2.2 Wiener Method

This section is concerned with the fundamental Wiener filtering theory, which fully exploits the second-order statistics of the signals through the optimization of the classical mean-squared error criterion.

2.2.1 Linear Filtering

We try to estimate the desired signal sample, $x(t)$, by applying a real-valued linear filter to the observation signal vector, $\mathbf{y}(t)$:

$$\begin{aligned} z(t) &= \sum_{l=1}^{L} h_l y(t+1-l) \\ &= \mathbf{h}^T \mathbf{y}(t), \end{aligned} \tag{2.5}$$

where $z(t)$ is the estimate of $x(t)$ and

$$\mathbf{h} = \begin{bmatrix} h_1 & h_2 & \cdots & h_L \end{bmatrix}^T \tag{2.6}$$

is a filter of length L (see Figure 2.1). This procedure is called single-channel signal enhancement in the time domain.

Using (2.2), we can express (2.5) as

$$\begin{aligned} z(t) &= \mathbf{h}^T\left[\mathbf{x}(t) + \mathbf{v}(t)\right] \\ &= x_{\text{fd}}(t) + v_{\text{rn}}(t), \end{aligned} \tag{2.7}$$

where

$$x_{\text{fd}}(t) = \mathbf{h}^T \mathbf{x}(t) \tag{2.8}$$

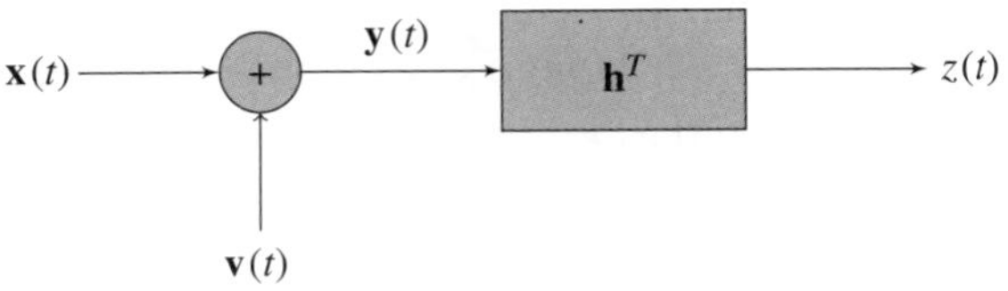

Figure 2.1 Block diagram of linear filtering in the time domain.

is the filtered desired signal and

$$v_{\text{rn}}(t) = \mathbf{h}^T \mathbf{v}(t) \tag{2.9}$$

is the residual noise.

Since the estimate of the desired signal at time t is the sum of two terms that are uncorrelated, the variance of $z(t)$ is

$$\begin{aligned} \sigma_z^2 &= E\left[z^2(t)\right] \\ &= \mathbf{h}^T \mathbf{R}_{\mathbf{y}} \mathbf{h} \\ &= \sigma_{x_{\text{fd}}}^2 + \sigma_{v_{\text{rn}}}^2, \end{aligned} \tag{2.10}$$

where

$$\sigma_{x_{\text{fd}}}^2 = \mathbf{h}^T \mathbf{R}_{\mathbf{x}} \mathbf{h} \tag{2.11}$$

is the variance of the filtered desired signal and

$$\sigma_{v_{\text{rn}}}^2 = \mathbf{h}^T \mathbf{R}_{\mathbf{v}} \mathbf{h} \tag{2.12}$$

is the variance of the residual noise. The variance of $z(t)$ is useful in some of the definitions of the performance measures.

2.2.2 Performance Measures

The first attempts to derive relevant and rigorous measures in the context of signal enhancement can be found in the literature [1, 5, 6]. These references are the main inspiration for the derivation of measures in the studied context throughout this work.

We are now ready to define the most important performance measures in the general context of signal enhancement described in Section 2.1. We can divide these measures into two distinct but related categories. The first category evaluates the noise reduction performance while the second one evaluates the distortion of the desired signal. We also discuss the very convenient mean-squared error criterion and show how it is related to the performance measures.

One of the most fundamental measures in all aspects of signal enhancement is the signal-to-noise ratio (SNR). The input SNR is a second-order measure, which quantifies the level of the noise present relative to the level of the desired signal. It is defined as

$$\begin{aligned} \text{iSNR} &= \frac{\text{tr}\left(\mathbf{R_x}\right)}{\text{tr}\left(\mathbf{R_v}\right)} \\ &= \frac{\sigma_x^2}{\sigma_v^2}, \end{aligned} \tag{2.13}$$

where $\text{tr}(\cdot)$ denotes the trace of a square matrix, and $\sigma_x^2 = E\left[x^2(t)\right]$ and $\sigma_v^2 = E\left[v^2(t)\right]$ are the variances of the desired and noise signals, respectively.

The output SNR helps quantify the level of the noise remaining in the filter output signal. The output SNR is obtained from (2.10):

$$\begin{aligned} \text{oSNR}\left(\mathbf{h}\right) &= \frac{\sigma_{x_{\text{fd}}}^2}{\sigma_{v_{\text{rn}}}^2} \\ &= \frac{\mathbf{h}^T\mathbf{R_x}\mathbf{h}}{\mathbf{h}^T\mathbf{R_v}\mathbf{h}}. \end{aligned} \tag{2.14}$$

Basically, (2.14) is the variance of the first signal (filtered desired signal) from the right-hand side of (2.10) over the variance of the second signal (filtered noise). The objective of the signal enhancement filter is to make the output SNR greater than the input SNR. Consequently, the quality of the filtered output signal, $z(t)$, is enhanced compared to the noisy signal, $y(t)$.

For a particular filter of length L:

$$\mathbf{i}_\text{i} = \left[\begin{array}{cccc} 1 & 0 & \cdots & 0 \end{array}\right]^T, \tag{2.15}$$

we have

$$\text{oSNR}\left(\mathbf{i}_\text{i}\right) = \text{iSNR}. \tag{2.16}$$

With the identity filter, $\mathbf{i}_\text{i}$, the SNR cannot be improved.

The noise reduction factor quantifies the amount of noise being rejected by the filter. This quantity is defined as the ratio of the power of the noise at the sensor over the power of the noise remaining at the filter output:

$$\xi_\text{n}\left(\mathbf{h}\right) = \frac{\sigma_v^2}{\mathbf{h}^T\mathbf{R_v}\mathbf{h}}. \tag{2.17}$$

The noise reduction factor is expected to be lower bounded by 1; otherwise, the filter amplifies the noise received at the sensor. The higher the value of the noise reduction factor, the more noise is rejected. While the output SNR is upper bounded, the noise reduction factor is not.

Since the noise is reduced by the filtering operation, so is, in general, the desired signal. This desired signal reduction (or cancellation) implies, in general, distortion. The desired signal reduction factor, the definition of which is somewhat similar to the noise reduction factor, is defined as the ratio of the variance of the desired signal at the sensor over the variance of the filtered desired signal:

$$\xi_{\mathrm{d}}(\mathbf{h}) = \frac{\sigma_x^2}{\mathbf{h}^T \mathbf{R}_{\mathbf{x}} \mathbf{h}}. \tag{2.18}$$

The closer the value of $\xi_{\mathrm{d}}(\mathbf{h})$ is to 1, the less distorted is the desired signal.

It is easy to verify that we have the following fundamental relation:

$$\frac{\mathrm{oSNR}(\mathbf{h})}{\mathrm{iSNR}} = \frac{\xi_{\mathrm{n}}(\mathbf{h})}{\xi_{\mathrm{d}}(\mathbf{h})}. \tag{2.19}$$

This expression indicates the equivalence between gain/loss in SNR and distortion (of both the desired and noise signals).

Another way to measure the distortion of the desired signal due to the filtering operation is via the desired signal distortion index, which is defined as the mean-squared error between the desired signal and the filtered desired signal, normalized by the variance of the desired signal:

$$\begin{aligned} \upsilon_{\mathrm{d}}(\mathbf{h}) &= \frac{E\left\{\left[x_{\mathrm{fd}}(t) - x(t)\right]^2\right\}}{E\left[x^2(t)\right]} \\ &= \frac{\left(\mathbf{h} - \mathbf{i}_{\mathrm{i}}\right)^T \mathbf{R}_{\mathbf{x}} \left(\mathbf{h} - \mathbf{i}_{\mathrm{i}}\right)}{\sigma_x^2}. \end{aligned} \tag{2.20}$$

The desired signal distortion index is close to 0 if there is no distortion and will be greater than 0 when distortion occurs.

Error criteria play a critical role in deriving optimal filters. The mean-squared error (MSE) [7] is, by far, the most practical one. We define the error signal between the estimated and desired signals as

$$\begin{aligned} e(t) &= z(t) - x(t) \\ &= x_{\mathrm{fd}}(t) + v_{\mathrm{rn}}(t) - x(t), \end{aligned} \tag{2.21}$$

which can be written as the sum of two uncorrelated error signals:

$$e(t) = e_{\mathrm{d}}(t) + e_{\mathrm{n}}(t), \tag{2.22}$$

where

$$\begin{aligned} e_{\mathrm{d}}(t) &= x_{\mathrm{fd}}(t) - x(t) \\ &= \left(\mathbf{h} - \mathbf{i}_{\mathrm{i}}\right)^T \mathbf{x}(t) \end{aligned} \tag{2.23}$$

is the desired signal distortion due to the filter and

$$\begin{aligned} e_{\mathrm{n}}(t) &= v_{\mathrm{rn}}(t) \\ &= \mathbf{h}^T \mathbf{v}(t) \end{aligned} \tag{2.24}$$

represents the residual noise. Therefore, the MSE criterion is

$$\begin{aligned} J(\mathbf{h}) &= E\left[e^2(t)\right] \\ &= \sigma_x^2 - 2\mathbf{h}^T \mathbf{R}_{\mathbf{x}} \mathbf{i}_{\mathrm{i}} + \mathbf{h}^T \mathbf{R}_{\mathbf{y}} \mathbf{h} \\ &= J_{\mathrm{d}}(\mathbf{h}) + J_{\mathrm{n}}(\mathbf{h}), \end{aligned} \tag{2.25}$$

where

$$\begin{aligned} J_{\mathrm{d}}(\mathbf{h}) &= E\left[e_{\mathrm{d}}^2(t)\right] \\ &= \left(\mathbf{h} - \mathbf{i}_{\mathrm{i}}\right)^T \mathbf{R}_{\mathbf{x}} \left(\mathbf{h} - \mathbf{i}_{\mathrm{i}}\right) \\ &= \sigma_x^2 \upsilon_{\mathrm{d}}(\mathbf{h}) \end{aligned} \tag{2.26}$$

and

$$\begin{aligned} J_{\mathrm{n}}(\mathbf{h}) &= E\left[e_{\mathrm{n}}^2(t)\right] \\ &= \mathbf{h}^T \mathbf{R}_{\mathbf{v}} \mathbf{h} \\ &= \frac{\sigma_v^2}{\xi_{\mathrm{n}}(\mathbf{h})}. \end{aligned} \tag{2.27}$$

We deduce that

$$J(\mathbf{h}) = \sigma_v^2 \left[\mathrm{iSNR} \times \upsilon_{\mathrm{d}}(\mathbf{h}) + \frac{1}{\xi_{\mathrm{n}}(\mathbf{h})}\right] \tag{2.28}$$

and

$$\begin{aligned} \frac{J_{\mathrm{d}}(\mathbf{h})}{J_{\mathrm{n}}(\mathbf{h})} &= \mathrm{iSNR} \times \xi_{\mathrm{n}}(\mathbf{h}) \times \upsilon_{\mathrm{d}}(\mathbf{h}) \\ &= \mathrm{oSNR}(\mathbf{h}) \times \xi_{\mathrm{d}}(\mathbf{h}) \times \upsilon_{\mathrm{d}}(\mathbf{h}). \end{aligned} \tag{2.29}$$

We observe how the MSEs are related to the different performance measures.

2.2.3 Optimal Filters

In this subsection, we derive the most important Wiener and Wiener-type filters that can help mitigate the level of the noise picked up by the sensor.

The Wiener filter is derived by taking the gradient of the MSE, $J(\mathbf{h})$ from Equation (2.25), with respect to $\mathbf{h}$ and equating the result to zero:

$$\mathbf{h}_{\mathrm{W}} = \mathbf{R}_{\mathbf{y}}^{-1} \mathbf{R}_{\mathbf{x}} \mathbf{i}_{\mathrm{i}}. \tag{2.30}$$

This optimal filter can also be expressed as

$$\mathbf{h}_{\mathrm{W}} = \left(\mathbf{I}_L - \mathbf{R}_{\mathbf{y}}^{-1}\mathbf{R}_{\mathbf{v}} \right) \mathbf{i}_{\mathrm{i}}, \tag{2.31}$$

where $\mathbf{I}_L$ is the identity matrix of size $L \times L$. The above formulation is more useful than (2.30) in practice, since it depends on the second-order statistics of the observation and noise signals. The correlation matrix $\mathbf{R}_{\mathbf{y}}$ can be immediately estimated from the observation signal while the other correlation matrix, $\mathbf{R}_{\mathbf{v}}$, is often known or can be indirectly estimated. In speech applications, for example, this matrix can be estimated during silences.

Let us define the normalized correlation matrices:

$$\begin{aligned}
\mathbf{\Gamma}_{\mathbf{v}} &= \frac{\mathbf{R}_{\mathbf{v}}}{\sigma_v^2}, \\
\mathbf{\Gamma}_{\mathbf{x}} &= \frac{\mathbf{R}_{\mathbf{x}}}{\sigma_x^2}, \\
\mathbf{\Gamma}_{\mathbf{y}} &= \frac{\mathbf{R}_{\mathbf{y}}}{\sigma_y^2}.
\end{aligned}$$

Another way to write the Wiener filter is

$$\begin{aligned}
\mathbf{h}_{\mathrm{W}} &= \left(\frac{\mathbf{I}_L}{\mathrm{iSNR}} + \mathbf{\Gamma}_{\mathbf{v}}^{-1}\mathbf{\Gamma}_{\mathbf{x}} \right)^{-1} \mathbf{\Gamma}_{\mathbf{v}}^{-1}\mathbf{\Gamma}_{\mathbf{x}}\mathbf{i}_{\mathrm{i}} \\
&= \rho^2(x, y)\mathbf{\Gamma}_{\mathbf{y}}^{-1}\mathbf{\Gamma}_{\mathbf{x}}\mathbf{i}_{\mathrm{i}} \\
&= \left[\mathbf{I}_L - \rho^2(v, y)\mathbf{\Gamma}_{\mathbf{y}}^{-1}\mathbf{\Gamma}_{\mathbf{v}} \right] \mathbf{i}_{\mathrm{i}},
\end{aligned} \tag{2.32}$$

where

$$\begin{aligned}
\rho^2(x, y) &= \frac{E^2\left[x(t)y(t)\right]}{\sigma_x^2\sigma_y^2} \\
&= \frac{\sigma_x^2}{\sigma_y^2} \\
&= \frac{\mathrm{iSNR}}{1 + \mathrm{iSNR}}
\end{aligned} \tag{2.33}$$

is the squared Pearson correlation coefficient (SPCC) between $x(t)$ and $y(t)$, and

$$\begin{aligned}
\rho^2(v, y) &= \frac{E^2\left[v(t)y(t)\right]}{\sigma_v^2\sigma_y^2} \\
&= \frac{\sigma_v^2}{\sigma_y^2} \\
&= \frac{1}{1 + \mathrm{iSNR}}
\end{aligned} \tag{2.34}$$

is the SPCC between $v(t)$ and $y(t)$. We can see from (2.32) that

$$\lim_{\mathrm{iSNR}\to\infty} \mathbf{h}_\mathrm{W} = \mathbf{i}_\mathrm{i}, \tag{2.35}$$

$$\lim_{\mathrm{iSNR}\to 0} \mathbf{h}_\mathrm{W} = \mathbf{0}, \tag{2.36}$$

where $\mathbf{0}$ is the zero vector. Clearly, the Wiener filter can have a disastrous effect at very low input SNRs since it may remove both noise *and* desired signals.

Hence, the estimate of the desired signal with the Wiener filter is

$$z_\mathrm{W}(t) = \mathbf{h}_\mathrm{W}^T \mathbf{y}(t). \tag{2.37}$$

We now describe a fundamental property, which was first shown by Chen et al. [6].

Property 2.2.1 With the optimal Wiener filter (2.30), the output SNR is always greater than or equal to the input SNR: $\mathrm{oSNR}\left(\mathbf{h}_\mathrm{W}\right) \geq \mathrm{iSNR}$.

Proof. There are different ways to show this property. Here, we do so with the help of the different SPCCs [8]. We recall that for any two zero-mean random variables $a(t)$ and $b(t)$, we have

$$0 \leq \rho^2(a, b) \leq 1. \tag{2.38}$$

It can be checked that

$$\rho^2(x, z) = \rho^2\left(x, x_\mathrm{fd}\right) \times \rho^2\left(x_\mathrm{fd}, z\right) \leq \rho^2\left(x_\mathrm{fd}, z\right), \tag{2.39}$$

where

$$\rho^2\left(x_\mathrm{fd}, z\right) = \frac{\mathrm{oSNR}\left(\mathbf{h}\right)}{1 + \mathrm{oSNR}\left(\mathbf{h}\right)}. \tag{2.40}$$

As a result, we have

$$\rho^2\left(x, z_\mathrm{W}\right) \leq \frac{\mathrm{oSNR}\left(\mathbf{h}_\mathrm{W}\right)}{1 + \mathrm{oSNR}\left(\mathbf{h}_\mathrm{W}\right)}. \tag{2.41}$$

Let us evaluate the SPCC between $y(t)$ and $z_\mathrm{W}(t)$:

$$\begin{aligned}
\rho^2(y, z_\mathrm{W}) &= \frac{\left(\mathbf{i}_\mathrm{i}^T \mathbf{R}_\mathbf{y} \mathbf{h}_\mathrm{W}\right)^2}{\sigma_y^2 \mathbf{h}_\mathrm{W}^T \mathbf{R}_\mathbf{y} \mathbf{h}_\mathrm{W}} \\
&= \frac{\sigma_x^2}{\sigma_y^2} \times \frac{\sigma_x^2}{\mathbf{i}_\mathrm{i}^T \mathbf{R}_\mathbf{x} \mathbf{h}_\mathrm{W}} \\
&= \frac{\rho^2(x, y)}{\rho^2(x, z_\mathrm{W})}.
\end{aligned}$$

Therefore,

$$\rho^2(x,y) = \rho^2(y,z_W) \times \rho^2(x,z_W) \le \rho^2(x,z_W). \tag{2.42}$$

Substituting (2.33) and (2.41) into (2.42), we get

$$\frac{\text{iSNR}}{1+\text{iSNR}} \le \frac{\text{oSNR}\left(\mathbf{h}_W\right)}{1+\text{oSNR}\left(\mathbf{h}_W\right)},$$

which we can slightly rearrange to give:

$$\frac{1}{1+\frac{1}{\text{iSNR}}} \le \frac{1}{1+\frac{1}{\text{oSNR}\left(\mathbf{h}_W\right)}},$$

which implies that

$$\frac{1}{\text{iSNR}} \ge \frac{1}{\text{oSNR}\left(\mathbf{h}_W\right)}.$$

Consequently, we have

$$\text{oSNR}\left(\mathbf{h}_W\right) \ge \text{iSNR}.$$

■

The minimum MSE (MMSE) is obtained by replacing (2.30) in (2.25):

$$\begin{aligned} J\left(\mathbf{h}_W\right) &= \sigma_x^2 - \mathbf{i}_i^T \mathbf{R}_\mathbf{x} \mathbf{R}_\mathbf{y}^{-1} \mathbf{R}_\mathbf{x} \mathbf{i}_i \\ &= \sigma_v^2 - \mathbf{i}_i^T \mathbf{R}_\mathbf{v} \mathbf{R}_\mathbf{y}^{-1} \mathbf{R}_\mathbf{v} \mathbf{i}_i, \end{aligned} \tag{2.43}$$

which can be rewritten as

$$\begin{aligned} J\left(\mathbf{h}_W\right) &= \sigma_x^2 \left[1 - \rho^2(x,z_W)\right] \\ &= \sigma_v^2 \left[1 - \rho^2(v,y-z_W)\right]. \end{aligned} \tag{2.44}$$

Clearly, we always have

$$J\left(\mathbf{h}_W\right) \le J\left(\mathbf{h}\right), \ \forall \mathbf{h} \tag{2.45}$$

and, in particular,

$$J\left(\mathbf{h}_W\right) \le J\left(\mathbf{i}_i\right) = \sigma_v^2. \tag{2.46}$$

The different performance measures with the Wiener filter are

$$\text{oSNR}\left(\mathbf{h}_W\right) = \frac{\mathbf{i}_i^T \mathbf{R}_\mathbf{x} \mathbf{R}_\mathbf{y}^{-1} \mathbf{R}_\mathbf{x} \mathbf{R}_\mathbf{y}^{-1} \mathbf{R}_\mathbf{x} \mathbf{i}_i}{\mathbf{i}_i^T \mathbf{R}_\mathbf{x} \mathbf{R}_\mathbf{y}^{-1} \mathbf{R}_\mathbf{v} \mathbf{R}_\mathbf{y}^{-1} \mathbf{R}_\mathbf{x} \mathbf{i}_i} \ge \text{iSNR}, \tag{2.47}$$

$$\xi_{\mathrm{n}}\left(\mathbf{h}_{\mathrm{W}}\right) = \frac{\sigma_v^2}{\mathbf{i}_{\mathrm{i}}^T\mathbf{R}_{\mathbf{x}}\mathbf{R}_{\mathbf{y}}^{-1}\mathbf{R}_{\mathbf{v}}\mathbf{R}_{\mathbf{y}}^{-1}\mathbf{R}_{\mathbf{x}}\mathbf{i}_{\mathrm{i}}} \geq 1, \tag{2.48}$$

$$\xi_{\mathrm{d}}\left(\mathbf{h}_{\mathrm{W}}\right) = \frac{\sigma_x^2}{\mathbf{i}_{\mathrm{i}}^T\mathbf{R}_{\mathbf{x}}\mathbf{R}_{\mathbf{y}}^{-1}\mathbf{R}_{\mathbf{x}}\mathbf{R}_{\mathbf{y}}^{-1}\mathbf{R}_{\mathbf{x}}\mathbf{i}_{\mathrm{i}}} \geq 1, \tag{2.49}$$

$$\upsilon_{\mathrm{d}}\left(\mathbf{h}_{\mathrm{W}}\right) = \frac{\left(\mathbf{R}_{\mathbf{y}}^{-1}\mathbf{R}_{\mathbf{x}}\mathbf{i}_{\mathrm{i}} - \mathbf{i}_{\mathrm{i}}\right)^T \mathbf{R}_{\mathbf{x}} \left(\mathbf{R}_{\mathbf{y}}^{-1}\mathbf{R}_{\mathbf{x}}\mathbf{i}_{\mathrm{i}} - \mathbf{i}_{\mathrm{i}}\right)}{\sigma_x^2} \leq 1. \tag{2.50}$$

Example 2.2.1 Suppose that the desired signal is a harmonic random process:

$$x(t) = A\cos\left(2\pi f_0 t + \phi\right),$$

with fixed amplitude A and frequency f_0, and random phase ϕ, uniformly distributed on the interval from 0 to 2π. This signal needs to be recovered from the noisy observation $y(t) = x(t) + v(t)$, where $v(t)$ is additive white Gaussian noise – in other words, $v(t) \sim \mathcal{N}\left(0, \sigma_v^2\right)$ – that is uncorrelated with $x(t)$.

The input SNR is

$$\mathrm{iSNR} = 10\log\frac{A^2/2}{\sigma_v^2} \quad (\mathrm{dB}).$$

The correlation matrix of $\mathbf{v}(t)$ is $\mathbf{R}_{\mathbf{v}} = \sigma_v^2\mathbf{I}_L$, and the elements of the correlation matrix of $\mathbf{x}(t)$ are $\left[\mathbf{R}_{\mathbf{x}}\right]_{i,j} = \frac{1}{2}A^2\cos\left[2\pi f_0(i-j)\right]$. Since the desired and noise signals are uncorrelated, the correlation matrix of the observation signal vector $\mathbf{y}(t)$ is $\mathbf{R}_{\mathbf{y}} = \mathbf{R}_{\mathbf{x}}+\mathbf{R}_{\mathbf{v}}$. The optimal filter $\mathbf{h}_{\mathrm{W}}$ is obtained from (2.30). The output SNR and the MMSE are obtained by substituting $\mathbf{h}_{\mathrm{W}}$ into (2.14) and (2.25), respectively.

To demonstrate the performance of the Wiener filter, we choose $A = 0.5, f_0 = 0.1$, and $\sigma_v^2 = 0.3$. The input SNR is −3.80 dB. Figure 2.2 shows the effect of the filter length, L, on the gain in SNR, $\mathcal{G}(\mathbf{h}_{\mathrm{W}}) = \mathrm{oSNR}\left(\mathbf{h}_{\mathrm{W}}\right)/\mathrm{iSNR}$, and the MMSE, $J(\mathbf{h}_{\mathrm{W}})$. As the length of the filter increases, the Wiener filter better enhances the harmonic signal, in terms of higher gain in SNR and lower MMSE. If we choose a fixed filter length, $L = 30$, and change σ_v^2 so that iSNR varies from 0 to 20 dB, then Figure 2.3 shows plots of the output SNR and the MMSE as a function of the input SNR. Figure 2.4 shows plots of the noise reduction factor, $\xi_{\mathrm{n}}\left(\mathbf{h}_{\mathrm{W}}\right)$, the desired signal reduction factor, $\xi_{\mathrm{d}}\left(\mathbf{h}_{\mathrm{W}}\right)$, and the desired signal distortion index, $\upsilon_{\mathrm{d}}\left(\mathbf{h}_{\mathrm{W}}\right)$, as a function of the input SNR. Figure 2.5 shows a realization of the noise corrupted and filtered sinusoidal signals for iSNR = 0 dB. ■

The objective of the Wiener filter is to minimize the MSE; therefore, it leads to the MMSE. However, this optimal filter is inflexible since it is not possible to compromise between desired signal distortion and noise reduction. It is instructive to observe that the MSE as given in (2.25) is the sum of two other MSEs. One depends on the desired signal distortion while the other depends on the noise reduction. Instead of minimizing the MSE with respect to $\mathbf{h}$ as already done to find the Wiener filter, we can instead minimize the distortion-based MSE subject to the constraint that the noise-reduction-based MSE is equal to some desired value. Mathematically, this is equivalent to

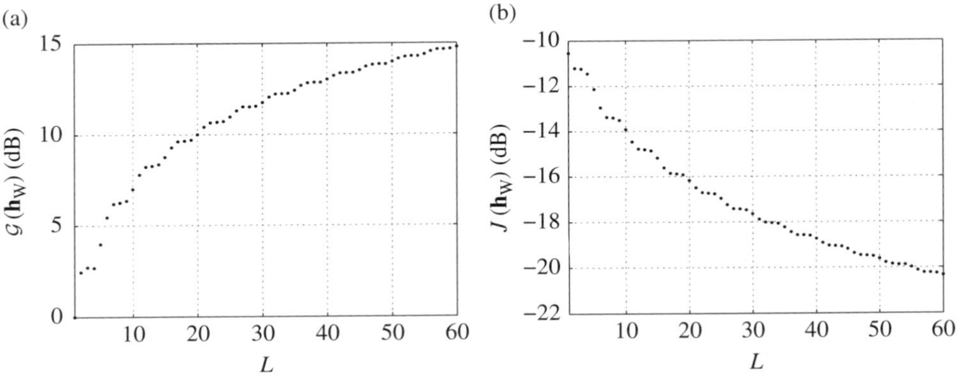

Figure 2.2 (a) The gain in SNR and (b) the MMSE of the Wiener filter as a function of the filter length.

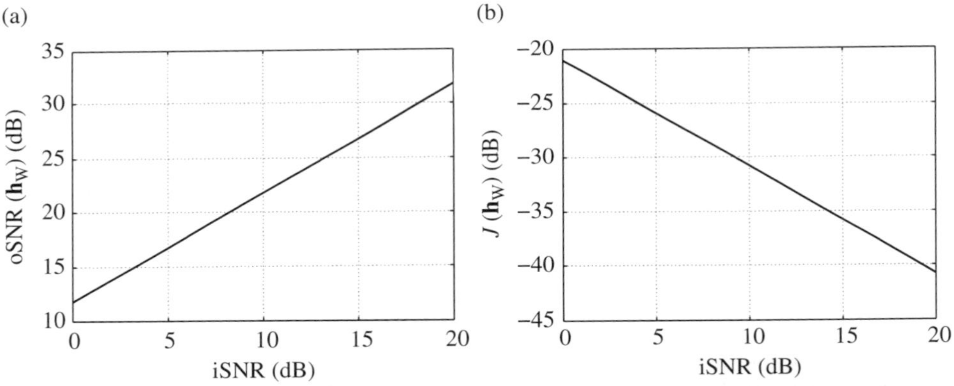

Figure 2.3 (a) The output SNR and (b) the MMSE of the Wiener filter as a function of the input SNR.

$$\min_{\mathbf{h}} J_{\mathrm{d}}(\mathbf{h}) \quad \text{subject to} \quad J_{\mathrm{n}}(\mathbf{h}) = \aleph\sigma_v^2, \tag{2.51}$$

where $0 < \aleph < 1$ to ensure that we have some noise reduction. If we use a Lagrange multiplier, μ, to adjoin the constraint to the cost function, (2.51) can be rewritten as

$$\mathbf{h}_{\mathrm{T},\mu} = \arg\min_{\mathbf{h}} \mathcal{L}(\mathbf{h}, \mu), \tag{2.52}$$

with

$$\mathcal{L}(\mathbf{h}, \mu) = J_{\mathrm{d}}(\mathbf{h}) + \mu\left[J_{\mathrm{n}}(\mathbf{h}) - \aleph\sigma_v^2\right] \tag{2.53}$$

and $\mu \geq 0$. From (2.52), we easily derive the tradeoff filter:

$$\begin{aligned} \mathbf{h}_{\mathrm{T},\mu} &= \left(\mathbf{R}_{\mathbf{x}} + \mu\mathbf{R}_{\mathbf{v}}\right)^{-1}\mathbf{R}_{\mathbf{x}}\mathbf{i}_{\mathrm{i}} \\ &= \left[\mathbf{R}_{\mathbf{y}} + (\mu - 1)\mathbf{R}_{\mathbf{v}}\right]^{-1}\left(\mathbf{R}_{\mathbf{y}} - \mathbf{R}_{\mathbf{v}}\right)\mathbf{i}_{\mathrm{i}}, \end{aligned} \tag{2.54}$$

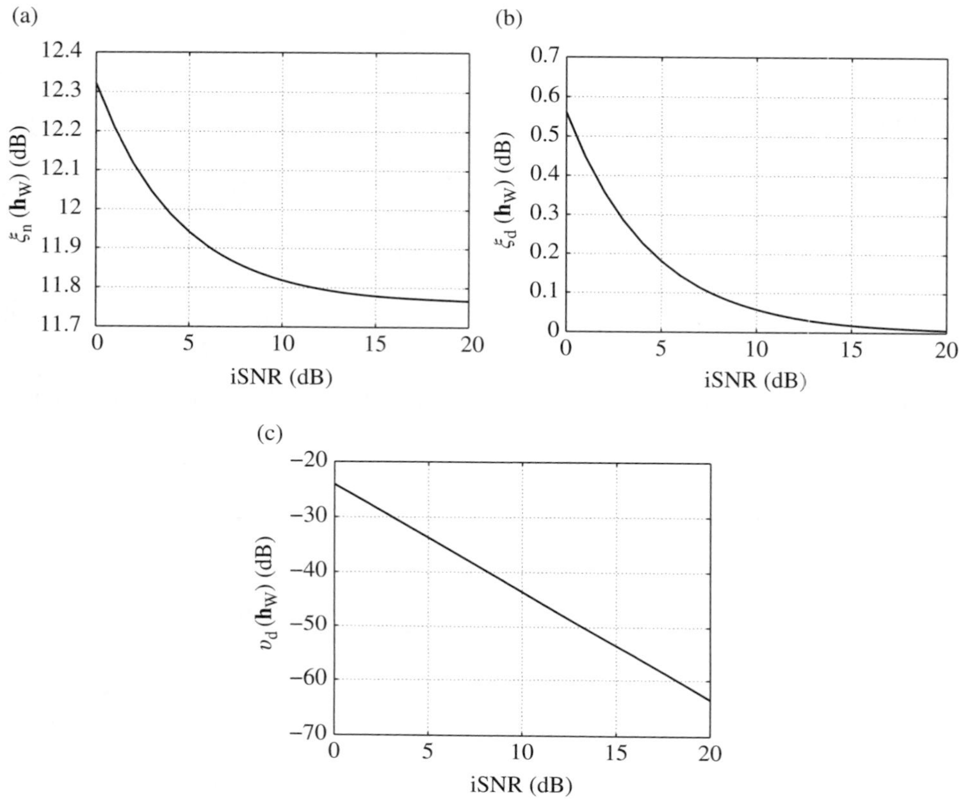

Figure 2.4 (a) The noise reduction factor, (b) the desired signal reduction factor, and (c) the desired signal distortion index of the Wiener filter as a function of the input SNR.

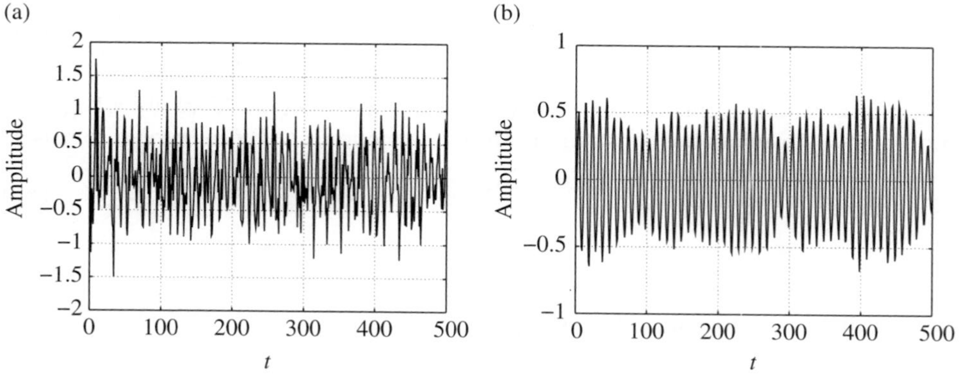

Figure 2.5 Example of (a) noise-corrupted and (b) Wiener-filtered sinusoidal signals.

where the Lagrange multiplier, μ, satisfies $J_{\text{n}}\left(\mathbf{h}_{\text{T},\mu}\right) = \aleph\sigma_v^2$, which implies that

$$\xi_{\text{n}}\left(\mathbf{h}_{\text{T},\mu}\right) = \frac{1}{\aleph} > 1. \tag{2.55}$$

In practice it is not easy to determine the optimal μ. Therefore, when this parameter is chosen in a heuristic way, we can see that for

- $\mu = 1$, $\mathbf{h}_{\text{T},1} = \mathbf{h}_{\text{W}}$, which is the Wiener filter
- $\mu = 0$, $\mathbf{h}_{\text{T},0} = \mathbf{i}_{\text{i}}$, which is the identity filter
- $\mu > 1$ results in a filter with low residual noise at the expense of high desired signal distortion
- $\mu < 1$ results in a filter with low desired signal distortion and small amount of noise reduction.

We are now ready to give a fundamental property about the tradeoff filter.

Property 2.2.2 With the tradeoff filter given in (2.54), the output SNR is always greater than or equal to the input SNR: $\text{oSNR}\left(\mathbf{h}_{\text{T},\mu}\right) \geq \text{iSNR}$, $\forall\mu \geq 0$.

Proof. The SPCC between $x(t)$ and $x(t) + \sqrt{\mu}v(t)$ is

$$\begin{aligned}\rho^2\left(x, x + \sqrt{\mu}v\right) &= \frac{\sigma_x^4}{\sigma_x^2\left(\sigma_x^2 + \mu\sigma_v^2\right)} \\ &= \frac{\text{iSNR}}{\mu + \text{iSNR}}.\end{aligned}$$

The SPCC between $x(t)$ and $\mathbf{h}_{\text{T},\mu}^T\mathbf{x}(t) + \sqrt{\mu}\mathbf{h}_{\text{T},\mu}^T\mathbf{v}(t)$ is

$$\begin{aligned}\rho^2\left(x, \mathbf{h}_{\text{T},\mu}^T\mathbf{x} + \sqrt{\mu}\mathbf{h}_{\text{T},\mu}^T\mathbf{v}\right) &= \frac{\left(\mathbf{h}_{\text{T},\mu}^T\mathbf{R}_{\mathbf{x}}\mathbf{i}_{\text{i}}\right)^2}{\sigma_x^2\mathbf{h}_{\text{T},\mu}^T\left(\mathbf{R}_{\mathbf{x}} + \mu\mathbf{R}_{\mathbf{v}}\right)\mathbf{h}_{\text{T},\mu}} \\ &= \frac{\mathbf{h}_{\text{T},\mu}^T\mathbf{R}_{\mathbf{x}}\mathbf{i}_{\text{i}}}{\sigma_x^2}.\end{aligned}$$

Another way to write the same SPCC is the following:

$$\begin{aligned}\rho^2\left(x, \mathbf{h}_{\text{T},\mu}^T\mathbf{x} + \sqrt{\mu}\mathbf{h}_{\text{T},\mu}^T\mathbf{v}\right) &= \frac{\left(\mathbf{h}_{\text{T},\mu}^T\mathbf{R}_{\mathbf{x}}\mathbf{i}_{\text{i}}\right)^2}{\sigma_x^2\mathbf{h}_{\text{T},\mu}^T\mathbf{R}_{\mathbf{x}}\mathbf{h}_{\text{T},\mu}} \times \frac{\text{oSNR}\left(\mathbf{h}_{\text{T},\mu}\right)}{\mu + \text{oSNR}\left(\mathbf{h}_{\text{T},\mu}\right)} \\ &= \rho^2\left(x, \mathbf{h}_{\text{T},\mu}^T\mathbf{x}\right) \times \\ &\quad\ \rho^2\left(\mathbf{h}_{\text{T},\mu}^T\mathbf{x}, \mathbf{h}_{\text{T},\mu}^T\mathbf{x} + \sqrt{\mu}\mathbf{h}_{\text{T},\mu}^T\mathbf{v}\right) \\ &\leq \frac{\text{oSNR}\left(\mathbf{h}_{\text{T},\mu}\right)}{\mu + \text{oSNR}\left(\mathbf{h}_{\text{T},\mu}\right)}.\end{aligned}$$

Now, let us evaluate the SPCC between $x(t) + \sqrt{\mu}v(t)$ and $\mathbf{h}_{\mathrm{T},\mu}^T \mathbf{x}(t) + \sqrt{\mu}\mathbf{h}_{\mathrm{T},\mu}^T \mathbf{v}(t)$:

$$\rho^2\left(x + \sqrt{\mu}v, \mathbf{h}_{\mathrm{T},\mu}^T \mathbf{x} + \sqrt{\mu}\mathbf{h}_{\mathrm{T},\mu}^T \mathbf{v}\right) = \frac{\left[\mathbf{h}_{\mathrm{T},\mu}^T\left(\mathbf{R}_{\mathbf{x}} + \mu\mathbf{R}_{\mathbf{v}}\right)\mathbf{i}_{\mathrm{i}}\right]^2}{\left(\sigma_x^2 + \mu\sigma_v^2\right)\mathbf{h}_{\mathrm{T},\mu}^T\left(\mathbf{R}_{\mathbf{x}} + \mu\mathbf{R}_{\mathbf{v}}\right)\mathbf{h}_{\mathrm{T},\mu}}$$
$$= \frac{\sigma_x^2}{\sigma_x^2 + \mu\sigma_v^2} \times \frac{\sigma_x^2}{\mathbf{h}_{\mathrm{T},\mu}^T\mathbf{R}_{\mathbf{x}}\mathbf{i}_{\mathrm{i}}}$$
$$= \frac{\rho^2\left(x, x + \sqrt{\mu}v\right)}{\rho^2\left(x, \mathbf{h}_{\mathrm{T},\mu}^T \mathbf{x} + \sqrt{\mu}\mathbf{h}_{\mathrm{T},\mu}^T \mathbf{v}\right)}.$$

Therefore,

$$\rho^2\left(x, x + \sqrt{\mu}v\right) = \frac{\mathrm{iSNR}}{\mu + \mathrm{iSNR}}$$
$$= \rho^2\left(x + \sqrt{\mu}v, \mathbf{h}_{\mathrm{T},\mu}^T \mathbf{x} + \sqrt{\mu}\mathbf{h}_{\mathrm{T},\mu}^T \mathbf{v}\right) \times \rho^2\left(x, \mathbf{h}_{\mathrm{T},\mu}^T \mathbf{x} + \sqrt{\mu}\mathbf{h}_{\mathrm{T},\mu}^T \mathbf{v}\right)$$
$$\leq \rho^2\left(x, \mathbf{h}_{\mathrm{T},\mu}^T \mathbf{x} + \sqrt{\mu}\mathbf{h}_{\mathrm{T},\mu}^T \mathbf{v}\right)$$
$$\leq \frac{\mathrm{oSNR}\left(\mathbf{h}_{\mathrm{T},\mu}\right)}{\mu + \mathrm{oSNR}\left(\mathbf{h}_{\mathrm{T},\mu}\right)}.$$

As a result,

$$\mathrm{oSNR}\left(\mathbf{h}_{\mathrm{T},\mu}\right) \geq \mathrm{iSNR}.$$

■

Example 2.2.2 Consider a desired signal, $x(t)$, with the autocorrelation sequence:

$$E\left[x(t)x(t')\right] = \alpha^{|t-t'|}, \quad -1 < \alpha < 1,$$

which is corrupted by additive white Gaussian noise $v(t) \sim \mathcal{N}\left(0, \sigma_v^2\right)$ that is uncorrelated with $x(t)$. The desired signal needs to be recovered from the noisy observation $y(t) = x(t) + v(t)$.

The input SNR is

$$\mathrm{iSNR} = 10 \log \frac{1}{\sigma_v^2} \quad (\mathrm{dB}).$$

The correlation matrix of $\mathbf{v}(t)$ is $\mathbf{R}_{\mathbf{v}} = \sigma_v^2 \mathbf{I}_L$, and the elements of the correlation matrix of $\mathbf{x}(t)$ are $\left[\mathbf{R}_{\mathbf{x}}\right]_{i,j} = \alpha^{|i-j|}$. Since the desired signal and the noise signal are uncorrelated, the

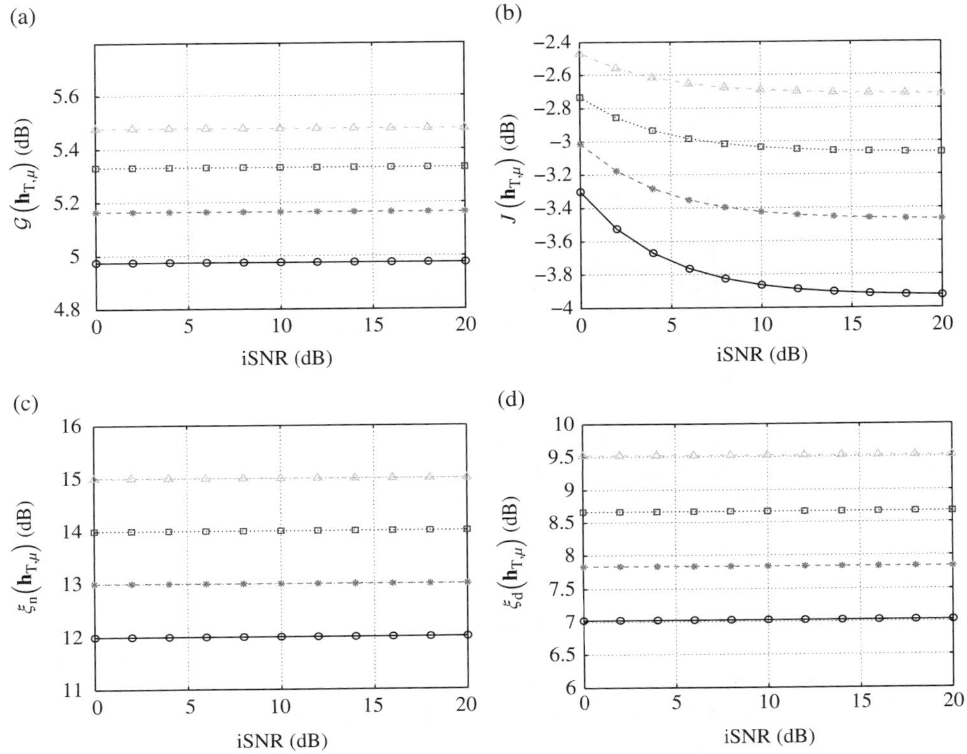

Figure 2.6 (a) The gain in SNR, (b) the MSE, (c) the noise reduction factor, and (d) the desired signal reduction factor of the tradeoff filter as a function of the input SNR for several values of $\aleph$: $\aleph = -12$ dB (solid line with circles), $\aleph = -13$ dB (dashed line with asterisks), $\aleph = -14$ dB (dotted line with squares), and $\aleph = -15$ dB (dash-dot line with triangles).

correlation matrix of observation signal vector $\mathbf{y}(t)$ is $\mathbf{R}_\mathbf{y} = \mathbf{R}_\mathbf{x} + \mathbf{R}_\mathbf{v}$. The tradeoff filter $\mathbf{h}_{T,\mu}$ is obtained from (2.54), where the Lagrange multiplier, μ, satisfies $J_n(\mathbf{h}_{T,\mu}) = \aleph\sigma_v^2$.

To demonstrate the performance of the tradeoff filter, we choose $\alpha = 0.8$, a filter length $L = 30$, and several values of $\aleph$. Figure 2.6 shows plots of the gain in SNR, $\mathcal{G}(\mathbf{h}_{T,\mu})$, the MSE, $J(\mathbf{h}_{T,\mu})$, the noise reduction factor, $\xi_n(\mathbf{h}_{T,\mu})$, and the desired signal reduction factor, $\xi_d(\mathbf{h}_{T,\mu})$, as a function of the input SNR, for several values of $\aleph$. Figure 2.6c shows that the tradeoff filter satisfies $\xi_n(\mathbf{h}_{T,\mu}) = -10\log(\aleph)$ dB. ■

Both Wiener and tradeoff filters always distort the desired signal since $\xi_d(\mathbf{h}_{T,\mu}) \neq 1$, $\forall \mu \geq 0$. It is fair to ask if it is possible to derive a distortionless filter that can mitigate the level of the noise. The answer is positive so long as the desired signal correlation matrix is rank deficient. Let us assume that rank $(\mathbf{R}_\mathbf{x}) = P \leq L$. Using the well-known eigenvalue decomposition [9], the desired signal correlation matrix can be diagonalized as

$$\mathbf{Q}_\mathbf{x}^T \mathbf{R}_\mathbf{x} \mathbf{Q}_\mathbf{x} = \mathbf{\Lambda}_\mathbf{x}, \tag{2.56}$$

where

$$\mathbf{Q}_\mathbf{x} = \begin{bmatrix} \mathbf{q}_{\mathbf{x},1} & \mathbf{q}_{\mathbf{x},2} & \cdots & \mathbf{q}_{\mathbf{x},L} \end{bmatrix} \tag{2.57}$$

is an orthogonal matrix (in other words, $\mathbf{Q}_\mathbf{x}^T\mathbf{Q}_\mathbf{x} = \mathbf{Q}_\mathbf{x}\mathbf{Q}_\mathbf{x}^T = \mathbf{I}_L$) and

$$\mathbf{\Lambda}_\mathbf{x} = \text{diag}\left(\lambda_{\mathbf{x},1}, \lambda_{\mathbf{x},2}, \ldots, \lambda_{\mathbf{x},L}\right) \tag{2.58}$$

is a diagonal matrix. The orthonormal vectors $\mathbf{q}_{\mathbf{x},1}, \mathbf{q}_{\mathbf{x},2}, \ldots, \mathbf{q}_{\mathbf{x},L}$ are the eigenvectors corresponding, respectively, to the eigenvalues $\lambda_{\mathbf{x},1}, \lambda_{\mathbf{x},2}, \ldots, \lambda_{\mathbf{x},L}$ of the matrix $\mathbf{R}_\mathbf{x}$, where $\lambda_{\mathbf{x},1} \geq \lambda_{\mathbf{x},2} \geq \cdots \geq \lambda_{\mathbf{x},P} > 0$ and $\lambda_{\mathbf{x},P+1} = \lambda_{\mathbf{x},P+2} = \cdots = \lambda_{\mathbf{x},L} = 0$. Let

$$\mathbf{Q}_\mathbf{x} = \begin{bmatrix} \mathbf{Q}'_\mathbf{x} & \mathbf{Q}''_\mathbf{x} \end{bmatrix}, \tag{2.59}$$

where the $L \times P$ matrix $\mathbf{Q}'_\mathbf{x}$ contains the eigenvectors corresponding to the nonzero eigenvalues of $\mathbf{R}_\mathbf{x}$ and the $L\times(L-P)$ matrix $\mathbf{Q}''_\mathbf{x}$ contains the eigenvectors corresponding to the null eigenvalues of $\mathbf{R}_\mathbf{x}$. It can be verified that

$$\mathbf{I}_L = \mathbf{Q}'_\mathbf{x}\mathbf{Q}'^T_\mathbf{x} + \mathbf{Q}''_\mathbf{x}\mathbf{Q}''^T_\mathbf{x}. \tag{2.60}$$

Notice that $\mathbf{Q}'_\mathbf{x}\mathbf{Q}'^T_\mathbf{x}$ and $\mathbf{Q}''_\mathbf{x}\mathbf{Q}''^T_\mathbf{x}$ are two orthogonal projection matrices of rank P and $L - P$, respectively. Hence, $\mathbf{Q}'_\mathbf{x}\mathbf{Q}'^T_\mathbf{x}$ is the orthogonal projector onto the desired signal subspace (where all the energy of the desired signal is concentrated) or the range of $\mathbf{R}_\mathbf{x}$. $\mathbf{Q}''_\mathbf{x}\mathbf{Q}''^T_\mathbf{x}$ is the orthogonal projector onto the null subspace of $\mathbf{R}_\mathbf{x}$. Using (2.60), we can write the desired signal vector as

$$\begin{aligned} \mathbf{x}(t) &= \mathbf{Q}_\mathbf{x}\mathbf{Q}_\mathbf{x}^T\mathbf{x}(t) \\ &= \mathbf{Q}'_\mathbf{x}\mathbf{Q}'^T_\mathbf{x}\mathbf{x}(t). \end{aligned} \tag{2.61}$$

We deduce from (2.61) that the distortionless constraint is

$$\mathbf{h}^T\mathbf{Q}'_\mathbf{x} = \mathbf{i}_\mathrm{i}^T\mathbf{Q}'_\mathbf{x}, \tag{2.62}$$

since, in this case,

$$\begin{aligned} \mathbf{h}^T\mathbf{x}(t) &= \mathbf{h}^T\mathbf{Q}'_\mathbf{x}\mathbf{Q}'^T_\mathbf{x}\mathbf{x}(t) \\ &= \mathbf{i}_\mathrm{i}^T\mathbf{Q}'_\mathbf{x}\mathbf{Q}'^T_\mathbf{x}\mathbf{x}(t) \\ &= x(t). \end{aligned} \tag{2.63}$$

Now, from the minimization of the criterion:

$$\min_{\mathbf{h}} J_\mathrm{n}(\mathbf{h}) \quad \text{subject to} \quad \mathbf{h}^T\mathbf{Q}'_\mathbf{x} = \mathbf{i}_\mathrm{i}^T\mathbf{Q}'_\mathbf{x}, \tag{2.64}$$

we find the minimum variance distortionless response (MVDR) filter:

$$\mathbf{h}_\mathrm{MVDR} = \mathbf{R}_\mathbf{v}^{-1}\mathbf{Q}'_\mathbf{x}\left(\mathbf{Q}'^T_\mathbf{x}\mathbf{R}_\mathbf{v}^{-1}\mathbf{Q}'_\mathbf{x}\right)^{-1}\mathbf{Q}'^T_\mathbf{x}\mathbf{i}_\mathrm{i}. \tag{2.65}$$

It can be shown that Equation 2.65 can also be expressed as

$$\mathbf{h}_{\text{MVDR}} = \mathbf{R}_{\mathbf{y}}^{-1}\mathbf{Q}'_{\mathbf{x}}\left(\mathbf{Q}'^{T}_{\mathbf{x}}\mathbf{R}_{\mathbf{y}}^{-1}\mathbf{Q}'_{\mathbf{x}}\right)^{-1}\mathbf{Q}'^{T}_{\mathbf{x}}\mathbf{i}_{\text{i}}. \tag{2.66}$$

It can be verified that, indeed, $J_{\text{d}}\left(\mathbf{h}_{\text{MVDR}}\right) = 0$. Of course, for $P = L$, the MVDR filter degenerates to the identity filter: $\mathbf{h}_{\text{MVDR}} = \mathbf{i}_{\text{i}}$. As a consequence, we can state that the higher the dimension of the nullspace of $\mathbf{R}_{\mathbf{x}}$, the more the MVDR filter is efficient in terms of noise reduction. The best scenario corresponds to $P = 1$. For a white noise signal – in other words, for $\mathbf{R}_{\mathbf{v}} = \sigma_v^2\mathbf{I}_L$ – the MVDR filter simplifies to

$$\mathbf{h}_{\text{MVDR}} = \mathbf{Q}'_{\mathbf{x}}\mathbf{Q}'^{T}_{\mathbf{x}}\mathbf{i}_{\text{i}}, \tag{2.67}$$

which is the minimum-norm solution of (2.62).

Property 2.2.3 With the MVDR filter given in Equation 2.65, the output SNR is always greater than or equal to the input SNR: $\text{oSNR}\left(\mathbf{h}_{\text{MVDR}}\right) \geq \text{iSNR}$.

Example 2.2.3 Consider a desired signal that is a sum of harmonic random processes:

$$x(t) = \sum_{k=1}^{K} A_k \cos\left(2\pi f_k t + \phi_k\right),$$

with fixed amplitudes $\{A_k\}$ and frequencies $\{f_k\}$, and independent and identically distributed (IID) random phases $\{\phi_k\}$, uniformly distributed on the interval from 0 to 2π. This signal needs to be recovered from the noisy observation $y(t) = x(t) + v(t)$, where $v(t)$ is additive white Gaussian noise, $v(t) \sim \mathcal{N}\left(0, \sigma_v^2\right)$, which is uncorrelated with $x(t)$.

The input SNR is

$$\text{iSNR} = 10\log\frac{\sum_{k=1}^{K} A_k^2}{2\sigma_v^2} \quad \text{(dB)}.$$

The correlation matrix of $\mathbf{v}(t)$ is $\mathbf{R}_{\mathbf{v}} = \sigma_v^2\mathbf{I}_L$ and the elements of the correlation matrix of $\mathbf{x}(t)$ are $\left[\mathbf{R}_{\mathbf{x}}\right]_{i,j} = \frac{1}{2}\sum_{k=1}^{K} A_k^2\cos\left[2\pi f_k(i-j)\right]$. The rank of this matrix is $\text{rank}\left(\mathbf{R}_{\mathbf{x}}\right) = 2K$. The MVDR filter, $\mathbf{h}_{\text{MVDR}}$, for the case of white noise is obtained from (2.67).

To demonstrate the performance of the MVDR filter, we choose $A_k = 0.5$ ($k = 1, \ldots, K$), $f_k = 0.05\,k$ ($k = 1, \ldots, K$), a filter length of $L = 30$, and several values of K. The dimension of the nullspace of $\mathbf{R}_{\mathbf{x}}$ is $L - 2K$. Figure 2.7 shows plots of the gain in SNR, $\mathcal{G}\left(\mathbf{h}_{\text{MVDR}}\right)$, the MSE, $J\left(\mathbf{h}_{\text{MVDR}}\right)$, the noise reduction factor, $\xi_{\text{n}}\left(\mathbf{h}_{\text{MVDR}}\right)$, and the desired signal reduction factor, $\xi_{\text{d}}\left(\mathbf{h}_{\text{MVDR}}\right)$, as a function of the input SNR, for several values of K. Clearly, the desired signal reduction factor is zero, and the higher the dimension of the nullspace of $\mathbf{R}_{\mathbf{x}}$ (smaller K), the higher the noise reduction factor. ■

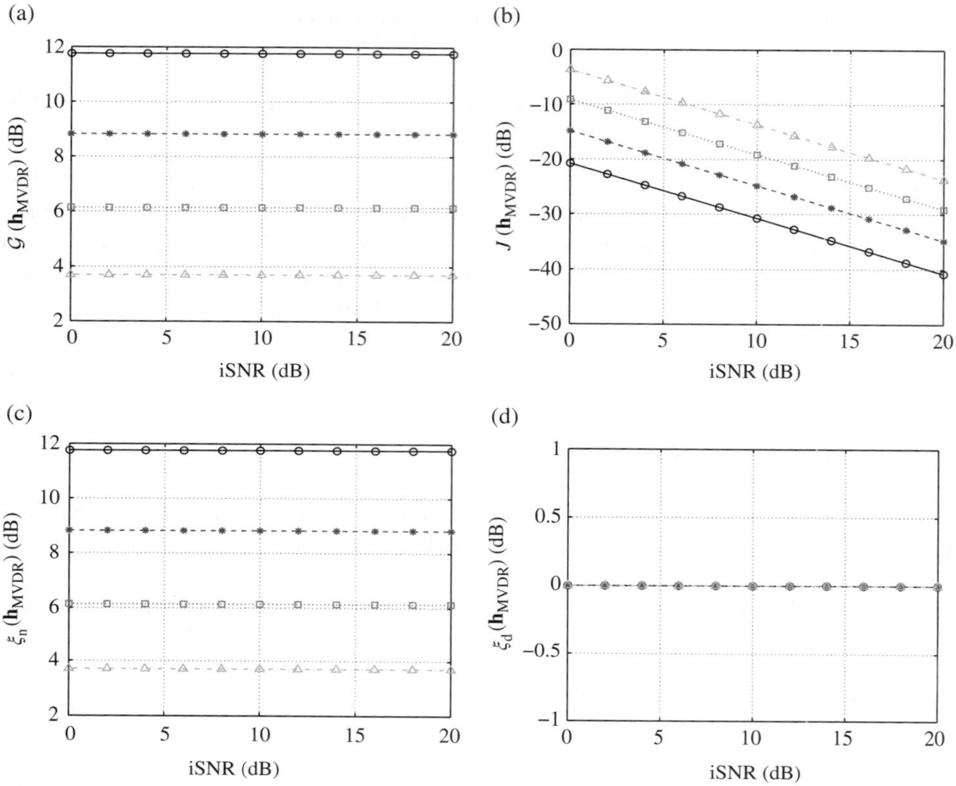

Figure 2.7 (a) The gain in SNR, (b) the MSE, (c) the noise reduction factor, and (d) the desired signal reduction factor of the MVDR filter as a function of the input SNR for several desired signals with different values of K: $K = 1$ (solid line with circles), $K = 2$ (dashed line with asterisks), $K = 4$ (dotted line with squares), and $K = 8$ (dash-dot line with triangles).

With the eigenvalue decomposition of $\mathbf{R}_\mathbf{x}$, the correlation matrix of the observation signal vector can be written as

$$\mathbf{R}_\mathbf{y} = \mathbf{Q}'_\mathbf{x}\mathbf{\Lambda}'_\mathbf{x}\mathbf{Q}'^T_\mathbf{x} + \mathbf{R}_\mathbf{v}, \tag{2.68}$$

where

$$\mathbf{\Lambda}'_\mathbf{x} = \text{diag}\left(\lambda_{\mathbf{x},1}, \lambda_{\mathbf{x},2}, \ldots, \lambda_{\mathbf{x},P}\right). \tag{2.69}$$

Determining the inverse of $\mathbf{R}_\mathbf{y}$ from (2.68) with the Woodbury's identity, we get

$$\mathbf{R}_\mathbf{y}^{-1} = \mathbf{R}_\mathbf{v}^{-1} - \mathbf{R}_\mathbf{v}^{-1}\mathbf{Q}'_\mathbf{x}\left(\mathbf{\Lambda}'^{-1}_\mathbf{x} + \mathbf{Q}'^T_\mathbf{x}\mathbf{R}_\mathbf{v}^{-1}\mathbf{Q}'_\mathbf{x}\right)^{-1}\mathbf{Q}'^T_\mathbf{x}\mathbf{R}_\mathbf{v}^{-1}. \tag{2.70}$$

Substituting (2.70) into (2.30), leads to another useful formulation of the Wiener filter:

$$\mathbf{h}_\mathrm{W} = \mathbf{R}_\mathbf{v}^{-1}\mathbf{Q}'_\mathbf{x}\left(\mathbf{\Lambda}'^{-1}_\mathbf{x} + \mathbf{Q}'^T_\mathbf{x}\mathbf{R}_\mathbf{v}^{-1}\mathbf{Q}'_\mathbf{x}\right)^{-1}\mathbf{Q}'^T_\mathbf{x}\mathbf{i}_\mathrm{i}. \tag{2.71}$$

This formulation shows how the MVDR and Wiener filters are closely related. In the same way, we can express the tradeoff filter as

$$\mathbf{h}_{\mathrm{T},\mu} = \mathbf{R}_{\mathbf{v}}^{-1}\mathbf{Q}_{\mathbf{x}}' \left(\mu\mathbf{\Lambda}_{\mathbf{x}}'^{-1} + \mathbf{Q}_{\mathbf{x}}'^{T}\mathbf{R}_{\mathbf{v}}^{-1}\mathbf{Q}_{\mathbf{x}}'\right)^{-1} \mathbf{Q}_{\mathbf{x}}'^{T}\mathbf{i}_{\mathrm{i}}. \tag{2.72}$$

This filter is strictly equivalent to the tradeoff filter given in (2.54), except for $\mu = 0$, where the two give different results when $\mathbf{R}_{\mathbf{x}}$ is not full rank; the one in (2.54) leads to the identity filter while the one in (2.72) leads to the MVDR filter. In fact, the filter given in (2.54) is not defined for $\mu = 0$ and when $\mathbf{R}_{\mathbf{x}}$ is not full rank.

So far, we have shown how to exploit the MSE criterion to derive all kinds of useful optimal filters. However, we can also exploit the definition of the output SNR to derive the so-called maximum SNR filter. Let us denote by λ_1 the maximum eigenvalue of the matrix $\mathbf{R}_{\mathbf{v}}^{-1}\mathbf{R}_{\mathbf{x}}$ and by $\mathbf{t}_1$ the corresponding eigenvector. The maximum SNR filter, $\mathbf{h}_{\mathrm{max}}$, is obtained by maximizing the output SNR as given in (2.14), from which we recognize the generalized Rayleigh quotient [9]. It is well known that this quotient is maximized with the eigenvector corresponding to the maximum eigenvalue of $\mathbf{R}_{\mathbf{v}}^{-1}\mathbf{R}_{\mathbf{x}}$. Therefore, we have

$$\mathbf{h}_{\mathrm{max}} = \varsigma\mathbf{t}_1, \tag{2.73}$$

where $\varsigma \neq 0$ is an arbitrary real number. We deduce that

$$\mathrm{oSNR}\left(\mathbf{h}_{\mathrm{max}}\right) = \lambda_1. \tag{2.74}$$

Clearly, we always have

$$\mathrm{oSNR}\left(\mathbf{h}_{\mathrm{max}}\right) \geq \mathrm{iSNR} \tag{2.75}$$

and

$$\mathrm{oSNR}\left(\mathbf{h}_{\mathrm{max}}\right) \geq \mathrm{oSNR}\left(\mathbf{h}\right), \ \forall\mathbf{h}. \tag{2.76}$$

While the maximum SNR filter maximizes the output SNR, it leads to large distortions of the desired signal.

Let us consider the very particular case of a matrix $\mathbf{R}_{\mathbf{v}}^{-1}\mathbf{R}_{\mathbf{x}}$ that has a maximum eigenvalue λ_1 with multiplicity $P \leq L$. We denote by $\mathbf{t}_1, \mathbf{t}_2, \ldots, \mathbf{t}_P$ the corresponding eigenvectors. It is not hard to see that the maximum SNR filter is now

$$\mathbf{h}_{\mathrm{max}} = \sum_{p=1}^{P} \varsigma_p\mathbf{t}_p, \tag{2.77}$$

since

$$\mathrm{oSNR}\left(\mathbf{h}_{\mathrm{max}}\right) = \lambda_1, \tag{2.78}$$

where ς_p, $p = 1, 2, \ldots, P$ are real numbers with at least one of them different from 0.

Table 2.1 Optimal linear filters for single-channel signal enhancement in the time domain.

Filter	
Wiener	$\mathbf{h}_{\mathrm{W}} = \mathbf{R}_{\mathbf{y}}^{-1}\mathbf{R}_{\mathbf{x}}\mathbf{i}_{\mathrm{i}}$
Tradeoff	$\mathbf{h}_{\mathrm{T},\mu} = \left(\mathbf{R}_{\mathbf{x}} + \mu\mathbf{R}_{\mathbf{v}}\right)^{-1}\mathbf{R}_{\mathbf{x}}\mathbf{i}_{\mathrm{i}},\ \mu \geq 0$
MVDR	$\mathbf{h}_{\mathrm{MVDR}} = \mathbf{R}_{\mathbf{v}}^{-1}\mathbf{Q}'_{\mathbf{x}}\left(\mathbf{Q}'^{T}_{\mathbf{x}}\mathbf{R}_{\mathbf{v}}^{-1}\mathbf{Q}'_{\mathbf{x}}\right)^{-1}\mathbf{Q}'^{T}_{\mathbf{x}}\mathbf{i}_{\mathrm{i}}$
Maximum SNR	$\mathbf{h}_{\max} = \varsigma\mathbf{t}_1,\ \varsigma \neq 0$

To summarize the performance of all the optimal filters derived in this subsection, we can state that for $\mu < 1$,

$$\mathrm{oSNR}\left(\mathbf{h}_{\max}\right) \geq \mathrm{oSNR}\left(\mathbf{h}_{\mathrm{W}}\right) \geq \mathrm{oSNR}\left(\mathbf{h}_{\mathrm{T},\mu}\right) \geq \mathrm{oSNR}\left(\mathbf{h}_{\mathrm{MVDR}}\right), \tag{2.79}$$

and for $\mu > 1$,

$$\mathrm{oSNR}\left(\mathbf{h}_{\max}\right) \geq \mathrm{oSNR}\left(\mathbf{h}_{\mathrm{T},\mu}\right) \geq \mathrm{oSNR}\left(\mathbf{h}_{\mathrm{W}}\right) \geq \mathrm{oSNR}\left(\mathbf{h}_{\mathrm{MVDR}}\right). \tag{2.80}$$

In Table 2.1, we summarize all the optimal filters described in this subsection.

Example 2.2.4 Consider a desired signal consisting of four harmonic random processes:

$$x(t) = \sum_{k=1}^{4} A_k \cos\left(2\pi f_k t + \phi_k\right),$$

with fixed amplitudes $\left\{A_k\right\}$ and frequencies $\left\{f_k\right\}$, and IID random phases $\left\{\phi_k\right\}$, uniformly distributed on the interval from 0 to 2π. This signal needs to be recovered from the noisy observation $y(t) = x(t) + v(t)$, where $v(t)$ is additive white Gaussian noise, $v(t) \sim \mathcal{N}\left(0, \sigma_v^2\right)$, which is uncorrelated with $x(t)$.

The input SNR is

$$\mathrm{iSNR} = 10 \log \frac{\sum_{k=1}^{4} A_k^2}{2\sigma_v^2} \quad \text{(dB)}.$$

The correlation matrix of $\mathbf{v}(t)$ is $\mathbf{R}_{\mathbf{v}} = \sigma_v^2\mathbf{I}_L$ and the elements of the correlation matrix of $\mathbf{x}(t)$ are $\left[\mathbf{R}_{\mathbf{x}}\right]_{i,j} = \frac{1}{2}\sum_{k=1}^{4} A_k^2 \cos\left[2\pi f_k(i-j)\right]$. The rank of this matrix is $\mathrm{rank}\left(\mathbf{R}_{\mathbf{x}}\right) = 8$.

To demonstrate the performances of the optimal filters, we choose $A_k = 0.5$ $(k = 1, \ldots, 4)$, $f_k = 0.1 + 0.03(k-1)$ $(k = 1, \ldots, 4)$, and a filter length of $L = 20$. The value of ς in (2.73) is chosen to minimize the MSE. Substituting (2.73) into (2.25) we have

$$J\left(\mathbf{h}_{\max}\right) = \sigma_x^2 - 2\varsigma\mathbf{t}_1^T\mathbf{R}_{\mathbf{x}}\mathbf{i}_{\mathrm{i}} + \varsigma^2\mathbf{t}_1^T\mathbf{R}_{\mathbf{y}}\mathbf{t}_1.$$

Taking the derivative of the MSE with respect to ς and equating the result to zero, we get

$$\varsigma = \frac{\mathbf{t}_1^T \mathbf{R}_\mathbf{x} \mathbf{i}_\mathrm{i}}{\mathbf{t}_1^T \mathbf{R}_\mathbf{y} \mathbf{t}_1}.$$

Figures 2.8 and 2.9 show plots of the gain in SNR, $\mathcal{G}(\mathbf{h})$, the MSE, $J(\mathbf{h})$, the noise reduction factor, $\xi_\mathrm{n}(\mathbf{h})$, and the desired signal reduction factor, $\xi_\mathrm{d}(\mathbf{h})$, as a function of the input SNR, for all the optimal filters derived in this subsection: the maximum SNR, Wiener, MVDR, and tradeoff filters. In Figure 2.8 the Lagrange multiplier of the tradeoff filter is $\mu = 0.5$, whereas in Figure 2.9, $\mu = 5$. Clearly, (2.79) is satisfied in Figure 2.8a, whereas (2.80) is satisfied in Figure 2.9a. Specifically, the maximum oSNR is obtained with $\mathbf{h}_\mathrm{max}$, the minimum oSNR is obtained with $\mathbf{h}_\mathrm{MVDR}$, and oSNR is larger when applying the Wiener filter than the tradeoff filter if $\mu < 1$, while the opposite is true if $\mu > 1$. Furthermore, the MSE is minimal for the Wiener filter, and the desired

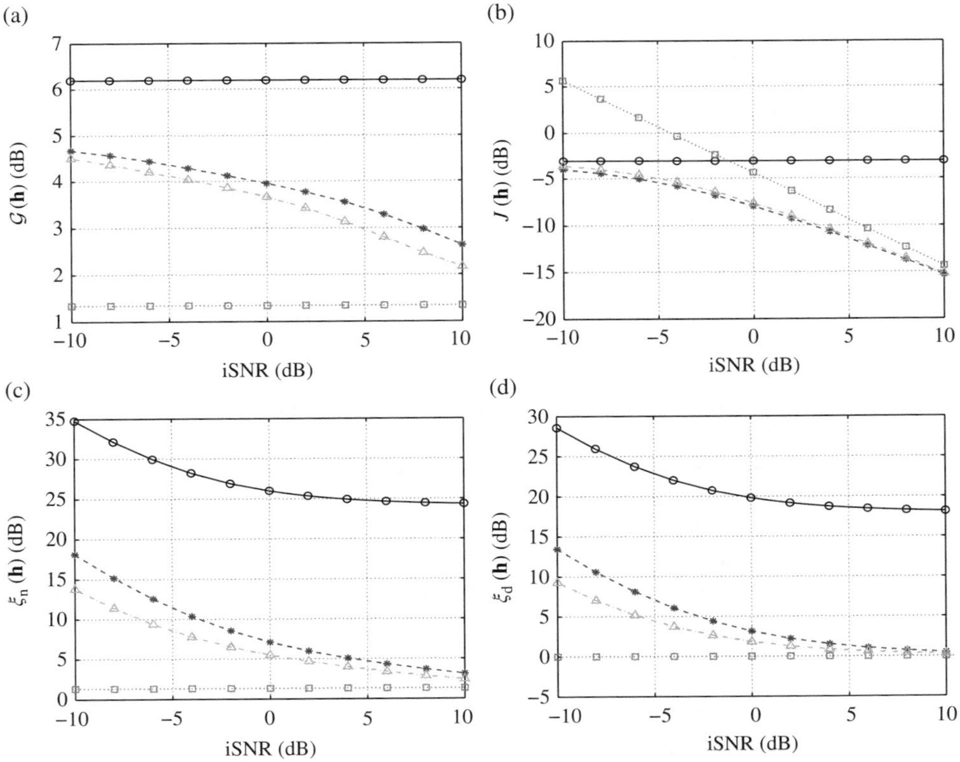

Figure 2.8 (a) The gain in SNR, (b) the MSE, (c) the noise reduction factor, and (d) the desired signal reduction factor as a function of the input SNR for different optimal filters: $\mathbf{h}_\mathrm{max}$ (solid line with circles), $\mathbf{h}_\mathrm{W}$ (dashed line with asterisks), $\mathbf{h}_\mathrm{MVDR}$ (dotted line with squares), and $\mathbf{h}_{\mathrm{T},\mu}$ with $\mu = 0.5$ (dash-dot line with triangles).

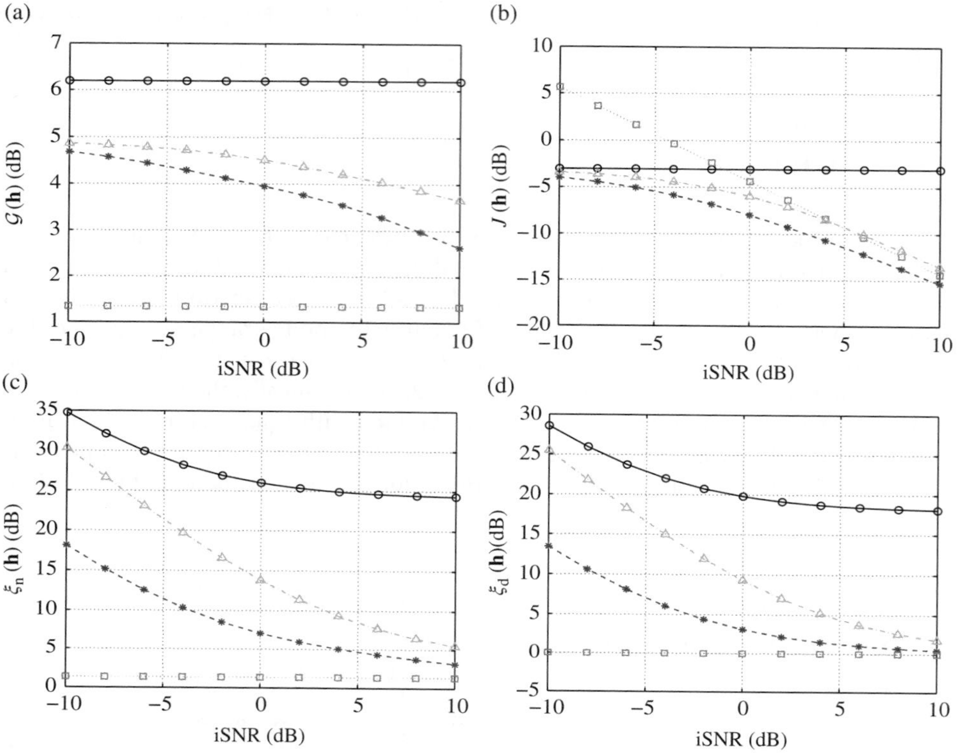

Figure 2.9 (a) The gain in SNR, (b) the MSE, (c) the noise reduction factor, and (d) the desired signal reduction factor as a function of the input SNR for different optimal filters: $\mathbf{h}_{\text{max}}$ (solid line with circles), $\mathbf{h}_{\text{W}}$ (dashed line with asterisks), $\mathbf{h}_{\text{MVDR}}$ (dotted line with squares), and $\mathbf{h}_{\text{T},\mu}$ with $\mu = 5$ (dash-dot line with triangles).

signal reduction factor is 0 dB for the MVDR filter, since the desired signal correlation matrix is rank deficient.

Suppose that we wish to design a tradeoff filter, $\mathbf{h}_{\text{T},\mu}$, that satisfies $\xi_{\text{n}}\left(\mathbf{h}_{\text{T},\mu}\right) = 10$ dB. A plot of the Lagrange multiplier, μ, that satisfies this constraint is shown in Figure 2.10. Plots of the gain in SNR, the MSE, the noise reduction factor, and the desired signal reduction factor, under this constraint, are shown in Figure 2.11. In this scenario, $\mu < 1$ for iSNR < -3.61 dB, and $\mu > 1$ for iSNR > -3.61 dB. Hence, oSNR $\left(\mathbf{h}_{\text{W}}\right) \geq$ oSNR $\left(\mathbf{h}_{\text{T},\mu}\right)$ for iSNR < -3.61 dB, and oSNR $\left(\mathbf{h}_{\text{W}}\right) \leq$ oSNR $\left(\mathbf{h}_{\text{T},\mu}\right)$ for iSNR > -3.61 dB. ■

2.3 Spectral Method

In this section, we give a spectral perspective of the single-channel signal enhancement problem and try to unify several known approaches. To this end, we use the joint diagonalization technique.

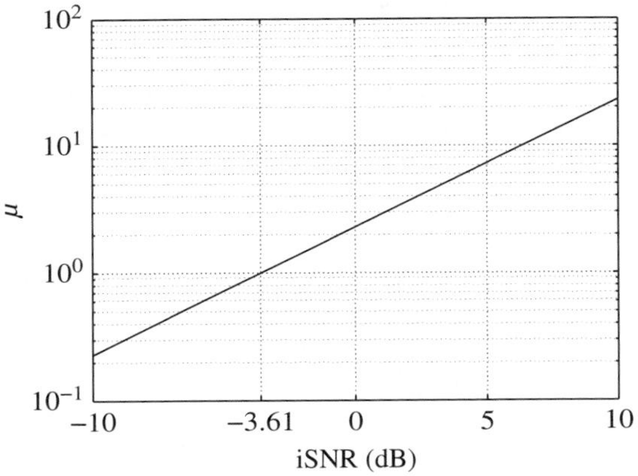

Figure 2.10 The Lagrange multiplier, μ, of the tradeoff filter, $\mathbf{h}_{T,\mu}$, as a function of the input SNR, which yields a constant noise reduction factor $\xi_n\left(\mathbf{h}_{T,\mu}\right) = 10$ dB.

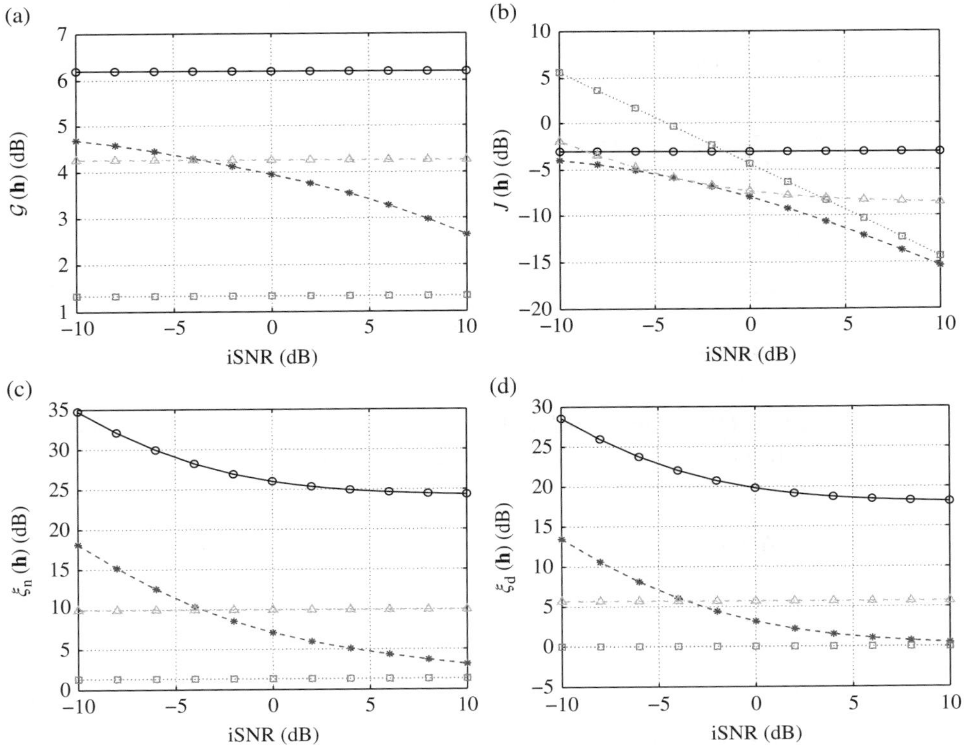

Figure 2.11 (a) The gain in SNR, (b) the MSE, (c) the noise reduction factor, and (d) the desired signal reduction factor as a function of the input SNR for different optimal filters: $\mathbf{h}_{max}$ (solid line with circles), $\mathbf{h}_W$ (dashed line with asterisks), $\mathbf{h}_{MVDR}$ (dotted line with squares), and $\mathbf{h}_{T,\mu}$ with μ that satisfies $\xi_n\left(\mathbf{h}_{T,\mu}\right) = 10$ dB (dash-dot line with triangles).

2.3.1 Joint Diagonalization and Reformulation of the Problem

The use of the joint diagonalization is going to be very useful in the rest of this chapter and will help us reformulate the original time-domain problem to the spectral domain, so we now briefly explain how it works. The two symmetric matrices $\mathbf{R_x}$ and $\mathbf{R_v}$ can be jointly diagonalized as follows [10]:

$$\mathbf{T}^T\mathbf{R_x}\mathbf{T} = \mathbf{\Lambda}, \tag{2.81}$$

$$\mathbf{T}^T\mathbf{R_v}\mathbf{T} = \mathbf{I}_L, \tag{2.82}$$

where $\mathbf{T}$ is a full-rank square matrix (of size $L \times L$) and $\mathbf{\Lambda}$ is a diagonal matrix the main elements of which are real and nonnegative. The procedure for jointly diagonalizing $\mathbf{R_x}$ and $\mathbf{R_v}$ consists of two steps:

i) Calculate $\mathbf{\Lambda}$ and $\mathbf{T}'$, the eigenvalue and eigenvector matrices, respectively, of $\mathbf{R_v}^{-1}\mathbf{R_x}$:

$$\mathbf{R_v}^{-1}\mathbf{R_x}\mathbf{T}' = \mathbf{T}'\mathbf{\Lambda}. \tag{2.83}$$

ii) Normalize the eigenvectors of $\mathbf{R_v}^{-1}\mathbf{R_x}$ such that (2.82) is satisfied. Denoting by $\mathbf{t}'_l$, $l = 1, \ldots, L$ the eigenvectors of $\mathbf{R_v}^{-1}\mathbf{R_x}$, we need to find the constants c_l such that $\mathbf{t}_l = c_l\mathbf{t}'_l$ satisfy $\mathbf{t}_l^T\mathbf{R_v}\mathbf{t}_l = 1$. Hence,

$$c_l = \frac{1}{\sqrt{\mathbf{t}_l'^T\mathbf{R_v}\mathbf{t}'_l}}, \; l = 1, \ldots, L. \tag{2.84}$$

Thus, we have

$$\mathbf{T} = \mathbf{T}'\mathbf{C}, \tag{2.85}$$

where $\mathbf{C}$ is a diagonal normalization matrix with the elements $\{c_1, \ldots, c_L\}$ on its main diagonal.

The eigenvalues of $\mathbf{R_v}^{-1}\mathbf{R_x}$ can be ordered as $\lambda_1 \geq \lambda_2 \geq \cdots \geq \lambda_L \geq 0$. We also denote by $\mathbf{t}_1, \mathbf{t}_2, \ldots, \mathbf{t}_L$, the corresponding eigenvectors. Therefore, the noisy signal correlation matrix can also be diagonalized as

$$\mathbf{T}^T\mathbf{R_y}\mathbf{T} = \mathbf{\Lambda} + \mathbf{I}_L. \tag{2.86}$$

By left multiplying both sides of (2.2) by $\mathbf{t}_l^T$, we get the lth spectral mode of the noisy signal:

$$\begin{aligned} y(t;l) &= \mathbf{t}_l^T\mathbf{y}(t) \\ &= x(t;l) + v(t;l), \end{aligned} \tag{2.87}$$

where $x(t;l) = \mathbf{t}_l^T\mathbf{x}(t)$ and $v(t;l) = \mathbf{t}_l^T\mathbf{v}(t)$ are the lth spectral modes of the desired and noise signals, respectively. We deduce that the variance of $y(t;l)$ is

$$\begin{aligned} \sigma_y^2(;l) &= E\left[y^2(t;l)\right] \\ &= \mathbf{t}_l^T\mathbf{R_y}\mathbf{t}_l \end{aligned} \tag{2.88}$$

$$= \lambda_l + 1$$
$$= \sigma_x^2(;l) + \sigma_v^2(;l),$$

where $\sigma_x^2(;l) = \lambda_l$ and $\sigma_v^2(;l) = 1$ are the variances of $x(t;l)$ and $v(t;l)$, respectively. Now, we consider (2.87) as our new signal model and, therefore, our aim is to recover $x(t;l)$ given $y(t;l)$.

We end this part by giving two useful properties.

Property 2.3.1 Let $\lambda_1 \geq \lambda_2 \geq \cdots \geq \lambda_L \geq 0$. We have

$$\frac{\sum_{i=1}^{L} \alpha_i^2 \lambda_i}{\sum_{i=1}^{L} \alpha_i^2} \leq \frac{\sum_{i=1}^{L-1} \alpha_i^2 \lambda_i}{\sum_{i=1}^{L-1} \alpha_i^2} \leq \cdots \leq \frac{\sum_{i=1}^{2} \alpha_i^2 \lambda_i}{\sum_{i=1}^{2} \alpha_i^2} \leq \lambda_1, \tag{2.89}$$

where α_i, $i = 1, 2, \ldots, L$ are arbitrary real numbers with at least one of them different from 0.

Proof. These inequalities can be easily shown by induction. ■

Property 2.3.2 Let $\lambda_1 \geq \lambda_2 \geq \cdots \geq \lambda_L \geq 0$. We have

$$\lambda_L \leq \frac{\sum_{i=1}^{2} \beta_{L+1-i}^2 \lambda_{L+1-i}}{\sum_{i=1}^{2} \beta_{L+1-i}^2} \leq \cdots \leq$$
$$\frac{\sum_{i=1}^{L-1} \beta_{L+1-i}^2 \lambda_{L+1-i}}{\sum_{i=1}^{L-1} \beta_{L+1-i}^2} \leq \frac{\sum_{i=1}^{L} \beta_{L+1-i}^2 \lambda_{L+1-i}}{\sum_{i=1}^{L} \beta_{L+1-i}^2}, \tag{2.90}$$

where β_{L+1-i}, $i = 1, 2, \ldots, L$ are arbitrary real numbers with at least one of them different from 0.

Proof. These inequalities can be easily shown by induction. ■

2.3.2 Noise Reduction with Gains

The simplest way to perform noise reduction is, as shown in Figure 2.12, by applying a real-valued gain, $h(;l)$, to the observation, $y(t;l)$:

$$z(t;l) = h(;l)y(t;l) \tag{2.91}$$
$$= x_{\text{fd}}(t;l) + v_{\text{fn}}(t;l),$$

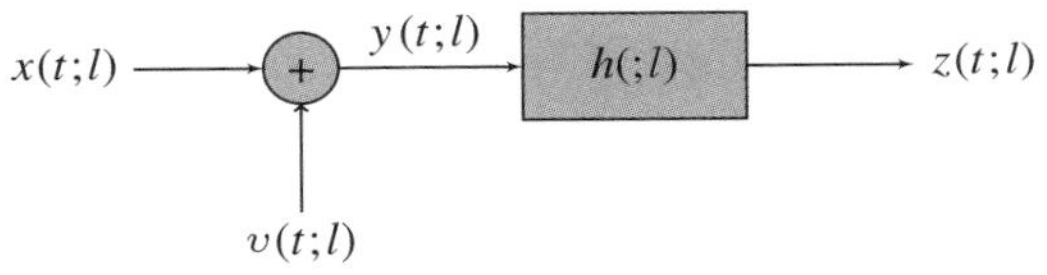

Figure 2.12 Block diagram of noise reduction with gains.

where $z(t;l)$ can be either the estimate of $x(t;l)$ or $v(t;l)$,

$$x_{\rm fd}(t;l) = h(;l)x(t;l) \tag{2.92}$$

is the filtered desired signal, and

$$v_{\rm fn}(t;l) = h(;l)v(t;l) \tag{2.93}$$

is the filtered noise. If $z(t;l)$ is the estimate of $v(t;l)$, then the estimate of $x(t;l)$ is

$$\hat{x}(t;l) = y(t;l) - z(t;l). \tag{2.94}$$

The technique in (2.91) is equivalent to the frequency-domain methods, where a gain is also used at each frequency to reduce noise (see Chapter 3). The variance of $z(t;l)$ is then

$$\begin{aligned} \sigma_z^2(;l) &= h^2(;l)\sigma_y^2(;l) \\ &= \sigma_{x_{\rm fd}}^2(;l) + \sigma_{v_{\rm fn}}^2(;l), \end{aligned} \tag{2.95}$$

where

$$\begin{aligned} \sigma_{x_{\rm fd}}^2(;l) &= h^2(;l)\sigma_x^2(;l) \\ &= h^2(;l)\lambda_l \end{aligned} \tag{2.96}$$

and

$$\begin{aligned} \sigma_{v_{\rm fn}}^2(;l) &= h^2(;l)\sigma_v^2(;l) \\ &= h^2(;l) \end{aligned} \tag{2.97}$$

are the variances of $x_{\rm fd}(t;l)$ and $v_{\rm fn}(t;l)$, respectively.

Eventually, the vector of length L:

$$\mathbf{z}(t) = \begin{bmatrix} z_1(t) & z_2(t) & \cdots & z_L(t) \end{bmatrix}^T, \tag{2.98}$$

which can be either the estimate of $\mathbf{x}(t)$ or $\mathbf{v}(t)$, is obtained as follows (see Figure 2.13):

$$\begin{aligned} \mathbf{z}(t) &= \mathbf{B}\,\mathrm{diag}\,[\mathbf{h}(;)]\,\mathbf{T}^T\mathbf{y}(t) \\ &= \left[\sum_{l=1}^{L} h(;l)\mathbf{b}_l\mathbf{t}_l^T\right]\mathbf{y}(t), \end{aligned} \tag{2.99}$$

where

$$\begin{aligned} \mathbf{B} &= \mathbf{T}^{-T} \\ &= \begin{bmatrix} \mathbf{b}_1 & \mathbf{b}_2 & \cdots & \mathbf{b}_L \end{bmatrix} \end{aligned} \tag{2.100}$$

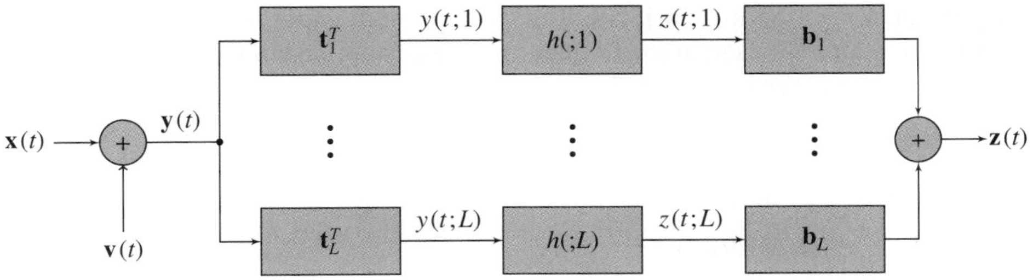

Figure 2.13 Block diagram of spectral mode linear filtering.

and diag $[\mathbf{h}(;)]$ is an $L \times L$ diagonal matrix whose diagonal elements are equal to the components of the vector of length L:

$$\mathbf{h}(;) = \begin{bmatrix} h(;1) & h(;2) & \cdots & h(;L) \end{bmatrix}^T, \tag{2.101}$$

which contains all the spectral mode gains.

2.3.3 Performance Measures

In this subsection, we consider that the filter $\mathbf{h}(;)$ is for the estimation of $\mathbf{T}^T\mathbf{x}(t)$; in other words, $\mathbf{z}(t)$ is an estimate of $\mathbf{x}(t)$.

We define the lth spectral mode input SNR:

$$\begin{aligned} \text{iSNR}(;l) &= \frac{\sigma_x^2(;l)}{\sigma_v^2(;l)} \\ &= \lambda_l \end{aligned} \tag{2.102}$$

and the fullmode input SNR:

$$\begin{aligned} \text{iSNR}(;) &= \frac{\sum_{l=1}^{L} \sigma_x^2(;l)}{\sum_{l=1}^{L} \sigma_v^2(;l)} \\ &= \frac{\text{tr}\,(\boldsymbol{\Lambda})}{L} \\ &= \frac{\text{tr}\left(\mathbf{R}_{\mathbf{v}}^{-1}\mathbf{R}_{\mathbf{x}}\right)}{L}. \end{aligned} \tag{2.103}$$

It can be seen that

$$\text{iSNR}(;L) \leq \text{iSNR}(;) \leq \text{iSNR}(;1). \tag{2.104}$$

In words, the fullmode input SNR can never exceed the maximum spectral mode input SNR and can never go below the minimum spectral mode input SNR. Notice that the definition of the fullmode input SNR is slightly different from the conventional definition of the input SNR, as set out in Equation 2.13. However, for white noise, it is easy to check that $\mathrm{iSNR}(;) = \mathrm{iSNR}$. The spectral mode input SNR is similar to the narrowband input SNR used in the frequency-domain approaches (see Chapter 3).

The lth spectral mode output SNR is

$$\mathrm{oSNR}\left[h(;l)\right] = \frac{\sigma^2_{x_{\mathrm{fd}}}(;l)}{\sigma^2_{v_{\mathrm{fn}}}(;l)} = \mathrm{iSNR}(;l). \tag{2.105}$$

Therefore, the spectral mode output SNR cannot be improved. The fullmode output SNR is

$$\mathrm{oSNR}\left[\mathbf{h}(;)\right] = \frac{\sum_{l=1}^{L} \sigma^2_{x_{\mathrm{fd}}}(;l)}{\sum_{l=1}^{L} \sigma^2_{v_{\mathrm{fn}}}(;l)} = \frac{\mathbf{h}^T(;)\mathbf{\Lambda}\mathbf{h}(;)}{\mathbf{h}^T(;)\mathbf{h}(;)}. \tag{2.106}$$

Clearly, we always have

$$\mathrm{iSNR}(;L) \le \mathrm{oSNR}\left[\mathbf{h}(;)\right] \le \mathrm{iSNR}(;1),\ \forall \mathbf{h}(;). \tag{2.107}$$

Therefore, our aim is to find the L spectral mode gains, $h(;l)$, $l = 1, 2, \ldots, L$, in such a way that the fullmode output SNR is greater than the fullmode input SNR; in other words, $\mathrm{oSNR}\left[\mathbf{h}(;)\right] > \mathrm{iSNR}(;)$.

It is not hard to show that the lth spectral mode and fullmode noise reduction factors are, respectively,

$$\xi_{\mathrm{n}}\left[h(;l)\right] = \frac{1}{h^2(;l)} \tag{2.108}$$

and

$$\xi_{\mathrm{n}}\left[\mathbf{h}(;)\right] = \frac{L}{\mathbf{h}^T(;)\mathbf{h}(;)}. \tag{2.109}$$

For optimal spectral mode gains, we should have $\xi_{\mathrm{n}}\left[h(;l)\right] \ge 1$ and $\xi_{\mathrm{n}}\left[\mathbf{h}(;)\right] \ge 1$. Large values of the noise reduction factors imply good noise reduction.

In the same way, we define the lth spectral mode and fullmode desired signal reduction factors as, respectively,

$$\xi_{\mathrm{d}}\left[h(;l)\right] = \frac{1}{h^2(;l)} \tag{2.110}$$

and

$$\xi_{\mathrm{d}}\left[\mathbf{h}(;)\right] = \frac{\mathrm{tr}\left(\mathbf{\Lambda}\right)}{\mathbf{h}^T(;)\mathbf{\Lambda}\mathbf{h}(;)}. \tag{2.111}$$

For optimal spectral mode gains, we should have $\xi_{\rm d}\left[h(;l)\right] \geq 1$ and $\xi_{\rm d}\left[\mathbf{h}(;)\right] \geq 1$. The closer are the values of the desired signal reduction factors to one, the less distorted is the desired signal.

We deduce the following important relation:

$$\frac{\text{oSNR}\left[\mathbf{h}(;)\right]}{\text{iSNR}(;)} = \frac{\xi_{\rm n}\left[\mathbf{h}(;)\right]}{\xi_{\rm d}\left[\mathbf{h}(;)\right]}. \tag{2.112}$$

It is easy to derive the lth spectral mode desired signal distortion index:

$$\upsilon_{\rm d}\left[h(;l)\right] = \left[h(;l) - 1\right]^2 \tag{2.113}$$

and the fullmode desired signal distortion index:

$$\upsilon_{\rm d}\left[\mathbf{h}(;)\right] = \frac{\sum_{l=1}^{L} \lambda_l \left[h(;l) - 1\right]^2}{\text{tr}\left(\mathbf{\Lambda}\right)}. \tag{2.114}$$

For completeness, we define the fullmode MSE for any filter $\mathbf{h}(;)$ as

$$\begin{aligned} J\left[\mathbf{h}(;)\right] &= E\left\{ \left[\mathbf{T}^T\mathbf{x}(t) - \text{diag}\left[\mathbf{h}(;)\right]\mathbf{T}^T\mathbf{y}(t)\right]^T \right. \\ &\quad \left. \times \left[\mathbf{T}^T\mathbf{x}(t) - \text{diag}\left[\mathbf{h}(;)\right]\mathbf{T}^T\mathbf{y}(t)\right] \right\} \\ &= \text{tr}\left(\mathbf{T}^T\mathbf{R}_{\mathbf{x}}\mathbf{T}\right) - 2\text{tr}\left\{\text{diag}\left[\mathbf{h}(;)\right]\mathbf{T}^T\mathbf{R}_{\mathbf{x}}\mathbf{T}\right\} \\ &\quad + \text{tr}\left\{\text{diag}\left[\mathbf{h}(;)\right]\mathbf{T}^T\mathbf{R}_{\mathbf{y}}\mathbf{T}\text{diag}\left[\mathbf{h}(;)\right]\right\} \\ &= \text{tr}\left(\mathbf{\Lambda}\right) - 2\mathbf{h}^T(;)\mathbf{\Lambda}\mathbf{1} + \mathbf{h}^T(;)\mathbf{\Lambda}\mathbf{h}(;) + \mathbf{h}^T(;)\mathbf{h}(;) \\ &= \left[\mathbf{1} - \mathbf{h}(;)\right]^T \mathbf{\Lambda}\left[\mathbf{1} - \mathbf{h}(;)\right] + \mathbf{h}^T(;)\mathbf{h}(;) \\ &= \text{tr}\left(\mathbf{\Lambda}\right)\upsilon_{\rm d}\left[\mathbf{h}(;)\right] + \frac{L}{\xi_{\rm n}\left[\mathbf{h}(;)\right]} \\ &= J_{\rm d}\left[\mathbf{h}(;)\right] + J_{\rm n}\left[\mathbf{h}(;)\right], \end{aligned} \tag{2.115}$$

where $\mathbf{1}$ is a vector of length L with all its elements equal to 1, which is also the fullmode identity filter since, with it, the fullmode output SNR is equal to the fullmode input SNR. This definition of the fullmode MSE is clearly connected to all the fullmode performance measures.

2.3.4 Determination of the Gains from the Fullmode Output SNR

There are two approaches to find the gains from the fullmode output SNR in order to perform noise reduction. The first one considers the largest spectral mode input SNRs. In this case, we get the estimate of the desired signal directly. The second method considers the smallest spectral mode input SNRs. As a result, we get the estimate of the noise signal, from which we deduce the estimate of the desired signal.

2.3.4.1 Maximization of the Fullmode Output SNR

The filter $\mathbf{h}(;)$ that maximizes the fullmode output SNR given in (2.106) is simply the eigenvector corresponding to the maximum eigenvalue of the matrix $\mathbf{\Lambda}$. Since this matrix is diagonal, its maximum eigenvalue is its largest diagonal element, λ_1. As a consequence, the maximum SNR filter is

$$\mathbf{h}_{\alpha_1}(;) = \alpha_1 \mathbf{i}_1, \tag{2.116}$$

where $\alpha_1 \neq 0$ is an arbitrary real number and $\mathbf{i}_1$ is the first column of $\mathbf{I}_L$. Equivalently, we can write (2.116) as

$$\begin{cases} h_{\alpha_1}(;1) = \alpha_1 \\ h(;i) = 0, \ i = 2, 3, \dots, L \end{cases}. \tag{2.117}$$

With (2.116), we get the maximum possible fullmode output SNR, which is

$$\text{oSNR}\left[\mathbf{h}_{\alpha_1}(;)\right] = \lambda_1 \geq \text{iSNR}(;). \tag{2.118}$$

As a result,

$$\text{oSNR}\left[\mathbf{h}_{\alpha_1}(;)\right] \geq \text{oSNR}\left[\mathbf{h}(;)\right], \ \forall \mathbf{h}(;). \tag{2.119}$$

We deduce that the estimate of the desired signal is

$$\begin{cases} \widehat{x}_{\alpha_1}(t;1) = h_{\alpha_1}(;1) y(t;1) \\ \widehat{x}(t;i) = 0, \ i = 2, 3, \dots, L \end{cases} \tag{2.120}$$

and the estimate of $\mathbf{x}(t)$ is

$$\widehat{\mathbf{x}}_{\alpha_1}(t) = \left[h_{\alpha_1}(;1)\mathbf{b}_1\mathbf{t}_1^T\right]\mathbf{y}(t) \tag{2.121}$$

as illustrated in Figure 2.14.

Now, we need to determine α_1. There are at least two ways to find this parameter. The first one is from the MSE between $x(t;1)$ and $\widehat{x}_{\alpha_1}(t;1)$:

$$J\left[h_{\alpha_1}(;1)\right] = E\left\{\left[x(t;1) - h_{\alpha_1}(;1)y(t;1)\right]^2\right\}. \tag{2.122}$$

The second possibility is to use the desired signal distortion-based MSE:

$$J_\text{d}\left[h_{\alpha_1}(;1)\right] = E\left\{\left[x(t;1) - h_{\alpha_1}(;1)x(t;1)\right]^2\right\}. \tag{2.123}$$

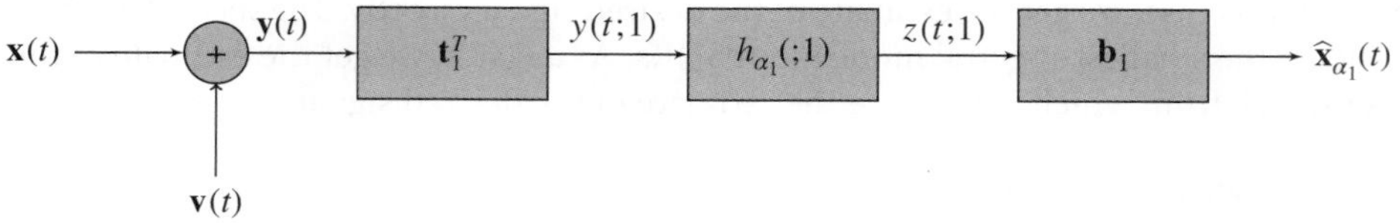

Figure 2.14 Estimation of the desired signal using the maximum SNR filter.

The minimization of $J\left[h_{\alpha_1}(;1)\right]$ leads to the Wiener gain at the spectral mode 1:

$$h_{\mathrm{W}}(;1) = \frac{\mathrm{iSNR}(;1)}{1 + \mathrm{iSNR}(;1)}, \tag{2.124}$$

while the minimization of $J_{\mathrm{d}}\left[h_{\alpha_1}(;1)\right]$ gives the unitary gain at the spectral mode 1:

$$h_{\mathrm{U}}(;1) = 1. \tag{2.125}$$

Notice that $h_{\mathrm{W}}(;l),\ l = 1, 2, \ldots, L$ resembles the Wiener gain approach in frequency-domain noise reduction, which will be discussed in the next chapter. It is obvious that $0 \leq h_{\mathrm{W}}(;1) \leq 1$.

Even though this method maximizes the fullmode output SNR, it is expected to introduce a large amount of distortion to the desired signal, since all its spectral modes are set to 0 except at 1. A much better approach when we deal with broadband signals is to form the filter from a linear combination of the eigenvectors corresponding to the $P(\leq L)$ largest eigenvalues of $\mathbf{\Lambda}$:

$$\mathbf{h}_{\alpha_{1:P}}(;) = \sum_{p=1}^{P} \alpha_p \mathbf{i}_p, \tag{2.126}$$

where $\alpha_p,\ p = 1, 2, \ldots, P$ are arbitrary real numbers with at least one of them different from 0, and $\mathbf{i}_p$ is the pth column of $\mathbf{I}_L$. We can also express (2.126) as

$$\begin{cases} h_{\alpha_p}(;p) = \alpha_p,\ p = 1, 2, \ldots, P \\ h(;i) = 0,\ i = P+1, P+2, \ldots, L \end{cases}. \tag{2.127}$$

Hence, the estimate of the desired signal is

$$\begin{cases} \widehat{x}_{\alpha_p}(t;p) = h_{\alpha_p}(;p)y(t;p),\ p = 1, 2, \ldots, P \\ \widehat{x}(t;i) = 0,\ i = P+1, P+2, \ldots, L \end{cases} \tag{2.128}$$

and the estimate of $\mathbf{x}(t)$ is

$$\widehat{\mathbf{x}}_{\alpha_{1:P}}(t) = \left[\sum_{p=1}^{P} h_{\alpha_p}(;p)\mathbf{b}_p\mathbf{t}_p^T\right]\mathbf{y}(t). \tag{2.129}$$

To find the various α_p, we can optimize either $J\left[h_{\alpha_p}(;p)\right]$ or $J_{\mathrm{d}}\left[h_{\alpha_p}(;p)\right]$. The first one leads to the Wiener gains at the spectral modes $p,\ p = 1, 2, \ldots, P$:

$$h_{\mathrm{W}}(;p) = \frac{\mathrm{iSNR}(;p)}{1 + \mathrm{iSNR}(;p)},\ p = 1, 2, \ldots, P, \tag{2.130}$$

while the second one gives the unitary gains at the spectral modes $p,\ p = 1, 2, \ldots, P$:

$$h_{\mathrm{U}}(;p) = 1,\ p = 1, 2, \ldots, P. \tag{2.131}$$

The filters (of length L) corresponding to (2.130) and (2.131) are, respectively,

$$\mathbf{h}_{\mathrm{W},P}(;) = \begin{bmatrix} h_{\mathrm{W}}(;1) & \cdots & h_{\mathrm{W}}(;P) & 0 \cdots 0 \end{bmatrix}^T \tag{2.132}$$

and

$$\mathbf{h}_{\mathrm{U},P}(;) = \begin{bmatrix} 1 & \cdots & 1 & 0 \cdots 0 \end{bmatrix}^T. \tag{2.133}$$

For $P = L$, $\mathbf{h}_{\mathrm{W},L}(;)$ corresponds to the classical Wiener approach and $\mathbf{h}_{\mathrm{U},L}(;)$ is the identity filter, which does not affect the observations. Indeed, it is easy to check that

$$\widehat{\mathbf{x}}_{\mathrm{W}}(t) = \mathbf{R}_{\mathbf{x}}\mathbf{R}_{\mathbf{y}}^{-1}\mathbf{y}(t), \tag{2.134}$$

$$\widehat{\mathbf{x}}_{\mathrm{U}}(t) = \mathbf{y}(t). \tag{2.135}$$

Clearly, $\mathbf{h}_{\mathrm{U},P}(;)$ corresponds to the ideal binary mask [11], since the spectral mode observation signals with the P largest spectral mode input SNRs are not affected while the $L - P$ others with the smallest spectral mode input SNRs are set to 0. This is also equivalent to the subspace approach, where the desired signal-plus-noise subspace is not processed while the (dominant) noise subspace is cancelled [12–14]. The corresponding estimator is

$$\widehat{\mathbf{x}}_{\mathrm{U},1:P}(t) = \left(\sum_{p=1}^{P} \mathbf{b}_p \mathbf{t}_p^T \right) \mathbf{y}(t). \tag{2.136}$$

We should always have

$$\mathrm{oSNR}\left[\mathbf{h}_{\mathrm{U},P}(;)\right] \leq \mathrm{oSNR}\left[\mathbf{h}_{\mathrm{W},P}(;)\right]. \tag{2.137}$$

From Property 2.3.1, we deduce that

$$\begin{aligned} \mathrm{iSNR}(;) &\leq \mathrm{oSNR}\left[\mathbf{h}_{\mathrm{W},L}(;)\right] \leq \mathrm{oSNR}\left[\mathbf{h}_{\mathrm{W},L-1}(;)\right] \\ &\leq \cdots \leq \mathrm{oSNR}\left[\mathbf{h}_{\mathrm{W},1}(;)\right] = \lambda_1 \end{aligned} \tag{2.138}$$

and

$$\begin{aligned} \mathrm{iSNR}(;) &= \mathrm{oSNR}\left[\mathbf{h}_{\mathrm{U},L}(;)\right] \leq \mathrm{oSNR}\left[\mathbf{h}_{\mathrm{U},L-1}(;)\right] \\ &\leq \cdots \leq \mathrm{oSNR}\left[\mathbf{h}_{\mathrm{U},1}(;)\right] = \lambda_1. \end{aligned} \tag{2.139}$$

Example 2.3.1 Consider a desired signal consisting of five harmonic random processes:

$$x(t) = \sum_{k=1}^{5} A_k \cos\left(2\pi f_k t + \phi_k\right),$$

with fixed amplitudes $\{A_k\}$ and frequencies $\{f_k\}$, and IID random phases $\{\phi_k\}$, uniformly distributed on the interval from 0 to 2π. This signal needs to be recovered

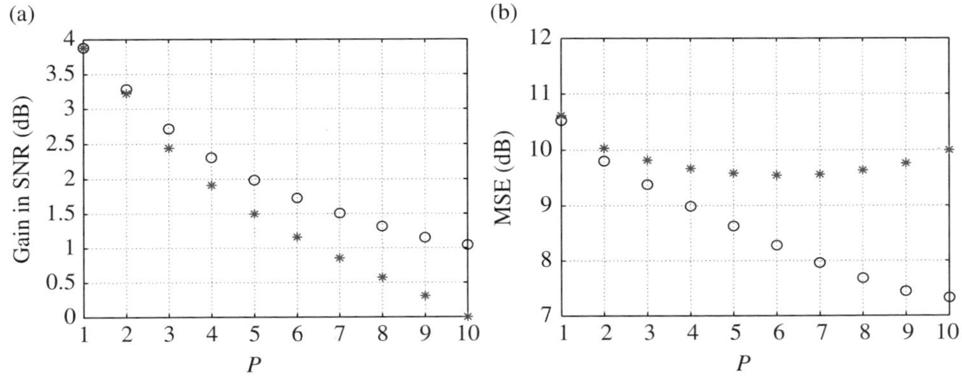

Figure 2.15 (a) The fullmode gain in SNR, $\mathcal{G}\left[\mathbf{h}_{W,P}(;)\right]$ (circles) and $\mathcal{G}\left[\mathbf{h}_{U,P}(;)\right]$ (asterisks), as a function of P; (b) the fullmode MSE, $J\left[\mathbf{h}_{W,P}(;)\right]$ (circles) and $J\left[\mathbf{h}_{U,P}(;)\right]$ (asterisks), as a function of P.

from the noisy observation $y(t) = x(t) + v(t)$, where $v(t)$ is colored noise that is uncorrelated with $x(t)$, whose correlation matrix is $\left[\mathbf{R}_\mathbf{v}\right]_{i,j} = \sigma_v^2 \alpha^{|i-j|}$ $(-1 < \alpha < 1)$.

To demonstrate the performances of the spectral mode gains $\mathbf{h}_{W,P}(;)$ and $\mathbf{h}_{U,P}(;)$, we choose $A_k = 0.5/k$ $(k = 1, \ldots, 5)$, $f_k = 0.05 + 0.1(k-1)$ $(k = 1, \ldots, 5)$, $\sigma^2 = 0.1$, $\alpha = 0.5$, and $L = 10$. The two symmetric matrices $\mathbf{R}_\mathbf{x}$ and $\mathbf{R}_\mathbf{v}$ are jointly diagonalized as in (2.81) and (2.82). The fullmode input SNR is $\text{iSNR}(;) = 10 \log \left[\text{tr}\left(\mathbf{\Lambda}\right)/L\right] = 1.44$ dB.

Figure 2.15 shows plots of the fullmode gain in SNR and MSE for the two spectral mode gains, $\mathbf{h}_{W,P}(;)$ and $\mathbf{h}_{U,P}(;)$, as a function of P. Figure 2.15a verifies that both $\text{oSNR}\left[\mathbf{h}_{W,P}(;)\right]$ and $\text{oSNR}\left[\mathbf{h}_{U,P}(;)\right]$ are decreasing functions of P, and that $\text{oSNR}\left[\mathbf{h}_{U,P}(;)\right] \leq \text{oSNR}\left[\mathbf{h}_{W,P}(;)\right]$. Figure 2.15b shows that the fullmode MSE for the Wiener gains, $J\left[\mathbf{h}_{W,P}(;)\right]$, is a decreasing function of P, and that $J\left[\mathbf{h}_{W,P}(;)\right] \leq J\left[\mathbf{h}_{U,P}(;)\right]$. ■

2.3.4.2 Minimization of the Fullmode Output SNR

It is clear that the filter $\mathbf{h}(;)$ that minimizes the fullmode output SNR given in (2.106) is the eigenvector corresponding to the minimum eigenvalue of the matrix $\mathbf{\Lambda}$, which is λ_L. Therefore, the minimum SNR filter is

$$\mathbf{h}_{\beta_L}(;) = \beta_L \mathbf{i}_L, \tag{2.140}$$

where $\beta_L \neq 0$ is an arbitrary real number and $\mathbf{i}_L$ is the Lth column of $\mathbf{I}_L$. Equivalently, we can write (2.140) as

$$\begin{cases} h(;i) = 0, \ i = 1, 2, \ldots, L-1 \\ h_{\beta_L}(;L) = \beta_L \end{cases}. \tag{2.141}$$

With (2.140), we get the minimum possible fullmode output SNR, which is

$$\text{oSNR}\left[\mathbf{h}_{\beta_L}(;)\right] = \lambda_L \leq \text{iSNR}(;). \tag{2.142}$$

As a result,

$$\text{oSNR}\left[\mathbf{h}_{\beta_L}(;)\right] \leq \text{oSNR}\left[\mathbf{h}(;)\right], \ \forall \mathbf{h}(;). \tag{2.143}$$

We deduce that the estimates of the noise and desired signals are, respectively,

$$\begin{cases} \widehat{v}(t;i) = 0, \ i = 1,2,\ldots,L-1 \\ \widehat{v}_{\beta_L}(t;L) = h_{\beta_L}(;L)y(t;L) \end{cases} \tag{2.144}$$

and

$$\begin{cases} \widehat{x}(t;i) = y(t;i), \ i = 1,2,\ldots,L-1 \\ \widehat{x}_{\beta_L}(t;L) = h'_{\beta_L}(;L)y(t;L) \end{cases}, \tag{2.145}$$

where

$$h'_{\beta_L}(;L) = 1 - h_{\beta_L}(;L) \tag{2.146}$$

is the equivalent gain for the estimation of $x(t;L)$. The equivalent filter is then

$$\mathbf{h}'_{\beta_L}(;) = \mathbf{1} - \mathbf{h}_{\beta_L}(;). \tag{2.147}$$

The fullmode output SNR corresponding to $\mathbf{h}'_{\beta_L}(;)$ is

$$\text{oSNR}\left[\mathbf{h}'_{\beta_L}(;)\right] = \frac{\mathbf{h}'^T_{\beta_L}(;)\mathbf{\Lambda}\mathbf{h}'_{\beta_L}(;)}{\mathbf{h}'^T_{\beta_L}(;)\mathbf{h}'_{\beta_L}(;)}. \tag{2.148}$$

It can be shown that $\text{oSNR}\left[\mathbf{h}'_{\beta_L}(;)\right] \geq \text{iSNR}(;)$ if and only if $\left[1 - h_{\beta_L}(;L)\right]^2 \leq 1$. We also see that the estimate of $\mathbf{v}(t)$ is

$$\widehat{\mathbf{v}}_{\beta_L}(t) = \left[h_{\beta_L}(;L)\mathbf{b}_L\mathbf{t}_L^T\right]\mathbf{y}(t). \tag{2.149}$$

The MSE between $v(t;L)$ and $\widehat{v}_{\beta_L}(t;L)$ is

$$\begin{aligned} J\left[h_{\beta_L}(;L)\right] &= E\left\{\left[v(t;L) - h_{\beta_L}(;L)y(t;L)\right]^2\right\} \\ &= h^2_{\beta_L}(;L)\lambda_L + \left[1 - h_{\beta_L}(;L)\right]^2 \\ &= J_\text{d}\left[h_{\beta_L}(;L)\right] + J_\text{n}\left[h_{\beta_L}(;L)\right]. \end{aligned} \tag{2.150}$$

From the previous expression, we see that there are at least two ways to find $h_{\beta_L}(;L)$ or $h'_{\beta_L}(;L)$. The minimization of $J\left[h_{\beta_L}(;L)\right]$ and using the relation (2.146) lead to

$$\begin{aligned} h'_\text{W}(;L) &= \frac{\text{iSNR}(;L)}{1 + \text{iSNR}(;L)} \\ &= h_\text{W}(;L), \end{aligned} \tag{2.151}$$

which is the Wiener gain at the spectral mode L. The minimization of the power of the residual noise, $J_{\rm n}\left[h_{\beta_L}(;L)\right]$ and using the relation (2.146) give

$$h'_{\rm N}(;L) = 0, \tag{2.152}$$

which is the null gain at the spectral mode L.

Obviously, the approach presented above is not meaningful for broadband signals, since only one spectral mode is processed while all the others are not affected at all. This is far from enough as far as noise reduction is concerned, even though very little distortion is expected. A more practical approach is to form the filter from a linear combination of the eigenvectors corresponding to the $Q(\leq L)$ smallest eigenvalues of $\mathbf{\Lambda}$:

$$\mathbf{h}_{\beta_{L-Q+1:L}}(;) = \sum_{q=1}^{Q} \beta_{L-Q+q}\mathbf{i}_{L-Q+q}, \tag{2.153}$$

where β_{L-Q+q}, $q = 1, 2, \ldots, Q$ are arbitrary real numbers with at least one of them different from 0 and $\mathbf{i}_{L-Q+q}$ is the $(L-Q+q)$th column of $\mathbf{I}_L$. Therefore, the equivalent filter for the estimation of the desired signal at the different spectral modes is

$$\mathbf{h}'_{\beta_{L-Q+1:L}}(;) = \mathbf{1} - \mathbf{h}_{\beta_{L-Q+1:L}}(;). \tag{2.154}$$

We can also express (2.154) as

$$\begin{cases} h'(;i) = 1, \ i = 1, 2, \ldots, L-Q \\ h'_{\beta_{L-Q+q}}(;L-Q+q) = 1 - \beta_{L-Q+q}, \ q = 1, 2, \ldots, Q \end{cases}. \tag{2.155}$$

Hence, the estimate of the desired signal is

$$\begin{cases} \widehat{x}(t;i) = y(t;i), \ i = 1, 2, \ldots, L-Q \\ \widehat{x}_{\beta_{L-Q+q}}(t;L-Q+q) = h'_{\beta_{L-Q+q}}(;L-Q+q)y(t;L-Q+q) \\ \qquad q = 1, 2, \ldots, Q \end{cases}. \tag{2.156}$$

Following the same steps as above, we deduce the two filters of interest:

$$\mathbf{h}'_{\rm W,Q}(;) = \begin{bmatrix} 1 & \cdots & 1 & h_{\rm W}(;L-Q+1) & \cdots & h_{\rm W}(;L) \end{bmatrix}^T \tag{2.157}$$

and

$$\mathbf{h}'_{\rm N,Q} = \begin{bmatrix} 1 & \cdots & 1 & 0 & \cdots & 0 \end{bmatrix}^T. \tag{2.158}$$

For $Q = L$, $\mathbf{h}'_{{\rm W},L}(;) = \mathbf{h}_{{\rm W},L}(;)$ corresponds to the classical Wiener approach and $\mathbf{h}'_{{\rm N},L}(;) = \mathbf{0}$ is the null filter, which completely cancels the observations. The filter $\mathbf{h}'_{\rm W,Q}(;)$ can be seen as a combination of the ideal binary mask and Wiener, where the observations with large spectral mode input SNRs are not affected while the ones with small spectral mode input SNRs are processed with Wiener gains. The filter $\mathbf{h}'_{\mathrm{N},Q}(;)$ is, obviously, the ideal binary mask. The estimator corresponding to $\mathbf{h}'_{\mathrm{W},Q}(;)$ is then

$$\widehat{\mathbf{x}}_{\mathrm{W},Q}(t) = \left(\sum_{i=1}^{L-Q} \mathbf{b}_i \mathbf{t}_i^T + \sum_{q=1}^{Q} \frac{\lambda_{L-Q+q}}{1+\lambda_{L-Q+q}} \mathbf{b}_{L-Q+q} \mathbf{t}_{L-Q+q}^T \right) \mathbf{y}(t). \tag{2.159}$$

We should always have

$$\mathrm{oSNR}\left[\mathbf{h}'_{\mathrm{N},Q}(;)\right] \geq \mathrm{oSNR}\left[\mathbf{h}'_{\mathrm{W},Q}(;)\right]. \tag{2.160}$$

We can also deduce that

$$\mathrm{oSNR}\left[\mathbf{h}'_{\mathrm{N},L-1}(;)\right] \geq \cdots \geq \mathrm{oSNR}\left[\mathbf{h}'_{\mathrm{N},1}(;)\right] \geq \mathrm{iSNR}(;). \tag{2.161}$$

Example 2.3.2 Returning to Example 2.3.1, Figure 2.16 shows plots of the fullmode gain in SNR and MSE for the two spectral mode gains, $\mathbf{h}'_{\mathrm{W},Q}(;)$ and $\mathbf{h}'_{\mathrm{N},Q}(;)$, as a function of Q. Figure 2.16a demonstrates that $\mathrm{oSNR}\left[\mathbf{h}'_{\mathrm{N},Q}(;)\right]$ is an increasing function of Q and that $\mathrm{oSNR}\left[\mathbf{h}'_{\mathrm{N},Q}(;)\right] \geq \mathrm{oSNR}\left[\mathbf{h}'_{\mathrm{W},Q}(;)\right]$. Figure 2.16b shows that the fullmode MSE for the Wiener gains, $J\left[\mathbf{h}'_{\mathrm{W},Q}(;)\right]$, is a decreasing function of Q and that $J\left[\mathbf{h}'_{\mathrm{W},Q}(;)\right] \leq J\left[\mathbf{h}'_{\mathrm{N},Q}(;)\right]$. ■

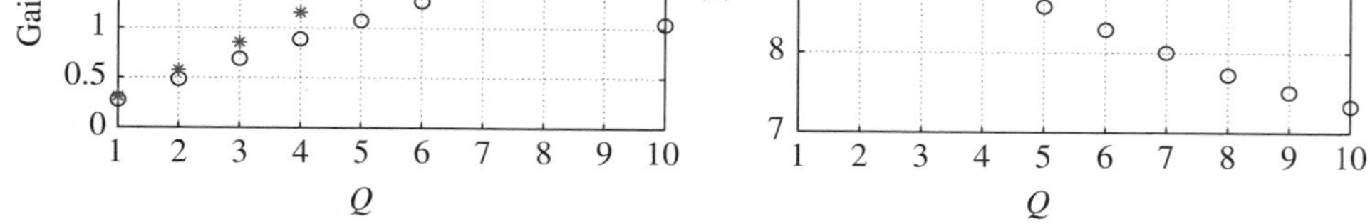

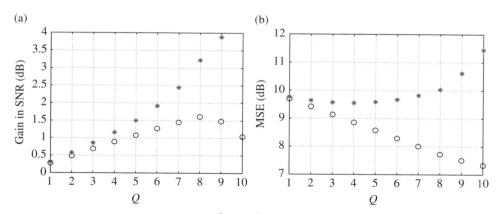

Figure 2.16 (a) The fullmode gain in SNR, $\mathcal{G}\left[\mathbf{h}'_{\mathrm{W},Q}(;)\right]$ (circles) and $\mathcal{G}\left[\mathbf{h}'_{\mathrm{N},Q}(;)\right]$ (asterisks), as a function of Q. (b) Plots of the fullmode MSE, $J\left[\mathbf{h}'_{\mathrm{W},Q}(;)\right]$ (circles) and $J\left[\mathbf{h}'_{\mathrm{N},Q}(;)\right]$ (asterisks), as a function of Q.

Problems

2.1 Show that the MSEs, $J(\mathbf{h})$, $J_{\mathrm{d}}(\mathbf{h})$, and $J_{\mathrm{n}}(\mathbf{h})$, are related to the different performance measures by

$$J(\mathbf{h}) = \sigma_v^2 \left[\mathrm{iSNR} \times v_{\mathrm{d}}(\mathbf{h}) + \frac{1}{\xi_{\mathrm{n}}(\mathbf{h})} \right]$$

and

$$\begin{aligned} \frac{J_{\mathrm{d}}(\mathbf{h})}{J_{\mathrm{n}}(\mathbf{h})} &= \mathrm{iSNR} \times \xi_{\mathrm{n}}(\mathbf{h}) \times v_{\mathrm{d}}(\mathbf{h}) \\ &= \mathrm{oSNR}(\mathbf{h}) \times \xi_{\mathrm{d}}(\mathbf{h}) \times v_{\mathrm{d}}(\mathbf{h}). \end{aligned}$$

2.2 Show that taking the gradient of the MSE:

$$J(\mathbf{h}) = \sigma_x^2 - 2\mathbf{h}^T \mathbf{R}_{\mathbf{x}} \mathbf{i}_{\mathrm{i}} + \mathbf{h}^T \mathbf{R}_{\mathbf{y}} \mathbf{h},$$

with respect to $\mathbf{h}$ and equating the result to zero yields the Wiener filter:

$$\mathbf{h}_{\mathrm{W}} = \mathbf{R}_{\mathbf{y}}^{-1} \mathbf{R}_{\mathbf{x}} \mathbf{i}_{\mathrm{i}}.$$

2.3 Show that the Wiener filter can be expressed as

$$\mathbf{h}_{\mathrm{W}} = \left[\mathbf{I}_L - \rho^2(v, y) \boldsymbol{\Gamma}_{\mathbf{y}}^{-1} \boldsymbol{\Gamma}_{\mathbf{v}} \right] \mathbf{i}_{\mathrm{i}}.$$

2.4 Prove that $z_{\mathrm{W}}(t)$, the estimate of the desired signal with the Wiener filter, satisfies

$$\rho^2 \left(x, z_{\mathrm{W}} \right) \leq \frac{\mathrm{oSNR}\left(\mathbf{h}_{\mathrm{W}}\right)}{1 + \mathrm{oSNR}\left(\mathbf{h}_{\mathrm{W}}\right)}.$$

2.5 Prove that with the optimal Wiener filter, the output SNR is always greater than or equal to the input SNR; in other words, $\mathrm{oSNR}\left(\mathbf{h}_{\mathrm{W}}\right) \geq \mathrm{iSNR}$.

2.6 Show that the MMSE can be expressed as

$$\begin{aligned} J\left(\mathbf{h}_{\mathrm{W}}\right) &= \sigma_x^2 \left[1 - \rho^2(x, z_{\mathrm{W}}) \right] \\ &= \sigma_v^2 \left[1 - \rho^2(v, y - z_{\mathrm{W}}) \right]. \end{aligned}$$

2.7 Show that the performance measures with the Wiener filter are

$$\mathrm{oSNR}\left(\mathbf{h}_{\mathrm{W}}\right) = \frac{\mathbf{i}_{\mathrm{i}}^T \mathbf{R}_{\mathbf{x}} \mathbf{R}_{\mathbf{y}}^{-1} \mathbf{R}_{\mathbf{x}} \mathbf{R}_{\mathbf{y}}^{-1} \mathbf{R}_{\mathbf{x}} \mathbf{i}_{\mathrm{i}}}{\mathbf{i}_{\mathrm{i}}^T \mathbf{R}_{\mathbf{x}} \mathbf{R}_{\mathbf{y}}^{-1} \mathbf{R}_{\mathbf{v}} \mathbf{R}_{\mathbf{y}}^{-1} \mathbf{R}_{\mathbf{x}} \mathbf{i}_{\mathrm{i}}},$$

$$\xi_{\mathrm{n}}\left(\mathbf{h}_{\mathrm{W}}\right) = \frac{\sigma_v^2}{\mathbf{i}_{\mathrm{i}}^T \mathbf{R}_{\mathbf{x}} \mathbf{R}_{\mathbf{y}}^{-1} \mathbf{R}_{\mathbf{v}} \mathbf{R}_{\mathbf{y}}^{-1} \mathbf{R}_{\mathbf{x}} \mathbf{i}_{\mathrm{i}}},$$

$$\xi_{\mathrm{d}}\left(\mathbf{h}_{\mathrm{W}}\right) = \frac{\sigma_x^2}{\mathbf{i}_{\mathrm{i}}^T \mathbf{R}_{\mathbf{x}} \mathbf{R}_{\mathbf{y}}^{-1} \mathbf{R}_{\mathbf{x}} \mathbf{R}_{\mathbf{y}}^{-1} \mathbf{R}_{\mathbf{x}} \mathbf{i}_{\mathrm{i}}},$$

$$\upsilon_{\mathrm{d}}\left(\mathbf{h}_{\mathrm{W}}\right) = \frac{\left(\mathbf{R}_{\mathbf{y}}^{-1}\mathbf{R}_{\mathbf{x}}\mathbf{i}_{\mathrm{i}} - \mathbf{i}_{\mathrm{i}}\right)^T \mathbf{R}_{\mathbf{x}} \left(\mathbf{R}_{\mathbf{y}}^{-1}\mathbf{R}_{\mathbf{x}}\mathbf{i}_{\mathrm{i}} - \mathbf{i}_{\mathrm{i}}\right)}{\sigma_x^2}.$$

2.8 Assume an harmonic random process:

$$x(t) = A\cos\left(2\pi f_0 t + \phi\right),$$

with fixed amplitude A and frequency f_0, and random phase ϕ, uniformly distributed on the interval from 0 to 2π. Show that the elements of the correlation matrix of $\mathbf{x}(t)$ are $\left[\mathbf{R}_{\mathbf{x}}\right]_{i,j} = \frac{1}{2}A^2\cos\left[2\pi f_0(i-j)\right]$.

2.9 Show that the tradeoff filter $\mathbf{h}_{\mathrm{T},\mu}$ can be expressed as

$$\mathbf{h}_{\mathrm{T},\mu} = \left[\mathbf{R}_{\mathbf{y}} + (\mu - 1)\mathbf{R}_{\mathbf{v}}\right]^{-1}\left(\mathbf{R}_{\mathbf{y}} - \mathbf{R}_{\mathbf{v}}\right)\mathbf{i}_{\mathrm{i}},$$

where μ is a Lagrange multiplier that satisfies $J_{\mathrm{n}}\left(\mathbf{h}_{\mathrm{T},\mu}\right) = \aleph\sigma_v^2$.

2.10 Show that the squared Pearson correlation coefficient between $x(t)$ and $x(t) + \sqrt{\mu}v(t)$ is

$$\rho^2\left(x, x + \sqrt{\mu}v\right) = \frac{\mathrm{iSNR}}{\mu + \mathrm{iSNR}}.$$

2.11 Show that the squared Pearson correlation coefficient between $x(t)$ and $\mathbf{h}_{\mathrm{T},\mu}^T\mathbf{x}(t) + \sqrt{\mu}\mathbf{h}_{\mathrm{T},\mu}^T\mathbf{v}(t)$ satisfies

$$\rho^2\left(x, \mathbf{h}_{\mathrm{T},\mu}^T\mathbf{x} + \sqrt{\mu}\mathbf{h}_{\mathrm{T},\mu}^T\mathbf{v}\right) \le \frac{\mathrm{oSNR}\left(\mathbf{h}_{\mathrm{T},\mu}\right)}{\mu + \mathrm{oSNR}\left(\mathbf{h}_{\mathrm{T},\mu}\right)}.$$

2.12 Show that the squared Pearson correlation coefficient between $x(t) + \sqrt{\mu}v(t)$ and $\mathbf{h}_{\mathrm{T},\mu}^T\mathbf{x}(t) + \sqrt{\mu}\mathbf{h}_{\mathrm{T},\mu}^T\mathbf{v}(t)$ satisfies

$$\rho^2\left(x + \sqrt{\mu}v, \mathbf{h}_{\mathrm{T},\mu}^T\mathbf{x} + \sqrt{\mu}\mathbf{h}_{\mathrm{T},\mu}^T\mathbf{v}\right) = \frac{\rho^2\left(x, x + \sqrt{\mu}v\right)}{\rho^2\left(x, \mathbf{h}_{\mathrm{T},\mu}^T\mathbf{x} + \sqrt{\mu}\mathbf{h}_{\mathrm{T},\mu}^T\mathbf{v}\right)}.$$

2.13 Show that the squared Pearson correlation coefficient between $x(t)$ and $x(t) + \sqrt{\mu}v(t)$ satisfies

$$\rho^2\left(x, x + \sqrt{\mu}v\right) \le \frac{\mathrm{oSNR}\left(\mathbf{h}_{\mathrm{T},\mu}\right)}{\mu + \mathrm{oSNR}\left(\mathbf{h}_{\mathrm{T},\mu}\right)}.$$

2.14 Show that with the tradeoff filter, the output SNR is always greater than or equal to the input SNR; in other words, $\text{oSNR}\left(\mathbf{h}_{\text{T},\mu}\right) \geq \text{iSNR}, \ \forall \mu \geq 0$.

2.15 Let us denote the matrix $\mathbf{Q}'_{\mathbf{x}}$ containing the eigenvectors corresponding to the nonzero eigenvalues of $\mathbf{R}_{\mathbf{x}}$. Show that the distortionless constraint $\mathbf{h}^T\mathbf{x}(t) = x(t)$ can be expressed as

$$\mathbf{h}^T\mathbf{Q}'_{\mathbf{x}} = \mathbf{i}_{\text{i}}^T\mathbf{Q}'_{\mathbf{x}}.$$

2.16 Prove that the MVDR filter is

$$\mathbf{h}_{\text{MVDR}} = \mathbf{R}_{\mathbf{v}}^{-1}\mathbf{Q}'_{\mathbf{x}}\left(\mathbf{Q}'^T_{\mathbf{x}}\mathbf{R}_{\mathbf{v}}^{-1}\mathbf{Q}'_{\mathbf{x}}\right)^{-1}\mathbf{Q}'^T_{\mathbf{x}}\mathbf{i}_{\text{i}}.$$

2.17 Show that the MVDR filter can be expressed as

$$\mathbf{h}_{\text{MVDR}} = \mathbf{R}_{\mathbf{y}}^{-1}\mathbf{Q}'_{\mathbf{x}}\left(\mathbf{Q}'^T_{\mathbf{x}}\mathbf{R}_{\mathbf{y}}^{-1}\mathbf{Q}'_{\mathbf{x}}\right)^{-1}\mathbf{Q}'^T_{\mathbf{x}}\mathbf{i}_{\text{i}}.$$

2.18 Verify that the MVDR filter satisfies $J_{\text{d}}\left(\mathbf{h}_{\text{MVDR}}\right) = 0$.

2.19 Show that with the MVDR filter, the output SNR is always greater than or equal to the input SNR; in other words, $\text{oSNR}\left(\mathbf{h}_{\text{MVDR}}\right) \geq \text{iSNR}$.

2.20 Show that the inverse of $\mathbf{R}_{\mathbf{y}}$ can be expressed as

$$\mathbf{R}_{\mathbf{y}}^{-1} = \mathbf{R}_{\mathbf{v}}^{-1} - \mathbf{R}_{\mathbf{v}}^{-1}\mathbf{Q}'_{\mathbf{x}}\left(\mathbf{\Lambda}'^{-1}_{\mathbf{x}} + \mathbf{Q}'^T_{\mathbf{x}}\mathbf{R}_{\mathbf{v}}^{-1}\mathbf{Q}'_{\mathbf{x}}\right)^{-1}\mathbf{Q}'^T_{\mathbf{x}}\mathbf{R}_{\mathbf{v}}^{-1}.$$

2.21 Show that the Wiener filter $\mathbf{h}_{\text{W}}$ can be written as

$$\mathbf{h}_{\text{W}} = \mathbf{R}_{\mathbf{v}}^{-1}\mathbf{Q}'_{\mathbf{x}}\left(\mathbf{\Lambda}'^{-1}_{\mathbf{x}} + \mathbf{Q}'^T_{\mathbf{x}}\mathbf{R}_{\mathbf{v}}^{-1}\mathbf{Q}'_{\mathbf{x}}\right)^{-1}\mathbf{Q}'^T_{\mathbf{x}}\mathbf{i}_{\text{i}}.$$

2.22 Show that the tradeoff filter $\mathbf{h}_{\text{T},\mu}$ can be expressed as

$$\mathbf{h}_{\text{T},\mu} = \mathbf{R}_{\mathbf{v}}^{-1}\mathbf{Q}'_{\mathbf{x}}\left(\mu\mathbf{\Lambda}'^{-1}_{\mathbf{x}} + \mathbf{Q}'^T_{\mathbf{x}}\mathbf{R}_{\mathbf{v}}^{-1}\mathbf{Q}'_{\mathbf{x}}\right)^{-1}\mathbf{Q}'^T_{\mathbf{x}}\mathbf{i}_{\text{i}}.$$

2.23 Let λ_1 denote the maximum eigenvalue of the matrix $\mathbf{R}_{\mathbf{v}}^{-1}\mathbf{R}_{\mathbf{x}}$, and let $\mathbf{t}_1$ denote the corresponding eigenvector.

a) Show that the filter that maximizes the output SNR is given by

$$\mathbf{h}_{\max} = \varsigma\mathbf{t}_1,$$

where $\varsigma \neq 0$ is an arbitrary real number.

b) Show that the MSE obtained with the maximum SNR filter is

$$J\left(\mathbf{h}_{\max}\right) = \sigma_x^2 - 2\varsigma\mathbf{t}_1^T\mathbf{R}_{\mathbf{x}}\mathbf{i}_{\text{i}} + \varsigma^2\mathbf{t}_1^T\mathbf{R}_{\mathbf{y}}\mathbf{t}_1.$$

c) Show that the maximum SNR filter, $\mathbf{h}_{\max}$, that minimizes the MSE is given by

$$\mathbf{h}_{\max} = \frac{\mathbf{t}_1^T \mathbf{R}_\mathbf{x} \mathbf{i}_\mathrm{i}}{\mathbf{t}_1^T \mathbf{R}_\mathbf{y} \mathbf{t}_1} \mathbf{t}_1.$$

2.24 Assume that the matrix $\mathbf{R}_\mathbf{v}^{-1}\mathbf{R}_\mathbf{x}$ has a maximum eigenvalue λ_1 with multiplicity $P \leq L$, and denote by $\mathbf{t}_1, \mathbf{t}_2, \ldots, \mathbf{t}_P$ the corresponding eigenvectors. Show that the maximum SNR filter $\mathbf{h}_{\max}$ is given by

$$\mathbf{h}_{\max} = \sum_{p=1}^{P} \varsigma_p \mathbf{t}_p,$$

where ς_p, $p = 1, 2, \ldots, P$ are real numbers with at least one of them different from 0.

2.25 Show that the output SNRs of the optimal filters are related by

a) for $\mu < 1$,

$$\mathrm{oSNR}\left(\mathbf{h}_{\max}\right) \geq \mathrm{oSNR}\left(\mathbf{h}_\mathrm{W}\right) \geq \mathrm{oSNR}\left(\mathbf{h}_{\mathrm{T},\mu}\right) \geq \mathrm{oSNR}\left(\mathbf{h}_\mathrm{MVDR}\right),$$

b) and for $\mu > 1$,

$$\mathrm{oSNR}\left(\mathbf{h}_{\max}\right) \geq \mathrm{oSNR}\left(\mathbf{h}_{\mathrm{T},\mu}\right) \geq \mathrm{oSNR}\left(\mathbf{h}_\mathrm{W}\right) \geq \mathrm{oSNR}\left(\mathbf{h}_\mathrm{MVDR}\right).$$

2.26 Prove that the fullmode input SNR can never exceed the maximum spectral mode input SNR and can never go below the minimum spectral mode input SNR; in other words,

$$\mathrm{iSNR}(;L) \leq \mathrm{iSNR}(;) \leq \mathrm{iSNR}(;1).$$

2.27 Prove that the fullmode output SNR can never exceed the maximum spectral mode input SNR and can never go below the minimum spectral mode input SNR; in other words,

$$\mathrm{iSNR}(;L) \leq \mathrm{oSNR}\left[\mathbf{h}(;)\right] \leq \mathrm{iSNR}(;1), \ \forall \mathbf{h}(;).$$

2.28 Show that the fullmode desired signal distortion index, $\upsilon_\mathrm{d}\left[\mathbf{h}(;)\right]$, can be expressed as

$$\upsilon_\mathrm{d}\left[\mathbf{h}(;)\right] = \frac{\sum_{l=1}^{L} \lambda_l \left[h(;l) - 1\right]^2}{\mathrm{tr}\left(\mathbf{\Lambda}\right)}.$$

2.29 Show that the fullmode MSE is given by

$$J\left[\mathbf{h}(;)\right] = \left[\mathbf{1} - \mathbf{h}(;)\right]^T \mathbf{\Lambda} \left[\mathbf{1} - \mathbf{h}(;)\right] + \mathbf{h}^T(;)\mathbf{h}(;).$$

2.30 Show that the fullmode output SNR obtained with the Wiener gains at the spectral modes $p,\ p = 1, 2, \dots, P$, i.e., $\text{oSNR}\left[\mathbf{h}_{\text{W},P}(;)\right]$, satisfies

$$\text{oSNR}\left[\mathbf{h}_{\text{W},L}(;)\right] \leq \text{oSNR}\left[\mathbf{h}_{\text{W},P}(;)\right] \leq \text{oSNR}\left[\mathbf{h}_{\text{W},1}(;)\right],$$

for all $P, 1 \leq P \leq L$.

2.31 Show that the fullmode output SNR corresponding to $\mathbf{h}'_{\beta_L}(;)$ is

$$\text{oSNR}\left[\mathbf{h}'_{\beta_L}(;)\right] = \frac{\mathbf{h}'^T_{\beta_L}(;)\mathbf{\Lambda}\mathbf{h}'_{\beta_L}(;)}{\mathbf{h}'^T_{\beta_L}(;)\mathbf{h}'_{\beta_L}(;)}.$$

2.32 Show that the estimator corresponding to $\mathbf{h}'_{\text{W},Q}(;)$ is

$$\widehat{\mathbf{x}}_{\text{W},Q}(t) = \left(\sum_{i=1}^{L-Q} \mathbf{b}_i\mathbf{t}_i^T + \sum_{q=1}^{Q} \frac{\lambda_{L-Q+q}}{1+\lambda_{L-Q+q}}\mathbf{b}_{L-Q+q}\mathbf{t}_{L-Q+q}^T\right)\mathbf{y}(t).$$

2.33 Show that:

a) the fullmode output SNR corresponding to $\mathbf{h}'_{\text{N},Q}(;)$ is not smaller than that corresponding to $\mathbf{h}'_{\text{W},Q}(;)$; in other words,

$$\text{oSNR}\left[\mathbf{h}'_{\text{N},Q}(;)\right] \geq \text{oSNR}\left[\mathbf{h}'_{\text{W},Q}(;)\right],$$

b) the fullmode output SNR corresponding to $\mathbf{h}'_{\text{N},Q}(;)$ is a decreasing function of Q for $1 \leq Q \leq L-1$; in other words,

$$\text{oSNR}\left[\mathbf{h}'_{\text{N},L-1}(;)\right] \geq \cdots \geq \text{oSNR}\left[\mathbf{h}'_{\text{N},1}(;)\right] \geq \text{iSNR}(;).$$

References

1 J. Benesty, J. Chen, Y. Huang, and I. Cohen, *Noise Reduction in Speech Processing.* Berlin, Germany: Springer-Verlag, 2009.

2 P. Vary and R. Martin, *Digital Speech Transmission: Enhancement, Coding and Error Concealment.* Chichester, England: John Wiley & Sons Ltd, 2006.

3 P. Loizou, *Speech Enhancement: Theory and Practice.* Boca Raton, FL: CRC Press, 2007.

4 J. Benesty and J. Chen, *Optimal Time-domain Noise Reduction Filters – A Theoretical Study.* Springer Briefs in Electrical and Computer Engineering, 2011.

5 J. Benesty, J. Chen, Y. Huang, and S. Doclo, “Study of the Wiener filter for noise reduction,” in *Speech Enhancement,* J. Benesty, S. Makino, and J. Chen (eds). Berlin, Germany: Springer-Verlag, 2005.

6 J. Chen, J. Benesty, Y. Huang, and S. Doclo, “New insights into the noise reduction Wiener filter,” *IEEE Trans. Audio, Speech, Language Process.*, vol. 14, pp. 1218–1234, Jul. 2006.

7 S. Haykin, *Adaptive Filter Theory*, 4th edn. Upper Saddle River, NJ: Prentice-Hall, 2002.
8 J. Benesty, J. Chen, and Y. Huang, "On the importance of the Pearson correlation coefficient in noise reduction," *IEEE Trans. Audio, Speech, Language Process.*, vol. 16, pp. 757–765, May 2008.
9 G. H. Golub and C. F. van Loan, *Matrix Computations*, 3rd edn. Baltimore, MD: The Johns Hopkins University Press, 1996.
10 J. Franklin, *Matrix Theory*. Englewood Cliffs, NJ: Prentice-Hall, 1968.
11 D. Wang, "On ideal binary mask as the computational goal of auditory scene analysis," in *Speech Separation by Humans and Machines*, Pierre Divenyi (ed). Boston, MA: Kluwer, 2005.
12 M. Dendrinos, S. Bakamidis, and G. Carayannis, "Speech enhancement from noise: a regenerative approach," *Speech Commun.*, vol. 10, pp. 45–57, Jan. 1991.
13 Y. Ephraim and H. Van Trees, "A signal subspace approach for speech enhancement," *IEEE Trans. Speech Audio Process.*, vol. 3, pp. 251–266, Jul. 1995.
14 S. H. Jensen, P. C. Hansen, S. D. Hansen, and J. A. Sørensen, "Reduction of broad-band noise in speech by truncated QSVD," *IEEE Trans. Speech Audio Process.*, vol. 3, pp. 439–448, Nov. 1995.

3

Single-Channel Signal Enhancement in the Frequency Domain

In this chapter, we continue our investigation of the single-channel signal enhancement problem but, this time, in the frequency domain. In many respects, the frequency-domain approach is equivalent to the spectral method discussed in the previous chapter. The advantages of the frequency-domain technique are twofold. First, it is very flexible, in the sense that the observation signal at each frequency can be processed independently of the others. Second, thanks to the fast Fourier transform, all algorithms can be implemented very efficiently. We start by formulating the problem. We then explain how to perform noise reduction with just simple gains. We give all relevant performance measures. We also derive all kinds of optimal gains and show how we can compromise between distortion of the desired signal and reduction of the additive noise. Finally, we explain how these gains can be implemented in the short-time Fourier transform domain.

3.1 Signal Model and Problem Formulation

We recall from Chapter 2 that the observation signal in the time domain is

$$y(t) = x(t) + v(t), \tag{3.1}$$

where $x(t)$ and $v(t)$ are the desired and noise signals, respectively. In the frequency domain, (3.1) can be written as [1]:

$$Y(f) = X(f) + V(f), \tag{3.2}$$

where $Y(f)$, $X(f)$, and $V(f)$ are the frequency-domain representations of $y(t)$, $x(t)$, and $v(t)$, respectively, at the frequency index f. Obviously, $X(f)$ and $V(f)$ are, respectively, the desired and noise signals in the frequency domain. Since the zero-mean signals $X(f)$ and $V(f)$ are assumed to be uncorrelated, the variance of $Y(f)$ is

$$\begin{aligned} \phi_Y(f) &= E\left[|Y(f)|^2\right] \\ &= \phi_X(f) + \phi_V(f), \end{aligned} \tag{3.3}$$

where $\phi_X(f) = E\left[|X(f)|^2\right]$ and $\phi_V(f) = E\left[|V(f)|^2\right]$ are the variances of $X(f)$ and $V(f)$, respectively.

Fundamentals of Signal Enhancement and Array Signal Processing, First Edition.
Jacob Benesty, Israel Cohen, and Jingdong Chen.

Companion website: www.wiley.com/go/benesty/arraysignalprocessing

The objective of single-channel noise reduction in the frequency domain is then to find an estimate of $X(f)$ from $Y(f)$.

3.2 Noise Reduction with Gains

An estimate of $X(f)$ can be obtained by multiplying $Y(f)$ with a complex gain, $H(f)$, as illustrated in Figure 3.1:

$$\begin{aligned} Z(f) &= H(f)Y(f) \\ &= H(f)\left[X(f) + V(f)\right] \\ &= X_{\mathrm{fd}}(f) + V_{\mathrm{rn}}(f), \end{aligned} \tag{3.4}$$

where $Z(f)$ is the frequency-domain representation of the signal $z(t)$,

$$X_{\mathrm{fd}}(f) = H(f)X(f) \tag{3.5}$$

is the filtered desired signal, and

$$V_{\mathrm{rn}}(f) = H(f)V(f) \tag{3.6}$$

is the residual noise. The variance of $Z(f)$ can then be written as

$$\begin{aligned} \phi_Z(f) &= E\left[|Z(f)|^2\right] \\ &= |H(f)|^2 \phi_Y(f) \\ &= \phi_{X_{\mathrm{fd}}}(f) + \phi_{V_{\mathrm{rn}}}(f), \end{aligned} \tag{3.7}$$

where

$$\phi_{X_{\mathrm{fd}}}(f) = |H(f)|^2 \phi_X(f) \tag{3.8}$$

is the variance of the filtered desired signal and

$$\phi_{V_{\mathrm{rn}}}(f) = |H(f)|^2 \phi_V(f) \tag{3.9}$$

is the variance of the residual noise.

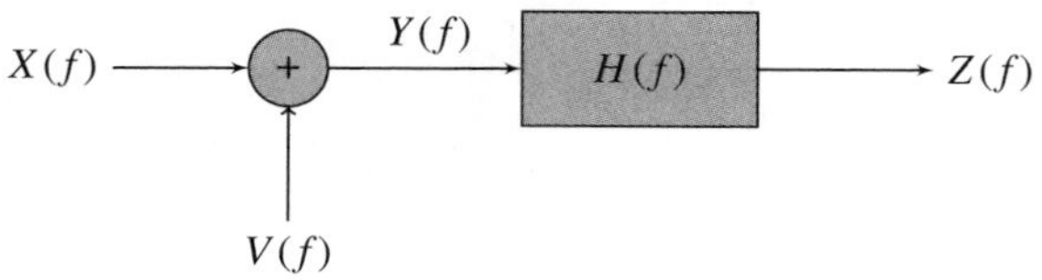

Figure 3.1 Block diagram of noise reduction with gains in the frequency domain.

3.3 Performance Measures

In this section, we discuss both the narrowband and broadband performance measures.

In a similar way to the spectral mode input SNR, we define the narrowband input SNR as

$$\mathrm{iSNR}(f) = \frac{\phi_X(f)}{\phi_V(f)}. \tag{3.10}$$

The broadband input SNR is obtained by simply integrating over all frequencies the numerator and denominator of iSNR(f). We get

$$\mathrm{iSNR} = \frac{\int_f \phi_X(f)df}{\int_f \phi_V(f)df}. \tag{3.11}$$

After noise reduction with the frequency-domain model given in (3.4), the narrowband output SNR can be written as

$$\begin{aligned} \mathrm{oSNR}\left[H(f)\right] &= \frac{\phi_{X_{\mathrm{fd}}}(f)}{\phi_{V_{\mathrm{rn}}}(f)} \\ &= \mathrm{iSNR}(f). \end{aligned} \tag{3.12}$$

It is important to observe that the narrowband output SNR is not influenced by $H(f)$. We deduce that the broadband output SNR is

$$\begin{aligned} \mathrm{oSNR}\left(H\right) &= \frac{\int_f \phi_{X_{\mathrm{fd}}}(f)df}{\int_f \phi_{V_{\mathrm{rn}}}(f)df} \\ &= \frac{\int_f |H(f)|^2 \phi_X(f)df}{\int_f |H(f)|^2 \phi_V(f)df}. \end{aligned} \tag{3.13}$$

It is essential to find the complex gains $H(f)$ at all frequencies in such a way that oSNR(H) > iSNR.

Other important measures in noise reduction are the noise reduction factors. We define the narrowband and broadband noise reduction factors as, respectively,

$$\xi_{\mathrm{n}}\left[H(f)\right] = \frac{1}{|H(f)|^2} \tag{3.14}$$

and

$$\xi_{\mathrm{n}}\left(H\right) = \frac{\int_f \phi_V(f)df}{\int_f |H(f)|^2 \phi_V(f)df}. \tag{3.15}$$

The larger the values of the noise reduction factors, the more the noise is reduced.

In the same way, we define the narrowband and broadband desired signal reduction factors as, respectively,

$$\xi_{\mathrm{d}}\left[H(f)\right] = \frac{1}{|H(f)|^2} \tag{3.16}$$

and

$$\xi_{\mathrm{d}}(H) = \frac{\int_f \phi_X(f)df}{\int_f |H(f)|^2 \phi_X(f)df}. \tag{3.17}$$

The larger the values of the desired signal reduction factors, the more the desired signal is distorted.

We always have

$$\frac{\mathrm{oSNR}(H)}{\mathrm{iSNR}} = \frac{\xi_{\mathrm{n}}(H)}{\xi_{\mathrm{d}}(H)}. \tag{3.18}$$

This means that the gain in SNR comes with the distortion of the desired and noise signals.

Another way to quantify distortion is via the narrowband desired signal distortion index:

$$\begin{aligned} \upsilon_{\mathrm{d}}\left[H(f)\right] &= \frac{E\left[|H(f)X(f) - X(f)|^2\right]}{\phi_X(f)} \\ &= |1 - H(f)|^2 \end{aligned} \tag{3.19}$$

and the broadband desired signal distortion index:

$$\begin{aligned} \upsilon_{\mathrm{d}}(H) &= \frac{\int_f E\left[|H(f)X(f) - X(f)|^2\right] df}{\int_f \phi_X(f)df} \\ &= \frac{\int_f |1 - H(f)|^2 \phi_X(f)df}{\int_f \phi_X(f)df} \\ &= \frac{\int_f \upsilon_{\mathrm{d}}\left[H(f)\right] \phi_X(f)df}{\int_f \phi_X(f)df}. \end{aligned} \tag{3.20}$$

The desired signal distortion index has a lower bound of 0 and an upper bound of 1 for optimal gains.

We define the error signal between the estimated and desired signals at frequency f as

$$\begin{aligned} \mathcal{E}(f) &= Z(f) - X(f) \\ &= H(f)Y(f) - X(f). \end{aligned} \tag{3.21}$$

This error can also be put into the form:

$$\mathcal{E}(f) = \mathcal{E}_{\mathrm{d}}(f) + \mathcal{E}_{\mathrm{n}}(f), \tag{3.22}$$

where

$$\mathcal{E}_{\mathrm{d}}(f) = \left[H(f) - 1\right] X(f) \tag{3.23}$$

is the desired signal distortion due to the complex gain and

$$\mathcal{E}_{\mathrm{n}}(f) = H(f)V(f) \tag{3.24}$$

represents the residual noise. The narrowband MSE criterion is then

$$\begin{aligned} J\left[H(f)\right] &= E\left[|\mathcal{E}(f)|^2\right] \\ &= \phi_X(f) + |H(f)|^2 \phi_Y(f) - H(f)\phi_X(f) - H^*(f)\phi_X(f) \\ &= |1 - H(f)|^2 \phi_X(f) + |H(f)|^2 \phi_V(f), \end{aligned} \tag{3.25}$$

where the superscript * is the complex-conjugate operator. The narrowband MSE is also

$$\begin{aligned} J\left[H(f)\right] &= E\left[|\mathcal{E}_{\mathrm{d}}(f)|^2\right] + E\left[|\mathcal{E}_{\mathrm{n}}(f)|^2\right] \\ &= J_{\mathrm{d}}\left[H(f)\right] + J_{\mathrm{n}}\left[H(f)\right], \end{aligned} \tag{3.26}$$

where

$$J_{\mathrm{d}}\left[H(f)\right] = \phi_X(f)\upsilon_{\mathrm{d}}\left[H(f)\right] \tag{3.27}$$

and

$$J_{\mathrm{n}}\left[H(f)\right] = \frac{\phi_V(f)}{\xi_{\mathrm{n}}\left[H(f)\right]}. \tag{3.28}$$

We deduce that

$$J\left[H(f)\right] = \phi_V(f)\left\{\mathrm{iSNR}(f) \times \upsilon_{\mathrm{d}}\left[H(f)\right] + \frac{1}{\xi_{\mathrm{n}}\left[H(f)\right]}\right\} \tag{3.29}$$

and

$$\begin{aligned} \frac{J_{\mathrm{d}}\left[H(f)\right]}{J_{\mathrm{n}}\left[H(f)\right]} &= \mathrm{iSNR}(f) \times \xi_{\mathrm{n}}\left[H(f)\right] \times \upsilon_{\mathrm{d}}\left[H(f)\right] \\ &= \mathrm{oSNR}\left[H(f)\right] \times \xi_{\mathrm{d}}\left[H(f)\right] \times \upsilon_{\mathrm{d}}\left[H(f)\right], \end{aligned} \tag{3.30}$$

showing how the narrowband MSEs are related to the different narrowband performance measures.

The extension of the narrowband MSE to the broadband case is straightforward. We define the broadband MSE criterion as

$$\begin{aligned} J(H) &= \int_f J\left[H(f)\right] df \\ &= \int_f |1 - H(f)|^2 \phi_X(f) df + \int_f |H(f)|^2 \phi_V(f) df \\ &= J_{\rm d}(H) + J_{\rm n}(H), \end{aligned} \tag{3.31}$$

where

$$J_{\rm d}(H) = \upsilon_{\rm d}(H) \int_f \phi_X(f) df \tag{3.32}$$

and

$$J_{\rm n}(H) = \frac{\int_f \phi_V(f) df}{\xi_{\rm n}(H)}. \tag{3.33}$$

These expressions show how the broadband MSEs are fundamentally equivalent to the broadband performance measures.

3.4 Optimal Gains

Now, we focus our attention on the derivation and analysis of some important gains for noise reduction. Taking the gradient of $J\left[H(f)\right]$ (from Equation 3.25) with respect to $H^*(f)$ and equating the result to 0 leads to

$$-E\left\{Y^*(f)\left[X(f) - H_{\rm W}(f)Y(f)\right]\right\} = 0. \tag{3.34}$$

Hence,

$$\phi_Y(f) H_{\rm W}(f) = \phi_{XY}(f), \tag{3.35}$$

where

$$\begin{aligned} \phi_{XY}(f) &= E\left[X(f)Y^*(f)\right] \\ &= \phi_X(f) \end{aligned} \tag{3.36}$$

is the the cross-correlation between $X(f)$ and $Y(f)$, which simplifies to the variance of $X(f)$ in this particular model. Therefore, the optimal Wiener gain can be put into the following forms:

$$H_W(f) = \frac{\phi_X(f)}{\phi_Y(f)} = 1 - \frac{\phi_V(f)}{\phi_Y(f)} = \frac{\text{iSNR}(f)}{1 + \text{iSNR}(f)}. \tag{3.37}$$

We observe that this gain is always real, positive, and smaller than one. Another way to write the Wiener gain is with the magnitude squared coherence functions (MSCFs). Indeed, it is easy to see that

$$H_W(f) = \left|\rho\left[X(f), Y(f)\right]\right|^2 = 1 - \left|\rho\left[V(f), Y(f)\right]\right|^2, \tag{3.38}$$

where

$$\left|\rho\left[X(f), Y(f)\right]\right|^2 = \frac{\left|E\left[X(f)Y^*(f)\right]\right|^2}{E\left[|X(f)|^2\right]E\left[|Y(f)|^2\right]} = \frac{|\phi_{XY}(f)|^2}{\phi_X(f)\phi_Y(f)} = \frac{\text{iSNR}(f)}{1 + \text{iSNR}(f)} \tag{3.39}$$

is the MSCF between $X(f)$ and $Y(f)$, and

$$\left|\rho\left[V(f), Y(f)\right]\right|^2 = \frac{|\phi_{VY}(f)|^2}{\phi_V(f)\phi_Y(f)} = \frac{1}{1 + \text{iSNR}(f)} \tag{3.40}$$

is the MSCF between $V(f)$ and $Y(f)$. When the level of the noise is high at frequency f, $\left|\rho\left[V(f), Y(f)\right]\right|^2 \approx 1$, then $H_W(f)$ is close to 0 since there is a large amount of noise that needs to be removed. When the level of the noise is low at frequency f, $\left|\rho\left[V(f), Y(f)\right]\right|^2 \approx 0$, then $H_W(f)$ is close to 1 and this gain is not going to greatly affect the signals since there is little noise that needs to be removed.

Now, let us define the complex number[1]:

$$\varrho\left[X(f), V(f)\right] = \rho\left[X(f), Y(f)\right] + \jmath\rho\left[V(f), Y(f)\right] = \cos\theta(f) + \jmath\sin\theta(f), \tag{3.41}$$

1 Notice that both $\rho\left[X(f), Y(f)\right]$ and $\rho\left[V(f), Y(f)\right]$ are real numbers.

where $\jmath = \sqrt{-1}$ is the imaginary unit and $\theta(f)$ is the phase of $\varrho\left[X(f), V(f)\right]$ whose modulus is equal to 1. On the complex plane, $\varrho\left[X(f), V(f)\right]$ is on the unit circle. Since $0 \leq \rho\left[X(f), Y(f)\right] \leq 1$ and $0 \leq \rho\left[V(f), Y(f)\right] \leq 1$, therefore $0 \leq \theta(f) \leq \frac{\pi}{2}$. We can then rewrite the Wiener gain as a function of the angle $\theta(f)$:

$$\begin{aligned} H_{\mathrm{W}}(f) &= \cos^2 \theta(f) \\ &= 1 - \sin^2 \theta(f). \end{aligned} \tag{3.42}$$

Hence,

$$\lim_{\theta(f)\to 0} H_{\mathrm{W}}(f) = 1, \tag{3.43}$$

$$\lim_{\theta(f)\to \frac{\pi}{2}} H_{\mathrm{W}}(f) = 0. \tag{3.44}$$

The MMSE is obtained by replacing (3.37) in (3.25):

$$\begin{aligned} J\left[H_{\mathrm{W}}(f)\right] &= \phi_X(f) - \frac{\phi_X^2(f)}{\phi_Y(f)} \\ &= \phi_V(f) - \frac{\phi_V^2(f)}{\phi_Y(f)}, \end{aligned} \tag{3.45}$$

which can be rewritten as

$$\begin{aligned} J\left[H_{\mathrm{W}}(f)\right] &= \phi_X(f)\left\{1 - \left|\rho\left[X(f), Y(f)\right]\right|^2\right\} \\ &= \phi_V(f)\left\{1 - \left|\rho\left[V(f), Y(f)\right]\right|^2\right\} \\ &= H_{\mathrm{W}}(f)\phi_V(f) \\ &= \left[1 - H_{\mathrm{W}}(f)\right]\phi_X(f). \end{aligned} \tag{3.46}$$

We deduce all the narrowband performance measures with the Wiener gain:

$$\xi_{\mathrm{n}}\left[H_{\mathrm{W}}(f)\right] = \frac{1}{\cos^4 \theta(f)} \geq 1, \tag{3.47}$$

$$\xi_{\mathrm{d}}\left[H_{\mathrm{W}}(f)\right] = \frac{1}{\cos^4 \theta(f)} \geq 1, \tag{3.48}$$

$$\upsilon_{\mathrm{d}}\left[H_{\mathrm{W}}(f)\right] = \sin^4 \theta(f) \leq 1. \tag{3.49}$$

We recall that the narrowband output SNR is equal to the narrowband input SNR.

Figure 3.2 shows plots of the optimal Wiener gain, $H_{\mathrm{W}}(f)$, the angle, $\theta(f)$, the narrowband noise reduction factor, $\xi_{\mathrm{n}}\left[H_{\mathrm{W}}(f)\right]$, and the narrowband desired signal distortion index, $\upsilon_{\mathrm{d}}\left[H_{\mathrm{W}}(f)\right]$, as a function of the narrowband input SNR. As the input SNR increases, the Wiener gain increases, since there is less noise to suppress. As a result, both the noise reduction factor and the desired signal distortion index decrease.

We now give a fundamental property about the broadband output SNR with the Wiener gain.

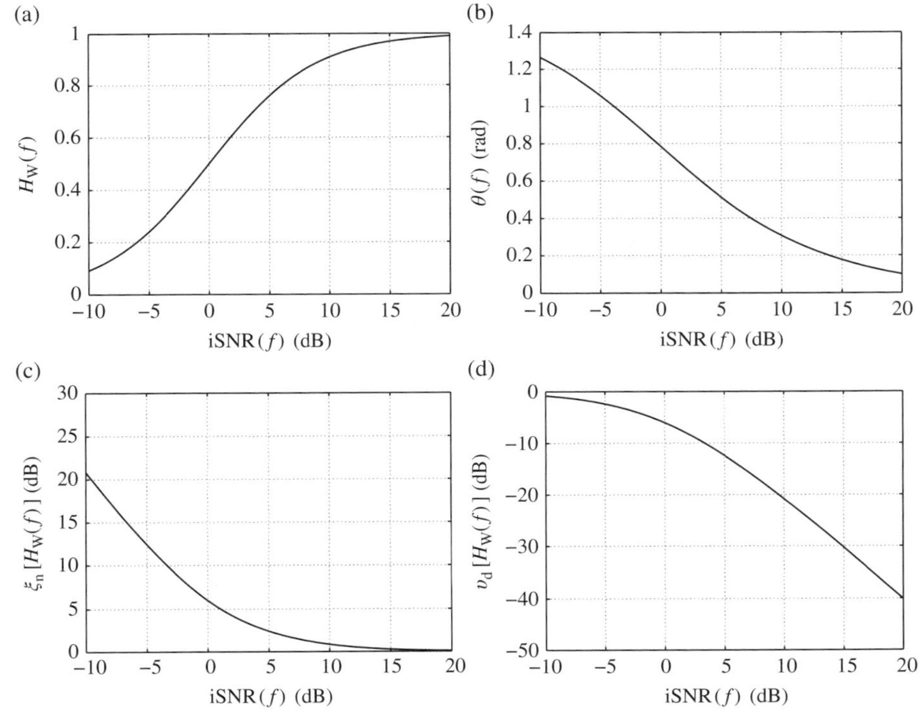

Figure 3.2 (a) The optimal Wiener gain, (b) the angle, (c) the narrowband noise reduction factor, and (d) the narrowband desired signal distortion index as a function of the narrowband input SNR.

Property 3.4.1 With the optimal Wiener gain given in (3.37), the broadband output SNR is always greater than or equal to the broadband input SNR; in other words, $\text{oSNR}\left(H_W\right) \geq \text{iSNR}$.

Proof. The broadband MSCF, which is equivalent to the SPCC, between the two zero-mean random variables $A(f)$ and $B(f)$, which are the frequency-domain representations of the time-domain real signals $a(t)$ and $b(t)$, is defined as

$$
\begin{aligned}
|\rho(A,B)|^2 &= \frac{\left|E\left[\int_f A(f)B^*(f)df\right]\right|^2}{E\left[\int_f |A(f)|^2 df\right] E\left[\int_f |B(f)|^2 df\right]} \\
&= \frac{\left|\int_f \phi_{AB}(f)df\right|^2}{\left[\int_f \phi_A(f)df\right]\left[\int_f \phi_B(f)df\right]} \\
&= \frac{E^2\left[a(t)b(t)\right]}{\sigma_a^2\sigma_b^2} \\
&= \rho^2(a,b).
\end{aligned}
\tag{3.50}
$$

Let us evaluate the broadband MSCF between $Y(f)$ and $Z_W(f) = H_W(f)Y(f)$:

$$\left|\rho\left(Y, Z_W\right)\right|^2 = \frac{\left[\int_f H_W(f)\phi_Y(f)df\right]^2}{\left[\int_f \phi_Y(f)df\right]\left[\int_f H_W^2(f)\phi_Y(f)df\right]}$$
$$= \frac{\int_f \phi_X(f)df}{\int_f \phi_Y(f)df} \times \frac{\int_f \phi_X(f)df}{\int_f H_W(f)\phi_X(f)df}$$
$$= \frac{|\rho(X, Y)|^2}{\left|\rho\left(X, Z_W\right)\right|^2}.$$

Therefore,

$$|\rho(X, Y)|^2 = \left|\rho\left(Y, Z_W\right)\right|^2 \times \left|\rho\left(X, Z_W\right)\right|^2 \leq \left|\rho\left(X, Z_W\right)\right|^2. \tag{3.51}$$

On the other hand, it can be shown that

$$|\rho(X, Y)|^2 = \frac{\text{iSNR}}{1 + \text{iSNR}}$$

and

$$\left|\rho\left(X, Z_W\right)\right|^2 \leq \frac{\text{oSNR}\left(H_W\right)}{1 + \text{oSNR}\left(H_W\right)}.$$

Substituting the two previous expressions into (3.51), we obtain

$$\frac{\text{iSNR}}{1 + \text{iSNR}} \leq \frac{\text{oSNR}\left(H_W\right)}{1 + \text{oSNR}\left(H_W\right)}.$$

As a result, we have

$$\text{oSNR}\left(H_W\right) \geq \text{iSNR}.$$

■

Example 3.4.1 Consider a desired signal, $X(f)$, with the variance:

$$\phi_X(f) = \begin{cases} \alpha, & |f| \leq \frac{1}{4} \\ 0, & \frac{1}{4} \leq |f| \leq \frac{1}{2} \end{cases},$$

which is corrupted with additive noise, $V(f)$, with the variance:

$$\phi_V(f) = \beta\left(1 - 2|f|\right), \ -\frac{1}{2} \leq |f| \leq \frac{1}{2}.$$

The desired signal is uncorrelated with the noise, and needs to be recovered from the noisy observation, $Y(f) = X(f) + V(f)$.

The narrowband input SNR is

$$\begin{aligned}\text{iSNR}(f) &= \frac{\phi_X(f)}{\phi_V(f)} \\ &= \begin{cases} \frac{\alpha}{\beta}\left(1 - 2|f|\right)^{-1}, & |f| \leq \frac{1}{4} \\ 0, & \frac{1}{4} \leq |f| \leq \frac{1}{2} \end{cases}\end{aligned}$$

and the broadband input SNR is

$$\begin{aligned}\text{iSNR} &= \frac{\int_f \phi_X(f)df}{\int_f \phi_V(f)df} \\ &= \frac{\alpha}{\beta}.\end{aligned}$$

The optimal Wiener gain is given by

$$\begin{aligned}H_{\text{W}}(f) &= \frac{\text{iSNR}(f)}{1 + \text{iSNR}(f)} \\ &= \begin{cases} \frac{\alpha}{\beta}\left(\frac{\alpha}{\beta} + 1 - 2|f|\right)^{-1}, & |f| \leq \frac{1}{4} \\ 0, & \frac{1}{4} \leq |f| \leq \frac{1}{2} \end{cases}.\end{aligned}$$

The broadband output SNR, $\text{oSNR}\left(H_{\text{W}}\right)$, is computed using (3.13), and the broadband gain in SNR is obtained using $\mathcal{G}\left(H_{\text{W}}\right) = \text{oSNR}\left(H_{\text{W}}\right)/\text{iSNR}$. Figure 3.3 shows plots of the broadband gain in SNR, the broadband MSE, $J\left(H_{\text{W}}\right)$, the broadband noise reduction factor, $\xi_{\text{n}}\left(H_{\text{W}}\right)$, and the broadband desired signal reduction factor, $\xi_{\text{d}}\left(H_{\text{W}}\right)$, as a function of the broadband input SNR. As the input SNR increases, the less the noise that needs to be suppressed, and the less the distortion that is introduced into the filtered desired signal. ■

Example 3.4.2 Suppose that the desired signal is a harmonic pulse of T samples:

$$x(t) = \begin{cases} A\sin\left(2\pi f_0 t + \phi\right), & 0 \leq t \leq T-1 \\ 0, & t < 0,\ t \geq T \end{cases},$$

with fixed amplitude A and frequency f_0, and random phase ϕ, uniformly distributed on the interval from 0 to 2π. This signal needs to be recovered from the noisy observation, $y(t) = x(t) + v(t)$, where $v(t)$ is additive white Gaussian noise; in other words, $v(t) \sim \mathcal{N}\left(0, \sigma_v^2\right)$, which is uncorrelated with $x(t)$.

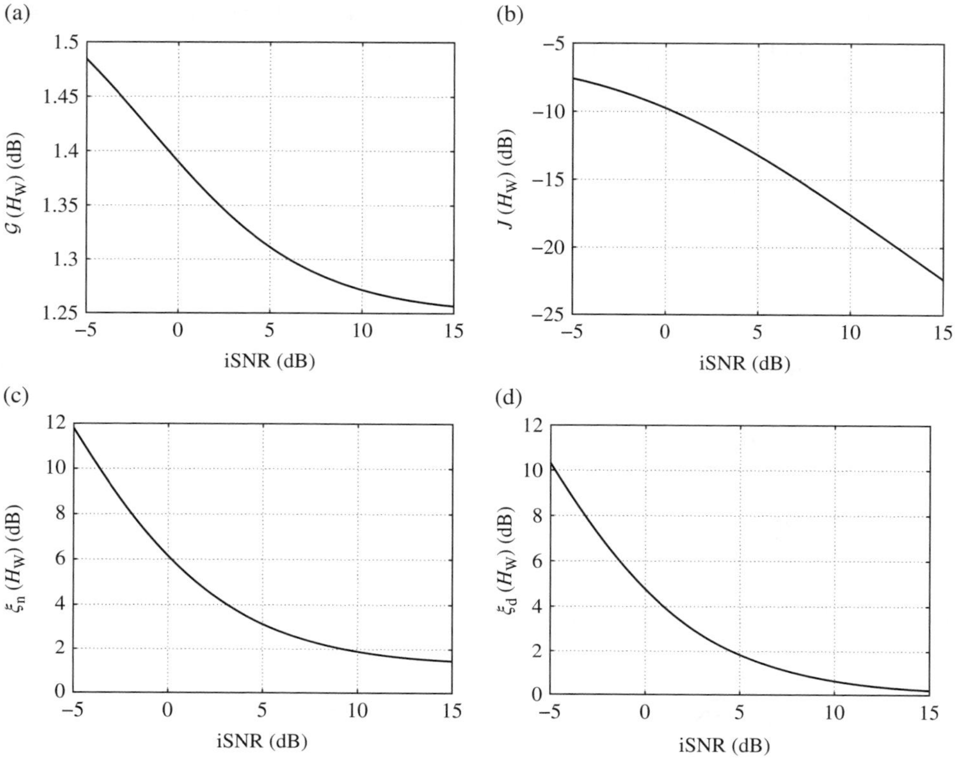

Figure 3.3 (a) The broadband gain in SNR, (b) the broadband MSE, (c) the broadband noise reduction factor, and (d) the broadband desired signal reduction factor of the Wiener gain as a function of the broadband input SNR.

The frequency-domain representation of the desired signal is given by

$$
\begin{aligned}
X(f) &= \sum_{t=-\infty}^{\infty} x(t) e^{\jmath 2\pi f t} \\
&= \sum_{t=0}^{T-1} A \sin\left(2\pi f_0 t + \phi\right) e^{\jmath 2\pi f t} \\
&= \frac{A}{2\jmath} e^{\jmath\phi + \jmath\pi(f+f_0)(T-1)} D_T\left[\pi\left(f + f_0\right)\right] \\
&\quad + \frac{A}{2\jmath} e^{-\jmath\phi + \jmath\pi(f-f_0)(T-1)} D_T\left[\pi\left(f - f_0\right)\right],
\end{aligned}
$$

where the function $D_T(x)$ is the Dirichlet kernel, defined as

$$
D_T(x) = \frac{\sin(Tx)}{\sin(x)}.
$$

Hence, the variance of $X(f)$ is

$$\phi_X(f) = \frac{A^2}{4} D_T^2 \left[\pi \left(f + f_0\right)\right] + \frac{A^2}{4} D_T^2 \left[\pi \left(f - f_0\right)\right].$$

The frequency-domain representation of the noise signal is

$$V(f) = \sum_{t=0}^{T-1} v(t) e^{j2\pi ft}.$$

Hence, the variance of $V(f)$ is $\phi_V(f) = T\sigma_v^2$. The narrowband input SNR is

$$\begin{aligned} \text{iSNR}(f) &= \frac{\phi_X(f)}{\phi_V(f)} \\ &= \frac{A^2}{4T\sigma_v^2} D_T^2 \left[\pi \left(f + f_0\right)\right] + \frac{A^2}{4T\sigma_v^2} D_T^2 \left[\pi \left(f - f_0\right)\right] \end{aligned}$$

and the broadband input SNR is

$$\begin{aligned} \text{iSNR} &= \frac{\int_f \phi_X(f) df}{\int_f \phi_V(f) df} \\ &= \frac{\sum_t E\left[|x(t)|^2\right]}{\sum_t E\left[|v(t)|^2\right]} \\ &= \frac{A^2}{2\sigma_v^2}, \end{aligned}$$

where we have used Parseval's identity. The optimal Wiener gain is obtained from (3.37).

To demonstrate the performance of the Wiener gain, we choose $A = 0.5, f_0 = 0.1$, and $T = 500$. Figure 3.4 shows plots of the broadband gain in SNR, the broadband MSE, $J\left(H_{\text{W}}\right)$, the broadband noise reduction factor, $\xi_{\text{n}}\left(H_{\text{W}}\right)$, and the broadband desired signal reduction factor, $\xi_{\text{d}}\left(H_{\text{W}}\right)$, as a function of the broadband input SNR. Figure 3.5 shows a realization of the noise corrupted and filtered sinusoidal signals for iSNR = 0 dB. ■

An important gain can be designed by minimizing the desired signal-distortion-based MSE with the noise-reductionraint that the noise-reduction-based MSE is equal to a positive number smaller than the level of the original noise. This optimization problem can be translated mathematically as

$$\min_{H(f)} J_{\text{d}}\left[H(f)\right] \quad \text{subject to} \quad J_{\text{n}}\left[H(f)\right] = \aleph \phi_V(f), \tag{3.52}$$

where

$$J_{\text{d}}\left[H(f)\right] = |1 - H(f)|^2 \phi_X(f), \tag{3.53}$$

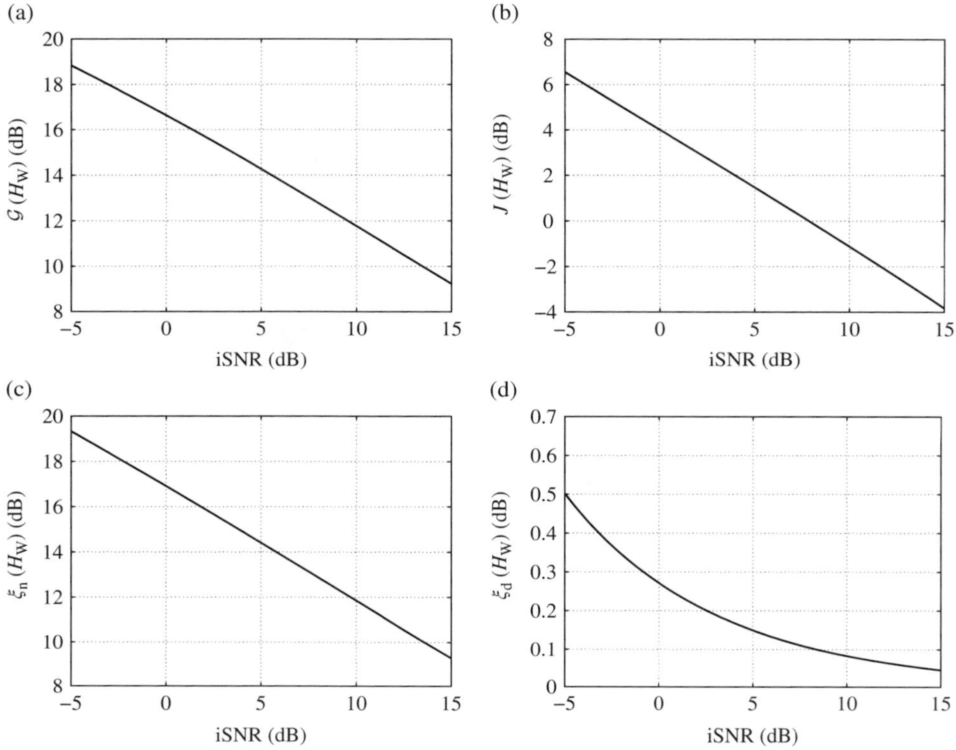

Figure 3.4 (a) The broadband gain in SNR, (b) the broadband MSE, (c) the broadband noise reduction factor, and (d) the broadband desired signal reduction factor of the Wiener gain as a function of the broadband input SNR.

$$J_n\left[H(f)\right] = |H(f)|^2\,\phi_V(f), \tag{3.54}$$

and $0 < \aleph < 1$ to ensure that we have some noise reduction at frequency f. If we use a Lagrange multiplier, $\mu(f) \geq 0$, to adjoin the constraint to the cost function, we easily find the tradeoff gain:

$$\begin{aligned} H_{T,\mu}(f) &= \frac{\phi_X(f)}{\phi_X(f) + \mu(f)\phi_V(f)} \\ &= \frac{\phi_Y(f) - \phi_V(f)}{\phi_Y(f) + \left[\mu(f) - 1\right]\phi_V(f)} \\ &= \frac{\text{iSNR}(f)}{\mu(f) + \text{iSNR}(f)}. \end{aligned} \tag{3.55}$$

This gain can be seen as a Wiener gain with an adjustable input noise level $\mu(f)\phi_V(f)$. Obviously, the particular case of $\mu(f) = 1$ corresponds to the Wiener gain.

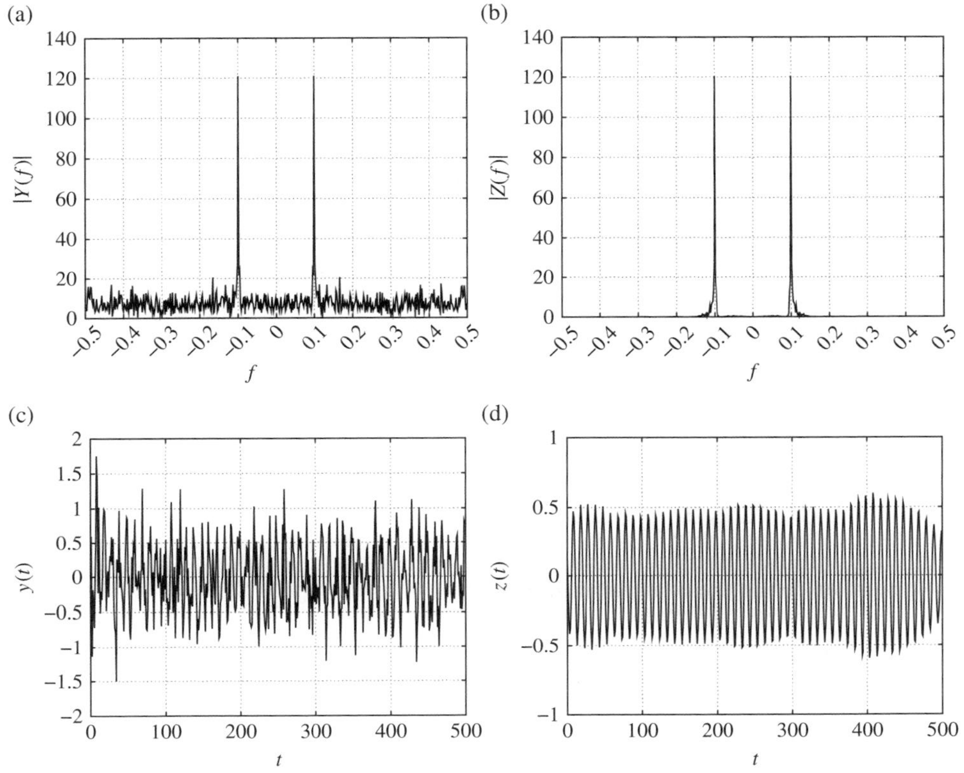

Figure 3.5 Example of noise corrupted and Wiener filtered sinusoidal signals for iSNR = 0 dB. (a) Magnitude of frequency-domain observation signal, $|Y(f)|$, (b) magnitude of frequency-domain estimated signal, $|Z(f)|$, (c) time-domain observation signal, $y(t)$, and (d) time-domain estimated signal, $z(t)$.

We can also find the optimal $\mu(f)$ corresponding to a given value of $\aleph$. Substituting $H_{\mathrm{T},\mu}(f)$ from (3.55) into the constraint in (3.52), we get

$$\begin{aligned} J_{\mathrm{n}}\left[H_{\mathrm{T},\mu}(f)\right] &= \left|H_{\mathrm{T},\mu}(f)\right|^2 \phi_V(f) \\ &= \aleph \phi_V(f). \end{aligned} \tag{3.56}$$

From the previous expression, we easily find that

$$\mu(f) = \mathrm{iSNR}(f) \frac{1-\sqrt{\aleph}}{\sqrt{\aleph}} \tag{3.57}$$

and the tradeoff simplifies to a constant gain:

$$H_{\mathrm{T},\aleph} = \sqrt{\aleph}. \tag{3.58}$$

In the rest, we assume that $\mu(f)$ is a constant, so it does not depend on frequency and we can drop the variable f. Usually, the value of μ is given by design.

The MSCF between the two signals $X(f)$ and $X(f) + \sqrt{\mu}V(f)$ at frequency f is

$$\left|\rho\left[X(f), X(f) + \sqrt{\mu}V(f)\right]\right|^2 = \frac{\text{iSNR}(f)}{\mu + \text{iSNR}(f)}. \tag{3.59}$$

The MSCF between the two signals $V(f)$ and $X(f) + \sqrt{\mu}V(f)$ at frequency f is

$$\left|\rho\left[V(f), X(f) + \sqrt{\mu}V(f)\right]\right|^2 = \frac{\mu}{\mu + \text{iSNR}(f)}. \tag{3.60}$$

Therefore, we can write the tradeoff gain as a function of these two MSCFs:

$$\begin{aligned} H_{\text{T},\mu}(f) &= \left|\rho\left[X(f), X(f) + \sqrt{\mu}V(f)\right]\right|^2 \\ &= 1 - \left|\rho\left[V(f), X(f) + \sqrt{\mu}V(f)\right]\right|^2. \end{aligned} \tag{3.61}$$

Now, let us define the complex number[2]:

$$\begin{aligned} \varrho_\mu\left[X(f), V(f)\right] &= \rho\left[X(f), X(f) + \sqrt{\mu}V(f)\right] \\ &\quad + \jmath\rho\left[V(f), X(f) + \sqrt{\mu}V(f)\right] \\ &= \cos\theta_\mu(f) + \jmath \sin\theta_\mu(f), \end{aligned} \tag{3.62}$$

where $\theta_\mu(f)$ is the phase of $\varrho_\mu\left[X(f), V(f)\right]$ whose modulus is equal to 1. On the complex plane, $\varrho_\mu\left[X(f), V(f)\right]$ is on the unit circle. Since $0 \leq \rho\left[X(f), X(f) + \sqrt{\mu}V(f)\right] \leq 1$ and $0 \leq \rho\left[V(f), X(f) + \sqrt{\mu}V(f)\right] \leq 1$, therefore $0 \leq \theta_\mu(f) \leq \frac{\pi}{2}$. We can then rewrite the tradeoff gain as a function of the angle $\theta_\mu(f)$:

$$\begin{aligned} H_{\text{T},\mu}(f) &= \cos^2\theta_\mu(f) \\ &= 1 - \sin^2\theta_\mu(f). \end{aligned} \tag{3.63}$$

We deduce all the narrowband performance measures with the tradeoff gain:

$$\text{oSNR}\left[H_{\text{T},\mu}(f)\right] = \text{iSNR}(f), \tag{3.64}$$

$$\xi_\text{n}\left[H_{\text{T},\mu}(f)\right] = \frac{1}{\cos^4\theta_\mu(f)} \geq 1, \tag{3.65}$$

$$\xi_\text{d}\left[H_{\text{T},\mu}(f)\right] = \frac{1}{\cos^4\theta_\mu(f)} \geq 1, \tag{3.66}$$

$$\upsilon_\text{d}\left[H_{\text{T},\mu}(f)\right] = \sin^4\theta_\mu(f) \leq 1. \tag{3.67}$$

2 Notice that both $\rho\left[X(f), X(f) + \sqrt{\mu}V(f)\right]$ and $\rho\left[V(f), X(f) + \sqrt{\mu}V(f)\right]$ are real numbers.

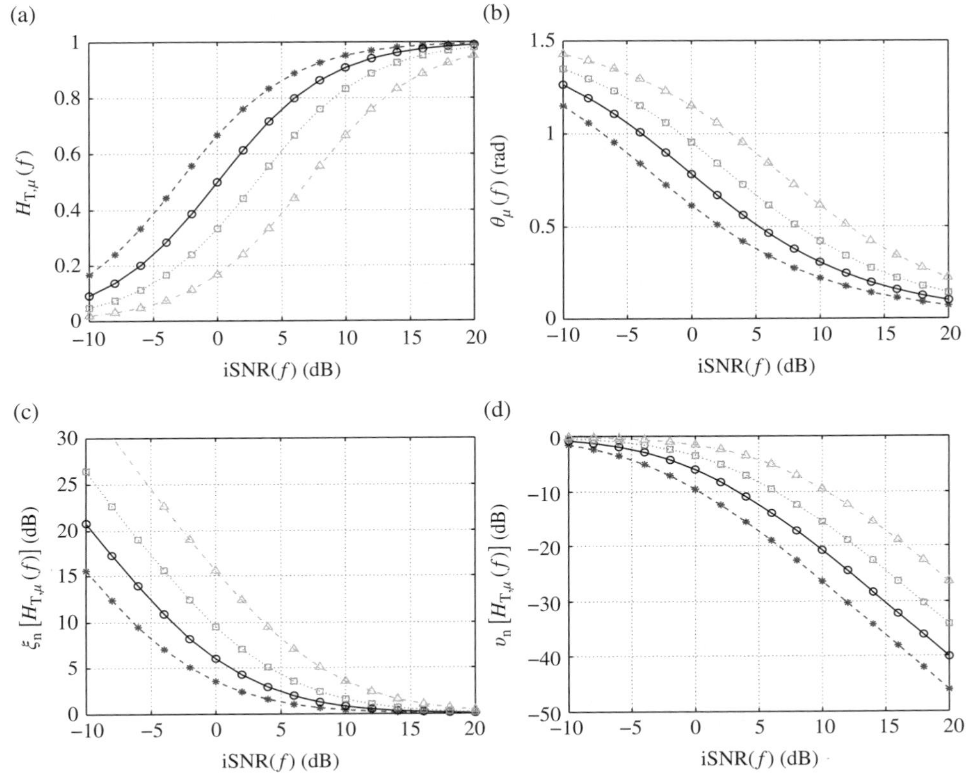

Figure 3.6 (a) The tradedoff gain, (b) the angle, (c) the narrowband noise reduction factor, and (d) the narrowband desired signal distortion index as a function of the narrowband input SNR for different values of μ: $\mu = 0.5$ (dashed line with asterisks), $\mu = 1$ (solid line with circles), $\mu = 2$ (dotted line with squares), and $\mu = 5$ (dash-dot line with triangles).

Figure 3.6 shows plots of the tradedoff gain, $H_{T,\mu}(f)$, the angle, $\theta_\mu(f)$, the narrowband noise reduction factor, $\xi_n\left[H_{T,\mu}(f)\right]$, and the narrowband desired signal distortion index, $\upsilon_d\left[H_{T,\mu}(f)\right]$, as a function of the narrowband input SNR for different values of μ. For a given input SNR, the higher the value of μ, the lower the tradedoff gain. Hence, both the noise reduction factor and the desired signal distortion index monotonically increase as a function of μ.

We give the following fundamental property about the broadband output SNR with the tradeoff gain.

Property 3.4.2 With the tradeoff gain given in (3.55), the broadband output SNR is always greater than or equal to the broadband input SNR; in other words, $\text{oSNR}\left(H_{T,\mu}\right) \geq \text{iSNR}, \forall \mu \geq 0$.

Proof. The broadband MSCF between the two variables $X(f)$ and $X(f) + \sqrt{\mu}V(f)$ is

$$\begin{aligned}
\left|\rho\left(X, X + \sqrt{\mu}V\right)\right|^2 &= \frac{\left[\int_f \phi_X(f)df\right]^2}{\left[\int_f \phi_X(f)df\right]\left[\int_f \phi_X(f)df + \mu \int_f \phi_V(f)df\right]} \\
&= \frac{\text{iSNR}}{\mu + \text{iSNR}}.
\end{aligned}$$

The broadband MSCF between the two variables $X(f)$ and $H_{\text{T},\mu}(f)X(f) + \sqrt{\mu}H_{\text{T},\mu}(f)$ $V(f)$ is

$$\begin{aligned}
&\left|\rho\left(X, H_{\text{T},\mu}X + \sqrt{\mu}H_{\text{T},\mu}V\right)\right|^2 \\
&= \frac{\left[\int_f H_{\text{T},\mu}(f)\phi_X(f)df\right]^2}{\left[\int_f \phi_X(f)df\right]\left[\int_f H^2_{\text{T},\mu}(f)\phi_X(f)df + \mu \int_f H^2_{\text{T},\mu}(f)\phi_V(f)df\right]} \\
&= \frac{\int_f H_{\text{T},\mu}(f)\phi_X(f)df}{\int_f \phi_X(f)df}.
\end{aligned}$$

Another way to write the same broadband MSCF is as follows:

$$\begin{aligned}
\left|\rho\left(X, H_{\text{T},\mu}X + \sqrt{\mu}H_{\text{T},\mu}V\right)\right|^2 &= \frac{\left[\int_f H_{\text{T},\mu}(f)\phi_X(f)df\right]^2}{\left[\int_f \phi_X(f)df\right]\left[\int_f H^2_{\text{T},\mu}(f)\phi_X(f)df\right]} \\
&\quad \times \frac{\text{oSNR}\left(H_{\text{T},\mu}\right)}{\mu + \text{oSNR}\left(H_{\text{T},\mu}\right)} \\
&= \left|\rho\left(X, H_{\text{T},\mu}X\right)\right|^2 \\
&\quad \times \left|\rho\left(H_{\text{T},\mu}X, H_{\text{T},\mu}X + \sqrt{\mu}H_{\text{T},\mu}V\right)\right|^2 \\
&\le \frac{\text{oSNR}\left(H_{\text{T},\mu}\right)}{\mu + \text{oSNR}\left(H_{\text{T},\mu}\right)}.
\end{aligned}$$

Now, let us evaluate the broadband MSCF between the two variables $X(f) + \sqrt{\mu}V(f)$ and $H_{\text{T},\mu}(f)X(f) + \sqrt{\mu}H_{\text{T},\mu}(f)V(f)$:

$$\begin{aligned}
\left|\rho\left(X + \sqrt{\mu}V, H_{\text{T},\mu}X + \sqrt{\mu}H_{\text{T},\mu}V\right)\right|^2 &= \frac{\int_f \phi_X(f)df}{\int_f \phi_X(f)df + \mu \int_f \phi_V(f)df} \\
&\quad \times \frac{\int_f \phi_X(f)df}{\int_f H_{\text{T},\mu}(f)\phi_X(f)df} \\
&= \frac{\left|\rho\left(X, X + \sqrt{\mu}V\right)\right|^2}{\left|\rho\left(X, H_{\text{T},\mu}X + \sqrt{\mu}H_{\text{T},\mu}V\right)\right|^2}.
\end{aligned}$$

Therefore,

$$\begin{aligned}\left|\rho\left(X, X+\sqrt{\mu}V\right)\right|^2 &= \frac{\text{iSNR}}{\mu+\text{iSNR}} \\ &= \left|\rho\left(X+\sqrt{\mu}V, H_{\text{T},\mu}X+\sqrt{\mu}H_{\text{T},\mu}V\right)\right|^2 \\ &\quad \times \left|\rho\left(X, H_{\text{T},\mu}X+\sqrt{\mu}H_{\text{T},\mu}V\right)\right|^2 \\ &\leq \left|\rho\left(X, H_{\text{T},\mu}X+\sqrt{\mu}H_{\text{T},\mu}V\right)\right|^2 \\ &\leq \frac{\text{oSNR}\left(H_{\text{T},\mu}\right)}{\mu+\text{oSNR}\left(H_{\text{T},\mu}\right)}.\end{aligned}$$

As a result, we have

$$\text{oSNR}\left(H_{\text{T},\mu}\right) \geq \text{iSNR}.$$

■

Example 3.4.3 Returning to Example 3.4.1, Figure 3.7 shows plots of the broadband gain in SNR, $\mathcal{G}\left(H_{\text{T},\mu}\right)$, the broadband MSE, $J\left(H_{\text{T},\mu}\right)$, the broadband noise reduction factor, $\xi_{\text{n}}\left(H_{\text{T},\mu}\right)$, and the broadband desired signal reduction factor, $\xi_{\text{d}}\left(H_{\text{T},\mu}\right)$, as a function of the broadband input SNR for different values of μ. Figure 3.8 shows similar plots for the signals in Example 3.4.2.

For a given broadband input SNR, the higher the value of μ, the higher the broadband SNR gain and noise reduction, but at the expense of higher broadband desired signal reduction. ■

Some applications may need aggressive noise reduction while others may require minimal desired signal distortion (and so less aggressive noise reduction). An easy way to control the compromise between noise reduction and signal distortion is via the parametric Wiener gain[3] [2, 3]:

$$H_{\mu_1,\mu_2}(f) = \left[1 - \sin^{\mu_1}\theta(f)\right]^{\mu_2}, \tag{3.68}$$

where μ_1 and μ_2 are two positive parameters that allow for control of this compromise. For $(\mu_1, \mu_2) = (2, 1)$, we get the Wiener gain developed previously. Taking $(\mu_1, \mu_2) = (2, 1/2)$, leads to

$$\begin{aligned}H_{\text{pow}}(f) &= \sqrt{1-\sin^2\theta(f)} \\ &= \cos\theta(f),\end{aligned} \tag{3.69}$$

3 There is nothing optimal about the parametric Wiener gain but, for convenience of presentation, we included it in this section.

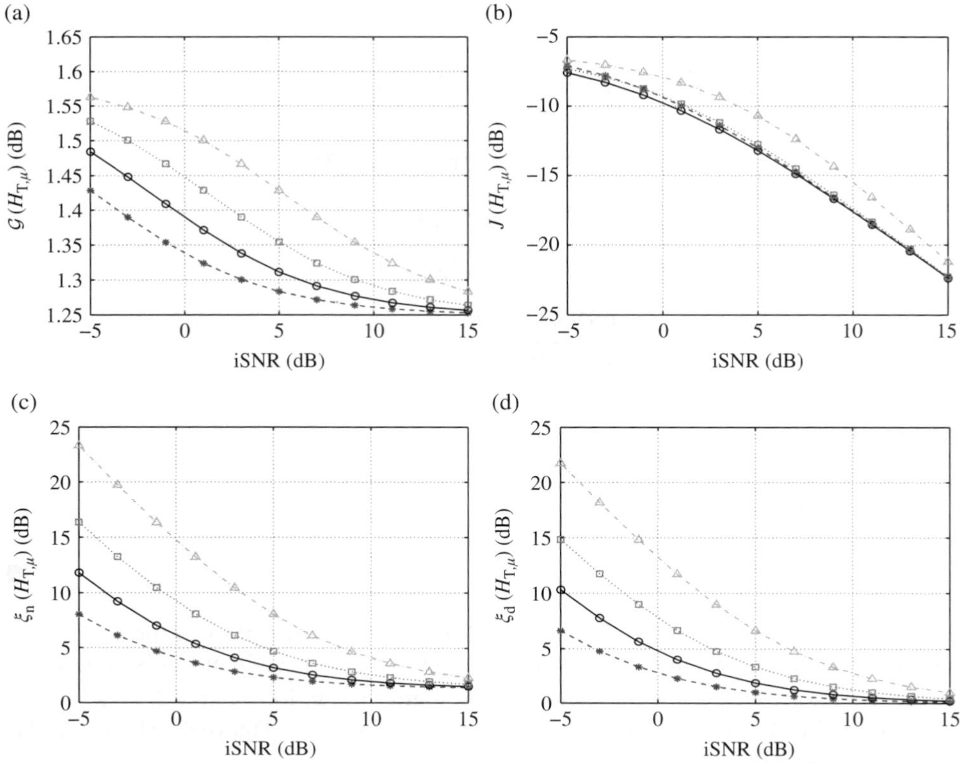

Figure 3.7 (a) The broadband gain in SNR, (b) the broadband MSE, (c) the broadband noise reduction factor, and (d) the broadband desired signal reduction factor of the tradeoff gain as a function of the broadband input SNR for different values of μ: $\mu = 0.5$ (dashed line with asterisks), $\mu = 1$ (solid line with circles), $\mu = 2$ (dotted line with squares), and $\mu = 5$ (dash-dot line with triangles).

which is the power subtraction method [2–6]. The pair $(\mu_1, \mu_2) = (1, 1)$ gives the magnitude subtraction method [7–11]:

$$\begin{aligned} H_{\rm mag}(f) &= 1 - \sin\theta(f) \\ &= 1 - \sqrt{1 - \cos^2\theta(f)}. \end{aligned} \tag{3.70}$$

We can verify that the narrowband noise reduction factors for the power subtraction and magnitude subtraction methods are

$$\xi_{\rm n}\left[H_{\rm pow}(f)\right] = \frac{1}{\cos^2\theta(f)}, \tag{3.71}$$

$$\xi_{\rm n}\left[H_{\rm mag}(f)\right] = \frac{1}{\left[1 - \sin\theta(f)\right]^2}, \tag{3.72}$$

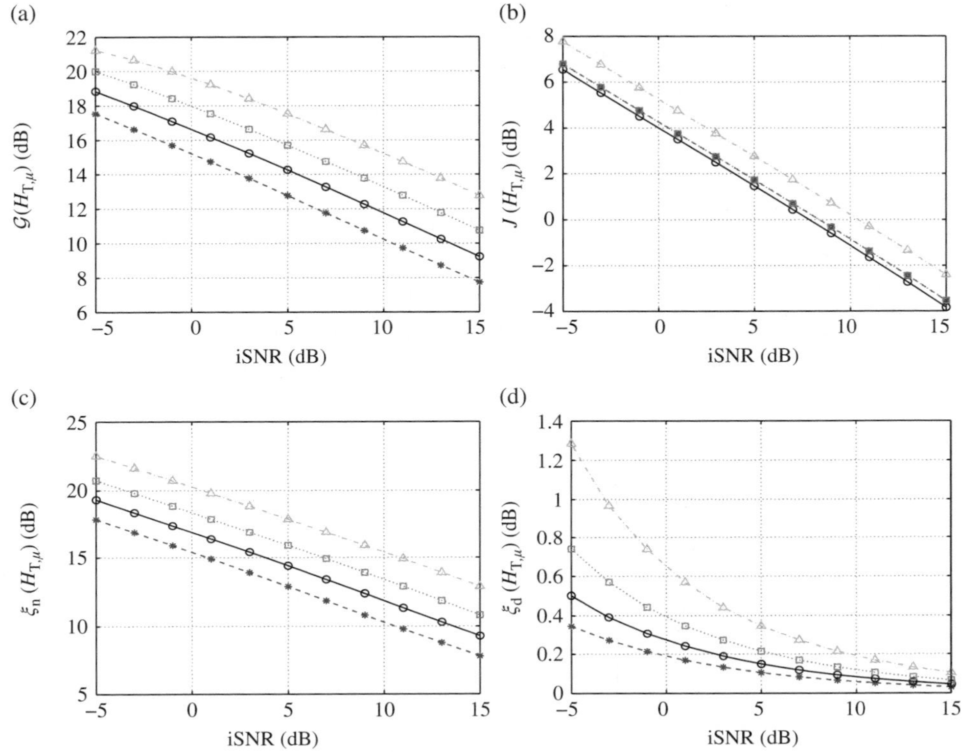

Figure 3.8 (a) The broadband gain in SNR, (b) the broadband MSE, (c) the broadband noise reduction factor, and (d) the broadband desired signal reduction factor of the tradeoff gain as a function of the broadband input SNR for different values of μ: $\mu = 0.5$ (dashed line with asterisks), $\mu = 1$ (solid line with circles), $\mu = 2$ (dotted line with squares), and $\mu = 5$ (dash-dot line with triangles).

and the corresponding narrowband desired signal distortion indices are

$$\upsilon_d \left[H_{pow}(f)\right] = \left[1 - \cos\theta(f)\right]^2, \tag{3.73}$$

$$\upsilon_d \left[H_{mag}(f)\right] = \sin^2\theta(f). \tag{3.74}$$

We can also easily check that

$$\xi_n \left[H_{mag}(f)\right] \geq \xi_n \left[H_W(f)\right] \geq \xi_n \left[H_{pow}(f)\right], \tag{3.75}$$

$$\upsilon_d \left[H_{pow}(f)\right] \leq \upsilon_d \left[H_W(f)\right] \leq \upsilon_d \left[H_{mag}(f)\right]. \tag{3.76}$$

These two inequalities are very important from a practical point of view. They show that, of the three methods, magnitude subtraction is the most aggressive as far as noise reduction is concerned, a very well-known fact in the literature [12] but, at the same time, it is the one that will likely distort the desired signal the most. The smoothest approach is power subtraction while the Wiener gain is between the two others in

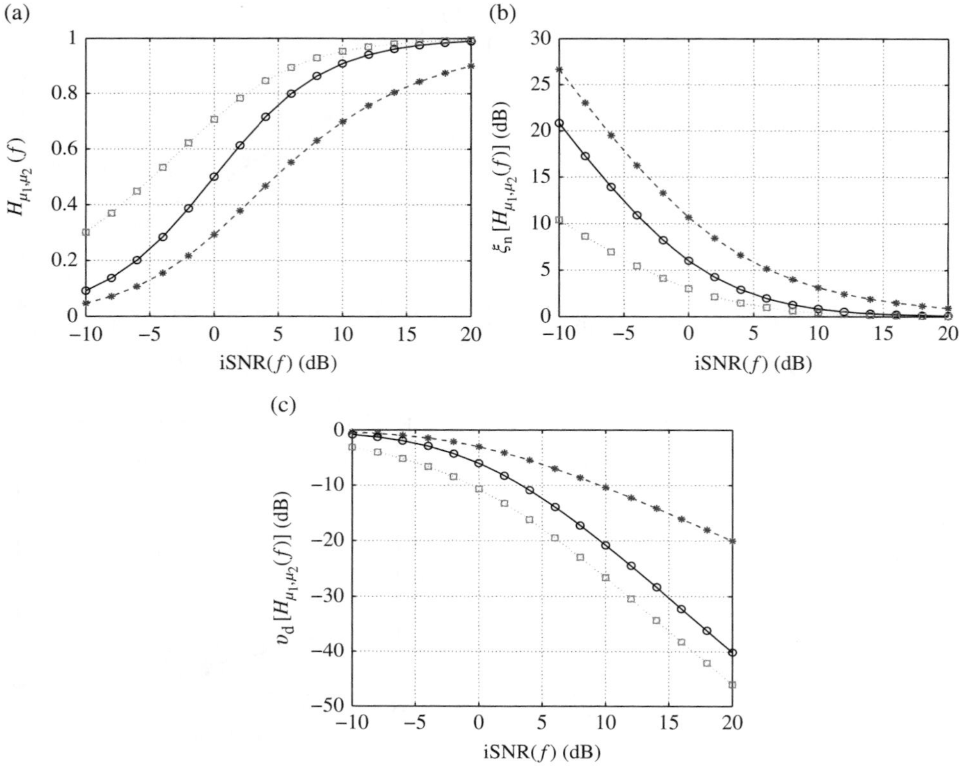

Figure 3.9 (a) The parametric Wiener gain, (b) the narrowband noise reduction factor, and (c) the narrowband desired signal distortion index as a function of the narrowband input SNR for different values of (μ_1, μ_2): magnitude subtraction with $(\mu_1, \mu_2) = (1, 1)$ (dashed line with asterisks), Wiener gain with $(\mu_1, \mu_2) = (2, 1)$ (solid line with circles), and power subtraction with $(\mu_1, \mu_2) = (2, 1/2)$ (dotted line with squares).

terms of desired signal distortion and noise reduction. Several other variants of these algorithms can be found in the literature [13–15].

Figure 3.9 shows plots of the parametric Wiener gain, $H_{\mu_1,\mu_2}(f)$, the narrowband noise reduction factor, $\xi_n\left[H_{\mu_1,\mu_2}(f)\right]$, and the narrowband desired signal distortion index, $\upsilon_d\left[H_{\mu_1,\mu_2}(f)\right]$, as a function of the narrowband input SNR for different values of the pair (μ_1, μ_2). For a given input SNR, $H_W(f)$ is larger than $H_{mag}(f)$ and smaller than $H_{pow}(f)$. Hence the magnitude subtraction method is associated with higher noise reduction and desired signal distortion than the Wiener method, while the power subtraction method is associated with less noise reduction and desired signal distortion than the Wiener method.

Example 3.4.4 Returning to Example 3.4.1, Figure 3.10 shows plots of the broadband gain in SNR, $\mathcal{G}\left(H_{\mu_1,\mu_2}\right)$, the broadband MSE, $J\left(H_{\mu_1,\mu_2}\right)$, the broadband noise reduction factor, $\xi_n\left(H_{\mu_1,\mu_2}\right)$, and the broadband desired signal reduction factor, $\xi_d\left(H_{\mu_1,\mu_2}\right)$, as a function of the broadband input SNR for different values of (μ_1, μ_2). Figure 3.11 shows similar plots for the signals in Example 3.4.2.

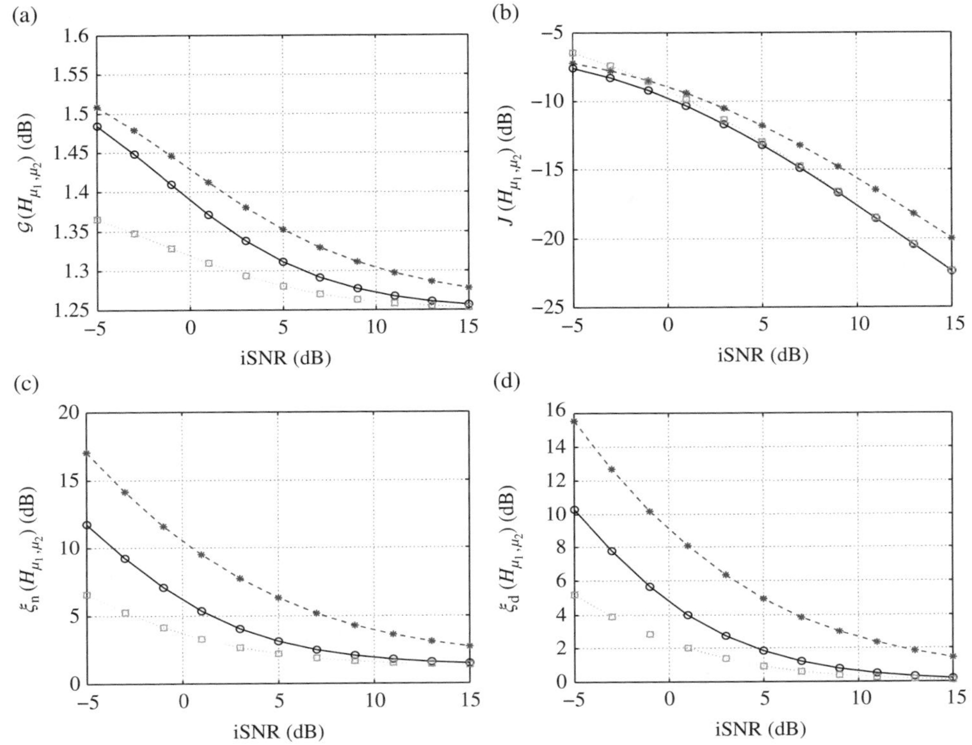

Figure 3.10 (a) The broadband gain in SNR, (b) the broadband MSE, (c) the broadband noise reduction factor, and (d) the broadband desired signal reduction factor of the parametric Wiener gain as a function of the broadband input SNR for different values of (μ_1, μ_2): magnitude subtraction with $(\mu_1, \mu_2) = (1, 1)$ (dashed line with asterisks), Wiener gain with $(\mu_1, \mu_2) = (2, 1)$ (solid line with circles), and power subtraction with $(\mu_1, \mu_2) = (2, 1/2)$ (dotted line with squares).

For a given broadband input SNR, the magnitude subtraction method is associated with higher broadband SNR gain and noise reduction than the Wiener method, but at the expense of higher broadband desired signal reduction. On the other hand, the power subtraction method is associated with lower broadband desired signal reduction than the Wiener method, but at the expense of lower broadband SNR gain and noise reduction. ■

3.5 Constraint Wiener Gains

In this section, we slightly change the notation for convenience. From the previous section, we know that the traditional way to estimate the desired signal, $X(f)$, is by applying a gain, $H_X(f)$, to the observation, $Y(f)$:

$$\hat{X}(f) = Y(f)H_X(f). \tag{3.77}$$

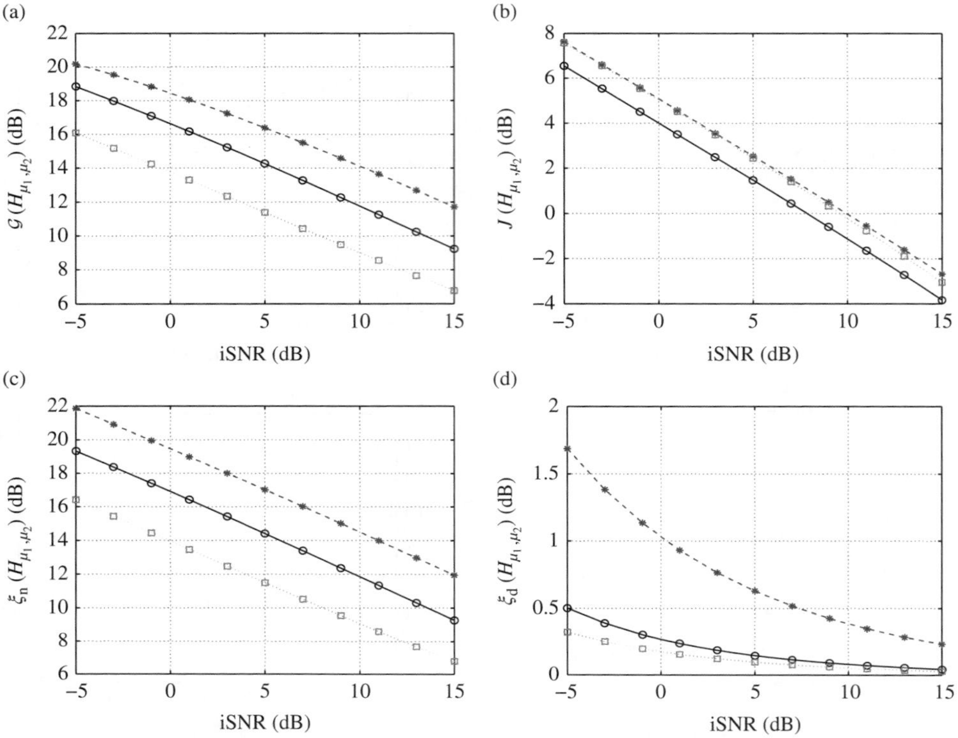

Figure 3.11 (a) The broadband gain in SNR, (b) the broadband MSE, (c) the broadband noise reduction factor, and (d) the broadband desired signal reduction factor of the parametric Wiener gain as a function of the broadband input SNR for different values of (μ_1, μ_2): magnitude subtraction with $(\mu_1, \mu_2) = (1, 1)$ (dashed line with asterisks), Wiener gain with $(\mu_1, \mu_2) = (2, 1)$ (solid line with circles), and power subtraction with $(\mu_1, \mu_2) = (2, 1/2)$ (dotted line with squares).

One reasonable way to find this gain is via the MSE criterion given by

$$J_X\left[H_X(f)\right] = E\left[\left|\widehat{X}(f) - X(f)\right|^2\right]. \tag{3.78}$$

The minimization of the last expression leads to the conventional Wiener gain given in (3.37), which we now denote by $H_{X,\mathrm{W}}(f)$. Therefore, the optimal estimate (in the MMSE sense) of $X(f)$ and the MMSE are, respectively,

$$\widehat{X}_{\mathrm{W}}(f) = H_{X,\mathrm{W}}(f)Y(f) \tag{3.79}$$

and

$$J_X\left[H_{X,\mathrm{W}}(f)\right] = \phi_X(f) - \phi_{\widehat{X}_{\mathrm{W}}}(f), \tag{3.80}$$

where

$$\phi_{\widehat{X}_{\mathrm{W}}}(f) = \frac{\phi_X^2(f)}{\phi_Y(f)} \tag{3.81}$$

is the variance of $\widehat{X}_{\mathrm{W}}(f)$.

Alternatively, we can also estimate the noise signal, $V(f)$, by applying a gain, $H_V(f)$, to the observation, $Y(f)$:

$$\widehat{V}(f) = Y(f)H_V(f). \tag{3.82}$$

By using the MSE criterion:

$$J_V\left[H_V(f)\right] = E\left[\left|\widehat{V}(f) - V(f)\right|^2\right], \tag{3.83}$$

we easily find that the optimal gain and estimator are, respectively,

$$\begin{aligned} H_{V,\mathrm{W}}(f) &= \frac{\phi_V(f)}{\phi_Y(f)} \\ &= \frac{1}{1 + \mathrm{iSNR}(f)} \end{aligned} \tag{3.84}$$

and

$$\widehat{V}_{\mathrm{W}}(f) = H_{V,\mathrm{W}}(f)Y(f). \tag{3.85}$$

We also find that the MMSE is

$$J_V\left[H_{V,\mathrm{W}}(f)\right] = \phi_V(f) - \phi_{\widehat{V}_{\mathrm{W}}}(f), \tag{3.86}$$

where

$$\phi_{\widehat{V}_{\mathrm{W}}}(f) = \frac{\phi_V^2(f)}{\phi_Y(f)} \tag{3.87}$$

is the variance of $\widehat{V}_{\mathrm{W}}(f)$. Now that we have the optimal estimate of $V(f)$, we can estimate $X(f)$ as follows:

$$\begin{aligned} \widehat{X}_{\mathrm{W},2}(f) &= Y(f) - \widehat{V}_{\mathrm{W}}(f) \\ &= \widehat{X}_{\mathrm{W}}(f). \end{aligned} \tag{3.88}$$

Obviously, the two methods are strictly equivalent here. It is easy to show that

$$\begin{aligned} J_X\left[H_{X,\mathrm{W}}(f)\right] &= J_V\left[H_{V,\mathrm{W}}(f)\right] \\ &= E\left[\widehat{X}_{\mathrm{W}}(f)\widehat{V}_{\mathrm{W}}^*(f)\right] \\ &= \frac{\phi_X(f)\phi_V(f)}{\phi_Y(f)}, \end{aligned} \tag{3.89}$$

which is the conditional variance of $X(f)$ given $Y(f)$ or the conditional variance of $V(f)$ given $Y(f)$. Also, it is interesting to observe that the sum of the estimated desired and noise signals is equal to the observation; that is,

$$\widehat{X}_{\mathrm{W}}(f) + \widehat{V}_{\mathrm{W}}(f) = Y(f), \tag{3.90}$$

which is equivalent to

$$H_{X,\mathrm{W}}(f) + H_{V,\mathrm{W}}(f) = 1, \tag{3.91}$$

assuming that $Y(f) \neq 0$. Then we can state that the Wiener gains are derived in such a way that (3.90) is verified. However, the sum of the variances of the estimated desired and noise signals is not equal to the variance of the observation:

$$\begin{aligned} \phi_{\widehat{X}_{\mathrm{W}}}(f) + \phi_{\widehat{V}_{\mathrm{W}}}(f) &= \frac{\phi_X^2(f) + \phi_V^2(f)}{\phi_Y(f)} \\ &\geq \phi_Y(f). \end{aligned} \tag{3.92}$$

This is due to the fact that $\widehat{X}_{\mathrm{W}}(f)$ and $\widehat{V}_{\mathrm{W}}(f)$ are correlated, as shown in (3.89). At first glance, this may come as a surprise to some readers since $X(f)$ and $V(f)$ are uncorrelated but this result actually makes sense.

Let us define the MSE criterion:

$$\begin{aligned} J\left[H_X(f), H_V(f)\right] &= J_X\left[H_X(f)\right] + J_V\left[H_V(f)\right] \\ &= E\left[\left|H_X(f)Y(f) - X(f)\right|^2\right] \\ &\quad + E\left[\left|H_V(f)Y(f) - V(f)\right|^2\right]. \end{aligned} \tag{3.93}$$

The minimization of $J\left[H_X(f), H_V(f)\right]$ without any constraint or with the constraint that $\widehat{X}(f) + \widehat{V}(f) = Y(f)$ – in other words that $H_X(f) + H_V(f) = 1$ – leads to $H_{X,\mathrm{W}}(f)$ and $H_{V,\mathrm{W}}(f)$. Another interesting possibility is to minimize $J\left[H_X(f), H_V(f)\right]$ with the constraint that the sum of the variances of the estimated desired and noise signals is equal to the variance of the observation:

$$\phi_{\widehat{X}}(f) + \phi_{\widehat{V}}(f) = \phi_Y(f) \tag{3.94}$$

or, equivalently,

$$\left|H_X(f)\right|^2 + \left|H_V(f)\right|^2 = 1. \tag{3.95}$$

By using the Lagrange multiplier technique, we easily find that the constraint Wiener gains for the estimation of the desired and noise signals are, respectively,

$$\begin{aligned} H_{X,\mathrm{cW}}(f) &= \frac{\phi_X(f)}{\sqrt{\phi_X^2(f) + \phi_V^2(f)}} \\ &= \sqrt{\frac{\mathrm{iSNR}^2(f)}{1 + \mathrm{iSNR}^2(f)}} \end{aligned} \tag{3.96}$$

and

$$H_{V,\mathrm{cW}}(f) = \frac{\phi_V(f)}{\sqrt{\phi_X^2(f) + \phi_V^2(f)}} = \sqrt{\frac{1}{1 + \mathrm{iSNR}^2(f)}}. \quad (3.97)$$

Then, we deduce two different estimators for $X(f)$:

$$\widehat{X}_{\mathrm{cW}}(f) = H_{X,\mathrm{cW}}(f)Y(f) \quad (3.98)$$

and

$$\begin{aligned}\widehat{X}_{\mathrm{cW},2}(f) &= Y(f) - \widehat{V}_{\mathrm{cW}}(f) \\ &= Y(f) - H_{V,\mathrm{cW}}(f)Y(f) \\ &= H_{X,\mathrm{cW},2}(f)Y(f),\end{aligned} \quad (3.99)$$

where

$$\widehat{V}_{\mathrm{cW}}(f) = H_{V,\mathrm{cW}}(f)Y(f) \quad (3.100)$$

and

$$H_{X,\mathrm{cW},2}(f) = 1 - H_{V,\mathrm{cW}}(f). \quad (3.101)$$

Now, contrary to the conventional Wiener approach, $\widehat{X}_{\mathrm{cW}}(f) \neq \widehat{X}_{\mathrm{cW},2}(f)$. It can be verified that

$$E\left[\widehat{X}_{\mathrm{cW}}(f)\widehat{V}_{\mathrm{cW}}^*(f)\right] \geq E\left[\widehat{X}_{\mathrm{W}}(f)\widehat{V}_{\mathrm{W}}^*(f)\right] \quad (3.102)$$

and

$$H_{X,\mathrm{cW},2}(f) \leq H_{X,\mathrm{W}}(f) \leq H_{X,\mathrm{cW}}(f). \quad (3.103)$$

As a consequence, we can state that $\widehat{X}_{\mathrm{cW}}(f)$ [resp. $\widehat{X}_{\mathrm{cW},2}(f)$] is more (resp. less) noisy but less (resp. more) distorted than $\widehat{X}_{\mathrm{W}}(f) = \widehat{X}_{\mathrm{W},2}(f)$.

Figure 3.12 shows plots of the constraint gains, $H_{X,\mathrm{cW}}(f)$ and $H_{X,\mathrm{cW},2}(f)$, the narrowband noise reduction factor, and the narrowband desired signal distortion index as a function of the narrowband input SNR. The plots for the Wiener gain, $H_{X,\mathrm{W}}(f)$, are also included as a reference. For a given input SNR, the Wiener gain is larger than $H_{X,\mathrm{cW},2}(f)$ and smaller than $H_{X,\mathrm{cW}}(f)$. Hence, $H_{X,\mathrm{cW},2}(f)$ is associated with higher noise reduction and desired signal distortion than the Wiener gain, while $H_{X,\mathrm{cW}}(f)$ is associated with less noise reduction and desired signal distortion than the Wiener gain.

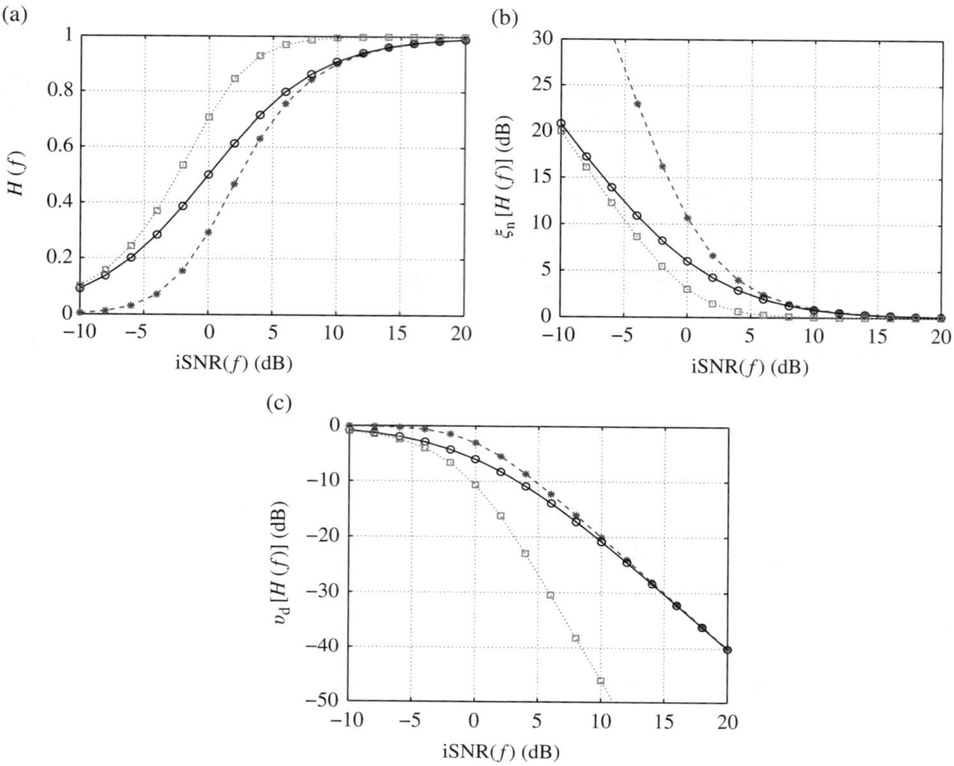

Figure 3.12 (a) The gain, (b) the narrowband noise reduction factor, and (c) the narrowband desired signal distortion index as a function of the narrowband input SNR for different optimal gains: $H_{X,\mathrm{cW},2}(f)$ (dashed line with asterisks), $H_{X,\mathrm{W}}(f)$ (solid line with circles), and $H_{X,\mathrm{cW}}(f)$ (dotted line with squares).

Example 3.5.1 Returning to Example 3.4.1, Figure 3.13 shows plots of the broadband gain in SNR, $\mathcal{G}(H)$, the broadband MSE, $J(H)$, the broadband noise reduction factor, $\xi_n(H)$, and the broadband desired signal reduction factor, $\xi_d(H)$, as a function of the broadband input SNR for different optimal gains: $H_{X,\mathrm{cW},2}(f)$, $H_{X,\mathrm{W}}(f)$, and $H_{X,\mathrm{cW}}(f)$. Figure 3.14 shows similar plots for the signals in Example 3.4.2.

For a given broadband input SNR, $H_{X,\mathrm{cW},2}(f)$ is associated with higher broadband SNR gain and noise reduction than $H_{X,\mathrm{cW}}(f)$, but at the expense of higher broadband desired signal reduction. ■

In Table 3.1, we summarize all the optimal gains studied in this section and the previous one.

3.6 Implementation with the Short-time Fourier Transform

In this section, we show how to implement the different gains in the short-time Fourier transform (STFT) domain.

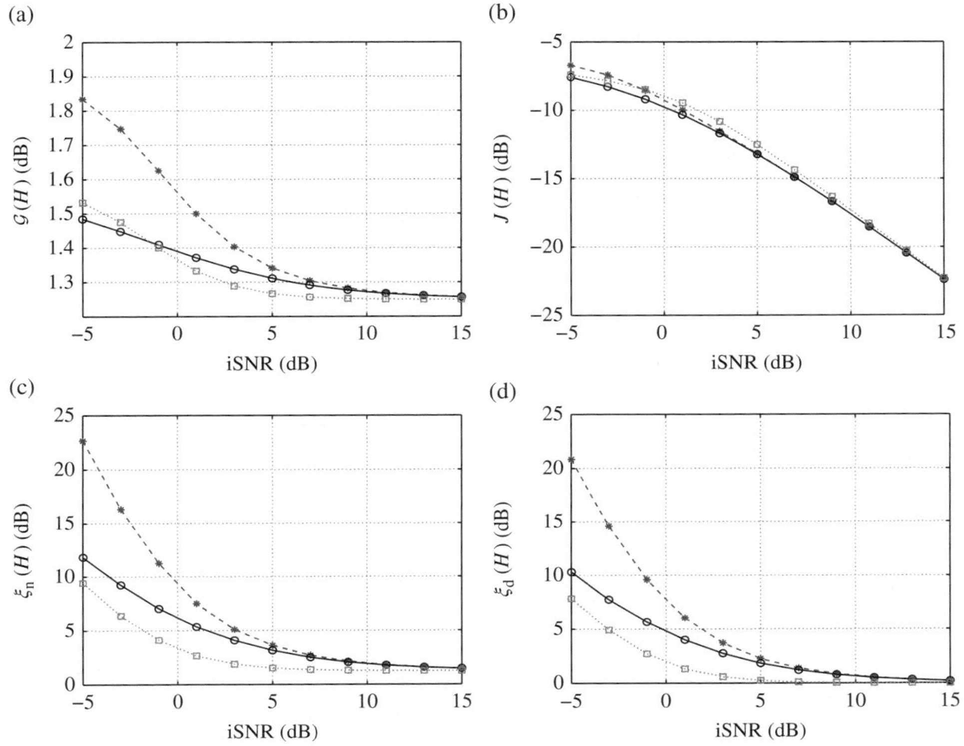

Figure 3.13 (a) The broadband gain in SNR, (b) the broadband MSE, (c) the broadband noise reduction factor, and (d) the broadband desired signal reduction factor as a function of the broadband input SNR for different optimal gains: $H_{X,\text{cW},2}(f)$ (dashed line with asterisks), $H_{X,\text{W}}(f)$ (solid line with circles), and $H_{X,\text{cW}}(f)$ (dotted line with squares).

The signal model given in (3.1) can be put into a vector form by considering the L most recent successive time samples:

$$\mathbf{y}(t) = \mathbf{x}(t) + \mathbf{v}(t), \tag{3.104}$$

where

$$\mathbf{y}(t) = \begin{bmatrix} y(t) & y(t-1) & \cdots & y(t-L+1) \end{bmatrix}^T \tag{3.105}$$

is a vector of length L, and $\mathbf{x}(t)$ and $\mathbf{v}(t)$ are defined in a similar way to $\mathbf{y}(t)$ from (3.105). A short-time segment of the measured signal – that is, $\mathbf{y}(t)$ – is multiplied with an analysis window of length L:

$$\mathbf{g} = \begin{bmatrix} g(0) & g(1) & \cdots & g(L-1) \end{bmatrix}^T \tag{3.106}$$

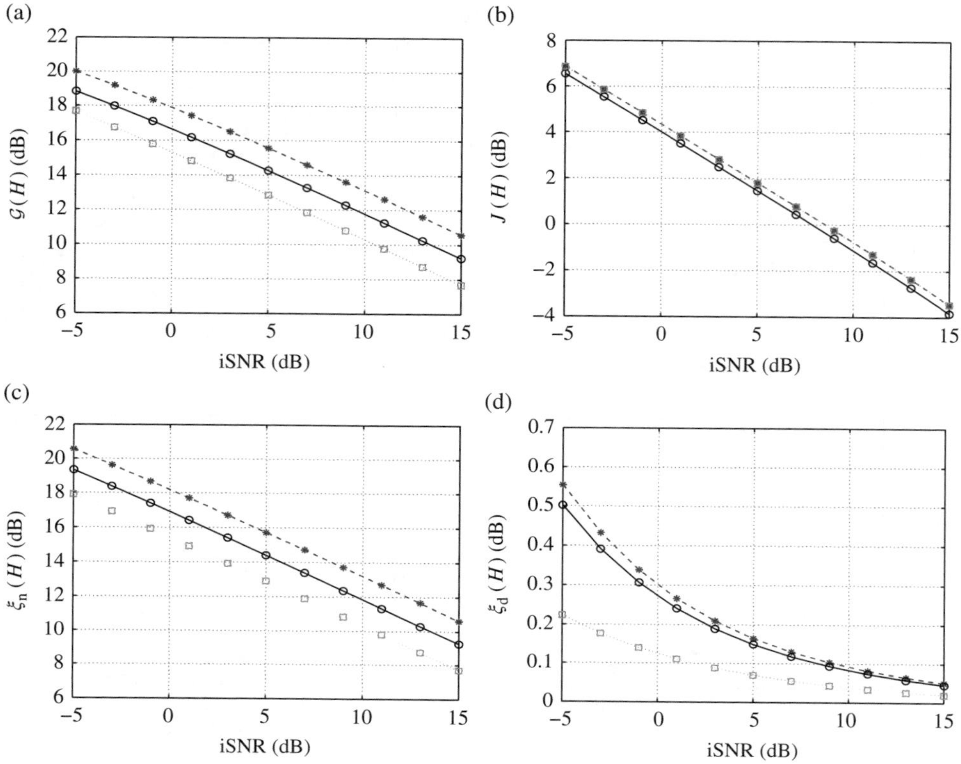

Figure 3.14 (a) The broadband gain in SNR, (b) the broadband MSE, (c) the broadband noise reduction factor, and (d) the broadband desired signal reduction factor as a function of the broadband input SNR for different optimal gains: $H_{X,\mathrm{cW},2}(f)$ (dashed line with asterisks), $H_{X,\mathrm{W}}(f)$ (solid line with circles), and $H_{X,\mathrm{cW}}(f)$ (dotted line with squares).

and transformed into the frequency domain by using the discrete Fourier transform (DFT). Let $\mathbf{W}$ denote the DFT matrix of size $L \times L$, with

$$[\mathbf{W}]_{i,j} = \exp\left(-\frac{j2\pi ij}{L}\right),\ i,j = 0,\dots,L-1. \tag{3.107}$$

Then, the STFT representation of the measured signal is defined as [16]:

$$\mathbf{Y}(t) = \mathbf{W}\mathrm{diag}(\mathbf{g})\,\mathbf{y}(t), \tag{3.108}$$

where

$$\mathbf{Y}(t) = \begin{bmatrix} Y(t,0) & Y(t,1) & \cdots & Y(t,L-1) \end{bmatrix}^T. \tag{3.109}$$

In practice, the STFT representation is decimated in time by a factor R $(1 \le R \le L)$ [17]:

$$\begin{aligned}\mathbf{Y}(rR) &= \mathbf{Y}(t)\big|_{t=rR} \\ &= \begin{bmatrix} Y(rR,0) & Y(rR,1) & \cdots & Y(rR,L-1) \end{bmatrix}^T,\ r \in \mathbb{Z}.\end{aligned} \tag{3.110}$$

Table 3.1 Optimal gains for single-channel signal enhancement in the frequency domain.

Gain	
Wiener	$H_{\mathrm{W}}(f) = \dfrac{\mathrm{iSNR}(f)}{1+\mathrm{iSNR}(f)}$
Tradeoff	$H_{\mathrm{T},\mu}(f) = \dfrac{\mathrm{iSNR}(f)}{\mu+\mathrm{iSNR}(f)}, \ \mu \geq 0$
Parametric Wiener	$H_{\mu_1,\mu_2}(f) = \left[1 - \sin^{\mu_1} \theta(f)\right]^{\mu_2}, \ \mu_1, \mu_2 \geq 0$
Constraint Wiener	$H_{X,\mathrm{cW}}(f) = \sqrt{\dfrac{\mathrm{iSNR}^2(f)}{1+\mathrm{iSNR}^2(f)}}$
	$H_{X,\mathrm{cW},2}(f) = 1 - \sqrt{\dfrac{1}{1+\mathrm{iSNR}^2(f)}}$

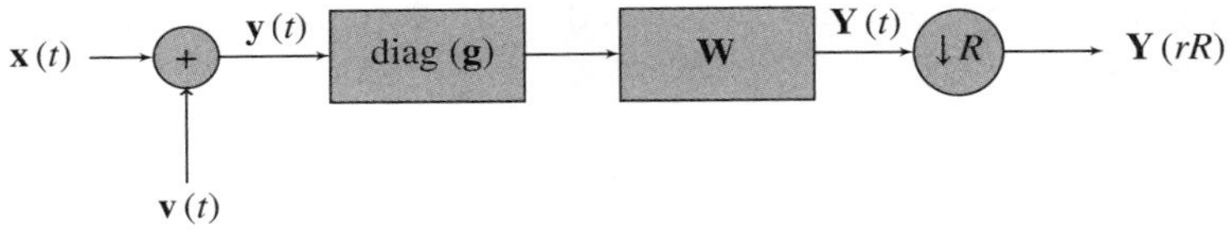

Figure 3.15 STFT representation of the measured signal.

Figure 3.15 shows the STFT representation of the measured signal. Therefore, in the STFT domain, (3.1) can be written as

$$Y(rR,k) = X(rR,k) + V(rR,k), \tag{3.111}$$

where $k = 0, \dots, L-1$ denotes the frequency index, and $X(rR,k)$ and $V(rR,k)$ are the STFT representations of $x(t)$ and $v(t)$, respectively. Since the zero-mean signals $X(rR,k)$ and $V(rR,k)$ are assumed to be uncorrelated, the variance of $Y(rR,k)$ is

$$\begin{aligned} \phi_Y(rR,k) &= E\left[|Y(rR,k)|^2\right] \\ &= \phi_X(rR,k) + \phi_V(rR,k), \end{aligned} \tag{3.112}$$

where $\phi_X(rR,k) = E\left[|X(rR,k)|^2\right]$ and $\phi_V(rR,k) = E\left[|V(rR,k)|^2\right]$ are the variances of $X(rR,k)$ and $V(rR,k)$, respectively.

An estimate of $X(rR,k)$ can be obtained by multiplying $Y(rR,k)$ with a gain $H(rR,k)$, as illustrated in Figure 3.16:

$$\begin{aligned} Z(rR,k) &= H(rR,k)Y(rR,k) \\ &= H(rR,k)\left[X(rR,k) + V(rR,k)\right] \\ &= X_{\mathrm{fd}}(rR,k) + V_{\mathrm{rn}}(rR,k), \end{aligned} \tag{3.113}$$

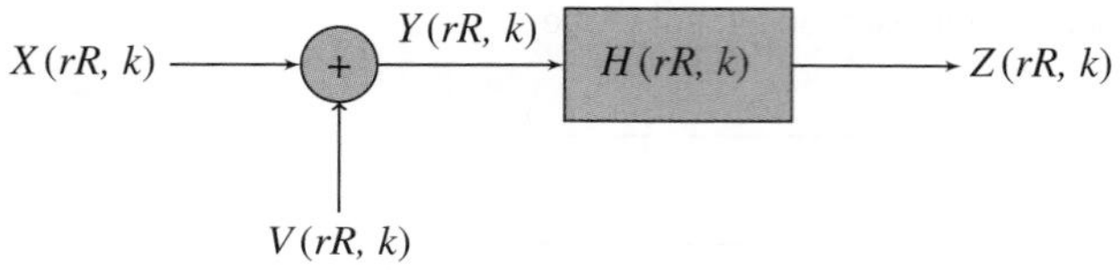

Figure 3.16 Block diagram of noise reduction in the STFT domain.

where $Z(rR, k)$ is the STFT representation of the signal $z(t)$,

$$X_{\rm fd}(rR, k) = H(rR, k)X(rR, k) \tag{3.114}$$

is the filtered desired signal, and

$$V_{\rm rn}(rR, k) = H(rR, k)V(rR, k) \tag{3.115}$$

is the residual noise.

A short-time segment of $z(t)$ can be reconstructed in the time domain by applying the inverse DFT to the vector:

$$\mathbf{Z}(rR) = \begin{bmatrix} Z(rR, 0) & Z(rR, 1) & \cdots & Z(rR, L-1) \end{bmatrix}^T \tag{3.116}$$

and multiplying the result with a synthesis window of length L:

$$\widetilde{\mathbf{g}} = \begin{bmatrix} \widetilde{g}(0) & \widetilde{g}(1) & \cdots & \widetilde{g}(L-1) \end{bmatrix}^T. \tag{3.117}$$

That is,

$$\mathbf{z}(rR) = \text{diag}\left(\widetilde{\mathbf{g}}\right) \mathbf{W}^H \mathbf{Z}(rR), \tag{3.118}$$

where the superscript ^{H} denotes conjugate-transpose of a vector or a matrix. The estimate $z(t)$ of the desired signal can be reconstructed in the time domain by the overlap-add method [18]; in other words, summing the values at time t of all the short-time segments that overlap at time t:

$$z(t) = \sum_r \mathbf{i}_{rR-t+1}^T \mathbf{z}(rR), \tag{3.119}$$

where $\mathbf{i}_i$ $(1 \leq i \leq L)$ is the ith column of $\mathbf{I}_L$ and the summation is over integer values of r in the range $\frac{t}{R} \leq r \leq \frac{t+L-1}{R}$. The inverse STFT is illustrated in Figure 3.17.

The synthesis window $\widetilde{\mathbf{g}}$ must satisfy a condition for exact reconstruction of $x(t)$ when $H(rR, k) = 1$ and $V(rR, k) = 0$ for all (r, k) [17]. Specifically, from (3.108) we have

$$\mathbf{X}(rR) = \mathbf{W} \,\text{diag}(\mathbf{g})\, \mathbf{x}(rR). \tag{3.120}$$

For exact reconstruction of $z(t) = x(t)$ using (3.118) and (3.119), we require

$$x(t) = \sum_r \mathbf{i}_{rR-t+1}^T \,\text{diag}\left(\widetilde{\mathbf{g}}\right) \mathbf{W}^H \mathbf{X}(rR). \tag{3.121}$$

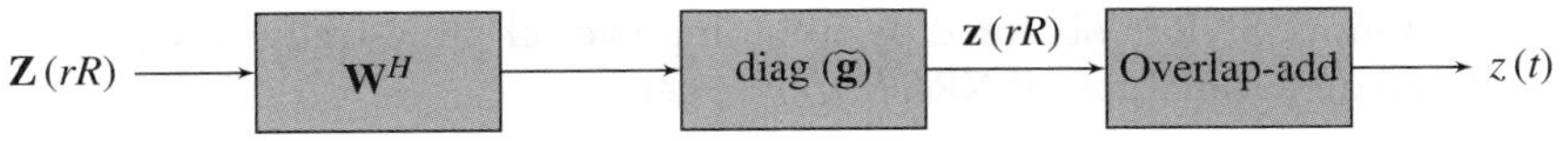

Figure 3.17 Block diagram of the inverse STFT.

Substituting (3.120) into (3.121), we get

$$x(t) = \sum_r \mathbf{i}_{rR-t+1}^T \operatorname{diag}\left(\widetilde{\mathbf{g}}\right) \operatorname{diag}\left(\mathbf{g}\right) \mathbf{x}(rR), \tag{3.122}$$

for all signals $x(t)$ and for all t. Therefore, the condition for exact reconstruction is

$$\sum_r \widetilde{g}(\ell + rR) g(\ell + rR) = 1, \forall \ell \in \{0, \ldots, R-1\}. \tag{3.123}$$

Property 3.6.1 For a given analysis window $\mathbf{g}$ of length $L > R$, there are infinite solutions $\widetilde{\mathbf{g}}$ that satisfy (3.123). A synthesis window of a minimal norm that satisfies (3.123) is given by [17]

$$\widetilde{g}(\ell) = \frac{g(\ell)}{\sum_r g^2(\ell + rR)}, \quad \ell = 0, \ldots, L-1. \tag{3.124}$$

Proof. Define

$$\mathbf{g}_\ell = \begin{bmatrix} \cdots & g(\ell - R) & g(\ell) & g(\ell + R) & \cdots \end{bmatrix}^T,$$
$$\widetilde{\mathbf{g}}_\ell = \begin{bmatrix} \cdots & \widetilde{g}(\ell - R) & \widetilde{g}(\ell) & \widetilde{g}(\ell + R) & \cdots \end{bmatrix}^T.$$

Then condition (3.123) can be written as

$$\mathbf{g}_\ell^T \widetilde{\mathbf{g}}_\ell = 1, \ \forall \ell \in \{0, \ldots, R-1\}. \tag{3.125}$$

The minimum-norm solution to this equation is the pseudo inverse of $\mathbf{g}_\ell$:

$$\widetilde{\mathbf{g}}_\ell = \mathbf{g}_\ell \left(\mathbf{g}_\ell^T \mathbf{g}_\ell\right)^{-1}, \tag{3.126}$$

which is equivalent to (3.124). ■

In a similar way to the frequency-domain input SNR, we define the narrowband input SNR as

$$\text{iSNR}(rR, k) = \frac{\phi_X(rR, k)}{\phi_V(rR, k)}. \tag{3.127}$$

The optimal gains, summarized in Table 3.1, are employed in the STFT domain by replacing $\text{iSNR}(f)$ with $\text{iSNR}(rR, k)$.

The broadband input SNR is obtained by summing over all time-frequency indices the numerator and denominator of iSNR(rR, k). We get

$$\text{iSNR} = \frac{\sum_{r,k} \phi_X(rR, k)}{\sum_{r,k} \phi_V(rR, k)}. \tag{3.128}$$

Similarly, the broadband output SNR is

$$\begin{aligned} \text{oSNR}(H) &= \frac{\sum_{r,k} \phi_{X_{\text{fd}}}(rR, k)}{\sum_{r,k} \phi_{V_{\text{rn}}}(rR, k)} \\ &= \frac{\sum_{r,k} |H(rR, k)|^2 \phi_X(rR, k)}{\sum_{r,k} |H(rR, k)|^2 \phi_V(rR, k)}, \end{aligned} \tag{3.129}$$

the broadband noise reduction and desired signal reduction factors are, respectively,

$$\xi_{\text{n}}(H) = \frac{\sum_{r,k} \phi_V(rR, k)}{\sum_{r,k} |H(rR, k)|^2 \phi_V(rR, k)} \tag{3.130}$$

and

$$\xi_{\text{d}}(H) = \frac{\sum_{r,k} \phi_X(rR, k)}{\sum_{r,k} |H(rR, k)|^2 \phi_X(rR, k)}, \tag{3.131}$$

and the broadband MSE is defined as

$$\begin{aligned} J(H) &= \sum_{r,k} J\left[H(rR, k)\right] \\ &= \sum_{r,k} |1 - H(rR, k)|^2 \phi_X(rR, k) + \sum_{r,k} |H(rR, k)|^2 \phi_V(rR, k). \end{aligned} \tag{3.132}$$

Example 3.6.1 Consider a speech signal, $x(t)$, sampled at 16 kHz, that is corrupted with uncorrelated additive white Gaussian noise, $v(t) \sim \mathcal{N}\left(0, \sigma_v^2\right)$. The observed signal, $y(t)$, given by $y(t) = x(t) + v(t)$, is transformed into the STFT domain, multiplied at each time-frequency bin by a spectral gain $H(rR, k)$, and transformed back into the time domain using (3.118) and (3.119).

To demonstrate noise reduction in the STFT domain, we choose a Hamming window of length $L = 512$ as the analysis window, a decimation factor $R = L/4 = 128$, and the Wiener gain in the STFT domain:

$$H_{\text{W}}(rR, k) = \frac{\text{iSNR}(rR, k)}{1 + \text{iSNR}(rR, k)}. \tag{3.133}$$

An estimate for the noise variance $\widehat{\phi}_V(rR, k)$ can be simply obtained by averaging past spectral power values of the noisy measurement during speech inactivity:

$$\widehat{\phi}_V(rR, k) = \begin{cases} \alpha \widehat{\phi}_V\left[(r-1)R, k\right] + (1-\alpha)\,|Y(rR, k)|^2, & X(rR, k) = 0 \\ \widehat{\phi}_V\left[(r-1)R, k\right], & X(rR, k) \neq 0 \end{cases}, \tag{3.134}$$

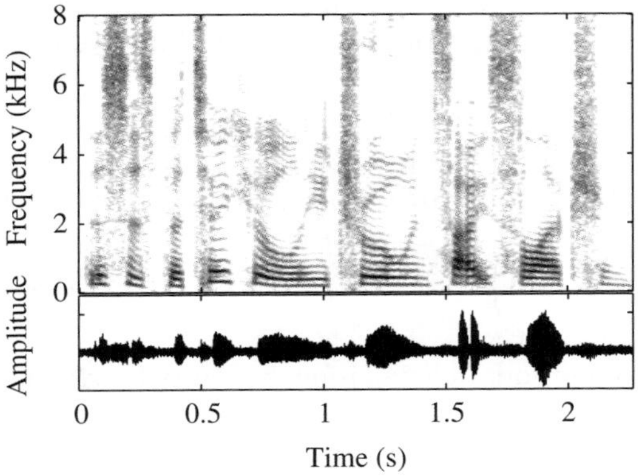

Figure 3.18 Speech spectrogram and waveform of a clean speech signal, $x(t)$: "This is particularly true in site selection."

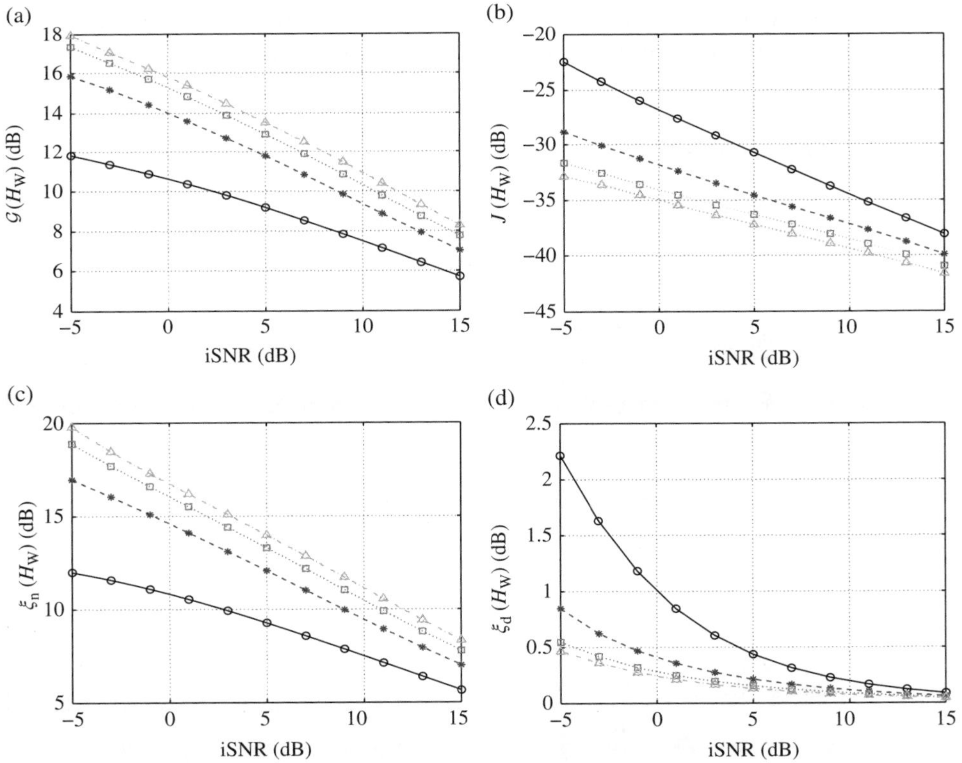

Figure 3.19 (a) The broadband gain in SNR, (b) the broadband MSE, (c) the broadband noise reduction factor, and (d) the broadband desired signal reduction factor of the Wiener gain as a function of the broadband input SNR for different oversubtraction factors β: $\beta = 1$ (solid line with circles), $\beta = 2$ (dashed line with asterisks), $\beta = 3$ (dotted line with squares), and $\beta = 4$ (dash-dot line with triangles).

where α ($0 < \alpha < 1$) denotes a smoothing parameter. This method requires a voice activity detector, but there are alternative and more efficient methods that are based on minimum statistics [19, 20].

Finding an estimate for $\phi_X(rR, k)$ is a much more challenging problem [21, 22]. In this example, for simplicity, we smooth $|Y(rR, k)|^2$ in both time and frequency axes and subtract an estimate of the noise that is multiplied by an oversubtraction factor β ($\beta \geq 1$):

$$\widehat{\phi}_X(rR, k) = \max\left\{\widehat{\phi}_Y(rR, k) - \beta\widehat{\phi}_V(rR, k), 0\right\}, \tag{3.135}$$

where $\widehat{\phi}_Y(rR, k)$ is obtained as a two-dimensional convolution between $|Y(rR, k)|^2$ and a smoothing window $w(rR, k)$. Here, the smoothing window is a two-dimensional Hamming window of size 3×11, normalized to $\sum_{r,k} w(rR, k) = 1$.

Figure 3.18 shows the spectrogram (magnitude of the STFT representation) and waveform of the clean speech signal, $x(t)$. Figure 3.19 shows plots of the broadband gain in SNR, the broadband MSE, $J\left(H_{\rm W}\right)$, the broadband noise reduction factor, $\xi_{\rm n}\left(H_{\rm W}\right)$, and the broadband desired signal reduction factor, $\xi_{\rm d}\left(H_{\rm W}\right)$, as a function of the broadband input SNR for different values of the oversubtraction factor β. Figure 3.20 shows a realization of the noise corrupted and filtered speech signals for different values

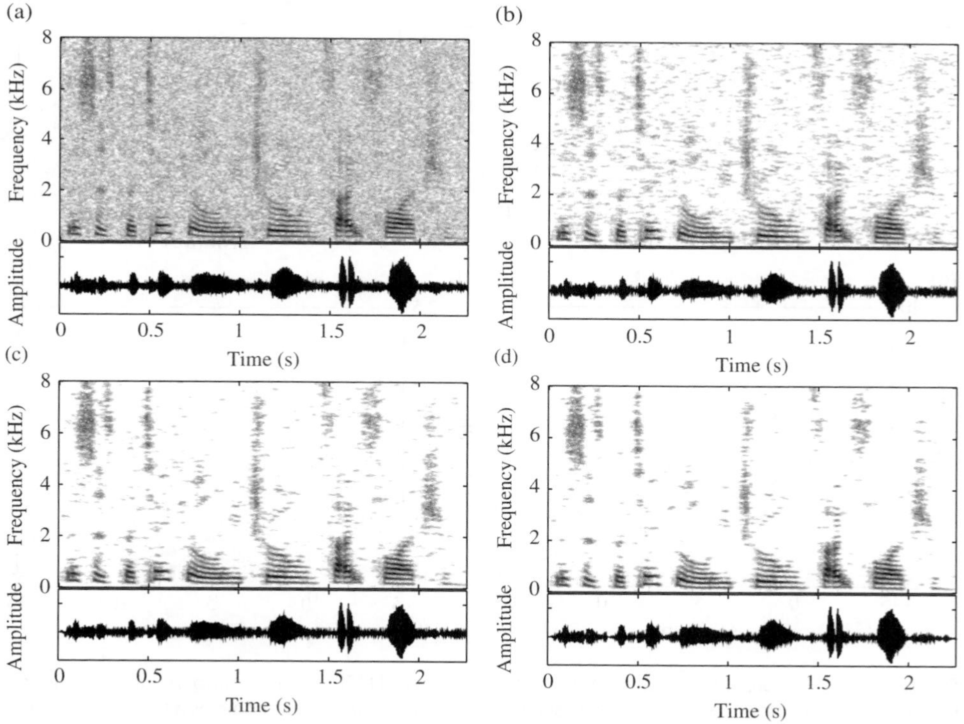

Figure 3.20 Speech spectrograms and waveforms of (a) noisy speech signal, $y(t)$, (b) filtered signal, $z(t)$, using an oversubtraction factor, $\beta = 1$, (c) filtered signal, $z(t)$, using an oversubtraction factor, $\beta = 2$, and (d) filtered signal, $z(t)$, using an oversubtraction factor, $\beta = 3$.

of β. For larger values of β, there is less residual musical noise, but at the expense of larger distortion of weak speech components.

Note that more useful algorithms for enhancing noisy speech signals in the STFT domain are presented in [1, 23, 24]. ■

Problems

3.1 Show that the narrowband MSE is given by

$$J\left[H(f)\right] = |1 - H(f)|^2 \phi_X(f) + |H(f)|^2 \phi_V(f).$$

3.2 Show that the narrowband MSE is related to the different narrowband performance measures by

$$J\left[H(f)\right] = \phi_V(f)\left\{ \text{iSNR}(f) \times \upsilon_{\text{d}}\left[H(f)\right] + \frac{1}{\xi_{\text{n}}\left[H(f)\right]} \right\}.$$

3.3 Show that the narrowband MSEs $J_{\text{d}}\left[H(f)\right]$ and $J_{\text{n}}\left[H(f)\right]$ are related to the different narrowband performance measures by

$$\begin{aligned} \frac{J_{\text{d}}\left[H(f)\right]}{J_{\text{n}}\left[H(f)\right]} &= \text{iSNR}(f) \times \xi_{\text{n}}\left[H(f)\right] \times \upsilon_{\text{d}}\left[H(f)\right] \\ &= \text{oSNR}\left[H(f)\right] \times \xi_{\text{d}}\left[H(f)\right] \times \upsilon_{\text{d}}\left[H(f)\right]. \end{aligned}$$

3.4 Show that the Wiener gain is is given by

$$H_{\text{W}}(f) = \frac{\text{iSNR}(f)}{1 + \text{iSNR}(f)}.$$

3.5 Show that the Wiener gain is equal to the MSCF between $X(f)$ and $Y(f)$; in other words:

$$H_{\text{W}}(f) = \left|\rho\left[X(f), Y(f)\right]\right|^2.$$

3.6 Show that the MMSE can be expressed as

$$J\left[H_{\text{W}}(f)\right] = \left[1 - H_{\text{W}}(f)\right] \phi_X(f).$$

3.7 Show that with the Wiener gain, the broadband output SNR is always greater than or equal to the broadband input SNR: $\text{oSNR}\left(H_{\text{W}}\right) \geq \text{iSNR}$.

3.8 Consider a desired signal, $X(f)$, with the variance:

$$\phi_X(f) = \begin{cases} \alpha, & |f| \leq \frac{1}{4} \\ 0, & \frac{1}{4} \leq |f| \leq \frac{1}{2} \end{cases},$$

which is corrupted with additive noise, $V(f)$, with the variance:

$$\phi_V(f) = \beta\left(1 - 2|f|\right), \ -\frac{1}{2} \leq |f| \leq \frac{1}{2}.$$

a) Show that the broadband input SNR is

$$\text{iSNR} = \frac{\alpha}{\beta}.$$

b) Show that the optimal Wiener gain is given by

$$H_W(f) = \begin{cases} \frac{\alpha}{\beta}\left(\frac{\alpha}{\beta} + 1 - 2|f|\right)^{-1}, & |f| \leq \frac{1}{4} \\ 0, & \frac{1}{4} \leq |f| \leq \frac{1}{2} \end{cases}.$$

3.9 Consider a narrowband desired signal, $X(f)$, with the variance:

$$\phi_X(f) = \begin{cases} \alpha, & |f - f_0| \leq \beta \\ 0, & \text{otherwise} \end{cases},$$

where $\beta \ll f_0$ and $\beta + f_0 < \frac{1}{2}$. The desired signal is corrupted with additive noise, $V(f)$, with the variance:

$$\phi_V(f) = N_0,$$

which is uncorrelated with $X(f)$.

a) Compute the broadband input SNR.
b) Show that the optimal Wiener gain is given by

$$H_W(f) = \begin{cases} \frac{\alpha}{\alpha + N_0}, & |f - f_0| \leq \beta \\ 0, & \text{otherwise} \end{cases}.$$

c) Show that the MMSE is given by

$$J\left[H_W(f)\right] = \frac{2N_0\alpha\beta}{\alpha + N_0}.$$

d) Show how the MMSE changes if the variance of the noise is

$$\phi_V(f) = \begin{cases} N_0, & |f - f_0| \leq \beta \\ 0, & \text{otherwise} \end{cases}.$$

3.10 Consider a harmonic pulse of T samples:

$$x(t) = \begin{cases} A \sin\left(2\pi f_0 t + \phi\right), & 0 \leq t \leq T-1 \\ 0, & t < 0,\ t \geq T \end{cases},$$

with fixed amplitude A and frequency f_0, and random phase ϕ, uniformly distributed on the interval from 0 to 2π.

a) Show that the variance of $X(f)$ is

$$\phi_X(f) = \frac{A^2}{4} D_T^2 \left[\pi \left(f + f_0\right)\right] + \frac{A^2}{4} D_T^2 \left[\pi \left(f - f_0\right)\right].$$

b) Assume that $x(t)$ is corrupted with additive white Gaussian noise $v(t) \sim \mathcal{N}\left(0, \sigma_v^2\right)$, that is uncorrelated with $x(t)$. Using Parseval's identity, show that the broadband input SNR is

$$\text{iSNR} = \frac{A^2}{2\sigma_v^2}.$$

3.11 Show that the tradeoff gain, at frequency f, is equal to the MSCF between the two signals $X(f)$ and $X(f) + \sqrt{\mu} V(f)$:

$$H_{\text{T},\mu}(f) = \left|\rho\left[X(f), X(f) + \sqrt{\mu} V(f)\right]\right|^2.$$

3.12 Show that the broadband MSCF between the two variables $X(f)$ and $X(f) + \sqrt{\mu} V(f)$ is

$$\left|\rho\left(X, X + \sqrt{\mu} V\right)\right|^2 = \frac{\text{iSNR}}{\mu + \text{iSNR}}.$$

3.13 Show that the broadband MSCF between the two variables $X(f)$ and $H_{\text{T},\mu}(f) X(f) + \sqrt{\mu} H_{\text{T},\mu}(f) V(f)$ satisfies

$$\left|\rho\left(X, H_{\text{T},\mu} X + \sqrt{\mu} H_{\text{T},\mu} V\right)\right|^2 \leq \frac{\text{oSNR}\left(H_{\text{T},\mu}\right)}{\mu + \text{oSNR}\left(H_{\text{T},\mu}\right)}.$$

3.14 Show that the broadband MSCF between the two variables $X(f)$ and $X(f) + \sqrt{\mu} V(f)$ satisfies

$$\left|\rho\left(X, X + \sqrt{\mu} V\right)\right|^2 \leq \frac{\text{oSNR}\left(H_{\text{T},\mu}\right)}{\mu + \text{oSNR}\left(H_{\text{T},\mu}\right)}.$$

3.15 Show that with the tradeoff gain, the broadband output SNR is always greater than or equal to the broadband input SNR: $\text{oSNR}\left(H_{\text{T},\mu}\right) \geq \text{iSNR}, \ \forall \mu \geq 0$.

3.16 Prove that the different performance measures obtained with the Wiener gain and the power subtraction and magnitude subtraction methods are related by

$$\xi_{\text{n}}\left[H_{\text{mag}}(f)\right] \geq \xi_{\text{n}}\left[H_{\text{W}}(f)\right] \geq \xi_{\text{n}}\left[H_{\text{pow}}(f)\right],$$
$$\upsilon_{\text{d}}\left[H_{\text{pow}}(f)\right] \leq \upsilon_{\text{d}}\left[H_{\text{W}}(f)\right] \leq \upsilon_{\text{d}}\left[H_{\text{mag}}(f)\right].$$

3.17 Prove that the MMSE obtained with $H_{X,\text{W}}(f)$ is the same as that obtained with $H_{V,\text{W}}(f)$:

$$J_X\left[H_{X,\text{W}}(f)\right] = J_V\left[H_{V,\text{W}}(f)\right].$$

3.18 Show that $\widehat{X}_{\text{W}}(f) + \widehat{V}_{\text{W}}(f) = Y(f)$, but $\phi_{\widehat{X}_{\text{W}}}(f) + \phi_{\widehat{V}_{\text{W}}}(f) \geq \phi_Y(f)$. Explain this result.

3.19 Show that minimization of $J\left[H_X(f), H_V(f)\right]$ with the constraint that $\phi_{\widehat{X}}(f) + \phi_{\widehat{V}}(f) = \phi_Y(f)$ yields

$$H_{X,\text{cW}}(f) = \sqrt{\frac{\text{iSNR}^2(f)}{1 + \text{iSNR}^2(f)}}.$$

3.20 Show that $\widehat{X}_{\text{cW}}(f)$ is more noisy but less distorted than $\widehat{X}_{\text{W}}(f)$, and that $\widehat{X}_{\text{cW},2}(f)$ is less noisy but more distorted than $\widehat{X}_{\text{W}}(f)$:

$$H_{X,\text{cW},2}(f) \leq H_{X,\text{W}}(f) \leq H_{X,\text{cW}}(f).$$

3.21 Let $\mathbf{g}$ and $\widetilde{\mathbf{g}}$ be, respectively, analysis and synthesis windows of the STFT. Show that the condition for exact reconstruction with the inverse STFT is

$$\sum_r \widetilde{g}(\ell + rR)g(\ell + rR) = 1, \forall \ell \in \{0, \ldots, R-1\}.$$

3.22 Show that for a given analysis window $\mathbf{g}$ of length $L > R$, the synthesis window of a minimal norm that satisfies the condition of exact reconstruction is given by

$$\widetilde{g}(\ell) = \frac{g(\ell)}{\sum_r g^2(\ell + rR)}, \ \ell = 0, \ldots, L-1.$$

References

1 J. Benesty, J. Chen, Y. Huang, and I. Cohen, *Noise Reduction in Speech Processing.* Berlin, Germany: Springer-Verlag, 2009.

2 W. Etter and G. S. Moschytz, "Noise reduction by noise-adaptive spectral magnitude expansion," *J. Audio Eng. Soc.*, vol. 42, pp. 341–349, May 1994.

3 J. S. Lim and A. V. Oppenheim, "Enhancement and bandwidth compression of noisy speech," *Proc. IEEE*, vol. 67, pp. 1586–1604, Dec. 1979.

4 Y. Ephraim and D. Malah, "Speech enhancement using a minimum mean-square error short-time spectral amplitude estimator," *IEEE Trans. Acoust., Speech, Signal Process.*, vol. ASSP-32, pp. 1109–1121, Dec. 1984.

5 R. J. McAulay and M. L. Malpass, "Speech enhancement using a soft-decision noise suppression filter," *IEEE Trans. Acoust., Speech, Signal Process.*, vol. ASSP-28, pp. 137–145, Apr. 1980.

6 M. M. Sondhi, C. E. Schmidt, and L. R. Rabiner, "Improving the quality of a noisy speech signal," *Bell Syst. Techn. J.*, vol. 60, pp. 1847–1859, Oct. 1981.

7 M. Berouti, R. Schwartz, and J. Makhoul, "Enhancement of speech corrupted by acoustic noise," in *Proc. IEEE ICASSP*, 1979, pp. 208–211.

8 S. F. Boll, "Suppression of acoustic noise in speech using spectral subtraction," *IEEE Trans. Acoust., Speech, Signal Process.*, vol. ASSP-27, pp. 113–120, Apr. 1979.

9 M. R. Schroeder, "Apparatus for suppressing noise and distortion in communication signals," US Patent No. 3,180,936, filed Dec. 1, 1960, issued Apr. 27, 1965.

10 M. R. Schroeder, "Processing of communication signals to reduce effects of noise," US Patent No. 3,403,224, filed May 28, 1965, issued Sept. 24, 1968.

11 M. R. Weiss, E. Aschkenasy, and T. W. Parsons, "Processing speech signals to attenuate interference," in *Proc. IEEE Symposium on Speech Recognition*, 1974, pp. 292–295.

12 E. J. Diethorn, "Subband noise reduction methods for speech enhancement," in *Audio Signal Processing for Next-Generation Multimedia Communication Systems*, Y. Huang and J. Benesty, (eds), Boston, MA, USA: Kluwer, 2004.

13 J. H. L. Hansen, "Speech enhancement employing adaptive boundary detection and morphological based spectral constraints," in *Proc. IEEE ICASSP*, 1991, pp. 901–904.

14 Y. Lu and P. C. Loizou, "A geometric approach to spectral subtraction," *Speech Communication*, vol. 50, pp. 453–466, Jun. 2008.

15 B. L. Sim, Y. C. Tong, J. S. Chang, and C. T. Tan, "A parametric formulation of the generalized spectral subtraction method," *IEEE Trans. Speech, Audio Process.*, vol. 6, pp. 328–337, Jul. 1998.

16 J. Wexler and S. Raz, "Discrete Gabor expansions," *Speech Process.*, vol. 21, pp. 207–220, Nov. 1990.

17 S. Qian and D. Chen, "Discrete Gabor transform," *IEEE Trans. Signal Process.*, vol. 41, pp. 2429–2438, Jul. 1993.

18 R. E. Crochiere and L. R. Rabiner, *Multirate Digital Signal Processing*. Englewood Cliffs, NJ: Prentice-Hall, 1983.

19 R. Martin, "Noise power spectral density estimation based on optimal smoothing and minimum statistics," *IEEE Trans. Speech and Audio Process.*, vol. 9, pp. 504–512, Jul. 2001.

20 I. Cohen, "Noise spectrum estimation in adverse environments: improved minima controlled recursive averaging," *IEEE Trans. Speech and Audio Process.*, vol. 11, pp. 466–475, Sep. 2003.

21 I. Cohen, "Relaxed statistical model for speech enhancement and a priori SNR estimation," *IEEE Trans. Speech and Audio Process.*, vol. 13, pp. 870–881, Sep. 2005.

22 I. Cohen, "Speech spectral modeling and enhancement based on autoregressive conditional heteroscedasticity models," *Signal Process.*, vol. 86, pp. 698–709, Apr. 2006.

23 I. Cohen and B. Berdugo, "Speech enhancement for non-stationary noise environments," *Signal Process.*, vol. 81, pp. 2403–2418, Nov. 2001.

24 I. Cohen and S. Gannot, "Spectral enhancement methods," in J. Benesty, M. M. Sondhi and Y. Huang (eds), *Springer Handbook of Speech Processing*, Springer, 2008.

4

Multichannel Signal Enhancement in the Time Domain

The time-domain multichannel signal enhancement problem is an important generalization of the single-channel case described in Chapter 2. The fundamental difference is that now we take the spatial information into account thanks to the multiple sensors, which are in different positions in the space. Each sensor has its own perspective on the desired and noise signals. This rich diversity can be exploited in order to derive much better filters than those in the single-channel scenario in terms of reduction of the additive noise and distortion of the desired signal. In other words, we have much more flexibility to compromise between noise reduction and distortion of the desired signal thanks to the space-time processing. In this chapter, we explore two different, although roughly equivalent, apparent avenues and derive many useful optimal filters for signal enhancement in a variety of contexts.

4.1 Signal Model and Problem Formulation

We consider the conventional signal model in which an array of M sensors with an arbitrary geometry captures a convolved desired source signal in some noise field. The received signals, at the discrete-time index t, are expressed as [1–3]:

$$\begin{aligned} y_m(t) &= g_m(t) * x(t) + v_m(t) \\ &= x_m(t) + v_m(t),\ m = 1, 2, \dots, M, \end{aligned} \tag{4.1}$$

where $g_m(t)$ is the impulse response from location of the unknown desired source, $x(t)$, to the mth sensor, $*$ stands for linear convolution, and $v_m(t)$ is the additive noise at sensor m. We assume that the signals $x_m(t) = g_m(t) * x(t)$ and $v_m(t)$ are uncorrelated, zero mean, stationary, real, and broadband. By definition the convolved signals, $x_m(t)$, $m = 1, 2, \dots, M$, are coherent across the array while the noise terms, $v_m(t)$, $m = 1, 2, \dots, M$, are typically only partially coherent across the array. The signal model given in (4.1) corresponds to the multichannel signal enhancement (or noise reduction) problem.

By processing the data in blocks of L successive time samples, the signal model given in (4.1) can be put into a vector form as

$$\mathbf{y}_m(t) = \mathbf{x}_m(t) + \mathbf{v}_m(t),\ m = 1, 2, \dots, M, \tag{4.2}$$

Fundamentals of Signal Enhancement and Array Signal Processing, First Edition.
Jacob Benesty, Israel Cohen, and Jingdong Chen.

Companion website: www.wiley.com/go/benesty/arraysignalprocessing

where

$$\mathbf{y}_m(t) = \left[\ y_m(t) \quad y_m(t-1) \quad \cdots \quad y_m(t-L+1)\ \right]^T \tag{4.3}$$

is a vector of length L, and $\mathbf{x}_m(t)$ and $\mathbf{v}_m(t)$ are defined similarly to $\mathbf{y}_m(t)$ from (4.3). It is more convenient to concatenate the M vectors $\mathbf{y}_m(t)$, $m = 1, 2, \ldots, M$, together as

$$\begin{aligned} \underline{\mathbf{y}}(t) &= \left[\ \mathbf{y}_1^T(t) \quad \mathbf{y}_2^T(t) \quad \cdots \quad \mathbf{y}_M^T(t)\ \right]^T \\ &= \underline{\mathbf{x}}(t) + \underline{\mathbf{v}}(t), \end{aligned} \tag{4.4}$$

where the vectors $\underline{\mathbf{x}}(t)$ and $\underline{\mathbf{v}}(t)$ of length ML are defined in a similar way to $\underline{\mathbf{y}}(t)$. Since $x_m(t)$ and $v_m(t)$ are uncorrelated by assumption, the correlation matrix (of size $ML \times ML$) of the observations is

$$\begin{aligned} \mathbf{R}_{\underline{\mathbf{y}}} &= E\left[\underline{\mathbf{y}}(t)\underline{\mathbf{y}}^T(t)\right] \\ &= \mathbf{R}_{\underline{\mathbf{x}}} + \mathbf{R}_{\underline{\mathbf{v}}}, \end{aligned} \tag{4.5}$$

where $\mathbf{R}_{\underline{\mathbf{x}}} = E\left[\underline{\mathbf{x}}(t)\underline{\mathbf{x}}^T(t)\right]$ and $\mathbf{R}_{\underline{\mathbf{v}}} = E\left[\underline{\mathbf{v}}(t)\underline{\mathbf{v}}^T(t)\right]$ are the correlation matrices of $\underline{\mathbf{x}}(t)$ and $\underline{\mathbf{v}}(t)$, respectively. From now on, unless stated otherwise, it is assumed that rank $\left(\mathbf{R}_{\underline{\mathbf{x}}}\right) = P < ML$ while rank $\left(\mathbf{R}_{\underline{\mathbf{v}}}\right) = ML$. In other words, $\mathbf{R}_{\underline{\mathbf{x}}}$ is rank deficient while $\mathbf{R}_{\underline{\mathbf{v}}}$ is full rank.

In this chapter, we consider the first sensor as the reference, so everything will be defined with respect to this sensor. In this case, the desired signal is the whole vector $\mathbf{x}_1(t)$ of length L. Our problem then may be stated as follows: given M mixtures of two uncorrelated signals $x_m(t)$ and $v_m(t)$, our aim is to preserve $\mathbf{x}_1(t)$ while minimizing the contribution of the noise signal vector, $\underline{\mathbf{v}}(t)$, at the array output. To achieve this goal, we develop two different approaches in the next two sections, which are fundamentally equivalent. The two methods give different perspectives on how things work. In the last section, we explore the case where the noise correlation matrix does not have full rank.

4.2 Conventional Method

In this section, we develop some important optimal filtering matrices for multichannel noise reduction in the time domain. In order to unify these different algorithms, we propose to use the joint diagonalization technique, which seems to be a natural thing to do to tackle this fundamental problem.

4.2.1 Joint Diagonalization

Since $\mathbf{R}_{\underline{\mathbf{v}}}$ has full rank, the two symmetric matrices $\mathbf{R}_{\underline{\mathbf{x}}}$ and $\mathbf{R}_{\underline{\mathbf{v}}}$ can be jointly diagonalized as follows [4]:

$$\underline{\mathbf{T}}^T\mathbf{R}_{\underline{\mathbf{x}}}\underline{\mathbf{T}} = \underline{\boldsymbol{\Lambda}}, \tag{4.6}$$

$$\underline{\mathbf{T}}^T\mathbf{R}_{\underline{\mathbf{v}}}\underline{\mathbf{T}} = \mathbf{I}_{ML}, \tag{4.7}$$

where $\underline{\mathbf{T}}$ is a full-rank square matrix (of size $ML \times ML$), $\underline{\mathbf{\Lambda}}$ is a diagonal matrix the main elements of which are real and nonnegative, and $\mathbf{I}_{ML}$ is the $ML \times ML$ identity matrix. Furthermore, $\underline{\mathbf{\Lambda}}$ and $\underline{\mathbf{T}}$ are the eigenvalue and eigenvector matrices, respectively, of $\mathbf{R}_{\underline{\mathbf{v}}}^{-1}\mathbf{R}_{\underline{\mathbf{x}}}$; that is,

$$\mathbf{R}_{\underline{\mathbf{v}}}^{-1}\mathbf{R}_{\underline{\mathbf{x}}}\underline{\mathbf{T}} = \underline{\mathbf{T}}\,\underline{\mathbf{\Lambda}}. \tag{4.8}$$

The eigenvalues of $\mathbf{R}_{\underline{\mathbf{v}}}^{-1}\mathbf{R}_{\underline{\mathbf{x}}}$ can be ordered as $\underline{\lambda}_1 \geq \underline{\lambda}_2 \geq \cdots \geq \underline{\lambda}_P > \underline{\lambda}_{P+1} = \cdots = \underline{\lambda}_{ML} = 0$. We denote by $\underline{\mathbf{t}}_1, \underline{\mathbf{t}}_2, \ldots, \underline{\mathbf{t}}_{ML}$, the corresponding eigenvectors. Therefore, the noisy signal correlation matrix can also be diagonalized as

$$\underline{\mathbf{T}}^T\mathbf{R}_{\underline{\mathbf{y}}}\underline{\mathbf{T}} = \underline{\mathbf{\Lambda}} + \mathbf{I}_{ML}. \tag{4.9}$$

It can be verified from (4.6) and (4.7) that

$$\underline{\mathbf{t}}_i^T\underline{\mathbf{x}}(t) = 0,\ i = P+1, P+2, \ldots, ML \tag{4.10}$$

and

$$\mathbf{R}_{\underline{\mathbf{v}}}^{-1} = \sum_{i=1}^{ML} \underline{\mathbf{t}}_i\underline{\mathbf{t}}_i^T. \tag{4.11}$$

4.2.2 Linear Filtering

Since we want to estimate the desired signal vector, $\mathbf{x}_1(t)$, of length L, from the observation signal vector, $\underline{\mathbf{y}}(t)$, of length ML, a real-valued rectangular filtering matrix, $\underline{\mathbf{H}}$, of size $L \times ML$ should be used, as follows (see Figure 4.1):

$$\begin{aligned}\mathbf{z}(t) &= \underline{\mathbf{H}}\,\underline{\mathbf{y}}(t) \\ &= \mathbf{x}_{\mathrm{fd}}(t) + \mathbf{v}_{\mathrm{rn}}(t),\end{aligned} \tag{4.12}$$

where $\mathbf{z}(t)$, a vector of length L, is the estimate of $\mathbf{x}_1(t)$,

$$\mathbf{x}_{\mathrm{fd}}(t) = \underline{\mathbf{H}}\,\underline{\mathbf{x}}(t) \tag{4.13}$$

is the filtered desired signal, and

$$\mathbf{v}_{\mathrm{rn}}(t) = \underline{\mathbf{H}}\,\underline{\mathbf{v}}(t) \tag{4.14}$$

is the residual noise. This procedure is called the multichannel signal enhancement problem in the time domain.

We can always express $\underline{\mathbf{H}}$ as

$$\underline{\mathbf{H}} = \underline{\mathbf{A}}\,\underline{\mathbf{T}}^T, \tag{4.15}$$

where $\underline{\mathbf{A}}$ is the transformed rectangular filtering matrix, also of size $L \times ML$. Instead of manipulating $\underline{\mathbf{H}}$ directly, we can, equivalently, manipulate $\underline{\mathbf{A}}$, since $\underline{\mathbf{T}}$ (or $\underline{\mathbf{T}}^T$) is a

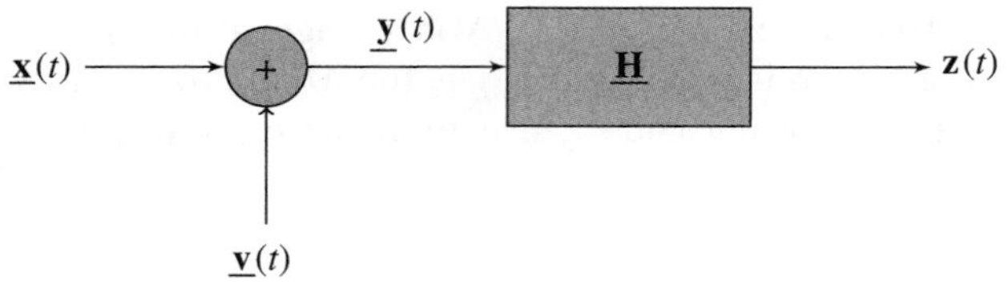

Figure 4.1 Block diagram of multichannel linear filtering in the time domain.

full-rank square matrix. So when $\underline{\mathbf{A}}$ is estimated, we can easily find $\underline{\mathbf{H}}$ from (4.15). In this section, we will mostly work with $\underline{\mathbf{A}}$ for convenience. Consequently, we can write (4.12) as

$$\mathbf{z}(t) = \underline{\mathbf{A}}\,\underline{\mathbf{T}}^T\underline{\mathbf{y}}(t). \tag{4.16}$$

We deduce that the correlation matrix of $\mathbf{z}(t)$ is

$$\begin{aligned}\mathbf{R}_{\mathbf{z}} &= E\left[\mathbf{z}(t)\mathbf{z}^T(t)\right] \\ &= \underline{\mathbf{A}}\left(\underline{\mathbf{\Lambda}} + \mathbf{I}_{ML}\right)\underline{\mathbf{A}}^T \\ &= \mathbf{R}_{\mathbf{x}_{\mathrm{fd}}} + \mathbf{R}_{\mathbf{v}_{\mathrm{rn}}},\end{aligned} \tag{4.17}$$

where

$$\mathbf{R}_{\mathbf{x}_{\mathrm{fd}}} = \underline{\mathbf{A}}\,\underline{\mathbf{\Lambda}}\,\underline{\mathbf{A}}^T \tag{4.18}$$

is the correlation matrix of the filtered desired signal and

$$\mathbf{R}_{\mathbf{v}_{\mathrm{rn}}} = \underline{\mathbf{A}}\,\underline{\mathbf{A}}^T \tag{4.19}$$

is the correlation matrix of the residual noise.

4.2.3 Performance Measures

In this subsection, we define some fundamental measures that fit well in the multiple sensor case and with a linear filtering matrix. We recall that sensor 1 is the reference, so all measures are derived with respect to this sensor.

The input SNR is defined as

$$\mathrm{iSNR} = \frac{\mathrm{tr}\left(\mathbf{R}_{\mathbf{x}_1}\right)}{\mathrm{tr}\left(\mathbf{R}_{\mathbf{v}_1}\right)}, \tag{4.20}$$

where $\mathbf{R}_{\mathbf{x}_1} = E\left[\mathbf{x}_1(t)\mathbf{x}_1^T(t)\right]$ and $\mathbf{R}_{\mathbf{v}_1} = E\left[\mathbf{v}_1(t)\mathbf{v}_1^T(t)\right]$ are the correlation matrices of $\mathbf{x}_1(t)$ and $\mathbf{v}_1(t)$, respectively. This definition of the input SNR is straightforwardly obtained from the correlation matrix of $\mathbf{y}_1(t)$, which is $\mathbf{R}_{\mathbf{y}_1} = \mathbf{R}_{\mathbf{x}_1} + \mathbf{R}_{\mathbf{v}_1}$.

The output SNR is easily derived from (4.17):

$$\begin{aligned} \text{oSNR}\left(\underline{\mathbf{H}}\right) &= \frac{\text{tr}\left(\mathbf{R}_{\mathbf{x}_{\text{fd}}}\right)}{\text{tr}\left(\mathbf{R}_{\mathbf{v}_{\text{rn}}}\right)} \\ &= \frac{\text{tr}\left(\underline{\mathbf{A}}\,\underline{\mathbf{\Lambda}}\,\underline{\mathbf{A}}^T\right)}{\text{tr}\left(\underline{\mathbf{A}}\,\underline{\mathbf{A}}^T\right)} \\ &= \text{oSNR}\left(\underline{\mathbf{A}}\right). \end{aligned} \tag{4.21}$$

It is clear that we always have

$$\text{oSNR}\left(\underline{\mathbf{A}}\right) \leq \underline{\lambda}_1, \tag{4.22}$$

showing how the output SNR is always upper bounded as long as $\mathbf{R}_{\underline{\mathbf{v}}}$ has full rank.

The noise reduction factor is given by

$$\begin{aligned} \xi_{\text{n}}\left(\underline{\mathbf{H}}\right) &= \frac{\text{tr}\left(\mathbf{R}_{\mathbf{v}_1}\right)}{\text{tr}\left(\underline{\mathbf{A}}\,\underline{\mathbf{A}}^T\right)} \\ &= \xi_{\text{n}}\left(\underline{\mathbf{A}}\right). \end{aligned} \tag{4.23}$$

For optimal filtering matrices, we should have $\xi_{\text{n}}\left(\underline{\mathbf{A}}\right) \geq 1$.

Since the desired signal may be distorted by the filtering matrix, we define the desired signal reduction factor as

$$\begin{aligned} \xi_{\text{d}}\left(\underline{\mathbf{H}}\right) &= \frac{\text{tr}\left(\mathbf{R}_{\mathbf{x}_1}\right)}{\text{tr}\left(\underline{\mathbf{A}}\,\underline{\mathbf{\Lambda}}\,\underline{\mathbf{A}}^T\right)} \\ &= \xi_{\text{d}}\left(\underline{\mathbf{A}}\right). \end{aligned} \tag{4.24}$$

For optimal filtering matrices, we generally have $\xi_{\text{d}}\left(\underline{\mathbf{A}}\right) \geq 1$. The closer the value of $\xi_{\text{d}}\left(\underline{\mathbf{A}}\right)$ is to 1, the less distorted is the desired signal.

Obviously, we have the fundamental relationship:

$$\frac{\text{oSNR}\left(\underline{\mathbf{A}}\right)}{\text{iSNR}} = \frac{\xi_{\text{n}}\left(\underline{\mathbf{A}}\right)}{\xi_{\text{d}}\left(\underline{\mathbf{A}}\right)}, \tag{4.25}$$

which, basically, states that nothing comes for free.

We can also evaluate distortion via the desired signal distortion index:

$$\begin{aligned} \upsilon_{\text{d}}\left(\underline{\mathbf{H}}\right) &= \frac{E\left\{\left[\mathbf{x}_{\text{fd}}(t) - \mathbf{x}_1(t)\right]^T\left[\mathbf{x}_{\text{fd}}(t) - \mathbf{x}_1(t)\right]\right\}}{\text{tr}\left(\mathbf{R}_{\mathbf{x}_1}\right)} \\ &= \upsilon_{\text{d}}\left(\underline{\mathbf{A}}\right). \end{aligned} \tag{4.26}$$

For optimal filtering matrices, we should have $\upsilon_{\text{d}}\left(\underline{\mathbf{A}}\right) \leq 1$.

We define the error signal vector between the estimated and desired signals as

$$\begin{aligned} \mathbf{e}(t) &= \mathbf{z}(t) - \mathbf{x}_1(t) \\ &= \underline{\mathbf{A}}\,\underline{\mathbf{T}}^T \underline{\mathbf{y}}(t) - \mathbf{x}_1(t) \\ &= \mathbf{e}_{\mathrm{d}}(t) + \mathbf{e}_{\mathrm{n}}(t), \end{aligned} \tag{4.27}$$

where

$$\begin{aligned} \mathbf{e}_{\mathrm{d}}(t) &= \mathbf{x}_{\mathrm{fd}}(t) - \mathbf{x}_1(t) \\ &= \left(\underline{\mathbf{A}}\,\underline{\mathbf{T}}^T - \underline{\mathbf{I}}_{\mathrm{i}}\right) \underline{\mathbf{x}}(t) \end{aligned} \tag{4.28}$$

is the desired signal distortion due to the filtering matrix with

$$\underline{\mathbf{I}}_{\mathrm{i}} = \begin{bmatrix} \mathbf{I}_L & \mathbf{0}_{L\times(M-1)L} \end{bmatrix} \tag{4.29}$$

being the identity filtering matrix of size $L \times ML$, $\underline{\mathbf{I}}_{\mathrm{i}}\underline{\mathbf{x}}(t) = \mathbf{x}_1(t)$, and

$$\begin{aligned} \mathbf{e}_{\mathrm{n}}(t) &= \mathbf{v}_{\mathrm{rn}}(t) \\ &= \underline{\mathbf{A}}\,\underline{\mathbf{T}}^T \underline{\mathbf{v}}(t) \end{aligned} \tag{4.30}$$

is the residual noise. We deduce that the MSE criterion is

$$\begin{aligned} J\left(\underline{\mathbf{A}}\right) &= \mathrm{tr}\left\{E\left[\mathbf{e}(t)\mathbf{e}^T(t)\right]\right\} \\ &= \mathrm{tr}\left[\mathbf{R}_{\mathbf{x}_1} - 2\underline{\mathbf{A}}\,\underline{\mathbf{T}}^T \mathbf{R}_{\underline{\mathbf{x}}} \underline{\mathbf{I}}_{\mathrm{i}}^T + \underline{\mathbf{A}}\left(\underline{\mathbf{\Lambda}} + \mathbf{I}_{ML}\right)\underline{\mathbf{A}}^T\right] \\ &= J_{\mathrm{d}}\left(\underline{\mathbf{A}}\right) + J_{\mathrm{n}}\left(\underline{\mathbf{A}}\right), \end{aligned} \tag{4.31}$$

where

$$\begin{aligned} J_{\mathrm{d}}\left(\underline{\mathbf{A}}\right) &= \mathrm{tr}\left\{E\left[\mathbf{e}_{\mathrm{d}}(t)\mathbf{e}_{\mathrm{d}}^T(t)\right]\right\} \\ &= \mathrm{tr}\left(\mathbf{R}_{\mathbf{x}_1} - 2\underline{\mathbf{A}}\,\underline{\mathbf{T}}^T \mathbf{R}_{\underline{\mathbf{x}}} \underline{\mathbf{I}}_{\mathrm{i}}^T + \underline{\mathbf{A}}\,\underline{\mathbf{\Lambda}}\,\underline{\mathbf{A}}^T\right) \\ &= \mathrm{tr}\left(\mathbf{R}_{\mathbf{x}_1}\right) \upsilon_{\mathrm{d}}\left(\underline{\mathbf{A}}\right) \end{aligned} \tag{4.32}$$

and

$$\begin{aligned} J_{\mathrm{n}}\left(\underline{\mathbf{A}}\right) &= \mathrm{tr}\left\{E\left[\mathbf{e}_{\mathrm{n}}(t)\mathbf{e}_{\mathrm{n}}^T(t)\right]\right\} \\ &= \mathrm{tr}\left(\underline{\mathbf{A}}\,\underline{\mathbf{A}}^T\right) \\ &= \frac{\mathrm{tr}\left(\mathbf{R}_{\mathbf{v}_1}\right)}{\xi_{\mathrm{n}}\left(\underline{\mathbf{A}}\right)}. \end{aligned} \tag{4.33}$$

As a result, we have

$$\frac{J_{\mathrm{d}}\left(\underline{\mathbf{A}}\right)}{J_{\mathrm{n}}\left(\underline{\mathbf{A}}\right)} = \mathrm{iSNR} \times \xi_{\mathrm{n}}\left(\underline{\mathbf{A}}\right) \times \upsilon_{\mathrm{d}}\left(\underline{\mathbf{A}}\right) \tag{4.34}$$
$$= \mathrm{oSNR}\left(\underline{\mathbf{A}}\right) \times \xi_{\mathrm{d}}\left(\underline{\mathbf{A}}\right) \times \upsilon_{\mathrm{d}}\left(\underline{\mathbf{A}}\right),$$

showing how the different performance measures are related to the MSEs.

4.2.4 Optimal Filtering Matrices

From the different MSEs, we now show how to derive different optimal filtering matrices for multichannel signal enhancement and how to compromise between noise reduction and desired signal distortion in a very flexible way.

The Wiener filtering matrix is derived from the minimization of the MSE criterion, $J\left(\underline{\mathbf{A}}\right)$. From this optimization, we obtain

$$\underline{\mathbf{A}}_{\mathrm{W}} = \underline{\mathbf{I}}_{\mathrm{i}}\mathbf{R}_{\underline{\mathbf{x}}}\underline{\mathbf{T}}\left(\underline{\mathbf{\Lambda}} + \mathbf{I}_{ML}\right)^{-1} \tag{4.35}$$
$$= \underline{\mathbf{I}}_{\mathrm{i}}\underline{\mathbf{T}}^{-T}\underline{\mathbf{\Lambda}}\left(\underline{\mathbf{\Lambda}} + \mathbf{I}_{ML}\right)^{-1}.$$

We deduce that the Wiener filtering matrix is

$$\underline{\mathbf{H}}_{\mathrm{W}} = \underline{\mathbf{A}}_{\mathrm{W}}\underline{\mathbf{T}}^{T} \tag{4.36}$$
$$= \underline{\mathbf{I}}_{\mathrm{i}}\mathbf{R}_{\underline{\mathbf{x}}}\sum_{i=1}^{ML}\frac{\underline{\mathbf{t}}_i\underline{\mathbf{t}}_i^T}{1+\underline{\lambda}_i}$$
$$= \underline{\mathbf{I}}_{\mathrm{i}}\mathbf{R}_{\underline{\mathbf{y}}}\sum_{i=1}^{ML}\frac{\underline{\lambda}_i}{1+\underline{\lambda}_i}\underline{\mathbf{t}}_i\underline{\mathbf{t}}_i^T.$$

Obviously, we can also express $\underline{\mathbf{H}}_{\mathrm{W}}$ as

$$\underline{\mathbf{H}}_{\mathrm{W}} = \underline{\mathbf{I}}_{\mathrm{i}}\mathbf{R}_{\underline{\mathbf{x}}}\mathbf{R}_{\underline{\mathbf{y}}}^{-1}. \tag{4.37}$$

Property 4.2.1 With the optimal Wiener filtering matrix given in (4.37), the output SNR is always greater than or equal to the input SNR: $\mathrm{oSNR}\left(\underline{\mathbf{H}}_{\mathrm{W}}\right) \geq \mathrm{iSNR}$.

Example 4.2.1 Consider an array of M sensors located on a line with a uniform spacing d, as shown in Figure 4.2. Such an array is known as a uniform linear array (ULA). Suppose that a desired signal impinges on the ULA from the broadside direction ($\theta = 90°$), and that an interference impinges on the ULA from the endfire direction ($\theta = 0°$). Assume that the desired signal is a harmonic random process:

$$x(t) = A\cos\left(2\pi f_0 t + \phi\right),$$

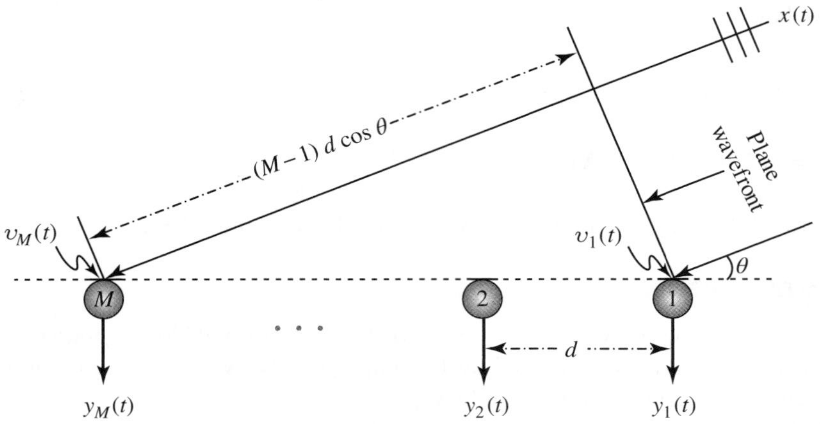

Figure 4.2 Illustration of a uniform linear array for signal capture in the farfield.

with fixed amplitude A and frequency f_0, and random phase ϕ, uniformly distributed on the interval from 0 to 2π. Assume that the interference $u(t)$ is white Gaussian noise, $u(t) \sim \mathcal{N}\left(0, \sigma_u^2\right)$, which is uncorrelated with $x(t)$. In addition, the sensors contain thermal white Gaussian noise, $w_m(t) \sim \mathcal{N}\left(0, \sigma_w^2\right)$, the signals of which are mutually uncorrelated. The desired signal needs to be recovered from the noisy received signals, $y_m(t) = x_m(t) + v_m(t)$, $m = 1, \ldots, M$, where $v_m(t) = u_m(t) + w_m(t)$, $m = 1, \ldots, M$ are the interference-plus-noise signals.

Since the desired source is in the broadside direction and the interference source is in the endfire direction, we have for $i = 2, \ldots, M$:

$$x_i(t) = x_1(t), \tag{4.38}$$

$$u_i(t) = u_1\left(t - \tau_i\right), \tag{4.39}$$

where

$$\tau_i = \frac{(i-1)d}{cT_s} \tag{4.40}$$

is the relative time delay in samples between the ith sensor and the first sensor for an endfire source, c is the speed of wave propagation, and T_s is the sampling interval. Assuming that the sampling interval satisfies $T_s = \frac{d}{c}$, then the delay $\tau_i = i - 1$ becomes an integer and, therefore, (4.38) and (4.39) can be written as

$$\left[\underline{\mathbf{x}}(t)\right]_{l+(m-1)L} = \left[\underline{\mathbf{x}}(t)\right]_l, \tag{4.41}$$

$$\left[\underline{\mathbf{u}}(t)\right]_{l+(m-1)L} = \left[\underline{\mathbf{u}}(t)\right]_{l+m-1}, \tag{4.42}$$

for $l = 1, \ldots, L$, $m = 1, \ldots, M$, and $l + m - 1 \leq L$. Hence, the correlation matrix of $\underline{\mathbf{x}}(t)$ is

$$\mathbf{R}_{\underline{\mathbf{x}}} = \mathbf{1}_M \otimes \mathbf{R}_{\mathbf{x}_1},$$

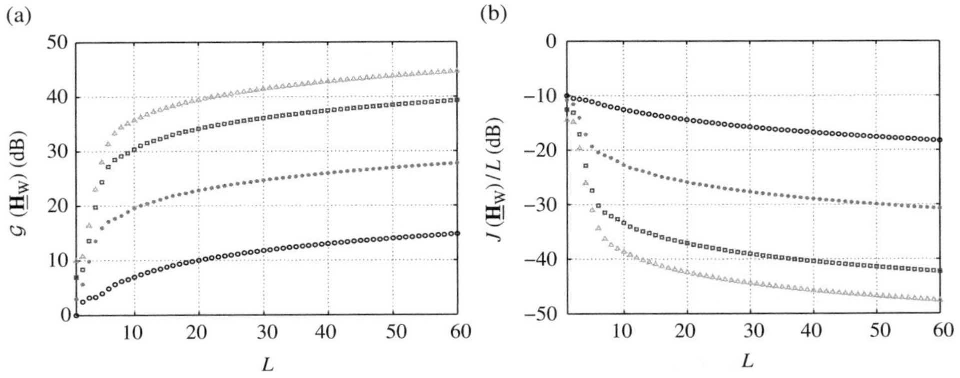

Figure 4.3 (a) The gain in SNR and (b) the MMSE per sample of the Wiener filtering matrix as a function of the filter length, L, for different numbers of sensors, M: $M = 1$ (circles), $M = 2$ (asterisks), $M = 5$ (squares), and $M = 10$ (triangles).

where $\otimes$ is the Kronecker product, $\mathbf{1}_M$ is an $M \times M$ matrix of all ones, and the elements of the correlation matrix of $\mathbf{x}_1(t)$ are $\left[\mathbf{R}_{\mathbf{x}_1}\right]_{i,j} = \frac{1}{2}A^2 \cos\left[2\pi f_0(i-j)\right]$. The correlation matrix of $\underline{\mathbf{v}}(t)$ is $\mathbf{R}_{\underline{\mathbf{v}}} = \mathbf{R}_{\underline{\mathbf{u}}} + \sigma_w^2 \mathbf{I}_{LM}$, where the elements of the $LM \times LM$ matrix $\mathbf{R}_{\underline{\mathbf{u}}}$ are

$$\left[\mathbf{R}_{\underline{\mathbf{u}}}\right]_{i+(m_1-1)L, j+(m_2-1)L} = \sigma_u^2 \, \delta\left(i + m_1 - j - m_2\right),$$
$$i, j = 1, \ldots, L, \; m_1, m_2 = 1, \ldots, M.$$

The input SNR is

$$\text{iSNR} = 10 \log \frac{A^2/2}{\sigma_u^2 + \sigma_w^2} \quad \text{(dB)}.$$

The optimal filter $\underline{\mathbf{H}}_W$ is obtained from (4.36).

To demonstrate the performance of the Wiener filtering matrix, we choose $A = 0.5$, $f_0 = 0.1$, $\sigma_u^2 = 0.5$, and $\sigma_w^2 = 0.01\sigma_u^2$. The input SNR is -6.06 dB. Figure 4.3 shows the effect of the filter length, L, and the number of sensors, M, on the gain in SNR, $\mathcal{G}\left(\underline{\mathbf{H}}_W\right) = \text{oSNR}\left(\underline{\mathbf{H}}_W\right)/\text{iSNR}$, and the MMSE per sample, $J\left(\underline{\mathbf{H}}_W\right)/L$. As the length of the filter increases, or as the number of sensors increases, the Wiener filtering matrix better enhances the harmonic signal, in terms of higher gain in SNR and lower MMSE per sample. If we choose a fixed filter length, $L = 30$, and change σ_u^2 so that iSNR varies from -5 to 15 dB, then Figure 4.4 shows plots of the gain in SNR, the MMSE, the noise reduction factor, and the desired signal reduction factor, as a function of the input SNR for different numbers of sensors, M. For a given input SNR, as the number of sensors increases, the gain in SNR and the noise reduction factor increase, while the MMSE and the desired signal reduction factor decrease.

Figure 4.5 shows a realization of the noise corrupted signal received at the first sensor, $y_1(t)$, and filtered signals for iSNR $= -5$ dB and different numbers of sensors. The filtered signal, $z(t)$, is obtained by taking, at each t, the first element of $\mathbf{z}(t) = \underline{\mathbf{H}}_W \, \underline{\mathbf{y}}(t)$.

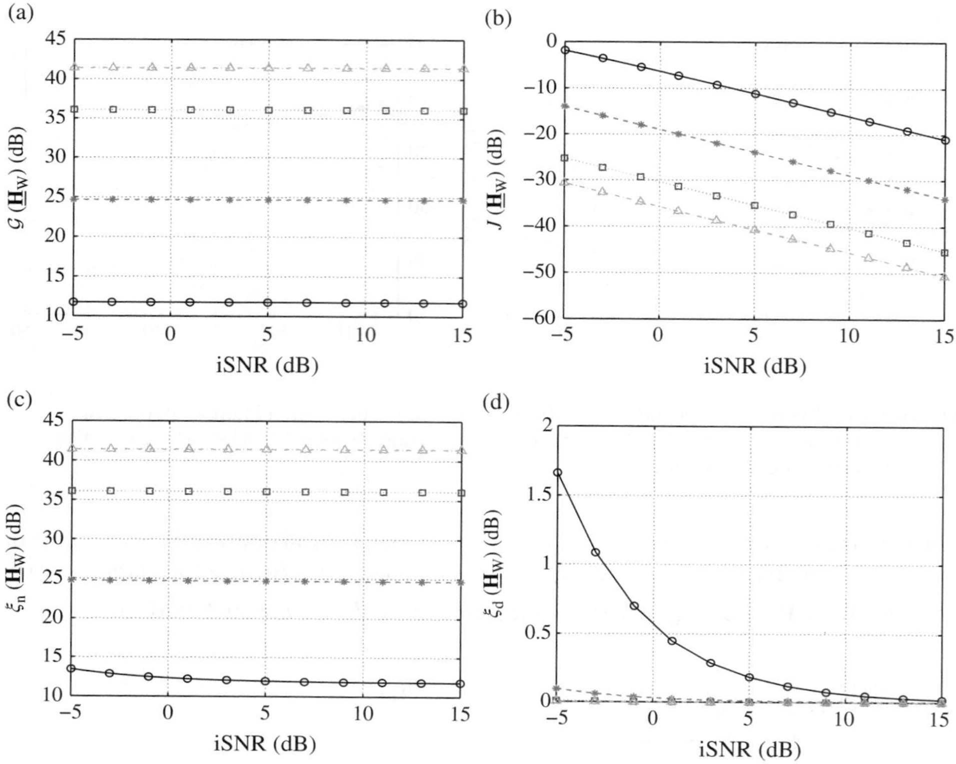

Figure 4.4 (a) The gain in SNR, (b) the MMSE, (c) the noise reduction factor, and (d) the desired signal reduction factor of the Wiener filtering matrix as a function of the input SNR for different numbers of sensors, M: $M = 1$ (solid line with circles), $M = 2$ (dashed line with asterisks), $M = 5$ (dotted line with squares), and $M = 10$ (dash-dot line with triangles).

Obviously, as the number of sensors increases, the Wiener filtering matrix better enhances the harmonic signal. ■

From the formulation given in (4.36), we propose a variable span (VS) Wiener filtering matrix [5, 6]:

$$\underline{\mathbf{H}}_{\mathrm{W},Q} = \underline{\mathbf{I}}_{\mathrm{i}}\mathbf{R}_{\underline{\mathbf{x}}} \sum_{q=1}^{Q} \frac{\underline{\mathbf{t}}_q \underline{\mathbf{t}}_q^T}{1 + \underline{\lambda}_q}, \tag{4.43}$$

where $1 \leq Q \leq ML$. We see that $\underline{\mathbf{H}}_{\mathrm{W},ML} = \underline{\mathbf{H}}_{\mathrm{W}}$ and, for $Q = 1$, we obtain the maximum SNR filtering matrix with minimum MSE:

$$\underline{\mathbf{H}}_{\max,1} = \underline{\mathbf{I}}_{\mathrm{i}}\mathbf{R}_{\underline{\mathbf{x}}} \frac{\underline{\mathbf{t}}_1 \underline{\mathbf{t}}_1^T}{1 + \underline{\lambda}_1}, \tag{4.44}$$

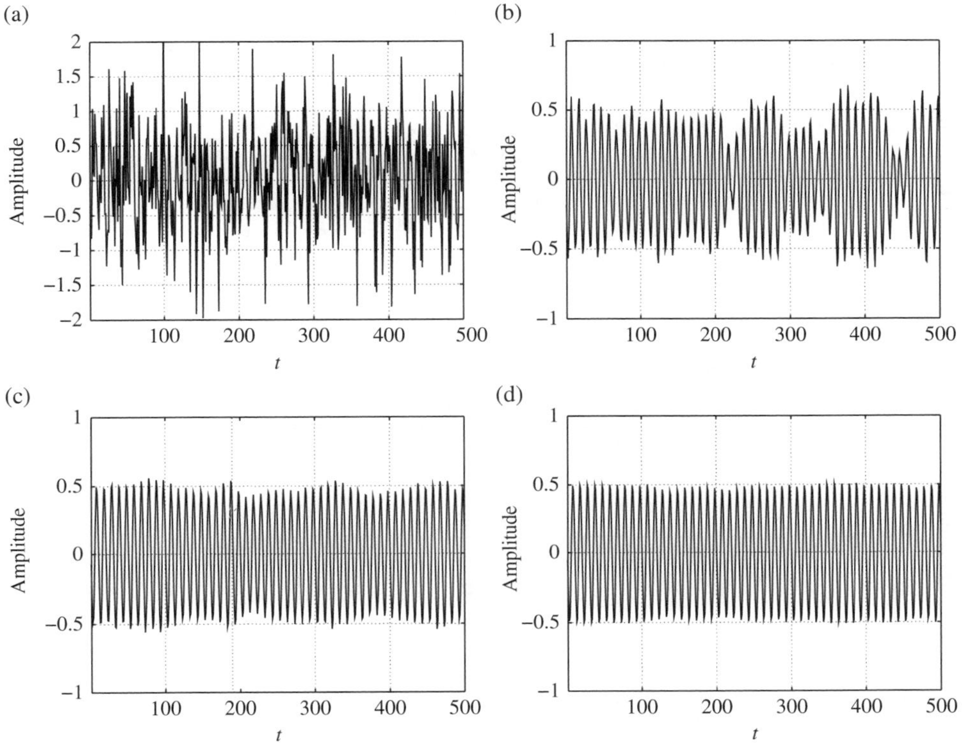

Figure 4.5 Example of noise corrupted and filtered sinusoidal signals for different numbers of sensors, M: (a) noise corrupted signal received at the first sensor, $y_1(t)$ (iSNR = −5 dB), and filtered signals for (b) $M = 1$ [$\mathrm{oSNR}\left(\underline{\mathbf{H}}_{\mathrm{W}}\right) = 6.76$ dB], (c) $M = 2$ [$\mathrm{oSNR}\left(\underline{\mathbf{H}}_{\mathrm{W}}\right) = 19.68$ dB], and (d) $M = 5$ [$\mathrm{oSNR}\left(\underline{\mathbf{H}}_{\mathrm{W}}\right) = 31.09$ dB].

since

$$\mathrm{oSNR}\left(\underline{\mathbf{H}}_{\max,1}\right) = \underline{\lambda}_1. \tag{4.45}$$

Example 4.2.2 Returning to Example 4.2.1, we now assume a desired signal, $x(t)$, with the autocorrelation sequence:

$$E\left[x(t)x(t')\right] = \alpha^{|t-t'|}, \ -1 < \alpha < 1.$$

The desired signal needs to be recovered from the noisy observation, $\underline{\mathbf{y}}(t) = \underline{\mathbf{x}}(t) + \underline{\mathbf{v}}(t)$. Since the desired source is at the broadside direction, the correlation matrix of $\underline{\mathbf{x}}(t)$ is

$$\mathbf{R}_{\underline{\mathbf{x}}} = \mathbf{1}_M \otimes \mathbf{R}_{\mathbf{x}_1},$$

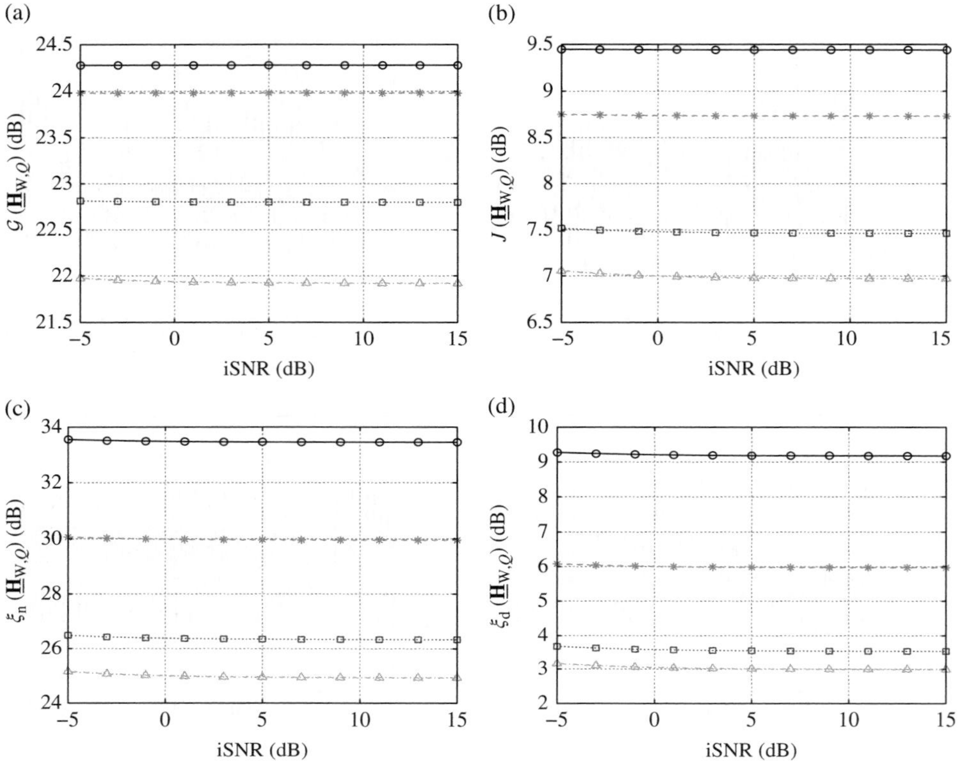

Figure 4.6 (a) The gain in SNR, (b) the MSE, (c) the noise reduction factor, and (d) the desired signal reduction factor of the VS Wiener filtering matrix as a function of the input SNR for several values of Q: $Q = 1$ (solid line with circles), $Q = 2$ (dashed line with asterisks), $Q = 5$ (dotted line with squares), and $Q = 9$ (dash-dot line with triangles).

where $\left[\mathbf{R}_{\mathbf{x}_1}\right]_{i,j} = \alpha^{|i-j|}$. The input SNR is

$$\text{iSNR} = 10 \log \frac{1}{\sigma_u^2 + \sigma_w^2} \quad \text{(dB)}.$$

The optimal filter $\underline{\mathbf{H}}_{\text{W},Q}$ is obtained from (4.43).

To demonstrate the performance of the VS Wiener filtering matrix, we choose $\alpha = 0.8$, $L = 10$, and $M = 5$. Figure 4.6 shows plots of the gain in SNR, $\mathcal{G}\left(\underline{\mathbf{H}}_{\text{W},Q}\right)$, the MSE, $J\left(\underline{\mathbf{H}}_{\text{W},Q}\right)$, the noise reduction factor, $\xi_{\text{n}}\left(\underline{\mathbf{H}}_{\text{W},Q}\right)$, and the desired signal reduction factor, $\xi_{\text{d}}\left(\underline{\mathbf{H}}_{\text{W},Q}\right)$, as a function of the input SNR for several values of Q. For a given input SNR, the higher the value of Q, the lower are the MSE and the desired signal reduction factor, but at the expense of lower gain in SNR and lower noise reduction factor. ■

We can also try to minimize the distortion-based MSE. Taking the gradient of $J_{\rm d}\left(\underline{\mathbf{A}}\right)$ with respect to $\underline{\mathbf{A}}$ and equating the result to zero, we get

$$\underline{\mathbf{A}}\,\underline{\boldsymbol{\Lambda}} = \underline{\mathbf{I}}_{\rm i}\mathbf{R}_{\underline{\mathbf{x}}}\underline{\mathbf{T}}. \tag{4.46}$$

Since $\underline{\boldsymbol{\Lambda}}$ is not invertible, we can take its pseudo-inverse. Then, the solution to (4.46) is

$$\underline{\mathbf{A}}_{\rm MVDR} = \underline{\mathbf{I}}_{\rm i}\mathbf{R}_{\underline{\mathbf{x}}}\underline{\mathbf{T}}\,\underline{\boldsymbol{\Lambda}}'^{-1}, \tag{4.47}$$

where

$$\underline{\boldsymbol{\Lambda}}'^{-1} = {\rm diag}\left(\underline{\lambda}_1^{-1}, \underline{\lambda}_2^{-1}, \ldots, \underline{\lambda}_P^{-1}, 0, \ldots, 0\right). \tag{4.48}$$

Therefore, the MVDR filtering matrix is

$$\begin{aligned}\underline{\mathbf{H}}_{\rm MVDR} &= \underline{\mathbf{A}}_{\rm MVDR}\underline{\mathbf{T}}^T \\ &= \underline{\mathbf{I}}_{\rm i}\mathbf{R}_{\underline{\mathbf{x}}}\sum_{p=1}^{P}\frac{\underline{\mathbf{t}}_p\underline{\mathbf{t}}_p^T}{\underline{\lambda}_p} \\ &= \underline{\mathbf{I}}_{\rm i}\mathbf{R}_{\underline{\mathbf{v}}}\sum_{p=1}^{P}\underline{\mathbf{t}}_p\underline{\mathbf{t}}_p^T.\end{aligned} \tag{4.49}$$

Now, let us show that (4.49) is the MVDR filtering matrix. With $\underline{\mathbf{H}}_{\rm MVDR}$, the filtered desired signal vector is

$$\begin{aligned}\mathbf{x}_{\rm fd}(t) &= \underline{\mathbf{I}}_{\rm i}\mathbf{R}_{\underline{\mathbf{v}}}\sum_{p=1}^{P}\underline{\mathbf{t}}_p\underline{\mathbf{t}}_p^T\underline{\mathbf{x}}(t) \\ &= \underline{\mathbf{I}}_{\rm i}\left(\mathbf{I}_{ML} - \mathbf{R}_{\underline{\mathbf{v}}}\sum_{i=P+1}^{ML}\underline{\mathbf{t}}_i\underline{\mathbf{t}}_i^T\right)\underline{\mathbf{x}}(t) \\ &= \mathbf{x}_1(t) - \underline{\mathbf{I}}_{\rm i}\mathbf{R}_{\underline{\mathbf{v}}}\sum_{i=P+1}^{ML}\underline{\mathbf{t}}_i\underline{\mathbf{t}}_i^T\underline{\mathbf{x}}(t) \\ &= \mathbf{x}_1(t),\end{aligned} \tag{4.50}$$

where, in the previous expression, we have used (4.10) and (4.11). Then, it is clear that

$$\upsilon_{\rm d}\left(\underline{\mathbf{H}}_{\rm MVDR}\right) = 0, \tag{4.51}$$

proving that, indeed, $\underline{\mathbf{H}}_{\rm MVDR}$ is the MVDR filtering matrix.

Property 4.2.2 With the MVDR filtering matrix given in (4.49), the output SNR is always greater than or equal to the input SNR: ${\rm oSNR}\left(\underline{\mathbf{H}}_{\rm MVDR}\right) \geq {\rm iSNR}$.

From the MVDR filtering matrix, we can derive the controlled distortion (CD) filtering matrix:

$$\underline{\mathbf{H}}_{\mathrm{CD},P'} = \underline{\mathbf{I}}_{\mathrm{i}} \mathbf{R}_{\underline{\mathbf{x}}} \sum_{p'=1}^{P'} \frac{\underline{\mathbf{t}}_{p'} \underline{\mathbf{t}}_{p'}^T}{\underline{\lambda}_{p'}}, \tag{4.52}$$

where $1 \leq P' \leq P$. We observe that $\underline{\mathbf{H}}_{\mathrm{CD},P} = \underline{\mathbf{H}}_{\mathrm{MVDR}}$ and, for $P' = 1$, we obtain the maximum SNR filtering matrix with minimum distortion:

$$\underline{\mathbf{H}}_{\max,0} = \underline{\mathbf{I}}_{\mathrm{i}} \mathbf{R}_{\underline{\mathbf{x}}} \frac{\underline{\mathbf{t}}_1 \underline{\mathbf{t}}_1^T}{\underline{\lambda}_1}, \tag{4.53}$$

since

$$\mathrm{oSNR}\left(\underline{\mathbf{H}}_{\max,0}\right) = \underline{\lambda}_1. \tag{4.54}$$

Example 4.2.3 Returning to Example 4.2.2, we now employ the CD filtering matrix, $\underline{\mathbf{H}}_{\mathrm{CD},P'}$, given in (4.52). Figure 4.7 shows plots of the gain in SNR, $\mathcal{G}\left(\underline{\mathbf{H}}_{\mathrm{CD},P'}\right)$, the MSE, $J\left(\underline{\mathbf{H}}_{\mathrm{CD},P'}\right)$, the noise reduction factor, $\xi_{\mathrm{n}}\left(\underline{\mathbf{H}}_{\mathrm{CD},P'}\right)$, and the desired signal reduction factor, $\xi_{\mathrm{d}}\left(\underline{\mathbf{H}}_{\mathrm{CD},P'}\right)$, as a function of the input SNR for several values of P'. For a given input SNR, the higher the value of P', the lower are the MSE and the desired signal reduction factor, but at the expense of lower gain in SNR and a lower noise reduction factor. ■

Another practical approach that can give a compromise between noise reduction and desired signal distortion is the tradeoff filtering matrix, which is obtained from:

$$\min_{\underline{\mathbf{A}}} J_{\mathrm{d}}\left(\underline{\mathbf{A}}\right) \quad \text{subject to} \quad J_{\mathrm{n}}\left(\underline{\mathbf{A}}\right) = \aleph \mathrm{tr}\left(\mathbf{R}_{\mathbf{v}_1}\right), \tag{4.55}$$

where $0 < \aleph < 1$ to ensure that filtering achieves some degree of noise reduction. We find that the optimal filtering matrix is

$$\underline{\mathbf{H}}_{\mathrm{T},\mu} = \underline{\mathbf{I}}_{\mathrm{i}} \mathbf{R}_{\underline{\mathbf{x}}} \sum_{i=1}^{ML} \frac{\underline{\mathbf{t}}_i \underline{\mathbf{t}}_i^T}{\mu + \underline{\lambda}_i}, \tag{4.56}$$

where $\mu \geq 0$ is a Lagrange multiplier. For $\mu = 1$, we get the Wiener filtering matrix.

Property 4.2.3 With the tradeoff filtering matrix given in (4.56), the output SNR is always greater than or equal to the input SNR: $\mathrm{oSNR}\left(\underline{\mathbf{H}}_{\mathrm{T},\mu}\right) \geq \mathrm{iSNR}, \forall \mu \geq 0$.

Example 4.2.4 Returning to Example 4.2.2, we now employ the tradeoff filtering matrix, $\underline{\mathbf{H}}_{\mathrm{T},\mu}$, given in (4.56). Figure 4.8 shows plots of the gain in SNR, $\mathcal{G}\left(\underline{\mathbf{H}}_{\mathrm{T},\mu}\right)$,

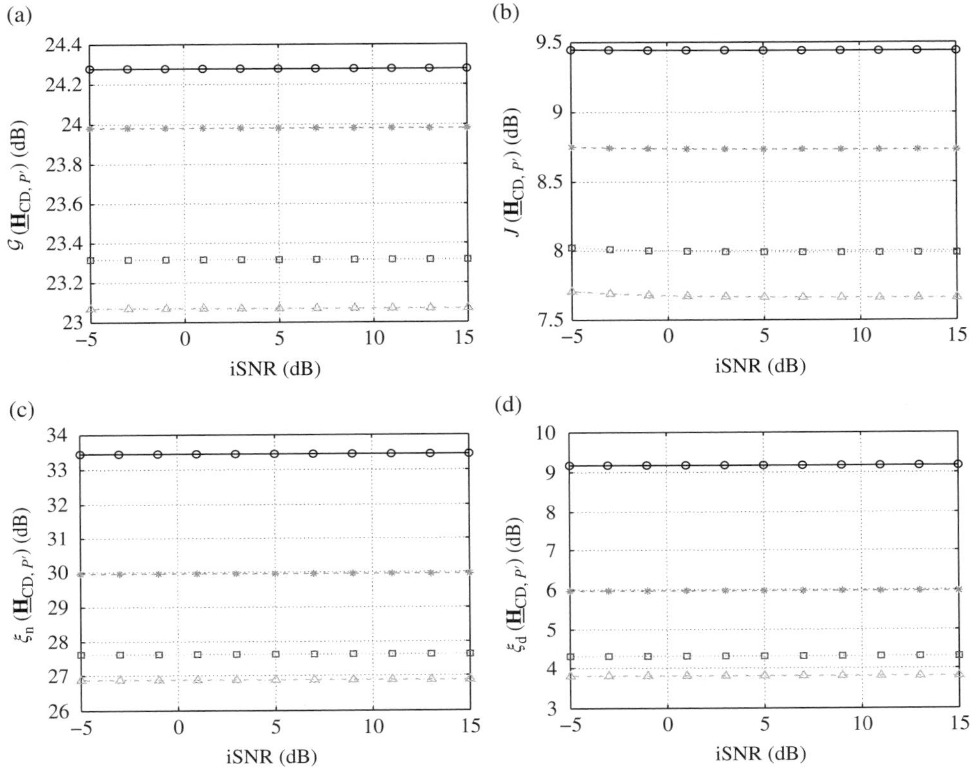

Figure 4.7 (a) The gain in SNR, (b) the MSE, (c) the noise reduction factor, and (d) the desired signal reduction factor of the CD filtering matrix as a function of the input SNR for several values of P': $P' = 1$ (solid line with circles), $P' = 2$ (dashed line with asterisks), $P' = 3$ (dotted line with squares), and $P' = 4$ (dash-dot line with triangles).

the MSE, $J\left(\underline{\mathbf{H}}_{\mathrm{T},\mu}\right)$, the noise reduction factor, $\xi_{\mathrm{n}}\left(\underline{\mathbf{H}}_{\mathrm{T},\mu}\right)$, and the desired signal reduction factor, $\xi_{\mathrm{d}}\left(\underline{\mathbf{H}}_{\mathrm{T},\mu}\right)$, as a function of the input SNR for several values of μ. For a given input SNR, the higher the value of μ, the higher are the gain in SNR and the noise reduction factor, but at the expense of a higher desired signal reduction factor. ■

From what we have seen so far, we can propose a very general subspace (GS) noise reduction filtering matrix [7]:

$$\underline{\mathbf{H}}_{\mu,Q} = \underline{\mathbf{I}}_{\mathrm{i}}\mathbf{R}_{\underline{\mathbf{x}}} \sum_{q=1}^{Q} \frac{\underline{\mathbf{t}}_q \underline{\mathbf{t}}_q^T}{\mu + \underline{\lambda}_q}, \tag{4.57}$$

where $1 \leq Q \leq ML$. This form encompasses most known optimal filtering matrices. Indeed, it is clear that

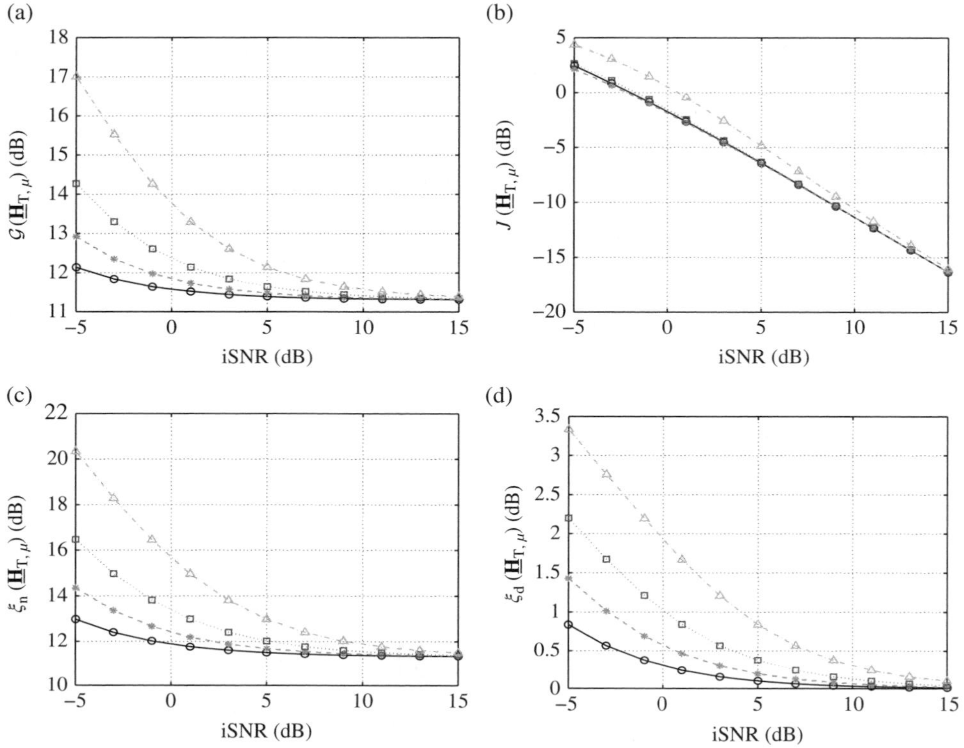

Figure 4.8 (a) The gain in SNR, (b) the MSE, (c) the noise reduction factor, and (d) the desired signal reduction factor of the tradeoff filtering matrix as a function of the input SNR for several values of μ: $\mu = 0.5$ (solid line with circles), $\mu = 1$ (dashed line with asterisks), $\mu = 2$ (dotted line with squares), and $\mu = 5$ (dash-dot line with triangles).

- $\underline{\mathbf{H}}_{1,ML} = \underline{\mathbf{H}}_{\mathrm{W}}$
- $\underline{\mathbf{H}}_{1,Q} = \underline{\mathbf{H}}_{\mathrm{W},Q}$
- $\underline{\mathbf{H}}_{1,1} = \underline{\mathbf{H}}_{\mathrm{max},1}$
- $\underline{\mathbf{H}}_{0,P} = \underline{\mathbf{H}}_{\mathrm{MVDR}}$
- $\underline{\mathbf{H}}_{0,P'} = \underline{\mathbf{H}}_{\mathrm{CD},P'}$
- $\underline{\mathbf{H}}_{0,1} = \underline{\mathbf{H}}_{\mathrm{max},0}$
- $\underline{\mathbf{H}}_{\mu,ML} = \underline{\mathbf{H}}_{\mathrm{T},\mu}$.

In Table 4.1, we present all the optimal filtering matrices developed in this subsection, showing how they are closely related.

4.3 Spectral Method

In this section, we show how to exploit the spectrum of each one of the sensors' signals. Thanks to this formulation, we better see the effect of the spatial information, which plays a critical role in multichannel signal enhancement when compared to the single-channel case.

Table 4.1 Optimal linear filtering matrices for multichannel signal enhancement in the time domain.

Filter	
Wiener	$\underline{\mathbf{H}}_{\mathrm{W}} = \underline{\mathbf{I}}_{\mathrm{i}} \mathbf{R}_{\underline{\mathbf{x}}} \sum_{i=1}^{ML} \frac{\underline{\mathbf{t}}_i \underline{\mathbf{t}}_i^T}{1 + \underline{\lambda}_i}$
VS Wiener	$\underline{\mathbf{H}}_{\mathrm{W},Q} = \underline{\mathbf{I}}_{\mathrm{i}} \mathbf{R}_{\underline{\mathbf{x}}} \sum_{q=1}^{Q} \frac{\underline{\mathbf{t}}_q \underline{\mathbf{t}}_q^T}{1 + \underline{\lambda}_q}, \ 1 \leq Q \leq ML$
MVDR	$\underline{\mathbf{H}}_{\mathrm{MVDR}} = \underline{\mathbf{I}}_{\mathrm{i}} \mathbf{R}_{\underline{\mathbf{x}}} \sum_{p=1}^{P} \frac{\underline{\mathbf{t}}_p \underline{\mathbf{t}}_p^T}{\underline{\lambda}_p}$
CD	$\underline{\mathbf{H}}_{\mathrm{CD},P'} = \underline{\mathbf{I}}_{\mathrm{i}} \mathbf{R}_{\underline{\mathbf{x}}} \sum_{p'=1}^{P'} \frac{\underline{\mathbf{t}}_{p'} \underline{\mathbf{t}}_{p'}^T}{\underline{\lambda}_{p'}}, \ 1 \leq P' \leq P$
Maximum SNR	$\underline{\mathbf{H}}_{\max,\mu} = \underline{\mathbf{I}}_{\mathrm{i}} \mathbf{R}_{\underline{\mathbf{x}}} \frac{\underline{\mathbf{t}}_1 \underline{\mathbf{t}}_1^T}{\mu + \underline{\lambda}_1}, \ \mu \geq 0$
Tradeoff	$\underline{\mathbf{H}}_{\mathrm{T},\mu} = \underline{\mathbf{I}}_{\mathrm{i}} \mathbf{R}_{\underline{\mathbf{x}}} \sum_{i=1}^{ML} \frac{\underline{\mathbf{t}}_i \underline{\mathbf{t}}_i^T}{\mu + \underline{\lambda}_i}, \ \mu \geq 0$
GS	$\underline{\mathbf{H}}_{\mu,Q} = \underline{\mathbf{I}}_{\mathrm{i}} \mathbf{R}_{\underline{\mathbf{x}}} \sum_{q=1}^{Q} \frac{\underline{\mathbf{t}}_q \underline{\mathbf{t}}_q^T}{\mu + \underline{\lambda}_q}, \ \mu \geq 0, \ 1 \leq Q \leq ML$

4.3.1 Temporal Joint Diagonalization and Reformulation of the Problem

We recall that the temporal correlation matrix (of size $L \times L$) of the mth sensor signal can be written as

$$\begin{aligned} \mathbf{R}_{\mathbf{y}_m} &= E\left[\mathbf{y}_m(t)\mathbf{y}_m^T(t)\right] \\ &= \mathbf{R}_{\mathbf{x}_m} + \mathbf{R}_{\mathbf{v}_m}, \end{aligned} \tag{4.58}$$

where $\mathbf{R}_{\mathbf{x}_m} = E\left[\mathbf{x}_m(t)\mathbf{x}_m^T(t)\right]$ and $\mathbf{R}_{\mathbf{v}_m} = E\left[\mathbf{v}_m(t)\mathbf{v}_m^T(t)\right]$ are the temporal correlation matrices of $\mathbf{x}_m(t)$ and $\mathbf{v}_m(t)$, respectively. The noise temporal correlation matrix, $\mathbf{R}_{\mathbf{v}_m}$, is assumed to be full rank: $\mathrm{rank}\left(\mathbf{R}_{\mathbf{v}_m}\right) = L$.

The joint diagonalization [4] of the two symmetric matrices $\mathbf{R}_{\mathbf{x}_m}$ and $\mathbf{R}_{\mathbf{v}_m}$ is

$$\mathbf{T}_m^T \mathbf{R}_{\mathbf{x}_m} \mathbf{T}_m = \mathbf{\Lambda}_m, \tag{4.59}$$

$$\mathbf{T}_m^T \mathbf{R}_{\mathbf{v}_m} \mathbf{T}_m = \mathbf{I}_L, \tag{4.60}$$

where

$$\mathbf{T}_m = \left[\begin{array}{cccc} \mathbf{t}_{m,1} & \mathbf{t}_{m,2} & \cdots & \mathbf{t}_{m,L} \end{array}\right] \tag{4.61}$$

is a full-rank square matrix (of size $L \times L$),

$$\mathbf{\Lambda}_m = \mathrm{diag}\left(\lambda_{m,1}, \lambda_{m,2}, \ldots, \lambda_{m,L}\right) \tag{4.62}$$

is a diagonal matrix with $\lambda_{m,1} \geq \lambda_{m,2} \geq \cdots \geq \lambda_{m,L} \geq 0$, and $\mathbf{I}_L$ is the $L \times L$ identity matrix. Also, we have

$$\mathbf{R}_{\mathbf{v}_m}^{-1}\mathbf{R}_{\mathbf{x}_m}\mathbf{T}_m = \mathbf{T}_m\mathbf{\Lambda}_m. \tag{4.63}$$

In other words, $\mathbf{\Lambda}_m$ and $\mathbf{T}_m$ are the eigenvalue and eigenvector matrices of $\mathbf{R}_{\mathbf{v}_m}^{-1}\mathbf{R}_{\mathbf{x}_m}$, respectively. The mth noisy signal temporal correlation matrix is then diagonalized as

$$\mathbf{T}_m^T\mathbf{R}_{\mathbf{y}_m}\mathbf{T}_m = \mathbf{\Lambda}_m + \mathbf{I}_L. \tag{4.64}$$

By left multiplying both sides of (4.2) by $\mathbf{t}_{m,l}^T$, we get the lth spectral mode of the mth sensor signal:

$$\begin{aligned} y_m(t;l) &= \mathbf{t}_{m,l}^T\mathbf{y}_m(t) \\ &= x_m(t;l) + v_m(t;l), \end{aligned} \tag{4.65}$$

where $x_m(t;l) = \mathbf{t}_{m,l}^T\mathbf{x}_m(t)$ and $v_m(t;l) = \mathbf{t}_{m,l}^T\mathbf{v}_m(t)$ are the lth spectral modes of the convolved desired and noise signals, respectively. We deduce that the variance of $y_m(t;l)$ is

$$\begin{aligned} \sigma_{y_m}^2(;l) &= E\left[y_m^2(t;l)\right] \\ &= \mathbf{t}_{m,l}^T\mathbf{R}_{\mathbf{y}_m}\mathbf{t}_{m,l} \\ &= \lambda_{m,l} + 1 \\ &= \sigma_{x_m}^2(;l) + \sigma_{v_m}^2(;l), \end{aligned} \tag{4.66}$$

where $\sigma_{x_m}^2(;l) = \lambda_{m,l}$ and $\sigma_{v_m}^2(;l) = 1$ are the variances of $x_m(t;l)$ and $v_m(t;l)$, respectively. Now, (4.65) can be considered as the new signal model and, as such, the aim is to recover the desired signal, $x_1(t;l)$, given $y_m(t;l)$, $m = 1,2,\ldots,M$. When the elements $x_1(t;l)$, $l = 1,2,\ldots,L$ are correctly estimated, it is easy to determine the estimate of $\mathbf{x}_1(t)$ by premultiplying the vector of length L containing the estimates of $x_1(t;l)$, $l = 1,2,\ldots,L$ by $\mathbf{T}_1^{-T}$.

4.3.2 Spatial Joint Diagonalization

We will again the joint diagonalization, but this time on the spatial correlation matrices. Let us define the stacked vector of length M:

$$\begin{aligned} \mathbf{y}(t;l) &= \begin{bmatrix} y_1(t;l) & y_2(t;l) & \cdots & y_M(t;l) \end{bmatrix}^T \\ &= \mathbf{x}(t;l) + \mathbf{v}(t;l), \end{aligned} \tag{4.67}$$

where $\mathbf{x}(t;l)$ and $\mathbf{v}(t;l)$ are defined in a similar way to $\mathbf{y}(t;l)$. The vector $\mathbf{y}(t;l)$ is the spatial information of the observations at the lth spectral mode. The spatial correlation matrix (of size $M \times M$) of $\mathbf{y}(t;l)$ is then

$$\begin{aligned}\mathbf{R}_{\mathbf{y}}(;l) &= E\left[\mathbf{y}(t;l)\mathbf{y}^T(t;l)\right] \\ &= \mathbf{R}_{\mathbf{x}}(;l) + \mathbf{R}_{\mathbf{v}}(;l),\end{aligned} \tag{4.68}$$

where $\mathbf{R}_{\mathbf{x}}(;l) = E\left[\mathbf{x}(t;l)\mathbf{x}^T(t;l)\right]$ and $\mathbf{R}_{\mathbf{v}}(;l) = E\left[\mathbf{v}(t;l)\mathbf{v}^T(t;l)\right]$ are the spatial correlation matrices of $\mathbf{x}(t;l)$ and $\mathbf{v}(t;l)$, respectively. Since the convolved desired signal is coherent at the sensors, the matrix $\mathbf{R}_{\mathbf{x}}(;l)$ should be rank deficient. We assume that this rank is equal to $P_l \leq M$. Since the noise is only partially coherent, $\mathbf{R}_{\mathbf{v}}(;l)$ is full rank.

The two spatial correlation matrices $\mathbf{R}_{\mathbf{x}}(;l)$ and $\mathbf{R}_{\mathbf{v}}(;l)$ can also be jointly diagonalized as

$$\mathbf{S}^T(;l)\mathbf{R}_{\mathbf{x}}(;l)\mathbf{S}(;l) = \mathbf{\Omega}(;l), \tag{4.69}$$

$$\mathbf{S}^T(;l)\mathbf{R}_{\mathbf{v}}(;l)\mathbf{S}(;l) = \mathbf{I}_M, \tag{4.70}$$

where

$$\begin{aligned}\mathbf{S}(;l) &= \begin{bmatrix} \mathbf{s}_1(;l) & \mathbf{s}_2(;l) & \cdots & \mathbf{s}_M(;l) \end{bmatrix} \\ &= \begin{bmatrix} \mathbf{S}'(;l) & \mathbf{S}''(;l) \end{bmatrix}\end{aligned} \tag{4.71}$$

is a full-rank square matrix (of size $M \times M$), $\mathbf{S}'(;l)$ and $\mathbf{S}''(;l)$ contain the first P_l and last $M - P_l$ columns of $\mathbf{S}(;l)$, respectively,

$$\mathbf{\Omega}(;l) = \text{diag}\left[\omega_1(;l), \omega_2(;l), \dots, \omega_M(;l)\right], \tag{4.72}$$

with $\omega_1(;l) \geq \omega_2(;l) \geq \cdots \geq \omega_{P_l}(;l) > \omega_{P_l+1}(;l) = \cdots = \omega_M(;l) = 0$, and $\mathbf{I}_M$ is the $M \times M$ identity matrix. Obviously, $\mathbf{\Omega}(;l)$ and $\mathbf{S}(;l)$ are the eigenvalue and eigenvector matrices of $\mathbf{R}_{\mathbf{v}}^{-1}(;l)\mathbf{R}_{\mathbf{x}}(;l)$, respectively:

$$\mathbf{R}_{\mathbf{v}}^{-1}(;l)\mathbf{R}_{\mathbf{x}}(;l)\mathbf{S}(;l) = \mathbf{S}(;l)\mathbf{\Omega}(;l) \tag{4.73}$$

and

$$\mathbf{S}^T(;l)\mathbf{R}_{\mathbf{y}}(;l)\mathbf{S}(;l) = \mathbf{\Omega}(;l) + \mathbf{I}_M. \tag{4.74}$$

Furthermore, it can be verified from (4.69) and (4.70) that

$$\mathbf{S}''^T(;l)\mathbf{x}(t;l) = \mathbf{0} \tag{4.75}$$

and

$$\mathbf{R}_{\mathbf{v}}^{-1}(;l) = \mathbf{S}'(;l)\mathbf{S}'^T(;l) + \mathbf{S}''(;l)\mathbf{S}''^T(;l). \tag{4.76}$$

4.3.3 Spatial Linear Filtering

The conventional way to perform multichannel noise reduction is to apply a filter to the spatial observation signal vector, as illustrated in Figure 4.9:

$$\begin{aligned} z(t;l) &= \mathbf{h}^T(;l)\mathbf{y}(t;l) \\ &= \mathbf{h}^T(;l)\left[\mathbf{x}(t;l) + \mathbf{v}(t;l)\right] \\ &= x_{\mathrm{fd}}(t;l) + v_{\mathrm{rn}}(t;l), \end{aligned} \tag{4.77}$$

where $z(t;l)$ is the estimate of $x_1(t;l)$,

$$\mathbf{h}(;l) = \begin{bmatrix} h_1(;l) & h_2(;l) & \cdots & h_M(;l) \end{bmatrix}^T \tag{4.78}$$

is a spatial linear filter of length M at the lth spectral mode, $x_{\mathrm{fd}}(t;l) = \mathbf{h}^T(;l)\mathbf{x}(t;l)$ is the filtered desired signal, and $v_{\mathrm{rn}}(t;l) = \mathbf{h}^T(;l)\mathbf{v}(t;l)$ is the residual noise. We deduce that the variance of $z(t;l)$ is

$$\begin{aligned} \sigma_z^2(;l) &= E\left[z^2(t;l)\right] \\ &= \mathbf{h}^T(;l)\mathbf{R}_{\mathbf{y}}(;l)\mathbf{h}(;l) \\ &= \sigma_{x_{\mathrm{fd}}}^2(;l) + \sigma_{v_{\mathrm{rn}}}^2(;l), \end{aligned} \tag{4.79}$$

where

$$\sigma_{x_{\mathrm{fd}}}^2(;l) = \mathbf{h}^T(;l)\mathbf{R}_{\mathbf{x}}(;l)\mathbf{h}(;l) \tag{4.80}$$

and

$$\sigma_{v_{\mathrm{rn}}}^2(;l) = \mathbf{h}^T(;l)\mathbf{R}_{\mathbf{v}}(;l)\mathbf{h}(;l) \tag{4.81}$$

are the variances of $x_{\mathrm{fd}}(t;l)$ and $v_{\mathrm{rn}}(t;l)$, respectively.

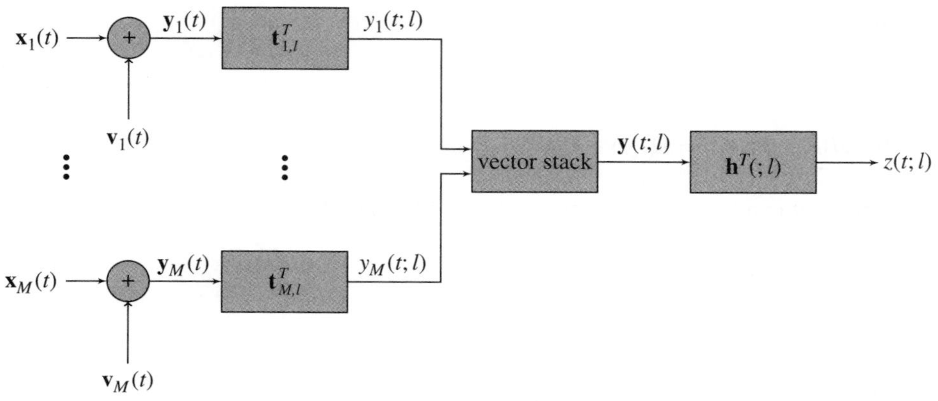

Figure 4.9 Block diagram of multichannel spatial linear filtering.

4.3.4 Performance Measures

Performance measures are an important area of the signal enhancement problem. Without them, there is no way to know how a filter really performs. In this section, we explain the most intuitive ones, as we did in Section 4.2.3. We start by deriving measures related to the noise reduction. Then we discuss the evaluation of the desired signal distortion. Finally, we present the MSE criterion, which is very convenient to use in signal enhancement applications.

4.3.4.1 Noise Reduction

Since the first sensor is the reference, we define the lth spectral mode input SNR as

$$\begin{aligned} \text{iSNR}(;l) &= \frac{\sigma_{x_1}^2(;l)}{\sigma_{v_1}^2(;l)} \\ &= \lambda_{1,l}. \end{aligned} \tag{4.82}$$

We deduce that the fullmode input SNR is

$$\begin{aligned} \text{iSNR}(;) &= \frac{\sum_{l=1}^{L} \sigma_{x_1}^2(;l)}{\sum_{l=1}^{L} \sigma_{v_1}^2(;l)} \\ &= \frac{\text{tr}\left(\mathbf{\Lambda}_1\right)}{L} \\ &= \frac{\text{tr}\left(\mathbf{R}_{\mathbf{v}_1}^{-1}\mathbf{R}_{\mathbf{x}_1}\right)}{L}. \end{aligned} \tag{4.83}$$

It is clear that

$$\text{iSNR}(;L) \leq \text{iSNR}(;) \leq \text{iSNR}(;1). \tag{4.84}$$

In other words, the fullmode input SNR can never exceed the maximum spectral mode input SNR and can never go below the minimum spectral mode input SNR.

From (4.79), we define the lth spectral mode output SNR:

$$\text{oSNR}\left[\mathbf{h}(;l)\right] = \frac{\mathbf{h}^T(;l)\mathbf{R}_{\mathbf{x}}(;l)\mathbf{h}(;l)}{\mathbf{h}^T(;l)\mathbf{R}_{\mathbf{v}}(;l)\mathbf{h}(;l)} \tag{4.85}$$

and it can be verified that it is upper bounded:

$$\text{oSNR}\left[\mathbf{h}(;l)\right] \leq \omega_1(;l). \tag{4.86}$$

The fullmode output SNR is

$$\text{oSNR}\left[\mathbf{h}(;:)\right] = \frac{\sum_{l=1}^{L} \mathbf{h}^T(;l)\mathbf{R}_{\mathbf{x}}(;l)\mathbf{h}(;l)}{\sum_{l=1}^{L} \mathbf{h}^T(;l)\mathbf{R}_{\mathbf{v}}(;l)\mathbf{h}(;l)}. \tag{4.87}$$

It can be shown that

$$\text{oSNR}\left[\mathbf{h}(;:)\right] \leq \max_{l} \omega_1(;l). \tag{4.88}$$

Then, the objective is to find the spatial filters, $\mathbf{h}(;l)$, $l = 1, 2, \ldots, L$, in such a way that $\text{oSNR}\left[\mathbf{h}(;l)\right] \geq \text{iSNR}(;l)$ and $\text{oSNR}\left[\mathbf{h}(;:)\right] > \text{iSNR}(;)$.

The lth spectral mode and fullmode noise reduction factors are given by, respectively,

$$\xi_{\text{n}}\left[\mathbf{h}(;l)\right] = \frac{1}{\mathbf{h}^T(;l)\mathbf{R}_{\mathbf{v}}(;l)\mathbf{h}(;l)} \tag{4.89}$$

and

$$\xi_{\text{n}}\left[\mathbf{h}(;:)\right] = \frac{L}{\sum_{l=1}^{L} \mathbf{h}^T(;l)\mathbf{R}_{\mathbf{v}}(;l)\mathbf{h}(;l)}. \tag{4.90}$$

For optimal filters, the noise reduction factors are greater than 1. The higher the value of ξ_{n}, the more noise reduction there is.

4.3.4.2 Desired Signal Distortion

The distortion of the desired signal vector can be measured with the lth spectral mode desired signal reduction factor:

$$\xi_{\text{d}}\left[\mathbf{h}(;l)\right] = \frac{\lambda_{1,l}}{\mathbf{h}^T(;l)\mathbf{R}_{\mathbf{x}}(;l)\mathbf{h}(;l)} \tag{4.91}$$

and the fullmode desired signal reduction factor:

$$\xi_{\text{d}}\left[\mathbf{h}(;:)\right] = \frac{\text{tr}\left(\mathbf{\Lambda}_1\right)}{\sum_{l=1}^{L} \mathbf{h}^T(;l)\mathbf{R}_{\mathbf{x}}(;l)\mathbf{h}(;l)}. \tag{4.92}$$

For optimal filters, the desired signal reduction factors are greater than 1. The higher the value of ξ_{d}, the more the distortion of the desired signal.

It is obvious that we always have

$$\frac{\text{oSNR}\left[\mathbf{h}(;l)\right]}{\text{iSNR}(;l)} = \frac{\xi_{\text{n}}\left[\mathbf{h}(;l)\right]}{\xi_{\text{d}}\left[\mathbf{h}(;l)\right]}, \tag{4.93}$$

$$\frac{\text{oSNR}\left[\mathbf{h}(;:)\right]}{\text{iSNR}(;)} = \frac{\xi_{\text{n}}\left[\mathbf{h}(;:)\right]}{\xi_{\text{d}}\left[\mathbf{h}(;:)\right]}. \tag{4.94}$$

Another way to measure distortion is via the lth spectral mode desired signal distortion index:

$$\upsilon_{\text{d}}\left[\mathbf{h}(;l)\right] = \frac{E\left\{\left[x_{\text{fd}}(t;l) - x_1(t;l)\right]^2\right\}}{\lambda_{1,l}}. \tag{4.95}$$

We deduce that the fullmode desired signal distortion index is

$$\upsilon_{\rm d}\left[\mathbf{h}(;:)\right] = \frac{\sum_{l=1}^{L} E\left\{\left[x_{\rm fd}(t;l) - x_1(t;l)\right]^2\right\}}{\text{tr}\left(\mathbf{\Lambda}_1\right)}. \tag{4.96}$$

This index should be lower than 1 for optimal filters; the higher its value, the more distorted the desired signal.

4.3.4.3 MSE Criterion

The error signal between the estimated and desired signals is

$$\begin{aligned} e(t;l) &= z(t;l) - x_1(t;l) \\ &= \mathbf{h}^T(;l)\mathbf{y}(t;l) - x_1(t;l) \\ &= e_{\rm d}(t;l) + e_{\rm n}(t;l), \end{aligned} \tag{4.97}$$

where

$$e_{\rm d}(t;l) = x_{\rm fd}(t;l) - x_1(t;l) \tag{4.98}$$

and

$$e_{\rm n}(t;l) = v_{\rm rn}(t;l) \tag{4.99}$$

are the errors quantifying distortion and residual noise, respectively. We deduce that the MSE criterion is

$$\begin{aligned} J\left[\mathbf{h}(;l)\right] &= E\left[e^2(t;l)\right] \\ &= \lambda_{1,l} + \mathbf{h}^T(;l)\mathbf{R}_{\mathbf{y}}(;l)\mathbf{h}(;l) - 2\mathbf{h}^T(;l)\mathbf{R}_{\mathbf{x}}(;l)\mathbf{i}_{\rm i}, \end{aligned} \tag{4.100}$$

where $\mathbf{i}_{\rm i}$, the identity filter, is the first column of $\mathbf{I}_M$. Since $E\left[e_{\rm d}(t;l)e_{\rm n}(t;l)\right] = 0, J\left[\mathbf{h}(;l)\right]$ can also be expressed as

$$\begin{aligned} J\left[\mathbf{h}(;l)\right] &= E\left[e_{\rm d}^2(t;l)\right] + E\left[e_{\rm n}^2(t;l)\right] \\ &= J_{\rm d}\left[\mathbf{h}(;l)\right] + J_{\rm n}\left[\mathbf{h}(;l)\right], \end{aligned} \tag{4.101}$$

where

$$\begin{aligned} J_{\rm d}\left[\mathbf{h}(;l)\right] &= \lambda_{1,l} + \mathbf{h}^T(;l)\mathbf{R}_{\mathbf{x}}(;l)\mathbf{h}(;l) - 2\mathbf{h}^T(;l)\mathbf{R}_{\mathbf{x}}(;l)\mathbf{i}_{\rm i} \\ &= \lambda_{1,l}\upsilon_{\rm d}\left[\mathbf{h}(;l)\right] \end{aligned} \tag{4.102}$$

and

$$\begin{aligned} J_{\rm n}\left[\mathbf{h}(;l)\right] &= \mathbf{h}^T(;l)\mathbf{R}_{\mathbf{v}}(;l)\mathbf{h}(;l) \\ &= \frac{1}{\xi_{\rm n}\left[\mathbf{h}(;l)\right]}. \end{aligned} \tag{4.103}$$

Finally, we have

$$\frac{J_{\mathrm{d}}\left[\mathbf{h}(;l)\right]}{J_{\mathrm{n}}\left[\mathbf{h}(;l)\right]} = \mathrm{iSNR}(;l) \times \xi_{\mathrm{n}}\left[\mathbf{h}(;l)\right] \times \upsilon_{\mathrm{d}}\left[\mathbf{h}(;l)\right] \tag{4.104}$$
$$= \mathrm{oSNR}\left[\mathbf{h}(;l)\right] \times \xi_{\mathrm{d}}\left[\mathbf{h}(;l)\right] \times \upsilon_{\mathrm{d}}\left[\mathbf{h}(;l)\right],$$

showing how the MSEs are related to the most fundamental lth spectral mode performance measures.

4.3.5 Optimal Filters

In this subsection, we derive some fundamental filters that can help reduce the level of the noise. We will see how these optimal filters are very closely related, thanks to the joint diagonalization formulation.

4.3.5.1 Maximum SNR

In the definition of the lth spectral mode output SNR (see Equation 4.85), we recognize the generalized Rayleigh quotient [4]. It is well known that this quotient is maximized with the maximum eigenvector, $\mathbf{s}_1(;l)$, of the matrix $\mathbf{R}_{\mathbf{v}}^{-1}(;l)\mathbf{R}_{\mathbf{x}}(;l)$. Therefore, the maximum SNR filter is

$$\mathbf{h}_{\max}(;l) = \varsigma(;l)\mathbf{s}_1(;l), \tag{4.105}$$

where $\varsigma(;l) \neq 0$ is an arbitrary real number. With this filter, we have the maximum possible lth spectral mode output SNR:

$$\mathrm{oSNR}\left[\mathbf{h}_{\max}(;l)\right] = \omega_1(;l). \tag{4.106}$$

Clearly,

$$\mathrm{oSNR}\left[\mathbf{h}_{\max}(;l)\right] \geq \mathrm{iSNR}(;l) \tag{4.107}$$

and

$$0 \leq \mathrm{oSNR}\left[\mathbf{h}(;l)\right] \leq \mathrm{oSNR}\left[\mathbf{h}_{\max}(;l)\right], \ \forall \mathbf{h}(;l). \tag{4.108}$$

Now, we need to determine $\varsigma(;l)$. The best way to do so is by minimizing distortion. Substituting (4.105) into the distortion-based MSE, we get

$$J_{\mathrm{d}}\left[\mathbf{h}_{\max}(;l)\right] = \lambda_{1,l} + \varsigma^2(;l)\omega_1(;l) - 2\varsigma(;l)\mathbf{s}_1^T(;l)\mathbf{R}_{\mathbf{x}}(;l)\mathbf{i}_{\mathrm{i}} \tag{4.109}$$

and, minimizing the previous expression with respect to $\varsigma(;l)$, we find

$$\varsigma(;l) = \frac{\mathbf{s}_1^T(;l)\mathbf{R}_{\mathbf{x}}(;l)\mathbf{i}_{\mathrm{i}}}{\omega_1(;l)}. \tag{4.110}$$

Plugging this optimal value into (4.105), we obtain the optimal maximum SNR filter with minimum desired signal distortion:

$$\mathbf{h}_{\max}(;l) = \frac{\mathbf{s}_1(;l)\mathbf{s}_1^T(;l)}{\omega_1(;l)}\mathbf{R}_{\mathbf{x}}(;l)\mathbf{i}_{\mathrm{i}} \\ = \mathbf{s}_1(;l)\mathbf{s}_1^T(;l)\mathbf{R}_{\mathbf{v}}(;l)\mathbf{i}_{\mathrm{i}}. \tag{4.111}$$

Example 4.3.1 Consider a ULA of M sensors, as shown in Figure 4.2. Suppose that a desired signal impinges on the ULA from the broadside direction ($\theta = 90°$), and that interference impinges on the ULA from the endfire direction ($\theta = 0°$). Assume that the desired signal is a sum of harmonic random processes:

$$x(t) = \sum_{k=1}^{K} A_k \cos\left(2\pi f_k t + \phi_k\right),$$

with fixed amplitudes $\{A_k\}$ and frequencies $\{f_k\}$, and IID random phases $\{\phi_k\}$, uniformly distributed on the interval from 0 to 2π. Assume that the interference $u(t)$ is colored noise that is uncorrelated with $x(t)$, with the autocorrelation sequence:

$$E\left[u(t)u(t')\right] = \sigma_u^2 \alpha^{|t-t'|}, \ -1 < \alpha < 1.$$

In addition, the sensors contain thermal white Gaussian noise, $w_m(t) \sim \mathcal{N}\left(0, \sigma_w^2\right)$, the signals of which are mutually uncorrelated. The desired signal needs to be recovered from the noisy received signals, $y_m(t) = x_m(t) + v_m(t)$, $m = 1, \dots, M$, where $v_m(t) = u_m(t) + w_m(t)$, $m = 1, \dots, M$ are the interference-plus-noise signals.

As in Example 4.2.1, we choose a sampling interval T_s that satisfies $T_s = \frac{d}{c}$. Hence, we have for $i = 2, \dots, M$:

$$x_i(t) = x_1(t), \tag{4.112}$$

$$u_i(t) = u_1\left(t - i + 1\right). \tag{4.113}$$

Therefore, the correlation matrix of $\mathbf{x}_i(t)$ is

$$\mathbf{R}_{\mathbf{x}_i} = \mathbf{R}_{\mathbf{x}_1},$$

where the elements of the $L \times L$ correlation matrix of $\mathbf{x}_1(t)$ are $\left[\mathbf{R}_{\mathbf{x}_1}\right]_{i,j} = \frac{1}{2}\sum_{k=1}^{K} A_k^2 \cos\left[2\pi f_k(i-j)\right]$. The correlation matrix of $\mathbf{v}_m(t)$ is

$$\mathbf{R}_{\mathbf{v}_m} = \mathbf{R}_{\mathbf{u}_1} + \sigma_w^2 \mathbf{I}_L,$$

where the elements of the $L \times L$ correlation matrix of $\mathbf{u}_1(t)$ are $\left[\mathbf{R}_{\mathbf{u}_1}\right]_{i,j} = \sigma_u^2 \alpha^{|i-j|}$. The fullmode input SNR is

$$\mathrm{iSNR}(;) = 10 \log \frac{\sum_{k=1}^{K} A_k^2}{2(\sigma_u^2 + \sigma_w^2)} \quad \text{(dB)}.$$

After joint diagonalization of $\mathbf{R}_{\mathbf{x}_m}$ and $\mathbf{R}_{\mathbf{v}_m}$ for all $m = 1, \ldots, M$ using (4.59) and (4.60), we compute the $M \times M$ matrices $\mathbf{R}_{\mathbf{x}}(;l)$ and $\mathbf{R}_{\mathbf{v}}(;l)$ for all $l = 1, \ldots, L$:

$$\left[\mathbf{R}_{\mathbf{x}}(;l)\right]_{m_1,m_2} = \mathbf{t}_{m_1,l}^T \mathbf{R}_{\mathbf{x}_{m_1},\mathbf{x}_{m_2}} \mathbf{t}_{m_2,l},$$
$$\left[\mathbf{R}_{\mathbf{v}}(;l)\right]_{m_1,m_2} = \mathbf{t}_{m_1,l}^T \mathbf{R}_{\mathbf{v}_{m_1},\mathbf{v}_{m_2}} \mathbf{t}_{m_2,l},$$

where $\mathbf{R}_{\mathbf{x}_{m_1},\mathbf{x}_{m_2}} = E\left[\mathbf{x}_{m_1}(t)\mathbf{x}_{m_2}^T(t)\right]$ and $\mathbf{R}_{\mathbf{v}_{m_1},\mathbf{v}_{m_2}} = E\left[\mathbf{v}_{m_1}(t)\mathbf{v}_{m_2}^T(t)\right]$ are $L \times L$ spatiotemporal correlation matrices of the desired and noise signals, respectively. In this example, since the desired source is in the broadside direction and the interference source is in the endfire direction, we have

$$\mathbf{R}_{\mathbf{x}_{m_1},\mathbf{x}_{m_2}} = \mathbf{R}_{\mathbf{x}_1},$$
$$\left[\mathbf{R}_{\mathbf{v}_{m_1},\mathbf{v}_{m_2}}\right]_{i,j} = \sigma_u^2 \alpha^{|i-j+m_1-m_2|} + \sigma_w^2 \delta\left(m_1 - m_2\right) \delta\left(i-j\right).$$

After joint diagonalization of $\mathbf{R}_{\mathbf{x}}(;l)$ and $\mathbf{R}_{\mathbf{v}}(;l)$ for all $l = 1, \ldots, L$ using (4.69) and (4.70), the optimal maximum SNR filter $\mathbf{h}_{\max}(;l)$ is obtained from (4.111).

To demonstrate the performance of the maximum SNR filter, we choose $A_k = 0.5/k$, $k = 1, \ldots, 5$, $f_k = 0.05 + 0.1(k-1)$, $k = 1, \ldots, 5$, $\alpha = 0.5$, $L = 10$, $\sigma_w^2 = 0.01\sigma_u^2$, and several values of M. Figure 4.10 shows plots of the fullmode gain in SNR, $\mathcal{G}\left[\mathbf{h}_{\max}(;:)\right]$, the fullmode MSE, $J\left[\mathbf{h}_{\max}(;:)\right]$, the fullmode noise reduction factor, $\xi_{\mathrm{n}}\left[\mathbf{h}_{\max}(;:)\right]$, and the fullmode desired signal reduction factor, $\xi_{\mathrm{d}}\left[\mathbf{h}_{\max}(;:)\right]$, as a function of the fullmode input SNR for different numbers of sensors, M. For a given fullmode input SNR, as the number of sensors increases, the fullmode gain in SNR and the fullmode noise reduction factor increase, while the fullmode MSE decreases. ■

4.3.5.2 Wiener

Taking the gradient of the MSE criterion, $J\left[\mathbf{h}(;l)\right]$, with respect to $\mathbf{h}(;l)$ and equating the result to zero, we find the Wiener filter:

$$\mathbf{h}_{\mathrm{W}}(;l) = \mathbf{R}_{\mathbf{y}}^{-1}(;l)\mathbf{R}_{\mathbf{x}}(;l)\mathbf{i}_{\mathrm{i}}, \tag{4.114}$$

which we can also express as

$$\begin{aligned}\mathbf{h}_{\mathrm{W}}(;l) &= \sum_{m=1}^{M} \frac{\mathbf{s}_m(;l)\mathbf{s}_m^T(;l)}{1+\omega_m(;l)} \mathbf{R}_{\mathbf{x}}(;l)\mathbf{i}_{\mathrm{i}} \\ &= \sum_{m=1}^{M} \frac{\omega_m(;l)}{1+\omega_m(;l)} \mathbf{s}_m(;l)\mathbf{s}_m^T(;l)\mathbf{R}_{\mathbf{v}}(;l)\mathbf{i}_{\mathrm{i}}.\end{aligned} \tag{4.115}$$

From this formulation, we see clearly how $\mathbf{h}_{\mathrm{W}}(;l)$ and $\mathbf{h}_{\max}(;l)$ are related. Besides a (slight) different weighting factor, $\mathbf{h}_{\mathrm{W}}(;l)$ considers all directions where the desired signal and noise are present, while $\mathbf{h}_{\max}(;l)$ relies only on the direction where the maximum of the desired signal energy is found.

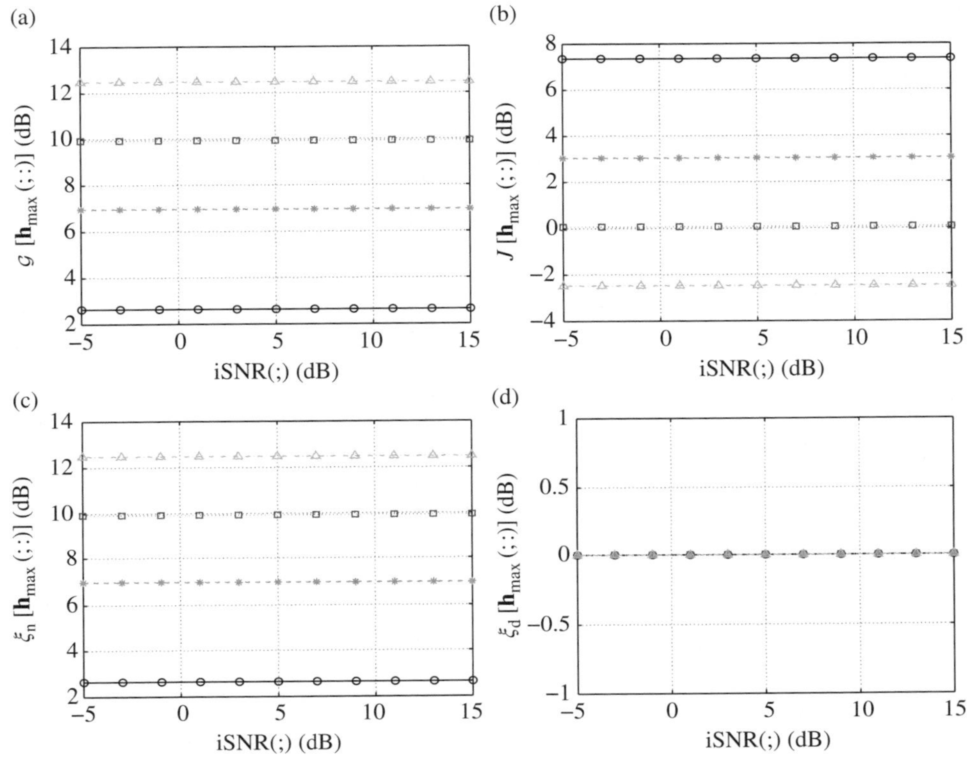

Figure 4.10 (a) The fullmode gain in SNR, (b) the fullmode MSE, (c) the fullmode noise reduction factor, and (d) the fullmode desired signal reduction factor of the maximum SNR filter as a function of the fullmode input SNR for different numbers of sensors, M: $M = 2$ (solid line with circles), $M = 5$ (dashed line with asterisks), $M = 10$ (dotted line with squares), and $M = 20$ (dash-dot line with triangles).

Obviously, we have

$$\mathrm{oSNR}\left[\mathbf{h}_{\mathrm{W}}(;l)\right] \leq \mathrm{oSNR}\left[\mathbf{h}_{\max}(;l)\right] \tag{4.116}$$

and, in general,

$$\upsilon_{\mathrm{d}}\left[\mathbf{h}_{\mathrm{W}}(;l)\right] \leq \upsilon_{\mathrm{d}}\left[\mathbf{h}_{\max}(;l)\right]. \tag{4.117}$$

Example 4.3.2 Returning to Example 4.3.1, we now employ the Wiener filter, $\mathbf{h}_{\mathrm{W}}(;l)$, given in (4.114). Figure 4.11 shows plots of the fullmode gain in SNR, $\mathcal{G}\left[\mathbf{h}_{\mathrm{W}}(;:)\right]$, the fullmode MSE, $J\left[\mathbf{h}_{\mathrm{W}}(;:)\right]$, the fullmode noise reduction factor, $\xi_{\mathrm{n}}\left[\mathbf{h}_{\mathrm{W}}(;:)\right]$, and the fullmode desired signal reduction factor, $\xi_{\mathrm{d}}\left[\mathbf{h}_{\mathrm{W}}(;:)\right]$, as a function of the fullmode input SNR for different numbers of sensors, M. For a given fullmode input SNR, as the number of sensors increases, the fullmode gain in SNR and the fullmode noise reduction factor increase, while the fullmode MSE and the fullmode desired signal reduction factor decrease. ■

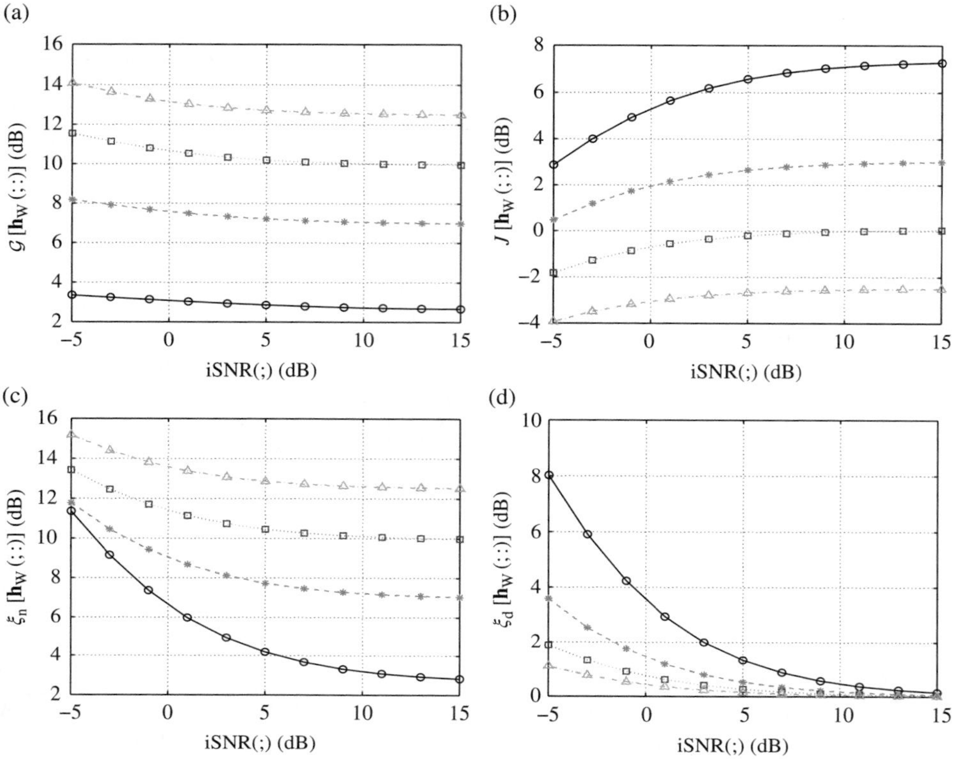

Figure 4.11 (a) The fullmode gain in SNR, (b) the fullmode MSE, (c) the fullmode noise reduction factor, and (d) the fullmode desired signal reduction factor of the Wiener filter as a function of the fullmode input SNR for different numbers of sensors, M: $M = 2$ (solid line with circles), $M = 5$ (dashed line with asterisks), $M = 10$ (dotted line with squares), and $M = 20$ (dash-dot line with triangles).

4.3.5.3 MVDR

Here, we show how to derive a filter that does not distort the desired signal. This approach exploits the nullspace of $\mathbf{R}_{\mathbf{x}}(;l)$.

By using (4.75) and (4.76), the filtered desired signal can be expressed as

$$\begin{aligned} x_{\mathrm{fd}}(t;l) &= \mathbf{h}^T(;l)\mathbf{x}(t;l) \\ &= \mathbf{h}^T(;l)\mathbf{R}_{\mathbf{v}}(;l)\mathbf{R}_{\mathbf{v}}^{-1}(;l)\mathbf{x}(t;l) \\ &= \mathbf{h}^T(;l)\mathbf{R}_{\mathbf{v}}(;l)\left[\mathbf{S}'(;l)\mathbf{S}'^T(;l) + \mathbf{S}''(;l)\mathbf{S}''^T(;l)\right]\mathbf{x}(t;l) \\ &= \mathbf{h}^T(;l)\mathbf{R}_{\mathbf{v}}(;l)\mathbf{S}'(;l)\mathbf{S}'^T(;l)\mathbf{x}(t;l). \end{aligned} \tag{4.118}$$

We see from (4.118) that in order to completely recover the desired signal, $x_1(t;l)$, we must use the constraint:

$$\mathbf{h}^T(;l)\mathbf{R}_{\mathbf{v}}(;l)\mathbf{S}'(;l) = \mathbf{i}_{\mathrm{i}}^T\mathbf{R}_{\mathbf{v}}(;l)\mathbf{S}'(;l). \tag{4.119}$$

Therefore, it is desired to minimize the residual noise subject to (4.119):

$$\min_{\mathbf{h}(;l)} \mathbf{h}^T(;l)\mathbf{R}_{\mathbf{v}}(;l)\mathbf{h}(;l) \quad \text{subject to}$$
$$\mathbf{h}^T(;l)\mathbf{R}_{\mathbf{v}}(;l)\mathbf{S}'(;l) = \mathbf{i}_{\mathrm{i}}^T\mathbf{R}_{\mathbf{v}}(;l)\mathbf{S}'(;l), \tag{4.120}$$

from which we find the MVDR filter:

$$\begin{aligned}\mathbf{h}_{\mathrm{MVDR}}(;l) &= \mathbf{S}'(;l)\mathbf{S}'^T(;l)\mathbf{R}_{\mathbf{v}}(;l)\mathbf{i}_{\mathrm{i}} \\ &= \sum_{p=1}^{P_l} \mathbf{s}_p(;l)\mathbf{s}_p^T(;l)\mathbf{R}_{\mathbf{v}}(;l)\mathbf{i}_{\mathrm{i}}.\end{aligned} \tag{4.121}$$

Another formulation of this filter is

$$\mathbf{h}_{\mathrm{MVDR}}(;l) = \sum_{p=1}^{P_l} \frac{\mathbf{s}_p(;l)\mathbf{s}_p^T(;l)}{\omega_p(;l)}\mathbf{R}_{\mathbf{x}}(;l)\mathbf{i}_{\mathrm{i}}. \tag{4.122}$$

For $P_l = M$, the MVDR filter degenerates to the identity filter: $\mathbf{h}_{\mathrm{MVDR}}(;l) = \mathbf{i}_{\mathrm{i}}$, which does not affect the observations. For $P_l = 1$, the MVDR filters becomes the maximum SNR filter. Therefore, we can state that the higher the dimension of the nullspace of $\mathbf{R}_{\mathbf{x}}(;l)$, the more noise reduction there is using the MVDR filter.

The lth spectral mode output SNR of the MVDR filter is

$$\begin{aligned}\mathrm{oSNR}\left[\mathbf{h}_{\mathrm{MVDR}}(;l)\right] &= \frac{\lambda_{1,l}}{\mathbf{i}_{\mathrm{i}}^T\mathbf{R}_{\mathbf{v}}(;l)\mathbf{S}'(;l)\mathbf{S}'^T(;l)\mathbf{R}_{\mathbf{v}}(;l)\mathbf{i}_{\mathrm{i}}} \\ &= \frac{\lambda_{1,l}}{1-\mathbf{i}_{\mathrm{i}}^T\mathbf{R}_{\mathbf{v}}(;l)\mathbf{S}''(;l)\mathbf{S}''^T(;l)\mathbf{R}_{\mathbf{v}}(;l)\mathbf{i}_{\mathrm{i}}}.\end{aligned} \tag{4.123}$$

As a result,

$$\mathrm{oSNR}\left[\mathbf{h}_{\mathrm{MVDR}}(;l)\right] \geq \lambda_{1,l} = \mathrm{iSNR}\,(;l)\,. \tag{4.124}$$

We always have

$$\mathrm{oSNR}\left[\mathbf{h}_{\mathrm{MVDR}}(;l)\right] \leq \mathrm{oSNR}\left[\mathbf{h}_{\mathrm{W}}(;l)\right]. \tag{4.125}$$

4.3.5.4 Controlled Distortion

From the obvious relationship between the MVDR and maximum SNR filters, we can deduce a whole class of controlled distortion (CD) filters:

$$\mathbf{h}_{\mathrm{CD},P_l'}(;l) = \sum_{p'=1}^{P_l'} \mathbf{s}_{p'}(;l)\mathbf{s}_{p'}^T(;l)\mathbf{R}_{\mathbf{v}}(;l)\mathbf{i}_{\mathrm{i}}, \tag{4.126}$$

where $1 \leq P_l' \leq P_l$. We observe that $\mathbf{h}_{\text{CD},1}(;l) = \mathbf{h}_{\max}(;l)$ and $\mathbf{h}_{\text{CD},P_l}(;l) = \mathbf{h}_{\text{MVDR}}(;l)$. Also, we always have

$$\text{oSNR}\left[\mathbf{h}_{\text{CD},P_l}(;l)\right] \leq \text{oSNR}\left[\mathbf{h}_{\text{CD},P_l-1}(;l)\right] \leq \cdots \leq \text{oSNR}\left[\mathbf{h}_{\text{CD},1}(;l)\right] \tag{4.127}$$

and

$$0 = \upsilon_{\text{d}}\left[\mathbf{h}_{\text{CD},P_l}(;l)\right] \leq \upsilon_{\text{d}}\left[\mathbf{h}_{\text{CD},P_l-1}(;l)\right] \leq \cdots \leq \upsilon_{\text{d}}\left[\mathbf{h}_{\text{CD},1}(;l)\right]. \tag{4.128}$$

Example 4.3.3 Returning to Example 4.3.1, the desired signal now impinges on the ULA from $\theta_0 = 60°$, rather than from the broadside direction. The interference and thermal noise remain the same as in Example 4.3.1. We assume that the desired signal is the sum of three harmonic random processes:

$$x(t) = \sum_{k=1}^{3} A_k \cos\left(2\pi f_k t + \phi_k\right),$$

with $A_k = 0.5/k$, $k = 1, \ldots, 3$, $f_k = 0.05 + 0.1(k-1)$, $k = 1, \ldots, 3$, and IID random phases $\{\phi_k\}$, uniformly distributed on the interval from 0 to 2π.

The desired signal at sensor m is a delayed version of the desired signal at the first sensor:

$$x_m(t) = x_1\left(t - \tau_m\right), \tag{4.129}$$

where

$$\begin{aligned}\tau_m &= \frac{(m-1)d\cos\theta_0}{cT_\text{s}} \\ &= (m-1)\cos\theta_0,\ m = 1, 2, \ldots, M\end{aligned} \tag{4.130}$$

is the relative time delay in samples (not necessarily an integer number) between the mth sensor and the first one. Therefore, the elements of the $L \times L$ spatiotemporal correlation matrix $\mathbf{R}_{\mathbf{x}_{m_1},\mathbf{x}_{m_2}} = E\left[\mathbf{x}_{m_1}(t)\mathbf{x}_{m_2}^T(t)\right]$ are

$$\left[\mathbf{R}_{\mathbf{x}_{m_1},\mathbf{x}_{m_2}}\right]_{i,j} = \frac{1}{2}\sum_{k=1}^{3} A_k^2 \cos\left[2\pi f_k(i - j + \tau_{m_1} - \tau_{m_2})\right].$$

The fullmode input SNR is

$$\text{iSNR}(;) = 10\log\frac{\sum_{k=1}^{3} A_k^2}{2(\sigma_u^2 + \sigma_w^2)} \quad \text{(dB)}.$$

After joint diagonalization of $\mathbf{R}_\mathbf{x}(;l)$ and $\mathbf{R}_\mathbf{v}(;l)$ for all $l = 1, \ldots, L$ using (4.69) and (4.70), the controlled distortion filter, $\mathbf{h}_{\text{CD},P_l'}(;l)$, is obtained from (4.126).

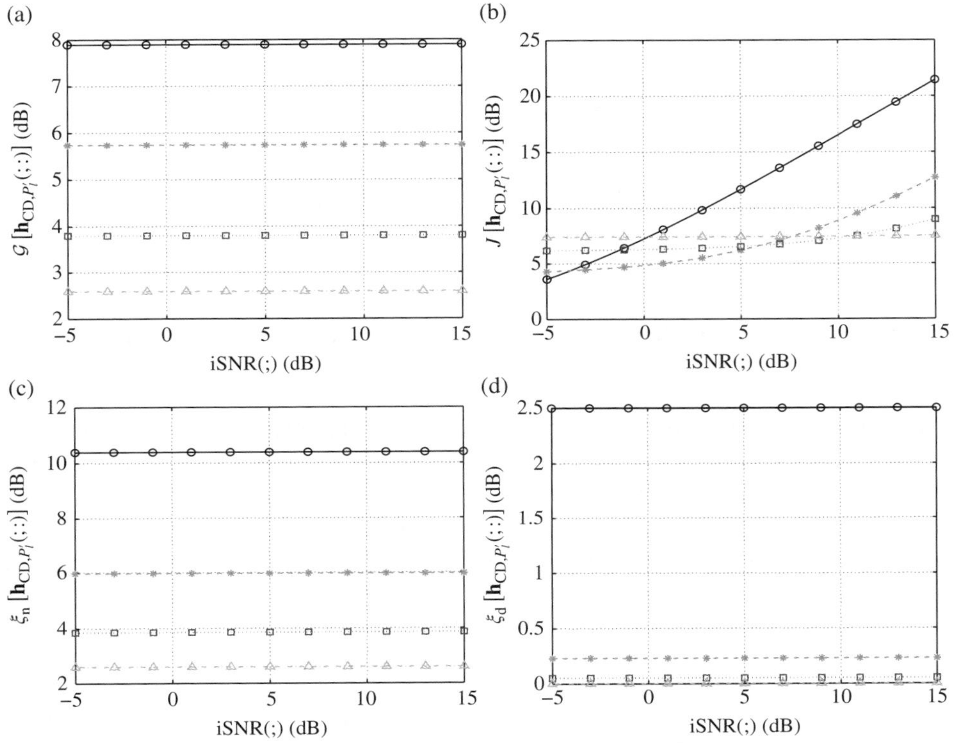

Figure 4.12 (a) The fullmode gain in SNR, (b) the fullmode MSE, (c) the fullmode noise reduction factor, and (d) the fullmode desired signal reduction factor of the controlled distortion filter as a function of the fullmode input SNR for several values of P'_l: $P'_l = 1$ (solid line with circles), $P'_l = 2$ (dashed line with asterisks), $P'_l = 3$ (dotted line with squares), and $P'_l = 4$ (dash-dot line with triangles).

Figure 4.12 shows plots of the fullmode gain in SNR, $\mathcal{G}\left[\mathbf{h}_{\mathrm{CD},P'_l}(;:)\right]$, the fullmode MSE, $J\left[\mathbf{h}_{\mathrm{CD},P'_l}(;:)\right]$, the fullmode noise reduction factor, $\xi_{\mathrm{n}}\left[\mathbf{h}_{\mathrm{CD},P'_l}(;:)\right]$, and the fullmode desired signal reduction factor, $\xi_{\mathrm{d}}\left[\mathbf{h}_{\mathrm{CD},P'_l}(;:)\right]$, as a function of the fullmode input SNR for several values of P'_l. For a given fullmode input SNR, as the value of P'_l increases, the fullmode desired signal reduction factor decreases, at the expense of lower fullmode gain in SNR and a lower fullmode noise reduction factor. ■

4.3.5.5 Tradeoff

Another way to compromise between noise reduction and desired signal distortion is to minimize the desired signal distortion index with the constraint that the noise reduction factor is equal to a positive value that is greater than 1. Mathematically, this is equivalent to

$$\min_{\mathbf{h}(;l)} J_{\mathrm{d}}\left[\mathbf{h}(;l)\right] \quad \text{subject to} \quad J_{\mathrm{n}}\left[\mathbf{h}(;l)\right] = \aleph, \tag{4.131}$$

where $0 < \aleph < 1$. We easily find the tradeoff filter:

$$\mathbf{h}_{\mathrm{T},\mu}(;l) = \sum_{m=1}^{M} \frac{\omega_m(;l)}{\mu + \omega_m(;l)} \mathbf{s}_m(;l)\mathbf{s}_m^T(;l)\mathbf{R}_{\mathbf{v}}(;l)\mathbf{i}_{\mathrm{i}}, \tag{4.132}$$

where $\mu \geq 0$ is a Lagrange multiplier. We observe that $\mathbf{h}_{\mathrm{T},1}(;l) = \mathbf{h}_{\mathrm{W}}(;l)$. We have

$$\mathrm{oSNR}\left[\mathbf{h}_{\mathrm{T},\mu}(;l)\right] \geq \mathrm{iSNR}(;l),\ \forall \mu \geq 0. \tag{4.133}$$

For μ greater (resp. less) than 1, the tradeoff filter reduces more (resp. less) noise than the Wiener filter but introduces more (resp. less) distortion.

Example 4.3.4 Returning to Example 4.3.3, we now employ the tradeoff filter, $\mathbf{h}_{\mathrm{T},\mu}(;l)$, given in (4.132). Figure 4.13 shows plots of the fullmode gain in SNR, $\mathcal{G}\left[\mathbf{h}_{\mathrm{T},\mu}(;:)\right]$, the fullmode MSE, $J\left[\mathbf{h}_{\mathrm{T},\mu}(;:)\right]$, the fullmode noise reduction factor, $\xi_{\mathrm{n}}\left[\mathbf{h}_{\mathrm{T},\mu}(;:)\right]$, and the fullmode desired signal reduction factor, $\xi_{\mathrm{d}}\left[\mathbf{h}_{\mathrm{T},\mu}(;:)\right]$, as a function of the fullmode input SNR for several values of μ. For a given fullmode input SNR, as the value of μ increases, the fullmode gain in SNR and the fullmode noise reduction factor increase, at the expense of a higher fullmode desired signal reduction factor. ■

4.3.5.6 General Subspace

Let $f\left[\omega_m(;l)\right]$ be a function of $\omega_m(;l)$, where $0 \leq f\left[\omega_m(;l)\right] \leq 1$. A general subspace (GS) approach for multichannel noise reduction is

$$\mathbf{h}_{\mathrm{GS}}(;l) = \sum_{m'=1}^{M'} f\left[\omega_{m'}(;l)\right] \mathbf{s}_{m'}(;l)\mathbf{s}_{m'}^T(;l)\mathbf{R}_{\mathbf{v}}(;l)\mathbf{i}_{\mathrm{i}}, \tag{4.134}$$

where $1 \leq M' \leq M$. It can be verified that the general form given in (4.134) encompasses all the filters derived in the previous subsections. Obviously, many other filters can be derived too.

In Table 4.2, we summarize all the filters described in this subsection, showing how they are closely related.

4.4 Case of a Rank Deficient Noise Correlation Matrix

So far, for both the single-channel and multichannel noise reduction problems, we have assumed that the noise correlation matrix has full rank. What happens if this is not the case? Many good optimal linear filters, such as the Wiener filter, are guaranteed to behave well only if the noise correlation matrix is full rank. However, in some applications, it may be that the noise is narrowband [8]. In such situations, classical linear filters for noise reduction may not function correctly. In this section, we show how to derive some very efficient linear filters in the particular case of a rank deficient noise correlation matrix.

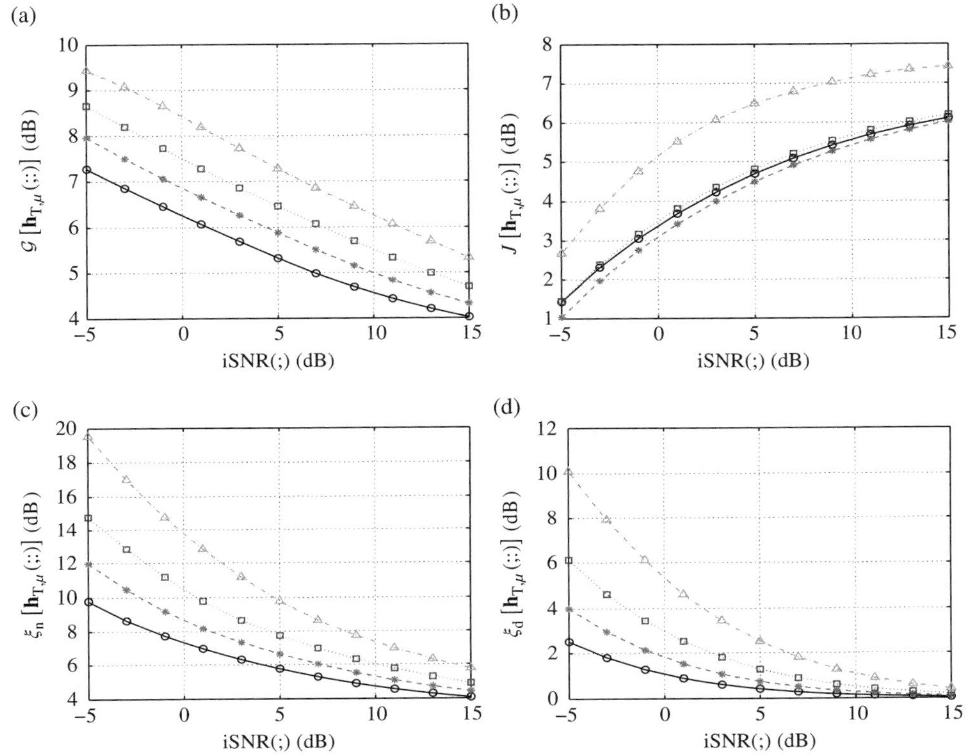

Figure 4.13 (a) The fullmode gain in SNR, (b) the fullmode MSE, (c) the fullmode noise reduction factor, and (d) the fullmode desired signal reduction factor of the tradeoff filter as a function of the fullmode input SNR for several values of μ: $\mu = 0.5$ (solid line with circles), $\mu = 1$ (dashed line with asterisks), $\mu = 2$ (dotted line with squares), and $\mu = 5$ (dash-dot line with triangles).

4.4.1 Eigenvalue Decompositions

From here on, it is assumed that rank $\left(\mathbf{R}_{\underline{\mathbf{x}}}\right) = P < ML$ while rank $\left(\mathbf{R}_{\underline{\mathbf{v}}}\right) = Q < ML$. Using the convenient eigenvalue decomposition [9], the noise correlation matrix can be diagonalized as

$$\mathbf{Q}_{\underline{\mathbf{v}}}^T \mathbf{R}_{\underline{\mathbf{v}}} \mathbf{Q}_{\underline{\mathbf{v}}} = \mathbf{\Lambda}_{\underline{\mathbf{v}}}, \tag{4.135}$$

where

$$\mathbf{Q}_{\underline{\mathbf{v}}} = \begin{bmatrix} \mathbf{q}_{\underline{\mathbf{v}},1} & \mathbf{q}_{\underline{\mathbf{v}},2} & \cdots & \mathbf{q}_{\underline{\mathbf{v}},ML} \end{bmatrix} \tag{4.136}$$

is an orthogonal matrix: $\mathbf{Q}_{\underline{\mathbf{v}}}^T \mathbf{Q}_{\underline{\mathbf{v}}} = \mathbf{Q}_{\underline{\mathbf{v}}} \mathbf{Q}_{\underline{\mathbf{v}}}^T = \mathbf{I}_{ML}$ and

$$\mathbf{\Lambda}_{\underline{\mathbf{v}}} = \text{diag}\left(\lambda_{\underline{\mathbf{v}},1}, \lambda_{\underline{\mathbf{v}},2}, \ldots, \lambda_{\underline{\mathbf{v}},ML}\right) \tag{4.137}$$

Table 4.2 Optimal linear filters for multichannel signal enhancement in the spectral domain.

Filter	
Maximum SNR	$\mathbf{h}_{\max}(;l) = \mathbf{s}_1(;l)\mathbf{s}_1^T(;l)\mathbf{R}_{\mathbf{v}}(;l)\mathbf{i}_{\mathrm{i}}$
Wiener	$\mathbf{h}_{\mathrm{W}}(;l) = \sum_{m=1}^{M} \frac{\omega_m(;l)}{1+\omega_m(;l)}\mathbf{s}_m(;l)\mathbf{s}_m^T(;l)\mathbf{R}_{\mathbf{v}}(;l)\mathbf{i}_{\mathrm{i}}$
MVDR	$\mathbf{h}_{\mathrm{MVDR}}(;l) = \sum_{p=1}^{P_l} \mathbf{s}_p(;l)\mathbf{s}_p^T(;l)\mathbf{R}_{\mathbf{v}}(;l)\mathbf{i}_{\mathrm{i}}$
CD	$\mathbf{h}_{\mathrm{CD},P_l'}(;l) = \sum_{p'=1}^{P_l'} \mathbf{s}_{p'}(;l)\mathbf{s}_{p'}^T(;l)\mathbf{R}_{\mathbf{v}}(;l)\mathbf{i}_{\mathrm{i}}$
Tradeoff	$\mathbf{h}_{\mathrm{T},\mu}(;l) = \sum_{m=1}^{M} \frac{\omega_m(;l)}{\mu+\omega_m(;l)}\mathbf{s}_m(;l)\mathbf{s}_m^T(;l)\mathbf{R}_{\mathbf{v}}(;l)\mathbf{i}_{\mathrm{i}}$
GS	$\mathbf{h}_{\mathrm{GS}}(;l) = \sum_{m'=1}^{M'} f\left[\omega_{m'}(;l)\right]\mathbf{s}_{m'}(;l)\mathbf{s}_{m'}^T(;l)\mathbf{R}_{\mathbf{v}}(;l)\mathbf{i}_{\mathrm{i}}$

is a diagonal matrix. The orthonormal vectors $\mathbf{q}_{\underline{\mathbf{v}},1}, \mathbf{q}_{\underline{\mathbf{v}},2}, \ldots, \mathbf{q}_{\underline{\mathbf{v}},ML}$ are the eigenvectors corresponding, respectively, to the eigenvalues $\lambda_{\underline{\mathbf{v}},1}, \lambda_{\underline{\mathbf{v}},2}, \ldots, \lambda_{\underline{\mathbf{v}},ML}$ of the matrix $\mathbf{R}_{\underline{\mathbf{v}}}$, where $\lambda_{\underline{\mathbf{v}},1} \geq \lambda_{\underline{\mathbf{v}},2} \geq \cdots \geq \lambda_{\underline{\mathbf{v}},Q} > 0$ and $\lambda_{\underline{\mathbf{v}},Q+1} = \lambda_{\underline{\mathbf{v}},Q+2} = \cdots = \lambda_{\underline{\mathbf{v}},ML} = 0$. In the same way, the desired signal correlation matrix can be diagonalized as

$$\mathbf{Q}_{\underline{\mathbf{x}}}^T \mathbf{R}_{\underline{\mathbf{x}}} \mathbf{Q}_{\underline{\mathbf{x}}} = \boldsymbol{\Lambda}_{\underline{\mathbf{x}}}, \tag{4.138}$$

where the orthogonal and diagonal matrices $\mathbf{Q}_{\underline{\mathbf{x}}}$ and $\boldsymbol{\Lambda}_{\underline{\mathbf{x}}}$ are defined in a similar way to $\mathbf{Q}_{\underline{\mathbf{v}}}$ and $\boldsymbol{\Lambda}_{\underline{\mathbf{v}}}$, respectively, with $\lambda_{\underline{\mathbf{x}},1} \geq \lambda_{\underline{\mathbf{x}},2} \geq \cdots \geq \lambda_{\underline{\mathbf{x}},P} > 0$ and $\lambda_{\underline{\mathbf{x}},P+1} = \lambda_{\underline{\mathbf{x}},P+2} = \cdots = \lambda_{\underline{\mathbf{x}},ML} = 0$. The two previous decompositions will be extensively used in the rest of this section for the purpose of deriving optimal linear filters.

4.4.2 Maximization of the Output SNR

In this part, we show how to exploit the nullspace of the noise correlation matrix in order to derive noise reduction filters. Thanks to this nullspace, we can completely cancel the noise; this leads to infinite noise reduction filtering matrices.

Let

$$\mathbf{Q}_{\underline{\mathbf{v}}}'' = \begin{bmatrix} \mathbf{q}_{\underline{\mathbf{v}},Q+1} & \mathbf{q}_{\underline{\mathbf{v}},Q+2} & \cdots & \mathbf{q}_{\underline{\mathbf{v}},ML} \end{bmatrix} \tag{4.139}$$

be the matrix of size $ML \times (ML - Q)$ containing the eigenvectors corresponding to the null eigenvalues of $\mathbf{R}_{\underline{\mathbf{v}}}$. We are interested in the linear filtering matrices of the form:

$$\underline{\mathbf{H}} = \underline{\mathbf{A}}\mathbf{Q}_{\underline{\mathbf{v}}}''^T, \tag{4.140}$$

where $\underline{\mathbf{A}} \neq \mathbf{0}$ is a matrix of size $L \times (ML - Q)$. Since $\underline{\mathbf{R}}_{\mathbf{v}}\mathbf{Q}''_{\underline{\mathbf{v}}} = \mathbf{0}$ and assuming that $\mathbf{R}_{\underline{\mathbf{x}}}\mathbf{Q}''_{\underline{\mathbf{v}}} \neq \mathbf{0}$, which is reasonable since $\mathbf{R}_{\underline{\mathbf{x}}}$ and $\mathbf{R}_{\underline{\mathbf{v}}}$ cannot be diagonalized by the same orthogonal matrix unless one of the two signals is white, we have

$$\begin{aligned} \text{oSNR}\left(\underline{\mathbf{H}}\right) &= \frac{\text{tr}\left(\underline{\mathbf{H}}\mathbf{R}_{\underline{\mathbf{x}}}\underline{\mathbf{H}}^T\right)}{\text{tr}\left(\underline{\mathbf{H}}\mathbf{R}_{\underline{\mathbf{v}}}\underline{\mathbf{H}}^T\right)} \\ &= \frac{\text{tr}\left(\underline{\mathbf{A}}\mathbf{Q}''^T_{\underline{\mathbf{v}}}\mathbf{R}_{\underline{\mathbf{x}}}\mathbf{Q}''_{\underline{\mathbf{v}}}\underline{\mathbf{A}}^T\right)}{\text{tr}\left(\underline{\mathbf{A}}\mathbf{Q}''^T_{\underline{\mathbf{v}}}\mathbf{R}_{\underline{\mathbf{v}}}\mathbf{Q}''_{\underline{\mathbf{v}}}\underline{\mathbf{A}}^T\right)} \\ &= \infty. \end{aligned} \tag{4.141}$$

As a consequence, the estimate of $\mathbf{x}_1(t)$ is

$$\begin{aligned} \widehat{\mathbf{x}}_1(t) &= \underline{\mathbf{H}}\,\underline{\mathbf{y}}(t) \\ &= \underline{\mathbf{A}}\mathbf{Q}''^T_{\underline{\mathbf{v}}}\underline{\mathbf{x}}(t) + \underline{\mathbf{A}}\mathbf{Q}''^T_{\underline{\mathbf{v}}}\underline{\mathbf{v}}(t) \\ &= \underline{\mathbf{A}}\mathbf{Q}''^T_{\underline{\mathbf{v}}}\underline{\mathbf{x}}(t). \end{aligned} \tag{4.142}$$

We observe from the previous expression that this approach completely cancels the noise. Now, we need to find $\underline{\mathbf{A}}$. The best way to find this matrix is by minimizing the distortion of the estimated desired signal.

The distortion-based MSE is

$$\begin{aligned} J_{\text{d}}\left(\underline{\mathbf{A}}\right) &= E\left\{\left[\widehat{\mathbf{x}}_1(t) - \mathbf{x}_1(t)\right]^T\left[\widehat{\mathbf{x}}_1(t) - \mathbf{x}_1(t)\right]\right\} \\ &= \text{tr}\left(\mathbf{R}_{\mathbf{x}_1} - 2\underline{\mathbf{A}}\mathbf{Q}''^T_{\underline{\mathbf{v}}}\mathbf{R}_{\underline{\mathbf{x}}}\underline{\mathbf{I}}_{\text{i}}^T + \underline{\mathbf{A}}\mathbf{Q}''^T_{\underline{\mathbf{v}}}\mathbf{R}_{\underline{\mathbf{x}}}\mathbf{Q}''_{\underline{\mathbf{v}}}\underline{\mathbf{A}}^T\right). \end{aligned} \tag{4.143}$$

From the minimization of $J_{\text{d}}\left(\underline{\mathbf{A}}\right)$ and (4.140), we easily deduce the optimal infinite noise reduction filtering matrix:

$$\underline{\mathbf{H}}_{\infty} = \underline{\mathbf{I}}_{\text{i}}\mathbf{R}_{\underline{\mathbf{x}}}\mathbf{Q}''_{\underline{\mathbf{v}}}\left(\mathbf{Q}''^T_{\underline{\mathbf{v}}}\mathbf{R}_{\underline{\mathbf{x}}}\mathbf{Q}''_{\underline{\mathbf{v}}}\right)^{-1}\mathbf{Q}''^T_{\underline{\mathbf{v}}}. \tag{4.144}$$

It is very important to see from (4.144) that for $\underline{\mathbf{H}}_{\infty}$ to be unique, we must have $P \geq ML - Q$. It is worth noticing that the larger the dimension of the nullspace of $\mathbf{R}_{\underline{\mathbf{v}}}$: $ML - Q$, the less distorted the desired signal, since the the number of columns of $\underline{\mathbf{A}}$ and hence the number of degrees of freedom to minimize distortion are larger. The worst case is when $Q = ML - 1$. Indeed, while the output SNR is still equal to infinity, distortion may be very high, since $\underline{\mathbf{A}}$ simplifies to a vector, which may not be enough to reduce this distortion.

Let us consider the particular case: $P = ML - Q$. We can always express the estimate of $\mathbf{x}_1(t)$ as

$$\begin{aligned}\widehat{\mathbf{x}}_1(t) &= \underline{\mathbf{A}}\mathbf{Q}_{\underline{\mathbf{v}}}^{\prime\prime T}\underline{\mathbf{x}}(t) \\ &= \underline{\mathbf{A}}\mathbf{Q}_{\underline{\mathbf{v}}}^{\prime\prime T}\mathbf{Q}_{\underline{\mathbf{x}}}^{\prime}\mathbf{Q}_{\underline{\mathbf{x}}}^{\prime T}\underline{\mathbf{x}}(t),\end{aligned} \tag{4.145}$$

where

$$\mathbf{Q}_{\underline{\mathbf{x}}}^{\prime} = \begin{bmatrix} \mathbf{q}_{\underline{\mathbf{x}},1} & \mathbf{q}_{\underline{\mathbf{x}},2} & \cdots & \mathbf{q}_{\underline{\mathbf{x}},P} \end{bmatrix} \tag{4.146}$$

is the matrix of size $ML \times P$ containing the eigenvectors corresponding to the nonnull eigenvalues of $\mathbf{R}_{\underline{\mathbf{x}}}$. We see from (4.145) that in order to recover the desired signal, $\mathbf{x}_1(t)$, we must have

$$\underline{\mathbf{A}}\mathbf{Q}_{\underline{\mathbf{v}}}^{\prime\prime T}\mathbf{Q}_{\underline{\mathbf{x}}}^{\prime} = \underline{\mathbf{I}}_{\mathrm{i}}\mathbf{Q}_{\underline{\mathbf{x}}}^{\prime}. \tag{4.147}$$

Therefore, since $\mathbf{Q}_{\underline{\mathbf{v}}}^{\prime\prime T}\mathbf{Q}_{\underline{\mathbf{x}}}^{\prime}$ is a square invertible matrix, we have a unique solution for (4.147). As a result, the optimal filtering matrix for this particular case is

$$\underline{\mathbf{H}}_{\infty} = \underline{\mathbf{I}}_{\mathrm{i}}\mathbf{Q}_{\underline{\mathbf{x}}}^{\prime}\left(\mathbf{Q}_{\underline{\mathbf{v}}}^{\prime\prime T}\mathbf{Q}_{\underline{\mathbf{x}}}^{\prime}\right)^{-1}\mathbf{Q}_{\underline{\mathbf{v}}}^{\prime\prime T}. \tag{4.148}$$

This filter perfectly recovers the desired signal: it completely cancels the noise without any distortion.

Another interesting scenario is when $P < ML - Q$. One reasonable approach is to take the minimum-norm solution of (4.147). Therefore, the filtering matrix becomes

$$\underline{\mathbf{H}}_{\infty} = \underline{\mathbf{I}}_{\mathrm{i}}\mathbf{Q}_{\underline{\mathbf{x}}}^{\prime}\left(\mathbf{Q}_{\underline{\mathbf{x}}}^{\prime T}\mathbf{Q}_{\underline{\mathbf{v}}}^{\prime\prime}\mathbf{Q}_{\underline{\mathbf{v}}}^{\prime\prime T}\mathbf{Q}_{\underline{\mathbf{x}}}^{\prime}\right)^{-1}\mathbf{Q}_{\underline{\mathbf{x}}}^{\prime T}\mathbf{Q}_{\underline{\mathbf{v}}}^{\prime\prime}\mathbf{Q}_{\underline{\mathbf{v}}}^{\prime\prime T}. \tag{4.149}$$

This filtering matrix cancels the noise but some distortion of the desired signal is expected.

4.4.3 Minimization of the Output SNR

The approach in this subsection involves two successive stages. We get an estimate of the noise in the first stage. Then, this estimate is used in the second stage by subtracting it from the observation signal. This will lead to the estimation of the desired signal [10]. Also, we exploit the nullspace of the desired signal correlation matrix. We will see that, thanks to this nullspace, we can derive distortionless filtering matrices.

Let

$$\mathbf{Q}_{\underline{\mathbf{x}}}^{\prime\prime} = \begin{bmatrix} \mathbf{q}_{\underline{\mathbf{x}},P+1} & \underline{\mathbf{q}}_{\mathbf{x},P+2} & \cdots & \mathbf{q}_{\underline{\mathbf{x}},ML} \end{bmatrix} \tag{4.150}$$

be the matrix of size $ML \times (ML - P)$ containing the eigenvectors corresponding to the null eigenvalues of $\mathbf{R}_{\underline{\mathbf{x}}}$. In this part, we are interested in the filtering matrices of the form:

$$\underline{\mathbf{H}}_{\mathbf{v}_1} = \underline{\mathbf{A}}_{\mathbf{v}_1}\mathbf{Q}_{\underline{\mathbf{x}}}^{\prime\prime T}, \tag{4.151}$$

where $\underline{\mathbf{A}}_{\mathbf{v}_1} \neq \mathbf{0}$ is a matrix of size $L \times (ML - P)$. We assume that $\underline{\mathbf{R}}_{\mathbf{v}}\underline{\mathbf{Q}}''_{\mathbf{x}} \neq \mathbf{0}$. Therefore, we have

$$\begin{aligned} \text{oSNR}\left(\underline{\mathbf{H}}_{\mathbf{v}_1}\right) &= \frac{\text{tr}\left(\underline{\mathbf{H}}_{\mathbf{v}_1}\underline{\mathbf{R}}_{\mathbf{x}}\underline{\mathbf{H}}_{\mathbf{v}_1}^T\right)}{\text{tr}\left(\underline{\mathbf{H}}_{\mathbf{v}_1}\underline{\mathbf{R}}_{\mathbf{v}}\underline{\mathbf{H}}_{\mathbf{v}_1}^T\right)} \\ &= \frac{\text{tr}\left(\underline{\mathbf{A}}_{\mathbf{v}_1}\underline{\mathbf{Q}}''^T_{\mathbf{x}}\underline{\mathbf{R}}_{\mathbf{x}}\underline{\mathbf{Q}}''_{\mathbf{x}}\underline{\mathbf{A}}_{\mathbf{v}_1}^T\right)}{\text{tr}\left(\underline{\mathbf{A}}_{\mathbf{v}_1}\underline{\mathbf{Q}}''^T_{\mathbf{x}}\underline{\mathbf{R}}_{\mathbf{v}}\underline{\mathbf{Q}}''_{\mathbf{x}}\underline{\mathbf{A}}_{\mathbf{v}_1}^T\right)} \\ &= 0, \end{aligned} \tag{4.152}$$

since $\underline{\mathbf{R}}_{\mathbf{x}}\underline{\mathbf{Q}}''_{\mathbf{x}} = \mathbf{0}$. As a consequence, the estimate of $\mathbf{v}_1(t)$ is

$$\begin{aligned} \widehat{\mathbf{v}}_1(t) &= \underline{\mathbf{H}}_{\mathbf{v}_1}\underline{\mathbf{y}}(t) \\ &= \underline{\mathbf{A}}_{\mathbf{v}_1}\underline{\mathbf{Q}}''^T_{\mathbf{x}}\underline{\mathbf{x}}(t) + \underline{\mathbf{A}}_{\mathbf{v}_1}\underline{\mathbf{Q}}''^T_{\mathbf{x}}\underline{\mathbf{v}}(t) \\ &= \underline{\mathbf{A}}_{\mathbf{v}_1}\underline{\mathbf{Q}}''^T_{\mathbf{x}}\underline{\mathbf{v}}(t), \end{aligned} \tag{4.153}$$

from which we deduce the estimate of $\mathbf{x}_1(t)$:

$$\begin{aligned} \widehat{\mathbf{x}}_1(t) &= \mathbf{y}_1(t) - \widehat{\mathbf{v}}_1(t) \\ &= \mathbf{x}_1(t) + \mathbf{v}_1(t) - \underline{\mathbf{A}}_{\mathbf{v}_1}\underline{\mathbf{Q}}''^T_{\mathbf{x}}\underline{\mathbf{v}}(t) \\ &= \underline{\mathbf{H}}\,\underline{\mathbf{y}}(t), \end{aligned} \tag{4.154}$$

where

$$\underline{\mathbf{H}} = \underline{\mathbf{I}}_{\text{i}} - \underline{\mathbf{A}}_{\mathbf{v}_1}\underline{\mathbf{Q}}''^T_{\mathbf{x}} \tag{4.155}$$

is the equivalent filtering matrix for the estimation of $\mathbf{x}_1(t)$. The output SNR with this filtering matrix is then

$$\text{oSNR}\left(\underline{\mathbf{H}}\right) = \frac{\text{tr}\left(\underline{\mathbf{H}}\underline{\mathbf{R}}_{\mathbf{x}}\underline{\mathbf{H}}^T\right)}{\text{tr}\left(\underline{\mathbf{H}}\underline{\mathbf{R}}_{\mathbf{v}}\underline{\mathbf{H}}^T\right)}. \tag{4.156}$$

It is important to notice that the estimator given in (4.154) does not introduce any distortion to the desired signal, since it is not filtered at all.

Now, we need to determine $\underline{\mathbf{A}}_{\mathbf{v}_1}$. The best way to do so is from the MSE of the residual noise:

$$\begin{aligned} J_{\text{n}}\left(\underline{\mathbf{H}}\right) &= \text{tr}\left(\underline{\mathbf{H}}\underline{\mathbf{R}}_{\mathbf{v}}\underline{\mathbf{H}}^T\right) \\ &= \text{tr}\left(\mathbf{R}_{\mathbf{v}_1} - 2\underline{\mathbf{A}}_{\mathbf{v}_1}\underline{\mathbf{Q}}''^T_{\mathbf{x}}\underline{\mathbf{R}}_{\mathbf{v}}\underline{\mathbf{I}}_{\text{i}}^T + \underline{\mathbf{A}}_{\mathbf{v}_1}\underline{\mathbf{Q}}''^T_{\mathbf{x}}\underline{\mathbf{R}}_{\mathbf{v}}\underline{\mathbf{Q}}''_{\mathbf{x}}\underline{\mathbf{A}}_{\mathbf{v}_1}^T\right) \\ &= J_{\text{n}}\left(\underline{\mathbf{A}}_{\mathbf{v}_1}\right). \end{aligned} \tag{4.157}$$

From the minimization of $J_{\mathrm{n}}\left(\underline{\mathbf{A}}_{\mathbf{v}_1}\right)$ and (4.151), we find the optimal distortionless filtering matrix:

$$\underline{\mathbf{H}}_{\mathrm{DL}} = \underline{\mathbf{I}}_{\mathrm{i}} - \underline{\mathbf{I}}_{\mathrm{i}}\mathbf{R}_{\underline{\mathbf{v}}}\mathbf{Q}''_{\underline{\mathbf{x}}}\left(\mathbf{Q}''^{T}_{\underline{\mathbf{x}}}\mathbf{R}_{\underline{\mathbf{v}}}\mathbf{Q}''_{\underline{\mathbf{x}}}\right)^{-1}\mathbf{Q}''^{T}_{\underline{\mathbf{x}}}. \tag{4.158}$$

It is very important to see from (4.158) that for $\underline{\mathbf{H}}_{\mathrm{DL}}$ to be unique, we must have $Q \geq ML - P$. For this filtering matrix to be useful as far as noise reduction is concerned, the number of columns of $\underline{\mathbf{A}}_{\mathbf{v}_1}$ should be large enough, which implies that the dimension of the nullspace of $\mathbf{R}_{\underline{\mathbf{x}}}$ should also be large.

Let us consider the particular case: $Q = ML - P$. We can always express the estimate of $\mathbf{v}_1(t)$ as

$$\hat{\mathbf{v}}_1(t) = \underline{\mathbf{A}}_{\mathbf{v}_1}\mathbf{Q}''^{T}_{\underline{\mathbf{x}}}\mathbf{Q}'_{\underline{\mathbf{v}}}\mathbf{Q}'^{T}_{\underline{\mathbf{v}}}\underline{\mathbf{v}}(t), \tag{4.159}$$

where

$$\mathbf{Q}'_{\underline{\mathbf{v}}} = \begin{bmatrix} \mathbf{q}_{\underline{\mathbf{v}},1} & \mathbf{q}_{\underline{\mathbf{v}},2} & \cdots & \mathbf{q}_{\underline{\mathbf{v}},Q} \end{bmatrix} \tag{4.160}$$

is the matrix of size $ML \times Q$ containing the eigenvectors corresponding to the nonnull eigenvalues of $\mathbf{R}_{\underline{\mathbf{v}}}$. We see from (4.159) that in order to recover the noise, $\mathbf{v}_1(t)$, we must have

$$\underline{\mathbf{A}}_{\mathbf{v}_1}\mathbf{Q}''^{T}_{\underline{\mathbf{x}}}\mathbf{Q}'_{\underline{\mathbf{v}}} = \underline{\mathbf{I}}_{\mathrm{i}}\mathbf{Q}'_{\underline{\mathbf{v}}}. \tag{4.161}$$

Therefore, since $\mathbf{Q}''^{T}_{\underline{\mathbf{x}}}\mathbf{Q}'_{\underline{\mathbf{v}}}$ is a square invertible matrix, we have a unique solution for (4.161). As a result, the optimal distortionless filtering matrix for this particular case is

$$\underline{\mathbf{H}}_{\mathrm{DL}} = \underline{\mathbf{I}}_{\mathrm{i}} - \underline{\mathbf{I}}_{\mathrm{i}}\mathbf{Q}'_{\underline{\mathbf{v}}}\left(\mathbf{Q}''^{T}_{\underline{\mathbf{x}}}\mathbf{Q}'_{\underline{\mathbf{v}}}\right)^{-1}\mathbf{Q}''^{T}_{\underline{\mathbf{x}}}. \tag{4.162}$$

This filtering matrix perfectly recovers the desired signal: it completely cancels the noise without any distortion.

Another interesting scenario is when $Q < ML - P$. One reasonable approach is to take the minimum-norm solution of (4.161). We then get another distortionless filtering matrix for this particular case:

$$\underline{\mathbf{H}}_{\mathrm{DL}} = \underline{\mathbf{I}}_{\mathrm{i}} - \underline{\mathbf{I}}_{\mathrm{i}}\mathbf{Q}'_{\underline{\mathbf{v}}}\left(\mathbf{Q}'^{T}_{\underline{\mathbf{v}}}\mathbf{Q}''_{\underline{\mathbf{x}}}\mathbf{Q}''^{T}_{\underline{\mathbf{x}}}\mathbf{Q}'_{\underline{\mathbf{v}}}\right)^{-1}\mathbf{Q}'^{T}_{\underline{\mathbf{v}}}\mathbf{Q}''_{\underline{\mathbf{x}}}\mathbf{Q}''_{\underline{\mathbf{x}}}. \tag{4.163}$$

Problems

4.1 Show that if two symmetric matrices $\mathbf{R}_{\underline{\mathbf{x}}}$ and $\mathbf{R}_{\underline{\mathbf{v}}}$ are jointly diagonalized:

$$\underline{\mathbf{T}}^{T}\mathbf{R}_{\underline{\mathbf{x}}}\underline{\mathbf{T}} = \underline{\mathbf{\Lambda}},$$

$$\underline{\mathbf{T}}^T \mathbf{R}_{\underline{\mathbf{v}}} \underline{\mathbf{T}} = \mathbf{I}_{ML},$$

then $\underline{\mathbf{\Lambda}}$ and $\underline{\mathbf{T}}$ are, respectively, the eigenvalue and eigenvector matrices of $\mathbf{R}_{\underline{\mathbf{v}}}^{-1} \mathbf{R}_{\underline{\mathbf{x}}}$:

$$\mathbf{R}_{\underline{\mathbf{v}}}^{-1} \mathbf{R}_{\underline{\mathbf{x}}} \underline{\mathbf{T}} = \underline{\mathbf{T}}\, \underline{\mathbf{\Lambda}}.$$

4.2 Denote by $\underline{\mathbf{t}}_1, \underline{\mathbf{t}}_2, \dots, \underline{\mathbf{t}}_{ML}$, the eigenvectors of $\mathbf{R}_{\underline{\mathbf{v}}}^{-1} \mathbf{R}_{\underline{\mathbf{x}}}$. Show that

$$\mathbf{R}_{\underline{\mathbf{v}}}^{-1} = \sum_{i=1}^{ML} \underline{\mathbf{t}}_i \underline{\mathbf{t}}_i^T.$$

4.3 Show that the MSE can be written as

$$J\left(\underline{\mathbf{A}}\right) = \text{tr}\left[\mathbf{R}_{\mathbf{x}_1} - 2\underline{\mathbf{A}}\, \underline{\mathbf{T}}^T \mathbf{R}_{\underline{\mathbf{x}}} \underline{\mathbf{I}}_{\text{i}}^T + \underline{\mathbf{A}} \left(\underline{\mathbf{\Lambda}} + \mathbf{I}_{ML}\right) \underline{\mathbf{A}}^T\right].$$

4.4 Show that the different performance measures are related to the MSEs by

$$\begin{aligned} \frac{J_{\text{d}}\left(\underline{\mathbf{A}}\right)}{J_{\text{n}}\left(\underline{\mathbf{A}}\right)} &= \text{iSNR} \times \xi_{\text{n}}\left(\underline{\mathbf{A}}\right) \times \upsilon_{\text{d}}\left(\underline{\mathbf{A}}\right) \\ &= \text{oSNR}\left(\underline{\mathbf{A}}\right) \times \xi_{\text{d}}\left(\underline{\mathbf{A}}\right) \times \upsilon_{\text{d}}\left(\underline{\mathbf{A}}\right). \end{aligned}$$

4.5 Show that the Wiener filtering matrix can be written as

$$\underline{\mathbf{H}}_{\text{W}} = \underline{\mathbf{I}}_{\text{i}} \mathbf{R}_{\underline{\mathbf{v}}} \sum_{i=1}^{ML} \frac{\underline{\lambda}_i}{1 + \underline{\lambda}_i} \underline{\mathbf{t}}_i \underline{\mathbf{t}}_i^T.$$

4.6 Prove that with the optimal Wiener filtering matrix, the output SNR is always greater than or equal to the input SNR: $\text{oSNR}\left(\underline{\mathbf{H}}_{\text{W}}\right) \geq \text{iSNR}$.

4.7 Consider a ULA of M sensors with interelement spacing d. Assume that the desired signal is a harmonic random process:

$$x(t) = A \cos\left(2\pi f_0 t + \phi\right),$$

with fixed amplitude A and frequency f_0, and random phase ϕ, uniformly distributed on the interval from 0 to 2π.

a) Suppose that the desired signal impinges on the ULA from the broadside direction. Compute the correlation matrix of $\underline{\mathbf{x}}(t)$.
b) Assume that the sampling interval satisfies $T_{\text{s}} = d/c$. Compute the correlation matrix of $\underline{\mathbf{x}}(t)$ when the desired signal impinges on the ULA from the endfire direction.
c) Compute the correlation matrix of $\underline{\mathbf{x}}(t)$ in the general case, when the desired signal impinges on the ULA from the direction θ.
d) Repeat the above for white Gaussian noise: $x(t) \sim \mathcal{N}\left(0, \sigma^2\right)$.

e) Repeat the above for a desired signal, $x(t)$, that has a autocorrelation sequence:

$$E\left[x(t)x(t')\right] = \alpha^{|t-t'|}, \ -1 < \alpha < 1.$$

4.8 Show that the maximum SNR filtering matrix with minimum MSE is given by

$$\underline{\mathbf{H}}_{\max,1} = \underline{\mathbf{I}}_{\mathrm{i}}\mathbf{R}_{\underline{\mathbf{x}}}\frac{\underline{\mathbf{t}}_1\underline{\mathbf{t}}_1^T}{1+\underline{\lambda}_1}.$$

4.9 Show that taking the gradient of $J_{\mathrm{d}}\left(\underline{\mathbf{A}}\right)$ with respect to $\underline{\mathbf{A}}$ and equating the result to zero yields the MVDR filtering matrix:

$$\underline{\mathbf{H}}_{\mathrm{MVDR}} = \underline{\mathbf{I}}_{\mathrm{i}}\mathbf{R}_{\underline{\mathbf{x}}}\sum_{p=1}^{P}\frac{\underline{\mathbf{t}}_p\underline{\mathbf{t}}_p^T}{\underline{\lambda}_p}.$$

4.10 Prove that with the MVDR filtering matrix, there is no distortion in the filtered signal:

$$v_{\mathrm{d}}\left(\underline{\mathbf{H}}_{\mathrm{MVDR}}\right) = 0. \tag{4.164}$$

4.11 Show that with the MVDR filtering matrix, the output SNR is always greater than or equal to the input SNR: $\mathrm{oSNR}\left(\underline{\mathbf{H}}_{\mathrm{MVDR}}\right) \geq \mathrm{iSNR}$.

4.12 Show that with the tradeoff filtering matrix $\underline{\mathbf{H}}_{\mathrm{T},\mu}$, the output SNR is always greater than or equal to the input SNR: $\mathrm{oSNR}\left(\underline{\mathbf{H}}_{\mathrm{T},\mu}\right) \geq \mathrm{iSNR}, \ \forall \mu \geq 0$.

4.13 Show that by jointly diagonalizing the two spatial correlation matrices $\mathbf{R}_{\mathbf{x}}(;l)$ and $\mathbf{R}_{\mathbf{v}}(;l)$, we obtain

$$\mathbf{S}''^T(;l)\mathbf{x}(t;l) = \mathbf{0}$$

and

$$\mathbf{R}_{\mathbf{v}}^{-1}(;l) = \mathbf{S}'(;l)\mathbf{S}'^T(;l) + \mathbf{S}''(;l)\mathbf{S}''^T(;l).$$

4.14 Show that the fullmode input SNR can never exceed the maximum spectral mode input SNR and can never go below the minimum spectral mode input SNR:

$$\mathrm{iSNR}(;L) \leq \mathrm{iSNR}(;) \leq \mathrm{iSNR}(;1).$$

4.15 Show that the MSE is given by

$$J\left[\mathbf{h}(;l)\right] = \lambda_{1,l} + \mathbf{h}^T(;l)\mathbf{R}_{\mathbf{y}}(;l)\mathbf{h}(;l) - 2\mathbf{h}^T(;l)\mathbf{R}_{\mathbf{x}}(;l)\mathbf{i}_{\mathrm{i}}.$$

4.16 Show that the MSEs are related to the lth spectral mode performance measures by

$$\frac{J_{\rm d}\left[\mathbf{h}(;l)\right]}{J_{\rm n}\left[\mathbf{h}(;l)\right]} = \mathrm{iSNR}(;l) \times \xi_{\rm n}\left[\mathbf{h}(;l)\right] \times \upsilon_{\rm d}\left[\mathbf{h}(;l)\right]$$
$$= \mathrm{oSNR}\left[\mathbf{h}(;l)\right] \times \xi_{\rm d}\left[\mathbf{h}(;l)\right] \times \upsilon_{\rm d}\left[\mathbf{h}(;l)\right].$$

4.17 Show that the optimal maximum SNR filter with minimum desired signal distortion is given by

$$\mathbf{h}_{\rm max}(;l) = \frac{\mathbf{s}_1(;l)\mathbf{s}_1^T(;l)}{\omega_1(;l)}\mathbf{R}_{\mathbf{x}}(;l)\mathbf{i}_{\rm i}$$
$$= \mathbf{s}_1(;l)\mathbf{s}_1^T(;l)\mathbf{R}_{\mathbf{v}}(;l)\mathbf{i}_{\rm i}.$$

4.18 Consider a desired signal that is a sum of harmonic random processes:

$$x(t) = \sum_{k=1}^{K} A_k \cos\left(2\pi f_k t + \phi_k\right),$$

with fixed amplitudes $\{A_k\}$ and frequencies $\{f_k\}$, and IID random phases $\{\phi_k\}$, uniformly distributed on the interval from 0 to 2π. Assume that the interference $u(t)$ is colored noise that is uncorrelated with $x(t)$, with the autocorrelation sequence:

$$E\left[u(t)u(t')\right] = \sigma_u^2\, \alpha^{|t-t'|},\ -1 < \alpha < 1.$$

a) Compute the fullmode input SNR.
b) Assume that the desired signal impinges on the ULA from the broadside direction. Describe the steps to find the correlation matrix $\mathbf{R}_{\mathbf{x}_m}$.
c) Assume that the desired signal impinges on the ULA from the direction θ. Describe the steps to find the correlation matrix $\mathbf{R}_{\mathbf{x}_m}$.
d) Assume that the sampling interval satisfies $T_{\rm s} = d/c$, and the interference impinges on the ULA from the endfire direction. Describe the steps to find the correlation matrix $\mathbf{R}_{\mathbf{u}_m}$.

4.19 Show that the Wiener filter can be expressed as

$$\mathbf{h}_{\rm W}(;l) = \sum_{m=1}^{M} \frac{\mathbf{s}_m(;l)\mathbf{s}_m^T(;l)}{1+\omega_m(;l)}\mathbf{R}_{\mathbf{x}}(;l)\mathbf{i}_{\rm i}$$
$$= \sum_{m=1}^{M} \frac{\omega_m(;l)}{1+\omega_m(;l)}\mathbf{s}_m(;l)\mathbf{s}_m^T(;l)\mathbf{R}_{\mathbf{v}}(;l)\mathbf{i}_{\rm i}.$$

4.20 Show that the MVDR filter can be expressed as

$$\begin{aligned}\mathbf{h}_{\mathrm{MVDR}}(;l) &= \mathbf{S}'(;l)\mathbf{S}'^T(;l)\mathbf{R}_{\mathbf{v}}(;l)\mathbf{i}_{\mathrm{i}} \\ &= \sum_{p=1}^{P_l} \mathbf{s}_p(;l)\mathbf{s}_p^T(;l)\mathbf{R}_{\mathbf{v}}(;l)\mathbf{i}_{\mathrm{i}}.\end{aligned}$$

4.21 Show that the MVDR filter can be expressed as

$$\mathbf{h}_{\mathrm{MVDR}}(;l) = \sum_{p=1}^{P_l} \frac{\mathbf{s}_p(;l)\mathbf{s}_p^T(;l)}{\omega_p(;l)}\mathbf{R}_{\mathbf{x}}(;l)\mathbf{i}_{\mathrm{i}}.$$

4.22 Show that the lth spectral mode output SNR of the MVDR filter is not larger than that of the Wiener filter:

$$\mathrm{oSNR}\left[\mathbf{h}_{\mathrm{MVDR}}(;l)\right] \leq \mathrm{oSNR}\left[\mathbf{h}_{\mathrm{W}}(;l)\right].$$

4.23 Show that with the controlled distortion filters, we have

$$\mathrm{oSNR}\left[\mathbf{h}_{\mathrm{CD},P_l}(;l)\right] \leq \mathrm{oSNR}\left[\mathbf{h}_{\mathrm{CD},P_l-1}(;l)\right] \leq \cdots \leq \mathrm{oSNR}\left[\mathbf{h}_{\mathrm{CD},1}(;l)\right]$$

and

$$0 = \upsilon_{\mathrm{d}}\left[\mathbf{h}_{\mathrm{CD},P_l}(;l)\right] \leq \upsilon_{\mathrm{d}}\left[\mathbf{h}_{\mathrm{CD},P_l-1}(;l)\right] \leq \cdots \leq \upsilon_{\mathrm{d}}\left[\mathbf{h}_{\mathrm{CD},1}(;l)\right].$$

4.24 Show that the tradeoff filter is given by

$$\mathbf{h}_{\mathrm{T},\mu}(;l) = \sum_{m=1}^{M} \frac{\omega_m(;l)}{\mu + \omega_m(;l)}\mathbf{s}_m(;l)\mathbf{s}_m^T(;l)\mathbf{R}_{\mathbf{v}}(;l)\mathbf{i}_{\mathrm{i}},$$

where $\mu \geq 0$ is a Lagrange multiplier.

4.25 Show that in the case of a rank deficient noise correlation matrix,

a) the distortion-based MSE can be written as

$$J_{\mathrm{d}}\left(\underline{\mathbf{A}}\right) = \mathrm{tr}\left(\mathbf{R}_{\mathbf{x}_1} - 2\underline{\mathbf{A}}\mathbf{Q}_{\underline{\mathbf{v}}}''^T\mathbf{R}_{\underline{\mathbf{x}}}\underline{\mathbf{I}}_{\mathrm{i}}^T + \underline{\mathbf{A}}\mathbf{Q}_{\underline{\mathbf{v}}}''^T\mathbf{R}_{\underline{\mathbf{x}}}\mathbf{Q}_{\underline{\mathbf{v}}}''\underline{\mathbf{A}}^T\right)$$

b) and the minimization of $J_{\mathrm{d}}\left(\underline{\mathbf{A}}\right)$ yields the optimal infinite noise reduction filtering matrix:

$$\underline{\mathbf{H}}_{\infty} = \underline{\mathbf{I}}_{\mathrm{i}}\mathbf{R}_{\underline{\mathbf{x}}}\mathbf{Q}_{\underline{\mathbf{v}}}''\left(\mathbf{Q}_{\underline{\mathbf{v}}}''^T\mathbf{R}_{\underline{\mathbf{x}}}\mathbf{Q}_{\underline{\mathbf{v}}}''\right)^{-1}\mathbf{Q}_{\underline{\mathbf{v}}}''^T.$$

4.26 Show that in the case of a rank deficient noise correlation matrix,

a) the MSE of the residual noise can be written as

$$J_{\mathrm{n}}\left(\underline{\mathbf{H}}\right) = \mathrm{tr}\left(\mathbf{R}_{\mathbf{v}_1} - 2\underline{\mathbf{A}}_{\mathbf{v}_1}\underline{\mathbf{Q}}_{\mathbf{x}}^{\prime\prime T}\underline{\mathbf{R}}_{\mathbf{v}}\underline{\mathbf{I}}_{\mathrm{i}}^{T} + \underline{\mathbf{A}}_{\mathbf{v}_1}\underline{\mathbf{Q}}_{\mathbf{x}}^{\prime\prime T}\underline{\mathbf{R}}_{\mathbf{v}}\underline{\mathbf{Q}}_{\mathbf{x}}^{\prime\prime}\underline{\mathbf{A}}_{\mathbf{v}_1}^{T}\right)$$

b) and the minimization of $J_{\mathrm{n}}\left(\underline{\mathbf{H}}\right)$ yields the optimal distortionless filtering matrix:

$$\underline{\mathbf{H}}_{\mathrm{DL}} = \underline{\mathbf{I}}_{\mathrm{i}} - \underline{\mathbf{I}}_{\mathrm{i}}\underline{\mathbf{R}}_{\mathbf{v}}\underline{\mathbf{Q}}_{\mathbf{x}}^{\prime\prime}\left(\underline{\mathbf{Q}}_{\mathbf{x}}^{\prime\prime T}\underline{\mathbf{R}}_{\mathbf{v}}\underline{\mathbf{Q}}_{\mathbf{x}}^{\prime\prime}\right)^{-1}\underline{\mathbf{Q}}_{\mathbf{x}}^{\prime\prime T}.$$

References

1 J. Benesty, J. Chen, and Y. Huang, *Microphone Array Signal Processing*. Berlin, Germany: Springer-Verlag, 2008.

2 M. Brandstein and D. B. Ward, Eds., *Microphone Arrays: Signal Processing Techniques and Applications*. Berlin, Germany: Springer-Verlag, 2001.

3 J. Benesty and J. Chen, *Optimal Time-Domain Noise Reduction Filters – A Theoretical Study*. Springer Briefs in Electrical and Computer Engineering, 2011.

4 J. N. Franklin, *Matrix Theory*. Englewood Cliffs, NJ: Prentice-Hall, 1968.

5 J. Benesty, M. G. Christensen, and J. R. Jensen, *Signal Enhancement with Variable Span Linear Filters*. Berlin, Germany: Springer-Verlag, 2016.

6 J. R. Jensen, J. Benesty, and M. G. Christensen, "Noise reduction with optimal variable span linear filters," *IEEE/ACM Trans. Audio, Speech, Language Process.*, vol. 24, pp. 631–644, Apr. 2016.

7 J. Benesty, J. R. Jensen, M. G. Christensen, and J. Chen, *Speech Enhancement – A Signal Subspace Perspective*. Academic Press, 2014.

8 P. C. Hansen and S. H. Jensen, "Prewhitening for rank-deficient noise in subspace methods for noise reduction," *IEEE Trans. Signal Process.*, vol. 53, pp. 3718–3726, Oct. 2005.

9 G. H. Golub and C. F. Van Loan, *Matrix Computations*, 3rd edn. Baltimore, MD: The Johns Hopkins University Press, 1996.

10 S. M. Nørholm, J. Benesty, J. R. Jensen, and M. G. Christensen, "Single-channel noise reduction using unified joint diagonalization and optimal filtering," *EURASIP J. Advances Signal Process.*, vol. 2014, pp. 37–48, 2014.

5

Multichannel Signal Enhancement in the Frequency Domain

Signal enhancement with multiple sensors or multichannel signal enhancement in the frequency domain is an important part of array signal processing. This chapter deals with this topic. As in the previous chapter, the spatial information is fully exploited here and in a much more obvious way. We explain the signal model and state the problem we wish to solve with the conventional linear filtering technique. We then derive the performance measures and show how to obtain the most well-known optimal linear filters. Finally, we give a signal subspace perspective, which can be an instructive alternative way to look at the problem at hand.

5.1 Signal Model and Problem Formulation

We consider the signal model of the previous chapter, i.e.,

$$\begin{aligned} y_m(t) &= g_m(t) * x(t) + v_m(t) \\ &= x_m(t) + v_m(t),\ m = 1, 2, \ldots, M, \end{aligned} \tag{5.1}$$

where $y_m(t)$, $x_m(t)$, and $v_m(t)$ are, respectively, the observation, convolved desired, and additive noise signals at the mth sensor, with M being the number of sensors and the array geometry is completely arbitrary. In the frequency domain, at the frequency index f, (5.1) can be expressed as [1–3]

$$\begin{aligned} Y_m(f) &= G_m(f)X(f) + V_m(f) \\ &= X_m(f) + V_m(f),\ m = 1, 2, \ldots, M, \end{aligned} \tag{5.2}$$

where $Y_m(f)$, $G_m(f)$, $X(f)$, $V_m(f)$, and $X_m(f) = G_m(f)X(f)$ are the frequency-domain representations of $y_m(t)$, $g_m(t)$, $x(t)$, $v_m(t)$, and $x_m(t) = g_m(t) * x(t)$, respectively. It is clear that $X_m(f)$ and $V_m(f)$, which are assumed to be zero mean and uncorrelated, are the frequency-domain desired and noise signals, respectively. Sensor 1 is the reference, so the objective of multichannel noise reduction in the frequency domain is to estimate the desired signal, $X_1(f)$, from the M observations $Y_m(f)$, $m = 1, 2, \ldots, M$, in the best possible way.

Fundamentals of Signal Enhancement and Array Signal Processing, First Edition.
Jacob Benesty, Israel Cohen, and Jingdong Chen.

Companion website: www.wiley.com/go/benesty/arraysignalprocessing

It is more convenient to write the M frequency-domain sensors' signals in a vector notation:

$$\begin{aligned}\mathbf{y}(f) &= \mathbf{g}(f)X(f) + \mathbf{v}(f) \\ &= \mathbf{x}(f) + \mathbf{v}(f) \\ &= \mathbf{d}(f)X_1(f) + \mathbf{v}(f),\end{aligned} \tag{5.3}$$

where

$$\begin{aligned}\mathbf{y}(f) &= \begin{bmatrix} Y_1(f) & Y_2(f) & \cdots & Y_M(f) \end{bmatrix}^T, \\ \mathbf{x}(f) &= \begin{bmatrix} X_1(f) & X_2(f) & \cdots & X_M(f) \end{bmatrix}^T \\ &= X(f)\mathbf{g}(f), \\ \mathbf{g}(f) &= \begin{bmatrix} G_1(f) & G_2(f) & \cdots & G_M(f) \end{bmatrix}^T, \\ \mathbf{v}(f) &= \begin{bmatrix} V_1(f) & V_2(f) & \cdots & V_M(f) \end{bmatrix}^T,\end{aligned}$$

and

$$\begin{aligned}\mathbf{d}(f) &= \begin{bmatrix} 1 & \frac{G_2(f)}{G_1(f)} & \cdots & \frac{G_M(f)}{G_1(f)} \end{bmatrix}^T \\ &= \frac{\mathbf{g}(f)}{G_1(f)}.\end{aligned} \tag{5.4}$$

Expression (5.3) depends explicitly on the desired signal, $X_1(f)$; as a result, (5.3) is the frequency-domain signal model for noise reduction. The vector $\mathbf{d}(f)$ can be seen as the steering vector for noise reduction [4] since the acoustic impulse responses ratios from the broadband source to the aperture convey information about the position of the source.

There is another useful way to write (5.3). First, it is easy to see that

$$X_m(f) = \gamma^*_{X_1X_m}(f)X_1(f),\ m = 1, 2, \ldots, M, \tag{5.5}$$

where

$$\begin{aligned}\gamma_{X_1X_m}(f) &= \frac{E\left[X_1(f)X_m^*(f)\right]}{E\left[|X_1(f)|^2\right]} \\ &= \frac{G_m^*(f)}{G_1^*(f)},\ m = 1, 2, \ldots, M\end{aligned} \tag{5.6}$$

is the partially normalized [with respect to $X_1(f)$] coherence function between $X_1(f)$ and $X_m(f)$. Using (5.5), we can rewrite (5.3) as

$$\mathbf{y}(f) = \boldsymbol{\gamma}^*_{X_1\mathbf{x}}(f)X_1(f) + \mathbf{v}(f), \tag{5.7}$$

where

$$\begin{aligned}\boldsymbol{\gamma}_{X_1\mathbf{x}}(f) &= \begin{bmatrix} 1 & \gamma_{X_1X_2}(f) & \cdots & \gamma_{X_1X_M}(f) \end{bmatrix}^T \\ &= \frac{E\left[X_1(f)\mathbf{x}^*(f)\right]}{E\left[|X_1(f)|^2\right]} \\ &= \mathbf{d}^*(f)\end{aligned} \tag{5.8}$$

is the partially normalized [with respect to $X_1(f)$] coherence vector (of length M) between $X_1(f)$ and $\mathbf{x}(f)$. In the rest, $\boldsymbol{\gamma}^*_{X_1\mathbf{x}}(f)$ and $\mathbf{d}(f)$ will be used interchangeably.

By definition, the signal $X_1(f)$ is completely coherent across all sensors [see eq. (5.5)]; however, $V_1(f)$ is usually partially coherent with the noise components, $V_m(f)$, at the other sensors. Therefore, any noise term $V_m(f)$ can be easily decomposed into two orthogonal components, i.e.,

$$V_m(f) = \gamma^*_{V_1V_m}(f)V_1(f) + V'_m(f),\ m = 1, 2, \dots, M, \tag{5.9}$$

where $\gamma_{V_1V_m}(f)$ is the partially normalized [with respect to $V_1(f)$] coherence function between $V_1(f)$ and $V_m(f)$ and

$$E\left[V_1^*(f)V'_m(f)\right] = 0,\ m = 1, 2, \dots, M. \tag{5.10}$$

The vector $\mathbf{v}(f)$ can then be written as the sum of two other vectors: one coherent with $V_1(f)$ and the other incoherent with $V_1(f)$, i.e.,

$$\mathbf{v}(f) = \boldsymbol{\gamma}^*_{V_1\mathbf{v}}(f)V_1(f) + \mathbf{v}'(f), \tag{5.11}$$

where

$$\boldsymbol{\gamma}_{V_1\mathbf{v}}(f) = \begin{bmatrix} 1 & \gamma_{V_1V_2}(f) & \cdots & \gamma_{V_1V_M}(f) \end{bmatrix}^T \tag{5.12}$$

is the partially normalized [with respect to $V_1(f)$] coherence vector (of length M) between $V_1(f)$ and $\mathbf{v}(f)$ and

$$\mathbf{v}'(f) = \begin{bmatrix} 0 & V'_2(f) & \cdots & V'_M(f) \end{bmatrix}^T.$$

If $V_1(f)$ is incoherent with $V_m(f)$, where $m \neq 1$, then $\gamma_{V_1V_m}(f) = 0$.

Another convenient way to write the sensors' signals vector is

$$\mathbf{y}(f) = \boldsymbol{\gamma}^*_{X_1\mathbf{x}}(f)X_1(f) + \boldsymbol{\gamma}^*_{V_1\mathbf{v}}(f)V_1(f) + \mathbf{v}'(f). \tag{5.13}$$

We see that $\mathbf{y}(f)$ is the sum of three mutual incoherent components. Therefore, the correlation matrix of $\mathbf{y}(f)$ is

$$\boldsymbol{\Phi}_{\mathbf{y}}(f) = E\left[\mathbf{y}(f)\mathbf{y}^H(f)\right] \tag{5.14}$$

$$= \phi_{X_1}(f)\mathbf{d}(f)\mathbf{d}^H(f) + \mathbf{\Phi}_{\mathbf{v}}(f)$$
$$= \phi_{X_1}(f)\boldsymbol{\gamma}^*_{X_1\mathbf{x}}(f)\boldsymbol{\gamma}^T_{X_1\mathbf{x}}(f) + \phi_{V_1}(f)\boldsymbol{\gamma}^*_{V_1\mathbf{v}}(f)\boldsymbol{\gamma}^T_{V_1\mathbf{v}}(f) + \mathbf{\Phi}_{\mathbf{v}'}(f),$$

where the superscript ^{H} is the conjugate-transpose operator, $\phi_{X_1}(f) = E\left[|X_1(f)|^2\right]$ and $\phi_{V_1}(f) = E\left[|V_1(f)|^2\right]$ are the variances of $X_1(f)$ and $V_1(f)$, respectively, and $\mathbf{\Phi}_{\mathbf{v}}(f) = E\left[\mathbf{v}(f)\mathbf{v}^H(f)\right]$ and $\mathbf{\Phi}_{\mathbf{v}'}(f) = E\left[\mathbf{v}'(f)\mathbf{v}'^H(f)\right]$ are the correlation matrices of $\mathbf{v}(f)$ and $\mathbf{v}'(f)$, respectively. The matrix $\mathbf{\Phi}_{\mathbf{y}}(f)$ is the sum of three other matrices: the first two are of rank equal to 1 and the last one (correlation matrix of the incoherent noise) is assumed to be of rank equal to $M - 1$.

5.2 Linear Filtering

In the frequency domain, conventional multichannel noise reduction is performed by applying a complex weight to the output of each sensor, at frequency f, and summing across the aperture (see Figure 5.1):

$$Z(f) = \sum_{m=1}^{M} H_m^*(f)Y_m(f) \tag{5.15}$$
$$= \mathbf{h}^H(f)\mathbf{y}(f),$$

where $Z(f)$ is the estimate of $X_1(f)$ and

$$\mathbf{h}(f) = \begin{bmatrix} H_1(f) & H_2(f) & \cdots & H_M(f) \end{bmatrix}^T \tag{5.16}$$

is a filter of length M containing all the complex gains applied to the sensors' outputs at frequency f.

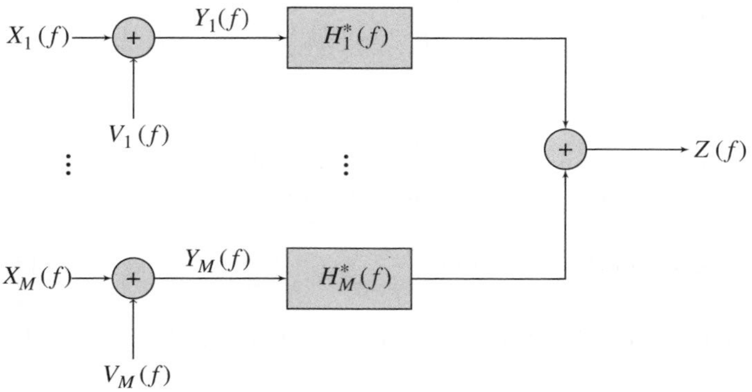

Figure 5.1 Block diagram of multichannel linear filtering in the frequency domain.

We can express (5.15) as a function of the steering vector, i.e.,

$$\begin{aligned} Z(f) &= \mathbf{h}^H(f)\left[\boldsymbol{\gamma}^*_{X_1\mathbf{x}}(f)X_1(f) + \mathbf{v}(f)\right] \\ &= X_{\rm fd}(f) + V_{\rm rn}(f), \end{aligned} \tag{5.17}$$

where

$$X_{\rm fd}(f) = X_1(f)\mathbf{h}^H(f)\boldsymbol{\gamma}^*_{X_1\mathbf{x}}(f) \tag{5.18}$$

is the filtered desired signal and

$$V_{\rm rn}(f) = \mathbf{h}^H(f)\mathbf{v}(f) \tag{5.19}$$

is the residual noise. This procedure is called the multichannel signal enhancement problem in the frequency domain.

The two terms on the right-hand side of (5.17) are incoherent. Hence, the variance of $Z(f)$ is also the sum of two variances:

$$\begin{aligned} \phi_Z(f) &= \mathbf{h}^H(f)\boldsymbol{\Phi}_{\mathbf{y}}(f)\mathbf{h}(f) \\ &= \phi_{X_{\rm fd}}(f) + \phi_{V_{\rm rn}}(f), \end{aligned} \tag{5.20}$$

where

$$\phi_{X_{\rm fd}}(f) = \phi_{X_1}(f)\left|\mathbf{h}^H(f)\boldsymbol{\gamma}^*_{X_1\mathbf{x}}(f)\right|^2, \tag{5.21}$$

$$\phi_{V_{\rm rn}}(f) = \mathbf{h}^H(f)\boldsymbol{\Phi}_{\mathbf{v}}(f)\mathbf{h}(f). \tag{5.22}$$

The different variances in (5.20) are important in the definitions of the performance measures.

5.3 Performance Measures

In the frequency domain, we must differentiate between the narrowband (i.e., single frequency) measures and the broadband (i.e., across the entire frequency range) measures. In this section, we define the most useful ones from the signal enhancement perspective. We recall that sensor 1 is our reference.

5.3.1 Input SNR

The input SNR gives an idea on the level of the noise as compared to the level of the desired signal at the reference sensor. From (5.2), it is obvious that the narrowband input SNR is

$$\text{iSNR}(f) = \frac{\phi_{X_1}(f)}{\phi_{V_1}(f)}. \tag{5.23}$$

From (5.23), we deduce the broadband input SNR:

$$\mathrm{iSNR} = \frac{\int_f \phi_{X_1}(f)df}{\int_f \phi_{V_1}(f)df}. \tag{5.24}$$

Notice that

$$\mathrm{iSNR} \neq \int_f \mathrm{iSNR}(f)df. \tag{5.25}$$

5.3.2 Output SNR

The output SNR quantifies the SNR after performing noise reduction. From (5.20), we deduce the narrowband output SNR:

$$\begin{aligned}\mathrm{oSNR}\left[\mathbf{h}(f)\right] &= \frac{\phi_{X_{\mathrm{fd}}}(f)}{\phi_{V_{\mathrm{rn}}}(f)} \\ &= \frac{\phi_{X_1}(f)\left|\mathbf{h}^H(f)\mathbf{d}(f)\right|^2}{\mathbf{h}^H(f)\mathbf{\Phi}_{\mathbf{v}}(f)\mathbf{h}(f)}\end{aligned} \tag{5.26}$$

and the broadband output SNR:

$$\mathrm{oSNR}\left(\mathbf{h}\right) = \frac{\int_f \phi_{X_1}(f)\left|\mathbf{h}^H(f)\mathbf{d}(f)\right|^2 df}{\int_f \mathbf{h}^H(f)\mathbf{\Phi}_{\mathbf{v}}(f)\mathbf{h}(f)df}. \tag{5.27}$$

It is clear that

$$\mathrm{oSNR}\left(\mathbf{h}\right) \neq \int_f \mathrm{oSNR}\left[\mathbf{h}(f)\right] df. \tag{5.28}$$

Assume that the matrix $\mathbf{\Phi}_{\mathbf{v}}(f)$ is nonsingular. In this case, for the two vectors $\mathbf{h}(f)$ and $\mathbf{d}(f)$, we have

$$\left|\mathbf{h}^H(f)\mathbf{d}(f)\right|^2 \leq \left[\mathbf{h}^H(f)\mathbf{\Phi}_{\mathbf{v}}(f)\mathbf{h}(f)\right]\left[\mathbf{d}^H(f)\mathbf{\Phi}_{\mathbf{v}}^{-1}(f)\mathbf{d}(f)\right], \tag{5.29}$$

with equality if and only if $\mathbf{h}(f) \propto \mathbf{\Phi}_{\mathbf{v}}^{-1}(f)\mathbf{d}(f)$. Using the inequality (5.29) in (5.26), we deduce an upper bound for the narrowband output SNR:

$$\mathrm{oSNR}\left[\mathbf{h}(f)\right] \leq \phi_{X_1}(f) \times \mathbf{d}^H(f)\mathbf{\Phi}_{\mathbf{v}}^{-1}(f)\mathbf{d}(f), \ \forall \mathbf{h}(f). \tag{5.30}$$

For the particular filter of length M:

$$\mathbf{h}(f) = \mathbf{i}_{\mathrm{i}} = \left[\begin{array}{cccc} 1 & 0 & \cdots & 0 \end{array}\right]^T, \tag{5.31}$$

we have

$$\text{oSNR}\left[\mathbf{i}_{\text{i}}(f)\right] = \text{iSNR}(f), \tag{5.32}$$

$$\text{oSNR}\left(\mathbf{i}_{\text{i}}\right) = \text{iSNR}. \tag{5.33}$$

With the identity filter, $\mathbf{i}_{\text{i}}$, the output SNRs cannot be improved and

$$\text{oSNR}\left[\mathbf{i}_{\text{i}}(f)\right] \leq \phi_{X_1}(f) \times \mathbf{d}^H(f)\boldsymbol{\Phi}_{\mathbf{v}}^{-1}(f)\mathbf{d}(f), \tag{5.34}$$

which implies that

$$\phi_{V_1}(f) \times \mathbf{d}^H(f)\boldsymbol{\Phi}_{\mathbf{v}}^{-1}(f)\mathbf{d}(f) \geq 1. \tag{5.35}$$

Our objective is then to find the filter, $\mathbf{h}(f)$, within the design constraints, in such a way that $\text{oSNR}\left[\mathbf{h}(f)\right] > \text{iSNR}(f)$. While the narrowband output SNR is important when we deal with narrowband and broadband signals, the broadband output SNR is even more important when we deal with broadband signals such as speech. Therefore, we also need to make sure finding $\mathbf{h}(f)$ in such a way that $\text{oSNR}(\mathbf{h}) > \text{iSNR}$.

5.3.3 Noise Rejection and Desired Signal Cancellation

The output SNR does not give any hint on the distortion of the desired signal introduced by the filtering process. Thus, this subsection introduces two measures which treat noise reduction and signal distortion individually.

The noise reduction factor or noise rejection factor quantifies the amount of noise being rejected by the filter. This quantity is defined as the ratio of the power of the noise at the reference sensor over the power of the noise remaining at the filter output. We provide the following definitions:

- the broadband noise reduction factor,

$$\xi_{\text{n}}(\mathbf{h}) = \frac{\int_f \phi_{V_1}(f)df}{\int_f \mathbf{h}^H(f)\boldsymbol{\Phi}_{\mathbf{v}}(f)\mathbf{h}(f)df} \tag{5.36}$$

- and the narrowband noise reduction factor,

$$\xi_{\text{n}}\left[\mathbf{h}(f)\right] = \frac{\phi_{V_1}(f)}{\mathbf{h}^H(f)\boldsymbol{\Phi}_{\mathbf{v}}(f)\mathbf{h}(f)}. \tag{5.37}$$

The broadband noise reduction factor is expected to be lower bounded by 1; otherwise, the filter amplifies the noise received at the senors. The higher the value of the noise reduction factor, the more the noise is rejected.

In practice, most filtering algorithms distort the desired signal. In order to quantify the level of this distortion, we define the desired signal reduction factor or desired signal cancellation factor as the ratio of the variance of the desired signal at the reference sensor over the variance of the filtered desired signal at the filter output. It is easy to deduce the following mathematical definitions:

- the broadband desired signal reduction factor,

$$\xi_{\mathrm{d}}(\mathbf{h}) = \frac{\int_f \phi_{X_1}(f)df}{\int_f \phi_{X_1}(f)\left|\mathbf{h}^H(f)\mathbf{d}(f)\right|^2 df} \tag{5.38}$$

- and the narrowband desired signal reduction factor,

$$\xi_{\mathrm{d}}\left[\mathbf{h}(f)\right] = \frac{1}{\left|\mathbf{h}^H(f)\mathbf{d}(f)\right|^2}. \tag{5.39}$$

Once again, note that

$$\xi_{\mathrm{n}}(\mathbf{h}) \neq \int_f \xi_{\mathrm{n}}\left[\mathbf{h}(f)\right] df, \tag{5.40}$$

$$\xi_{\mathrm{d}}(\mathbf{h}) \neq \int_f \xi_{\mathrm{d}}\left[\mathbf{h}(f)\right] df. \tag{5.41}$$

Another key observation is that the design of filters that do not cancel the broadband desired signal requires the constraint:

$$\mathbf{h}^H(f)\mathbf{d}(f) = 1. \tag{5.42}$$

Thus, the desired signal reduction factor is equal to 1 if there is no cancellation and expected to be greater than 1 when cancellation happens.

Lastly, by making the appropriate substitutions, one can derive the following relationships between the output and input SNRs, noise reduction factor, and desired signal reduction factor:

$$\frac{\mathrm{oSNR}(\mathbf{h})}{\mathrm{iSNR}} = \frac{\xi_{\mathrm{n}}(\mathbf{h})}{\xi_{\mathrm{d}}(\mathbf{h})}, \tag{5.43}$$

$$\frac{\mathrm{oSNR}\left[\mathbf{h}(f)\right]}{\mathrm{iSNR}(f)} = \frac{\xi_{\mathrm{n}}\left[\mathbf{h}(f)\right]}{\xi_{\mathrm{d}}\left[\mathbf{h}(f)\right]}. \tag{5.44}$$

5.3.4 Desired Signal Distortion Index

Another useful way to measure the distortion of the desired signal is via the desired signal distortion index, which is defined as the MSE between the desired signal and its estimate, normalized by the power of the desired signal. We have the following definitions:

- the broadband desired signal distortion index,

$$\upsilon_{\mathrm{d}}(\mathbf{h}) = \frac{\int_f \phi_{X_1}(f)\left|\mathbf{h}^H(f)\boldsymbol{\gamma}^*_{X_1\mathbf{x}}(f) - 1\right|^2 df}{\int_f \phi_{X_1}(f)df} \tag{5.45}$$

- and the narrowband desired signal distortion index,

$$\upsilon_{\rm d}\left[\mathbf{h}(f)\right] = \frac{E\left[\left|X_{\rm fd}(f) - X_1(f)\right|^2\right]}{\phi_{X_1}(f)} = \left|\mathbf{h}^H(f)\boldsymbol{\gamma}^*_{X_1\mathbf{x}}(f) - 1\right|^2. \quad (5.46)$$

It is interesting to point out that the broadband desired signal distortion index is a linear combination of the narrowband desired signal distortion indices as the denominator is simply a scaling factor, i.e.,

$$\upsilon_{\rm d}(\mathbf{h}) = \frac{\int_f \phi_{X_1}(f)\upsilon_{\rm d}\left[\mathbf{h}(f)\right] df}{\int_f \phi_{X_1}(f) df}. \quad (5.47)$$

The distortionless constraint implies that $\upsilon_{\rm d}\left[\mathbf{h}(f)\right] = 0,\ \forall f$.

5.3.5 MSE Criterion

We define the error signal between the estimated and desired signals at frequency f as

$$\begin{aligned}\mathcal{E}(f) &= Z(f) - X_1(f) \\ &= \mathbf{h}^H(f)\mathbf{y}(f) - X_1(f) \\ &= X_{\rm fd}(f) + V_{\rm rn}(f) - X_1(f).\end{aligned} \quad (5.48)$$

This error can also be expressed as

$$\mathcal{E}(f) = \mathcal{E}_{\rm d}(f) + \mathcal{E}_{\rm n}(f), \quad (5.49)$$

where

$$\mathcal{E}_{\rm d}(f) = \left[\mathbf{h}^H(f)\boldsymbol{\gamma}^*_{X_1\mathbf{x}}(f) - 1\right] X_1(f) \quad (5.50)$$

is the desired signal distortion due to the complex filter and

$$\mathcal{E}_{\rm n}(f) = \mathbf{h}^H(f)\mathbf{v}(f) \quad (5.51)$$

represents the residual noise. The error signals $\mathcal{E}_{\rm d}(f)$ and $\mathcal{E}_{\rm n}(f)$ are incoherent. The narrowband MSE is then

$$\begin{aligned}J\left[\mathbf{h}(f)\right] &= E\left[\left|\mathcal{E}(f)\right|^2\right] \\ &= \phi_{X_1}(f) + \mathbf{h}^H(f)\boldsymbol{\Phi}_{\mathbf{y}}(f)\mathbf{h}(f) - \phi_{X_1}(f)\mathbf{h}^H(f)\boldsymbol{\gamma}^*_{X_1\mathbf{x}}(f) \\ &\quad - \phi_{X_1}(f)\boldsymbol{\gamma}^T_{X_1\mathbf{x}}(f)\mathbf{h}(f),\end{aligned} \quad (5.52)$$

which can be rewritten as

$$J\left[\mathbf{h}(f)\right] = E\left[\left|\mathcal{E}_{\mathrm{d}}(f)\right|^2\right] + E\left[\left|\mathcal{E}_{\mathrm{n}}(f)\right|^2\right] = J_{\mathrm{d}}\left[\mathbf{h}(f)\right] + J_{\mathrm{n}}\left[\mathbf{h}(f)\right], \tag{5.53}$$

where

$$J_{\mathrm{d}}\left[\mathbf{h}(f)\right] = \phi_{X_1}(f)\left|\mathbf{h}^H(f)\boldsymbol{\gamma}_{X_1\mathbf{x}}^*(f) - 1\right|^2 = \phi_{X_1}(f)\upsilon_{\mathrm{d}}\left[\mathbf{h}(f)\right] \tag{5.54}$$

and

$$J_{\mathrm{n}}\left[\mathbf{h}(f)\right] = \mathbf{h}^H(f)\boldsymbol{\Phi}_{\mathbf{v}}(f)\mathbf{h}(f) = \frac{\phi_{V_1}(f)}{\xi_{\mathrm{n}}\left[\mathbf{h}(f)\right]}. \tag{5.55}$$

We deduce that

$$\frac{J_{\mathrm{d}}\left[\mathbf{h}(f)\right]}{J_{\mathrm{n}}\left[\mathbf{h}(f)\right]} = \mathrm{iSNR}(f) \times \xi_{\mathrm{n}}\left[\mathbf{h}(f)\right] \times \upsilon_{\mathrm{d}}\left[\mathbf{h}(f)\right] = \mathrm{oSNR}\left[\mathbf{h}(f)\right] \times \xi_{\mathrm{d}}\left[\mathbf{h}(f)\right] \times \upsilon_{\mathrm{d}}\left[\mathbf{h}(f)\right]. \tag{5.56}$$

We observe how the narrowband MSEs are related to the narrowband performance measures.

Sometimes, it is also important to examine the MSE from the broadband point of view. We define the broadband MSE as

$$J(\mathbf{h}) = \int_f J\left[\mathbf{h}(f)\right] df = \int_f J_{\mathrm{d}}\left[\mathbf{h}(f)\right] df + \int_f J_{\mathrm{n}}\left[\mathbf{h}(f)\right] df = J_{\mathrm{d}}(\mathbf{h}) + J_{\mathrm{n}}(\mathbf{h}). \tag{5.57}$$

It is easy to show the relations between the broadband MSEs and the broadband performance measures:

$$\frac{J_{\mathrm{d}}(\mathbf{h})}{J_{\mathrm{n}}(\mathbf{h})} = \mathrm{iSNR} \times \xi_{\mathrm{n}}(\mathbf{h}) \times \upsilon_{\mathrm{d}}(\mathbf{h}) = \mathrm{oSNR}(\mathbf{h}) \times \xi_{\mathrm{d}}(\mathbf{h}) \times \upsilon_{\mathrm{d}}(\mathbf{h}). \tag{5.58}$$

5.4 Optimal Filters

After our discussion on the performance measures and different error criteria, we now have all the necessary tools to begin our search for reliable and practical multichannel noise reduction filters. We start with the maximum SNR filter. Interestingly, this is the only optimal filter that is not derived from an MSE point of view. Nevertheless, it is strongly related to the other optimal filters.

5.4.1 Maximum SNR

Let us rewrite the narrowband output SNR:

$$\text{oSNR}\left[\mathbf{h}(f)\right] = \frac{\phi_{X_1}(f)\mathbf{h}^H(f)\boldsymbol{\gamma}_{X_1\mathbf{x}}^*(f)\boldsymbol{\gamma}_{X_1\mathbf{x}}^T(f)\mathbf{h}(f)}{\mathbf{h}^H(f)\boldsymbol{\Phi}_\mathbf{v}(f)\mathbf{h}(f)}. \tag{5.59}$$

The maximum SNR filter, $\mathbf{h}_{\max}(f)$, is obtained by maximizing the output SNR as given above. In (5.59), we recognize the generalized Rayleigh quotient [5]. It is well known that this quotient is maximized with the maximum eigenvector of the matrix $\phi_{X_1}(f)\boldsymbol{\Phi}_\mathbf{v}^{-1}(f)\boldsymbol{\gamma}_{X_1\mathbf{x}}^*(f)\boldsymbol{\gamma}_{X_1\mathbf{x}}^T(f)$. Let us denote by $\lambda_1(f)$ the maximum eigenvalue corresponding to this maximum eigenvector. Since the rank of the mentioned matrix is equal to 1, we have

$$\begin{aligned}\lambda_1(f) &= \text{tr}\left[\phi_{X_1}(f)\boldsymbol{\Phi}_\mathbf{v}^{-1}(f)\boldsymbol{\gamma}_{X_1\mathbf{x}}^*(f)\boldsymbol{\gamma}_{X_1\mathbf{x}}^T(f)\right] \\ &= \phi_{X_1}(f)\boldsymbol{\gamma}_{X_1\mathbf{x}}^T(f)\boldsymbol{\Phi}_\mathbf{v}^{-1}(f)\boldsymbol{\gamma}_{X_1\mathbf{x}}^*(f).\end{aligned} \tag{5.60}$$

As a result,

$$\begin{aligned}\text{oSNR}\left[\mathbf{h}_{\max}(f)\right] &= \lambda_1(f) \\ &= \text{oSNR}_{\max}(f),\end{aligned} \tag{5.61}$$

which corresponds to the maximum possible narrowband output SNR.

Obviously, we also have

$$\mathbf{h}_{\max}(f) = \varsigma(f)\boldsymbol{\Phi}_\mathbf{v}^{-1}(f)\boldsymbol{\gamma}_{X_1\mathbf{x}}^*(f), \tag{5.62}$$

where $\varsigma(f)$ is an arbitrary frequency-dependent complex number different from zero. While this factor has no effect on the narrowband output SNR, it has on the broadband output SNR and on the desired signal distortion. In fact, all the filters (except for the LCMV) derived in the rest of this section are equivalent up to this complex factor. These filters also try to find the respective complex factors at each frequency depending on what we optimize. It is important to understand that while the maximum SNR filter maximizes the narrowband output SNR, it certainly does not maximize the broadband output SNR whose value depends quite a lot on $\varsigma(f)$.

Let us denote by $\text{oSNR}_{\max}^{(m)}(f)$ the maximum narrowband output SNR of an array with m sensors. By virtue of the inclusion principle [5] for the matrix $\phi_{X_1}(f)\boldsymbol{\Phi}_\mathbf{v}^{-1}(f)\boldsymbol{\gamma}_{X_1\mathbf{x}}^*(f)\boldsymbol{\gamma}_{X_1\mathbf{x}}^T(f)$, we have

$$\begin{aligned}&\text{oSNR}_{\max}^{(M)}(f) \geq \text{oSNR}_{\max}^{(M-1)}(f) \geq \cdots \geq \text{oSNR}_{\max}^{(2)}(f) \geq \\ &\quad \text{oSNR}_{\max}^{(1)}(f) = \text{iSNR}(f).\end{aligned} \tag{5.63}$$

This shows that by increasing the number of sensors, we necessarily increase the narrowband output SNR. If there is one sensor only, the narrowband output SNR cannot be improved as expected.

Example 5.4.1 Consider a ULA of M sensors, as shown in Figure 4.2. Suppose that a desired signal impinges on the ULA from the direction θ_0, and that an interference impinges on the ULA from the endfire direction ($\theta = 0°$). Assume that the desired signal is a harmonic pulse of T samples:

$$x(t) = \begin{cases} A \sin\left(2\pi f_0 t + \phi\right), & 0 \leq t \leq T-1 \\ 0, & t < 0,\ t \geq T \end{cases},$$

with fixed amplitude A and frequency f_0, and random phase ϕ, uniformly distributed on the interval from 0 to 2π. Assume that the interference $u(t)$ is white Gaussian noise, i.e., $u(t) \sim \mathcal{N}\left(0, \sigma_u^2\right)$, uncorrelated with $x(t)$. In addition, the sensors contain thermal white Gaussian noise, $w_m(t) \sim \mathcal{N}\left(0, \sigma_w^2\right)$, that are mutually uncorrelated. The desired signal needs to be recovered from the noisy received signals, $y_m(t) = x_m(t) + v_m(t)$, $m = 1, \ldots, M$, where $v_m(t) = u_m(t) + w_m(t)$, $m = 1, \ldots, M$ are the interference-plus-noise signals.

As in Example 4.2.4, we choose a sampling interval T_s that satisfies $T_s = \frac{d}{c}$. Hence, the desired signal at sensor m is a delayed version of the desired signal at the first sensor:

$$x_m(t) = x_1\left(t - \tau_m\right),$$

where

$$\begin{aligned}\tau_m &= \frac{(m-1)d\cos\theta_0}{cT_s} \\ &= (m-1)\cos\theta_0,\ m = 1, 2, \ldots, M\end{aligned}$$

is the relative time delay in samples (not necessarily an integer number) between the mth sensor and the first one. The frequency-domain representation of the desired signal received at the first sensor is given by

$$\begin{aligned}X_1(f) &= \sum_{t=-\infty}^{\infty} x_1(t)e^{\jmath 2\pi ft} \\ &= \frac{A}{2\jmath}e^{\jmath\phi+\jmath\pi(f+f_0)(T-1)}D_T\left[\pi\left(f+f_0\right)\right] \\ &\quad + \frac{A}{2\jmath}e^{-\jmath\phi+\jmath\pi(f-f_0)(T-1)}D_T\left[\pi\left(f-f_0\right)\right],\end{aligned}$$

where

$$D_T(x) = \frac{\sin(Tx)}{\sin(x)}.$$

Therefore, the variance of $X_1(f)$ is

$$\phi_{X_1}(f) = \frac{A^2}{4} D_T^2 \left[\pi \left(f + f_0\right)\right] + \frac{A^2}{4} D_T^2 \left[\pi \left(f - f_0\right)\right].$$

Using the vector notation (5.3), we have

$$\mathbf{x}(f) = \mathbf{d}(f) X_1(f),$$
$$\mathbf{\Phi}_{\mathbf{x}}(f) = \phi_{X_1}(f) \mathbf{d}(f) \mathbf{d}^H(f),$$

where

$$\mathbf{d}(f) = \begin{bmatrix} 1 & e^{-\jmath 2\pi f \tau_2} & e^{-\jmath 2\pi f \tau_3} & \cdots & e^{-\jmath 2\pi f \tau_M} \end{bmatrix}^T.$$

The interference signal at sensor m is also a delayed version of the interference signal at the first sensor:

$$u_m(t) = u_1(t - m + 1).$$

The frequency-domain representation of the interference signal received at the first sensor is given by

$$U_1(f) = \sum_{t=0}^{T-1} u_1(t) e^{\jmath 2\pi f t}.$$

Hence, the variance of $U_1(f)$ is $\phi_{U_1}(f) = T\sigma_u^2$. Using the vector notation (5.13), we have

$$\mathbf{v}(f) = \boldsymbol{\gamma}_{U_1\mathbf{u}}^*(f) U_1(f) + \mathbf{w}(f),$$
$$\mathbf{\Phi}_{\mathbf{v}}(f) = \phi_{U_1}(f) \boldsymbol{\gamma}_{U_1\mathbf{u}}^*(f) \boldsymbol{\gamma}_{U_1\mathbf{u}}^T(f) + T\sigma_w^2 \mathbf{I}_M,$$

where

$$\boldsymbol{\gamma}_{U_1\mathbf{u}}^*(f) = \begin{bmatrix} 1 & e^{-\jmath 2\pi f} & e^{-\jmath 2\pi f 2} & \cdots & e^{-\jmath 2\pi f (M-1)} \end{bmatrix}^T$$

and $\mathbf{I}_M$ is the $M \times M$ identity matrix.

The narrowband and broadband input SNRs are, respectively,

$$\begin{aligned} \mathrm{iSNR}(f) &= \frac{\phi_{X_1}(f)}{\phi_{V_1}(f)} \\ &= \frac{A^2}{4T\left(\sigma_u^2 + \sigma_w^2\right)} D_T^2 \left[\pi \left(f + f_0\right)\right] \\ &\quad + \frac{A^2}{4T\left(\sigma_u^2 + \sigma_w^2\right)} D_T^2 \left[\pi \left(f - f_0\right)\right] \end{aligned}$$

and

$$\begin{aligned}\text{iSNR} &= \frac{\int_f \phi_{X_1}(f)df}{\int_f \phi_{V_1}(f)df} \\ &= \frac{\sum_t E\left[|x_1(t)|^2\right]}{\sum_t E\left[|v_1(t)|^2\right]} \\ &= \frac{A^2}{2\left(\sigma_u^2 + \sigma_w^2\right)},\end{aligned}$$

where we have used Parseval's identity. The maximum SNR filter, $\mathbf{h}_{\text{max}}(f)$, is obtained from (5.62). Using (5.61), we can write the narrowband gain in SNR as

$$\begin{aligned}\mathcal{G}\left[\mathbf{h}_{\text{max}}(f)\right] &= \frac{\text{oSNR}\left[\mathbf{h}_{\text{max}}(f)\right]}{\text{iSNR}(f)} \\ &= \mathbf{d}^H(f)\left[\frac{\sigma_u^2}{\sigma_u^2+\sigma_w^2}\boldsymbol{\gamma}_{U_1\mathbf{u}}^*(f)\boldsymbol{\gamma}_{U_1\mathbf{u}}^T(f) + \frac{\sigma_w^2}{\sigma_u^2+\sigma_w^2}\mathbf{I}_M\right]^{-1}\mathbf{d}(f).\end{aligned}$$

To demonstrate the performance of the maximum SNR filter, we choose $\sigma_w^2 = 0.01\sigma_u^2$. Figure 5.2 shows the effect of the number of sensors, M, on the narrowband gain in SNR, $\mathcal{G}\left[\mathbf{h}_{\text{max}}(f)\right]$, for different incidence angles of the desired signal and different frequencies. For a single sensor ($M = 1$), there is no narrowband gain in SNR. As the number of sensors increases, the narrowband gain in SNR increases. ■

5.4.2 Wiener

The Wiener filter is found by minimizing the narrowband MSE, $J\left[\mathbf{h}(f)\right]$ [eq. (5.52)]. We get

$$\mathbf{h}_{\text{W}}(f) = \phi_{X_1}(f)\boldsymbol{\Phi}_{\mathbf{y}}^{-1}(f)\boldsymbol{\gamma}_{X_1\mathbf{x}}^*(f). \tag{5.64}$$

Let

$$\boldsymbol{\Gamma}_{\mathbf{y}}(f) = \frac{\boldsymbol{\Phi}_{\mathbf{y}}(f)}{\phi_{Y_1}(f)} \tag{5.65}$$

be the pseudo-coherence matrix of the observations, we can rewrite (5.64) as

$$\begin{aligned}\mathbf{h}_{\text{W}}(f) &= \frac{\text{iSNR}(f)}{1+\text{iSNR}(f)}\boldsymbol{\Gamma}_{\mathbf{y}}^{-1}(f)\boldsymbol{\gamma}_{X_1\mathbf{x}}^*(f) \\ &= H_{\text{W}}(f)\boldsymbol{\Gamma}_{\mathbf{y}}^{-1}(f)\boldsymbol{\gamma}_{X_1\mathbf{x}}^*(f),\end{aligned} \tag{5.66}$$

where

$$H_{\text{W}}(f) = \frac{\text{iSNR}(f)}{1+\text{iSNR}(f)} \tag{5.67}$$

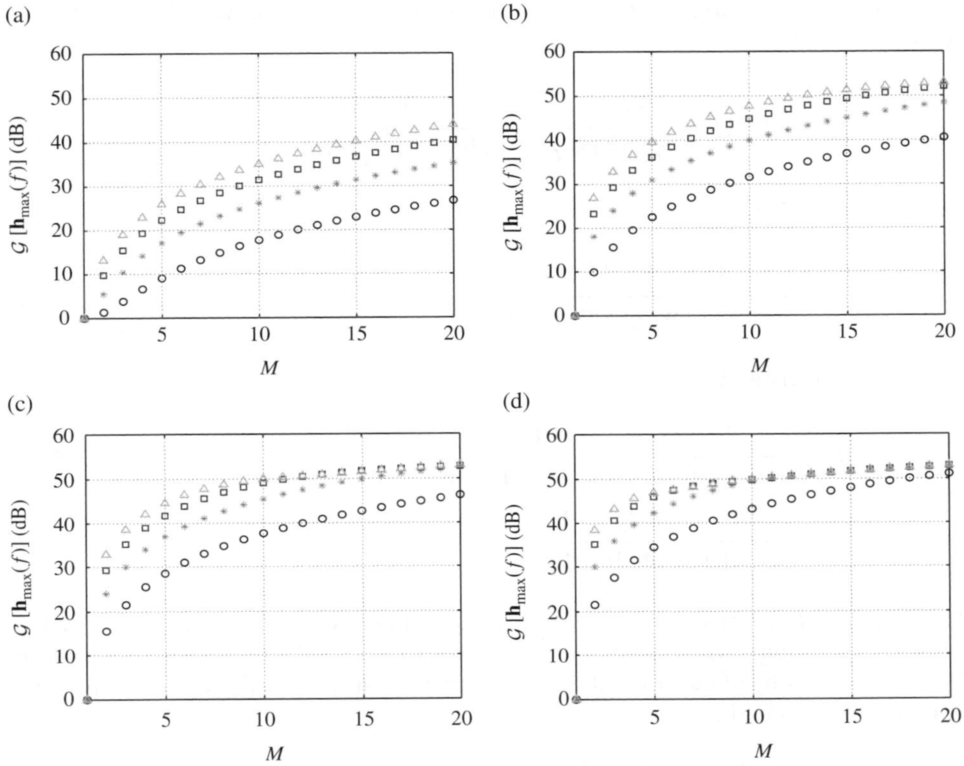

Figure 5.2 Narrowband gain in SNR of the maximum SNR filter as a function of the number of sensors, M, for different incidence angles of the desired signal and different frequencies: $\theta_0 = 30°$ (circles), $\theta_0 = 50°$ (asterisks), $\theta_0 = 70°$ (squares), and $\theta_0 = 90°$ (triangles); (a) $f = 0.01$, (b) $f = 0.05$, (c) $f = 0.1$, and (d) $f = 0.2$.

is the (single-channel) Wiener gain and $\mathbf{\Gamma}_{\mathbf{y}}^{-1}(f)\boldsymbol{\gamma}_{X_1\mathbf{x}}^{*}(f)$ is the spatial information vector. The decomposition in (5.66) is very useful; it shows separately the influence of the spectral and spatial processing on multichannel signal enhancement.

We can express (5.64) differently, i.e.,

$$\begin{aligned}\mathbf{h}_{\mathrm{W}}(f) &= \mathbf{\Phi}_{\mathbf{y}}^{-1}(f)E\left[\mathbf{x}(f)X_1^{*}(f)\right] \\ &= \mathbf{\Phi}_{\mathbf{y}}^{-1}(f)\mathbf{\Phi}_{\mathbf{x}}(f)\mathbf{i}_{\mathrm{i}} \\ &= \left[\mathbf{I}_M - \mathbf{\Phi}_{\mathbf{y}}^{-1}(f)\mathbf{\Phi}_{\mathbf{v}}(f)\right]\mathbf{i}_{\mathrm{i}}.\end{aligned} \tag{5.68}$$

In this form, the Wiener filter relies on the second-order statistics of the observation and noise signals.

We can write the general form of the Wiener filter in another way that will make it easier to compare to other optimal filters. We know that

$$\mathbf{\Phi}_{\mathbf{y}}(f) = \phi_{X_1}(f)\boldsymbol{\gamma}_{X_1\mathbf{x}}^{*}(f)\boldsymbol{\gamma}_{X_1\mathbf{x}}^{T}(f) + \mathbf{\Phi}_{\mathbf{v}}(f). \tag{5.69}$$

Determining the inverse of $\mathbf{\Phi}_{\mathbf{y}}(f)$ from the previous expression with the Woodbury's identity, we get

$$\mathbf{\Phi}_{\mathbf{y}}^{-1}(f) = \mathbf{\Phi}_{\mathbf{v}}^{-1}(f) - \frac{\mathbf{\Phi}_{\mathbf{v}}^{-1}(f)\boldsymbol{\gamma}_{X_1\mathbf{x}}^{*}(f)\boldsymbol{\gamma}_{X_1\mathbf{x}}^{T}(f)\mathbf{\Phi}_{\mathbf{v}}^{-1}(f)}{\phi_{X_1}^{-1}(f) + \boldsymbol{\gamma}_{X_1\mathbf{x}}^{T}(f)\mathbf{\Phi}_{\mathbf{v}}^{-1}(f)\boldsymbol{\gamma}_{X_1\mathbf{x}}^{*}(f)}. \tag{5.70}$$

Substituting (5.70) into (5.64) gives

$$\mathbf{h}_{\mathrm{W}}(f) = \frac{\phi_{X_1}(f)\mathbf{\Phi}_{\mathbf{v}}^{-1}(f)\boldsymbol{\gamma}_{X_1\mathbf{x}}^{*}(f)}{1 + \phi_{X_1}(f)\boldsymbol{\gamma}_{X_1\mathbf{x}}^{T}(f)\mathbf{\Phi}_{\mathbf{v}}^{-1}(f)\boldsymbol{\gamma}_{X_1\mathbf{x}}^{*}(f)}, \tag{5.71}$$

that we can rewrite as

$$\begin{aligned}\mathbf{h}_{\mathrm{W}}(f) &= \frac{\mathbf{\Phi}_{\mathbf{v}}^{-1}(f)\left[\mathbf{\Phi}_{\mathbf{y}}(f) - \mathbf{\Phi}_{\mathbf{v}}(f)\right]}{1 + \mathrm{tr}\left\{\mathbf{\Phi}_{\mathbf{v}}^{-1}(f)\left[\mathbf{\Phi}_{\mathbf{y}}(f) - \mathbf{\Phi}_{\mathbf{v}}(f)\right]\right\}}\mathbf{i}_{\mathrm{i}} \\ &= \frac{\mathbf{\Phi}_{\mathbf{v}}^{-1}(f)\mathbf{\Phi}_{\mathbf{y}}(f) - \mathbf{I}_M}{1 - M + \mathrm{tr}\left[\mathbf{\Phi}_{\mathbf{v}}^{-1}(f)\mathbf{\Phi}_{\mathbf{y}}(f)\right]}\mathbf{i}_{\mathrm{i}}.\end{aligned} \tag{5.72}$$

Comparing (5.68) with (5.72), we see that in the former, we invert the correlation matrix of the observations, while in the latter, we invert the correlation matrix of the noise. We can express $\mathbf{h}_{\mathrm{W}}(f)$ as a function of the narrowband input SNR and the pseudo-coherence matrices, i.e.,

$$\mathbf{h}_{\mathrm{W}}(f) = \frac{\left[1 + \mathrm{iSNR}(f)\right]\mathbf{\Gamma}_{\mathbf{v}}^{-1}(f)\mathbf{\Gamma}_{\mathbf{y}}(f) - \mathbf{I}_M}{1 - M + \left[1 + \mathrm{iSNR}(f)\right]\mathrm{tr}\left[\mathbf{\Gamma}_{\mathbf{v}}^{-1}(f)\mathbf{\Gamma}_{\mathbf{y}}(f)\right]}\mathbf{i}_{\mathrm{i}}, \tag{5.73}$$

where

$$\mathbf{\Gamma}_{\mathbf{v}}(f) = \frac{\mathbf{\Phi}_{\mathbf{v}}(f)}{\phi_{V_1}(f)}. \tag{5.74}$$

From (5.71), we deduce that the narrowband output SNR is

$$\begin{aligned}\mathrm{oSNR}\left[\mathbf{h}_{\mathrm{W}}(f)\right] &= \lambda_1(f) \\ &= \mathrm{tr}\left[\mathbf{\Phi}_{\mathbf{v}}^{-1}(f)\mathbf{\Phi}_{\mathbf{y}}(f)\right] - M\end{aligned} \tag{5.75}$$

and, obviously,

$$\mathrm{oSNR}\left[\mathbf{h}_{\mathrm{W}}(f)\right] \geq \mathrm{iSNR}(f), \tag{5.76}$$

since the Wiener filter maximizes the narrowband output SNR.

The desired signal distortion indices are

$$\upsilon_{\mathrm{d}}\left[\mathbf{h}_{\mathrm{W}}(f)\right] = \frac{1}{\left[1 + \lambda_1(f)\right]^2}, \tag{5.77}$$

$$v_{\mathrm{d}}\left(\mathbf{h}_{\mathrm{W}}\right) = \frac{\int_f \phi_{X_1}(f)\left[1+\lambda_1(f)\right]^{-2} df}{\int_f \phi_{X_1}(f)df}. \tag{5.78}$$

The higher the value of $\lambda_1(f)$ (and/or the number of sensors), the less the desired signal is distorted.

It is also easy to find the noise reduction factors:

$$\xi_{\mathrm{n}}\left[\mathbf{h}_{\mathrm{W}}(f)\right] = \frac{\left[1+\lambda_1(f)\right]^2}{\mathrm{iSNR}(f)\times\lambda_1(f)}, \tag{5.79}$$

$$\xi_{\mathrm{n}}\left(\mathbf{h}_{\mathrm{W}}\right) = \frac{\int_f \phi_{X_1}(f)\mathrm{iSNR}^{-1}(f)df}{\int_f \phi_{X_1}(f)\lambda_1(f)\left[1+\lambda_1(f)\right]^{-2} df}, \tag{5.80}$$

and the desired signal reduction factors:

$$\xi_{\mathrm{d}}\left[\mathbf{h}_{\mathrm{W}}(f)\right] = \frac{\left[1+\lambda_1(f)\right]^2}{\lambda_1^2(f)}, \tag{5.81}$$

$$\xi_{\mathrm{d}}\left(\mathbf{h}_{\mathrm{W}}\right) = \frac{\int_f \phi_{X_1}(f)df}{\int_f \phi_{X_1}(f)\lambda_1^2(f)\left[1+\lambda_1(f)\right]^{-2} df}. \tag{5.82}$$

The broadband output SNR of the Wiener filter is

$$\mathrm{oSNR}\left(\mathbf{h}_{\mathrm{W}}\right) = \frac{\displaystyle\int_f \phi_{X_1}(f)\frac{\lambda_1^2(f)}{\left[1+\lambda_1(f)\right]^2}df}{\displaystyle\int_f \phi_{X_1}(f)\frac{\lambda_1(f)}{\left[1+\lambda_1(f)\right]^2}df}. \tag{5.83}$$

Property 5.4.1 With the frequency-domain multichannel Wiener filter given in (5.64), the broadband output SNR is always greater than or equal to the broadband input SNR, i.e., $\mathrm{oSNR}\left(\mathbf{h}_{\mathrm{W}}\right) \geq \mathrm{iSNR}$.

Proof. See Subsection 5.4.4.

It is interesting to see that the two filters $\mathbf{h}_{\mathrm{W}}(f)$ and $\mathbf{h}_{\max}(f)$ differ only by a real-valued factor. Indeed, taking

$$\varsigma(f) = \frac{\phi_{X_1}(f)}{1+\lambda_1(f)} \tag{5.84}$$

in (5.62) (maximum SNR filter), we find (5.71) (Wiener filter). ■

Example 5.4.2 Returning to Example 5.4.1, we now employ the Wiener filter, $\mathbf{h}_{\mathrm{W}}(f)$, given in (5.64). To demonstrate the performance of the Wiener filter, we choose $A = 0.5$, $f_0 = 0.1$, $T = 500$, $\theta_0 = 70°$, and $\sigma_w^2 = 0.01\sigma_u^2$. Figure 5.3 shows plots of the broadband

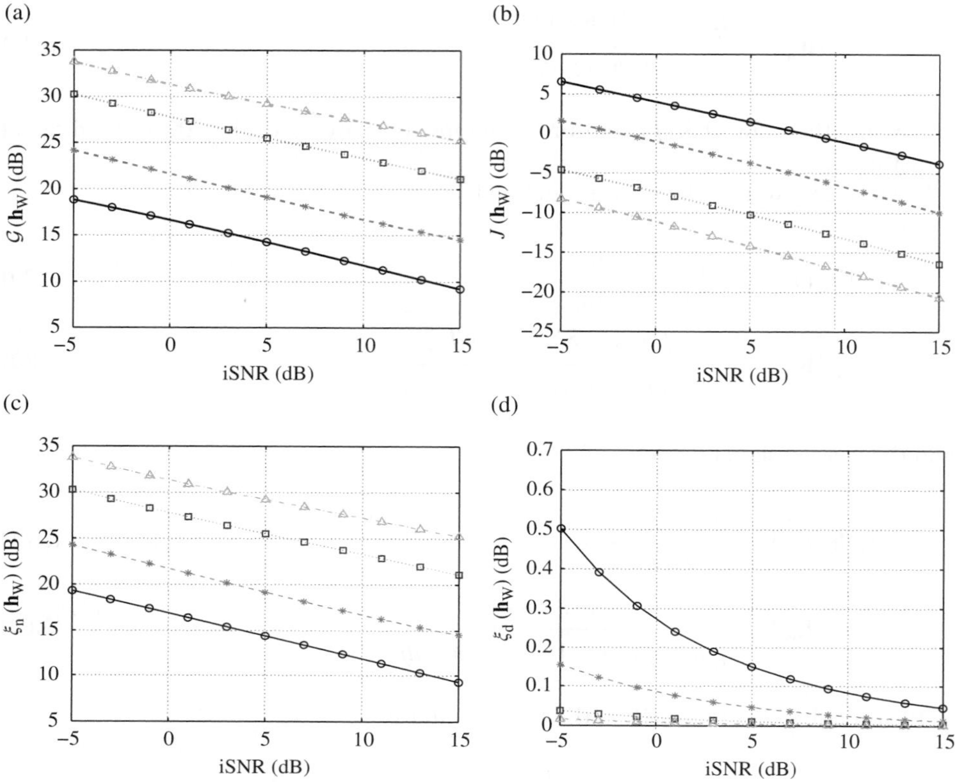

Figure 5.3 (a) The broadband gain in SNR, (b) the broadband MSE, (c) the broadband noise reduction factor, and (d) the broadband desired signal reduction factor of the Wiener filter as a function of the broadband input SNR, for different numbers of sensors, M: $M = 1$ (solid line with circles), $M = 2$ (dashed line with asterisks), $M = 5$ (dotted line with squares), and $M = 10$ (dash-dot line with triangles).

gain in SNR, $\mathcal{G}\left(\mathbf{h}_W\right)$, the broadband MSE, $J\left(\mathbf{h}_W\right)$, the broadband noise reduction factor, $\xi_n\left(\mathbf{h}_W\right)$, and the broadband desired signal reduction factor, $\xi_d\left(\mathbf{h}_W\right)$, as a function of the broadband input SNR, for different numbers of sensors. For a given broadband input SNR, as the number of sensors increases, the broadband gain in SNR and the broadband noise reduction factor increase, while the broadband MMSE and the broadband desired signal reduction factor decrease.

Figure 5.4 shows a realization of the frequency-domain noise corrupted signal received at the first sensor, $|Y_1(f)|$, and the error signals $\left|\mathcal{E}\left(f\right)\right| = |Z(f) - X_1(f)|$ for iSNR $= -5$ dB and different numbers of sensors. Figure 5.5 shows the corresponding time-domain observation signal at the first sensor, $y_1(t)$, and the time-domain estimated signals, $z(t)$. Obviously, as the number of sensors increases, the Wiener filter better enhances the desired signal. ■

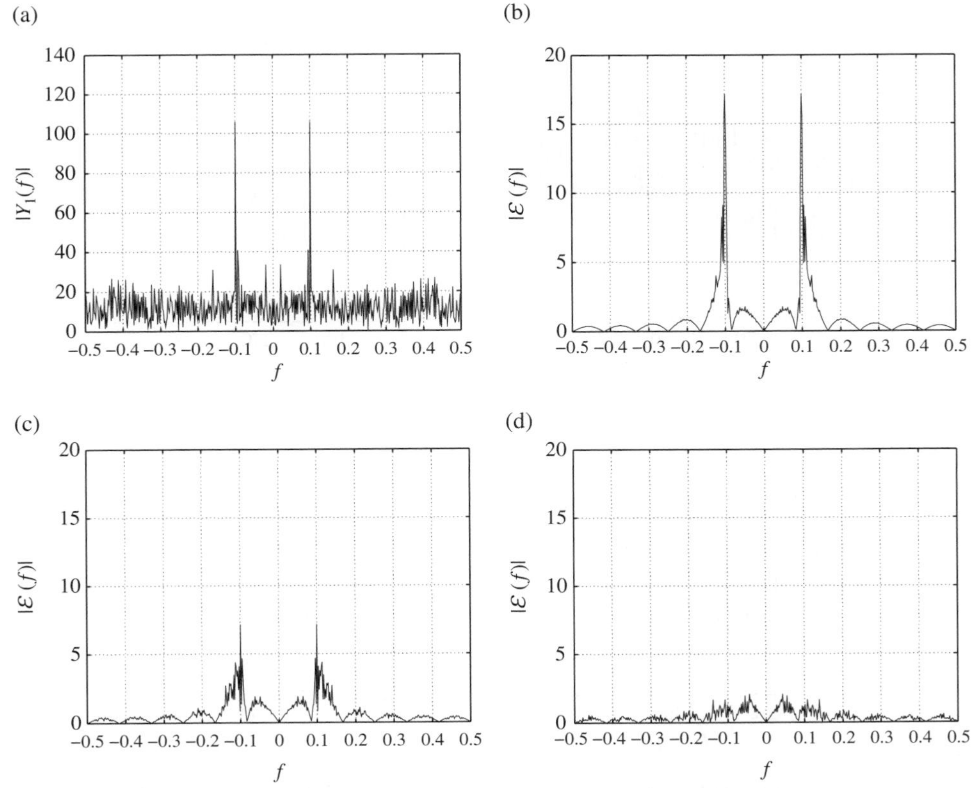

Figure 5.4 Example of frequency-domain noise corrupted and error signals of the Wiener filter for different numbers of sensors, M: (a) magnitude of the frequency-domain observation signal at the first sensor, $|Y_1(f)|$ (iSNR = −5 dB), and magnitude of the frequency-domain error signals, $|\mathcal{E}(f)| = |Z(f) - X_1(f)|$ for (b) $M = 1$ [oSNR $(\mathbf{h}_W) = 14.2$ dB], (c) $M = 2$ [oSNR $(\mathbf{h}_W) = 19.2$ dB], and (d) $M = 5$ [oSNR $(\mathbf{h}_W) = 25.5$ dB].

5.4.3 MVDR

The well-known MVDR filter proposed by Capon [6, 7] is easily derived by minimizing the narrowband MSE of the residual noise, $J_n\left[\mathbf{h}(f)\right]$, with the constraint that the desired signal is not distorted. Mathematically, this is equivalent to

$$\min_{\mathbf{h}(f)} \mathbf{h}^H(f)\mathbf{\Phi}_{\mathbf{v}}(f)\mathbf{h}(f) \quad \text{subject to} \quad \mathbf{h}^H(f)\boldsymbol{\gamma}^*_{X_1\mathbf{x}}(f) = 1, \tag{5.85}$$

for which the solution is

$$\mathbf{h}_{\text{MVDR}}(f) = \frac{\mathbf{\Phi}_{\mathbf{v}}^{-1}(f)\boldsymbol{\gamma}^*_{X_1\mathbf{x}}(f)}{\boldsymbol{\gamma}^T_{X_1\mathbf{x}}(f)\mathbf{\Phi}_{\mathbf{v}}^{-1}(f)\boldsymbol{\gamma}^*_{X_1\mathbf{x}}(f)}. \tag{5.86}$$

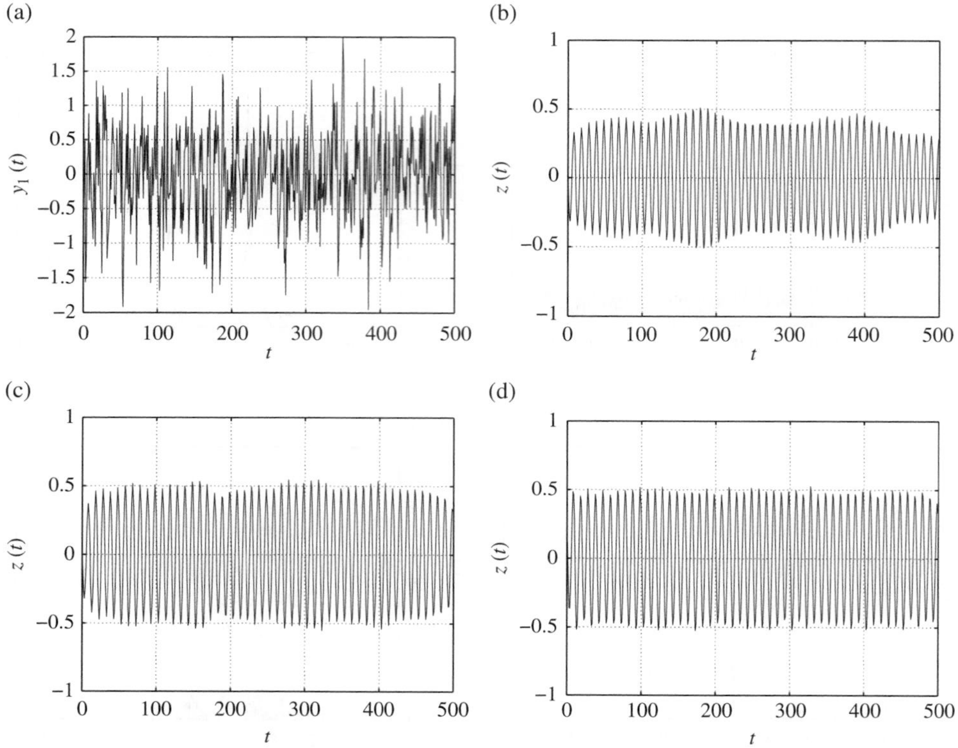

Figure 5.5 Example of time-domain noise corrupted and Wiener filtered sinusoidal signals for different numbers of sensors, M: (a) time-domain observation signal at the first sensor, $y_1(t)$ (iSNR = −5 dB), and time-domain estimated signal, $z(t)$, for (b) $M = 1$ [oSNR $\left(\mathbf{h}_\text{W}\right) = 14.2$ dB], (c) $M = 2$ [oSNR $\left(\mathbf{h}_\text{W}\right) = 19.2$ dB], and (d) $M = 5$ [oSNR $\left(\mathbf{h}_\text{W}\right) = 25.5$ dB].

Using the fact that $\mathbf{\Phi}_\mathbf{x}(f) = \phi_{X_1}(f)\boldsymbol{\gamma}^*_{X_1\mathbf{x}}(f)\boldsymbol{\gamma}^T_{X_1\mathbf{x}}(f)$, the explicit dependence of the above filter on the steering vector is eliminated to obtain the following forms:

$$\begin{aligned}\mathbf{h}_\text{MVDR}(f) &= \frac{\mathbf{\Phi}_\mathbf{v}^{-1}(f)\mathbf{\Phi}_\mathbf{x}(f)}{\lambda_1(f)}\mathbf{i}_\text{i} \\ &= \frac{\mathbf{\Phi}_\mathbf{v}^{-1}(f)\mathbf{\Phi}_\mathbf{y}(f) - \mathbf{I}_M}{\text{tr}\left[\mathbf{\Phi}_\mathbf{v}^{-1}(f)\mathbf{\Phi}_\mathbf{y}(f)\right] - M}\mathbf{i}_\text{i} \\ &= \frac{\left[1 + \text{iSNR}(f)\right]\mathbf{\Gamma}_\mathbf{v}^{-1}(f)\mathbf{\Gamma}_\mathbf{y}(f) - \mathbf{I}_M}{\left[1 + \text{iSNR}(f)\right]\text{tr}\left[\mathbf{\Gamma}_\mathbf{v}^{-1}(f)\mathbf{\Gamma}_\mathbf{y}(f)\right] - M}\mathbf{i}_\text{i}.\end{aligned} \tag{5.87}$$

Alternatively, we can also write the MVDR as

$$\mathbf{h}_\text{MVDR}(f) = \frac{\mathbf{\Phi}_\mathbf{y}^{-1}(f)\boldsymbol{\gamma}^*_{X_1\mathbf{x}}(f)}{\boldsymbol{\gamma}^T_{X_1\mathbf{x}}(f)\mathbf{\Phi}_\mathbf{y}^{-1}(f)\boldsymbol{\gamma}^*_{X_1\mathbf{x}}(f)} \tag{5.88}$$

$$= \frac{\mathbf{\Gamma}_{\mathbf{y}}^{-1}(f)\boldsymbol{\gamma}_{X_1\mathbf{x}}^*(f)}{\boldsymbol{\gamma}_{X_1\mathbf{x}}^T(f)\mathbf{\Gamma}_{\mathbf{y}}^{-1}(f)\boldsymbol{\gamma}_{X_1\mathbf{x}}^*(f)}.$$

Taking

$$\varsigma(f) = \frac{\phi_{X_1}(f)}{\lambda_1(f)} \tag{5.89}$$

in (5.62) (maximum SNR filter), we find (5.86) (MVDR filter), showing how the maximum SNR and MVDR filters are equivalent up to a real-valued factor.

The Wiener and MVDR filters are simply related as follows:

$$\mathbf{h}_{\mathrm{W}}(f) = C_{\mathrm{W}}(f)\mathbf{h}_{\mathrm{MVDR}}(f), \tag{5.90}$$

where

$$\begin{aligned} C_{\mathrm{W}}(f) &= \mathbf{h}_{\mathrm{W}}^H(f)\boldsymbol{\gamma}_{X_1\mathbf{x}}^*(f) \\ &= \frac{\lambda_1(f)}{1+\lambda_1(f)} \end{aligned} \tag{5.91}$$

can be seen as a single-channel frequency-domain Wiener gain. In fact, any filter of the form:

$$\mathbf{h}(f) = C(f)\mathbf{h}_{\mathrm{MVDR}}(f), \tag{5.92}$$

where $C(f)$ is a real number, with $0 < C(f) < 1$, removes more noise than the MVDR filter at the price of some desired signal distortion, which is

$$\xi_{\mathrm{d}}\left[\mathbf{h}(f)\right] = \frac{1}{C^2(f)} \tag{5.93}$$

or

$$\upsilon_{\mathrm{d}}\left[\mathbf{h}(f)\right] = \left[C(f)-1\right]^2. \tag{5.94}$$

It can be verified that we always have

$$\mathrm{oSNR}\left[\mathbf{h}_{\mathrm{MVDR}}(f)\right] = \mathrm{oSNR}\left[\mathbf{h}_{\mathrm{W}}(f)\right], \tag{5.95}$$

$$\upsilon_{\mathrm{d}}\left[\mathbf{h}_{\mathrm{MVDR}}(f)\right] = 0, \tag{5.96}$$

$$\xi_{\mathrm{d}}\left[\mathbf{h}_{\mathrm{MVDR}}(f)\right] = 1, \tag{5.97}$$

and

$$\xi_{\mathrm{n}}\left[\mathbf{h}_{\mathrm{MVDR}}(f)\right] \leq \xi_{\mathrm{n}}\left[\mathbf{h}_{\mathrm{W}}(f)\right], \tag{5.98}$$

$$\xi_{\mathrm{n}}\left(\mathbf{h}_{\mathrm{MVDR}}\right) \leq \xi_{\mathrm{n}}\left(\mathbf{h}_{\mathrm{W}}\right). \tag{5.99}$$

The MVDR filter rejects the maximum level of noise allowable without distorting the desired signal at each frequency.

While the narrowband output SNRs of the Wiener and MVDR are strictly equal, their broadband output SNRs are not. The broadband output SNR of the MVDR is

$$\text{oSNR}\left(\mathbf{h}_{\text{MVDR}}\right) = \frac{\int_f \phi_{X_1}(f)df}{\int_f \phi_{X_1}(f)\lambda_1^{-1}(f)df} \tag{5.100}$$

and

$$\text{oSNR}\left(\mathbf{h}_{\text{MVDR}}\right) \leq \text{oSNR}\left(\mathbf{h}_{\text{W}}\right). \tag{5.101}$$

Property 5.4.2 With the frequency-domain MVDR filter given in (5.86), the broadband output SNR is always greater than or equal to the broadband input SNR, i.e., $\text{oSNR}\left(\mathbf{h}_{\text{MVDR}}\right) \geq \text{iSNR}$.

Proof. See Subsection 5.4.4. ■

Example 5.4.3 Returning to Example 5.4.2, we now employ the MVDR filter, $\mathbf{h}_{\text{MVDR}}(f)$, given in (5.86). Figure 5.6 shows plots of the broadband gain in SNR, $\mathcal{G}\left(\mathbf{h}_{\text{MVDR}}\right)$, the broadband MSE, $J\left(\mathbf{h}_{\text{MVDR}}\right)$, the broadband noise reduction factor, $\xi_{\text{n}}\left(\mathbf{h}_{\text{MVDR}}\right)$, and the broadband desired signal reduction factor, $\xi_{\text{d}}\left(\mathbf{h}_{\text{MVDR}}\right)$, as a function of the broadband input SNR, for different numbers of sensors. For a given broadband input SNR, as the number of sensors increases, the broadband gain in SNR and the broadband noise reduction factor increase, while the broadband MSE decreases. ■

5.4.4 Tradeoff

As we have learned from the previous subsections, not much flexibility is associated with the Wiener and MVDR filters in the sense that we do not know in advance by how much the narrowband output SNR will be improved. However, in many practical situations, we wish to control the compromise between noise reduction and desired signal distortion, and one possible way to do this is via the so-called tradeoff filter.

In the tradeoff approach, we minimize the narrowband desired signal distortion index with the constraint that the narrowband noise reduction factor is equal to a positive value that is greater than 1. Mathematically, this is equivalent to

$$\min_{\mathbf{h}(f)} J_{\text{d}}\left[\mathbf{h}(f)\right] \quad \text{subject to} \quad J_{\text{n}}\left[\mathbf{h}(f)\right] = \aleph\phi_{V_1}(f), \tag{5.102}$$

where $0 < \aleph < 1$ to ensure that we get some noise reduction. By using a Lagrange multiplier, $\mu > 0$, to adjoin the constraint to the cost function, we easily deduce the tradeoff filter:

$$\begin{aligned}\mathbf{h}_{\text{T},\mu}(f) &= \phi_{X_1}(f)\left[\mathbf{\Phi}_{\mathbf{x}}(f) + \mu\mathbf{\Phi}_{\mathbf{v}}(f)\right]^{-1}\boldsymbol{\gamma}_{X_1\mathbf{x}}^*(f) \\ &= \frac{\phi_{X_1}(f)\mathbf{\Phi}_{\mathbf{v}}^{-1}(f)\boldsymbol{\gamma}_{X_1\mathbf{x}}^*(f)}{\mu + \phi_{X_1}(f)\boldsymbol{\gamma}_{X_1\mathbf{x}}^T(f)\mathbf{\Phi}_{\mathbf{v}}^{-1}(f)\boldsymbol{\gamma}_{X_1\mathbf{x}}^*(f)}\end{aligned} \tag{5.103}$$

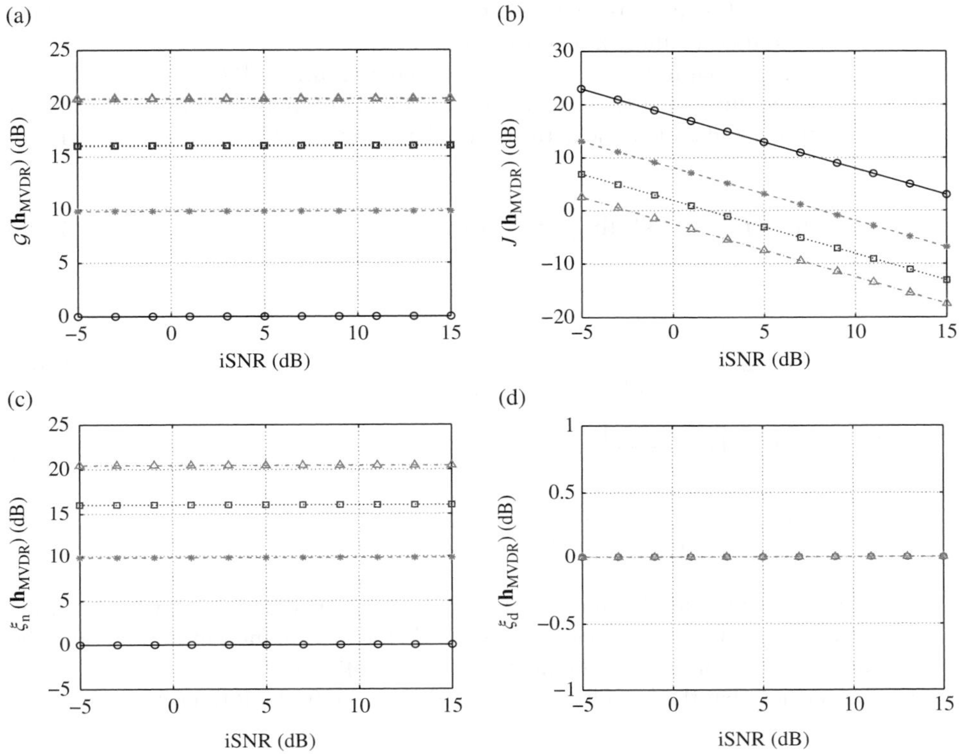

Figure 5.6 (a) The broadband gain in SNR, (b) the broadband MSE, (c) the broadband noise reduction factor, and (d) the broadband desired signal reduction factor of the MVDR filter as a function of the broadband input SNR, for different numbers of sensors, M: $M = 1$ (solid line with circles), $M = 2$ (dashed line with asterisks), $M = 5$ (dotted line with squares), and $M = 10$ (dash-dot line with triangles).

$$= \frac{\mathbf{\Phi}_{\mathbf{v}}^{-1}(f)\mathbf{\Phi}_{\mathbf{y}}(f) - \mathbf{I}_M}{\mu - M + \mathrm{tr}\left[\mathbf{\Phi}_{\mathbf{v}}^{-1}(f)\mathbf{\Phi}_{\mathbf{y}}(f)\right]}\mathbf{i}_{\mathrm{i}},$$

where the Lagrange multiplier, μ, satisfies

$$J_{\mathrm{n}}\left[\mathbf{h}_{\mathrm{T},\mu}(f)\right] = \aleph\phi_{V_1}(f). \tag{5.104}$$

However, in practice it is not easy to determine the optimal μ. Therefore, when this parameter is chosen in a heuristic way, we can see that for

- $\mu = 1$, $\mathbf{h}_{\mathrm{T},1}(f) = \mathbf{h}_{\mathrm{W}}(f)$, which is the Wiener filter;
- $\mu = 0$, $\mathbf{h}_{\mathrm{T},0}(f) = \mathbf{h}_{\mathrm{MVDR}}(f)$, which is the MVDR filter;
- $\mu > 1$, results in a filter with low residual noise at the expense of high desired signal distortion (as compared to Wiener); and
- $\mu < 1$, results in a filter with high residual noise and low desired signal distortion (as compared to Wiener).

Note that the MVDR cannot be derived from the first line of (5.103) since by taking $\mu = 0$, we have to invert a matrix that is not full rank.

It can be observed that the tradeoff, Wiener, and maximum SNR filters are equivalent up to a real-valued number. As a result, the narrowband output SNR of the tradeoff filter is independent of μ and is identical to the narrowband output SNR of the maximum SNR filter, i.e.,

$$\text{oSNR}\left[\mathbf{h}_{\text{T},\mu}(f)\right] = \text{oSNR}\left[\mathbf{h}_{\max}(f)\right], \ \forall \mu \geq 0. \tag{5.105}$$

We have

$$v_{\text{d}}\left[\mathbf{h}_{\text{T},\mu}(f)\right] = \left[\frac{\mu}{\mu + \lambda_1(f)}\right]^2, \tag{5.106}$$

$$\xi_{\text{d}}\left[\mathbf{h}_{\text{T},\mu}(f)\right] = \left[1 + \frac{\mu}{\lambda_1(f)}\right]^2, \tag{5.107}$$

$$\xi_{\text{n}}\left[\mathbf{h}_{\text{T},\mu}(f)\right] = \frac{\left[\mu + \lambda_1(f)\right]^2}{\text{iSNR}(f) \times \lambda_1(f)}. \tag{5.108}$$

The tradeoff filter is useful from several perspectives since it encompasses both the Wiener and MVDR filters. It is then useful to study the broadband output SNR and the broadband desired signal distortion index of the tradeoff filter.

It can be verified that the broadband output SNR of the tradeoff filter is

$$\text{oSNR}\left(\mathbf{h}_{\text{T},\mu}\right) = \frac{\displaystyle\int_f \phi_{X_1}(f) \frac{\lambda_1^2(f)}{\left[\mu + \lambda_1(f)\right]^2} df}{\displaystyle\int_f \phi_{X_1}(f) \frac{\lambda_1(f)}{\left[\mu + \lambda_1(f)\right]^2} df}. \tag{5.109}$$

We propose the following [8].

Property 5.4.3 The broadband output SNR of the tradeoff filter is an increasing function of the parameter μ.

Proof. We need to show that

$$\frac{d\text{oSNR}\left(\mathbf{h}_{\text{T},\mu}\right)}{d\mu} \geq 0. \tag{5.110}$$

The proof showing (5.110) is identical to the one given in [8]. But for completeness, we show it here again.

We have

$$\frac{d\text{oSNR}\left(\mathbf{h}_{\text{T},\mu}\right)}{d\mu} = 2\frac{\text{Num}(\mu)}{\text{Den}(\mu)}, \tag{5.111}$$

where

$$\text{Num}(\mu) = -\int_f \frac{\phi_{X_1}(f)\lambda_1(f)}{[\mu + \lambda_1(f)]^2} df \int_f \frac{\phi_{X_1}(f)\lambda_1^2(f)}{[\mu + \lambda_1(f)]^3} df + \int_f \frac{\phi_{X_1}(f)\lambda_1^2(f)}{[\mu + \lambda_1(f)]^2} df \int_f \frac{\phi_{X_1}(f)\lambda_1(f)}{[\mu + \lambda_1(f)]^3} df \quad (5.112)$$

and

$$\text{Den}(\mu) = \left\{ \int_f \frac{\phi_{X_1}(f)\lambda_1(f)}{[\mu + \lambda_1(f)]^2} df \right\}^2 . \quad (5.113)$$

We only focus on the numerator of the above derivative to see the variations of the broadband output SNR since the denominator is always positive. Multiplying and dividing by $\mu + \lambda_1(f)$, this numerator can be rewritten as

$$\begin{aligned}
\text{Num}(\mu) &= -\int_f \frac{\phi_{X_1}(f)\lambda_1(f)\,[\mu + \lambda_1(f)]}{[\mu + \lambda_1(f)]^3} df \int_f \frac{\phi_{X_1}(f)\lambda_1^2(f)}{[\mu + \lambda_1(f)]^3} df \\
&\quad + \int_f \frac{\phi_{X_1}(f)\lambda_1^2(f)\,[\mu + \lambda_1(f)]}{[\mu + \lambda_1(f)]^3} df \int_f \frac{\phi_{X_1}(f)\lambda_1(f)}{[\mu + \lambda_1(f)]^3} df \\
&= -\left\{ \int_f \frac{\phi_{X_1}(f)\lambda_1^2(f)}{[\mu + \lambda_1(f)]^3} df \right\}^2 \\
&\quad - \mu \int_f \frac{\phi_{X_1}(f)\lambda_1(f)}{[\mu + \lambda_1(f)]^3} df \int_f \frac{\phi_{X_1}(f)\lambda_1^2(f)}{[\mu + \lambda_1(f)]^3} df \\
&\quad + \int_f \frac{\phi_{X_1}(f)\lambda_1^3(f)}{[\mu + \lambda_1(f)]^3} df \int_f \frac{\phi_{X_1}(f)\lambda_1(f)}{[\mu + \lambda_1(f)]^3} df \\
&\quad + \mu \int_f \frac{\phi_{X_1}(f)\lambda_1(f)}{[\mu + \lambda_1(f)]^3} df \int_f \frac{\phi_{X_1}(f)\lambda_1^2(f)}{[\mu + \lambda_1(f)]^3} df \\
&= -\left\{ \int_f \frac{\phi_{X_1}(f)\lambda_1^2(f)}{[\mu + \lambda_1(f)]^3} df \right\}^2 \\
&\quad + \int_f \frac{\phi_{X_1}(f)\lambda_1^3(f)}{[\mu + \lambda_1(f)]^3} df \int_f \frac{\phi_{X_1}(f)\lambda_1(f)}{[\mu + \lambda_1(f)]^3} df .
\end{aligned} \quad (5.114)$$

As far as μ, $\lambda_1(f)$, and $\phi_{X_1}(f)$ are positive $\forall f$, we can use the Cauchy-Schwarz inequality:

$$\int_f \frac{\phi_{X_1}(f)\lambda_1^3(f)}{[\mu + \lambda_1(f)]^3} df \int_f \frac{\phi_{X_1}(f)\lambda_1(f)}{[\mu + \lambda_1(f)]^3} df$$

$$\geq \left\{ \int_f \sqrt{\frac{\phi_{X_1}(f)\lambda_1^3(f)}{[\mu + \lambda_1(f)]^3}} \sqrt{\frac{\phi_{X_1}(f)\lambda_1(f)}{[\mu + \lambda_1(f)]^3}} df \right\}^2$$

$$= \left\{ \int_f \frac{\phi_{X_1}(f)\lambda_1^2(f)}{[\mu + \lambda_1(f)]^3} df \right\}^2 . \tag{5.115}$$

Substituting (5.115) into (5.114), we conclude that

$$\frac{d\text{oSNR}\left(\mathbf{h}_{\text{T},\mu}\right)}{d\mu} \geq 0, \tag{5.116}$$

proving that the broadband output SNR is increasing with respect to μ. ■

From Property 5.4.3, we deduce that the MVDR filter gives the smallest broadband output SNR.

While the broadband output SNR is upper bounded, it is easy to see that the broadband noise reduction factor and broadband desired signal reduction factor are not. So when μ goes to infinity, so are $\xi_{\text{n}}\left(\mathbf{h}_{\text{T},\mu}\right)$ and $\xi_{\text{d}}\left(\mathbf{h}_{\text{T},\mu}\right)$.

The broadband desired signal distortion index is

$$\upsilon_{\text{d}}\left(\mathbf{h}_{\text{T},\mu}\right) = \frac{\int_f \phi_{X_1}(f) \frac{\mu^2}{[\mu + \lambda_1(f)]^2} df}{\int_f \phi_{X_1}(f) df}. \tag{5.117}$$

Property 5.4.4 The broadband desired signal distortion index of the tradeoff filter is an increasing function of the parameter μ.

Proof. It is straightforward to verify that

$$\frac{d\upsilon_{\text{d}}\left(\mathbf{h}_{\text{T},\mu}\right)}{d\mu} \geq 0, \tag{5.118}$$

which ends the proof. ■

It is clear that

$$0 \leq \upsilon_{\text{d}}\left(\mathbf{h}_{\text{T},\mu}\right) \leq 1, \ \forall \mu \geq 0. \tag{5.119}$$

Therefore, as μ increases, the broadband output SNR increases at the price of more distortion to the desired signal.

Property 5.4.5 With the frequency-domain tradeoff filter given in (5.103), the broadband output SNR is always greater than or equal to the broadband input SNR, i.e., $\text{oSNR}\left(\mathbf{h}_{\text{T},\mu}\right) \geq \text{iSNR}, \ \forall \mu \geq 0$.

Proof. We know that

$$\lambda_1(f) \geq \text{iSNR}(f), \tag{5.120}$$

which implies that

$$\int_f \phi_{V_1}(f) \frac{\text{iSNR}(f)}{\lambda_1(f)} df \leq \int_f \phi_{V_1}(f) df, \tag{5.121}$$

and, hence,

$$\begin{aligned} \text{oSNR}\left(\mathbf{h}_{\text{T},0}\right) &= \frac{\int_f \phi_{X_1}(f) df}{\int_f \phi_{V_1}(f) \frac{\text{iSNR}(f)}{\lambda_1(f)} df} \\ &\geq \frac{\int_f \phi_{X_1}(f) df}{\int_f \phi_{V_1}(f) df} = \text{iSNR}. \end{aligned} \tag{5.122}$$

But from Proposition 5.4.3, we have

$$\text{oSNR}\left(\mathbf{h}_{\text{T},\mu}\right) \geq \text{oSNR}\left(\mathbf{h}_{\text{T},0}\right), \ \forall \mu \geq 0. \tag{5.123}$$

As a result,

$$\text{oSNR}\left(\mathbf{h}_{\text{T},\mu}\right) \geq \text{iSNR}, \ \forall \mu \geq 0, \tag{5.124}$$

which ends the proof. ∎

From the previous results, we deduce that for $\mu \geq 1$,

$$\text{iSNR} \leq \text{oSNR}\left(\mathbf{h}_{\text{MVDR}}\right) \leq \text{oSNR}\left(\mathbf{h}_{\text{W}}\right) \leq \text{oSNR}\left(\mathbf{h}_{\text{T},\mu}\right), \tag{5.125}$$

$$0 = \upsilon_{\text{d}}\left(\mathbf{h}_{\text{MVDR}}\right) \leq \upsilon_{\text{d}}\left(\mathbf{h}_{\text{W}}\right) \leq \upsilon_{\text{d}}\left(\mathbf{h}_{\text{T},\mu}\right), \tag{5.126}$$

and for $0 \leq \mu \leq 1$,

$$\text{iSNR} \leq \text{oSNR}\left(\mathbf{h}_{\text{MVDR}}\right) \leq \text{oSNR}\left(\mathbf{h}_{\text{T},\mu}\right) \leq \text{oSNR}\left(\mathbf{h}_{\text{W}}\right), \tag{5.127}$$

$$0 = \upsilon_{\text{d}}\left(\mathbf{h}_{\text{MVDR}}\right) \leq \upsilon_{\text{d}}\left(\mathbf{h}_{\text{T},\mu}\right) \leq \upsilon_{\text{d}}\left(\mathbf{h}_{\text{W}}\right). \tag{5.128}$$

Example 5.4.4 Returning to Example 5.4.2, we now employ the tradeoff filter, $\mathbf{h}_{\text{T},\mu}(f)$, given in (5.103). We assume $M = 5$ sensors. Figure 5.7 shows plots of the broadband gain in SNR, $\mathcal{G}\left(\mathbf{h}_{\text{T},\mu}\right)$, the broadband desired signal distortion index, $\upsilon_{\text{d}}\left(\mathbf{h}_{\text{T},\mu}\right)$, the broadband noise reduction factor, $\xi_{\text{n}}\left(\mathbf{h}_{\text{T},\mu}\right)$, and the broadband desired signal reduction factor, $\xi_{\text{d}}\left(\mathbf{h}_{\text{T},\mu}\right)$, as a function of the broadband input SNR, for several values of μ. For a given broadband input SNR, the higher is the value of μ, the higher are the broadband gain in SNR and the broadband noise reduction factor, but at the expense of higher broadband desired signal distortion index and higher broadband desired signal reduction factor. ∎

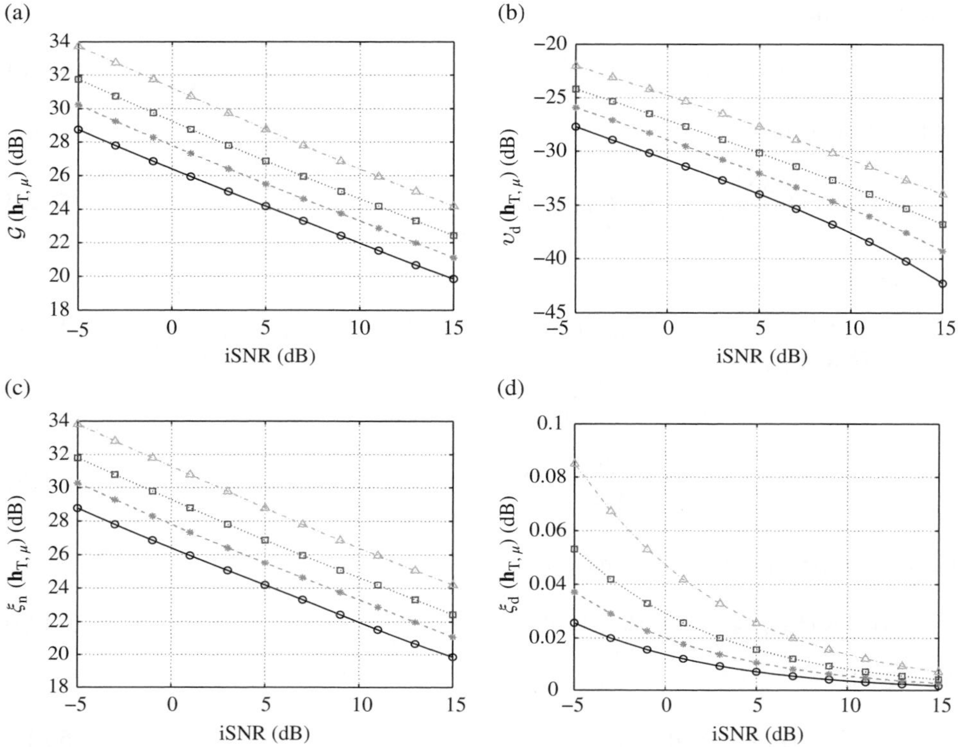

Figure 5.7 (a) The broadband gain in SNR, (b) the broadband desired signal distortion index, (c) the broadband noise reduction factor, and (d) the broadband desired signal reduction factor of the tradeoff filter as a function of the broadband input SNR, for several values of μ: $\mu = 0.5$ (solid line with circles), $\mu = 1$ (dashed line with asterisks), $\mu = 2$ (dotted line with squares), and $\mu = 5$ (dash-dot line with triangles).

5.4.5 LCMV

In the Wiener, MVDR, and tradeoff filters, we have fully exploited the structure of the desired signal vector, $\mathbf{x}(f)$. In this subsection, we are going to exploit as well the structure of the noise signal vector, $\mathbf{v}(f)$, in order to derive the linearly constrained minimum variance (LCMV) filter [9–12], which can handle more than one constraint. Our problem this time is the following. We wish to perfectly recover our desired signal, $X_1(f)$, and completely remove the coherent components, $\boldsymbol{\gamma}^*_{V_1\mathbf{v}}(f)V_1(f)$ [see eq. (5.11)]. Thus, the two constraints can be put together in a matrix form as

$$\mathbf{C}^H_{X_1V_1}(f)\mathbf{h}(f) = \begin{bmatrix} 1 \\ 0 \end{bmatrix}, \tag{5.129}$$

where

$$\mathbf{C}_{X_1V_1}(f) = \begin{bmatrix} \boldsymbol{\gamma}^*_{X_1\mathbf{x}}(f) & \boldsymbol{\gamma}^*_{V_1\mathbf{v}}(f) \end{bmatrix} \tag{5.130}$$

is our constraint matrix of size $M \times 2$. Then, our optimal filter is obtained by minimizing the energy at the filter output, with the constraints that the coherent noise components are cancelled and the desired signal is preserved, i.e.,

$$\mathbf{h}_{\text{LCMV}}(f) = \arg\min_{\mathbf{h}(f)} \mathbf{h}^H(f)\mathbf{\Phi}_{\mathbf{y}}(f)\mathbf{h}(f) \quad \text{subject to}$$
$$\mathbf{C}^H_{X_1V_1}(f)\mathbf{h}(f) = \begin{bmatrix} 1 \\ 0 \end{bmatrix}. \tag{5.131}$$

The solution to (5.131) is given by

$$\mathbf{h}_{\text{LCMV}}(f) = \mathbf{\Phi}_{\mathbf{y}}^{-1}(f)\mathbf{C}_{X_1V_1}(f)\left[\mathbf{C}^H_{X_1V_1}(f)\mathbf{\Phi}_{\mathbf{y}}^{-1}(f)\mathbf{C}_{X_1V_1}(f)\right]^{-1}\begin{bmatrix} 1 \\ 0 \end{bmatrix}. \tag{5.132}$$

We always have

$$\text{oSNR}\left(\mathbf{h}_{\text{LCMV}}\right) \leq \text{oSNR}\left(\mathbf{h}_{\text{MVDR}}\right), \tag{5.133}$$
$$\upsilon_{\text{d}}\left(\mathbf{h}_{\text{LCMV}}\right) = 0, \tag{5.134}$$
$$\xi_{\text{d}}\left(\mathbf{h}_{\text{LCMV}}\right) = 1, \tag{5.135}$$

and

$$\xi_{\text{n}}\left(\mathbf{h}_{\text{LCMV}}\right) \leq \xi_{\text{n}}\left(\mathbf{h}_{\text{MVDR}}\right) \leq \xi_{\text{n}}\left(\mathbf{h}_{\text{W}}\right). \tag{5.136}$$

The LCMV structure can be an useful solution in practical applications where the coherent noise is more problematic than the incoherent one.

Example 5.4.5 Returning to Example 5.4.2, we now employ the LCMV filter, $\mathbf{h}_{\text{LCMV}}(f)$, given in (5.132). We assume $M = 5$ sensors. Figure 5.8 shows plots of the broadband gain in SNR, $\mathcal{G}\left(\mathbf{h}_{\text{LCMV}}\right)$, the broadband MSE, $J\left(\mathbf{h}_{\text{LCMV}}\right)$, the broadband noise reduction factor, $\xi_{\text{n}}\left(\mathbf{h}_{\text{LCMV}}\right)$, and the broadband desired signal reduction factor, $\xi_{\text{d}}\left(\mathbf{h}_{\text{LCMV}}\right)$, as a function of the broadband input SNR, for several values of $\alpha = \sigma_w^2/\sigma_u^2$. For a given broadband input SNR, as the ratio between the coherent to incoherent noise increases (α decreases), the LCMV filter yields higher broadband gain in SNR and higher broadband noise reduction factor. ■

The LCMV filter shown above can, obviously, be extended to any number of linear constraints $M_{\text{c}} \leq M$. The constraint equation, which includes the distortionless constraint, can be expressed as

$$\mathbf{C}^H(f)\mathbf{h}(f) = \mathbf{i}_{\text{c}}, \tag{5.137}$$

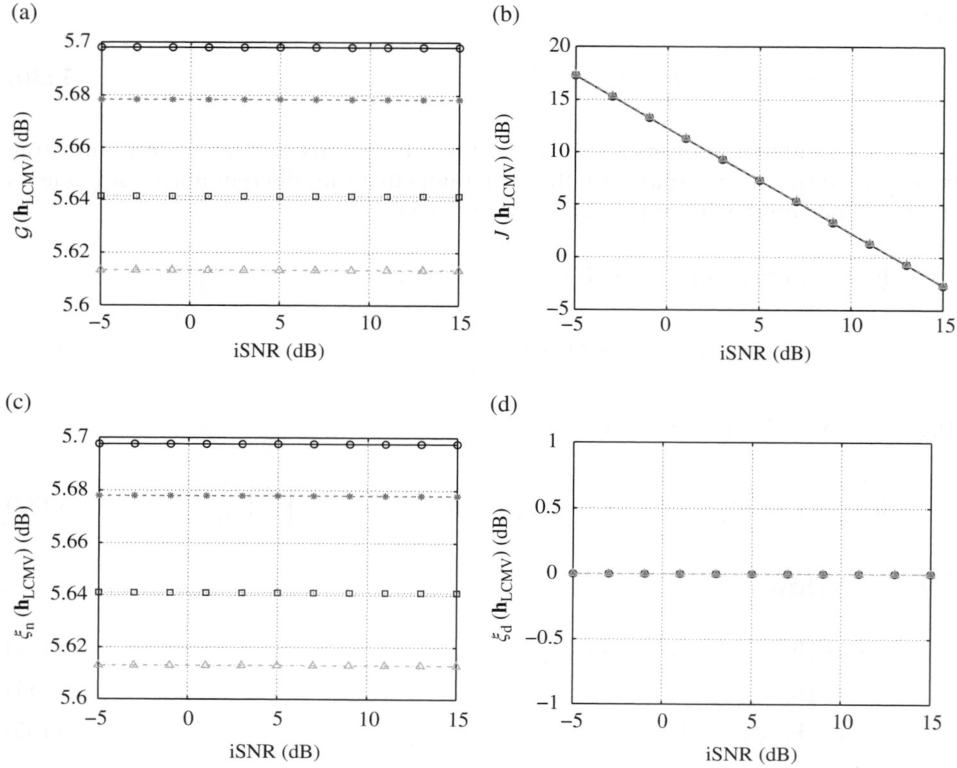

Figure 5.8 (a) The broadband gain in SNR, (b) the broadband MSE, (c) the broadband noise reduction factor, and (d) the broadband desired signal reduction factor of the LCMV filter as a function of the broadband input SNR, for several values of $\alpha = \sigma_w^2/\sigma_u^2$: $\alpha = 0.03$ (solid line with circles), $\alpha = 0.1$ (dashed line with asterisks), $\alpha = 0.3$ (dotted line with squares), and $\alpha = 1$ (dash-dot line with triangles).

where

$$\mathbf{C}(f) = \begin{bmatrix} \mathbf{d}(f) & \mathbf{c}_2(f) & \cdots & \mathbf{c}_{M_c}(f) \end{bmatrix} \tag{5.138}$$

is a matrix of size $M \times M_c$ whose M_c columns are linearly independent and $\mathbf{i}_c$ is a vector of length M_c whose first component is equal to 1 and the other components are some chosen real numbers to satisfy the constraints on the filter. Generally, these constraints are null ones where it is desired to completely cancel some interference sources. Following the same steps as above, we easily find the LCMV filter:

$$\mathbf{h}_{\text{LCMV}}(f) = \mathbf{\Phi}_{\mathbf{y}}^{-1}(f)\mathbf{C}(f)\left[\mathbf{C}^H(f)\mathbf{\Phi}_{\mathbf{y}}^{-1}(f)\mathbf{C}(f)\right]^{-1}\mathbf{i}_c. \tag{5.139}$$

For $M_c = 1$, $\mathbf{h}_{\text{LCMV}}(f)$ simplifies to $\mathbf{h}_{\text{MVDR}}(f)$.

In Table 5.1, we summarize all the optimal filters studied in this section.

Table 5.1 Optimal linear filters for multichannel signal enhancement in the frequency domain.

Filter	
Maximum SNR	$\mathbf{h}_{\max}(f) = \varsigma(f)\mathbf{\Phi}_{\mathbf{v}}^{-1}(f)\mathbf{d}(f),\ \varsigma(f) \neq 0$
Wiener	$\mathbf{h}_{\mathrm{W}}(f) = \dfrac{\mathbf{\Phi}_{\mathbf{v}}^{-1}(f)\mathbf{\Phi}_{\mathbf{y}}(f) - \mathbf{I}_M}{1 - M + \mathrm{tr}\left[\mathbf{\Phi}_{\mathbf{v}}^{-1}(f)\mathbf{\Phi}_{\mathbf{y}}(f)\right]}\mathbf{i}_{\mathrm{i}}$
MVDR	$\mathbf{h}_{\mathrm{MVDR}}(f) = \dfrac{\mathbf{\Phi}_{\mathbf{v}}^{-1}(f)\mathbf{\Phi}_{\mathbf{y}}(f) - \mathbf{I}_M}{\mathrm{tr}\left[\mathbf{\Phi}_{\mathbf{v}}^{-1}(f)\mathbf{\Phi}_{\mathbf{y}}(f)\right] - M}\mathbf{i}_{\mathrm{i}}$
Tradeoff	$\mathbf{h}_{\mathrm{T},\mu} = \dfrac{\mathbf{\Phi}_{\mathbf{v}}^{-1}(f)\mathbf{\Phi}_{\mathbf{y}}(f) - \mathbf{I}_M}{\mu - M + \mathrm{tr}\left[\mathbf{\Phi}_{\mathbf{v}}^{-1}(f)\mathbf{\Phi}_{\mathbf{y}}(f)\right]}\mathbf{i}_{\mathrm{i}},\ \mu \geq 0$
LCMV	$\mathbf{h}_{\mathrm{LCMV}}(f) = \mathbf{\Phi}_{\mathbf{y}}^{-1}(f)\mathbf{C}(f)\left[\mathbf{C}^H(f)\mathbf{\Phi}_{\mathbf{y}}^{-1}(f)\mathbf{C}(f)\right]^{-1}\mathbf{i}_{\mathrm{c}}$

5.5 Generalized Sidelobe Canceller Structure

The generalized sidelobe canceller (GSC) structure solves exactly the same problem as the LCMV approach by dividing the filter vector $\mathbf{h}_{\mathrm{LCMV}}(f)$ into two components operating on orthogonal subspaces [13–16]:

$$\mathbf{h}_{\mathrm{LCMV}}(f) = \mathbf{h}_{\mathrm{MN}}(f) - \mathbf{B}_{\mathrm{C}}(f)\mathbf{w}_{\mathrm{GSC}}(f), \tag{5.140}$$

where

$$\mathbf{h}_{\mathrm{MN}}(f) = \mathbf{C}(f)\left[\mathbf{C}^H(f)\mathbf{C}(f)\right]^{-1}\mathbf{i}_{\mathrm{c}} \tag{5.141}$$

is the minimum-norm solution of (5.137), $\mathbf{B}_{\mathrm{C}}(f)$ is the so-called blocking matrix that spans the nullspace of $\mathbf{C}^H(f)$, i.e.,

$$\mathbf{C}^H(f)\mathbf{B}_{\mathrm{C}}(f) = \mathbf{0}_{M_{\mathrm{c}}\times(M-M_{\mathrm{c}})}, \tag{5.142}$$

and $\mathbf{w}_{\mathrm{GSC}}(f)$ is a weighting vector derived as explained below. The size of $\mathbf{B}_{\mathrm{C}}(f)$ is $M \times (M - M_{\mathrm{c}})$, where $M - M_{\mathrm{c}}$ is the dimension of the nullspace of $\mathbf{C}^H(f)$. Therefore, the length of the vector $\mathbf{w}_{\mathrm{GSC}}(f)$ is $M - M_{\mathrm{c}}$. The blocking matrix is not unique and the most obvious choice is the following:

$$\mathbf{B}_{\mathrm{C}}(f) = \mathbf{P}_{\mathrm{C}}(f)\begin{bmatrix} \mathbf{I}_{M-M_{\mathrm{c}}} \\ \mathbf{0}_{M_{\mathrm{c}}\times(M-M_{\mathrm{c}})} \end{bmatrix}, \tag{5.143}$$

where

$$\mathbf{P}_{\mathrm{C}}(f) = \mathbf{I}_M - \mathbf{C}(f)\left[\mathbf{C}^H(f)\mathbf{C}(f)\right]^{-1}\mathbf{C}^H(f) \tag{5.144}$$

is a projection matrix whose rank is equal to $M - M_{\mathrm{c}}$ and $\mathbf{I}_{M-M_{\mathrm{c}}}$ is the $\left(M - M_{\mathrm{c}}\right) \times \left(M - M_{\mathrm{c}}\right)$ identity matrix.

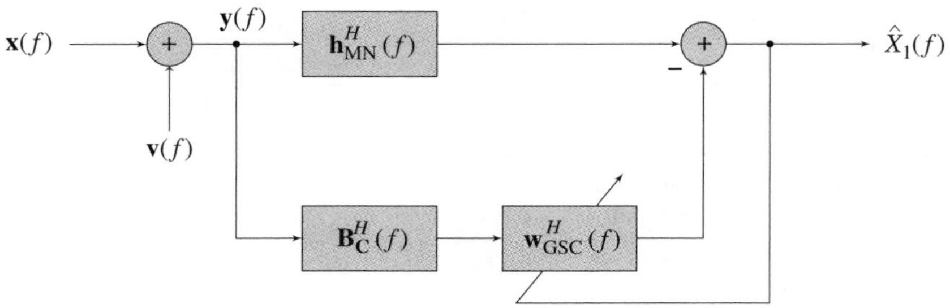

Figure 5.9 Block diagram of the generalized sidelobe canceller.

To obtain the filter $\mathbf{w}_{\mathrm{GSC}}(f)$, the GSC approach is used, which is formulated as the following unconstrained optimization problem:

$$\min_{\mathbf{w}(f)} \left[\mathbf{h}_{\mathrm{MN}}(f) - \mathbf{B}_{\mathbf{C}}(f)\mathbf{w}(f)\right]^H \mathbf{\Phi}_{\mathbf{y}}(f) \left[\mathbf{h}_{\mathrm{MN}}(f) - \mathbf{B}_{\mathbf{C}}(f)\mathbf{w}(f)\right]^H, \tag{5.145}$$

for which the solution is

$$\mathbf{w}_{\mathrm{GSC}}(f) = \left[\mathbf{B}_{\mathbf{C}}^H(f)\mathbf{\Phi}_{\mathbf{y}}(f)\mathbf{B}_{\mathbf{C}}(f)\right]^{-1} \mathbf{B}_{\mathbf{C}}^H(f)\mathbf{\Phi}_{\mathbf{y}}(f)\mathbf{h}_{\mathrm{MN}}(f). \tag{5.146}$$

Define the error signal, which is also the estimate of the desired signal, between the outputs of the two filters $\mathbf{h}_{\mathrm{MN}}(f)$ and $\mathbf{B}_{\mathbf{C}}(f)\mathbf{w}(f)$:

$$\widehat{X}_1(f) = \mathbf{h}_{\mathrm{MN}}^H(f)\mathbf{y}(f) - \mathbf{w}^H(f)\mathbf{B}_{\mathbf{C}}^H(f)\mathbf{y}(f). \tag{5.147}$$

It is easy to see that the minimization of $E\left[\left|\widehat{X}_1(f)\right|^2\right]$ with respect to $\mathbf{w}(f)$ is equivalent to (5.145). A block diagram of the GSC is illustrated in Figure 5.9.

Now, we need to check if indeed the two filters LCMV and GSC are equivalent, i.e.,

$$\begin{aligned} &\mathbf{i}_{\mathrm{c}}^T \left[\mathbf{C}^H(f)\mathbf{\Phi}_{\mathbf{y}}^{-1}(f)\mathbf{C}(f)\right]^{-1} \mathbf{C}^H(f)\mathbf{\Phi}_{\mathbf{y}}^{-1}(f) \\ &\quad = \mathbf{h}_{\mathrm{MN}}^H(f)\left\{\mathbf{I}_M - \mathbf{\Phi}_{\mathbf{y}}(f)\mathbf{B}_{\mathbf{C}}(f)\left[\mathbf{B}_{\mathbf{C}}^H(f)\mathbf{\Phi}_{\mathbf{y}}(f)\mathbf{B}_{\mathbf{C}}(f)\right]^{-1}\mathbf{B}_{\mathbf{C}}^H(f)\right\}. \end{aligned} \tag{5.148}$$

For that, we are going to follow the elegant proof given in [17]. The matrix in brackets in the second line of (5.148) can be rewritten as

$$\begin{aligned} &\mathbf{I}_M - \mathbf{\Phi}_{\mathbf{y}}(f)\mathbf{B}_{\mathbf{C}}(f)\left[\mathbf{B}_{\mathbf{C}}^H(f)\mathbf{\Phi}_{\mathbf{y}}(f)\mathbf{B}_{\mathbf{C}}(f)\right]^{-1}\mathbf{B}_{\mathbf{C}}^H(f) \\ &\quad = \mathbf{\Phi}_{\mathbf{y}}^{1/2}(f)\left[\mathbf{I}_M - \mathbf{P}_1(f)\right]\mathbf{\Phi}_{\mathbf{y}}^{-1/2}(f), \end{aligned} \tag{5.149}$$

where

$$\mathbf{P}_1(f) = \mathbf{\Phi}_{\mathbf{y}}^{1/2}(f)\mathbf{B}_{\mathbf{C}}(f)\left[\mathbf{B}_{\mathbf{C}}^H(f)\mathbf{\Phi}_{\mathbf{y}}(f)\mathbf{B}_{\mathbf{C}}(f)\right]^{-1}\mathbf{B}_{\mathbf{C}}^H(f)\mathbf{\Phi}_{\mathbf{y}}^{1/2}(f) \tag{5.150}$$

is a projection operator onto the subspace spanned by the columns of $\mathbf{\Phi}_{\mathbf{y}}^{1/2}(f)\mathbf{B}_{\mathbf{C}}(f)$. We have

$$\begin{aligned}\mathbf{B}_{\mathbf{C}}^{H}(f)\mathbf{C}(f) &= \mathbf{B}_{\mathbf{C}}^{H}(f)\mathbf{\Phi}_{\mathbf{y}}^{1/2}(f)\mathbf{\Phi}_{\mathbf{y}}^{-1/2}(f)\mathbf{C}(f) \\ &= \mathbf{0}_{(M-M_{\mathrm{c}})\times M_{\mathrm{c}}}.\end{aligned} \tag{5.151}$$

This implies that the rows of $\mathbf{B}_{\mathbf{C}}^{H}(f)$ are orthogonal to the columns of $\mathbf{C}(f)$ and the subspace spanned by the columns of $\mathbf{\Phi}_{\mathbf{y}}^{1/2}(f)\mathbf{B}_{\mathbf{C}}(f)$ is orthogonal to the subspace spanned by the columns of $\mathbf{\Phi}_{\mathbf{y}}^{-1/2}(f)\mathbf{C}(f)$. Since $\mathbf{B}_{\mathbf{C}}(f)$ has a rank equal to $M - M_{\mathrm{c}}$ where M_{c} is the rank of $\mathbf{C}(f)$, then the sum of the dimensions of the two subspaces is M and the subspaces are complementary. This means that

$$\mathbf{P}_1(f) + \mathbf{P}_2(f) = \mathbf{I}_M, \tag{5.152}$$

where

$$\mathbf{P}_2(f) = \mathbf{\Phi}_{\mathbf{y}}^{-1/2}(f)\mathbf{C}(f)\left[\mathbf{C}^{H}(f)\mathbf{\Phi}_{\mathbf{y}}^{-1}(f)\mathbf{C}(f)\right]^{-1}\mathbf{C}^{H}(f)\mathbf{\Phi}_{\mathbf{y}}^{-1/2}(f). \tag{5.153}$$

When this is substituted and the constraint $\mathbf{i}_{\mathrm{c}}^{T} = \mathbf{h}_{\mathrm{MN}}^{H}(f)\mathbf{C}(f)$ is applied, (5.148) becomes

$$\begin{aligned}&\mathbf{i}_{\mathrm{c}}^{T}\left[\mathbf{C}^{H}(f)\mathbf{\Phi}_{\mathbf{y}}^{-1}(f)\mathbf{C}(f)\right]^{-1}\mathbf{C}^{H}(f)\mathbf{\Phi}_{\mathbf{y}}^{-1}(f) \\ &= \mathbf{h}_{\mathrm{MN}}^{H}(f)\mathbf{\Phi}_{\mathbf{y}}^{1/2}(f)\mathbf{P}_2(f)\mathbf{\Phi}_{\mathbf{y}}^{-1/2}(f) \\ &= \mathbf{h}_{\mathrm{MN}}^{H}(f)\mathbf{\Phi}_{\mathbf{y}}^{1/2}(f)\left[\mathbf{I}_M - \mathbf{P}_1(f)\right]\mathbf{\Phi}_{\mathbf{y}}^{-1/2}(f) \\ &= \mathbf{h}_{\mathrm{MN}}^{H}(f)\left\{\mathbf{I}_M - \mathbf{\Phi}_{\mathbf{y}}(f)\mathbf{B}_{\mathbf{C}}(f)\left[\mathbf{B}_{\mathbf{C}}^{H}(f)\mathbf{\Phi}_{\mathbf{y}}(f)\mathbf{B}_{\mathbf{C}}(f)\right]^{-1}\mathbf{B}_{\mathbf{C}}^{H}(f)\right\}.\end{aligned} \tag{5.154}$$

Hence, the LCMV and GSC filters are strictly equivalent.

5.6 A Signal Subspace Perspective

In this section, we give a signal subspace perspective of some of the optimal filters derived in Section 5.4 by using the joint diagonalization.

5.6.1 Joint Diagonalization

The two Hermitian matrices $\mathbf{\Phi}_{\mathbf{x}}(f)$ and $\mathbf{\Phi}_{\mathbf{v}}(f)$ can be jointly diagonalized as follows [5]:

$$\mathbf{T}^{H}(f)\mathbf{\Phi}_{\mathbf{x}}(f)\mathbf{T}(f) = \mathbf{\Lambda}(f), \tag{5.155}$$

$$\mathbf{T}^{H}(f)\mathbf{\Phi}_{\mathbf{v}}(f)\mathbf{T}(f) = \mathbf{I}_M, \tag{5.156}$$

where

$$\mathbf{T}(f) = \begin{bmatrix} \mathbf{t}_1(f) & \mathbf{t}_2(f) & \cdots & \mathbf{t}_M(f) \end{bmatrix} \tag{5.157}$$

is a full-rank square matrix (of size $M \times M$),

$$\mathbf{t}_1(f) = \frac{\mathbf{\Phi}_{\mathbf{v}}^{-1}(f)\mathbf{d}(f)}{\sqrt{\mathbf{d}^H(f)\mathbf{\Phi}_{\mathbf{v}}^{-1}(f)\mathbf{d}(f)}} \tag{5.158}$$

is the first eigenvector of the matrix $\mathbf{\Phi}_{\mathbf{v}}^{-1}(f)\mathbf{\Phi}_{\mathbf{x}}(f)$,

$$\mathbf{\Lambda}(f) = \mathrm{diag}\left[\lambda_1(f), 0, \ldots, 0\right] \tag{5.159}$$

is a diagonal matrix (of size $M \times M$), and

$$\lambda_1(f) = \phi_{X_1}(f)\mathbf{d}^H(f)\mathbf{\Phi}_{\mathbf{v}}^{-1}(f)\mathbf{d}(f) \tag{5.160}$$

is the only nonnull eigenvalue of $\mathbf{\Phi}_{\mathbf{v}}^{-1}(f)\mathbf{\Phi}_{\mathbf{x}}(f)$, whose corresponding eigenvector is $\mathbf{t}_1(f)$. Also, the noisy signal correlation matrix can be diagonalized as

$$\mathbf{T}^H(f)\mathbf{\Phi}_{\mathbf{y}}(f)\mathbf{T}(f) = \mathbf{\Lambda}(f) + \mathbf{I}_M. \tag{5.161}$$

It can be checked from (5.155) that

$$\mathbf{t}_i^H(f)\mathbf{d}(f) = 0,\ i = 2, 3, \ldots, M. \tag{5.162}$$

5.6.2 Estimation of the Desired Signal

As explained in Section 5.2, the desired signal, $X_1(f)$, can be estimated with

$$Z(f) = \mathbf{h}^H(f)\mathbf{y}(f), \tag{5.163}$$

where $\mathbf{h}(f)$ is a complex-valued linear filter of length M. From (5.163), we easily find the variance of $Z(f)$:

$$\phi_Z(f) = \phi_{X_1}(f)\left|\mathbf{h}^H(f)\mathbf{d}(f)\right|^2 + \mathbf{h}^H(f)\mathbf{\Phi}_{\mathbf{v}}(f)\mathbf{h}(f), \tag{5.164}$$

from which we deduce the narrowband output SNR:

$$\mathrm{oSNR}\left[\mathbf{h}(f)\right] = \frac{\phi_{X_1}(f)\left|\mathbf{h}^H(f)\mathbf{d}(f)\right|^2}{\mathbf{h}^H(f)\mathbf{\Phi}_{\mathbf{v}}(f)\mathbf{h}(f)} \leq \lambda_1(f). \tag{5.165}$$

We define the narrowband white noise gain (WNG) as

$$\mathcal{W}\left[\mathbf{h}(f)\right] = \frac{\left|\mathbf{h}^H(f)\mathbf{d}(f)\right|^2}{\mathbf{h}^H(f)\mathbf{h}(f)}. \tag{5.166}$$

We will show next how to use this measure to derive a class of filters.

Let us define the matrix of size $M \times N$:

$$\mathbf{T}_{1:N}(f) = \begin{bmatrix} \mathbf{t}_1(f) & \mathbf{t}_2(f) & \cdots & \mathbf{t}_N(f) \end{bmatrix}, \quad (5.167)$$

with $1 \leq N \leq M$. We consider multichannel noise reduction filters that have the form:

$$\mathbf{h}_{1:N}(f) = \mathbf{T}_{1:N}(f)\mathbf{a}(f), \quad (5.168)$$

where

$$\mathbf{a}(f) = \begin{bmatrix} A_1(f) & A_2(f) & \cdots & A_N(f) \end{bmatrix}^T \neq \mathbf{0} \quad (5.169)$$

is a vector of length N. Then, the narrowband WNG can be expressed as

$$\begin{aligned} \mathcal{W}\left[\mathbf{h}_{1:N}(f)\right] &= \frac{\left|\mathbf{a}^H(f)\mathbf{T}_{1:N}^H(f)\mathbf{d}(f)\right|^2}{\mathbf{a}^H(f)\mathbf{T}_{1:N}^H(f)\mathbf{T}_{1:N}(f)\mathbf{a}(f)} \\ &= \mathcal{W}\left[\mathbf{a}(f)\right]. \end{aligned} \quad (5.170)$$

It is clear that the vector $\mathbf{a}(f)$ that maximizes $\mathcal{W}\left[\mathbf{a}(f)\right]$ is

$$\mathbf{a}(f) = \varsigma(f)\left[\mathbf{T}_{1:N}^H(f)\mathbf{T}_{1:N}(f)\right]^{-1}\mathbf{T}_{1:N}^H(f)\mathbf{d}(f), \quad (5.171)$$

where $\varsigma(f) \neq 0$ is an arbitrary complex number. As a result, the filter $\mathbf{h}_{1:N}(f)$ that maximizes $\mathcal{W}\left[\mathbf{h}_{1:N}(f)\right]$ is

$$\mathbf{h}_{1:N}(f) = \varsigma(f)\mathbf{P}_{\mathbf{T}_{1:N}}(f)\mathbf{d}(f), \quad (5.172)$$

where

$$\mathbf{P}_{\mathbf{T}_{1:N}}(f) = \mathbf{T}_{1:N}(f)\left[\mathbf{T}_{1:N}^H(f)\mathbf{T}_{1:N}(f)\right]^{-1}\mathbf{T}_{1:N}^H(f) \quad (5.173)$$

is a projection matrix whose rank is equal to N. With (5.172), the narrowband WNG is

$$\begin{aligned} \mathcal{W}\left[\mathbf{h}_{1:N}(f)\right] &= \mathbf{d}^H(f)\mathbf{P}_{\mathbf{T}_{1:N}}(f)\mathbf{d}(f) \\ &= \frac{\lambda_1(f)}{\phi_{X_1}(f)}\mathbf{i}^T\left[\mathbf{T}_{1:N}^H(f)\mathbf{T}_{1:N}(f)\right]^{-1}\mathbf{i}, \end{aligned} \quad (5.174)$$

where $\mathbf{i}$ is the first column of the $N \times N$ identity matrix, $\mathbf{I}_N$, with

$$\mathcal{W}\left[\mathbf{h}_{1:1}(f)\right] = \frac{\lambda_1(f)}{\phi_{X_1}(f)\mathbf{t}_1^H(f)\mathbf{t}_1(f)}, \quad (5.175)$$

$$\mathcal{W}\left[\mathbf{h}_{1:M}(f)\right] = \mathbf{d}^H(f)\mathbf{d}(f), \quad (5.176)$$

and

$$\mathcal{W}\left[\mathbf{h}_{1:1}(f)\right] \leq \mathcal{W}\left[\mathbf{h}_{1:2}(f)\right] \leq \cdots \leq \mathcal{W}\left[\mathbf{h}_{1:M}(f)\right]. \quad (5.177)$$

Example 5.6.1 Consider a ULA of $M = 12$ sensors. Suppose that a desired signal, $x(t)$, impinges on the ULA from the direction θ_x. Assume that the desired signal is a harmonic pulse of T samples:

$$x(t) = \begin{cases} A \sin\left(2\pi f_0 t + \phi\right), & 0 \leq t \leq T-1 \\ 0, & t < 0,\ t \geq T \end{cases},$$

with fixed amplitude A and frequency f_0, and random phase ϕ, uniformly distributed on the interval from 0 to 2π. Assume that the interference $u_m(t)$ is a diffuse noise uncorrelated with $x(t)$. In addition, the sensors contain thermal white Gaussian noise, $w_m(t) \sim \mathcal{N}\left(0, \sigma_w^2\right)$, that are mutually uncorrelated. The desired signal needs to be recovered from the noisy received signals, $y_m(t) = x_m(t) + v_m(t)$, $m = 1, \ldots, M$, where $v_m(t) = u_m(t) + w_m(t)$, $m = 1, \ldots, M$ are the interference-plus-noise signals.

As in Example 5.4.1, we choose a sampling interval T_s that satisfies $T_s = \frac{d}{c}$. Hence, the variance of $X_1(f)$ is

$$\phi_{X_1}(f) = \frac{A^2}{4} D_T^2 \left[\pi\left(f + f_0\right)\right] + \frac{A^2}{4} D_T^2 \left[\pi\left(f - f_0\right)\right]$$

and the correlation matrix of $\mathbf{x}(f)$ is

$$\mathbf{\Phi}_{\mathbf{x}}(f) = \phi_{X_1}(f)\mathbf{d}(f)\mathbf{d}^H(f),$$

where

$$\mathbf{d}(f) = \begin{bmatrix} 1 & e^{-\jmath 2\pi f \tau_{x,2}} & e^{-\jmath 2\pi f \tau_{x,3}} & \cdots & e^{-\jmath 2\pi f \tau_{x,M}} \end{bmatrix}^T,$$

$$\tau_{x,m} = (m-1)\cos\theta_x,\ m = 1, 2, \ldots, M.$$

The correlation matrix of $\mathbf{v}(f)$ is

$$\mathbf{\Phi}_{\mathbf{v}}(f) = T\sigma_u^2 \mathbf{\Gamma}_{\mathbf{u}}(f) + T\sigma_w^2 \mathbf{I}_M,$$

where

$$\begin{aligned}\left[\mathbf{\Gamma}_{\mathbf{u}}(f)\right]_{i,j} &= \frac{\sin\left(2\pi f |i-j|\right)}{2\pi f |i-j|} \\ &= \text{sinc}\left(2f|i-j|\right)\end{aligned}$$

is the normalized coherence between the ith and jth sensors for the diffuse noise.

The narrowband input SNR is

$$\begin{aligned}\text{iSNR}(f) &= \frac{\phi_{X_1}(f)}{\phi_{V_1}(f)} \\ &= \frac{A^2}{4T\left(\sigma_u^2 + \sigma_w^2\right)} D_T^2 \left[\pi\left(f + f_0\right)\right]\end{aligned}$$

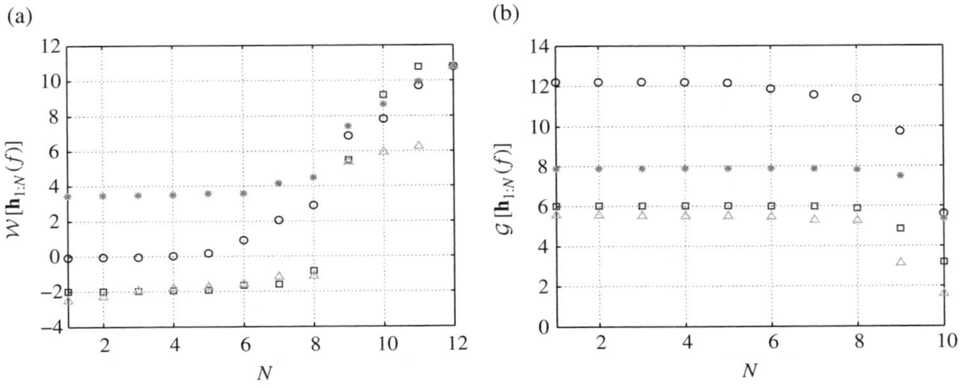

Figure 5.10 (a) Narrowband white noise gain and (b) narrowband gain in SNR of $\mathbf{h}_{1:N}(f)$ as a function of N for different incidence angles of the desired signal: $\theta_x = 0°$ (circles), $\theta_x = 30°$ (asterisks), $\theta_x = 60°$ (squares), and $\theta_x = 90°$ (triangles).

$$+ \frac{A^2}{4T\left(\sigma_u^2 + \sigma_w^2\right)} D_T^2\left[\pi\left(f - f_0\right)\right].$$

To demonstrate the performance of the filter $\mathbf{h}_{1:N}(f)$ in (5.172), we choose $A = 0.5$, $f_0 = 0.1$, $\sigma_w^2 = 0.01\sigma_u^2$, $\sigma_u^2 = 0.1$, and $T = 500$. Figure 5.10 shows the effect of N on the WNG, $\mathcal{W}\left[\mathbf{h}_{1:N}(f)\right]$, and on the narrowband gain in SNR, $\mathcal{G}\left[\mathbf{h}_{1:N}(f)\right]$, for different incidence angles of the desired signal and for $f = f_0$. For $N = 1$, the WNG is minimal. As N increases, the WNG increases. ■

Now, we need to determine $\varsigma(f)$. There are at least three different useful approaches for that.

The first idea to find $\varsigma(f)$ is from the distortionless constraint, i.e.,

$$\mathbf{h}_{1:N}^H(f)\mathbf{d}(f) = 1. \tag{5.178}$$

Substituting (5.172) into (5.178), we get

$$\varsigma(f) = \frac{1}{\mathbf{d}^H(f)\mathbf{P}_{\mathbf{T}_{1:N}}(f)\mathbf{d}(f)}. \tag{5.179}$$

Consequently, we obtain a class of distortionless filters:

$$\mathbf{h}_{1:N,\mathrm{DL}}(f) = \frac{\mathbf{P}_{\mathbf{T}_{1:N}}(f)\mathbf{d}(f)}{\mathbf{d}^H(f)\mathbf{P}_{\mathbf{T}_{1:N}}(f)\mathbf{d}(f)}, \tag{5.180}$$

with

$$\mathbf{h}_{1:1,\mathrm{DL}}(f) = \frac{\mathbf{t}_1(f)}{\mathbf{d}^H(f)\mathbf{t}_1(f)} \tag{5.181}$$

$$= \frac{\mathbf{\Phi}_{\mathbf{v}}^{-1}(f)\mathbf{d}(f)}{\mathbf{d}^H(f)\mathbf{\Phi}_{\mathbf{v}}^{-1}(f)\mathbf{d}(f)}$$
$$= \mathbf{h}_{\mathrm{MVDR}}(f)$$

being the MVDR filter (see Section 5.4) and

$$\mathbf{h}_{1:M,\mathrm{DL}}(f) = \frac{\mathbf{d}(f)}{\mathbf{d}^H(f)\mathbf{d}(f)} \quad (5.182)$$
$$= \mathbf{h}_{\mathrm{MN}}(f)$$

being the minimum-norm filter, which can be directly derived from (5.178). We should always have

$$\lambda_1(f) = \mathrm{oSNR}\left[\mathbf{h}_{1:1,\mathrm{DL}}(f)\right] \geq \mathrm{oSNR}\left[\mathbf{h}_{1:2,\mathrm{DL}}(f)\right] \geq \cdots \geq \mathrm{oSNR}\left[\mathbf{h}_{1:M,\mathrm{DL}}(f)\right]. \quad (5.183)$$

Example 5.6.2 Returning to Example 5.6.1, we assume the ULA contains $M = 6$ sensors, and that the desired signal impinges on the ULA from the direction $\theta_x = 120°$. Now, we employ the distortionless filter, $\mathbf{h}_{1:N,\mathrm{DL}}(f)$, given in (5.180). Figure 5.11 shows plots of the narrowband gain in SNR, $\mathcal{G}\left[\mathbf{h}_{1:N,\mathrm{DL}}(f)\right]$, the narrowband MSE, $J\left[\mathbf{h}_{1:N,\mathrm{DL}}(f)\right]$, the narrowband noise reduction factor, $\xi_{\mathrm{n}}\left[\mathbf{h}_{1:N,\mathrm{DL}}(f)\right]$, and the narrowband desired signal reduction factor, $\xi_{\mathrm{d}}\left[\mathbf{h}_{1:N,\mathrm{DL}}(f)\right]$, as a function of the narrowband input SNR, for $f = f_0$ and several values of N. For $N = 1$, the narrowband gain in SNR is maximal. As N increases, the narrowband gain in SNR and the narrowband noise reduction factor decrease, while the narrowband MSE increases. ■

Let us define the error signal between the estimated and desired signals at frequency f:

$$\mathcal{E}(f) = Z(f) - X_1(f) \quad (5.184)$$
$$= \mathbf{h}_{1:N}^H(f)\mathbf{y}(f) - X_1(f).$$

Then, the MSE is

$$J\left[\mathbf{h}_{1:N}(f)\right] = E\left[|\mathcal{E}(f)|^2\right] \quad (5.185)$$
$$= |\varsigma(f)|^2\,\mathbf{d}^H(f)\mathbf{P}_{\mathbf{T}_{1:N}}(f)\mathbf{\Phi}_{\mathbf{y}}(f)\mathbf{P}_{\mathbf{T}_{1:N}}(f)\mathbf{d}(f) + \phi_{X_1}(f)$$
$$- \varsigma^*(f)\phi_{X_1}(f)\mathbf{d}^H(f)\mathbf{P}_{\mathbf{T}_{1:N}}(f)\mathbf{d}(f)$$
$$- \varsigma(f)\phi_{X_1}(f)\mathbf{d}^H(f)\mathbf{P}_{\mathbf{T}_{1:N}}(f)\mathbf{d}(f).$$

The minimization of $J\left[\mathbf{h}_{1:N}(f)\right]$ with respect to $\varsigma^*(f)$ leads to

$$\varsigma(f) = \frac{\phi_{X_1}(f)\mathbf{d}^H(f)\mathbf{P}_{\mathbf{T}_{1:N}}(f)\mathbf{d}(f)}{\mathbf{d}^H(f)\mathbf{P}_{\mathbf{T}_{1:N}}(f)\mathbf{\Phi}_{\mathbf{y}}(f)\mathbf{P}_{\mathbf{T}_{1:N}}(f)\mathbf{d}(f)}. \quad (5.186)$$

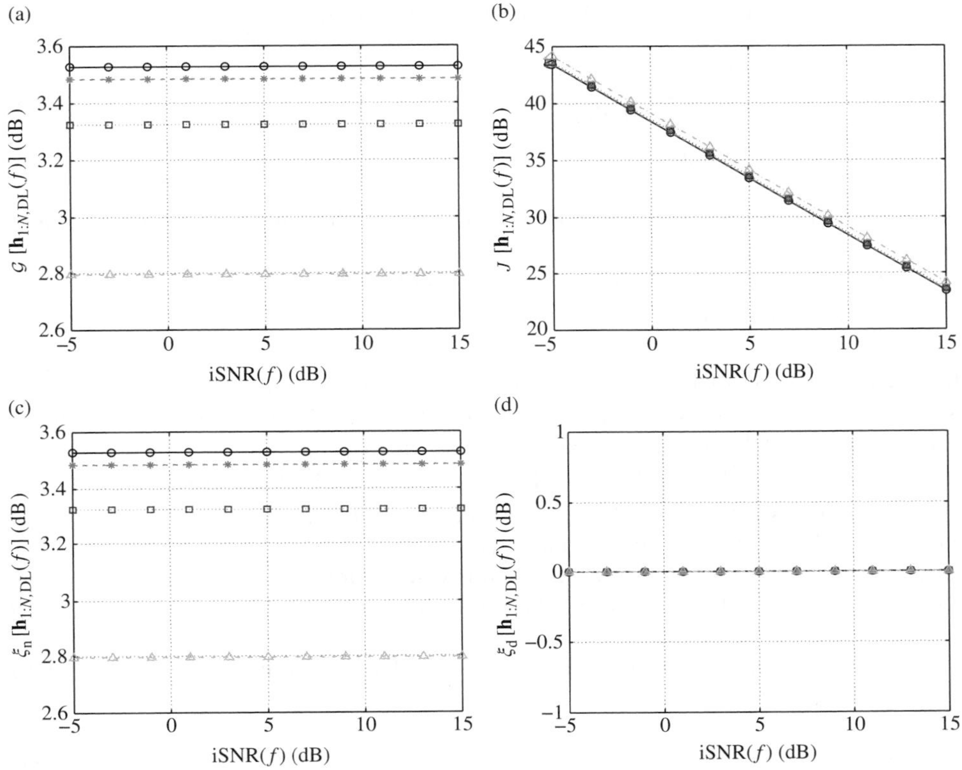

Figure 5.11 (a) The narrowband gain in SNR, (b) the narrowband MSE, (c) the narrowband noise reduction factor, and (d) the narrowband desired signal reduction factor of the distortionless filters as a function of the narrowband input SNR, for several values of N: $N = 1$ (solid line with circles), $N = 2$ (dashed line with asterisks), $N = 3$ (dotted line with squares), and $N = 4$ (dash-dot line with triangles).

We deduce a class of Wiener filters:

$$\mathbf{h}_{1:N,\mathrm{W}}(f) = \frac{\phi_{X_1}(f)\mathbf{P}_{\mathbf{T}_{1:N}}(f)\mathbf{d}(f)\mathbf{d}^H(f)\mathbf{P}_{\mathbf{T}_{1:N}}(f)\mathbf{d}(f)}{\mathbf{d}^H(f)\mathbf{P}_{\mathbf{T}_{1:N}}(f)\mathbf{\Phi}_{\mathbf{y}}(f)\mathbf{P}_{\mathbf{T}_{1:N}}(f)\mathbf{d}(f)}, \tag{5.187}$$

with

$$\begin{aligned}\mathbf{h}_{1:1,\mathrm{W}}(f) &= \frac{\phi_{X_1}(f)\mathbf{t}_1^H(f)\mathbf{d}(f)}{1+\lambda_1(f)}\mathbf{t}_1(f) \\ &= \phi_{X_1}(f)\mathbf{\Phi}_{\mathbf{y}}^{-1}(f)\mathbf{d}(f) \\ &= \frac{\lambda_1(f)}{1+\lambda_1(f)}\mathbf{h}_{\mathrm{MVDR}}(f) \\ &= \mathbf{h}_{\mathrm{W}}(f)\end{aligned} \tag{5.188}$$

being the classical multichannel Wiener filter (see Section 5.4) and

$$\begin{aligned}\mathbf{h}_{1:M,\mathrm{W}}(f) &= \frac{\phi_{X_1}(f)\mathbf{d}^H(f)\mathbf{d}(f)}{\mathbf{d}^H(f)\mathbf{\Phi}_\mathbf{y}(f)\mathbf{d}(f)}\mathbf{d}(f) \\ &= \mathbf{h}_{\mathrm{MN},2}(f)\end{aligned} \tag{5.189}$$

being another (distorted) form of the minimum-norm filter. We should always have

$$\begin{aligned}\lambda_1(f) = \mathrm{oSNR}\left[\mathbf{h}_{1:1,\mathrm{W}}(f)\right] \geq \mathrm{oSNR}\left[\mathbf{h}_{1:2,\mathrm{W}}(f)\right] \geq \\ \cdots \geq \mathrm{oSNR}\left[\mathbf{h}_{1:M,\mathrm{W}}(f)\right].\end{aligned} \tag{5.190}$$

Example 5.6.3 Returning to Example 5.6.2, we now employ the Wiener filter, $\mathbf{h}_{1:N,\mathrm{W}}(f)$, given in (5.187). Figure 5.12 shows plots of the narrowband gain in SNR, $\mathcal{G}\left[\mathbf{h}_{1:N,\mathrm{W}}(f)\right]$, the narrowband MSE, $J\left[\mathbf{h}_{1:N,\mathrm{W}}(f)\right]$, the narrowband noise reduction factor, $\xi_\mathrm{n}\left[\mathbf{h}_{1:N,\mathrm{W}}(f)\right]$, and the narrowband desired signal reduction factor, $\xi_\mathrm{d}\left[\mathbf{h}_{1:N,\mathrm{W}}(f)\right]$, as a function of the narrowband input SNR, for $f = f_0$ and several values of N. For $N = 1$, the narrowband gain in SNR is maximal. As N increases, the narrowband gain in SNR and the narrowband noise reduction factor decrease, while the narrowband MSE and the narrowband desired signal reduction factor increase. ■

We define the error signal between the estimated and reference microphone signals at frequency f as

$$\begin{aligned}\mathcal{E}'(f) &= Z(f) - Y_1(f) \\ &= \mathbf{h}_{1:N}^H(f)\mathbf{y}(f) - Y_1(f)\end{aligned} \tag{5.191}$$

and the corresponding MSE is

$$J'\left[\mathbf{h}_{1:N}(f)\right] = E\left[\left|\mathcal{E}'(f)\right|^2\right]. \tag{5.192}$$

The minimization of $J'\left[\mathbf{h}_{1:N}(f)\right]$ with respect to $\varsigma^*(f)$ gives

$$\varsigma(f) = \frac{\mathbf{d}^H(f)\mathbf{P}_{\mathbf{T}_{1:N}}(f)\mathbf{\Phi}_\mathbf{y}(f)\mathbf{i}_\mathrm{i}}{\mathbf{d}^H(f)\mathbf{P}_{\mathbf{T}_{1:N}}(f)\mathbf{\Phi}_\mathbf{y}(f)\mathbf{P}_{\mathbf{T}_{1:N}}(f)\mathbf{d}(f)}. \tag{5.193}$$

Therefore, we find a class of tradeoff filters:

$$\mathbf{h}_{1:N,\mathrm{T}}(f) = \frac{\mathbf{P}_{\mathbf{T}_{1:N}}(f)\mathbf{d}(f)\mathbf{d}^H(f)\mathbf{P}_{\mathbf{T}_{1:N}}(f)\mathbf{\Phi}_\mathbf{y}(f)\mathbf{i}_\mathrm{i}}{\mathbf{d}^H(f)\mathbf{P}_{\mathbf{T}_{1:N}}(f)\mathbf{\Phi}_\mathbf{y}(f)\mathbf{P}_{\mathbf{T}_{1:N}}(f)\mathbf{d}(f)}, \tag{5.194}$$

with

$$\begin{aligned}\lambda_1(f) = \mathrm{oSNR}\left[\mathbf{h}_{1:1,\mathrm{T}}(f)\right] \geq \mathrm{oSNR}\left[\mathbf{h}_{1:2,\mathrm{T}}(f)\right] \geq \\ \cdots \geq \mathrm{oSNR}\left[\mathbf{h}_{1:M,\mathrm{T}}(f)\right].\end{aligned} \tag{5.195}$$

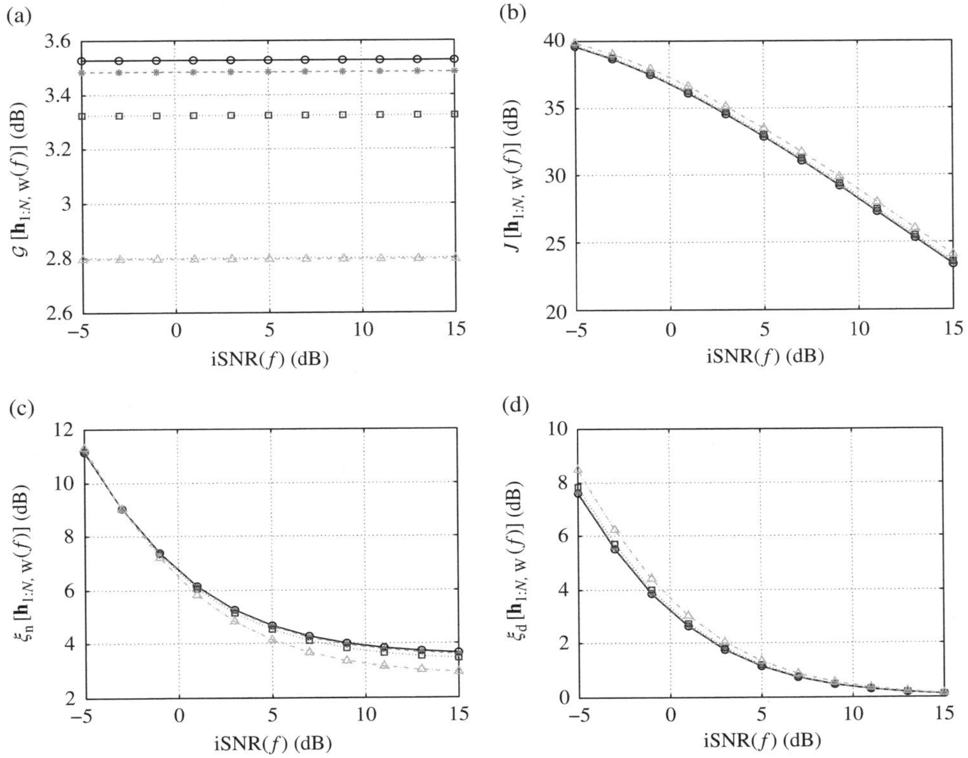

Figure 5.12 (a) The narrowband gain in SNR, (b) the narrowband MSE, (c) the narrowband noise reduction factor, and (d) the narrowband desired signal reduction factor of the Wiener filters as a function of the narrowband input SNR, for several values of N: $N = 1$ (solid line with circles), $N = 2$ (dashed line with asterisks), $N = 3$ (dotted line with squares), and $N = 4$ (dash-dot line with triangles).

Example 5.6.4 Returning to Example 5.6.2, we now employ the tradeoff filter, $\mathbf{h}_{1:N,\mathrm{T}}(f)$, given in (5.194). Figure 5.13 shows plots of the narrowband gain in SNR, $\mathcal{G}\left[\mathbf{h}_{1:N,\mathrm{T}}(f)\right]$, the narrowband MSE, $J\left[\mathbf{h}_{1:N,\mathrm{T}}(f)\right]$, the narrowband noise reduction factor, $\xi_{\mathrm{n}}\left[\mathbf{h}_{1:N,\mathrm{T}}(f)\right]$, and the narrowband desired signal reduction factor, $\xi_{\mathrm{d}}\left[\mathbf{h}_{1:N,\mathrm{T}}(f)\right]$, as a function of the narrowband input SNR, for $f = f_0$ and several values of N. For $N = 1$, the narrowband gain in SNR is maximal. As N increases, the narrowband gain in SNR and the narrowband noise reduction factor decrease, while the narrowband MSE increases. ■

5.7 Implementation with the STFT

In this section, we show how to implement the optimal filters in the STFT domain.

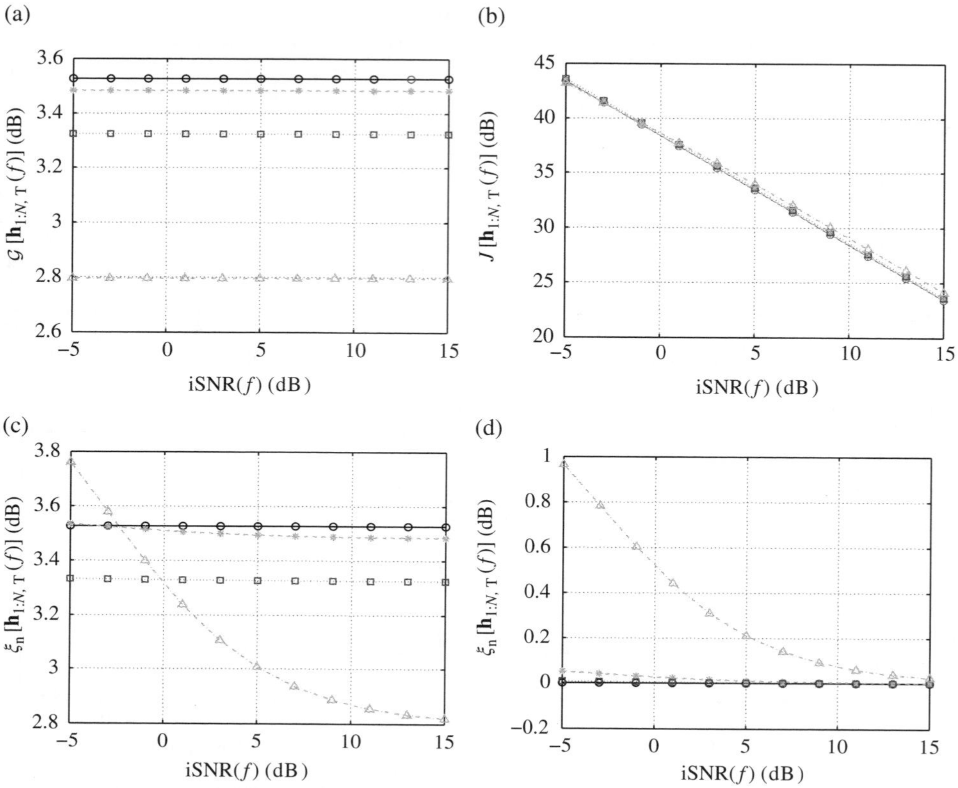

Figure 5.13 (a) The narrowband gain in SNR, (b) the narrowband MSE, (c) the narrowband noise reduction factor, and (d) the narrowband desired signal reduction factor of the tradeoff filters as a function of the narrowband input SNR, for several values of N: $N = 1$ (solid line with circles), $N = 2$ (dashed line with asterisks), $N = 3$ (dotted line with squares), and $N = 4$ (dash-dot line with triangles).

The signal model given in (5.1) can be put into a vector form by considering the L most recent successive time samples, i.e.,

$$\mathbf{y}_m(t) = \mathbf{x}_m(t) + \mathbf{v}_m(t), \ m = 1, 2, \ldots, M, \tag{5.196}$$

where

$$\mathbf{y}_m(t) = \begin{bmatrix} y_m(t) & y_m(t-1) & \cdots & y_m(t-L+1) \end{bmatrix}^T \tag{5.197}$$

is a vector of length L, and $\mathbf{x}_m(t)$ and $\mathbf{v}_m(t)$ are defined in a similar way to $\mathbf{y}_m(t)$ from (5.197). A short-time segment of the observation [i.e., $\mathbf{y}_m(t)$], is multiplied with an analysis window of length L:

$$\mathbf{g}_\mathrm{a} = \begin{bmatrix} g_\mathrm{a}(0) & g_\mathrm{a}(1) & \cdots & g_\mathrm{a}(L-1) \end{bmatrix}^T \tag{5.198}$$

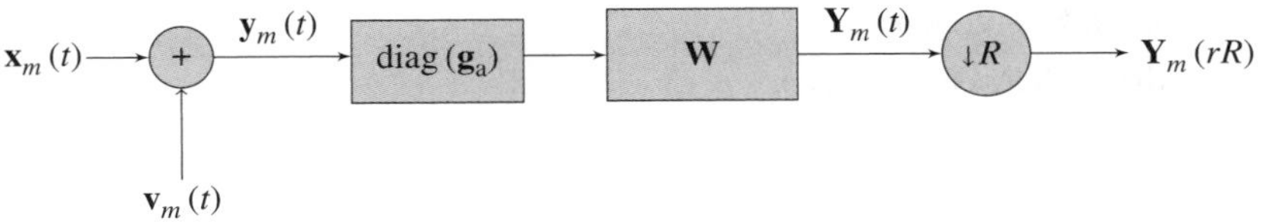

Figure 5.14 STFT representation of the measured signal at the mth sensor.

and transformed into the frequency domain by using the discrete Fourier transform (DFT). Let $\mathbf{W}$ denote the DFT matrix of size $L \times L$, with

$$[\mathbf{W}]_{i,j} = \exp\left(-\frac{\jmath 2\pi ij}{L}\right), \; i,j = 0,\dots,L-1. \tag{5.199}$$

Then, the STFT representation of the observation is defined as [18]

$$\mathbf{Y}_m(t) = \mathbf{W}\text{diag}\left(\mathbf{g}_\text{a}\right)\mathbf{y}_m(t), \tag{5.200}$$

where

$$\mathbf{Y}_m(t) = \left[\begin{array}{cccc} Y_m(t,0) & Y_m(t,1) & \cdots & Y_m(t,L-1) \end{array}\right]^T. \tag{5.201}$$

In practice, the STFT representation is decimated in time by a factor R ($1 \leq R \leq L$) [19]:

$$\begin{aligned} \mathbf{Y}_m(rR) &= \mathbf{Y}_m(t)\left.\right|_{t=rR} \\ &= \left[\begin{array}{cccc} Y_m(rR,0) & Y_m(rR,1) & \cdots & Y_m(rR,L-1) \end{array}\right]^T, \; r \in \mathbb{Z}. \end{aligned} \tag{5.202}$$

Figure 5.14 shows the STFT representation of the measured signal at the mth sensor. Therefore, in the STFT domain, (5.1) can be written as

$$Y_m(rR,k) = X_m(rR,k) + V_m(rR,k), \tag{5.203}$$

where $k = 0,\dots,L-1$ denotes the frequency index, and $X_m(rR,k)$ and $V_m(rR,k)$ are the STFT representations of $x_m(t)$ and $v_m(t)$, respectively. Assuming that L, the length of the analysis window $\mathbf{g}_\text{a}$, is sufficiently larger than the effective support of the acoustic impulse response $g_m(t)$ [20], we can apply the multiplicative transfer function (MTF) approximation [20] and write the convolved desired signal at the mth sensor as

$$X_m(rR,k) = G_m(k)X(rR,k), \tag{5.204}$$

where $X(rR,k)$ is the STFT representation of the desired signal, $x(t)$, and $G_m(k)$ is the DFT of $g_m(t)$.

Writing the M STFT representations of the sensors' signals in a vector notation, we have

$$\mathbf{y}(rR,k) = \mathbf{g}(k)X(rR,k) + \mathbf{v}(rR,k) \tag{5.205}$$

$$= \mathbf{x}(rR,k) + \mathbf{v}(rR,k)$$
$$= \mathbf{d}(k)X_1(rR,k) + \mathbf{v}(rR,k),$$

where

$$\mathbf{y}(rR,k) = \begin{bmatrix} Y_1(rR,k) & Y_2(rR,k) & \cdots & Y_M(rR,k) \end{bmatrix}^T,$$
$$\mathbf{x}(rR,k) = \begin{bmatrix} X_1(rR,k) & X_2(rR,k) & \cdots & X_M(rR,k) \end{bmatrix}^T$$
$$= X(rR,k)\mathbf{g}(k),$$
$$\mathbf{g}(k) = \begin{bmatrix} G_1(k) & G_2(k) & \cdots & G_M(k) \end{bmatrix}^T,$$
$$\mathbf{v}(rR,k) = \begin{bmatrix} V_1(rR,k) & V_2(rR,k) & \cdots & V_M(rR,k) \end{bmatrix}^T,$$

and

$$\mathbf{d}(k) = \begin{bmatrix} 1 & \frac{G_2(k)}{G_1(k)} & \cdots & \frac{G_M(k)}{G_1(k)} \end{bmatrix}^T \tag{5.206}$$
$$= \frac{\mathbf{g}(k)}{G_1(k)}.$$

The correlation matrix of $\mathbf{y}(rR,k)$ is

$$\mathbf{\Phi}_{\mathbf{y}}(rR,k) = E\left[\mathbf{y}(rR,k)\mathbf{y}^H(rR,k)\right] \tag{5.207}$$
$$= \phi_{X_1}(rR,k)\mathbf{d}(k)\mathbf{d}^H(k) + \mathbf{\Phi}_{\mathbf{v}}(rR,k),$$

where $\phi_{X_1}(rR,k) = E\left[|X_1(rR,k)|^2\right]$ is the variance of $X_1(rR,k)$ and $\mathbf{\Phi}_{\mathbf{v}}(rR,k) = E\left[\mathbf{v}(rR,k)\mathbf{v}^H(rR,k)\right]$ is the correlation matrix of $\mathbf{v}(f)$.

In the STFT domain, conventional multichannel noise reduction is performed by applying a complex weight to the output of each sensor, at time-frequency bin (rR,k), and summing across the aperture (see Figure 5.15):

$$Z(rR,k) = \sum_{m=1}^{M} H_m^*(rR,k)Y_m(rR,k) \tag{5.208}$$
$$= \mathbf{h}^H(rR,k)\mathbf{y}(rR,k),$$

where $Z(rR,k)$ is the estimate of $X_1(rR,k)$ and

$$\mathbf{h}(rR,k) = \begin{bmatrix} H_1(rR,k) & H_2(rR,k) & \cdots & H_M(rR,k) \end{bmatrix}^T \tag{5.209}$$

is a filter of length M containing all the complex gains applied to the sensors' outputs at time-frequency bin (rR,k).

We can express (5.208) as a function of the steering vector, i.e.,

$$Z(rR,k) = \mathbf{h}^H(rR,k)\left[\mathbf{d}(k)X_1(rR,k) + \mathbf{v}(rR,k)\right] \tag{5.210}$$

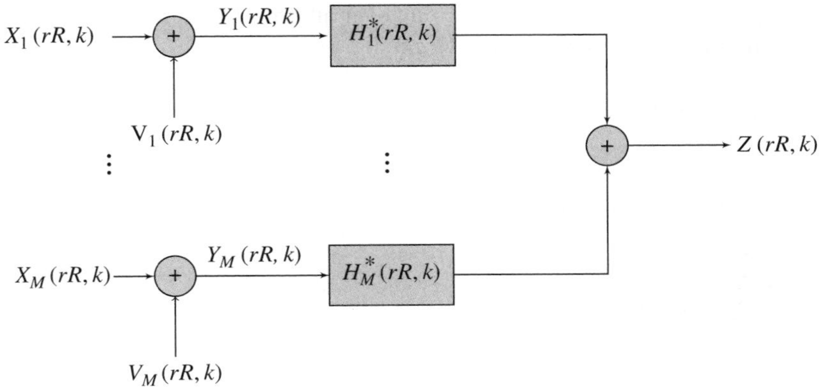

Figure 5.15 Block diagram of multichannel linear filtering in the STFT domain.

$$= X_{\mathrm{fd}}(rR,k) + V_{\mathrm{rn}}(rR,k),$$

where

$$X_{\mathrm{fd}}(rR,k) = X_1(rR,k)\mathbf{h}^H(rR,k)\mathbf{d}(k) \tag{5.211}$$

is the filtered desired signal and

$$V_{\mathrm{rn}}(rR,k) = \mathbf{h}^H(rR,k)\mathbf{v}(rR,k) \tag{5.212}$$

is the residual noise. This procedure is called multichannel signal enhancement in the STFT domain.

The two terms on the right-hand side of (5.210) are incoherent. Hence, the variance of $Z(rR,k)$ is the sum of two variances:

$$\begin{aligned}\phi_Z(rR,k) &= \mathbf{h}^H(rR,k)\mathbf{\Phi}_{\mathbf{y}}(rR,k)\mathbf{h}(rR,k) \\ &= \phi_{X_{\mathrm{fd}}}(rR,k) + \phi_{V_{\mathrm{rn}}}(rR,k),\end{aligned} \tag{5.213}$$

where

$$\phi_{X_{\mathrm{fd}}}(rR,k) = \phi_{X_1}(rR,k)\left|\mathbf{h}^H(rR,k)\mathbf{d}(k)\right|^2, \tag{5.214}$$

$$\phi_{V_{\mathrm{rn}}}(rR,k) = \mathbf{h}^H(rR,k)\mathbf{\Phi}_{\mathbf{v}}(rR,k)\mathbf{h}(rR,k). \tag{5.215}$$

In a similar way to the frequency-domain input SNR, we define the narrowband input SNR as

$$\mathrm{iSNR}(rR,k) = \frac{\phi_{X_1}(rR,k)}{\phi_{V_1}(rR,k)}. \tag{5.216}$$

The broadband input SNR is obtained by summing over all time-frequency indices the numerator and denominator of iSNR(rR, k). We get

$$\text{iSNR} = \frac{\sum_{r,k} \phi_{X_1}(rR, k)}{\sum_{r,k} \phi_{V_1}(rR, k)}. \tag{5.217}$$

Similarly, the broadband output SNR is

$$\begin{aligned} \text{oSNR}\left(\mathbf{h}\right) &= \frac{\sum_{r,k} \phi_{X_{\text{fd}}}(rR, k)}{\sum_{r,k} \phi_{V_{\text{rn}}}(rR, k)} \\ &= \frac{\sum_{r,k} \phi_{X_1}(rR, k) \left|\mathbf{h}^H(rR, k)\mathbf{d}(k)\right|^2}{\sum_{r,k} \mathbf{h}^H(rR, k)\mathbf{\Phi}_{\mathbf{v}}(rR, k)\mathbf{h}(rR, k)}, \end{aligned} \tag{5.218}$$

the broadband noise reduction and desired signal reduction factors are, respectively,

$$\xi_{\text{n}}\left(\mathbf{h}\right) = \frac{\sum_{r,k} \phi_{V_1}(rR, k)}{\sum_{r,k} \mathbf{h}^H(rR, k)\mathbf{\Phi}_{\mathbf{v}}(rR, k)\mathbf{h}(rR, k)} \tag{5.219}$$

and

$$\xi_{\text{d}}\left(\mathbf{h}\right) = \frac{\sum_{r,k} \phi_{X_1}(rR, k)}{\sum_{r,k} \phi_{X_1}(rR, k) \left|\mathbf{h}^H(rR, k)\mathbf{d}(k)\right|^2}, \tag{5.220}$$

the broadband desired signal distortion index is

$$\upsilon_{\text{d}}\left(\mathbf{h}\right) = \frac{\sum_{r,k} \phi_{X_1}(rR, k) \left|\mathbf{h}^H(rR, k)\mathbf{d}(k) - 1\right|^2}{\sum_{r,k} \phi_{X_1}(rR, k)}, \tag{5.221}$$

and the broadband MSE is defined as

$$\begin{aligned} J\left(\mathbf{h}\right) &= \sum_{r,k} J\left[\mathbf{h}(rR, k)\right] \\ &= \sum_{r,k} \left[\phi_{X_1}(rR, k)\left|\mathbf{h}^H(rR, k)\mathbf{d}(k) - 1\right|^2 \right. \\ &\quad \left. + \mathbf{h}^H(rR, k)\mathbf{\Phi}_{\mathbf{v}}(rR, k)\mathbf{h}(rR, k)\right]. \end{aligned} \tag{5.222}$$

The optimal filters, summarized in Table 5.1, are employed in the STFT domain by replacing $\mathbf{\Phi}_{\mathbf{y}}(f)$, $\mathbf{\Phi}_{\mathbf{v}}(f)$, and $\mathbf{d}(f)$ with $\mathbf{\Phi}_{\mathbf{y}}(rR, k)$, $\mathbf{\Phi}_{\mathbf{v}}(rR, k)$, and $\mathbf{d}(k)$, respectively.

Example 5.7.1 Consider a ULA of M sensors, as shown in Figure 4.2. Suppose that a desired speech signal, $x(t)$, impinges on the ULA from the direction θ_x, and that an interference $u(t)$ impinges on the ULA from the direction θ_u. Assume that the interference $u(t)$ is white Gaussian noise, i.e., $u(t) \sim \mathcal{N}\left(0, \sigma_u^2\right)$, uncorrelated with $x(t)$. In addition, the sensors contain thermal white Gaussian noise, $w_m(t) \sim \mathcal{N}\left(0, \sigma_w^2\right)$, that are mutually uncorrelated. The desired speech signal needs to be recovered from the noisy received signals, $y_m(t) = x_m(t) + v_m(t)$, $m = 1, \ldots, M$, where $v_m(t) = u_m(t) + w_m(t)$, $m = 1, \ldots, M$ are the interference-plus-noise signals.

Assume that the sampling frequency is 16 kHz, and that the sampling interval T_s satisfies $T_s = \frac{d}{c}$. We have

$$x_m(t) = x_1\left(t - \tau_{x,m}\right),$$
$$u_m(t) = u_1\left(t - \tau_{u,m}\right),$$

where

$$\tau_{x,m} = \frac{(m-1)d\cos\theta_x}{cT_s} = (m-1)\cos\theta_x,$$
$$\tau_{u,m} = \frac{(m-1)d\cos\theta_u}{cT_s} = (m-1)\cos\theta_u.$$

In the STFT domain, we obtain

$$\mathbf{x}(rR,k) = X_1(rR,k)\mathbf{d}(k),$$
$$\mathbf{u}(rR,k) = U_1(rR,k)\boldsymbol{\gamma}_{U_1\mathbf{u}}^*(k),$$

where

$$\mathbf{d}(k) = \begin{bmatrix} 1 & e^{-\jmath 2\pi k\tau_{x,2}/L} & e^{-\jmath 2\pi k\tau_{x,3}/L} & \cdots & e^{-\jmath 2\pi k\tau_{x,M}/L} \end{bmatrix}^T,$$
$$\boldsymbol{\gamma}_{U_1\mathbf{u}}^*(k) = \begin{bmatrix} 1 & e^{-\jmath 2\pi k\tau_{u,2}/L} & e^{-\jmath 2\pi k\tau_{u,3}/L} & \cdots & e^{-\jmath 2\pi k\tau_{u,M}/L} \end{bmatrix}^T.$$

To demonstrate noise reduction in the STFT domain, we choose $\theta_x = 70°$, $\theta_u = 20°$, $\sigma_w^2 = 0.1\sigma_u^2$, a Hamming window of length $L = 512$ as the analysis window, a decimation factor $R = L/4 = 128$, and the Wiener filter in the STFT domain:

$$\mathbf{h}_{\mathrm{W}}(rR,k) = \phi_{X_1}(rR,k)\left[\phi_{X_1}(rR,k)\mathbf{d}(k)\mathbf{d}^H(k) + \boldsymbol{\Phi}_{\mathbf{v}}(rR,k)\right]^{-1}\mathbf{d}(k). \tag{5.223}$$

An estimate for the correlation matrix of $\mathbf{v}(rR,k)$ can be obtained by averaging past cross-spectral power values of the noisy measurement during speech inactivity:

$$\widehat{\boldsymbol{\Phi}}_{\mathbf{v}}(rR,k) = \begin{cases} \alpha\widehat{\boldsymbol{\Phi}}_{\mathbf{v}}\left[(r-1)R,k\right] + (1-\alpha)\mathbf{y}(rR,k)\mathbf{y}^H(rR,k), & X(rR,k) = 0 \\ \widehat{\boldsymbol{\Phi}}_{\mathbf{v}}\left[(r-1)R,k\right], & X(rR,k) \neq 0 \end{cases}, \tag{5.224}$$

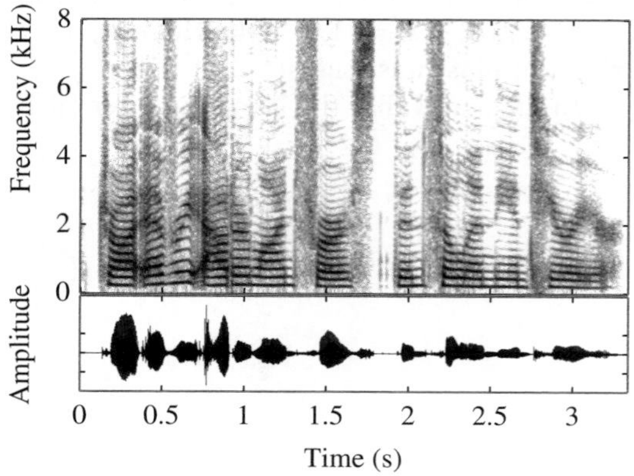

Figure 5.16 Speech spectrogram and waveform of a clean speech signal received at the first sensor, $x_1(t)$: "Draw every outer line first, then fill in the interior."

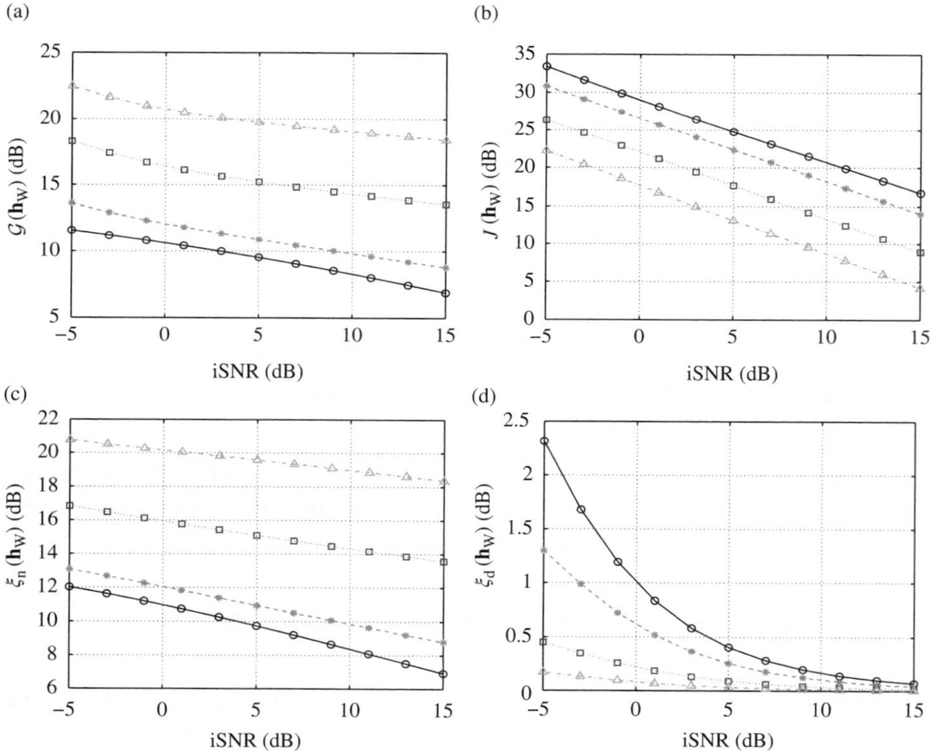

Figure 5.17 (a) The broadband gain in SNR, (b) the broadband MSE, (c) the broadband noise reduction factor, and (d) the broadband desired signal reduction factor of the Wiener filter as a function of the broadband input SNR, for different numbers of sensors, M: $M = 1$ (solid line with circles), $M = 2$ (dashed line with asterisks), $M = 5$ (dotted line with squares), and $M = 10$ (dash-dot line with triangles).

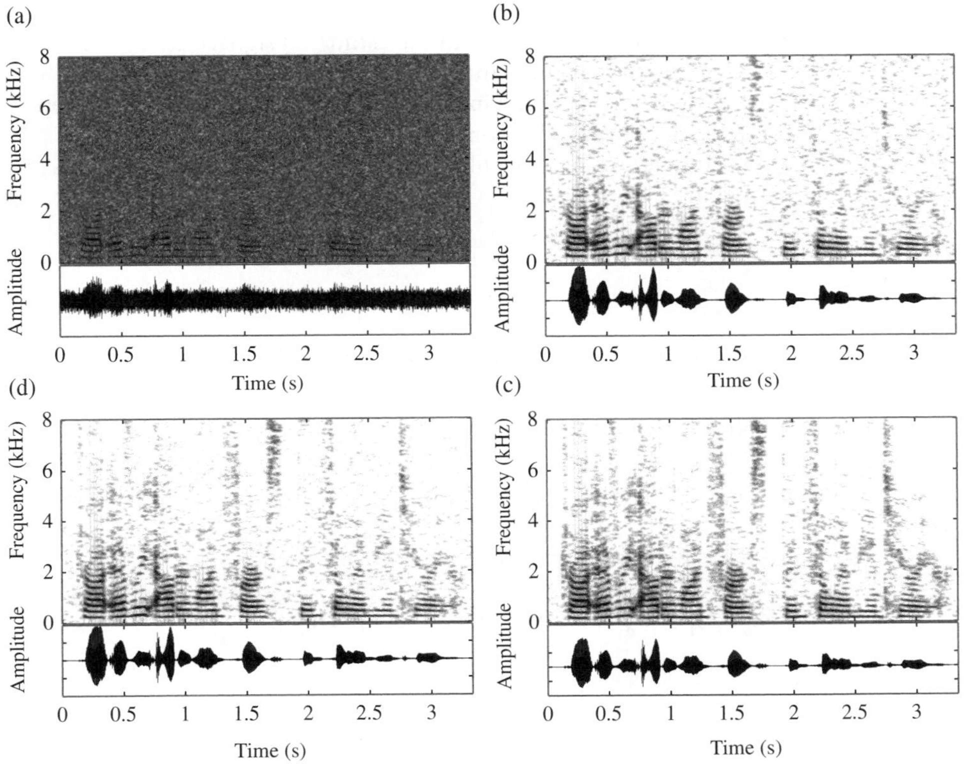

Figure 5.18 Speech spectrograms and waveforms of (a) noisy speech signal received at the first sensor, $y_1(t)$ (iSNR = −5 dB), and the estimated signal, $z(t)$, for (b) $M = 1$ [oSNR $(\mathbf{h}_{\mathrm{W}}) = 6.64$ dB], (c) $M = 2$ [oSNR $(\mathbf{h}_{\mathrm{W}}) = 8.72$ dB], and (d) $M = 5$ [oSNR $(\mathbf{h}_{\mathrm{W}}) = 13.34$ dB].

where α $(0 < \alpha < 1)$ denotes a smoothing parameter. This method requires a voice activity detector (VAD), but there are also alternative and more efficient methods that are based on minimum statistics [22, 23].

Finding an estimate for $\phi_{X_1}(rR, k)$ is a much more challenging problem [24, 25]. In this example, for simplicity, we smooth $|Y_1(rR, k)|^2$ in both time and frequency axes and subtract an estimate of the noise, i.e.,

$$\widehat{\phi}_{X_1}(rR, k) = \max\left\{\widehat{\phi}_{Y_1}(rR, k) - \widehat{\phi}_{V_1}(rR, k), 0\right\},$$

where $\widehat{\phi}_{Y_1}(rR, k)$ is obtained as a two-dimensional convolution between $|Y_1(rR, k)|^2$ and a smoothing window $w(rR, k)$. Here, the smoothing window is a two-dimensional Hamming window of size 3×11, normalized to $\sum_{r,k} w(rR, k) = 1$.

Figure 5.16 shows the spectrogram and waveform of the clean speech signal received at the first sensor, $x_1(t)$. Figure 5.17 shows plots of the broadband gain in SNR, $\mathcal{G}(\mathbf{h}_{\mathrm{W}})$, the broadband MSE, $J(\mathbf{h}_{\mathrm{W}})$, the broadband noise reduction factor, $\xi_{\mathrm{n}}(\mathbf{h}_{\mathrm{W}})$, and the broadband desired signal reduction factor, $\xi_{\mathrm{d}}(\mathbf{h}_{\mathrm{W}})$, as a function of the broadband input SNR, for different numbers of sensors, M. Figure 5.18 shows a realization of the observation signal at the first sensor, $y_1(t)$, and the estimated signals, $z(t)$, for different numbers of sensors, M. Clearly, as the number of sensors increases, the Wiener filter better enhances the desired speech signal in terms of higher SNR and noise reduction, and lower MSE and desired signal reduction.

Note that more useful algorithms for enhancing noisy speech signals in the STFT domain are presented in [2, 26, 27]. ■

Problems

5.1 Assume that the matrix $\mathbf{\Phi}_{\mathbf{v}}(f)$ is nonsingular. Show that

$$\left|\mathbf{h}^H(f)\mathbf{d}(f)\right|^2 \leq \left[\mathbf{h}^H(f)\mathbf{\Phi}_{\mathbf{v}}(f)\mathbf{h}(f)\right]\left[\mathbf{d}^H(f)\mathbf{\Phi}_{\mathbf{v}}^{-1}(f)\mathbf{d}(f)\right],$$

with equality if and only if $\mathbf{h}(f) \propto \mathbf{\Phi}_{\mathbf{v}}^{-1}(f)\mathbf{d}(f)$.

5.2 Show that the narrowband output SNR is upper bounded by

$$\mathrm{oSNR}\left[\mathbf{h}(f)\right] \leq \phi_{X_1}(f) \times \mathbf{d}^H(f)\mathbf{\Phi}_{\mathbf{v}}^{-1}(f)\mathbf{d}(f), \ \forall \mathbf{h}(f).$$

5.3 Show that

$$\mathrm{oSNR}\left[\mathbf{i}_{\mathrm{i}}(f)\right] \leq \phi_{X_1}(f) \times \mathbf{d}^H(f)\mathbf{\Phi}_{\mathbf{v}}^{-1}(f)\mathbf{d}(f).$$

5.4 Show that

$$\phi_{V_1}(f) \times \mathbf{d}^H(f)\mathbf{\Phi}_{\mathbf{v}}^{-1}(f)\mathbf{d}(f) \geq 1.$$

5.5 Show that the narrowband desired signal distortion index is given by

$$\upsilon_{\mathrm{d}}\left[\mathbf{h}(f)\right] = \left|\mathbf{h}^H(f)\boldsymbol{\gamma}_{X_1\mathbf{x}}^*(f) - 1\right|^2.$$

5.6 Show that the narrowband MSE can be written as

$$J\left[\mathbf{h}(f)\right] = \phi_{X_1}(f) + \mathbf{h}^H(f)\mathbf{\Phi}_{\mathbf{y}}(f)\mathbf{h}(f) - \phi_{X_1}(f)\mathbf{h}^H(f)\boldsymbol{\gamma}_{X_1\mathbf{x}}^*(f) - \phi_{X_1}(f)\boldsymbol{\gamma}_{X_1\mathbf{x}}^T(f)\mathbf{h}(f).$$

5.7 Show that the narrowband MSEs are related to the narrowband performance measures by

$$\begin{aligned}\frac{J_{\mathrm{d}}\left[\mathbf{h}(f)\right]}{J_{\mathrm{n}}\left[\mathbf{h}(f)\right]} &= \mathrm{iSNR}(f) \times \xi_{\mathrm{n}}\left[\mathbf{h}(f)\right] \times \upsilon_{\mathrm{d}}\left[\mathbf{h}(f)\right] \\ &= \mathrm{oSNR}\left[\mathbf{h}(f)\right] \times \xi_{\mathrm{d}}\left[\mathbf{h}(f)\right] \times \upsilon_{\mathrm{d}}\left[\mathbf{h}(f)\right].\end{aligned}$$

5.8 Show that the broadband MSEs are related to the broadband performance measures by

$$\frac{J_{\mathrm{d}}(\mathbf{h})}{J_{\mathrm{n}}(\mathbf{h})} = \mathrm{iSNR} \times \xi_{\mathrm{n}}(\mathbf{h}) \times \upsilon_{\mathrm{d}}(\mathbf{h})$$
$$= \mathrm{oSNR}(\mathbf{h}) \times \xi_{\mathrm{d}}(\mathbf{h}) \times \upsilon_{\mathrm{d}}(\mathbf{h}).$$

5.9 Show that the narrowband output SNR can be written as

$$\mathrm{oSNR}\left[\mathbf{h}(f)\right] = \frac{\phi_{X_1}(f)\mathbf{h}^H(f)\boldsymbol{\gamma}^*_{X_1\mathbf{x}}(f)\boldsymbol{\gamma}^T_{X_1\mathbf{x}}(f)\mathbf{h}(f)}{\mathbf{h}^H(f)\boldsymbol{\Phi}_{\mathbf{v}}(f)\mathbf{h}(f)}.$$

5.10 Show that the maximum eigenvalue corresponding to the eigenvector of the matrix $\phi_{X_1}(f)\boldsymbol{\Phi}^{-1}_{\mathbf{v}}(f)\boldsymbol{\gamma}^*_{X_1\mathbf{x}}(f)\boldsymbol{\gamma}^T_{X_1\mathbf{x}}(f)$ is given by

$$\lambda_1(f) = \phi_{X_1}(f)\boldsymbol{\gamma}^T_{X_1\mathbf{x}}(f)\boldsymbol{\Phi}^{-1}_{\mathbf{v}}(f)\boldsymbol{\gamma}^*_{X_1\mathbf{x}}(f).$$

5.11 Show that the maximum SNR filter is given by

$$\mathbf{h}_{\max}(f) = \varsigma(f)\boldsymbol{\Phi}^{-1}_{\mathbf{v}}(f)\boldsymbol{\gamma}^*_{X_1\mathbf{x}}(f),$$

where $\varsigma(f)$ is an arbitrary frequency-dependent complex number different from zero.

5.12 Denote by $\mathrm{oSNR}^{(m)}_{\max}(f)$ the maximum narrowband output SNR of an array with m sensors. Show that

$$\mathrm{oSNR}^{(M)}_{\max}(f) \geq \mathrm{oSNR}^{(M-1)}_{\max}(f) \geq \cdots \geq \mathrm{oSNR}^{(2)}_{\max}(f) \geq$$
$$\mathrm{oSNR}^{(1)}_{\max}(f) = \mathrm{iSNR}(f).$$

5.13 Consider a desired signal that is a harmonic pulse of T samples:

$$x(t) = \begin{cases} A\sin\left(2\pi f_0 t + \phi\right), & 0 \leq t \leq T-1 \\ 0, & t < 0,\ t \geq T \end{cases},$$

with fixed amplitude A and frequency f_0, and random phase ϕ, uniformly distributed on the interval from 0 to 2π. Suppose that the desired signal impinges on a ULA of M sensors from the direction θ_0.

a) Show that the variance of $\mathbf{x}(f)$ is

$$\boldsymbol{\Phi}_{\mathbf{x}}(f) = \phi_{X_1}(f)\mathbf{d}(f)\mathbf{d}^H(f),$$

where

$$\phi_{X_1}(f) = \frac{A^2}{4}D^2_T\left[\pi\left(f+f_0\right)\right] + \frac{A^2}{4}D^2_T\left[\pi\left(f-f_0\right)\right],$$
$$\mathbf{d}(f) = \left[\begin{array}{ccccc} 1 & e^{-\jmath 2\pi f\tau_2} & e^{-\jmath 2\pi f\tau_3} & \cdots & e^{-\jmath 2\pi f\tau_M} \end{array}\right]^T.$$

b) Assume an interference $u(t) \sim \mathcal{N}\left(0, \sigma_u^2\right)$ that impinges on the ULA from the endfire direction. Show that

$$\mathbf{\Phi}_{\mathbf{u}}(f) = \phi_{U_1}(f)\boldsymbol{\gamma}_{U_1\mathbf{u}}^*(f)\boldsymbol{\gamma}_{U_1\mathbf{u}}^T(f).$$

c) Compute the narrowband and broadband input SNRs.

d) Show that with the maximum SNR filter the narrowband gain in SNR can be written as

$$\begin{aligned}\mathcal{G}\left[\mathbf{h}_{\max}(f)\right] &= \frac{\text{oSNR}\left[\mathbf{h}_{\max}(f)\right]}{\text{iSNR}(f)} \\ &= \mathbf{d}^H(f)\left[\frac{\sigma_u^2}{\sigma_u^2+\sigma_w^2}\boldsymbol{\gamma}_{U_1\mathbf{u}}^*(f)\boldsymbol{\gamma}_{U_1\mathbf{u}}^T(f) + \frac{\sigma_w^2}{\sigma_u^2+\sigma_w^2}\mathbf{I}_M\right]^{-1}\mathbf{d}(f).\end{aligned}$$

5.14 Show that by minimizing the narrowband MSE, $J\left[\mathbf{h}(f)\right]$, we obtain the Wiener filter:

$$\mathbf{h}_{\text{W}}(f) = \phi_{X_1}(f)\mathbf{\Phi}_{\mathbf{y}}^{-1}(f)\boldsymbol{\gamma}_{X_1\mathbf{x}}^*(f).$$

5.15 Show that the Wiener filter can be expressed as

$$\mathbf{h}_{\text{W}}(f) = \frac{\text{iSNR}(f)}{1+\text{iSNR}(f)}\mathbf{\Gamma}_{\mathbf{y}}^{-1}(f)\boldsymbol{\gamma}_{X_1\mathbf{x}}^*(f).$$

5.16 Show that the Wiener filter can be written as

$$\mathbf{h}_{\text{W}}(f) = \left[\mathbf{I}_M - \mathbf{\Phi}_{\mathbf{y}}^{-1}(f)\mathbf{\Phi}_{\mathbf{v}}(f)\right]\mathbf{i}_{\text{i}}.$$

5.17 Show that the Wiener filter can be written as

$$\mathbf{h}_{\text{W}}(f) = \frac{\mathbf{\Phi}_{\mathbf{v}}^{-1}(f)\mathbf{\Phi}_{\mathbf{y}}(f) - \mathbf{I}_M}{1 - M + \text{tr}\left[\mathbf{\Phi}_{\mathbf{v}}^{-1}(f)\mathbf{\Phi}_{\mathbf{y}}(f)\right]}\mathbf{i}_{\text{i}}.$$

5.18 Show that with the Wiener filter, the narrowband output SNR is

$$\text{oSNR}\left[\mathbf{h}_{\text{W}}(f)\right] = \text{tr}\left[\mathbf{\Phi}_{\mathbf{v}}^{-1}(f)\mathbf{\Phi}_{\mathbf{y}}(f)\right] - M.$$

5.19 Show that with the Wiener filter, the broadband output SNR is always greater than or equal to the broadband input SNR, i.e., $\text{oSNR}\left(\mathbf{h}_{\text{W}}\right) \geq \text{iSNR}$.

5.20 Show that by minimizing the narrowband MSE of the residual noise, $J_{\text{n}}\left[\mathbf{h}(f)\right]$, with the constraint that the desired signal is not distorted yields

$$\mathbf{h}_{\text{MVDR}}(f) = \frac{\mathbf{\Phi}_{\mathbf{v}}^{-1}(f)\boldsymbol{\gamma}_{X_1\mathbf{x}}^*(f)}{\boldsymbol{\gamma}_{X_1\mathbf{x}}^T(f)\mathbf{\Phi}_{\mathbf{v}}^{-1}(f)\boldsymbol{\gamma}_{X_1\mathbf{x}}^*(f)}.$$

5.21 Show that the MVDR filter can be written as

$$\mathbf{h}_{\mathrm{MVDR}}(f) = \frac{\mathbf{\Phi}_{\mathbf{v}}^{-1}(f)\mathbf{\Phi}_{\mathbf{y}}(f) - \mathbf{I}_M}{\mathrm{tr}\left[\mathbf{\Phi}_{\mathbf{v}}^{-1}(f)\mathbf{\Phi}_{\mathbf{y}}(f)\right] - M}\mathbf{i}_{\mathrm{i}}.$$

5.22 Show that the MVDR filter can be written as

$$\mathbf{h}_{\mathrm{MVDR}}(f) = \frac{\left[1 + \mathrm{iSNR}(f)\right]\mathbf{\Gamma}_{\mathbf{v}}^{-1}(f)\mathbf{\Gamma}_{\mathbf{y}}(f) - \mathbf{I}_M}{\left[1 + \mathrm{iSNR}(f)\right]\mathrm{tr}\left[\mathbf{\Gamma}_{\mathbf{v}}^{-1}(f)\mathbf{\Gamma}_{\mathbf{y}}(f)\right] - M}\mathbf{i}_{\mathrm{i}}.$$

5.23 Show that the MVDR filter can be written as

$$\mathbf{h}_{\mathrm{MVDR}}(f) = \frac{\mathbf{\Phi}_{\mathbf{y}}^{-1}(f)\boldsymbol{\gamma}_{X_1\mathbf{x}}^{*}(f)}{\boldsymbol{\gamma}_{X_1\mathbf{x}}^{T}(f)\mathbf{\Phi}_{\mathbf{y}}^{-1}(f)\boldsymbol{\gamma}_{X_1\mathbf{x}}^{*}(f)}.$$

5.24 Show that the Wiener and MVDR filters are related by

$$\mathbf{h}_{\mathrm{W}}(f) = C(f)\mathbf{h}_{\mathrm{MVDR}}(f),$$

where $C(f)$ is a real number, with $0 < C(f) < 1$.

5.25 Show that

a) $\mathrm{oSNR}\left[\mathbf{h}_{\mathrm{MVDR}}(f)\right] = \mathrm{oSNR}\left[\mathbf{h}_{\mathrm{W}}(f)\right]$,
b) $\upsilon_{\mathrm{d}}\left[\mathbf{h}_{\mathrm{MVDR}}(f)\right] = 0$,
c) $\xi_{\mathrm{d}}\left[\mathbf{h}_{\mathrm{MVDR}}(f)\right] = 1$,
d) $\xi_{\mathrm{n}}\left[\mathbf{h}_{\mathrm{MVDR}}(f)\right] \leq \xi_{\mathrm{n}}\left[\mathbf{h}_{\mathrm{W}}(f)\right]$,
e) $\xi_{\mathrm{n}}\left(\mathbf{h}_{\mathrm{MVDR}}\right) \leq \xi_{\mathrm{n}}\left(\mathbf{h}_{\mathrm{W}}\right)$.

5.26 Show that with the MVDR filter, the broadband output SNR is always greater than or equal to the broadband input SNR, i.e., $\mathrm{oSNR}\left(\mathbf{h}_{\mathrm{MVDR}}\right) \geq \mathrm{iSNR}$.

5.27 Show that by minimizing the narrowband desired signal distortion index with the constraint that the narrowband noise reduction factor is equal to a positive value $\frac{1}{\aleph}$ we obtain the tradeoff filter:

$$\mathbf{h}_{\mathrm{T},\mu}(f) = \frac{\mathbf{\Phi}_{\mathbf{v}}^{-1}(f)\mathbf{\Phi}_{\mathbf{y}}(f) - \mathbf{I}_M}{\mu - M + \mathrm{tr}\left[\mathbf{\Phi}_{\mathbf{v}}^{-1}(f)\mathbf{\Phi}_{\mathbf{y}}(f)\right]}\mathbf{i}_{\mathrm{i}},$$

where μ is a Lagrange multiplier.

5.28 Show that for

a) $\mu = 1$, $\mathbf{h}_{\mathrm{T},1}(f) = \mathbf{h}_{\mathrm{W}}(f)$,
b) $\mu = 0$, $\mathbf{h}_{\mathrm{T},0}(f) = \mathbf{h}_{\mathrm{MVDR}}(f)$,
c) $\mu > 1$, $\mathbf{h}_{\mathrm{T},\mu}(f)$ results in a filter with low residual noise at the expense of high desired signal distortion (as compared to Wiener), and

d) $\mu < 1$, $\mathbf{h}_{\text{T},\mu}(f)$ results in a filter with high residual noise and low desired signal distortion (as compared to Wiener).

5.29 Show that the tradeoff, Wiener, and maximum SNR filters are equivalent up to a real-valued number.

5.30 Show that the narrowband output SNR of the tradeoff filter is independent of μ and is identical to the narrowband output SNR of the maximum SNR filter, i.e.,

$$\text{oSNR}\left[\mathbf{h}_{\text{T},\mu}(f)\right] = \text{oSNR}\left[\mathbf{h}_{\max}(f)\right], \ \forall \mu \geq 0.$$

5.31 Show that

$$\upsilon_{\text{d}}\left[\mathbf{h}_{\text{T},\mu}(f)\right] = \left[\frac{\mu}{\mu + \lambda_1(f)}\right]^2,$$

$$\xi_{\text{d}}\left[\mathbf{h}_{\text{T},\mu}(f)\right] = \left[1 + \frac{\mu}{\lambda_1(f)}\right]^2,$$

$$\xi_{\text{n}}\left[\mathbf{h}_{\text{T},\mu}(f)\right] = \frac{\left[\mu + \lambda_1(f)\right]^2}{\text{iSNR}(f) \times \lambda_1(f)}.$$

5.32 Show that the broadband output SNR of the tradeoff filter is an increasing function of the parameter μ.

5.33 Show that the broadband desired signal distortion index of the tradeoff filter is an increasing function of the parameter μ.

5.34 Show that with the tradeoff filter, the broadband output SNR is always greater than or equal to the broadband input SNR, i.e., $\text{oSNR}\left(\mathbf{h}_{\text{T},\mu}\right) \geq \text{iSNR}$, $\forall \mu \geq 0$.

5.35 Show that for $\mu \geq 1$,

$$\text{iSNR} \leq \text{oSNR}\left(\mathbf{h}_{\text{MVDR}}\right) \leq \text{oSNR}\left(\mathbf{h}_{\text{W}}\right) \leq \text{oSNR}\left(\mathbf{h}_{\text{T},\mu}\right),$$
$$0 = \upsilon_{\text{d}}\left(\mathbf{h}_{\text{MVDR}}\right) \leq \upsilon_{\text{d}}\left(\mathbf{h}_{\text{W}}\right) \leq \upsilon_{\text{d}}\left(\mathbf{h}_{\text{T},\mu}\right),$$

and for $0 \leq \mu \leq 1$,

$$\text{iSNR} \leq \text{oSNR}\left(\mathbf{h}_{\text{MVDR}}\right) \leq \text{oSNR}\left(\mathbf{h}_{\text{T},\mu}\right) \leq \text{oSNR}\left(\mathbf{h}_{\text{W}}\right),$$
$$0 = \upsilon_{\text{d}}\left(\mathbf{h}_{\text{MVDR}}\right) \leq \upsilon_{\text{d}}\left(\mathbf{h}_{\text{T},\mu}\right) \leq \upsilon_{\text{d}}\left(\mathbf{h}_{\text{W}}\right).$$

5.36 Show that by minimizing the energy at the filter output, with the constraints that the coherent noise components are cancelled and the desired signal is preserved, yields the LCMV filter:

$$\mathbf{h}_{\text{LCMV}}(f) = \mathbf{\Phi}_{\mathbf{y}}^{-1}(f)\mathbf{C}_{X_1V_1}(f)\left[\mathbf{C}_{X_1V_1}^H(f)\mathbf{\Phi}_{\mathbf{y}}^{-1}(f)\mathbf{C}_{X_1V_1}(f)\right]^{-1}\left[\begin{array}{c} 1 \\ 0 \end{array}\right].$$

5.37 Show that
a) $\text{oSNR}\left(\mathbf{h}_{\text{LCMV}}\right) \leq \text{oSNR}\left(\mathbf{h}_{\text{MVDR}}\right)$,
b) $\upsilon_{\text{d}}\left(\mathbf{h}_{\text{LCMV}}\right) = 0$,
c) $\xi_{\text{d}}\left(\mathbf{h}_{\text{LCMV}}\right) = 1$,
d) $\xi_{\text{n}}\left(\mathbf{h}_{\text{LCMV}}\right) \leq \xi_{\text{n}}\left(\mathbf{h}_{\text{MVDR}}\right) \leq \xi_{\text{n}}\left(\mathbf{h}_{\text{W}}\right)$.

5.38 Show that the LCMV filter can be split into two components operating on orthogonal subspaces:

$$\mathbf{h}_{\text{LCMV}}(f) = \mathbf{h}_{\text{MN}}(f) - \mathbf{B}_{\text{C}}(f)\mathbf{w}_{\text{GSC}}(f).$$

5.39 Show that the filter $\mathbf{w}_{\text{GSC}}(f)$ is given by

$$\mathbf{w}_{\text{GSC}}(f) = \left[\mathbf{B}_{\mathbf{C}}^H(f)\mathbf{\Phi}_{\mathbf{y}}(f)\mathbf{B}_{\text{C}}(f)\right]^{-1} \mathbf{B}_{\mathbf{C}}^H(f)\mathbf{\Phi}_{\mathbf{y}}(f)\mathbf{h}_{\text{MN}}(f).$$

5.40 Show that the minimization of $E\left[\left|\widehat{X}_1(f)\right|^2\right]$ with respect to $\mathbf{w}(f)$ yields the filter $\mathbf{w}_{\text{GSC}}(f)$.

5.41 Show that the two filters LCMV and GSC are equivalent, i.e.,

$$\begin{aligned}
&\mathbf{i}_{\text{c}}^T \left[\mathbf{C}^H(f)\mathbf{\Phi}_{\mathbf{y}}^{-1}(f)\mathbf{C}(f)\right]^{-1} \mathbf{C}^H(f)\mathbf{\Phi}_{\mathbf{y}}^{-1}(f) \\
&= \mathbf{h}_{\text{MN}}^H(f)\left\{\mathbf{I}_M - \mathbf{\Phi}_{\mathbf{y}}(f)\mathbf{B}_{\text{C}}(f)\left[\mathbf{B}_{\mathbf{C}}^H(f)\mathbf{\Phi}_{\mathbf{y}}(f)\mathbf{B}_{\text{C}}(f)\right]^{-1}\mathbf{B}_{\mathbf{C}}^H(f)\right\}.
\end{aligned}$$

5.42 Consider multichannel noise reduction filters that have the form $\mathbf{h}_{1:N}(f) = \mathbf{T}_{1:N}(f)\mathbf{a}(f)$. Show that
a) the narrowband WNG can be expressed as

$$\mathcal{W}\left[\mathbf{h}_{1:N}(f)\right] = \frac{\left|\mathbf{a}^H(f)\mathbf{T}_{1:N}^H(f)\mathbf{d}(f)\right|^2}{\mathbf{a}^H(f)\mathbf{T}_{1:N}^H(f)\mathbf{T}_{1:N}(f)\mathbf{a}(f)},$$

b) the vector $\mathbf{a}(f)$ that maximizes $\mathcal{W}\left[\mathbf{a}(f)\right]$ is

$$\mathbf{a}(f) = \varsigma(f)\left[\mathbf{T}_{1:N}^H(f)\mathbf{T}_{1:N}(f)\right]^{-1}\mathbf{T}_{1:N}^H(f)\mathbf{d}(f),$$

where $\varsigma(f) \neq 0$ is an arbitrary complex number,
c) the MSE is given by

$$\begin{aligned}
J\left[\mathbf{h}_{1:N}(f)\right] = &\left|\varsigma(f)\right|^2 \mathbf{d}^H(f)\mathbf{P}_{\mathbf{T}_{1:N}}(f)\mathbf{\Phi}_{\mathbf{y}}(f)\mathbf{P}_{\mathbf{T}_{1:N}}(f)\mathbf{d}(f) + \phi_{X_1}(f) \\
&- \varsigma^*(f)\phi_{X_1}(f)\mathbf{d}^H(f)\mathbf{P}_{\mathbf{T}_{1:N}}(f)\mathbf{d}(f) \\
&- \varsigma(f)\phi_{X_1}(f)\mathbf{d}^H(f)\mathbf{P}_{\mathbf{T}_{1:N}}(f)\mathbf{d}(f),
\end{aligned}$$

d) and the minimization of $J\left[\mathbf{h}_{1:N}(f)\right]$ leads to

$$\varsigma(f) = \frac{\phi_{X_1}(f)\mathbf{d}^H(f)\mathbf{P}_{\mathbf{T}_{1:N}}(f)\mathbf{d}(f)}{\mathbf{d}^H(f)\mathbf{P}_{\mathbf{T}_{1:N}}(f)\mathbf{\Phi}_{\mathbf{y}}(f)\mathbf{P}_{\mathbf{T}_{1:N}}(f)\mathbf{d}(f)}.$$

References

1 J. Benesty, J. Chen, and Y. Huang, *Microphone Array Signal Processing*. Berlin, Germany: Springer-Verlag, 2008.

2 J. Benesty, J. Chen, Y. Huang, and I. Cohen, *Noise Reduction in Speech Processing*. Berlin, Germany: Springer-Verlag, 2009.

3 J. Benesty, J. Chen, and E. Habets, *Speech Enhancement in the STFT Domain*. Springer Briefs in Electrical and Computer Engineering, 2011.

4 J. P. Dmochowski and J. Benesty, "Microphone arrays: fundamental concepts," in *Speech Processing in Modern Communication–Challenges and Perspectives*, I. Cohen, J. Benesty, and S. Gannot, Eds., Berlin, Germany: Springer-Verlag, 2008, Chapter 8, pp. 199–223, 2010.

5 J. N. Franklin, *Matrix Theory*. Englewood Cliffs, NJ: Prentice-Hall, 1968.

6 J. Capon, "High resolution frequency-wavenumber spectrum analysis," *Proc. IEEE*, vol. 57, pp. 1408–1418, Aug. 1969.

7 R. T. Lacoss, "Data adaptive spectral analysis methods," *Geophysics*, vol. 36, pp. 661–675, Aug. 1971.

8 M. Souden, J. Benesty, and S. Affes, "On the global output SNR of the parameterized frequency-domain multichannel noise reduction Wiener filter," *IEEE Signal Process. Lett.*, vol. 17, pp. 425–428, May 2010.

9 J. Benesty, J. Chen, Y. Huang, and J. Dmochowski, "On microphone-array beamforming from a MIMO acoustic signal processing perspective," *IEEE Trans. Audio, Speech, Language Process.*, vol. 15, pp. 1053–1065, Mar. 2007.

10 A. Booker and C. Y. Ong, "Multiple constraint adaptive filtering," *Geophysics*, vol. 36, pp. 498–509, June 1971.

11 O. Frost, "An algorithm for linearly constrained adaptive array processing," *Proc. IEEE*, vol. 60, pp. 926–935, Jan. 1972.

12 M. Er and A. Cantoni, "Derivative constraints for broad-band element space antenna array processors," *IEEE Trans. Acoust., Speech, Signal Process.*, vol. 31, pp. 1378–1393, Dec. 1983.

13 L. J. Griffiths and C. W. Jim, "An alternative approach to linearly constrained adaptive beamforming," *IEEE Trans. Antennas Propagat.*, vol. AP-30, pp. 27–34, Jan. 1982.

14 K. M. Buckley, "Broad-band beamforming and the generalized sidelobe canceller," *IEEE Trans. Acoust., Speech, Signal Process.*, vol. ASSP-34, pp. 1322–1323, Oct. 1986.

15 K. M. Buckley and L. J. Griffiths, "An adaptive generalized sidelobe canceller with derivative constraints," *IEEE Trans. Antennas Propagat.*, vol. AP-34, pp. 311–319, Mar. 1986.

16 S. Werner, J. A. Apolinário, Jr., and M. L. R. de Campos, "On the equivalence of RLS implementations of LCMV and GSC processors," *IEEE Signal Process. Lett.*, vol. 10, pp. 356–359, Dec. 2003.

17 B. R. Breed and J. Strauss, "A short proof of the equivalence of LCMV and GSC beamforming," *IEEE Signal Process. Lett.*, vol. 9, pp. 168–169, June 2002.
18 J. Wexler and S. Raz, "Discrete Gabor expansions," *Speech Process.*, vol. 21, pp. 207–220, Nov. 1990.
19 S. Qian and D. Chen, "Discrete Gabor transform," *IEEE Trans. Signal Process.*, vol. 41, pp. 2429–2438, July 1993.
20 Y. Avargel and I. Cohen, "On multiplicative transfer function approximation in the short-time Fourier transform domain," *IEEE Signal Process. Lett.*, vol. 14, pp. 337–340, May 2007.
21 R. E. Crochiere and L. R. Rabiner, *Multirate Digital Signal Processing.* Englewood Cliffs, New Jersey: Prentice-Hall, 1983.
22 R. Martin, "Noise power spectral density estimation based on optimal smoothing and minimum statistics," *IEEE Trans. Speech, Audio Process.*, vol. 9, pp. 504–512, July 2001.
23 I. Cohen, "Noise spectrum estimation in adverse environments: improved minima controlled recursive averaging," *IEEE Trans. Speech, Audio Process.*, vol. 11, pp. 466–475, Sept. 2003.
24 I. Cohen, "Relaxed statistical model for speech enhancement and a priori SNR estimation," *IEEE Trans. Speech, Audio Process.*, vol. 13, pp. 870–881, Sept. 2005.
25 I. Cohen, "Speech spectral modeling and enhancement based on autoregressive conditional heteroscedasticity models," *Signal Process.*, vol. 86, pp. 698–709, Apr. 2006.
26 I. Cohen and B. Berdugo, "Speech enhancement for non-stationary noise environments," *Signal Process.*, vol. 81, pp. 2403–2418, Nov. 2001.
27 I. Cohen and S. Gannot, "Spectral enhancement methods," in J. Benesty, M. M. Sondhi, and Y. Huang (Eds.), *Springer Handbook of Speech Processing*, Springer-Verlag, 2008, Part H, Chapter 44, pp. 873–901.

6

An Exhaustive Class of Linear Filters

This chapter is a generalization of the four previous chapters but presented in a more unified framework. Therefore the signal enhancement problem in the time or frequency domain, with one sensor or multiple sensors, looks very similar. Within this framework, we derive a very large class of well-known optimal linear filters as well as a category of filters whose output signal-to-interference-plus-noise ratios (SINRs) are between the conventional maximum SINR and Wiener filters. With this very flexible approach, any kind of filter can be designed in order to make a compromise, in a very precise manner, between interference-plus-noise reduction and desired signal distortion. This chapter will also serve as a bridge between the problem of noise reduction studied so far and the forthcoming chapters on beamforming.

6.1 Signal Model and Problem Formulation

We consider the very general signal model of an observed signal's vector of length M:

$$\begin{aligned}\mathbf{y} &= \begin{bmatrix} y_1 & y_2 & \cdots & y_M \end{bmatrix}^T \\ &= \mathbf{x} + \mathbf{v}_0 + \sum_{n=1}^{N} \mathbf{v}_n \\ &= \mathbf{x} + \mathbf{v}_0 + \mathbf{v},\end{aligned} \tag{6.1}$$

where $\mathbf{x}$ is the desired signal vector, $\mathbf{v}_0$ is the additive white noise signal vector, $\mathbf{v}_n$, $n = 1, 2, \ldots, N$ are N interferences, and $\mathbf{v} = \sum_{n=1}^{N} \mathbf{v}_n$. All vectors on the right-hand side of (6.1) are defined in a similar way to the noisy signal vector, $\mathbf{y}$. The entries of $\mathbf{y}$ can be, for example, the signals picked up by M sensors. All signals are considered to be random, complex, circular, zero mean, and stationary. Furthermore, the vectors $\mathbf{x}$, $\mathbf{v}_0$, and $\mathbf{v}_n$, $n = 1, 2, \ldots, N$ are assumed to be mutually uncorrelated: $E\left(\mathbf{x}\mathbf{v}_0^H\right) = E\left(\mathbf{x}\mathbf{v}_n^H\right) = E\left(\mathbf{v}_0\mathbf{v}_n^H\right) = E\left(\mathbf{v}_i\mathbf{v}_j^H\right) = \mathbf{0}$, $\forall i \neq j$, $i, j = 1, 2, \ldots, N$. In this context, the correlation matrix (of size $M \times M$) of the observations can be written as

Fundamentals of Signal Enhancement and Array Signal Processing, First Edition.
Jacob Benesty, Israel Cohen, and Jingdong Chen.

Companion website: www.wiley.com/go/benesty/arraysignalprocessing

$$\begin{aligned}\mathbf{\Phi}_{\mathbf{y}} &= E\left(\mathbf{y}\mathbf{y}^H\right) \\ &= \mathbf{\Phi}_{\mathbf{x}} + \mathbf{\Phi}_{\mathbf{v}_0} + \sum_{n=1}^{N} \mathbf{\Phi}_{\mathbf{v}_n} \\ &= \mathbf{\Phi}_{\mathbf{x}} + \mathbf{\Phi}_{\mathbf{v}_0} + \mathbf{\Phi}_{\mathbf{v}} \\ &= \mathbf{\Phi}_{\mathbf{x}} + \mathbf{\Phi}_{\mathrm{in}},\end{aligned} \tag{6.2}$$

where $\mathbf{\Phi}_{\mathbf{x}} = E\left(\mathbf{x}\mathbf{x}^H\right)$, $\mathbf{\Phi}_{\mathbf{v}_0} = E\left(\mathbf{v}_0\mathbf{v}_0^H\right)$, and $\mathbf{\Phi}_{\mathbf{v}_n} = E\left(\mathbf{v}_n\mathbf{v}_n^H\right)$ are the correlation matrices of $\mathbf{x}$, $\mathbf{v}_0$, and $\mathbf{v}_n$, respectively, $\mathbf{\Phi}_{\mathbf{v}} = \sum_{n=1}^{N} \mathbf{\Phi}_{\mathbf{v}_n}$, and $\mathbf{\Phi}_{\mathrm{in}} = \mathbf{\Phi}_{\mathbf{v}_0} + \mathbf{\Phi}_{\mathbf{v}}$ is the interference-plus-noise correlation matrix. Since $\mathbf{v}_0$ is assumed to be white, its correlation matrix simplifies to $\mathbf{\Phi}_{\mathbf{v}_0} = \phi_{v_0}\mathbf{I}_M$, where $\phi_{v_0} = E\left(|v_0|^2\right)$ is the variance of v_0, the first component of $\mathbf{v}_0$. In the rest of this chapter, the desired signal and interference correlation matrices are assumed to have the following ranks: rank $\left(\mathbf{\Phi}_{\mathbf{x}}\right) = R_x \leq M$ and rank $\left(\mathbf{\Phi}_{\mathbf{v}_n}\right) = R_{v_n} < M$. Let $R_v = \min\left(\sum_{n=1}^{N} R_{v_n}, M\right)$. We deduce that rank $\left(\mathbf{\Phi}_{\mathbf{v}}\right) = R_v$ and, obviously, rank $\left(\mathbf{\Phi}_{\mathrm{in}}\right) = M$. Then, the objective of signal enhancement (or noise reduction) is to estimate the first element of $\mathbf{x}$, x_1 (the desired signal sample), from the different second-order statistics available from (6.2) in the best possible way. This should be done in such a way that the noise and the interference are reduced as much as possible, with little or no distortion of the desired signal. The matrix $\mathbf{\Phi}_{\mathbf{y}}$ can be easily estimated from the observations, but $\mathbf{\Phi}_{\mathbf{v}_0}$ and $\mathbf{\Phi}_{\mathbf{v}}$ are more tricky to estimate. However, in many applications, it is still possible to get reliable estimates of these matrices [1, 2], which will be assumed here.

A very important particular case of the model described above is the conventional beamforming problem, which can be formulated as [3, 4]:

$$\mathbf{y} = \mathbf{d}x_1 + \mathbf{v}_0 + \mathbf{v}, \tag{6.3}$$

where $\mathbf{d}$ is the steering vector of length M, the first entry of which is equal to 1. This vector can be deterministic or random. In the former case, the desired signal correlation matrix is $\mathbf{\Phi}_{\mathbf{x}} = \phi_{x_1}\mathbf{d}\mathbf{d}^H$ (the rank of which is, indeed, equal to 1), where $\phi_{x_1} = E\left(|x_1|^2\right)$ is the variance of x_1. When the steering vector is random, the rank of $\mathbf{\Phi}_{\mathbf{x}}$ is no longer 1 [5].

Some decompositions of the different matrices are necessary in order to fully exploit the structure of the signals. Using the well-known eigenvalue decomposition [6], the desired signal correlation matrix can be diagonalized as

$$\mathbf{Q}_{\mathbf{x}}^H\mathbf{\Phi}_{\mathbf{x}}\mathbf{Q}_{\mathbf{x}} = \mathbf{\Lambda}_{\mathbf{x}}, \tag{6.4}$$

where

$$\mathbf{Q}_{\mathbf{x}} = \begin{bmatrix} \mathbf{q}_{\mathbf{x},1} & \mathbf{q}_{\mathbf{x},2} & \cdots & \mathbf{q}_{\mathbf{x},M} \end{bmatrix} \tag{6.5}$$

is a unitary matrix: $\mathbf{Q}_{\mathbf{x}}^H\mathbf{Q}_{\mathbf{x}} = \mathbf{Q}_{\mathbf{x}}\mathbf{Q}_{\mathbf{x}}^H = \mathbf{I}_M$, and

$$\mathbf{\Lambda}_{\mathbf{x}} = \mathrm{diag}\left(\lambda_{\mathbf{x},1}, \lambda_{\mathbf{x},2}, \ldots, \lambda_{\mathbf{x},M}\right) \tag{6.6}$$

is a diagonal matrix. The orthonormal vectors $\mathbf{q}_{\mathbf{x},1}, \mathbf{q}_{\mathbf{x},2}, \ldots, \mathbf{q}_{\mathbf{x},M}$ are the eigenvectors corresponding, respectively, to the eigenvalues $\lambda_{\mathbf{x},1}, \lambda_{\mathbf{x},2}, \ldots, \lambda_{\mathbf{x},M}$ of the matrix $\mathbf{\Phi}_{\mathbf{x}}$, where $\lambda_{\mathbf{x},1} \geq \lambda_{\mathbf{x},2} \geq \cdots \geq \lambda_{\mathbf{x},R_x} > 0$ and $\lambda_{\mathbf{x},R_x+1} = \lambda_{\mathbf{x},R_x+2} = \cdots = \lambda_{\mathbf{x},M} = 0$. In the same way, the nth interference correlation matrix can be diagonalized as

$$\mathbf{Q}_{\mathbf{v}_n}^H \mathbf{\Phi}_{\mathbf{v}_n} \mathbf{Q}_{\mathbf{v}_n} = \mathbf{\Lambda}_{\mathbf{v}_n}, \tag{6.7}$$

where the unitary and diagonal matrices $\mathbf{Q}_{\mathbf{v}_n}$ and $\mathbf{\Lambda}_{\mathbf{v}_n}$ are defined in a similar way to $\mathbf{Q}_{\mathbf{x}}$ and $\mathbf{\Lambda}_{\mathbf{x}}$, respectively, with $\lambda_{\mathbf{v}_n,1} \geq \lambda_{\mathbf{v}_n,2} \geq \cdots \geq \lambda_{\mathbf{v}_n,R_{v_n}} > 0$ and $\lambda_{\mathbf{v}_n,R_{v_n}+1} = \lambda_{\mathbf{v}_n,R_{v_n}+2} = \cdots = \lambda_{\mathbf{v}_n,M} = 0$. It may also be useful to diagonalize the matrix $\mathbf{\Phi}_{\mathbf{v}}$ as well; that is,

$$\mathbf{Q}_{\mathbf{v}}^H \mathbf{\Phi}_{\mathbf{v}} \mathbf{Q}_{\mathbf{v}} = \mathbf{\Lambda}_{\mathbf{v}}, \tag{6.8}$$

where $\mathbf{Q}_{\mathbf{v}}$ and $\mathbf{\Lambda}_{\mathbf{v}}$ are similarly defined to $\mathbf{Q}_{\mathbf{x}}$ and $\mathbf{\Lambda}_{\mathbf{x}}$, respectively, with $\lambda_{\mathbf{v},1} \geq \lambda_{\mathbf{v},2} \geq \cdots \geq \lambda_{\mathbf{v},R_v} > 0$ and $\lambda_{\mathbf{v},R_v+1} = \lambda_{\mathbf{v},R_v+2} = \cdots = \lambda_{\mathbf{v},M} = 0$. All these decompositions will be extensively used in the rest of this chapter.

6.2 Linear Filtering for Signal Enhancement

By far, the most convenient and practical way to perform signal enhancement – to estimate the desired signal, x_1 – is by applying a linear filter to the observation signal vector, $\mathbf{y}$, as illustrated in Figure 6.1:

$$\begin{aligned} z &= \mathbf{h}^H \mathbf{y} \\ &= \mathbf{h}^H \left(\mathbf{x} + \mathbf{v}_0 + \mathbf{v}\right) \\ &= x_{\text{fd}} + v_{\text{rn}} + v_{\text{ri}}, \end{aligned} \tag{6.9}$$

where z is the estimate of x_1 or the filter output signal,

$$\mathbf{h} = \begin{bmatrix} h_1 & h_2 & \cdots & h_M \end{bmatrix}^T \tag{6.10}$$

is a complex-valued filter of length M,

$$x_{\text{fd}} = \mathbf{h}^H \mathbf{x} \tag{6.11}$$

is the filtered desired signal,

$$v_{\text{rn}} = \mathbf{h}^H \mathbf{v}_0 \tag{6.12}$$

is the residual noise, and

$$v_{\text{ri}} = \mathbf{h}^H \mathbf{v} \tag{6.13}$$

is the residual interference. We deduce that the variance of z is

$$\begin{aligned} \phi_z &= E\left(|z|^2\right) \\ &= \phi_{x_{\text{fd}}} + \phi_{v_{\text{rn}}} + \phi_{v_{\text{ri}}}, \end{aligned} \tag{6.14}$$

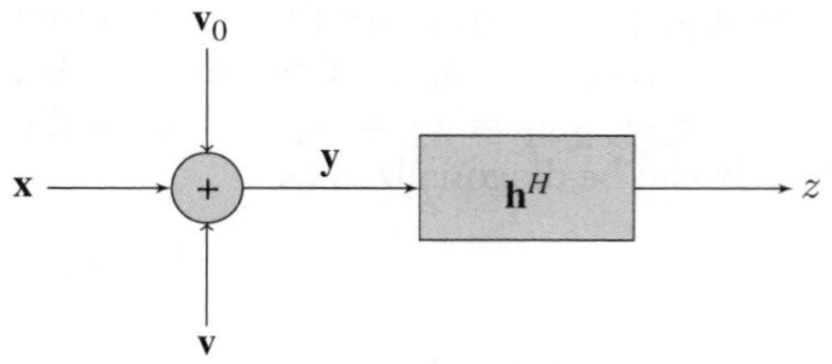

Figure 6.1 Block diagram of linear filtering.

where

$$\phi_{x_{\rm fd}} = \mathbf{h}^H \mathbf{\Phi}_{\mathbf{x}} \mathbf{h}, \tag{6.15}$$

$$\phi_{v_{\rm rn}} = \phi_{v_0} \mathbf{h}^H \mathbf{h}, \tag{6.16}$$

$$\phi_{v_{\rm ri}} = \mathbf{h}^H \mathbf{\Phi}_{\mathbf{v}} \mathbf{h}. \tag{6.17}$$

6.3 Performance Measures

Performance measures are not only useful for the derivation of different kinds of optimal filters but also for their evaluation. These measures can be divided into two distinct but related categories. The first category evaluates the noise reduction performance while the second evaluates the desired signal distortion. We will use, as before, the MSE criterion.

One of the most fundamental measures in our context is the signal-to-interference-plus-noise ratio (SINR). The input SINR is a second-order measure, which quantifies the level of the interference-plus-noise present relative to the level of the desired signal. By taking the first element of $\mathbf{y}$ as the reference, this measure is defined as

$$\text{iSINR} = \frac{\phi_{x_1}}{\phi_{v_0} + \phi_v}, \tag{6.18}$$

where ϕ_v is the variance of $v = \sum_{n=1}^{N} v_{n1}$, with v_{n1} being the first entry of $\mathbf{v}_n$. Another useful measure is the input signal-to-interference ratio (SIR):

$$\text{iSIR} = \frac{\phi_{x_1}}{\phi_v}. \tag{6.19}$$

The output SINR helps quantify the level of the interference-plus-noise remaining in the filter output signal. The output SINR is obtained from (6.14):

$$\text{oSINR}(\mathbf{h}) = \frac{\phi_{x_{\rm fd}}}{\phi_{v_{\rm rn}} + \phi_{v_{\rm ri}}} \tag{6.20}$$

$$= \frac{\mathbf{h}^H \mathbf{\Phi}_{\mathbf{x}} \mathbf{h}}{\mathbf{h}^H \mathbf{\Phi}_{\text{in}} \mathbf{h}}.$$

Basically, (6.20) is the variance of the first signal (filtered desired) from the right-hand side of (6.14) over the variance of the two other signals (residual interference-plus-noise). Since the matrix $\mathbf{\Phi}_{\text{in}}$ in the denominator of (6.20) is full rank, the output SINR is upper bounded. The objective of the signal enhancement filter is to make the output SINR greater than the input SINR. Consequently, the quality of the filter output signal may be enhanced compared to the noisy signal. It is straightforward to see that the output SIR is

$$\text{oSIR}(\mathbf{h}) = \frac{\mathbf{h}^H \mathbf{\Phi}_{\mathbf{x}} \mathbf{h}}{\mathbf{h}^H \mathbf{\Phi}_{\mathbf{v}} \mathbf{h}}. \tag{6.21}$$

Since the matrix $\mathbf{\Phi}_{\mathbf{v}}$ in the denominator of (6.21) may not be full rank, the output SIR may not be upper bounded.

For the particular filter of length M:

$$\mathbf{i}_{\text{i}} = \begin{bmatrix} 1 & 0 & \cdots & 0 \end{bmatrix}^T, \tag{6.22}$$

we have

$$\text{oSINR}\left(\mathbf{i}_{\text{i}}\right) = \text{iSINR}, \tag{6.23}$$

$$\text{oSIR}\left(\mathbf{i}_{\text{i}}\right) = \text{iSIR}. \tag{6.24}$$

With the identity filter, $\mathbf{i}_{\text{i}}$, neither the SINR nor the SIR can be improved.

Since the noise and interference are reduced by the filtering operation, so, in general, is the desired signal. This implies distortion, which we can measure with the desired signal distortion index, which is defined as the MSE between the desired signal and the filtered desired signal, normalized by the variance of the desired signal:

$$\upsilon(\mathbf{h}) = \frac{E\left(\left|x_{\text{fd}} - x_1\right|^2\right)}{E\left(\left|x_1\right|^2\right)} \tag{6.25}$$

$$= \frac{\left(\mathbf{h} - \mathbf{i}_{\text{i}}\right)^H \mathbf{\Phi}_{\mathbf{x}} \left(\mathbf{h} - \mathbf{i}_{\text{i}}\right)}{\phi_{x_1}}.$$

The desired signal distortion index is close to 0 if there is little distortion and greater than 0 when distortion occurs.

Error criteria play a critical role in deriving optimal filters. The MSE [7], as we already know, is, by far, the most practical one. We define the error signal between the estimated and desired signals as

$$e = z - x_1 \tag{6.26}$$

$$= x_{\text{fd}} + v_{\text{rn}} + v_{\text{ri}} - x_1,$$

which can be written as the sum of three mutually uncorrelated error signals:

$$e = e_{\rm d} + e_{\rm n} + e_{\rm i}, \tag{6.27}$$

where

$$\begin{aligned} e_{\rm d} &= x_{\rm fd} - x_1 \\ &= \left(\mathbf{h} - \mathbf{i}_{\rm i}\right)^H \mathbf{x} \end{aligned} \tag{6.28}$$

is the desired signal distortion due to the filter,

$$e_{\rm n} = v_{\rm rn} = \mathbf{h}^H \mathbf{v}_0 \tag{6.29}$$

is the residual noise, and

$$e_{\rm i} = v_{\rm ri} = \mathbf{h}^H \mathbf{v} \tag{6.30}$$

is the residual interference. Therefore, the MSE criterion is

$$\begin{aligned} J(\mathbf{h}) &= E\left(|e|^2\right) \\ &= \phi_{x_1} + \mathbf{h}^H \mathbf{\Phi}_{\mathbf{y}} \mathbf{h} - \mathbf{h}^H \mathbf{\Phi}_{\mathbf{x}} \mathbf{i}_{\rm i} - \mathbf{i}_{\rm i}^T \mathbf{\Phi}_{\mathbf{x}} \mathbf{h} \\ &= J_{\rm d}(\mathbf{h}) + J_{\rm n}(\mathbf{h}) + J_{\rm i}(\mathbf{h}), \end{aligned} \tag{6.31}$$

where

$$\begin{aligned} J_{\rm d}(\mathbf{h}) &= E\left(\left|e_{\rm d}\right|^2\right) \\ &= \left(\mathbf{h} - \mathbf{i}_{\rm i}\right)^H \mathbf{\Phi}_{\mathbf{x}} \left(\mathbf{h} - \mathbf{i}_{\rm i}\right) = \phi_{x_1} \upsilon(\mathbf{h}), \end{aligned} \tag{6.32}$$

$$J_{\rm n}(\mathbf{h}) = E\left(\left|e_{\rm n}\right|^2\right) = \phi_{v_0} \mathbf{h}^H \mathbf{h}, \tag{6.33}$$

$$J_{\rm i}(\mathbf{h}) = E\left(\left|e_{\rm i}\right|^2\right) = \mathbf{h}^H \mathbf{\Phi}_{\mathbf{v}} \mathbf{h}. \tag{6.34}$$

6.4 Optimal Filters

In this section, we derive a large class of well-known optimal linear filters by fully exploiting the structure of the signals, which was not really done before. To that end, the performance measures described in the previous section are of great help.

6.4.1 Wiener

The Wiener filter is derived from the MSE, $J(\mathbf{h})$, in (6.31), by taking its gradient with respect to $\mathbf{h}$ and equating the result to zero:

$$\mathbf{h}_{\rm W} = \mathbf{\Phi}_{\mathbf{y}}^{-1} \mathbf{\Phi}_{\mathbf{x}} \mathbf{i}_{\rm i}. \tag{6.35}$$

This optimal filter can also be expressed as

$$\mathbf{h}_{\mathrm{W}} = \left(\mathbf{I}_M - \boldsymbol{\Phi}_{\mathbf{y}}^{-1}\boldsymbol{\Phi}_{\mathrm{in}}\right)\mathbf{i}_{\mathrm{i}}. \tag{6.36}$$

The above formulation is more useful than (6.35) in practice since it depends on the second-order statistics of the observation and interference-plus-noise signals. The correlation matrix $\boldsymbol{\Phi}_{\mathbf{y}}$ can be estimated from the observation signal while the other correlation matrix, $\boldsymbol{\Phi}_{\mathrm{in}}$, is often known or can be indirectly estimated. In speech applications, for example, this matrix can be estimated during silences.

Let

$$\mathbf{Q}_{\mathbf{x}} = \begin{bmatrix} \mathbf{Q}'_{\mathbf{x}} & \mathbf{Q}''_{\mathbf{x}} \end{bmatrix}, \tag{6.37}$$

where the $M \times R_x$ matrix $\mathbf{Q}'_{\mathbf{x}}$ contains the eigenvectors corresponding to the nonzero eigenvalues of $\boldsymbol{\Phi}_{\mathbf{x}}$ and the $M \times (M - R_x)$ matrix $\mathbf{Q}''_{\mathbf{x}}$ contains the eigenvectors corresponding to the null eigenvalues of $\boldsymbol{\Phi}_{\mathbf{x}}$. It can be verified that

$$\mathbf{I}_M = \mathbf{Q}'_{\mathbf{x}}\mathbf{Q}'^{H}_{\mathbf{x}} + \mathbf{Q}''_{\mathbf{x}}\mathbf{Q}''^{H}_{\mathbf{x}}. \tag{6.38}$$

Notice that $\mathbf{Q}'_{\mathbf{x}}\mathbf{Q}'^{H}_{\mathbf{x}}$ and $\mathbf{Q}''_{\mathbf{x}}\mathbf{Q}''^{H}_{\mathbf{x}}$ are two orthogonal projection matrices of rank R_x and $M - R_x$, respectively. Hence $\mathbf{Q}'_{\mathbf{x}}\mathbf{Q}'^{H}_{\mathbf{x}}$ is the orthogonal projector onto the desired signal subspace (where all the energy of the desired signal is concentrated) or the range of $\boldsymbol{\Phi}_{\mathbf{x}}$, and $\mathbf{Q}''_{\mathbf{x}}\mathbf{Q}''^{H}_{\mathbf{x}}$ is the orthogonal projector onto the null subspace of $\boldsymbol{\Phi}_{\mathbf{x}}$. With the eigenvalue decomposition of $\boldsymbol{\Phi}_{\mathbf{x}}$, the correlation matrix of the observations' signal vector can be written as

$$\boldsymbol{\Phi}_{\mathbf{y}} = \mathbf{Q}'_{\mathbf{x}}\boldsymbol{\Lambda}'_{\mathbf{x}}\mathbf{Q}'^{H}_{\mathbf{x}} + \boldsymbol{\Phi}_{\mathrm{in}}, \tag{6.39}$$

where

$$\boldsymbol{\Lambda}'_{\mathbf{x}} = \mathrm{diag}\left(\lambda_{\mathbf{x},1}, \lambda_{\mathbf{x},2}, \ldots, \lambda_{\mathbf{x},R_x}\right) \tag{6.40}$$

is a diagonal matrix of size $R_x \times R_x$. Determining the inverse of $\boldsymbol{\Phi}_{\mathbf{y}}$ from (6.39) with the Woodbury identity, we get

$$\boldsymbol{\Phi}_{\mathbf{y}}^{-1} = \boldsymbol{\Phi}_{\mathrm{in}}^{-1} - \boldsymbol{\Phi}_{\mathrm{in}}^{-1}\mathbf{Q}'_{\mathbf{x}}\left(\boldsymbol{\Lambda}'^{-1}_{\mathbf{x}} + \mathbf{Q}'^{H}_{\mathbf{x}}\boldsymbol{\Phi}_{\mathrm{in}}^{-1}\mathbf{Q}'_{\mathbf{x}}\right)^{-1}\mathbf{Q}'^{H}_{\mathbf{x}}\boldsymbol{\Phi}_{\mathrm{in}}^{-1}. \tag{6.41}$$

Substituting (6.41) into (6.35), leads to another useful formulation of the Wiener filter:

$$\mathbf{h}_{\mathrm{W}} = \boldsymbol{\Phi}_{\mathrm{in}}^{-1}\mathbf{Q}'_{\mathbf{x}}\left(\boldsymbol{\Lambda}'^{-1}_{\mathbf{x}} + \mathbf{Q}'^{H}_{\mathbf{x}}\boldsymbol{\Phi}_{\mathrm{in}}^{-1}\mathbf{Q}'_{\mathbf{x}}\right)^{-1}\mathbf{Q}'^{H}_{\mathbf{x}}\mathbf{i}_{\mathrm{i}}. \tag{6.42}$$

It can be shown that with the optimal Wiener filter given in (6.35), the output SINR is always greater than or equal to the input SINR: $\mathrm{oSINR}\left(\mathbf{h}_{\mathrm{W}}\right) \geq \mathrm{iSINR}$ [8].

Example 6.4.1 Consider a ULA of M sensors, as shown in Figure 4.2. Suppose that a desired signal impinges on the ULA from the direction θ_x and that an interference impinges on the ULA from the direction θ_v. Assume that the desired signal received at the first sensor is a complex harmonic random process:

$$x_1(t) = A \exp\left(\jmath 2\pi f_0 t + \jmath\varphi\right),$$

with fixed amplitude A and frequency f_0, and random phase φ, uniformly distributed on the interval from 0 to 2π. Assume that the interference received at the first sensor, $v_1(t)$, is a random process with the autocorrelation sequence:

$$E\left[v_1(t)v_1(t')\right] = \alpha^{|t-t'|}, \ -1 < \alpha < 1.$$

In addition, the sensors contain thermal white Gaussian noise, with correlation matrix $\mathbf{\Phi}_{\mathbf{v}_0} = \phi_{v_0}\mathbf{I}_M$. The desired signal needs to be recovered from the noisy observation, $\mathbf{y}(t) = \mathbf{d}x_1(t) + \mathbf{v}_0 + \mathbf{v}$, where $\mathbf{d}$ is the steering vector of the desired signal.

Since the desired source impinges on the ULA from the direction θ_x, we have for $m = 2, \ldots, M$:

$$x_m(t) = x_1\left(t - \tau_{x,m}\right) = e^{-\jmath 2\pi f_0 \tau_{x,m}} x_1(t),$$

where

$$\tau_{x,m} = \frac{(m-1)d\cos\theta_x}{cT_s}$$

is the relative time delay in samples between the desired signals $x_m(t)$ and $x_1(t)$ received at the mth sensor and the first one, and T_s is the sampling interval. Hence, $\mathbf{x}(t) = \mathbf{d}x_1(t)$, where

$$\mathbf{d} = \begin{bmatrix} 1 & e^{-\jmath 2\pi f_0 \tau_{x,2}} & e^{-\jmath 2\pi f_0 \tau_{x,3}} & \cdots & e^{-\jmath 2\pi f_0 \tau_{x,M}} \end{bmatrix}^T.$$

Similarly,

$$u_m(t) = u_1\left(t - \tau_{v,m}\right),$$

where

$$\tau_{v,m} = \frac{(m-1)d\cos\theta_v}{cT_s}$$

is the relative time delay in samples between the interferences received at the mth sensor and the first one. Assuming that the sampling interval satisfies $T_s = \frac{d}{c}$, we have $\tau_{x,m} = (m-1)\cos\theta_x$ and $\tau_{v,m} = (m-1)\cos\theta_v$. The desired signal correlation matrix is $\mathbf{\Phi}_{\mathbf{x}} = \phi_{x_1}\mathbf{d}\mathbf{d}^H$, where $\phi_{x_1} = A^2$. The elements of the $M \times M$ correlation matrix of the interference are $\left[\mathbf{\Phi}_{\mathbf{v}}\right]_{i,j} = \alpha^{|\tau_{v,i} - \tau_{v,j}|}$.

The input SINR is

$$\text{iSINR} = 10\log\frac{A^2}{\phi_{v_0} + 1} \quad \text{(dB)}.$$

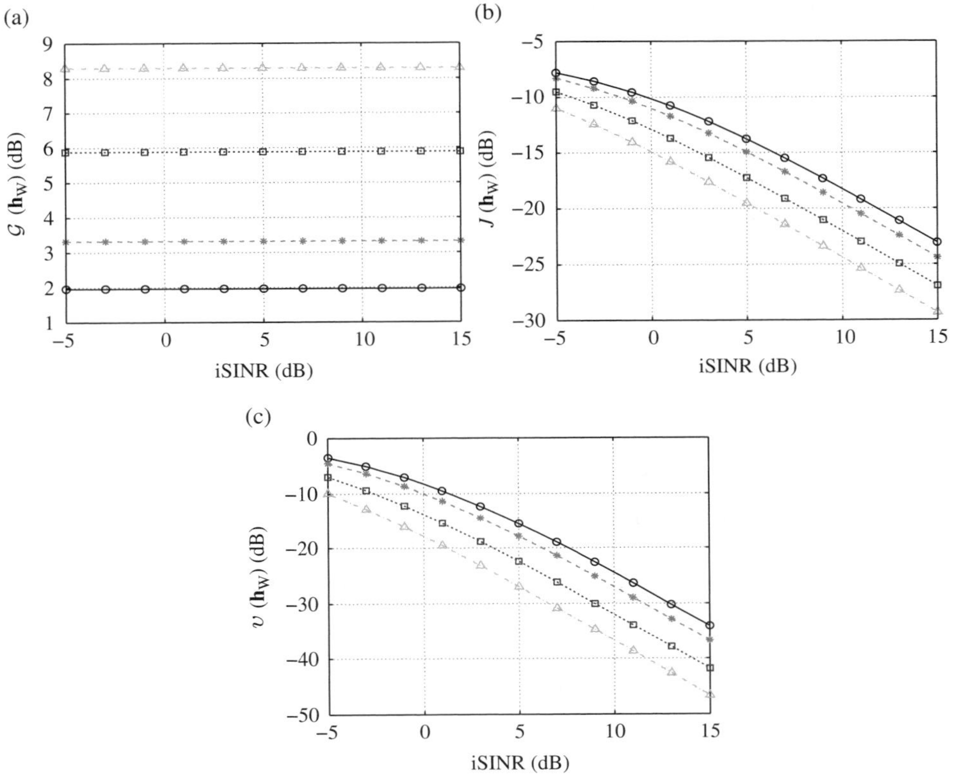

Figure 6.2 (a) The gain in SINR, (b) the MSE, and (c) the desired signal distortion index of the Wiener filter as a function of the input SINR for different numbers of sensors, M: $M = 10$ (solid line with circles), $M = 20$ (dashed line with asterisks), $M = 50$ (dotted line with squares), and $M = 100$ (dash-dot line with triangles).

The optimal filter $\mathbf{h}_W$ is obtained from (6.35).

To demonstrate the performance of the Wiener filter, we choose $f_0 = 0.1$, $\theta_x = 90°$, $\theta_v = 0°$, $\alpha = 0.9$, and $\phi_{v_0} = 0.1$. Figure 6.2 shows plots of the gain in SINR, $\mathcal{G}(\mathbf{h}_W) = \text{oSINR}(\mathbf{h}_W)/\text{iSINR}$, the MSE, $J(\mathbf{h}_W)$, and the desired signal distortion index, $\upsilon(\mathbf{h}_W)$, as a function of the input SINR for different numbers of sensors, M. The gain in SINR is always positive. For a given input SINR, as the number of sensors increases, the gain in SINR increases, while the MMSE and the desired signal distortion index decrease. ■

6.4.2 MVDR

In this subsection, we derive a distortionless filter, which is able to reduce the interference-plus-noise, by exploiting the nullspace of $\mathbf{\Phi}_\mathbf{x}$. Using (6.38), we can write the desired signal vector as

$$\begin{aligned}\mathbf{x} &= \mathbf{Q}_\mathbf{x}\mathbf{Q}_\mathbf{x}^H\mathbf{x} \\ &= \mathbf{Q}_\mathbf{x}'\mathbf{Q}_\mathbf{x}'^H\mathbf{x}.\end{aligned} \tag{6.43}$$

We deduce from (6.43) that the distortionless constraint is

$$\mathbf{h}^H \mathbf{Q}'_\mathbf{x} = \mathbf{i}_\mathrm{i}^T \mathbf{Q}'_\mathbf{x}, \tag{6.44}$$

since, in this case,

$$\begin{aligned} \mathbf{h}^H \mathbf{x} &= \mathbf{h}^H \mathbf{Q}'_\mathbf{x} \mathbf{Q}'^H_\mathbf{x} \mathbf{x} \\ &= \mathbf{i}_\mathrm{i}^T \mathbf{Q}'_\mathbf{x} \mathbf{Q}'^H_\mathbf{x} \mathbf{x} \\ &= x_1. \end{aligned} \tag{6.45}$$

Now, from the minimization of the criterion:

$$\min_{\mathbf{h}} \left[J_\mathrm{n}(\mathbf{h}) + J_\mathrm{i}(\mathbf{h}) \right] \quad \text{subject to} \quad \mathbf{h}^H \mathbf{Q}'_\mathbf{x} = \mathbf{i}_\mathrm{i}^T \mathbf{Q}'_\mathbf{x}, \tag{6.46}$$

which is the minimization of the residual interference-plus-noise subject to the distortionless constraint, we find the MVDR filter:

$$\mathbf{h}_\mathrm{MVDR} = \mathbf{\Phi}_\mathrm{in}^{-1} \mathbf{Q}'_\mathbf{x} \left(\mathbf{Q}'^H_\mathbf{x} \mathbf{\Phi}_\mathrm{in}^{-1} \mathbf{Q}'_\mathbf{x} \right)^{-1} \mathbf{Q}'^H_\mathbf{x} \mathbf{i}_\mathrm{i}. \tag{6.47}$$

It is interesting to compare this filter with the form of the Wiener filter given in (6.42). It can be shown that (6.47) can also be expressed as

$$\mathbf{h}_\mathrm{MVDR} = \mathbf{\Phi}_\mathbf{y}^{-1} \mathbf{Q}'_\mathbf{x} \left(\mathbf{Q}'^H_\mathbf{x} \mathbf{\Phi}_\mathbf{y}^{-1} \mathbf{Q}'_\mathbf{x} \right)^{-1} \mathbf{Q}'^H_\mathbf{x} \mathbf{i}_\mathrm{i}. \tag{6.48}$$

It can be verified that, indeed, $J_\mathrm{d}\left(\mathbf{h}_\mathrm{MVDR}\right) = 0$. Of course, for $R_x = M$, the MVDR filter simplifies to the identity filter: $\mathbf{h}_\mathrm{MVDR} = \mathbf{i}_\mathrm{i}$. As a consequence, we can state that the higher the dimension of the nullspace of $\mathbf{\Phi}_\mathbf{x}$, the more the MVDR filter is efficient in terms of noise reduction. The best scenario corresponds to $R_x = 1$, which is the form of the MVDR filter that is well known in the literature. The case $R_x > 1$ was discovered only recently [9, 10].

It can be shown that with the MVDR filter given in (6.47), the output SINR is always greater than or equal to the input SINR: $\mathrm{oSINR}(\mathbf{h}_\mathrm{MVDR}) \geq \mathrm{iSINR}$ [11].

Example 6.4.2 Returning to Example 6.4.1, we now employ the MVDR filter, $\mathbf{h}_\mathrm{MVDR}$, given in (6.47). Figure 6.3 shows plots of the gain in SINR, $\mathcal{G}\left(\mathbf{h}_\mathrm{MVDR}\right)$, and the MSE, $J\left(\mathbf{h}_\mathrm{MVDR}\right)$, as a function of the input SINR for different numbers of sensors, M. The desired signal distortion index, $\upsilon\left(\mathbf{h}_\mathrm{MVDR}\right)$, is zero. The gain in SINR is always positive. For a given input SINR, as the number of sensors increases, the gain in SINR increases, while the MSE decreases. ■

6.4.3 Tradeoff

We are now going to derive a filter that can compromise between interference-plus-noise reduction and desired signal distortion. For that, we need to minimize the distortion-based MSE subject to the constraint that the interference-plus-noise reduction-based MSE is equal to some desired value. Mathematically, this is equivalent to

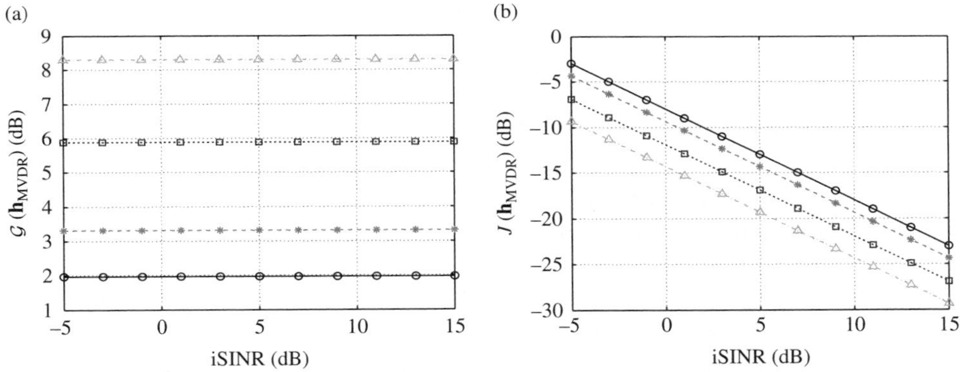

Figure 6.3 (a) The gain in SINR and (b) the MSE of the MVDR filter as a function of the input SINR for different numbers of sensors, M: $M = 10$ (solid line with circles), $M = 20$ (dashed line with asterisks), $M = 50$ (dotted line with squares), and $M = 100$ (dash-dot line with triangles).

$$\min_{\mathbf{h}} J_{\text{d}}(\mathbf{h}) \quad \text{subject to} \quad J_{\text{n}}(\mathbf{h}) + J_{\text{i}}(\mathbf{h}) = \aleph\left(\phi_{v_0} + \phi_v\right), \tag{6.49}$$

where $0 < \aleph < 1$ to ensure that we have some noise reduction. If we use a Lagrange multiplier, μ, to adjoin the constraint to the cost function, (6.49) can be rewritten as

$$\mathbf{h}_{\text{T},\mu} = \arg\min_{\mathbf{h}} \mathcal{L}(\mathbf{h}, \mu), \tag{6.50}$$

with

$$\mathcal{L}(\mathbf{h}, \mu) = J_{\text{d}}(\mathbf{h}) + \mu\left[J_{\text{n}}(\mathbf{h}) + J_{\text{i}}(\mathbf{h}) - \aleph\left(\phi_{v_0} + \phi_v\right)\right] \tag{6.51}$$

and $\mu > 0$. From (6.50), we easily derive the tradeoff filter:

$$\begin{aligned} \mathbf{h}_{\text{T},\mu} &= \left(\mathbf{\Phi}_{\mathbf{x}} + \mu\mathbf{\Phi}_{\text{in}}\right)^{-1}\mathbf{\Phi}_{\mathbf{x}}\mathbf{i}_{\text{i}} \\ &= \left[\mathbf{\Phi}_{\mathbf{y}} + (\mu - 1)\mathbf{\Phi}_{\text{in}}\right]^{-1}\left(\mathbf{\Phi}_{\mathbf{y}} - \mathbf{\Phi}_{\text{in}}\right)\mathbf{i}_{\text{i}}, \end{aligned} \tag{6.52}$$

where the Lagrange multiplier, μ, satisfies $J_{\text{n}}\left(\mathbf{h}_{\text{T},\mu}\right) + J_{\text{i}}\left(\mathbf{h}_{\text{T},\mu}\right) = \aleph\left(\phi_{v_0} + \phi_v\right)$.

In practice it is not easy to determine the optimal μ. Therefore, when this parameter is chosen in a heuristic way, we can see that for

- $\mu = 1$, $\mathbf{h}_{\text{T},1} = \mathbf{h}_{\text{W}}$, which is the Wiener filter
- $\mu = 0$ [if rank $\left(\mathbf{\Phi}_{\mathbf{x}}\right) = M$], $\mathbf{h}_{\text{T},0} = \mathbf{i}_{\text{i}}$, which is the identity filter
- $\mu > 1$ results in a filter with low residual interference-plus-noise at the expense of high desired signal distortion
- $\mu < 1$ results in a filter with low desired signal distortion and small amount of interference-plus-noise reduction.

It can be shown that with the tradeoff filter given in (6.52), the output SINR is always greater than or equal to the input SINR: $\text{oSINR}(\mathbf{h}_{\text{T},\mu}) \geq \text{iSINR}$, $\forall\mu \geq 0$ [1].

With the eigenvalue decomposition of $\mathbf{\Phi}_{\mathbf{x}}$ and the Woodbury identity, we can express the tradeoff filter as

$$\mathbf{h}_{\mathrm{T},\mu} = \mathbf{\Phi}_{\mathrm{in}}^{-1}\mathbf{Q}'_{\mathbf{x}} \left(\mu\mathbf{\Lambda}_{\mathbf{x}}^{\prime -1} + \mathbf{Q}_{\mathbf{x}}^{\prime H}\mathbf{\Phi}_{\mathrm{in}}^{-1}\mathbf{Q}'_{\mathbf{x}}\right)^{-1} \mathbf{Q}_{\mathbf{x}}^{\prime H}\mathbf{i}_{\mathrm{i}}. \tag{6.53}$$

This filter is strictly equivalent to the tradeoff filter given in (6.52), except for $\mu = 0$; indeed, the one in (6.52) is not defined while the one in (6.53) leads to the MVDR filter.

Example 6.4.3 Returning to Example 6.4.1, we now employ the tradeoff filter, $\mathbf{h}_{\mathrm{T},\mu}$, given in (6.52). Figures 6.4 and 6.5 show plots of the gain in SINR, $\mathcal{G}\left(\mathbf{h}_{\mathrm{T},\mu}\right)$, the MSE, $J\left(\mathbf{h}_{\mathrm{T},\mu}\right)$, and the desired signal distortion index, $\upsilon\left(\mathbf{h}_{\mathrm{T},\mu}\right)$, as a function of the input SINR for different numbers of sensors, M, for $\mu = 0.5$ and $\mu = 5$, respectively. The gain in SINR is always positive. For a given input SINR, as the number of sensors increases, the gain in SINR increases, while the MSE and the desired signal distortion index decrease. ■

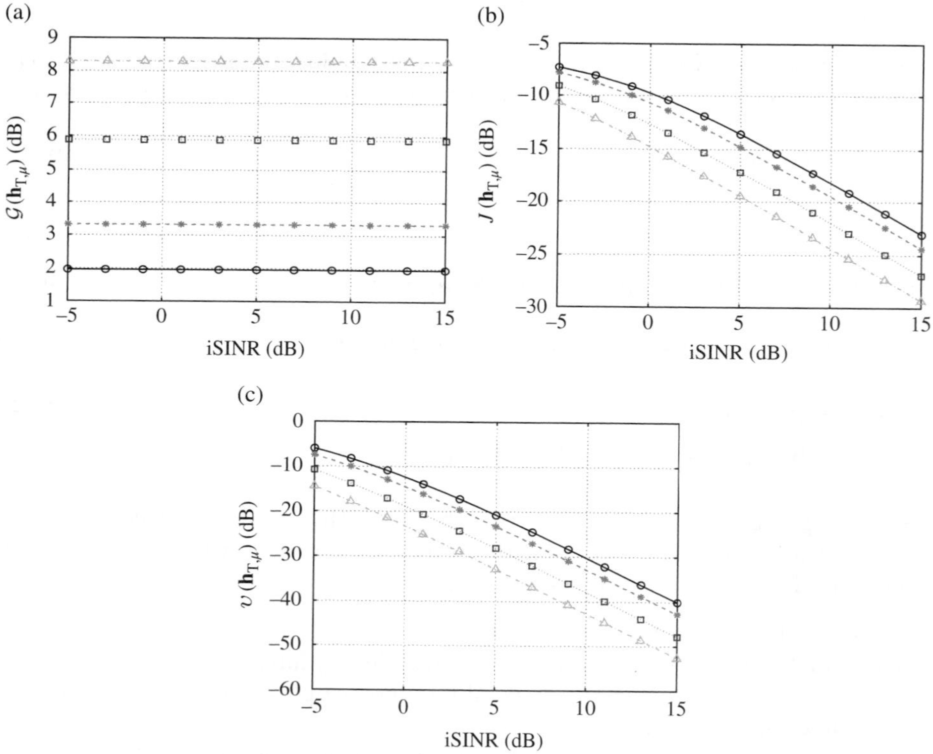

Figure 6.4 (a) The gain in SINR, (b) the MSE, and (c) the desired signal distortion index of the tradeoff filter as a function of the input SINR for different numbers of sensors, M, and $\mu = 0.5$: $M = 10$ (solid line with circles), $M = 20$ (dashed line with asterisks), $M = 50$ (dotted line with squares), and $M = 100$ (dash-dot line with triangles).

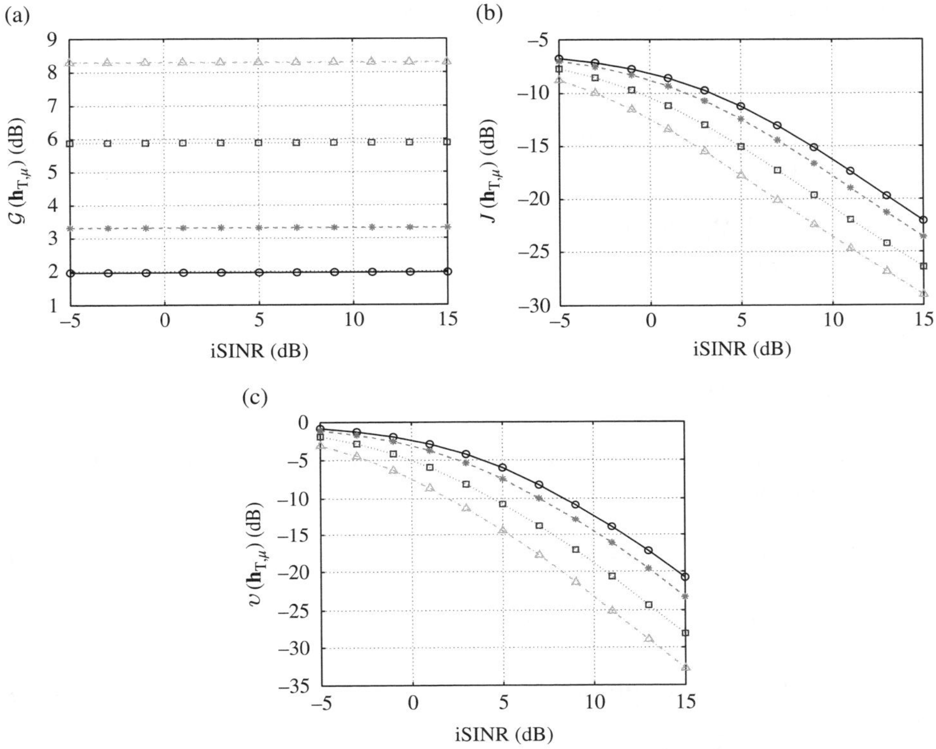

Figure 6.5 (a) The gain in SINR, (b) the MSE, and (c) the desired signal distortion index of the tradeoff filter as a function of the input SINR for different numbers of sensors, M, and $\mu = 5$: $M = 10$ (solid line with circles), $M = 20$ (dashed line with asterisks), $M = 50$ (dotted line with squares), and $M = 100$ (dash-dot line with triangles).

6.4.4 LCMV

In this approach, we would like to find a filter that completely cancels one interference, let's say $\mathbf{v}_1$, without any distortion to the desired signal, and attenuates as much as possible the rest of the interference-plus-noise signal.

Let

$$\mathbf{Q}_{\mathbf{v}_1} = \begin{bmatrix} \mathbf{Q}'_{\mathbf{v}_1} & \mathbf{Q}''_{\mathbf{v}_1} \end{bmatrix}, \tag{6.54}$$

where the $M \times R_{v_1}$ matrix $\mathbf{Q}'_{\mathbf{v}_1}$ contains the eigenvectors corresponding to the nonzero eigenvalues of $\mathbf{\Phi}_{\mathbf{v}_1}$ and the $M \times (M - R_{v_1})$ matrix $\mathbf{Q}''_{\mathbf{v}_1}$ contains the eigenvectors corresponding to the null eigenvalues of $\mathbf{\Phi}_{\mathbf{v}_1}$. We can write the interference $\mathbf{v}_1$ as

$$\begin{aligned} \mathbf{v}_1 &= \mathbf{Q}_{\mathbf{v}_1}\mathbf{Q}^H_{\mathbf{v}_1}\mathbf{v}_1 \\ &= \mathbf{Q}'_{\mathbf{v}_1}\mathbf{Q}'^H_{\mathbf{v}_1}\mathbf{v}_1. \end{aligned} \tag{6.55}$$

We deduce that the constraint to cancel this interference is

$$\mathbf{h}^H \mathbf{Q}'_{\mathbf{v}_1} = \mathbf{0}^T, \tag{6.56}$$

where $\mathbf{0}$ is the zero vector of length R_{v_1}. Combining this constraint with the distortionless one, we get

$$\begin{aligned} \mathbf{h}^H \mathbf{C}_{\mathbf{x}\mathbf{v}_1} &= \left[\begin{array}{cc} \mathbf{i}_{\mathrm{i}}^T \mathbf{Q}'_{\mathbf{x}} & \mathbf{0}^T \end{array} \right] \\ &= \mathbf{i}_{\mathrm{c}}^H, \end{aligned} \tag{6.57}$$

where

$$\mathbf{C}_{\mathbf{x}\mathbf{v}_1} = \left[\begin{array}{cc} \mathbf{Q}'_{\mathbf{x}} & \mathbf{Q}'_{\mathbf{v}_1} \end{array} \right] \tag{6.58}$$

is the constraint matrix of size $M \times (R_x + R_{v_1})$ and $\mathbf{i}_{\mathrm{c}}$ is a vector of length $R_x + R_{v_1}$. Now, the criterion to optimize is

$$\min_{\mathbf{h}} \left[J_{\mathrm{n}}(\mathbf{h}) + J_{\mathrm{i}}(\mathbf{h}) \right] \quad \text{subject to} \quad \mathbf{h}^H \mathbf{C}_{\mathbf{x}\mathbf{v}_1} = \mathbf{i}_{\mathrm{c}}^H, \tag{6.59}$$

which leads to the celebrated LCMV filter:

$$\mathbf{h}_{\mathrm{LCMV}} = \mathbf{\Phi}_{\mathrm{in}}^{-1} \mathbf{C}_{\mathbf{x}\mathbf{v}_1} \left(\mathbf{C}_{\mathbf{x}\mathbf{v}_1}^H \mathbf{\Phi}_{\mathrm{in}}^{-1} \mathbf{C}_{\mathbf{x}\mathbf{v}_1} \right)^{-1} \mathbf{i}_{\mathrm{c}}. \tag{6.60}$$

It is clear from (6.60) that for this filter to exist, we must have $M \geq R_x + R_{v_1}$. An equivalent way to express (6.60) is

$$\mathbf{h}_{\mathrm{LCMV}} = \mathbf{\Phi}_{\mathbf{y}}^{-1} \mathbf{C}_{\mathbf{x}\mathbf{v}_1} \left(\mathbf{C}_{\mathbf{x}\mathbf{v}_1}^H \mathbf{\Phi}_{\mathbf{y}}^{-1} \mathbf{C}_{\mathbf{x}\mathbf{v}_1} \right)^{-1} \mathbf{i}_{\mathrm{c}}. \tag{6.61}$$

While, with this filter, we can completely cancel the interference $\mathbf{v}_1$, there is no guarantee that the rest of the interference-plus-noise can be attenuated; in fact, it can even be amplified. This depends on how M is larger than $R_x + R_{v_1}$. As the difference of these two integers increases, so is the attenuation of the rest of the interference-plus-noise signal.

Example 6.4.4 Returning to Example 6.4.1, we now assume two uncorrelated complex harmonic random processes as interferences, $\mathbf{v}_1$ and $\mathbf{v}_2$, impinging on the ULA from the directions $\theta_{v_1} = 0°$ and $\theta_{v_2} = 45°$, respectively. We employ the LCMV filter, $\mathbf{h}_{\mathrm{LCMV}}$, given in (6.60). Figure 6.6 shows plots of the gain in SINR, $\mathcal{G}\left(\mathbf{h}_{\mathrm{LCMV}}\right)$, the MSE, $J\left(\mathbf{h}_{\mathrm{LCMV}}\right)$, and the desired signal distortion index, $\upsilon\left(\mathbf{h}_{\mathrm{LCMV}}\right)$, as a function of the input SINR for different numbers of sensors, M. The desired signal distortion index, $\upsilon\left(\mathbf{h}_{\mathrm{LCMV}}\right)$, is zero. The gain in SINR is always positive. For a given input SINR, as the number of sensors increases, the gain in SINR increases, while the MSE decreases. ■

The generalization of this approach to the cancellation of more than one interference is straightforward. Let's say that we want to cancel the two interferences $\mathbf{v}_1$ and $\mathbf{v}_2$. First, we take the correlation matrix of the signal $\mathbf{v}_1 + \mathbf{v}_2$. We perform the eigenvalue

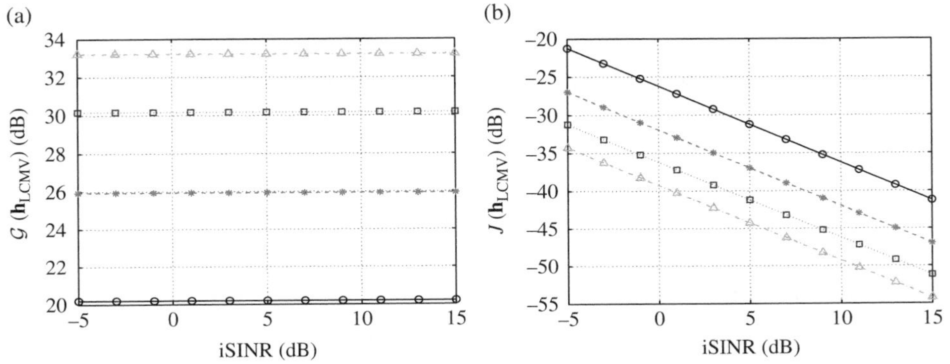

Figure 6.6 (a) The gain in SINR, (b) the MSE, and (c) the desired signal distortion index of the LCMV filter as a function of the input SINR for different numbers of sensors, M: $M = 10$ (solid line with circles), $M = 20$ (dashed line with asterisks), $M = 50$ (dotted line with squares), and $M = 100$ (dash-dot line with triangles).

decomposition of this matrix as we did for $\mathbf{\Phi}_{\mathbf{v}_1}$. Then, the derivation of the corresponding LCMV filter is as described above. The only thing that changes is the condition of the filter existing, which is now $M \geq R_x + R_{v_1} + R_{v_2}$.

Another interesting way to derive the LCMV filter is the following. Let us consider the filters that have the form:

$$\mathbf{h} = \mathbf{Q}''_{\mathbf{v}_1}\mathbf{a}, \tag{6.62}$$

where $\mathbf{a} \neq \mathbf{0}$ is a shorter filter of length $M - R_{v_1}$. It is easy to observe that

$$\mathbf{h}^H\mathbf{v}_1 = \mathbf{a}^H\mathbf{Q}''^H_{\mathbf{v}_1}\mathbf{v}_1 = \mathbf{0}, \tag{6.63}$$

since $\mathbf{Q}''^H_{\mathbf{v}_1}\mathbf{Q}'_{\mathbf{v}_1} = \mathbf{0}$. By its nature, the filter $\mathbf{h}$ in (6.62) cancels the interference. Substituting (6.62) into $J_{\text{n}}(\mathbf{h}) + J_{\text{i}}(\mathbf{h})$, we obtain

$$\begin{aligned} J_{\text{n}}(\mathbf{h}) + J_{\text{i}}(\mathbf{h}) &= \phi_{v_0}\mathbf{a}^H\mathbf{a} + \mathbf{a}^H\mathbf{Q}''^H_{\mathbf{v}_1}\mathbf{\Phi}_{\mathbf{v}}\mathbf{Q}''_{\mathbf{v}_1}\mathbf{a} \\ &= \mathbf{a}^H\mathbf{\Phi}'_{\text{in}}\mathbf{a} \\ &= J_{\text{n}}(\mathbf{a}) + J_{\text{i}}(\mathbf{a}), \end{aligned} \tag{6.64}$$

where

$$\mathbf{\Phi}'_{\text{in}} = \phi_{v_0}\mathbf{I}_{M-R_{v_1}} + \mathbf{Q}''^H_{\mathbf{v}_1}\mathbf{\Phi}_{\mathbf{v}}\mathbf{Q}''_{\mathbf{v}_1}, \tag{6.65}$$

with $\mathbf{I}_{M-R_{v_1}}$ being the $(M - R_{v_1}) \times (M - R_{v_1})$ identity matrix. Then, from the criterion:

$$\min_{\mathbf{a}} \left[J_{\text{n}}(\mathbf{a}) + J_{\text{i}}(\mathbf{a})\right] \quad \text{subject to} \quad \mathbf{a}^H\mathbf{Q}''^H_{\mathbf{v}_1}\mathbf{Q}'_{\mathbf{x}} = \mathbf{i}_{\text{i}}^T\mathbf{Q}'_{\mathbf{x}}, \tag{6.66}$$

we find that

$$\mathbf{a}_{\mathrm{LCMV}} = \mathbf{\Phi}_{\mathrm{in}}^{\prime -1} \mathbf{Q}_{\mathbf{v}_1}^{\prime\prime H} \mathbf{Q}_{\mathbf{x}}^{\prime} \left(\mathbf{Q}_{\mathbf{x}}^{\prime H} \mathbf{Q}_{\mathbf{v}_1}^{\prime\prime} \mathbf{\Phi}_{\mathrm{in}}^{\prime -1} \mathbf{Q}_{\mathbf{v}_1}^{\prime\prime H} \mathbf{Q}_{\mathbf{x}}^{\prime} \right)^{-1} \mathbf{Q}_{\mathbf{x}}^{\prime H} \mathbf{i}_{\mathrm{i}}. \tag{6.67}$$

As a result, another formulation of the LCMV filter is

$$\mathbf{h}_{\mathrm{LCMV}} = \mathbf{Q}_{\mathbf{v}_1}^{\prime\prime} \mathbf{\Phi}_{\mathrm{in}}^{\prime -1} \mathbf{Q}_{\mathbf{v}_1}^{\prime\prime H} \mathbf{Q}_{\mathbf{x}}^{\prime} \left(\mathbf{Q}_{\mathbf{x}}^{\prime H} \mathbf{Q}_{\mathbf{v}_1}^{\prime\prime} \mathbf{\Phi}_{\mathrm{in}}^{\prime -1} \mathbf{Q}_{\mathbf{v}_1}^{\prime\prime H} \mathbf{Q}_{\mathbf{x}}^{\prime} \right)^{-1} \mathbf{Q}_{\mathbf{x}}^{\prime H} \mathbf{i}_{\mathrm{i}}. \tag{6.68}$$

6.4.5 Maximum SINR

The maximum SINR filter is obtained by maximizing the output SINR as given in (6.20), from which we recognize the generalized Rayleigh quotient [6]. Since $\mathbf{\Phi}_{\mathrm{in}}$ is full rank, it is well known that this quotient is maximized with the eigenvector corresponding to the maximum eigenvalue of $\mathbf{\Phi}_{\mathrm{in}}^{-1}\mathbf{\Phi}_{\mathbf{x}}$. Let us denote by λ_1 the maximum eigenvalue of this matrix and by $\mathbf{t}_1$ the corresponding eigenvector. Therefore, we have

$$\mathbf{h}_{\mathrm{mSINR}} = \varsigma \mathbf{t}_1, \tag{6.69}$$

where $\varsigma \neq 0$ is an arbitrary complex number. We deduce that

$$\mathrm{oSINR}\left(\mathbf{h}_{\mathrm{mSINR}}\right) = \lambda_1. \tag{6.70}$$

Clearly, we always have

$$\mathrm{oSINR}\left(\mathbf{h}_{\mathrm{mSINR}}\right) \geq \mathrm{iSINR} \tag{6.71}$$

and

$$\mathrm{oSINR}\left(\mathbf{h}_{\mathrm{mSINR}}\right) \geq \mathrm{oSINR}\left(\mathbf{h}\right), \ \forall \mathbf{h}. \tag{6.72}$$

Now we need to determine ς. One possible way to find this parameter is by minimizing distortion. Substituting (6.69) into $J_{\mathrm{d}}\left(\mathbf{h}\right)$, we get

$$J_{\mathrm{d}}\left(\mathbf{h}_{\mathrm{mSINR}}\right) = \phi_{x_1} + \lambda_1 \left|\varsigma\right|^2 - \varsigma^* \mathbf{t}_1^H \mathbf{\Phi}_{\mathbf{x}} \mathbf{i}_{\mathrm{i}} - \varsigma \mathbf{i}_{\mathrm{i}}^T \mathbf{\Phi}_{\mathbf{x}} \mathbf{t}_1. \tag{6.73}$$

The minimization of the previous expression with respect to ς^* gives

$$\varsigma = \frac{\mathbf{t}_1^H \mathbf{\Phi}_{\mathbf{x}} \mathbf{i}_{\mathrm{i}}}{\lambda_1}. \tag{6.74}$$

We deduce that the maximum SINR filter with minimum distortion is

$$\begin{aligned} \mathbf{h}_{\mathrm{mSINR}} &= \frac{\mathbf{t}_1 \mathbf{t}_1^H \mathbf{\Phi}_{\mathbf{x}} \mathbf{i}_{\mathrm{i}}}{\lambda_1} \\ &= \mathbf{t}_1 \mathbf{t}_1^H \mathbf{\Phi}_{\mathrm{in}} \mathbf{i}_{\mathrm{i}}. \end{aligned} \tag{6.75}$$

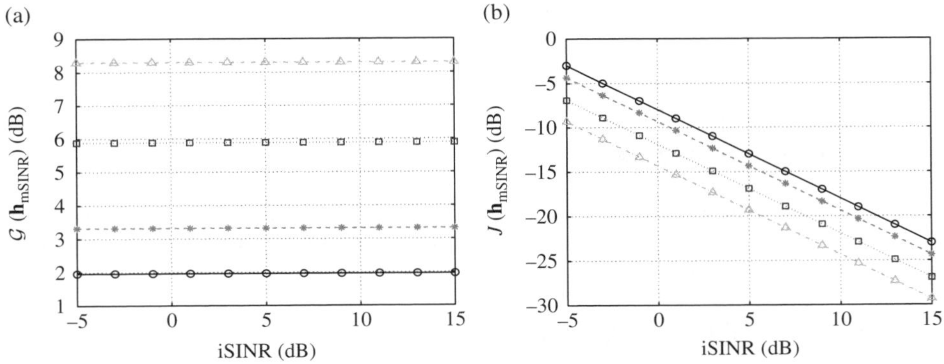

Figure 6.7 (a) The gain in SINR and (b) the MSE of the maximum SINR filter as a function of the input SINR for different numbers of sensors, M: $M = 10$ (solid line with circles), $M = 20$ (dashed line with asterisks), $M = 50$ (dotted line with squares), and $M = 100$ (dash-dot line with triangles).

Example 6.4.5 Returning to Example 6.4.1, we now employ the maximum SINR filter, $\mathbf{h}_{\mathrm{mSINR}}$, given in (6.75). Figure 6.7 shows plots of the gain in SINR, $\mathcal{G}\left(\mathbf{h}_{\mathrm{mSINR}}\right)$, and the MSE, $J\left(\mathbf{h}_{\mathrm{mSINR}}\right)$, as a function of the input SINR for different numbers of sensors, M. The desired signal distortion index, $\upsilon\left(\mathbf{h}_{\mathrm{mSINR}}\right)$, is zero. The gain in SINR is always positive. For a given input SINR, as the number of sensors increases, the gain in SINR increases, while the MSE decreases. ■

6.4.6 Maximum SIR

In the denominator of the output SIR appears the matrix $\mathbf{\Phi}_{\mathbf{v}}$, which can be either full rank or rank deficient. In the first case, it is easy to derive the maximum SIR filter, which is the eigenvector corresponding to the maximum eigenvalue of $\mathbf{\Phi}_{\mathbf{v}}^{-1}\mathbf{\Phi}_{\mathbf{x}}$. Fundamentally, this scenario is equivalent to what was done in the previous subsection for the maximization of the SINR. Therefore, we are only interested in the second case, where we assume that $\mathrm{rank}\left(\mathbf{\Phi}_{\mathbf{v}}\right) = R_v < M$.

Let

$$\mathbf{Q}_{\mathbf{v}} = \left[\begin{array}{cc} \mathbf{Q}_{\mathbf{v}}' & \mathbf{Q}_{\mathbf{v}}'' \end{array}\right], \tag{6.76}$$

where the $M \times R_v$ matrix $\mathbf{Q}_{\mathbf{v}}'$ contains the eigenvectors corresponding to the nonzero eigenvalues of $\mathbf{\Phi}_{\mathbf{v}}$ and the $M \times (M - R_v)$ matrix $\mathbf{Q}_{\mathbf{v}}''$ contains the eigenvectors corresponding to the null eigenvalues of $\mathbf{\Phi}_{\mathbf{v}}$. We are interested in the linear filters of the form:

$$\mathbf{h} = \mathbf{Q}_{\mathbf{v}}''\mathbf{a}, \tag{6.77}$$

where $\mathbf{a}$ is a vector of length $M - R_v$. Since $\mathbf{\Phi}_{\mathbf{v}}\mathbf{Q}_{\mathbf{v}}'' = \mathbf{0}$ and assuming that $\mathbf{\Phi}_{\mathbf{x}}\mathbf{Q}_{\mathbf{v}}'' \neq \mathbf{0}$, which is reasonable since $\mathbf{\Phi}_{\mathbf{x}}$ and $\mathbf{\Phi}_{\mathbf{v}}$ cannot be diagonalized by the same orthogonal matrix unless at least one of the two signals x_1 and v is white, we have

$$\mathrm{oSIR}\left(\mathbf{h}\right) = \mathrm{oSIR}\left(\mathbf{Q}_{\mathbf{v}}''\mathbf{a}\right) = \infty. \tag{6.78}$$

As a consequence, the estimate of x_1 is

$$\begin{aligned}\widehat{x}_1 &= \mathbf{h}^H\mathbf{y} \\ &= \mathbf{a}^H\mathbf{Q}_\mathbf{v}''^H\mathbf{x} + \mathbf{a}^H\mathbf{Q}_\mathbf{v}''^H\mathbf{v}_0 + \mathbf{a}^H\mathbf{Q}_\mathbf{v}''^H\mathbf{v} \\ &= \mathbf{a}^H\mathbf{Q}_\mathbf{v}''^H\mathbf{x} + \mathbf{a}^H\mathbf{Q}_\mathbf{v}''^H\mathbf{v}_0.\end{aligned} \tag{6.79}$$

We observe from the previous expression that this approach completely cancels the interference. Now, we need to find **a**. The best way to find this vector is by minimizing the MSE criterion. Substituting (6.77) into (6.31), we get

$$J(\mathbf{a}) = \phi_{x_1} + \mathbf{a}^H\mathbf{Q}_\mathbf{v}''^H\mathbf{\Phi}_\mathbf{y}\mathbf{Q}_\mathbf{v}''\mathbf{a} - \mathbf{a}^H\mathbf{Q}_\mathbf{v}''^H\mathbf{\Phi}_\mathbf{x}\mathbf{i}_\mathrm{i} - \mathbf{i}_\mathrm{i}^T\mathbf{\Phi}_\mathbf{x}\mathbf{Q}_\mathbf{v}''\mathbf{a}. \tag{6.80}$$

The minimization of the previous expression leads to

$$\mathbf{a}_\mathrm{mSIR} = \left(\mathbf{Q}_\mathbf{v}''^H\mathbf{\Phi}_\mathbf{y}\mathbf{Q}_\mathbf{v}''\right)^{-1}\mathbf{Q}_\mathbf{v}''^H\mathbf{\Phi}_\mathbf{x}\mathbf{i}_\mathrm{i}. \tag{6.81}$$

As a result, the maximum SIR filter with minimum MSE is

$$\mathbf{h}_\mathrm{mSIR} = \mathbf{Q}_\mathbf{v}''\left(\mathbf{Q}_\mathbf{v}''^H\mathbf{\Phi}_\mathbf{y}\mathbf{Q}_\mathbf{v}''\right)^{-1}\mathbf{Q}_\mathbf{v}''^H\mathbf{\Phi}_\mathbf{x}\mathbf{i}_\mathrm{i}. \tag{6.82}$$

Example 6.4.6 Returning to Example 6.4.4, we now employ the maximum SIR filter, $\mathbf{h}_\mathrm{mSIR}$, given in (6.82). Figure 6.8 shows plots of the gain in SINR, $\mathcal{G}\left(\mathbf{h}_\mathrm{mSIR}\right)$, the MSE, $J\left(\mathbf{h}_\mathrm{mSIR}\right)$, and the desired signal distortion index, $\upsilon\left(\mathbf{h}_\mathrm{mSIR}\right)$, as a function of the input SINR for different numbers of sensors, M. The gain in SINR is always positive. For a given input SINR, as the number of sensors increases, the gain in SINR increases, while the MSE and the desired signal distortion index decrease. ■

All the optimal filters derived in this section are summarized in Table 6.1.

6.5 Filling the Gap Between the Maximum SINR and Wiener Filters

In this section, we revisit the maximum SINR and Wiener filters. We show how they are related, and from this we derive a new class of filters.

The fact that the correlation matrix of the observation signal vector is the sum of the correlation matrices of the desired and interference-plus-noise signal vectors will make the analysis of potential filters easy if we jointly diagonalize these two matrices. Since $\mathbf{\Phi}_\mathrm{in}$ is full rank, the two Hermitian matrices $\mathbf{\Phi}_\mathbf{x}$ and $\mathbf{\Phi}_\mathrm{in}$ can indeed be jointly diagonalized, as follows [12]:

$$\mathbf{T}^H\mathbf{\Phi}_\mathbf{x}\mathbf{T} = \mathbf{\Lambda}, \tag{6.83}$$

$$\mathbf{T}^H\mathbf{\Phi}_\mathrm{in}\mathbf{T} = \mathbf{I}_M, \tag{6.84}$$

where $\mathbf{T}$ is a full-rank square matrix (of size $M \times M$) and $\mathbf{\Lambda}$ is a diagonal matrix whose main elements are real and nonnegative. Furthermore, $\mathbf{\Lambda}$ and $\mathbf{T}$ are the eigenvalue and eigenvector matrices, respectively, of $\mathbf{\Phi}_\mathrm{in}^{-1}\mathbf{\Phi}_\mathbf{x}$:

$$\mathbf{\Phi}_\mathrm{in}^{-1}\mathbf{\Phi}_\mathbf{x}\mathbf{T} = \mathbf{T}\mathbf{\Lambda}. \tag{6.85}$$

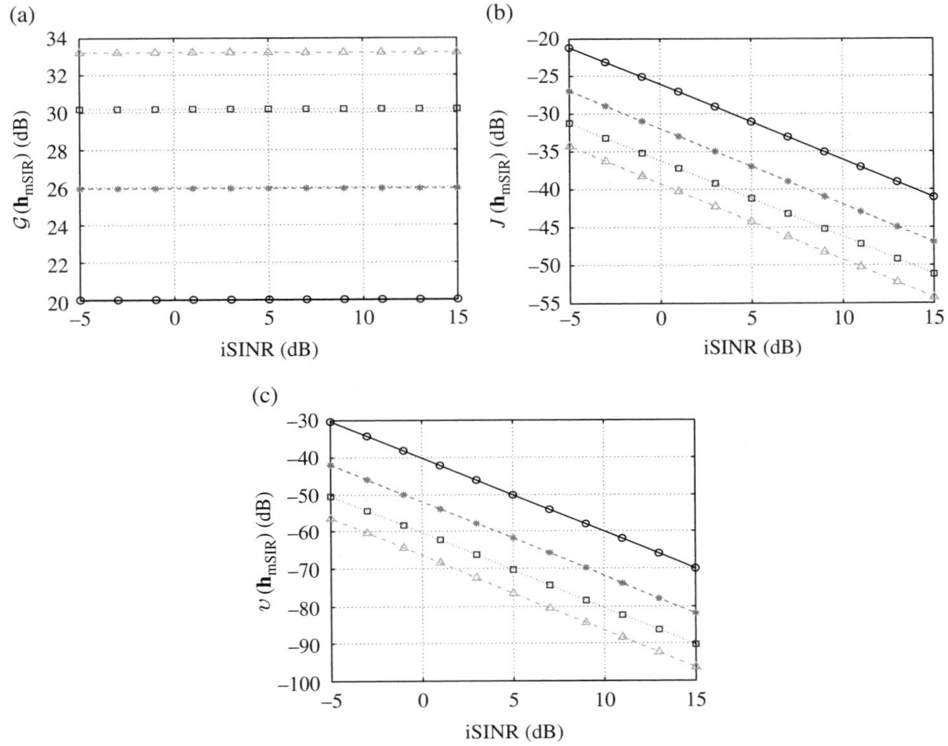

Figure 6.8 (a) The gain in SINR, (b) the MSE, and (c) the desired signal distortion index of the maximum SIR filter as a function of the input SINR for different numbers of sensors, M: $M = 10$ (solid line with circles), $M = 20$ (dashed line with asterisks), $M = 50$ (dotted line with squares), and $M = 100$ (dash-dot line with triangles).

Table 6.1 Optimal linear filters for signal enhancement.

Filter	
Wiener	$\mathbf{h}_{\rm W} = \boldsymbol{\Phi}_{\rm in}^{-1}\mathbf{Q}'_{\mathbf{x}}\left(\boldsymbol{\Lambda}'^{-1}_{\mathbf{x}} + \mathbf{Q}'^H_{\mathbf{x}}\boldsymbol{\Phi}_{\rm in}^{-1}\mathbf{Q}'_{\mathbf{x}}\right)^{-1}\mathbf{Q}'^H_{\mathbf{x}}\mathbf{i}_{\rm i}$
MVDR	$\mathbf{h}_{\rm MVDR} = \boldsymbol{\Phi}_{\rm in}^{-1}\mathbf{Q}'_{\mathbf{x}}\left(\mathbf{Q}'^H_{\mathbf{x}}\boldsymbol{\Phi}_{\rm in}^{-1}\mathbf{Q}'_{\mathbf{x}}\right)^{-1}\mathbf{Q}'^H_{\mathbf{x}}\mathbf{i}_{\rm i}$
Tradeoff	$\mathbf{h}_{{\rm T},\mu} = \boldsymbol{\Phi}_{\rm in}^{-1}\mathbf{Q}'_{\mathbf{x}}\left(\mu\boldsymbol{\Lambda}'^{-1}_{\mathbf{x}} + \mathbf{Q}'^H_{\mathbf{x}}\boldsymbol{\Phi}_{\rm in}^{-1}\mathbf{Q}'_{\mathbf{x}}\right)^{-1}\mathbf{Q}'^H_{\mathbf{x}}\mathbf{i}_{\rm i}$
LCMV	$\mathbf{h}_{\rm LCMV} = \mathbf{Q}''_{\mathbf{v}_1}\boldsymbol{\Phi}'^{-1}_{\rm in}\mathbf{Q}''^H_{\mathbf{v}_1}\mathbf{Q}'_{\mathbf{x}}\times$
	$\left(\mathbf{Q}'^H_{\mathbf{x}}\mathbf{Q}''_{\mathbf{v}_1}\boldsymbol{\Phi}'^{-1}_{\rm in}\mathbf{Q}''^H_{\mathbf{v}_1}\mathbf{Q}'_{\mathbf{x}}\right)^{-1}\mathbf{Q}'^H_{\mathbf{x}}\mathbf{i}_{\rm i}$
Maximum SINR	$\mathbf{h}_{\rm mSINR} = \mathbf{t}_1\mathbf{t}_1^H\boldsymbol{\Phi}_{\rm in}\mathbf{i}_{\rm i}$
Maximum SIR	$\mathbf{h}_{\rm mSIR} = \mathbf{Q}''_{\mathbf{v}}\left(\mathbf{Q}''^H_{\mathbf{v}}\boldsymbol{\Phi}_{\mathbf{y}}\mathbf{Q}''_{\mathbf{v}}\right)^{-1}\mathbf{Q}''^H_{\mathbf{v}}\boldsymbol{\Phi}_{\mathbf{x}}\mathbf{i}_{\rm i}$

Since the rank of the matrix $\boldsymbol{\Phi}_{\mathbf{x}}$ is equal to R_x, the eigenvalues of $\boldsymbol{\Phi}_{\rm in}^{-1}\boldsymbol{\Phi}_{\mathbf{x}}$ can be ordered as $\lambda_1 \geq \lambda_2 \geq \cdots \geq \lambda_{R_x} > \lambda_{R_x+1} = \cdots = \lambda_M = 0$. In other words, the last $M - R_x$ eigenvalues of the matrix product $\boldsymbol{\Phi}_{\rm in}^{-1}\boldsymbol{\Phi}_{\mathbf{x}}$ are exactly zero, while its first R_x

eigenvalues are positive, with λ_1 being the maximum eigenvalue. We also denote by $\mathbf{t}_1, \mathbf{t}_2, \ldots, \mathbf{t}_{R_x}, \mathbf{t}_{R_x+1}, \ldots, \mathbf{t}_M$, the corresponding eigenvectors. A consequence of this joint diagonalization is that the noisy signal correlation matrix can also be diagonalized as

$$\mathbf{T}^H \mathbf{\Phi}_{\mathbf{y}} \mathbf{T} = \mathbf{\Lambda} + \mathbf{I}_M. \tag{6.86}$$

It is always possible to write the filter $\mathbf{h}$ in a basis formed from the vectors $\mathbf{t}_m$, $m = 1, 2, \ldots, M$:

$$\mathbf{h} = \mathbf{T}\mathbf{a}, \tag{6.87}$$

where the components of

$$\mathbf{a} = \begin{bmatrix} a_1 & a_2 & \cdots & a_M \end{bmatrix}^T \tag{6.88}$$

are the coordinates of $\mathbf{h}$ in the new basis. Now, instead of estimating the coefficients of $\mathbf{h}$ as in conventional approaches, we can estimate, equivalently, the coordinates a_m, $m = 1, 2, \ldots, M$. When $\mathbf{a}$ is estimated, it is then straightforward to determine $\mathbf{h}$ from (6.87). From here onwards, we will refer to $\mathbf{h}$ and $\mathbf{a}$ as the direct and transformed filters, respectively. Both filters may be used interchangeably. Consequently, we can express (6.9) as

$$z = \mathbf{a}^H \mathbf{T}^H \mathbf{y}. \tag{6.89}$$

We deduce that the variance of z is

$$\phi_z = \mathbf{a}^H \mathbf{\Lambda} \mathbf{a} + \mathbf{a}^H \mathbf{a}. \tag{6.90}$$

The output SINR can then be expressed as

$$\begin{aligned} \text{oSINR}(\mathbf{a}) &= \frac{\mathbf{a}^H \mathbf{\Lambda} \mathbf{a}}{\mathbf{a}^H \mathbf{a}} \\ &= \frac{\sum_{i=1}^{R_x} |a_i|^2 \lambda_i}{\sum_{m=1}^{M} |a_m|^2} \end{aligned} \tag{6.91}$$

and, clearly,

$$\text{oSINR}(\mathbf{a}) \leq \lambda_1, \ \forall \mathbf{a}. \tag{6.92}$$

We define the transformed identity filter as

$$\mathbf{i}_{\mathbf{T}} = \mathbf{T}^{-1} \mathbf{i}_{\mathrm{i}}. \tag{6.93}$$

The particular filter $\mathbf{i}_{\mathbf{T}}$ does not affect the observations since $z = \mathbf{i}_{\mathbf{T}}^H \mathbf{T}^H \mathbf{y} = y_1$ and $\text{oSINR}(\mathbf{i}_{\mathbf{T}}) = \text{iSINR}$. As a result, we have

$$\text{iSINR} \leq \lambda_1. \tag{6.94}$$

In the same way, we can express the MSE criterion as

$$\begin{aligned} J(\mathbf{a}) &= E\left(\left|z - x_1\right|^2\right) \\ &= \phi_{x_1} - \mathbf{a}^H \mathbf{T}^H \mathbf{\Phi}_{\mathbf{x}} \mathbf{i}_{\mathrm{i}} - \mathbf{i}_{\mathrm{i}}^T \mathbf{\Phi}_{\mathbf{x}} \mathbf{T}\mathbf{a} + \mathbf{a}^H \left(\mathbf{\Lambda} + \mathbf{I}_M\right) \mathbf{a} \\ &= \phi_{x_1} - \mathbf{a}^H \mathbf{\Lambda} \mathbf{i}_{\mathbf{T}} - \mathbf{i}_{\mathbf{T}}^H \mathbf{\Lambda} \mathbf{a} + \mathbf{a}^H \left(\mathbf{\Lambda} + \mathbf{I}_M\right) \mathbf{a} \\ &= \left(\mathbf{a} - \mathbf{i}_{\mathbf{T}}\right)^H \mathbf{\Lambda} \left(\mathbf{a} - \mathbf{i}_{\mathbf{T}}\right) + \mathbf{a}^H \mathbf{a}. \end{aligned} \tag{6.95}$$

With the transformed identity filter, we have $J\left(\mathbf{i}_{\mathbf{T}}\right) = \mathbf{i}_{\mathbf{T}}^H \mathbf{i}_{\mathbf{T}}$.

We observe that the output SINR and MSE criteria are very different. A filter that gives a large output SINR does not necessarily imply a small MSE and a filter that leads to a small MSE does not mean that the output SINR is large (or strictly larger than the input SNR) [13]. Key questions that one may ask are:

- How are these two filters related?
- How to capture the best from the two criteria?

From (6.91), it is easy to see that the maximum SINR filter is

$$\mathbf{a}_{\mathrm{mSINR},2} = a_1 \mathbf{i}_{\mathrm{i}}, \tag{6.96}$$

where $a_1 \neq 0$ is an arbitrary complex number. Obviously, we have

$$\mathrm{oSINR}\left(\mathbf{a}_{\mathrm{mSINR},2}\right) = \lambda_1 \geq \mathrm{iSINR} \tag{6.97}$$

and

$$\mathrm{oSINR}\left(\mathbf{a}\right) \leq \mathrm{oSINR}\left(\mathbf{a}_{\mathrm{mSINR},2}\right), \ \forall \mathbf{a}. \tag{6.98}$$

We need to determine a_1. One reasonable way to find this parameter is from an MSE perspective. Indeed, substituting $\mathbf{a}_{\mathrm{mSINR},2}$ into the MSE criterion of (6.95) and minimizing $J\left(\mathbf{a}_{\mathrm{mSINR},2}\right)$ with respect to a_1, we easily get

$$\begin{aligned} a_1 &= \frac{\lambda_1}{1+\lambda_1} \mathbf{i}_{\mathrm{i}}^T \mathbf{i}_{\mathbf{T}} \\ &= \frac{\lambda_1}{1+\lambda_1} \mathbf{i}_{\mathrm{i}}^T \mathbf{T}^{-1} \mathbf{i}_{\mathrm{i}} \\ &= \frac{\lambda_1}{1+\lambda_1} \mathbf{t}_1^H \mathbf{\Phi}_{\mathrm{in}} \mathbf{i}_{\mathrm{i}}. \end{aligned} \tag{6.99}$$

As a result, the transformed and direct maximum SINR filters are, respectively,

$$\mathbf{a}_{\mathrm{mSINR},2} = \frac{\lambda_1}{1+\lambda_1} \mathbf{i}_{\mathrm{i}} \mathbf{i}_{\mathrm{i}}^T \mathbf{T}^{-1} \mathbf{i}_{\mathrm{i}} \tag{6.100}$$

and

$$\mathbf{h}_{\text{mSINR},2} = \frac{\lambda_1}{1+\lambda_1}\mathbf{t}_1\mathbf{t}_1^H\mathbf{\Phi}_{\text{in}}\mathbf{i}_{\text{i}}. \tag{6.101}$$

This filter is, obviously, very close to $\mathbf{h}_{\text{mSINR}}$; the two filters are equivalent up to a scaling factor. We then deduce that

$$\begin{aligned} J\left(\mathbf{a}_{\text{mSINR},2}\right) &= \mathbf{i}_{\mathbf{T}}^H\mathbf{\Lambda}\mathbf{i}_{\mathbf{T}} - \frac{\lambda_1^2}{1+\lambda_1}\left|\mathbf{i}_{\mathbf{T}}^H\mathbf{i}_{\text{i}}\right|^2 \\ &= \frac{\lambda_1}{1+\lambda_1}\left|\mathbf{i}_{\mathbf{T}}^H\mathbf{i}_{\text{i}}\right|^2 + \sum_{i=2}^{R_x}\lambda_i\left|\mathbf{i}_{\mathbf{T}}^H\mathbf{i}_i\right|^2, \end{aligned} \tag{6.102}$$

where $\mathbf{i}_i$ is the ith column of $\mathbf{I}_M$. It is not guaranteed that $J\left(\mathbf{a}_{\text{mSINR},2}\right) \le J\left(\mathbf{i}_{\mathbf{T}}\right)$.

Example 6.5.1 Returning to Example 6.4.1, we now assume that the desired signal impinges on the ULA from the direction $\theta_x = 70°$ and that the desired signal received at the first sensor is the sum of four complex harmonic random processes:

$$x_1(t) = A\sum_{k=1}^{4}\exp\left(\jmath 2\pi f_k t + \jmath\varphi_k\right),$$

with fixed amplitude A and frequencies $\left\{f_k = 0.1k\right\}$, and IID random phases $\left\{\varphi_k\right\}$, uniformly distributed on the interval from 0 to 2π.

The desired signal correlation matrix is

$$\mathbf{\Phi}_{\mathbf{x}} = A^2\sum_{k=1}^{4}\mathbf{d}_k\mathbf{d}_k^H$$

where

$$\mathbf{d}_k = \begin{bmatrix} 1 & e^{-\jmath 2\pi f_k\tau_{x,2}} & e^{-\jmath 2\pi f_k\tau_{x,3}} & \cdots & e^{-\jmath 2\pi f_k\tau_{x,M}} \end{bmatrix}^T.$$

The rank of $\mathbf{\Phi}_{\mathbf{x}}$ is $R_x = 4$. The input SINR is

$$\text{iSINR} = 10\log\frac{4A^2}{\phi_{v_0}+1} \quad \text{(dB)}.$$

The maximum SINR filter, $\mathbf{h}_{\text{mSINR},2}$, is obtained from (6.101).

Figure 6.9 shows plots of the gain in SINR, $\mathcal{G}\left(\mathbf{h}_{\text{mSINR},2}\right)$, the MSE, $J\left(\mathbf{h}_{\text{mSINR},2}\right)$, and the desired signal distortion index, $\upsilon\left(\mathbf{h}_{\text{mSINR},2}\right)$, as a function of the input SINR for different numbers of sensors, M. The gain in SINR is always positive. For a given input SINR, as the number of sensors increases, the gain in SINR increases. However, the three criteria – gain in SINR, MSE, and desired signal distortion index – are very different. A filter that gives a large gain in SINR does not necessarily imply a small MSE or a small

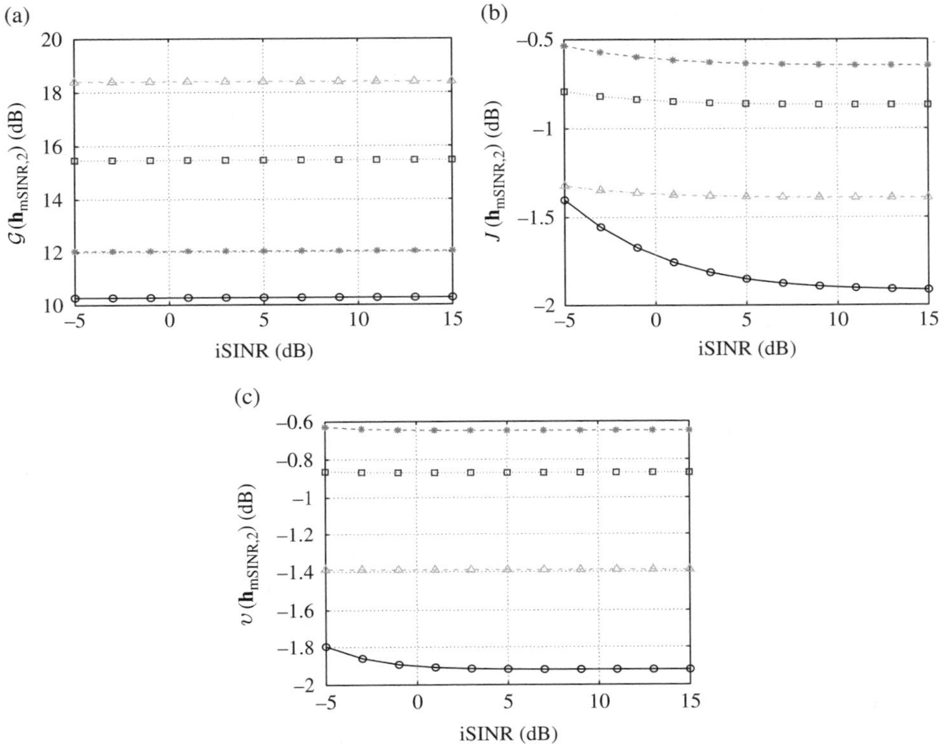

Figure 6.9 (a) The gain in SINR, (b) the MSE, and (c) the desired signal distortion index of the maximum SINR filter as a function of the input SINR for different numbers of sensors, M: $M = 10$ (solid line with circles), $M = 20$ (dashed line with asterisks), $M = 50$ (dotted line with squares), and $M = 100$ (dash-dot line with triangles).

desired signal distortion index, and a filter that leads to a small MSE or a small desired signal distortion index does not mean that the gain in SINR is large. ■

The classical Wiener filter is derived by minimizing $J(\mathbf{a})$ with respect to $\mathbf{a}$. We obtain

$$\begin{aligned}\mathbf{a}_{\mathrm{W}} &= \left(\mathbf{\Lambda} + \mathbf{I}_M\right)^{-1} \mathbf{\Lambda}\mathbf{i}_{\mathrm{T}} \\ &= \sum_{i=1}^{R_x} \frac{\lambda_i}{1+\lambda_i} \mathbf{i}_i \mathbf{i}_i^T \mathbf{T}^{-1} \mathbf{i}_{\mathrm{i}}.\end{aligned} \tag{6.103}$$

Therefore, another way to write the Wiener filter is

$$\mathbf{h}_{\mathrm{W}} = \sum_{i=1}^{R_x} \frac{\lambda_i}{1+\lambda_i} \mathbf{t}_i \mathbf{t}_i^H \mathbf{\Phi}_{\mathrm{in}} \mathbf{i}_{\mathrm{i}}. \tag{6.104}$$

It is of great interest to compare (6.101) to (6.104). For $R_x = 1$, we see that $\mathbf{h}_{\mathrm{mSINR},2} = \mathbf{h}_{\mathrm{W}}$, as we should expect. In the general case of $R_x \geq 1$, we can see clearly that the two filters are still closely connected. While the maximum SINR filter only takes into account the direction in which the energy of the desired signal is maximal, the Wiener filter takes into account the whole space in which the desired signal is present. This is the reason why the two filters work so differently in practice, even if they look similar. This result, although obvious, was never really shown in a such explicit way in the literature. Substituting (6.103) into $J(\mathbf{a})$, we find the MMSE:

$$\begin{aligned} J\left(\mathbf{a}_{\mathrm{W}}\right) &= \mathbf{i}_{\mathbf{T}}^H \boldsymbol{\Lambda} \mathbf{i}_{\mathbf{T}} - \sum_{i=1}^{R_x} \frac{\lambda_i^2}{1+\lambda_i} \left|\mathbf{i}_{\mathbf{T}}^H \mathbf{i}_i\right|^2 \\ &= \sum_{i=1}^{R_x} \frac{\lambda_i}{1+\lambda_i} \left|\mathbf{i}_{\mathbf{T}}^H \mathbf{i}_i\right|^2 \end{aligned} \quad (6.105)$$

and $J\left(\mathbf{a}_{\mathrm{W}}\right) \leq J\left(\mathbf{i}_{\mathbf{T}}\right)$. Obviously, we always have

$$J\left(\mathbf{a}_{\mathrm{W}}\right) \leq J(\mathbf{a}), \ \forall \mathbf{a}. \quad (6.106)$$

From this treatment, we can conclude that while the maximum SINR filter maximizes the output SINR, it may not improve its MSE compared to the variance of the noise. On the other hand, the Wiener filter can improve the output SINR while giving the minimum MSE.

Example 6.5.2 Returning to Example 6.5.1, we now employ the Wiener filter, $\mathbf{h}_{\mathrm{W}}$, given in (6.104). Figure 6.10 shows plots of the gain in SINR, $\mathcal{G}\left(\mathbf{h}_{\mathrm{W}}\right)$, the MSE, $J\left(\mathbf{h}_{\mathrm{W}}\right)$, and the desired signal distortion index, $v\left(\mathbf{h}_{\mathrm{W}}\right)$, as a function of the input SINR for different numbers of sensors, M. The gain in SINR is always positive. For a given input SINR, as the number of sensors increases, the gain in SINR increases, while the MSE and the desired signal distortion index decrease. ■

Now, we derive a class of filters that naturally fills the gap between the two fundamental filters studied above. For that purpose, let us first give the following property.

Property 6.5.1 Let $\lambda_1 \geq \lambda_2 \geq \cdots \geq \lambda_M \geq 0$. We have

$$\frac{\sum_{i=1}^{M} \left|a_i\right|^2 \lambda_i}{\sum_{i=1}^{M} \left|a_i\right|^2} \leq \frac{\sum_{i=1}^{M-1} \left|a_i\right|^2 \lambda_i}{\sum_{i=1}^{M-1} \left|a_i\right|^2} \leq \cdots \leq \frac{\sum_{i=1}^{2} \left|a_i\right|^2 \lambda_i}{\sum_{i=1}^{2} \left|a_i\right|^2} \leq \lambda_1, \quad (6.107)$$

where a_i, $i = 1, 2, \ldots, M$ are arbitrary complex numbers with at least one of them different from 0.

Proof. The previous inequalities can be easily shown by induction.

Property 6.5.1 suggests that we can define a class of filters that have the form:

$$\mathbf{a}_Q = \sum_{q=1}^{Q} \frac{\lambda_q}{1+\lambda_q} \mathbf{i}_q \mathbf{i}_q^T \mathbf{T}^{-1} \mathbf{i}_{\mathrm{i}} \quad (6.108)$$

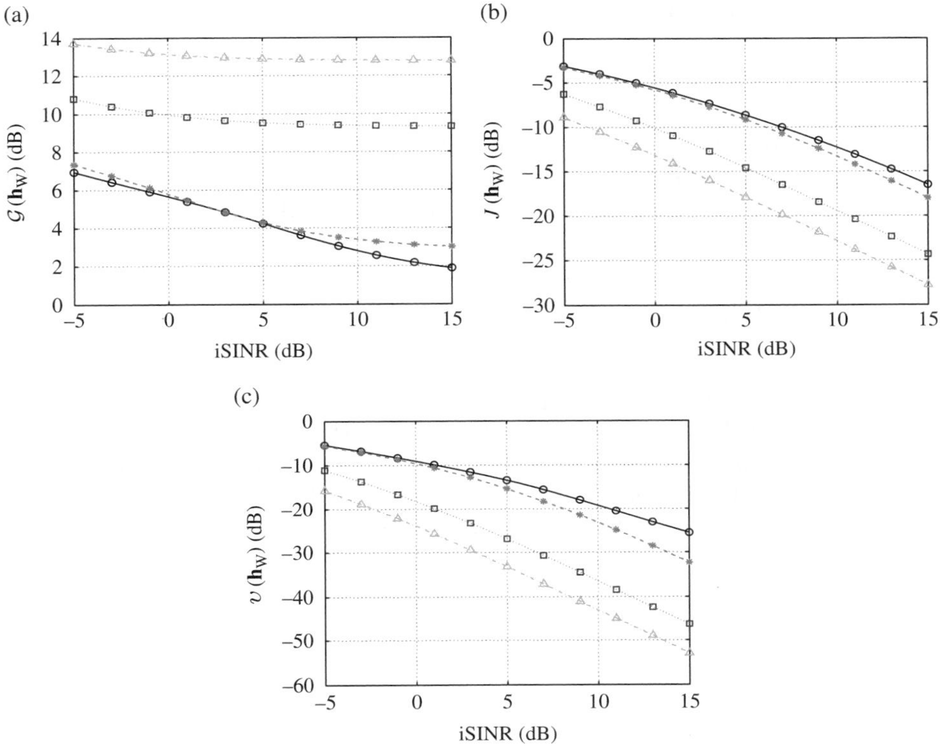

Figure 6.10 (a) The gain in SINR, (b) the MSE, and (c) the desired signal distortion index of the Wiener filter as a function of the input SINR for different numbers of sensors, M: $M = 10$ (solid line with circles), $M = 20$ (dashed line with asterisks), $M = 50$ (dotted line with squares), and $M = 100$ (dash-dot line with triangles).

or, equivalently,

$$\mathbf{h}_Q = \sum_{q=1}^{Q} \frac{\lambda_q}{1+\lambda_q} \mathbf{t}_q \mathbf{t}_q^H \mathbf{\Phi}_{\text{in}} \mathbf{i}_{\text{i}}, \tag{6.109}$$

where $1 \leq Q \leq R_x$. We see that $\mathbf{h}_1 = \mathbf{h}_{\text{mSINR},2}$ and $\mathbf{h}_{R_x} = \mathbf{h}_{\text{W}}$.

From Property 6.5.1, it is immediate obvious that

$$\text{iSNR} \leq \text{oSNR}\left(\mathbf{a}_{R_x}\right) \leq \text{oSNR}\left(\mathbf{a}_{R_x-1}\right) \leq \cdots \leq \text{oSNR}\left(\mathbf{a}_1\right) = \lambda_1. \tag{6.110}$$

It is straightforward to compute the MSE:

$$\begin{aligned} J\left(\mathbf{a}_Q\right) &= \mathbf{i}_{\mathbf{T}}^H \mathbf{\Lambda} \mathbf{i}_{\mathbf{T}} - \sum_{q=1}^{Q} \frac{\lambda_q^2}{1+\lambda_q} \left|\mathbf{i}_{\mathbf{T}}^H \mathbf{i}_q\right|^2 \\ &= \sum_{q=1}^{Q} \frac{\lambda_q}{1+\lambda_q} \left|\mathbf{i}_{\mathbf{T}}^H \mathbf{i}_q\right|^2 + \sum_{i=Q+1}^{R_x} \lambda_i \left|\mathbf{i}_{\mathbf{T}}^H \mathbf{i}_i\right|^2. \end{aligned} \tag{6.111}$$

We deduce from (6.111) that

$$J\left(\mathbf{a}_{R_x}\right) \leq J\left(\mathbf{a}_{R_x-1}\right) \leq \cdots \leq J\left(\mathbf{a}_1\right). \tag{6.112}$$

We see from (6.110) and (6.112) that the class of filters proposed here can give a compromise, in a very smooth way, between large values of the output SINR and small values of the MSE. This compromise depends, of course, on the rank of $\mathbf{\Phi}_{\mathbf{x}}$. For a value of R_x close to 1, the number of possibilities is very small but this is fine since, in this case, the two filters are very close to each other. For large values of R_x, we have many more options and this is desirable, since $\mathbf{h}_{\mathrm{mSINR},2}$ and $\mathbf{h}_{\mathrm{W}}$ now behave very differently. ■

Example 6.5.3 Returning to Example 6.5.1, we now employ the filter, $\mathbf{h}_Q$, given in (6.109). Figure 6.11 shows plots of the gain in SINR, $\mathcal{G}\left(\mathbf{h}_Q\right)$, the MSE, $J\left(\mathbf{h}_Q\right)$, and the desired signal distortion index, $\upsilon\left(\mathbf{h}_Q\right)$, as a function of the input SINR for $M = 20$ sensors and different values of Q. For $Q = 1$, $\mathbf{h}_Q$ reduces to $\mathbf{h}_{\mathrm{mSINR},2}$. For $Q = R_x = 4$, $\mathbf{h}_Q$ reduces to $\mathbf{h}_{\mathrm{W}}$. Clearly, the gain in SINR is always positive. For a given input SINR, as

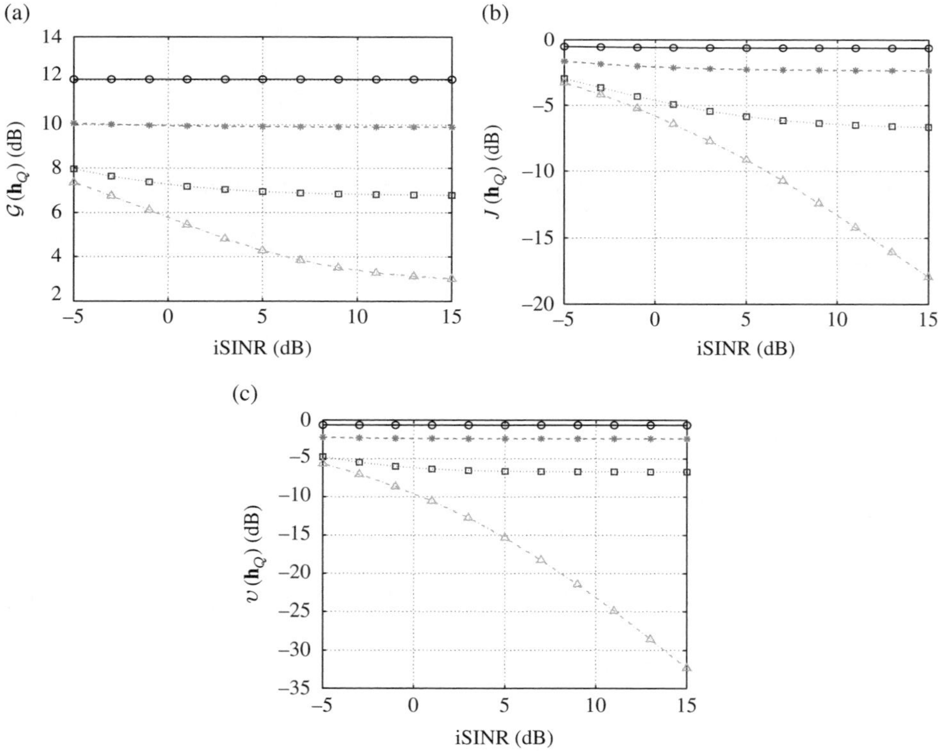

Figure 6.11 (a) The gain in SINR, (b) the MSE, and (c) the desired signal distortion index of the filter $\mathbf{h}_Q$ as a function of the input SINR for different values of Q: $Q = 1$ (solid line with circles), $Q = 2$ (dashed line with asterisks), $Q = 3$ (dotted line with squares), and $Q = 4$ (dash-dot line with triangles).

the value of Q decreases, the gain in SINR increases, at the expense of higher MSE and a higher desired signal distortion index. ■

Following the same steps as above, it is easy to derive another class of linear filters that fill the gap between the maximum SINR and MVDR filters:

$$\mathbf{h}'_Q = \sum_{q=1}^{Q} \mathbf{t}_q \mathbf{t}_q^H \mathbf{\Phi}_{\text{in}} \mathbf{i}_{\text{i}}, \tag{6.113}$$

where $1 \leq Q \leq R_x$. It is obvious that $\mathbf{h}'_1 = \mathbf{h}_{\text{mSINR}}$ and it can be shown that $\mathbf{h}'_{R_x} = \mathbf{h}_{\text{MVDR}}$.

Problems

6.1 Show that the Wiener filter can be expressed as

$$\mathbf{h}_{\text{W}} = \left(\mathbf{I}_M - \mathbf{\Phi}_{\mathbf{y}}^{-1}\mathbf{\Phi}_{\text{in}}\right) \mathbf{i}_{\text{i}}.$$

6.2 Using Woodbury's identity, show that

$$\mathbf{\Phi}_{\mathbf{y}}^{-1} = \mathbf{\Phi}_{\text{in}}^{-1} - \mathbf{\Phi}_{\text{in}}^{-1}\mathbf{Q}'_{\mathbf{x}} \left(\mathbf{\Lambda}'^{-1}_{\mathbf{x}} + \mathbf{Q}'^H_{\mathbf{x}}\mathbf{\Phi}_{\text{in}}^{-1}\mathbf{Q}'_{\mathbf{x}}\right)^{-1} \mathbf{Q}'^H_{\mathbf{x}}\mathbf{\Phi}_{\text{in}}^{-1}.$$

6.3 Show that the Wiener filter can be expressed as

$$\mathbf{h}_{\text{W}} = \mathbf{\Phi}_{\text{in}}^{-1}\mathbf{Q}'_{\mathbf{x}} \left(\mathbf{\Lambda}'^{-1}_{\mathbf{x}} + \mathbf{Q}'^H_{\mathbf{x}}\mathbf{\Phi}_{\text{in}}^{-1}\mathbf{Q}'_{\mathbf{x}}\right)^{-1} \mathbf{Q}'^H_{\mathbf{x}}\mathbf{i}_{\text{i}}.$$

6.4 Show that the MVDR filter is given by

$$\mathbf{h}_{\text{MVDR}} = \mathbf{\Phi}_{\text{in}}^{-1}\mathbf{Q}'_{\mathbf{x}} \left(\mathbf{Q}'^H_{\mathbf{x}}\mathbf{\Phi}_{\text{in}}^{-1}\mathbf{Q}'_{\mathbf{x}}\right)^{-1} \mathbf{Q}'^H_{\mathbf{x}}\mathbf{i}_{\text{i}}.$$

6.5 Show that the MVDR filter can be expressed as

$$\mathbf{h}_{\text{MVDR}} = \mathbf{\Phi}_{\mathbf{y}}^{-1}\mathbf{Q}'_{\mathbf{x}} \left(\mathbf{Q}'^H_{\mathbf{x}}\mathbf{\Phi}_{\mathbf{y}}^{-1}\mathbf{Q}'_{\mathbf{x}}\right)^{-1} \mathbf{Q}'^H_{\mathbf{x}}\mathbf{i}_{\text{i}}.$$

6.6 Show that the tradeoff filter is given by

$$\begin{aligned}\mathbf{h}_{\text{T},\mu} &= \left(\mathbf{\Phi}_{\mathbf{x}} + \mu\mathbf{\Phi}_{\text{in}}\right)^{-1} \mathbf{\Phi}_{\mathbf{x}}\mathbf{i}_{\text{i}} \\ &= \left[\mathbf{\Phi}_{\mathbf{y}} + (\mu - 1)\mathbf{\Phi}_{\text{in}}\right]^{-1} \left(\mathbf{\Phi}_{\mathbf{y}} - \mathbf{\Phi}_{\text{in}}\right) \mathbf{i}_{\text{i}},\end{aligned}$$

where μ is a Lagrange multiplier.

6.7 Show that the tradeoff filter can be expressed as

$$\mathbf{h}_{\text{T},\mu} = \mathbf{\Phi}_{\text{in}}^{-1}\mathbf{Q}'_{\mathbf{x}} \left(\mu\mathbf{\Lambda}'^{-1}_{\mathbf{x}} + \mathbf{Q}'^H_{\mathbf{x}}\mathbf{\Phi}_{\text{in}}^{-1}\mathbf{Q}'_{\mathbf{x}}\right)^{-1} \mathbf{Q}'^H_{\mathbf{x}}\mathbf{i}_{\text{i}}.$$

6.8 Show that the LCMV filter is given by

$$\mathbf{h}_{\text{LCMV}} = \mathbf{\Phi}_{\text{in}}^{-1}\mathbf{C}_{\mathbf{x}\mathbf{v}_1}\left(\mathbf{C}_{\mathbf{x}\mathbf{v}_1}^{H}\mathbf{\Phi}_{\text{in}}^{-1}\mathbf{C}_{\mathbf{x}\mathbf{v}_1}\right)^{-1}\mathbf{i}_{\text{c}}.$$

6.9 Show that the LCMV filter can be expressed as

$$\mathbf{h}_{\text{LCMV}} = \mathbf{\Phi}_{\mathbf{y}}^{-1}\mathbf{C}_{\mathbf{x}\mathbf{v}_1}\left(\mathbf{C}_{\mathbf{x}\mathbf{v}_1}^{H}\mathbf{\Phi}_{\mathbf{y}}^{-1}\mathbf{C}_{\mathbf{x}\mathbf{v}_1}\right)^{-1}\mathbf{i}_{\text{c}}.$$

6.10 Show that the LCMV filter can be expressed as

$$\mathbf{h}_{\text{LCMV}} = \mathbf{Q}_{\mathbf{v}_1}''\mathbf{\Phi}_{\text{in}}'^{-1}\mathbf{Q}_{\mathbf{v}_1}''^{H}\mathbf{Q}_{\mathbf{x}}'\left(\mathbf{Q}_{\mathbf{x}}'^{H}\mathbf{Q}_{\mathbf{v}_1}''\mathbf{\Phi}_{\text{in}}'^{-1}\mathbf{Q}_{\mathbf{v}_1}''^{H}\mathbf{Q}_{\mathbf{x}}'\right)^{-1}\mathbf{Q}_{\mathbf{x}}'^{H}\mathbf{i}_{\text{i}}.$$

6.11 Show that the maximum SINR filter with minimum distortion is given by

$$\begin{aligned}\mathbf{h}_{\text{mSINR}} &= \frac{\mathbf{t}_1\mathbf{t}_1^{H}\mathbf{\Phi}_{\mathbf{x}}\mathbf{i}_{\text{i}}}{\lambda_1} \\ &= \mathbf{t}_1\mathbf{t}_1^{H}\mathbf{\Phi}_{\text{in}}\mathbf{i}_{\text{i}}.\end{aligned}$$

6.12 Show that the maximum SIR filter with minimum MSE is given by

$$\mathbf{h}_{\text{mSIR}} = \mathbf{Q}_{\mathbf{v}}''\left(\mathbf{Q}_{\mathbf{v}}''^{H}\mathbf{\Phi}_{\mathbf{y}}\mathbf{Q}_{\mathbf{v}}''\right)^{-1}\mathbf{Q}_{\mathbf{v}}''^{H}\mathbf{\Phi}_{\mathbf{x}}\mathbf{i}_{\text{i}}.$$

6.13 Show that the output SINR can be expressed as

$$\begin{aligned}\text{oSINR}(\mathbf{a}) &= \frac{\mathbf{a}^{H}\mathbf{\Lambda}\mathbf{a}}{\mathbf{a}^{H}\mathbf{a}} \\ &= \frac{\sum_{i=1}^{R_x}|a_i|^2\lambda_i}{\sum_{m=1}^{M}|a_m|^2}.\end{aligned}$$

6.14 Show that the transformed identity filter, $\mathbf{i}_{\mathbf{T}}$, does not affect the observations: $z = \mathbf{i}_{\mathbf{T}}^{H}\mathbf{T}^{H}\mathbf{y} = y_1$ and $\text{oSINR}\left(\mathbf{i}_{\mathbf{T}}\right) = \text{iSINR}$.

6.15 Show that the MSE can be expressed as

$$J(\mathbf{a}) = \phi_{x_1} - \mathbf{a}^{H}\mathbf{\Lambda}\mathbf{i}_{\mathbf{T}} - \mathbf{i}_{\mathbf{T}}^{H}\mathbf{\Lambda}\mathbf{a} + \mathbf{a}^{H}\left(\mathbf{\Lambda} + \mathbf{I}_M\right)\mathbf{a}.$$

6.16 Show that the MSE can be expressed as

$$J(\mathbf{a}) = \left(\mathbf{a} - \mathbf{i}_{\mathbf{T}}\right)^{H}\mathbf{\Lambda}\left(\mathbf{a} - \mathbf{i}_{\mathbf{T}}\right) + \mathbf{a}^{H}\mathbf{a}.$$

6.17 Show that the maximum SINR filter with minimum MSE is given by

$$\mathbf{h}_{\text{mSINR},2} = \frac{\lambda_1}{1+\lambda_1}\mathbf{t}_1\mathbf{t}_1^{H}\mathbf{\Phi}_{\text{in}}\mathbf{i}_{\text{i}}.$$

6.18 Show that with the maximum SINR filter, $\mathbf{h}_{\text{mSINR},2}$, the minimum MSE is given by

$$J\left(\mathbf{h}_{\text{mSINR},2}\right) = \frac{\lambda_1}{1+\lambda_1}\left|\mathbf{i}_{\mathbf{T}}^H\mathbf{i}_{\mathrm{i}}\right|^2 + \sum_{i=2}^{R_x}\lambda_i\left|\mathbf{i}_{\mathbf{T}}^H\mathbf{i}_i\right|^2.$$

6.19 Show that the Wiener filter can be expressed as

$$\mathbf{h}_{\mathrm{W}} = \sum_{i=1}^{R_x}\frac{\lambda_i}{1+\lambda_i}\mathbf{t}_i\mathbf{t}_i^H\mathbf{\Phi}_{\text{in}}\mathbf{i}_{\mathrm{i}}.$$

6.20 Show that with the Wiener filter $\mathbf{h}_{\mathrm{W}}$, the MMSE is given by

$$\begin{aligned} J\left(\mathbf{h}_{\mathrm{W}}\right) &= \mathbf{i}_{\mathbf{T}}^H\mathbf{\Lambda}\mathbf{i}_{\mathbf{T}} - \sum_{i=1}^{R_x}\frac{\lambda_i^2}{1+\lambda_i}\left|\mathbf{i}_{\mathbf{T}}^H\mathbf{i}_i\right|^2 \\ &= \sum_{i=1}^{R_x}\frac{\lambda_i}{1+\lambda_i}\left|\mathbf{i}_{\mathbf{T}}^H\mathbf{i}_i\right|^2. \end{aligned}$$

6.21 Let $\lambda_1 \geq \lambda_2 \geq \cdots \geq \lambda_M \geq 0$ and let a_m, $m = 1, 2, \ldots, M$ denote arbitrary complex numbers with at least one of them different from 0. Prove that

$$\frac{\sum_{i=1}^{M}\left|a_i\right|^2\lambda_i}{\sum_{i=1}^{M}\left|a_i\right|^2} \leq \frac{\sum_{i=1}^{M-1}\left|a_i\right|^2\lambda_i}{\sum_{i=1}^{M-1}\left|a_i\right|^2} \leq \cdots \leq \frac{\sum_{i=1}^{2}\left|a_i\right|^2\lambda_i}{\sum_{i=1}^{2}\left|a_i\right|^2} \leq \lambda_1.$$

6.22 Show that the class of filters $\mathbf{a}_Q$ makes a compromise between large values of the output SINR and small values of the MSE; that is,

a) $\text{iSNR} \leq \text{oSNR}\left(\mathbf{a}_{R_x}\right) \leq \text{oSNR}\left(\mathbf{a}_{R_x-1}\right) \leq \cdots \leq \text{oSNR}\left(\mathbf{a}_1\right) = \lambda_1$,

b) $J\left(\mathbf{a}_{R_x}\right) \leq J\left(\mathbf{a}_{R_x-1}\right) \leq \cdots \leq J\left(\mathbf{a}_1\right)$.

6.23 Show that the class of linear filters $\mathbf{h}'_Q$ satisfies $\mathbf{h}'_1 = \mathbf{h}_{\text{mSINR}}$ and $\mathbf{h}'_{R_x} = \mathbf{h}_{\text{MVDR}}$.

References

1 J. Benesty, J. Chen, Y. Huang, and I. Cohen, *Noise Reduction in Speech Processing*. Berlin, Germany: Springer-Verlag, 2009.

2 J. Benesty and J. Chen, *Optimal Time-domain Noise Reduction Filters – A Theoretical Study*. Springer Briefs in Electrical and Computer Engineering, 2011.

3 P. Stoica and R. L. Moses, *Introduction to Spectral Analysis*. Englewood Cliffs, NJ: Prentice-Hall, 1997.

4 B. D. Van Veen and K. M. Buckley, "Beamforming: A versatile approach to spatial filtering," *IEEE Acoust., Speech, Signal Process. Mag.*, vol. 5, pp. 4–24, Apr. 1988.

5 S. Shahbazpanahi, A. B. Gershman, Z.-Q. Luo, and K. M. Wong, "Robust adaptive beamforming for general-rank signal models," *IEEE Trans. Signal Process.*, vol. 51, pp. 2257–2269, Sep. 2003.

6 G. H. Golub and C. F. Van Loan, *Matrix Computations*, 3rd edn. Baltimore, MD: The Johns Hopkins University Press, 1996.
7 S. Haykin, *Adaptive Filter Theory*, 4th edn. Upper Saddle River, NJ: Prentice-Hall, 2002.
8 J. Chen, J. Benesty, Y. Huang, and S. Doclo, "New insights into the noise reduction Wiener filter," *IEEE Trans. Audio, Speech, Language Process.*, vol. 14, pp. 1218–1234, Jul. 2006.
9 J. R. Jensen, J. Benesty, M. G. Christensen, and J. Chen, "A class of optimal rectangular filtering matrices for single-channel signal enhancement in the time domain," *IEEE Trans. Audio, Speech, Language Process.*, vol. 11, pp. 2595–2606, Dec. 2013.
10 J. Benesty, J. R. Jensen, M. G. Christensen, and J. Chen, *Speech Enhancement – A Signal Subspace Perspective*. Oxford, England: Academic Press, 2014.
11 S. M. Nørholm, J. Benesty, J. R. Jensen, and M. G. Christensen, "Single-channel noise reduction using unified joint diagonalization and optimal filtering," *EURASIP J. Advances Signal Process.*, vol. 2014, pp. 37–48.
12 J. N. Franklin, *Matrix Theory*. Englewood Cliffs, NJ: Prentice-Hall, 1968.
13 Y. Rong, Y. C. Eldar, and A. B. Gershman, "Performance tradeoffs among adaptive beamforming criteria," *IEEE J. Selected Topics Signal Process.*, vol. 1, pp. 651–659, Dec. 2007.

Part II

Array Signal Processing

7

Fixed Beamforming

A fixed beamformer is a spatial filter that has the ability to form a main beam in the direction of the desired signal and, possibly, place nulls in the directions of interferences without the knowledge of the data picked up by the array or the statistics of the desired and noise signals; as a consequence, the coefficients of this filter are fixed and do not depend on the changes of the wave propagation environment in which the array performs. However, fixed beamforming uses information about the location of the sensors in space and the directions of the desired and interference sources through the steering vectors. Therefore, the geometry of the array needs to be known. In this chapter, we derive and study a large class of fixed beamformers in tandem with uniform linear arrays (ULAs), where the sensors are located along a line with uniform spacing. In the rest of this text, only ULAs are considered. This simplifies the presentation of the main results. Generalization to other geometries is not difficult in general.

7.1 Signal Model and Problem Formulation

We consider a plane wave, in the farfield – that is, far enough from the array – that propagates in an anechoic environment at speed c[1], and impinges on a uniform linear sensor array consisting of M omnidirectional sensors. The distance between two successive sensors is equal to δ and the direction of the source signal to the array is parameterized by the azimuth angle θ (see Figure 7.1). In this context, the steering vector (of length M) is given by [1–3]:

$$\mathbf{d}\left(f, \cos\theta\right) = \begin{bmatrix} 1 & e^{-\jmath 2\pi f \tau_0 \cos\theta} & \cdots & e^{-\jmath (M-1) 2\pi f \tau_0 \cos\theta} \end{bmatrix}^T, \tag{7.1}$$

where $\jmath = \sqrt{-1}$ is the imaginary unit, $f > 0$ is the temporal frequency, and $\tau_0 = \delta/c$ is the delay between two successive sensors at the angle $\theta = 0$. We denote by $\omega = 2\pi f$ the angular frequency and by $\lambda = c/f$ the wavelength. Since $\cos\theta$ is an even function, so is $\mathbf{d}\left(f, \cos\theta\right)$. Therefore, the study is limited to angles $\theta \in [0, \pi]$.

1 For example, the speed of sound in the air is $c = 340$ m/s.

Fundamentals of Signal Enhancement and Array Signal Processing, First Edition.
Jacob Benesty, Israel Cohen, and Jingdong Chen.

Companion website: www.wiley.com/go/benesty/arraysignalprocessing

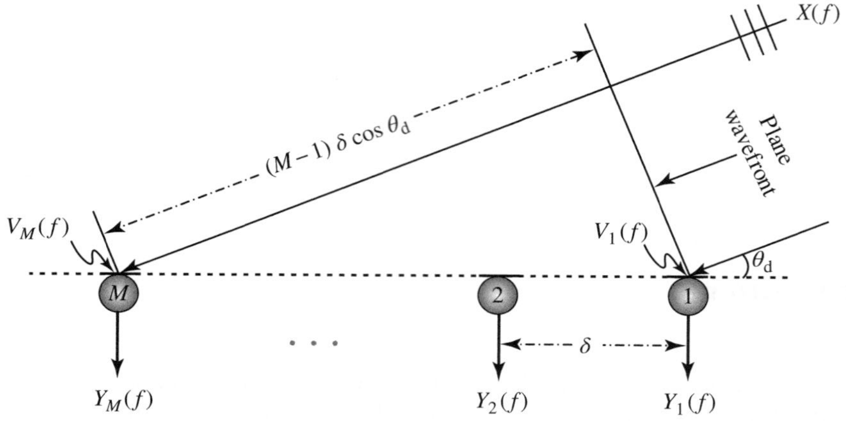

Figure 7.1 A uniform linear array with M sensors.

Assume that the desired signal propagates from the angle θ_{d}. Using the signal model of Chapter 5, the observation signal vector (of length M) is

$$\begin{aligned}\mathbf{y}(f) &= \begin{bmatrix} Y_1(f) & Y_2(f) & \cdots & Y_M(f) \end{bmatrix}^T \\ &= \mathbf{x}(f) + \mathbf{v}(f) \\ &= \mathbf{d}\left(f, \cos\theta_{\mathrm{d}}\right) X(f) + \mathbf{v}(f),\end{aligned} \tag{7.2}$$

where $Y_m(f)$ is the mth sensor signal, $\mathbf{x}(f) = \mathbf{d}\left(f, \cos\theta_{\mathrm{d}}\right) X(f)$, $X(f)$ is the desired signal, $\mathbf{d}\left(f, \cos\theta_{\mathrm{d}}\right)$ is the steering vector at $\theta = \theta_{\mathrm{d}}$ (direction of the desired source), and $\mathbf{v}(f)$ is the additive noise signal vector defined similarly to $\mathbf{y}(f)$. Then, the correlation matrix of $\mathbf{y}(f)$ is

$$\begin{aligned}\mathbf{\Phi}_{\mathbf{y}}(f) &= E\left[\mathbf{y}(f)\mathbf{y}^H(f)\right] \\ &= \phi_X(f)\mathbf{d}\left(f, \cos\theta_{\mathrm{d}}\right)\mathbf{d}^H\left(f, \cos\theta_{\mathrm{d}}\right) + \mathbf{\Phi}_{\mathbf{v}}(f),\end{aligned} \tag{7.3}$$

where $\phi_X(f)$ is the variance of $X(f)$ and $\mathbf{\Phi}_{\mathbf{v}}(f)$ is the correlation matrix of $\mathbf{v}(f)$.

Our objective in this chapter is to design beamformers, independent of the statistics of the signals, which are able to form a main beam in the direction of the desired signal, θ_{d}, in order to extract it undistorted while attenuating signals coming from other directions.

7.2 Linear Array Model

Usually, the array processing or beamforming is performed by applying a temporal filter to each sensor signal and summing the filtered signals. In the frequency domain, this is equivalent to adding a complex weight to the output of each sensor and summing across the aperture [4]:

$$\begin{aligned} Z(f) &= \sum_{m=1}^{M} H_m^*(f)Y_m(f) \\ &= \mathbf{h}^H(f)\mathbf{y}(f) \\ &= X_{\mathrm{fd}}(f) + V_{\mathrm{rn}}(f), \end{aligned} \tag{7.4}$$

where $Z(f)$ is the beamformer output signal,

$$\mathbf{h}(f) = \begin{bmatrix} H_1(f) & H_2(f) & \cdots & H_M(f) \end{bmatrix}^T \tag{7.5}$$

is the beamforming weight vector, which is suitable for performing spatial filtering at frequency f,

$$X_{\mathrm{fd}}(f) = X(f)\mathbf{h}^H(f)\mathbf{d}\left(f, \cos\theta_{\mathrm{d}}\right) \tag{7.6}$$

is the filtered desired signal, and

$$V_{\mathrm{rn}}(f) = \mathbf{h}^H(f)\mathbf{v}(f) \tag{7.7}$$

is the residual noise.

Since the two terms on the right-hand side of (7.4) are incoherent, the variance of $Z(f)$ is the sum of two variances:

$$\begin{aligned} \phi_Z(f) &= \mathbf{h}^H(f)\mathbf{\Phi}_{\mathbf{y}}(f)\mathbf{h}(f) \\ &= \phi_{X_{\mathrm{fd}}}(f) + \phi_{V_{\mathrm{rn}}}(f), \end{aligned} \tag{7.8}$$

where

$$\phi_{X_{\mathrm{fd}}}(f) = \phi_X(f)\left|\mathbf{h}^H(f)\mathbf{d}\left(f, \cos\theta_{\mathrm{d}}\right)\right|^2, \tag{7.9}$$

$$\phi_{V_{\mathrm{rn}}}(f) = \mathbf{h}^H(f)\mathbf{\Phi}_{\mathbf{v}}(f)\mathbf{h}(f). \tag{7.10}$$

In the context of fixed beamforming, the distortionless constraint is desired:

$$\mathbf{h}^H(f)\mathbf{d}\left(f, \cos\theta_{\mathrm{d}}\right) = 1, \tag{7.11}$$

meaning that any signal arriving along $\mathbf{d}\left(f, \cos\theta_{\mathrm{d}}\right)$ will pass through the beamformer undistorted. Consequently, all beamformers will be derived by taking (7.11) into account.

7.3 Performance Measures

In fixed beamforming, it is customary to focus on narrowband performance measures only. As in the previous chapters, the first sensor is considered the reference.

Each beamformer has a pattern of directional sensitivity: it has different sensitivities from sounds arriving from different directions. The beampattern or directivity pattern describes the sensitivity of the beamformer to a plane wave (source signal) impinging on the array from the direction θ. Mathematically, it is defined as

$$\mathcal{B}\left[\mathbf{h}(f), \cos\theta\right] = \mathbf{d}^H\left(f, \cos\theta\right)\mathbf{h}(f) \qquad (7.12)$$
$$= \sum_{m=1}^{M} H_m(f) e^{\jmath(m-1)2\pi f \tau_0 \cos\theta}.$$

Usually, $\left|\mathcal{B}\left[\mathbf{h}(f), \cos\theta\right]\right|^2$, which is the power pattern [5], is illustrated with a polar plot.

The (narrowband) input SNR is

$$\text{iSNR}(f) = \frac{\phi_X(f)}{\phi_{V_1}(f)}, \qquad (7.13)$$

where $\phi_{V_1}(f) = E\left[\left|V_1(f)\right|^2\right]$ is the variance of $V_1(f)$, which is the first element of $\mathbf{v}(f)$. The (narrowband) output SNR is defined as

$$\text{oSNR}\left[\mathbf{h}(f)\right] = \phi_X(f)\frac{\left|\mathbf{h}^H(f)\mathbf{d}\left(f, \cos\theta_{\text{d}}\right)\right|^2}{\mathbf{h}^H(f)\mathbf{\Phi}_{\mathbf{v}}(f)\mathbf{h}(f)} \qquad (7.14)$$
$$= \frac{\phi_X(f)}{\phi_{V_1}(f)} \times \frac{\left|\mathbf{h}^H(f)\mathbf{d}\left(f, \cos\theta_{\text{d}}\right)\right|^2}{\mathbf{h}^H(f)\mathbf{\Gamma}_{\mathbf{v}}(f)\mathbf{h}(f)},$$

where

$$\mathbf{\Gamma}_{\mathbf{v}}(f) = \frac{\mathbf{\Phi}_{\mathbf{v}}(f)}{\phi_{V_1}(f)} \qquad (7.15)$$

is the pseudo-coherence matrix of $\mathbf{v}(f)$. From the previous definitions of the SNRs, we deduce the array gain:

$$\mathcal{G}\left[\mathbf{h}(f)\right] = \frac{\text{oSNR}\left[\mathbf{h}(f)\right]}{\text{iSNR}(f)} \qquad (7.16)$$
$$= \frac{\left|\mathbf{h}^H(f)\mathbf{d}\left(f, \cos\theta_{\text{d}}\right)\right|^2}{\mathbf{h}^H(f)\mathbf{\Gamma}_{\mathbf{v}}(f)\mathbf{h}(f)}.$$

The most convenient way to evaluate the sensitivity of the array to some of its imperfections, such as sensor noise, is via the so-called (narrowband) white noise gain (WNG) (already used in Chapter 5), which is defined by taking $\mathbf{\Gamma}_{\mathbf{v}}(f) = \mathbf{I}_M$ in (7.16), where $\mathbf{I}_M$ is the $M \times M$ identity matrix:

$$\mathcal{W}\left[\mathbf{h}(f)\right] = \frac{\left|\mathbf{h}^H(f)\mathbf{d}\left(f, \cos\theta_{\text{d}}\right)\right|^2}{\mathbf{h}^H(f)\mathbf{h}(f)}. \qquad (7.17)$$

Using the Cauchy–Schwarz inequality,

$$\left|\mathbf{h}^H(f)\mathbf{d}\left(f,\cos\theta_{\rm d}\right)\right|^2 \leq \mathbf{h}^H(f)\mathbf{h}(f)\times\mathbf{d}^H\left(f,\cos\theta_{\rm d}\right)\mathbf{d}\left(f,\cos\theta_{\rm d}\right), \tag{7.18}$$

we easily deduce from (7.17) that

$$\mathcal{W}\left[\mathbf{h}(f)\right] \leq M,\ \forall\mathbf{h}(f). \tag{7.19}$$

As a result, the maximum WNG is

$$\mathcal{W}_{\max} = M, \tag{7.20}$$

which is frequency independent. Let

$$\cos\left[\mathbf{d}\left(f,\cos\theta_{\rm d}\right),\mathbf{h}(f)\right] = \frac{\mathbf{d}^H\left(f,\cos\theta_{\rm d}\right)\mathbf{h}(f)+\mathbf{h}^H(f)\mathbf{d}\left(f,\cos\theta_{\rm d}\right)}{2\left\|\mathbf{d}\left(f,\cos\theta_{\rm d}\right)\right\|_2\|\mathbf{h}(f)\|_2} \tag{7.21}$$

be the cosine of the angle between the two vectors $\mathbf{d}\left(f,\cos\theta_{\rm d}\right)$ and $\mathbf{h}(f)$, with $\|\cdot\|_2$ denoting the ℓ_2 norm. Assuming the distortionless constraint, we can rewrite the WNG as

$$\mathcal{W}\left[\mathbf{h}(f)\right] = \mathcal{W}_{\max}\cos^2\left[\mathbf{d}\left(f,\cos\theta_{\rm d}\right),\mathbf{h}(f)\right]. \tag{7.22}$$

Another important measure, which quantifies how the sensor array performs in the presence of reverberation, is the (narrowband) directivity factor (DF). Considering the spherically isotropic (diffuse) noise field, the DF is defined as [5]:

$$\begin{aligned}\mathcal{D}\left[\mathbf{h}(f)\right] &= \frac{\left|\mathcal{B}\left[\mathbf{h}(f),\cos\theta_{\rm d}\right]\right|^2}{\frac{1}{2}\int_0^{\pi}\left|\mathcal{B}\left[\mathbf{h}(f),\cos\theta\right]\right|^2\sin\theta d\theta} \\ &= \frac{\left|\mathbf{h}^H(f)\mathbf{d}\left(f,\cos\theta_{\rm d}\right)\right|^2}{\mathbf{h}^H(f)\mathbf{\Gamma}_{0,\pi}(f)\mathbf{h}(f)},\end{aligned} \tag{7.23}$$

where

$$\mathbf{\Gamma}_{0,\pi}(f) = \frac{1}{2}\int_0^{\pi}\mathbf{d}\left(f,\cos\theta\right)\mathbf{d}^H\left(f,\cos\theta\right)\sin\theta d\theta. \tag{7.24}$$

It can be verified that the elements of the $M\times M$ matrix $\mathbf{\Gamma}_{0,\pi}(f)$ are

$$\begin{aligned}\left[\mathbf{\Gamma}_{0,\pi}(f)\right]_{ij} &= \frac{\sin\left[2\pi f(j-i)\tau_0\right]}{2\pi f(j-i)\tau_0} \\ &= \text{sinc}\left[2\pi f(j-i)\tau_0\right],\end{aligned} \tag{7.25}$$

with $\left[\boldsymbol{\Gamma}_{0,\pi}(f)\right]_{mm} = 1,\ m = 1, 2, \ldots, M$. Again, by invoking the Cauchy–Schwarz inequality,

$$\left|\mathbf{h}^H(f)\mathbf{d}\left(f, \cos\theta_{\mathrm{d}}\right)\right|^2 \leq \mathbf{h}^H(f)\boldsymbol{\Gamma}_{0,\pi}(f)\mathbf{h}(f) \times \mathbf{d}^H\left(f, \cos\theta_{\mathrm{d}}\right)\boldsymbol{\Gamma}_{0,\pi}^{-1}(f)\mathbf{d}\left(f, \cos\theta_{\mathrm{d}}\right), \tag{7.26}$$

we find from (7.23) that

$$\mathcal{D}\left[\mathbf{h}(f)\right] \leq \mathbf{d}^H\left(f, \cos\theta_{\mathrm{d}}\right)\boldsymbol{\Gamma}_{0,\pi}^{-1}(f)\mathbf{d}\left(f, \cos\theta_{\mathrm{d}}\right),\ \forall \mathbf{h}(f). \tag{7.27}$$

As a result, the maximum DF is

$$\begin{aligned}\mathcal{D}_{\max}\left(f, \cos\theta_{\mathrm{d}}\right) &= \mathbf{d}^H\left(f, \cos\theta_{\mathrm{d}}\right)\boldsymbol{\Gamma}_{0,\pi}^{-1}(f)\mathbf{d}\left(f, \cos\theta_{\mathrm{d}}\right) \\ &= \mathrm{tr}\left[\boldsymbol{\Gamma}_{0,\pi}^{-1}(f)\mathbf{d}\left(f, \cos\theta_{\mathrm{d}}\right)\mathbf{d}^H\left(f, \cos\theta_{\mathrm{d}}\right)\right] \\ &\leq M\mathrm{tr}\left[\boldsymbol{\Gamma}_{0,\pi}^{-1}(f)\right],\end{aligned} \tag{7.28}$$

which is frequency and (desired signal) angle dependent. Let

$$\cos\left[\boldsymbol{\Gamma}_{0,\pi}^{-1/2}(f)\mathbf{d}\left(f, \cos\theta_{\mathrm{d}}\right), \boldsymbol{\Gamma}_{0,\pi}^{1/2}(f)\mathbf{h}(f)\right] = \frac{\mathbf{d}^H\left(f, \cos\theta_{\mathrm{d}}\right)\mathbf{h}(f) + \mathbf{h}^H(f)\mathbf{d}\left(f, \cos\theta_{\mathrm{d}}\right)}{2\left\|\boldsymbol{\Gamma}_{0,\pi}^{-1/2}(f)\mathbf{d}\left(f, \cos\theta_{\mathrm{d}}\right)\right\|_2\left\|\boldsymbol{\Gamma}_{0,\pi}^{1/2}(f)\mathbf{h}(f)\right\|_2} \tag{7.29}$$

be the cosine of the angle between the two vectors $\boldsymbol{\Gamma}_{0,\pi}^{-1/2}(f)\mathbf{d}\left(f, \cos\theta_{\mathrm{d}}\right)$ and $\boldsymbol{\Gamma}_{0,\pi}^{1/2}(f)\mathbf{h}(f)$. Assuming the distortionless constraint, we can express the DF as

$$\mathcal{D}\left[\mathbf{h}(f)\right] = \mathcal{D}_{\max}\left(f, \cos\theta_{\mathrm{d}}\right)\cos^2\left[\boldsymbol{\Gamma}_{0,\pi}^{-1/2}(f)\mathbf{d}\left(f, \cos\theta_{\mathrm{d}}\right), \boldsymbol{\Gamma}_{0,\pi}^{1/2}(f)\mathbf{h}(f)\right]. \tag{7.30}$$

7.4 Spatial Aliasing

We discuss here the spatial aliasing problem encountered in array processing; it is similar to the temporal aliasing that occurs when a continuous-time signal is sampled at a rate lower than twice its highest frequency.

Let θ_1 and θ_2 be two different angles; that is, $\theta_1 \neq \theta_2$. Spatial aliasing occurs when $\mathbf{d}\left(f, \cos\theta_1\right) = \mathbf{d}\left(f, \cos\theta_2\right)$, implying an ambiguity in source locations. Let

$$\cos\theta_1 = \frac{c}{f\delta} + \cos\theta_2 = \frac{\lambda}{\delta} + \cos\theta_2, \tag{7.31}$$

or, equivalently,

$$\frac{\delta}{\lambda} = \frac{1}{\cos\theta_1 - \cos\theta_2}. \tag{7.32}$$

It is straightforward to see that

$$e^{-\jmath(m-1)2\pi f\tau_0\cos\theta_1} = e^{-\jmath(m-1)2\pi f\tau_0\cos\theta_2},\ m = 1, 2, \ldots, M. \tag{7.33}$$

As a consequence,

$$\mathbf{d}\left(f, \cos\theta_1\right) = \mathbf{d}\left(f, \cos\theta_2\right), \tag{7.34}$$

meaning that spatial aliasing takes place.

Since $|\cos\theta| \leq 1$, we always have

$$\left|\cos\theta_1 - \cos\theta_2\right| \leq 2, \tag{7.35}$$

or, equivalently,

$$\frac{1}{\left|\cos\theta_1 - \cos\theta_2\right|} \geq \frac{1}{2}. \tag{7.36}$$

We conclude from (7.32) that to prevent aliasing, one needs to ensure that

$$\frac{\delta}{\lambda} < \frac{1}{2}, \tag{7.37}$$

which is the classical narrowband aliasing criterion.

7.5 Fixed Beamformers

In this section, we derive several useful fixed beamformers from the WNG and the DF, which can also be viewed as meaningful criteria as the MSE criterion and not only as performance measures.

7.5.1 Delay and Sum

The most well-known fixed beamformer is the so-called delay-and-sum (DS), which is derived by maximizing the WNG:

$$\min_{\mathbf{h}(f)} \mathbf{h}^H(f)\mathbf{h}(f) \quad \text{subject to} \quad \mathbf{h}^H(f)\mathbf{d}\left(f, \cos\theta_{\mathrm{d}}\right) = 1. \tag{7.38}$$

We easily get the optimal filter:

$$\mathbf{h}_{\mathrm{DS}}\left(f, \cos\theta_{\mathrm{d}}\right) = \frac{\mathbf{d}\left(f, \cos\theta_{\mathrm{d}}\right)}{\mathbf{d}^H\left(f, \cos\theta_{\mathrm{d}}\right)\mathbf{d}\left(f, \cos\theta_{\mathrm{d}}\right)} = \frac{\mathbf{d}\left(f, \cos\theta_{\mathrm{d}}\right)}{M}. \tag{7.39}$$

Therefore, with this beamformer, the WNG and the DF are, respectively,

$$\mathcal{W}\left[\mathbf{h}_{\mathrm{DS}}\left(f, \cos\theta_{\mathrm{d}}\right)\right] = M = \mathcal{W}_{\max} \tag{7.40}$$

and

$$\mathcal{D}\left[\mathbf{h}_{\mathrm{DS}}\left(f, \cos\theta_{\mathrm{d}}\right)\right] = \frac{M^2}{\mathbf{d}^H\left(f, \cos\theta_{\mathrm{d}}\right)\mathbf{\Gamma}_{0,\pi}(f)\mathbf{d}\left(f, \cos\theta_{\mathrm{d}}\right)}. \tag{7.41}$$

Since,

$$\mathbf{d}^H\left(f, \cos\theta_{\mathrm{d}}\right)\mathbf{\Gamma}_{0,\pi}(f)\mathbf{d}\left(f, \cos\theta_{\mathrm{d}}\right) \leq M\mathrm{tr}\left[\mathbf{\Gamma}_{0,\pi}(f)\right] = M^2, \tag{7.42}$$

we have $\mathcal{D}\left[\mathbf{h}_{\mathrm{DS}}\left(f, \cos\theta_{\mathrm{d}}\right)\right] \geq 1$. While the DS beamformer maximizes the WNG, it never amplifies the diffuse noise since $\mathcal{D}\left[\mathbf{h}_{\mathrm{DS}}\left(f, \cos\theta_{\mathrm{d}}\right)\right] \geq 1$.

We find that the beampattern is

$$\begin{aligned}\left|\mathcal{B}\left[\mathbf{h}_{\mathrm{DS}}\left(f, \cos\theta_{\mathrm{d}}\right), \cos\theta\right]\right|^2 &= \frac{1}{M^2}\left|\mathbf{d}^H\left(f, \cos\theta\right)\mathbf{d}\left(f, \cos\theta_{\mathrm{d}}\right)\right|^2 \\ &= \frac{1}{M^2}\left|\sum_{m=1}^{M} e^{\jmath(m-1)2\pi f\tau_0\left(\cos\theta-\cos\theta_{\mathrm{d}}\right)}\right|^2 \\ &= \frac{1}{M^2}\left|\frac{1-e^{\jmath M2\pi f\tau_0\left(\cos\theta-\cos\theta_{\mathrm{d}}\right)}}{1-e^{\jmath 2\pi f\tau_0\left(\cos\theta-\cos\theta_{\mathrm{d}}\right)}}\right|^2,\end{aligned} \tag{7.43}$$

with $\left|\mathcal{B}\left[\mathbf{h}_{\mathrm{DS}}\left(f, \cos\theta_{\mathrm{d}}\right), \cos\theta\right]\right|^2 \leq 1$. The beampattern of the DS beamformer is very frequency dependent.

Another interesting way to express (7.41) is [6]:

$$\begin{aligned}\mathcal{D}\left[\mathbf{h}_{\mathrm{DS}}\left(f, \cos\theta_{\mathrm{d}}\right)\right] &= \mathcal{D}_{\max}\left(f, \cos\theta_{\mathrm{d}}\right) \\ &\quad\times\cos^2\left[\mathbf{\Gamma}_{0,\pi}^{1/2}(f)\mathbf{d}\left(f, \cos\theta_{\mathrm{d}}\right), \mathbf{\Gamma}_{0,\pi}^{-1/2}(f)\mathbf{d}\left(f, \cos\theta_{\mathrm{d}}\right)\right],\end{aligned} \tag{7.44}$$

where

$$\cos\left[\mathbf{\Gamma}_{0,\pi}^{1/2}(f)\mathbf{d}\left(f, \cos\theta_{\mathrm{d}}\right), \mathbf{\Gamma}_{0,\pi}^{-1/2}(f)\mathbf{d}\left(f, \cos\theta_{\mathrm{d}}\right)\right] = \frac{\mathbf{d}^H\left(f, \cos\theta_{\mathrm{d}}\right)\mathbf{d}\left(f, \cos\theta_{\mathrm{d}}\right)}{\sqrt{\mathbf{d}^H\left(f, \cos\theta_{\mathrm{d}}\right)\mathbf{\Gamma}_{0,\pi}(f)\mathbf{d}\left(f, \cos\theta_{\mathrm{d}}\right)}\sqrt{\mathbf{d}^H\left(f, \cos\theta_{\mathrm{d}}\right)\mathbf{\Gamma}_{0,\pi}^{-1}(f)\mathbf{d}\left(f, \cos\theta_{\mathrm{d}}\right)}} \tag{7.45}$$

is the cosine of the angle between the two vectors $\mathbf{\Gamma}_{0,\pi}^{1/2}(f)\mathbf{d}\left(f,\cos\theta_{\mathrm{d}}\right)$ and $\mathbf{\Gamma}_{0,\pi}^{-1/2}(f)\mathbf{d}\left(f,\cos\theta_{\mathrm{d}}\right)$. Let $\sigma_1(f)$ and $\sigma_M(f)$ be the maximum and minimum eigenvalues of $\mathbf{\Gamma}_{0,\pi}(f)$, respectively. Using the Kantorovich inequality [7]:

$$\cos^2\left[\mathbf{\Gamma}_{0,\pi}^{1/2}(f)\mathbf{d}\left(f,\cos\theta_{\mathrm{d}}\right),\mathbf{\Gamma}_{0,\pi}^{-1/2}(f)\mathbf{d}\left(f,\cos\theta_{\mathrm{d}}\right)\right] \geq \frac{4\sigma_1(f)\sigma_M(f)}{\left[\sigma_1(f)+\sigma_M(f)\right]^2}, \quad (7.46)$$

we deduce that

$$\frac{4\sigma_1(f)\sigma_M(f)}{\left[\sigma_1(f)+\sigma_M(f)\right]^2} \leq \frac{\mathcal{D}\left[\mathbf{h}_{\mathrm{DS}}\left(f,\cos\theta_{\mathrm{d}}\right)\right]}{\mathcal{D}_{\max}\left(f,\cos\theta_{\mathrm{d}}\right)} \leq 1. \quad (7.47)$$

Example 7.5.1 Consider a ULA of M sensors, as shown in Figure 7.1. Suppose that a desired signal impinges on the ULA from the direction θ_{d}. Figure 7.2 shows plots of the WNG, $\mathcal{W}\left[\mathbf{h}_{\mathrm{DS}}\left(f,\cos\theta_{\mathrm{d}}\right)\right]$, as a function of frequency, for different numbers of sensors, M. Figure 7.3 shows plots of the DF, $\mathcal{D}\left[\mathbf{h}_{\mathrm{DS}}\left(f,\cos\theta_{\mathrm{d}}\right)\right]$, as a function of frequency, for different numbers of sensors, M, and several values of θ_{d} and δ. As the number of sensors increases, both the WNG and the DF of the DS beamformer increase. Figures 7.4–7.6 show beampatterns, $\left|\mathcal{B}\left[\mathbf{h}_{\mathrm{DS}}\left(f,\cos\theta_{\mathrm{d}}\right),\cos\theta\right]\right|$, for $M=8$, several values of θ_{d} and δ, and several frequencies. The main beam is in the direction of the desired signal, θ_{d}. As the frequency increases, the width of the main beam decreases. As δ/λ increases, we may observe spatial aliasing, as shown in Figure 7.6d. ■

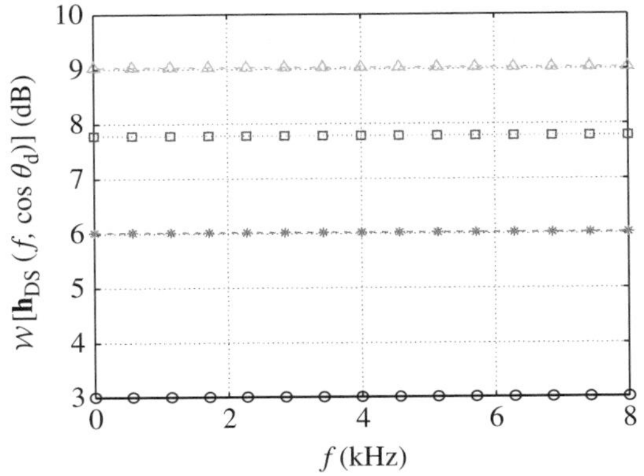

Figure 7.2 WNG of the DS beamformer as a function of frequency, for different numbers of sensors, M: $M=2$ (solid line with circles), $M=4$ (dashed line with asterisks), $M=6$ (dotted line with squares), and $M=8$ (dash-dot line with triangles).

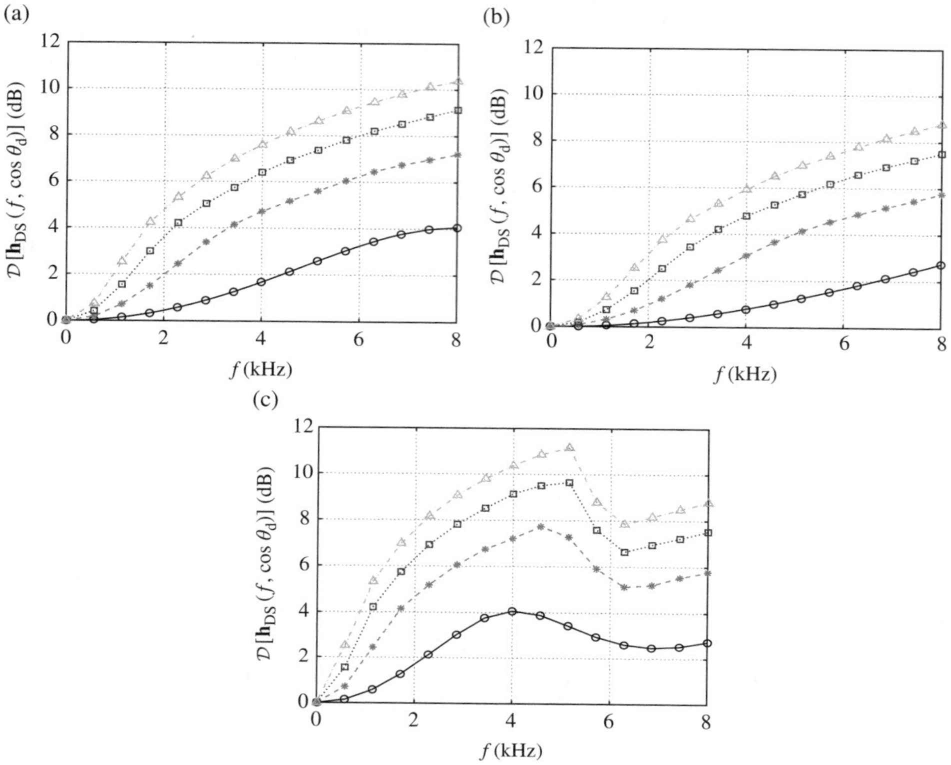

Figure 7.3 DF of the DS beamformer as a function of frequency, for different numbers of sensors, M, and several values of θ_{d} and δ: $M = 2$ (solid line with circles), $M = 4$ (dashed line with asterisks), $M = 6$ (dotted line with squares), and $M = 8$ (dash-dot line with triangles). (a) $\theta_{\mathrm{d}} = 90°$, $\delta = 3$ cm, (b) $\theta_{\mathrm{d}} = 0°$, $\delta = 1$ cm, and (c) $\theta_{\mathrm{d}} = 0°$, $\delta = 3$ cm.

7.5.2 Maximum DF

The maximum DF beamformer, as the name implies, maximizes the DF:

$$\min_{\mathbf{h}(f)} \mathbf{h}^H(f)\mathbf{\Gamma}_{0,\pi}(f)\mathbf{h}(f) \quad \text{subject to} \quad \mathbf{h}^H(f)\mathbf{d}\left(f, \cos\theta_{\mathrm{d}}\right) = 1. \tag{7.48}$$

Then, the maximum DF beamformer is

$$\mathbf{h}_{\mathrm{mDF}}\left(f, \cos\theta_{\mathrm{d}}\right) = \frac{\mathbf{\Gamma}_{0,\pi}^{-1}(f)\mathbf{d}\left(f, \cos\theta_{\mathrm{d}}\right)}{\mathbf{d}^H\left(f, \cos\theta_{\mathrm{d}}\right)\mathbf{\Gamma}_{0,\pi}^{-1}(f)\mathbf{d}\left(f, \cos\theta_{\mathrm{d}}\right)}. \tag{7.49}$$

We deduce that the WNG and the DF are, respectively,

$$\mathcal{W}\left[\mathbf{h}_{\mathrm{mDF}}\left(f, \cos\theta_{\mathrm{d}}\right)\right] = \frac{\left[\mathbf{d}^H\left(f, \cos\theta_{\mathrm{d}}\right)\mathbf{\Gamma}_{0,\pi}^{-1}(f)\mathbf{d}\left(f, \cos\theta_{\mathrm{d}}\right)\right]^2}{\mathbf{d}^H\left(f, \cos\theta_{\mathrm{d}}\right)\mathbf{\Gamma}_{0,\pi}^{-2}(f)\mathbf{d}\left(f, \cos\theta_{\mathrm{d}}\right)} \tag{7.50}$$

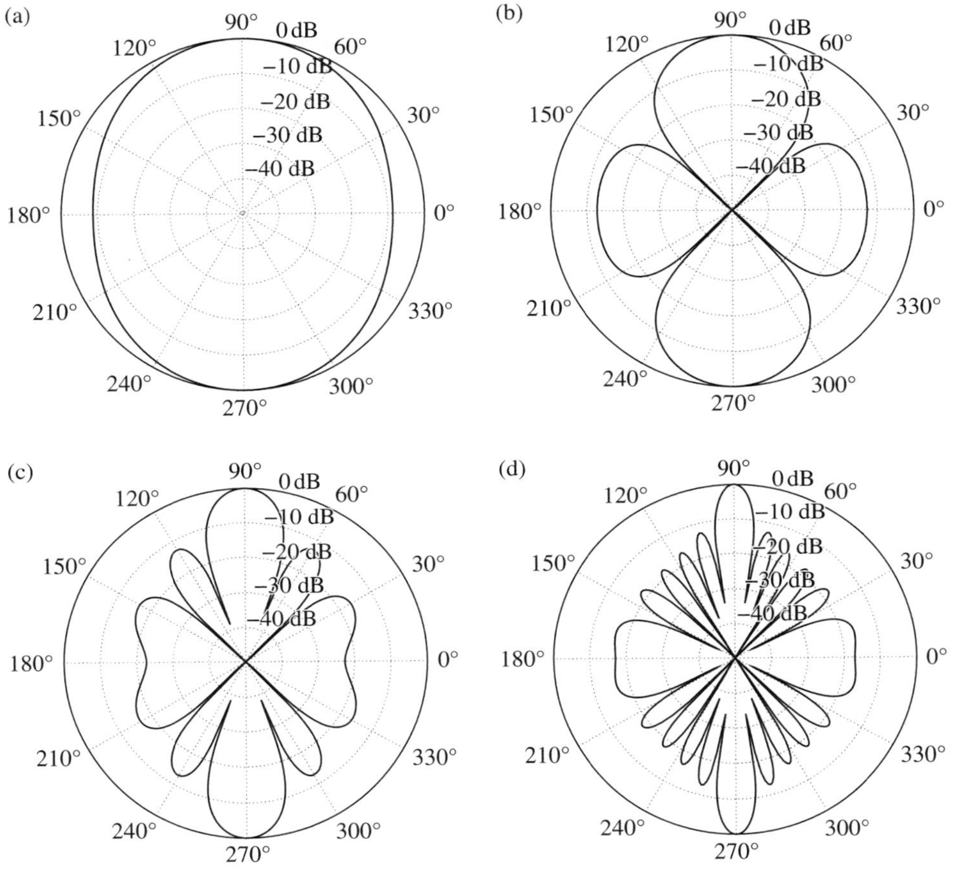

Figure 7.4 Beampatterns of the DS beamformer for several frequencies with $M = 8$, $\theta_d = 90°$, and $\delta = 3$ cm: (a) $f = 1$ kHz, (b) $f = 2$ kHz, (c) $f = 4$ kHz, and (d) $f = 8$ kHz.

and

$$\mathcal{D}\left[\mathbf{h}_{\text{mDF}}\left(f, \cos\theta_{\text{d}}\right)\right] = \mathbf{d}^H\left(f, \cos\theta_{\text{d}}\right)\boldsymbol{\Gamma}_{0,\pi}^{-1}(f)\mathbf{d}\left(f, \cos\theta_{\text{d}}\right) \\ = \mathcal{D}_{\max}\left(f, \cos\theta_{\text{d}}\right). \tag{7.51}$$

It is not hard to see that the beampattern is

$$\left|\mathcal{B}\left[\mathbf{h}_{\text{mDF}}\left(f, \cos\theta_{\text{d}}\right), \cos\theta\right]\right|^2 = \frac{\left|\mathbf{d}^H\left(f, \cos\theta\right)\boldsymbol{\Gamma}_{0,\pi}^{-1}(f)\mathbf{d}\left(f, \cos\theta_{\text{d}}\right)\right|^2}{\left[\mathbf{d}^H\left(f, \cos\theta_{\text{d}}\right)\boldsymbol{\Gamma}_{0,\pi}^{-1}(f)\mathbf{d}\left(f, \cos\theta_{\text{d}}\right)\right]^2}. \tag{7.52}$$

We can express the WNG as [6]:

$$\mathcal{W}\left[\mathbf{h}_{\text{mDF}}\left(f, \cos\theta_{\text{d}}\right)\right] = \mathcal{W}_{\max}\cos^2\left[\mathbf{d}\left(f, \cos\theta_{\text{d}}\right), \boldsymbol{\Gamma}_{0,\pi}^{-1}(f)\mathbf{d}\left(f, \cos\theta_{\text{d}}\right)\right], \tag{7.53}$$

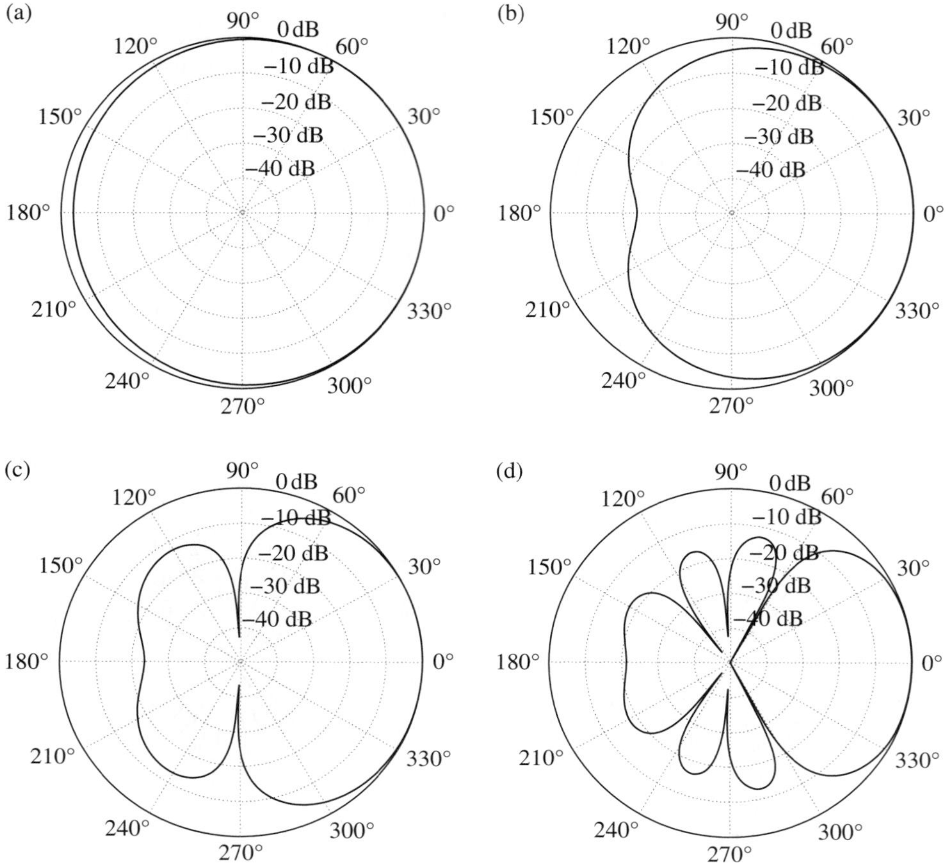

Figure 7.5 Beampatterns of the DS beamformer for several frequencies with $M = 8$, $\theta_d = 0°$, and $\delta = 1$ cm: (a) $f = 1$ kHz, (b) $f = 2$ kHz, (c) $f = 4$ kHz, and (d) $f = 8$ kHz.

where

$$\begin{aligned}\cos\left[\mathbf{d}\left(f,\cos\theta_{\mathrm{d}}\right),\boldsymbol{\Gamma}_{0,\pi}^{-1}(f)\mathbf{d}\left(f,\cos\theta_{\mathrm{d}}\right)\right] \\ = \frac{\mathbf{d}^H\left(f,\cos\theta_{\mathrm{d}}\right)\boldsymbol{\Gamma}_{0,\pi}^{-1}(f)\mathbf{d}\left(f,\cos\theta_{\mathrm{d}}\right)}{\sqrt{\mathbf{d}^H\left(f,\cos\theta_{\mathrm{d}}\right)\mathbf{d}\left(f,\cos\theta_{\mathrm{d}}\right)}\sqrt{\mathbf{d}^H\left(f,\cos\theta_{\mathrm{d}}\right)\boldsymbol{\Gamma}_{0,\pi}^{-2}(f)\mathbf{d}\left(f,\cos\theta_{\mathrm{d}}\right)}}\end{aligned} \tag{7.54}$$

is the cosine of the angle between the two vectors $\mathbf{d}\left(f,\cos\theta_{\mathrm{d}}\right)$ and $\boldsymbol{\Gamma}_{0,\pi}^{-1}(f)\left(f,\cos\theta_{\mathrm{d}}\right)$. Again, by invoking the Kantorovich inequality, we find that

$$\frac{4\sigma_1(f)\sigma_M(f)}{\left[\sigma_1(f)+\sigma_M(f)\right]^2} \leq \frac{\mathcal{W}\left[\mathbf{h}_{\mathrm{mDF}}\left(f,\cos\theta_{\mathrm{d}}\right)\right]}{\mathcal{W}_{\max}} \leq 1. \tag{7.55}$$

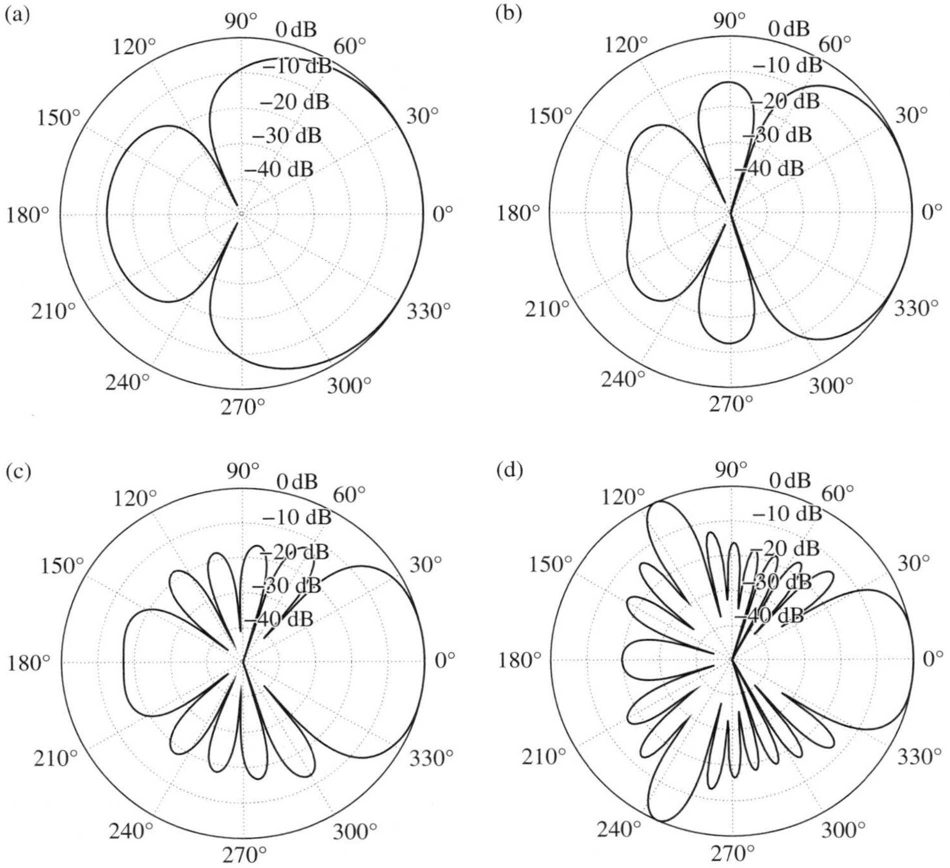

Figure 7.6 Beampatterns of the DS beamformer for several frequencies with $M = 8$, $\theta_d = 0°$, and $\delta = 3$ cm: (a) $f = 1$ kHz, (b) $f = 2$ kHz, (c) $f = 4$ kHz, and (d) $f = 8$ kHz.

We can see from (7.53) that the WNG may be smaller than 1, which implies white noise amplification.

It is interesting to observe that

$$\frac{1}{\mathbf{h}_{\text{mDF}}^H \left(f, \cos\theta_d\right) \mathbf{h}_{\text{DS}} \left(f, \cos\theta_d\right)} = \mathcal{W}_{\max} \tag{7.56}$$

and

$$\frac{1}{\mathbf{h}_{\text{mDF}}^H \left(f, \cos\theta_d\right) \mathbf{\Gamma}_{0,\pi}(f) \mathbf{h}_{\text{DS}} \left(f, \cos\theta_d\right)} = \mathcal{D}_{\max}\left(f, \cos\theta_d\right). \tag{7.57}$$

We also give the obvious relationship between the DS and maximum DF beamformers:

$$\mathcal{D}_{\max}\left(f, \cos\theta_d\right) \mathbf{\Gamma}_{0,\pi}(f) \mathbf{h}_{\text{mDF}}\left(f, \cos\theta_d\right) = \mathcal{W}_{\max} \mathbf{h}_{\text{DS}}\left(f, \cos\theta_d\right). \tag{7.58}$$

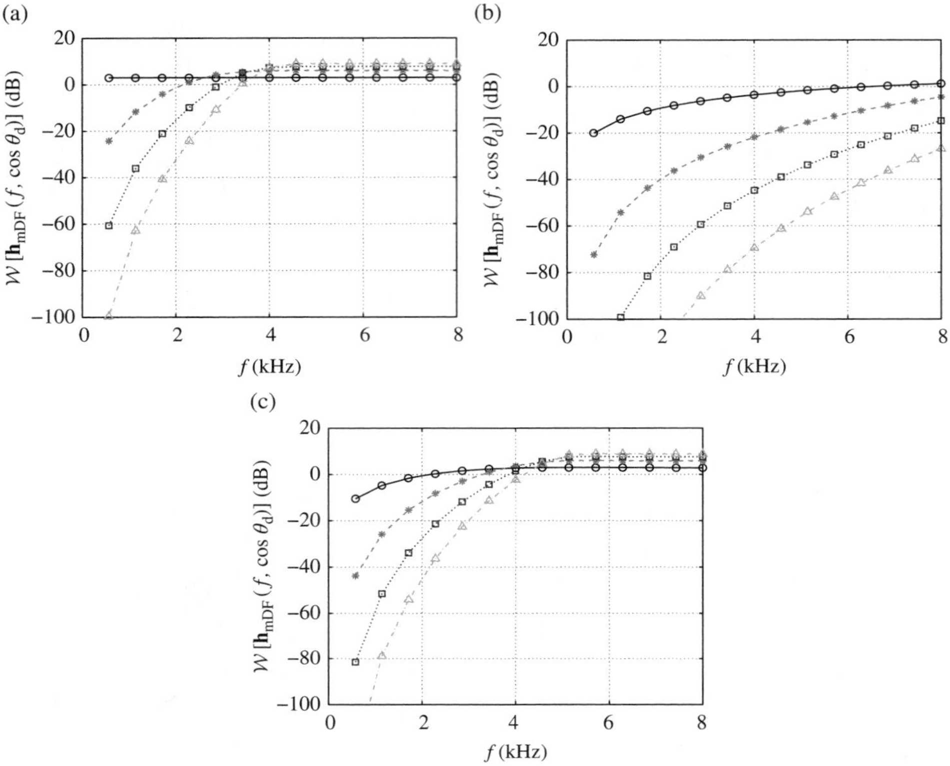

Figure 7.7 WNG of the maximum DF beamformer as a function of frequency, for different numbers of sensors, M, and several values of θ_d and δ: $M = 2$ (solid line with circles), $M = 4$ (dashed line with asterisks), $M = 6$ (dotted line with squares), and $M = 8$ (dash-dot line with triangles). (a) $\theta_d = 90°$, $\delta = 3$ cm, (b) $\theta_d = 0°$, $\delta = 1$ cm, and (c) $\theta_d = 0°$, $\delta = 3$ cm.

Example 7.5.2 Returning to Example 7.5.1, we now employ the maximum DF beamformer, $\mathbf{h}_{\text{mDF}}(f, \cos\theta_d)$, given in (7.49). Figure 7.7 shows plots of the WNG, $\mathcal{W}\left[\mathbf{h}_{\text{mDF}}(f, \cos\theta_d)\right]$, as a function of frequency, for different numbers of sensors, M, and several values of θ_d and δ. Figure 7.8 shows plots of the DF, $\mathcal{D}\left[\mathbf{h}_{\text{mDF}}(f, \cos\theta_d)\right]$, as a function of frequency, for different numbers of sensors, M, and several values of θ_d and δ. Compared to the DS beamformer, the maximum DF beamformer obtains higher DF, but lower WNG (cf. Figures 7.2 and 7.3). Generally, for high frequencies, as the number of sensors increases, both the DF and the WNG of the maximum DF beamformer increase. However, for low frequencies the WNG of the maximum DF beamformer is significantly lower than 0 dB, which implies that the maximum DF beamformer amplifies the white noise at low frequencies.

Figures 7.9–7.11 show beampatterns, $\left|\mathcal{B}\left[\mathbf{h}_{\text{mDF}}(f, \cos\theta_d), \cos\theta\right]\right|$, for $M = 8$, several values of θ_d and δ, and several frequencies. The main beam is in the direction of the desired signal, θ_d. As the frequency increases, the width of the main beam decreases. As δ/λ increases, we may observe spatial aliasing, as shown in Figure 7.11d. ■

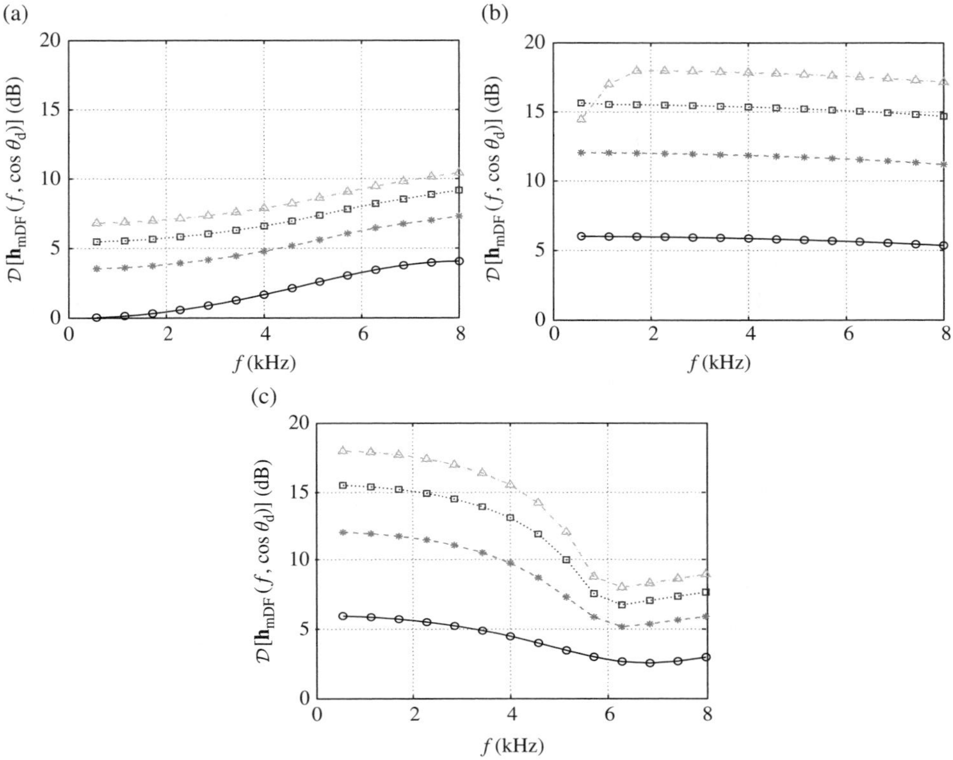

Figure 7.8 DF of the maximum DF beamformer as a function of frequency, for different numbers of sensors, M, and several values of θ_{d} and δ: $M = 2$ (solid line with circles), $M = 4$ (dashed line with asterisks), $M = 6$ (dotted line with squares), and $M = 8$ (dash-dot line with triangles). (a) $\theta_{\mathrm{d}} = 90°$, $\delta = 3$ cm, (b) $\theta_{\mathrm{d}} = 0°$, $\delta = 1$ cm, and (c) $\theta_{\mathrm{d}} = 0°$, $\delta = 3$ cm.

7.5.3 Superdirective

Let us evaluate the maximum DF, which is also the DF of the maximum DF beamformer, for $M = 2$. After simple algebraic manipulations, we find that

$$\mathcal{D}_{\max}\left(f, \cos\theta_{\mathrm{d}}\right) = 2\frac{1 - \mathrm{sinc}\left(2\pi f \tau_0\right)\cos\left(2\pi f \tau_0 \cos\theta_{\mathrm{d}}\right)}{1 - \mathrm{sinc}^2\left(2\pi f \tau_0\right)}. \tag{7.59}$$

Using in (7.59) the approximations:

$$\mathrm{sinc}\, x \approx 1 - \frac{x^2}{6}, \tag{7.60}$$

$$\cos x \approx 1 - \frac{x^2}{2}, \tag{7.61}$$

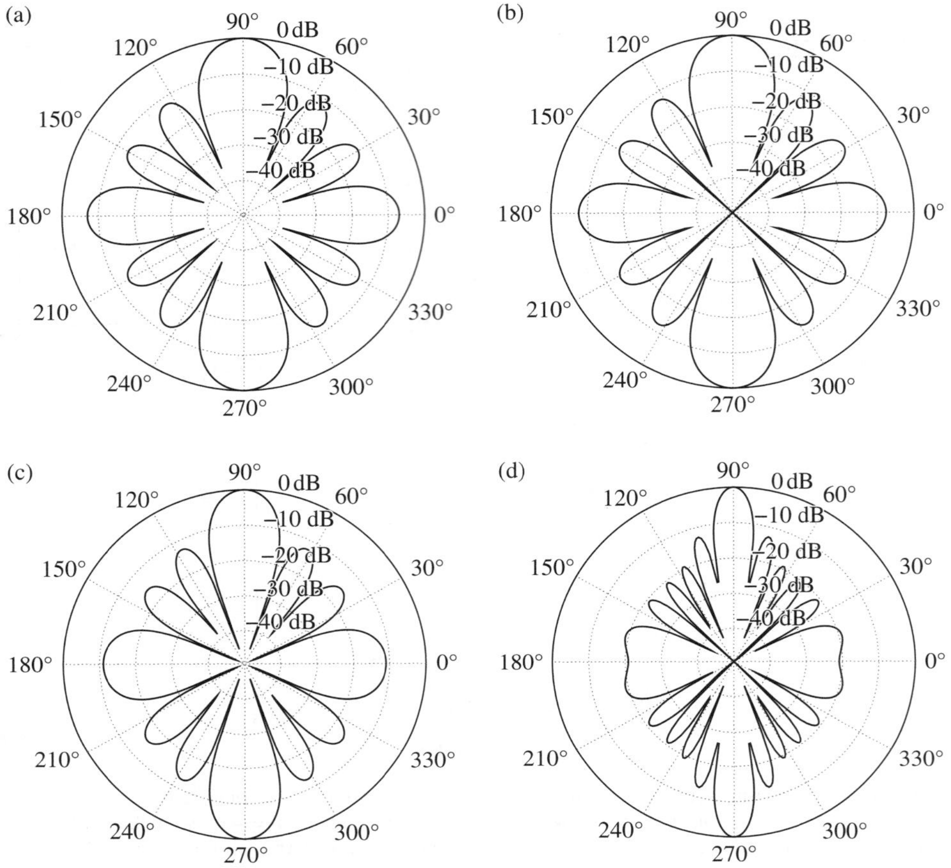

Figure 7.9 Beampatterns of the maximum DF beamformer for several frequencies with $M = 8$, $\theta_d = 90°$, and $\delta = 3$ cm: (a) $f = 1$ kHz, (b) $f = 2$ kHz, (c) $f = 4$ kHz, and (d) $f = 8$ kHz.

we obtain

$$\mathcal{D}_{\max}\left(f, \cos\theta_d\right) \approx \frac{6\left[6\cos^2\theta_d + 2 - \left(2\pi f \tau_0\right)^2 \cos^2\theta_d\right]}{12 - \left(2\pi f \tau_0\right)^2}. \tag{7.62}$$

First, for δ very small, the previous expression can be further approximated by

$$\mathcal{D}_{\max}\left(f, \cos\theta_d\right) \approx 3\cos^2\theta_d + 1, \tag{7.63}$$

which is frequency independent. Second, it is clear from (7.63) that the maximum DF is maximized for $\theta_d = 0$ or π (endfire direction). The minimum of the maximum DF is obtained for $\theta_d = \pi/2$ (broadside direction). From this simple example, we can conclude that the best arrays, as far the DF is concerned, are endfire arrays with a small interelement spacing. Broadside arrays do not perform very well in general. A deeper

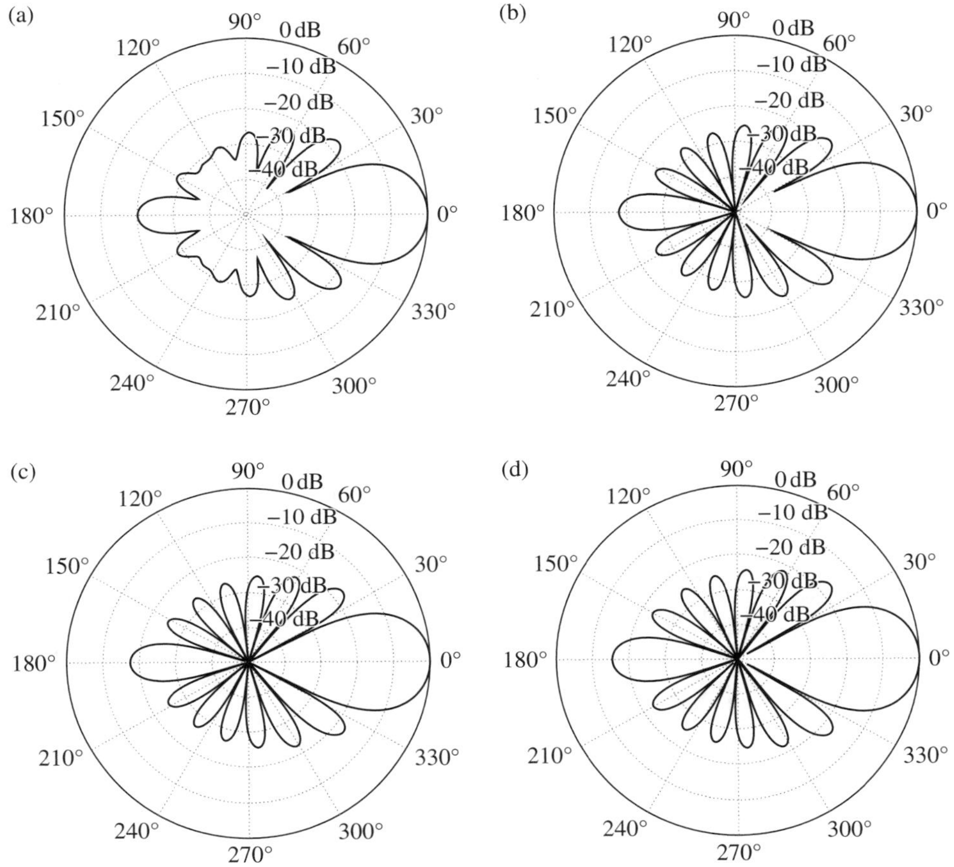

Figure 7.10 Beampatterns of the maximum DF beamformer for several frequencies with $M = 8$, $\theta_{\rm d} = 0°$, and $\delta = 1$ cm: (a) $f = 1$ kHz, (b) $f = 2$ kHz, (c) $f = 4$ kHz, and (d) $f = 8$ kHz.

study [8] draws the same conclusions. Other experimental studies show the benefits of endfire arrays [9]. In fact, it can be shown that [10]:

$$\lim_{\delta \to 0} \mathcal{D}_{\max}\left(f, 1\right) = M^2. \tag{7.64}$$

This high DF is called supergain in the literature.

The well-known superdirective beamformer is just a particular case of the maximum DF beamformer, where $\theta_{\rm d} = 0$ and δ is small. It is given by [11]:

$$\mathbf{h}_{\rm SD}(f) = \frac{\mathbf{\Gamma}_{0,\pi}^{-1}(f)\mathbf{d}\left(f, 1\right)}{\mathbf{d}^H\left(f, 1\right)\mathbf{\Gamma}_{0,\pi}^{-1}(f)\mathbf{d}\left(f, 1\right)}. \tag{7.65}$$

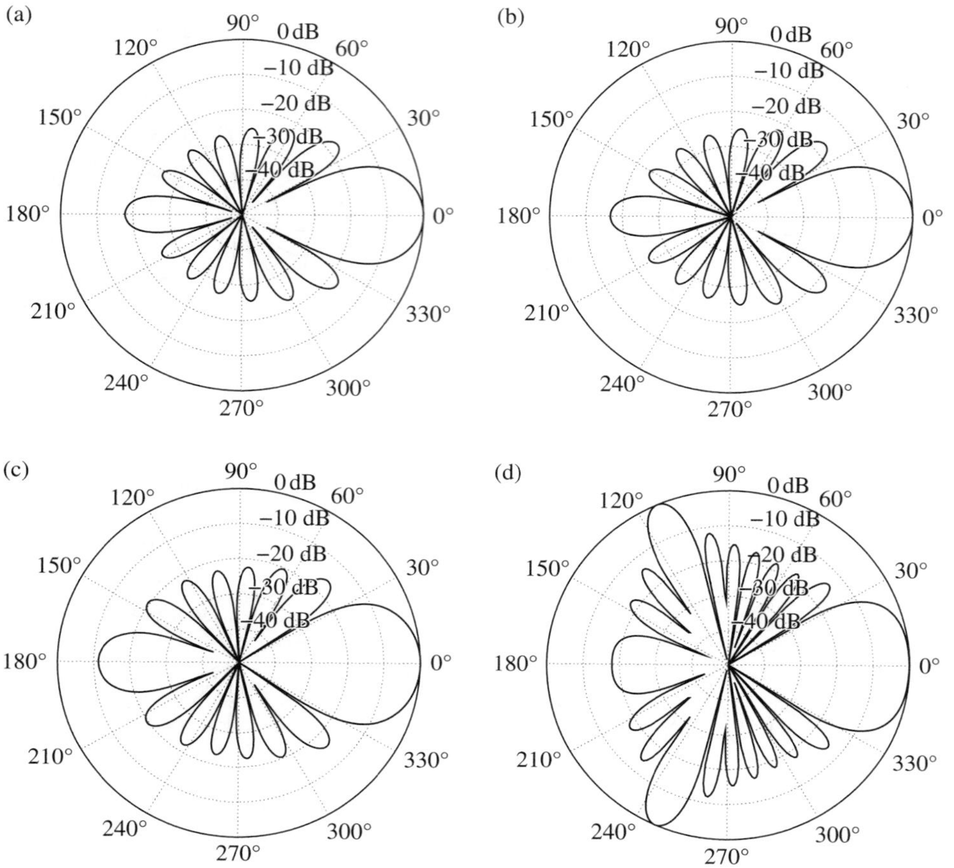

Figure 7.11 Beampatterns of the maximum DF beamformer for several frequencies with $M = 8$, $\theta_{\rm d} = 0°$, and $\delta = 3$ cm: (a) $f = 1$ kHz, (b) $f = 2$ kHz, (c) $f = 4$ kHz, and (d) $f = 8$ kHz.

While the superdirective beamformer maximizes the DF, it may amplify the white noise; in other words, the WNG may be smaller than 1, especially at low frequencies (see Figures 7.7b and 7.7c].

7.5.4 Robust Superdirective

It is well known that the superdirective beamformer is sensitive to spatially white noise, so it lacks robustness. In order to deal with this important problem, Cox et al. proposed maximizing the DF [11, 12]:

$$\mathcal{D}\left[\mathbf{h}(f)\right] = \frac{\left|\mathbf{h}^H(f)\mathbf{d}\left(f,1\right)\right|^2}{\mathbf{h}^H(f)\mathbf{\Gamma}_{0,\pi}(f)\mathbf{h}(f)}, \tag{7.66}$$

subject to a constraint on the WNG:

$$\mathcal{W}\left[\mathbf{h}(f)\right] = \frac{\left|\mathbf{h}^H(f)\mathbf{d}\left(f,1\right)\right|^2}{\mathbf{h}^H(f)\mathbf{h}(f)}. \tag{7.67}$$

This is equivalent to minimizing $1/\mathcal{D}\left[\mathbf{h}(f)\right]$ with a constraint on $1/\mathcal{W}\left[\mathbf{h}(f)\right]$; in other words, minimizing

$$\frac{1}{\mathcal{D}\left[\mathbf{h}(f)\right]} + \epsilon\frac{1}{\mathcal{W}\left[\mathbf{h}(f)\right]} = \frac{\mathbf{h}^H(f)\left[\mathbf{\Gamma}_{0,\pi}(f) + \epsilon\mathbf{I}_M\right]\mathbf{h}(f)}{\left|\mathbf{h}^H(f)\mathbf{d}\left(f,1\right)\right|^2}, \tag{7.68}$$

where $\epsilon \geq 0$ is a Lagrange multiplier. Using the distortionless constraint, we easily find that the optimal solution is

$$\mathbf{h}_{\mathrm{R},\epsilon}(f) = \frac{\left[\mathbf{\Gamma}_{0,\pi}(f) + \epsilon\mathbf{I}_M\right]^{-1}\mathbf{d}\left(f,1\right)}{\mathbf{d}^H\left(f,1\right)\left[\mathbf{\Gamma}_{0,\pi}(f) + \epsilon\mathbf{I}_M\right]^{-1}\mathbf{d}\left(f,1\right)}. \tag{7.69}$$

It is clear that (7.69) is a regularized version of (7.65), where ϵ is the regularization parameter. This parameter tries to find a good compromise between a supergain and white noise amplification. A small ϵ leads to a large DF and a low WNG, while a large ϵ gives a low DF and a large WNG. We have $\mathbf{h}_{\mathrm{R},0}(f) = \mathbf{h}_{\mathrm{SD}}(f)$ and $\mathbf{h}_{\mathrm{R},\infty}(f) = \mathbf{h}_{\mathrm{DS}}(f,1)$. In practice, since white noise amplification is much worse at low frequencies than at high frequencies, it is better to make ϵ frequency dependent. Therefore, (7.69) is rewritten as

$$\mathbf{h}_{\mathrm{R},\epsilon}(f) = \frac{\left[\mathbf{\Gamma}_{0,\pi}(f) + \epsilon(f)\mathbf{I}_M\right]^{-1}\mathbf{d}\left(f,1\right)}{\mathbf{d}^H\left(f,1\right)\left[\mathbf{\Gamma}_{0,\pi}(f) + \epsilon(f)\mathbf{I}_M\right]^{-1}\mathbf{d}\left(f,1\right)}. \tag{7.70}$$

An equivalent way to express (7.70) is

$$\mathbf{h}_{\mathrm{R},\alpha}(f) = \frac{\mathbf{\Gamma}_\alpha^{-1}(f)\mathbf{d}\left(f,1\right)}{\mathbf{d}^H\left(f,1\right)\mathbf{\Gamma}_\alpha^{-1}(f)\mathbf{d}\left(f,1\right)}, \tag{7.71}$$

where

$$\mathbf{\Gamma}_\alpha(f) = \left[1 - \alpha(f)\right]\mathbf{\Gamma}_{0,\pi}(f) + \alpha(f)\mathbf{I}_M, \tag{7.72}$$

with $\alpha(f)$ being a real number and $0 \leq \alpha(f) \leq 1$. It can be checked that the relationship between $\alpha(f)$ and $\epsilon(f)$ is

$$\epsilon(f) = \frac{\alpha(f)}{1 - \alpha(f)}. \tag{7.73}$$

The robust superdirective beamformer given in (7.71) may be preferable in practice to the equivalent form given in (7.70) since $\alpha(f)$ is set between 0 and 1 in the former while $\epsilon(f)$ is can be from 0 to ∞ in the latter. We find that the WNG and the DF are, respectively,

$$\mathcal{W}\left[\mathbf{h}_{\mathrm{R},\alpha}(f)\right] = \frac{\left[\mathbf{d}^H(f,1)\,\mathbf{\Gamma}_{\alpha}^{-1}(f)\mathbf{d}(f,1)\right]^2}{\mathbf{d}^H(f,1)\,\mathbf{\Gamma}_{\alpha}^{-2}(f)\mathbf{d}(f,1)} \tag{7.74}$$

and

$$\mathcal{D}\left[\mathbf{h}_{\mathrm{R},\alpha}(f)\right] = \frac{\left[\mathbf{d}^H(f,1)\,\mathbf{\Gamma}_{\alpha}^{-1}(f)\mathbf{d}(f,1)\right]^2}{\mathbf{d}^H(f,1)\,\mathbf{\Gamma}_{\alpha}^{-1}(f)\mathbf{\Gamma}_{0,\pi}(f)\mathbf{\Gamma}_{\alpha}^{-1}(f)\mathbf{d}(f,1)}. \tag{7.75}$$

Using the geometrical interpretation, the last two expressions become

$$\mathcal{W}\left[\mathbf{h}_{\mathrm{R},\alpha}(f)\right] = \mathcal{W}_{\max}\cos^2\left[\mathbf{d}(f,1),\mathbf{\Gamma}_{\alpha}^{-1}(f)\mathbf{d}(f,1)\right] \tag{7.76}$$

and

$$\begin{aligned}\mathcal{D}\left[\mathbf{h}_{\mathrm{R},\alpha}(f)\right] &= \mathcal{D}_{\max}(f,1) \\ &\quad\times\cos^2\left[\mathbf{\Gamma}_{0,\pi}^{1/2}(f)\mathbf{\Gamma}_{\alpha}^{-1}(f)\mathbf{d}(f,1),\mathbf{\Gamma}_{0,\pi}^{-1/2}(f)\mathbf{d}(f,1)\right].\end{aligned} \tag{7.77}$$

The beampattern of the robust superdirective beamformer is

$$\left|\mathcal{B}\left[\mathbf{h}_{\mathrm{R},\alpha}(f),\cos\theta\right]\right|^2 = \frac{\left|\mathbf{d}^H(f,\cos\theta)\,\mathbf{\Gamma}_{\alpha}^{-1}(f)\mathbf{d}(f,1)\right|^2}{\left[\mathbf{d}^H(f,\cos 1)\,\mathbf{\Gamma}_{\alpha}^{-1}(f)\mathbf{d}(f,1)\right]^2}. \tag{7.78}$$

Example 7.5.3 Returning to Example 7.5.1, we now use the robust superdirective beamformer, $\mathbf{h}_{\mathrm{R},\alpha}(f)$, given in (7.71). To demonstrate the performance of the robust superdirective beamformer, we choose $\delta = 1$ cm.

Figure 7.12 shows plots of the WNG, $\mathcal{W}\left[\mathbf{h}_{\mathrm{R},\alpha}(f)\right]$, as a function of frequency, for different numbers of sensors, M, and several values of α. Figure 7.13 shows plots of the WNG, $\mathcal{W}\left[\mathbf{h}_{\mathrm{R},\alpha}(f)\right]$, as a function of α, for different numbers of sensors, M, and several frequencies.

Figure 7.14 shows plots of the DF, $\mathcal{D}\left[\mathbf{h}_{\mathrm{R},\alpha}(f)\right]$, as a function of frequency, for different numbers of sensors, M, and several values of α. Figure 7.15 shows plots of the DF, $\mathcal{D}\left[\mathbf{h}_{\mathrm{R},\alpha}(f)\right]$, as a function of α, for different numbers of sensors, M, and several frequencies.

For given frequency and α, as the number of sensors increases, the DF of the robust superdirective beamformer increases. For given frequency and M, as the value of α increases, the WNG of the robust superdirective beamformer increases at the expense of a lower DF.

Figures 7.16–7.18 show beampatterns, $\left|\mathcal{B}\left[\mathbf{h}_{\mathrm{R},\alpha}(f),\cos\theta\right]\right|$, for $M = 8$, several values of α, and several frequencies. The main beam is in the direction of the desired signal: $\theta_{\mathrm{d}} = 0°$. For a given α, as the frequency increases, the width of the main beam decreases. For a given frequency, as the value of α increases, the width of the main beam increases (lower DF). ■

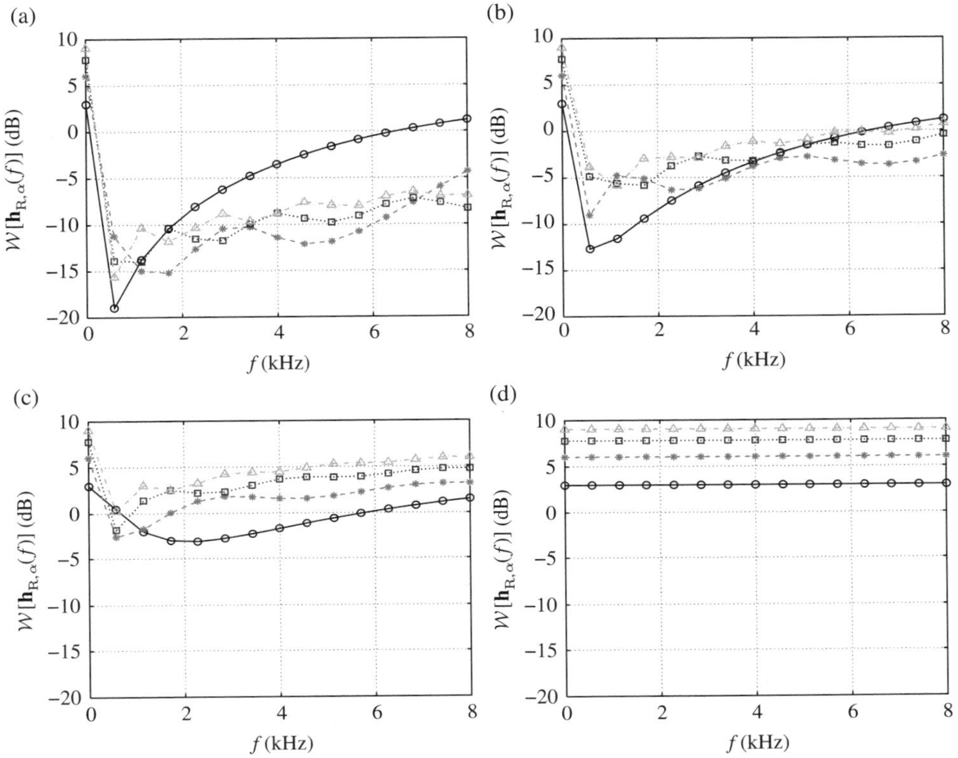

Figure 7.12 WNG of the robust superdirective beamformer as a function of frequency, for different numbers of sensors, M, and several values of α: $M = 2$ (solid line with circles), $M = 4$ (dashed line with asterisks), $M = 6$ (dotted line with squares), and $M = 8$ (dash-dot line with triangles). (a) $\alpha = 0.001$, (b) $\alpha = 0.01$, (c) $\alpha = 0.1$, and (d) $\alpha = 1$.

Let $\mathbf{A}$ and $\mathbf{B}$ be two invertible square matrices. If ϵ is small compared to $\mathbf{A}$ and $\mathbf{B}$, then [13]:

$$(\mathbf{A} + \epsilon\mathbf{B})^{-1} \approx \mathbf{A}^{-1} - \epsilon\mathbf{A}^{-1}\mathbf{B}\mathbf{A}^{-1}. \tag{7.79}$$

Using the previous approximation in $\boldsymbol{\Gamma}_\alpha(f)$, we get

$$\boldsymbol{\Gamma}_\alpha(f) \approx (1-\alpha)^{-2}\boldsymbol{\Gamma}_{0,\pi}^{-1}(f)\boldsymbol{\Gamma}_{\alpha^-}(f)\boldsymbol{\Gamma}_{0,\pi}^{-1}(f) \tag{7.80}$$

for $0 \leq \alpha \leq 0.5$, and

$$\boldsymbol{\Gamma}_\alpha(f) \approx -\alpha^{-2}\boldsymbol{\Gamma}_{\alpha^-}(f) \tag{7.81}$$

for $0.5 < \alpha \leq 1$, where

$$\boldsymbol{\Gamma}_{\alpha^-}(f) = \left[1 - \alpha(f)\right]\boldsymbol{\Gamma}_{0,\pi}(f) - \alpha(f)\mathbf{I}_M. \tag{7.82}$$

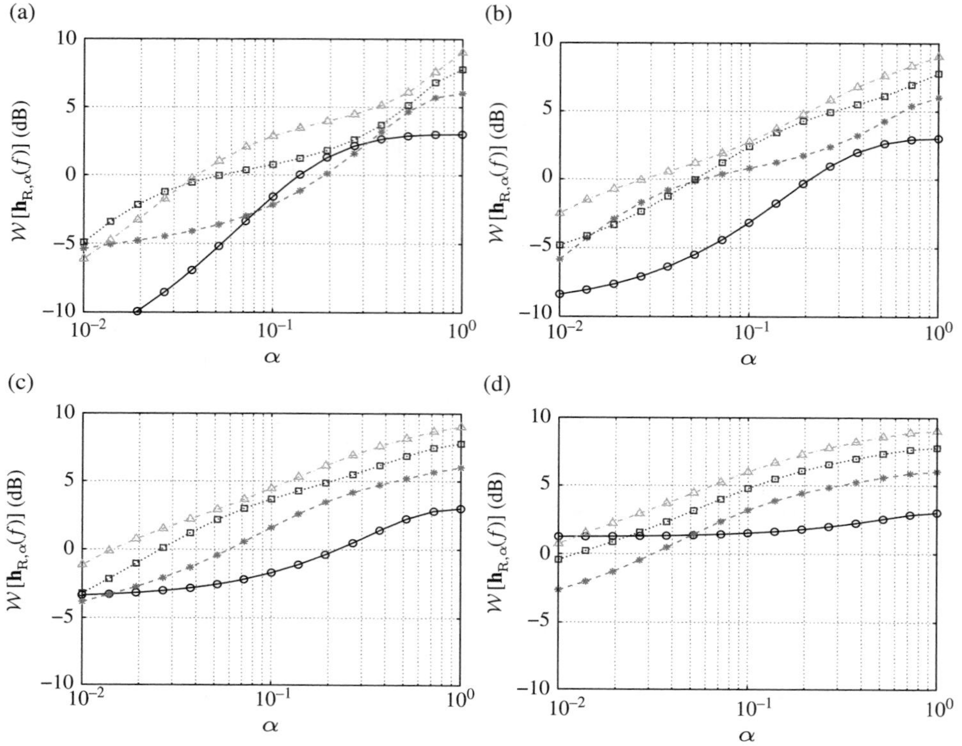

Figure 7.13 WNG of the robust superdirective beamformer as a function of α, for different numbers of sensors, M, and several frequencies: $M = 2$ (solid line with circles), $M = 4$ (dashed line with asterisks), $M = 6$ (dotted line with squares), and $M = 8$ (dash-dot line with triangles). (a) $f = 1$ kHz, (b) $f = 2$ kHz, (c) $f = 4$ kHz, and (d) $f = 8$ kHz.

As a result, the robust beamformer becomes

$$\mathbf{h}_{\mathrm{R},\alpha\leq 0.5}(f) = \frac{\mathbf{\Gamma}_{0,\pi}^{-1}(f)\mathbf{\Gamma}_{\alpha^-}(f)\mathbf{\Gamma}_{0,\pi}^{-1}(f)\mathbf{d}(f,1)}{\mathbf{d}^H(f,1)\mathbf{\Gamma}_{0,\pi}^{-1}(f)\mathbf{\Gamma}_{\alpha^-}(f)\mathbf{\Gamma}_{0,\pi}^{-1}(f)\mathbf{d}(f,1)}, \tag{7.83}$$

$$\mathbf{h}_{\mathrm{R},\alpha>0.5}(f) = \frac{\mathbf{\Gamma}_{\alpha^-}(f)\mathbf{d}(f,1)}{\mathbf{d}^H(f,1)\mathbf{\Gamma}_{\alpha^-}(f)\mathbf{d}(f,1)}. \tag{7.84}$$

We deduce that the WNG and the DF are, respectively,

$$\mathcal{W}\left[\mathbf{h}_{\mathrm{R},\alpha\leq 0.5}(f)\right] = \frac{\left[\mathbf{d}^H(f,1)\mathbf{\Gamma}_{0,\pi}^{-1}(f)\mathbf{\Gamma}_{\alpha^-}(f)\mathbf{\Gamma}_{0,\pi}^{-1}(f)\mathbf{d}(f,1)\right]^2}{\mathbf{d}^H(f,1)\mathbf{\Gamma}_{0,\pi}^{-1}(f)\mathbf{\Gamma}_{\alpha^-}(f)\mathbf{\Gamma}_{0,\pi}^{-2}(f)\mathbf{\Gamma}_{\alpha^-}(f)\mathbf{\Gamma}_{0,\pi}^{-1}(f)\mathbf{d}(f,1)}, \tag{7.85}$$

$$\mathcal{W}\left[\mathbf{h}_{\mathrm{R},\alpha>0.5}(f)\right] = \frac{\left[\mathbf{d}^H(f,1)\mathbf{\Gamma}_{\alpha^-}(f)\mathbf{d}(f,1)\right]^2}{\mathbf{d}^H(f,1)\mathbf{\Gamma}_{\alpha^-}^2(f)\mathbf{d}(f,1)}, \tag{7.86}$$

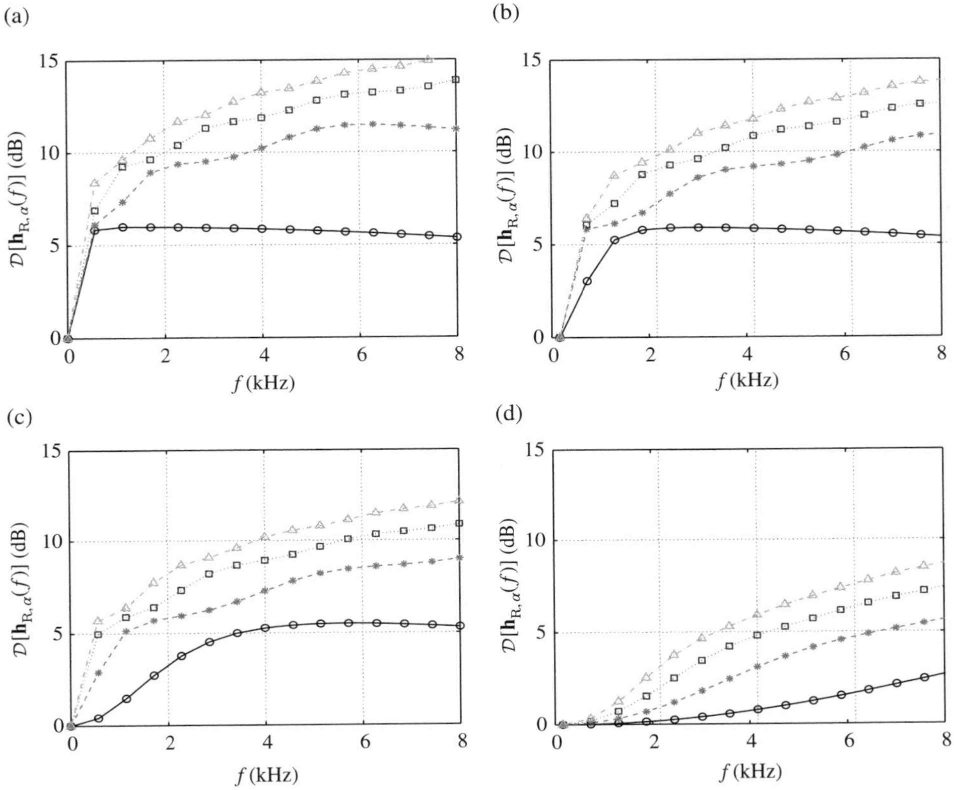

Figure 7.14 DF of the robust superdirective beamformer as a function of frequency, for different numbers of sensors, M, and several values of α: $M = 2$ (solid line with circles), $M = 4$ (dashed line with asterisks), $M = 6$ (dotted line with squares), and $M = 8$ (dash-dot line with triangles). (a) $\alpha = 0.001$, (b) $\alpha = 0.01$, (c) $\alpha = 0.1$, and (d) $\alpha = 1$.

and

$$\mathcal{D}\left[\mathbf{h}_{\mathrm{R},\alpha\leq 0.5}(f)\right] = \frac{\left[\mathbf{d}^H(f,1)\,\mathbf{\Gamma}_{0,\pi}^{-1}(f)\mathbf{\Gamma}_{\alpha^-}(f)\mathbf{\Gamma}_{0,\pi}^{-1}(f)\mathbf{d}(f,1)\right]^2}{\mathbf{d}^H(f,1)\,\mathbf{\Gamma}_{0,\pi}^{-1}(f)\mathbf{\Gamma}_{\alpha^-}(f)\mathbf{\Gamma}_{0,\pi}^{-1}(f)\mathbf{\Gamma}_{\alpha^-}(f)\mathbf{\Gamma}_{0,\pi}^{-1}(f)\mathbf{d}(f,1)}, \tag{7.87}$$

$$\mathcal{D}\left[\mathbf{h}_{\mathrm{R},\alpha>0.5}(f)\right] = \frac{\left[\mathbf{d}^H(f,1)\,\mathbf{\Gamma}_{\alpha^-}(f)\mathbf{d}(f,1)\right]^2}{\mathbf{d}^H(f,1)\,\mathbf{\Gamma}_{\alpha^-}(f)\mathbf{\Gamma}_{0,\pi}(f)\mathbf{\Gamma}_{\alpha^-}(f)\mathbf{d}(f,1)}. \tag{7.88}$$

The good thing about this approximation is that for a desired WNG or DF, we can find the corresponding value of $\alpha(f)$.

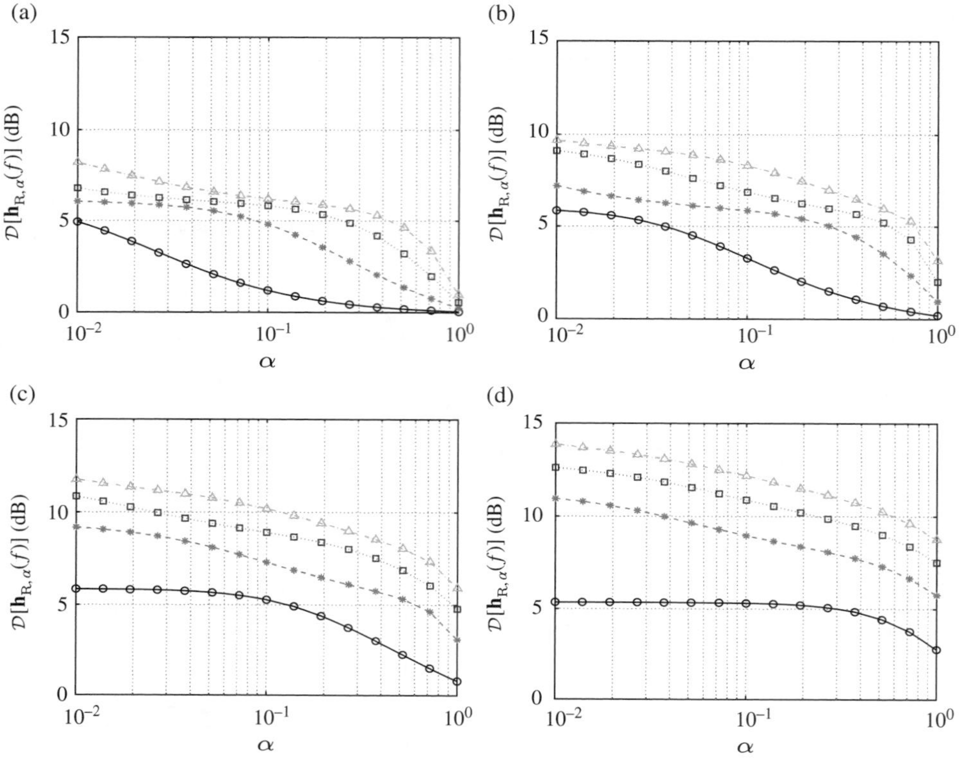

Figure 7.15 DF of the robust superdirective beamformer as a function of α, for different numbers of sensors, M, and several frequencies: $M = 2$ (solid line with circles), $M = 4$ (dashed line with asterisks), $M = 6$ (dotted line with squares), and $M = 8$ (dash-dot line with triangles). (a) $f = 1$ kHz, (b) $f = 2$ kHz, (c) $f = 4$ kHz, and (d) $f = 8$ kHz.

7.5.5 Null Steering

In this subsection, we assume that we have N sources, with $N < M$, impinging on the array from the directions $\theta_1 \neq \theta_2 \neq \cdots \neq \theta_N \neq \theta_{\mathrm{d}}$. These sources are considered as interferences that we would like to completely cancel; in other words we want to put nulls in the directions θ_n, $n = 1, 2, \ldots, N$, with a beamformer $\mathbf{h}(f)$, and, meanwhile, recover the desired source coming from the direction θ_{d}. Combining all these constraints together, we get the constraint equation:

$$\mathbf{C}^H \left(f, \theta_{\mathrm{d}}, \theta_{1:N}\right) \mathbf{h}(f) = \mathbf{i}_{\mathrm{c}}, \tag{7.89}$$

where

$$\mathbf{C}\left(f, \theta_{\mathrm{d}}, \theta_{1:N}\right) = \left[\begin{array}{cccc} \mathbf{d}\left(f, \theta_{\mathrm{d}}\right) & \mathbf{d}\left(f, \theta_1\right) & \cdots & \mathbf{d}\left(f, \theta_N\right) \end{array} \right] \tag{7.90}$$

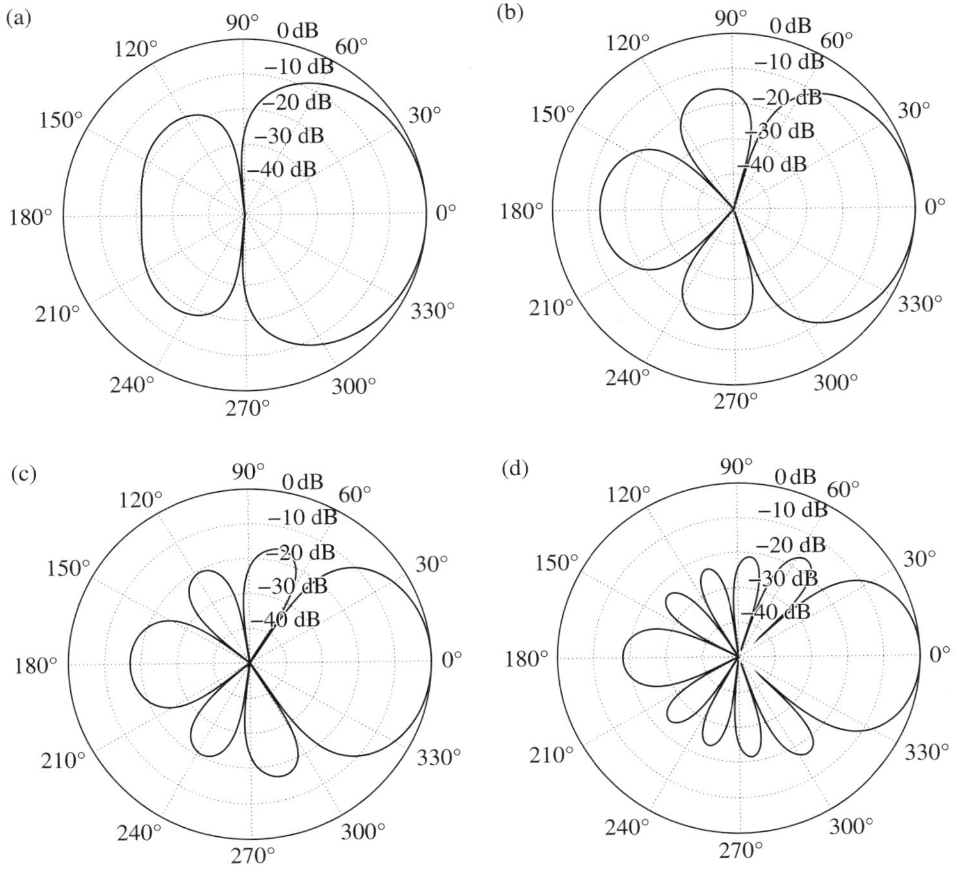

Figure 7.16 Beampatterns of the robust superdirective beamformer for $\alpha = 0.01$ and several frequencies: (a) $f = 1$ kHz, (b) $f = 2$ kHz, (c) $f = 4$ kHz, and (d) $f = 8$ kHz.

is the constraint matrix of size $M \times (N+1)$ whose $N+1$ columns are linearly independent and

$$\mathbf{i}_c = \begin{bmatrix} 1 & 0 & \cdots & 0 \end{bmatrix}^T \tag{7.91}$$

is a vector of length $N + 1$.

Depending on what it is desired, we have at least two different approaches to finding the optimal filter, which are based on the WNG and the DF as criteria. The first obvious beamformer is obtained by maximizing the WNG and by taking (7.89) into account:

$$\min_{\mathbf{h}(f)} \mathbf{h}^H(f)\mathbf{h}(f) \quad \text{subject to} \quad \mathbf{C}^H\left(f, \theta_d, \theta_{1:N}\right)\mathbf{h}(f) = \mathbf{i}_c. \tag{7.92}$$

From this criterion, we find the minimum-norm (MN) beamformer:

$$\mathbf{h}_{\text{MN}}\left(f, \cos\theta_d\right) = \mathbf{C}\left(f, \theta_d, \theta_{1:N}\right)\left[\mathbf{C}^H\left(f, \theta_d, \theta_{1:N}\right)\mathbf{C}\left(f, \theta_d, \theta_{1:N}\right)\right]^{-1}\mathbf{i}_c, \tag{7.93}$$

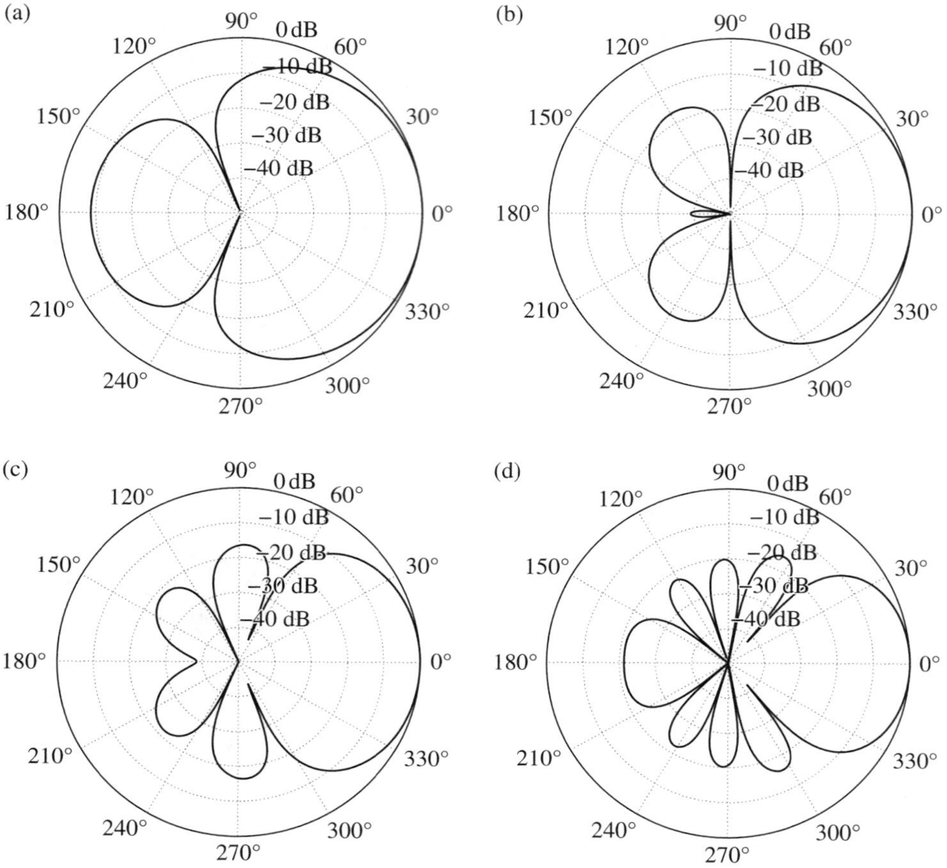

Figure 7.17 Beampatterns of the robust superdirective beamformer for $\alpha = 0.1$ and several frequencies: (a) $f = 1$ kHz, (b) $f = 2$ kHz, (c) $f = 4$ kHz, and (d) $f = 8$ kHz.

which is also the minimum-norm solution of (7.89). Clearly, we always have

$$\mathcal{W}\left[\mathbf{h}_{\mathrm{MN}}\left(f, \cos\theta_{\mathrm{d}}\right)\right] \leq \mathcal{W}\left[\mathbf{h}_{\mathrm{DS}}\left(f, \cos\theta_{\mathrm{d}}\right)\right]. \tag{7.94}$$

The other beamformer is obtained by maximizing the DF and by taking (7.89) into account:

$$\min_{\mathbf{h}(f)} \mathbf{h}^H(f)\boldsymbol{\Gamma}_{0,\pi}(f)\mathbf{h}(f) \quad \text{subject to} \quad \mathbf{C}^H\left(f, \theta_{\mathrm{d}}, \theta_{1:N}\right)\mathbf{h}(f) = \mathbf{i}_{\mathrm{c}}. \tag{7.95}$$

We then easily find the null-steering (NS) beamformer:

$$\begin{aligned}\mathbf{h}_{\mathrm{NS}}\left(f, \cos\theta_{\mathrm{d}}\right) = {} & \boldsymbol{\Gamma}_{0,\pi}^{-1}(f)\mathbf{C}\left(f, \theta_{\mathrm{d}}, \theta_{1:N}\right) \\ & \times \left[\mathbf{C}^H\left(f, \theta_{\mathrm{d}}, \theta_{1:N}\right)\boldsymbol{\Gamma}_{0,\pi}^{-1}(f)\mathbf{C}\left(f, \theta_{\mathrm{d}}, \theta_{1:N}\right)\right]^{-1}\mathbf{i}_{\mathrm{c}}.\end{aligned} \tag{7.96}$$

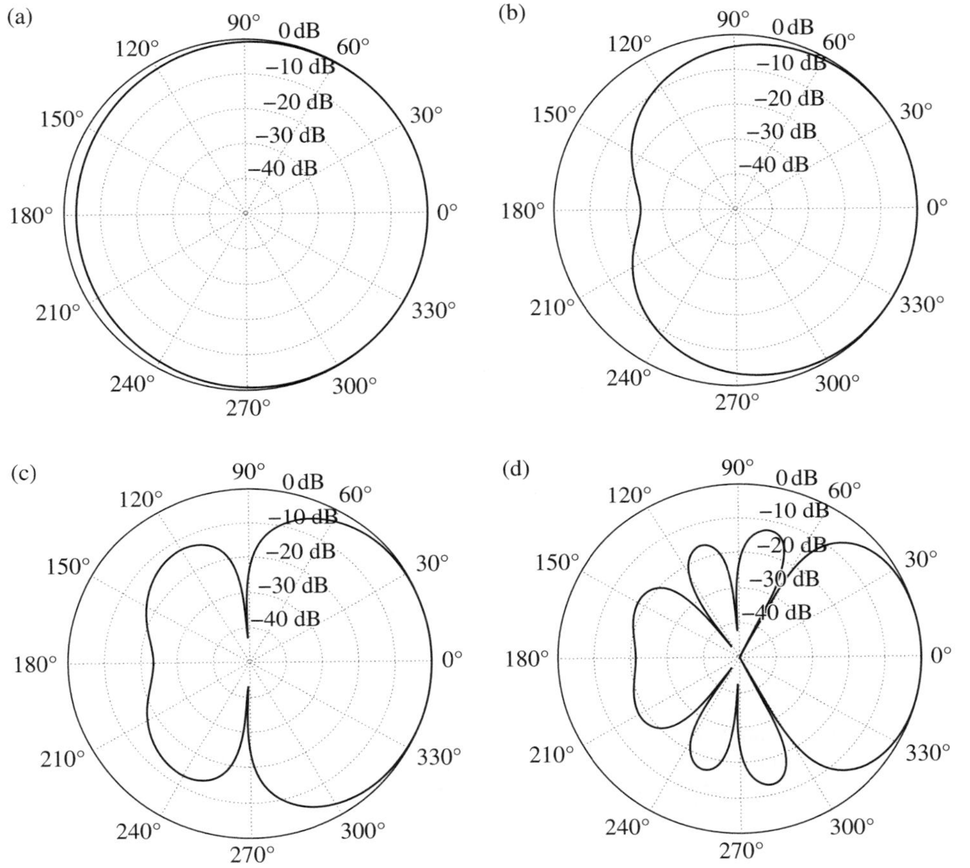

Figure 7.18 Beampatterns of the robust superdirective beamformer for $\alpha = 1$ and several frequencies: (a) $f = 1$ kHz, (b) $f = 2$ kHz, (c) $f = 4$ kHz, and (d) $f = 8$ kHz.

Obviously, we always have

$$\mathcal{D}\left[\mathbf{h}_{\text{NS}}\left(f, \cos\theta_{\text{d}}\right)\right] \leq \mathcal{D}\left[\mathbf{h}_{\text{mDF}}\left(f, \cos\theta_{\text{d}}\right)\right]. \tag{7.97}$$

A straightforward way to reach a compromise between the WNG and the DF with the null-steering approach is the following beamformer:

$$\begin{aligned}\mathbf{h}_{\alpha}\left(f, \cos\theta_{\text{d}}\right) &= \mathbf{\Gamma}_{\alpha}^{-1}(f)\mathbf{C}\left(f, \theta_{\text{d}}, \theta_{1:N}\right) \\ &\quad \times \left[\mathbf{C}^{H}\left(f, \theta_{\text{d}}, \theta_{1:N}\right)\mathbf{\Gamma}_{\alpha}^{-1}(f)\mathbf{C}\left(f, \theta_{\text{d}}, \theta_{1:N}\right)\right]^{-1}\mathbf{i}_{\text{c}},\end{aligned} \tag{7.98}$$

where $\mathbf{\Gamma}_{\alpha}(f)$ is defined in (7.72). We observe that $\mathbf{h}_{0}\left(f, \cos\theta_{\text{d}}\right) = \mathbf{h}_{\text{NS}}\left(f, \cos\theta_{\text{d}}\right)$ and $\mathbf{h}_{1}\left(f, \cos\theta_{\text{d}}\right) = \mathbf{h}_{\text{MN}}\left(f, \cos\theta_{\text{d}}\right)$.

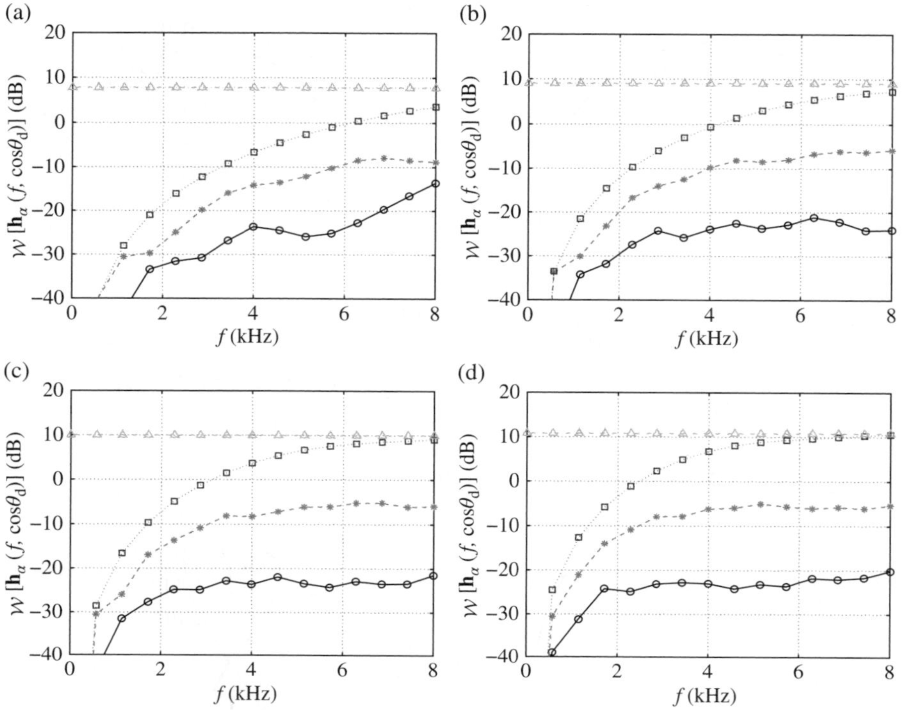

Figure 7.19 WNG of the MN/NS beamformer as a function of frequency, for different numbers of sensors, M, and several values of α: $\alpha = 1e-5$ (solid line with circles), $\alpha = 1e-3$ (dashed line with asterisks), and $\alpha = 1$ (dotted line with squares). (a) $M = 6$, (b) $M = 8$, (c) $M = 10$, and (d) $M = 12$. As a reference, $\mathcal{W}\left[\mathbf{h}_{\text{DS}}\left(f, \cos\theta_{\text{d}}\right)\right]$ is also plotted in the figures (dash-dot line with triangles).

Example 7.5.4 Returning to Example 7.5.1, we now employ the MN/NS beamformer, $\mathbf{h}_{\alpha}\left(f, \cos\theta_{\text{d}}\right)$, given in (7.98). To demonstrate the performance of the MN/NS beamformer, we choose $\theta_{\text{d}} = 0°$, $\theta_1 = 45°$, $\theta_2 = 90°$, and $\delta = 1$ cm. Figure 7.19 shows plots of the WNG, $\mathcal{W}\left[\mathbf{h}_{\alpha}\left(f, \cos\theta_{\text{d}}\right)\right]$, as a function of frequency, for different numbers of sensors, M, and several values of α. As a reference, the WNG of the DS beamformer, $\mathcal{W}\left[\mathbf{h}_{\text{DS}}\left(f, \cos\theta_{\text{d}}\right)\right]$, is also plotted. For given frequency and M, as the value of α increases, the WNG of the MN/NS beamformer increases, and is upper bounded by the WNG of the DS beamformer.

Figure 7.20 shows plots of the DF, $\mathcal{D}\left[\mathbf{h}_{\alpha}\left(f, \cos\theta_{\text{d}}\right)\right]$, as a function of frequency, for different numbers of sensors, M, and several values of α. As a reference, the DF of the maximum DF beamformer, $\mathcal{D}\left[\mathbf{h}_{\text{mDF}}\left(f, \cos\theta_{\text{d}}\right)\right]$, is also plotted. For given frequency and M, as the value of α increases, the DF of the MN/NS beamformer decreases, and is upper bounded by the WNG of the maximum DF beamformer.

Figures 7.21–7.23 show beampatterns, $\left|\mathcal{B}\left[\mathbf{h}_{\alpha}\left(f, \cos\theta_{\text{d}}\right), \cos\theta\right]\right|$, for $M = 8$, several values of α, and several frequencies. The main beam is in the direction of the desired signal, θ_{d}. Compared to the previous beamformers, here the width of the main beam is less sensitive to frequency. ■

In Table 7.1, we summarize all the fixed beamformers described in this section.

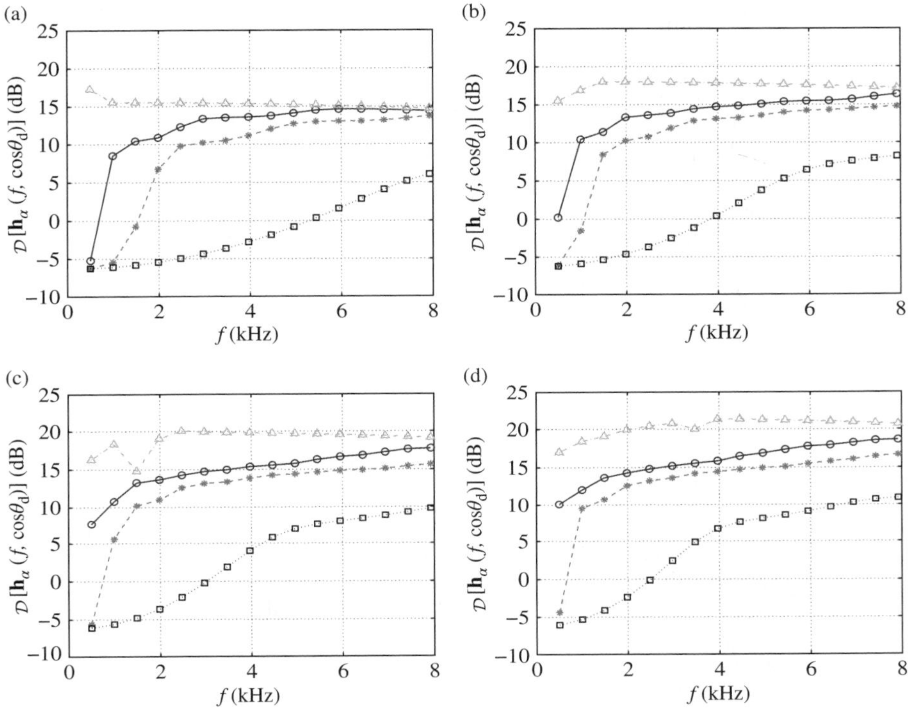

Figure 7.20 DF of the MN/NS beamformer as a function of frequency, for different numbers of sensors, M, and several values of α: $\alpha = 1e-5$ (solid line with circles), $\alpha = 1e-3$ (dashed line with asterisks), and $\alpha = 1$ (dotted line with squares). (a) $M = 6$, (b) $M = 8$, (c) $M = 10$, and (d) $M = 12$. As a reference, $\mathcal{D}\left[\mathbf{h}_{\mathrm{mDF}}\left(f, \cos\theta_{\mathrm{d}}\right)\right]$ is also plotted in the figures (dash-dot line with triangles).

7.6 A Signal Subspace Perspective

In this section, we give a signal subspace perspective of the superdirective beamformer by using joint diagonalization [14]. This approach leads to a class of beamformers that can better compromise between the WNG and the DF. It is assumed that $\theta_{\mathrm{d}} = 0$; that is, the desired signal is at the endfire.

7.6.1 Joint Diagonalization

The correlation matrix of $\mathbf{x}(f)$ is $\mathbf{\Phi}_{\mathbf{x}}(f) = \phi_X(f)\mathbf{d}\left(f,1\right)\mathbf{d}^H\left(f,1\right)$. Therefore, its pseudo-coherence matrix is

$$
\begin{aligned}
\mathbf{\Gamma}_{\mathbf{x}}(f) &= \frac{\mathbf{\Phi}_{\mathbf{x}}(f)}{\phi_X(f)} \\
&= \mathbf{d}\left(f,1\right)\mathbf{d}^H\left(f,1\right),
\end{aligned}
\tag{7.99}
$$

which does not depend on $X(f)$.

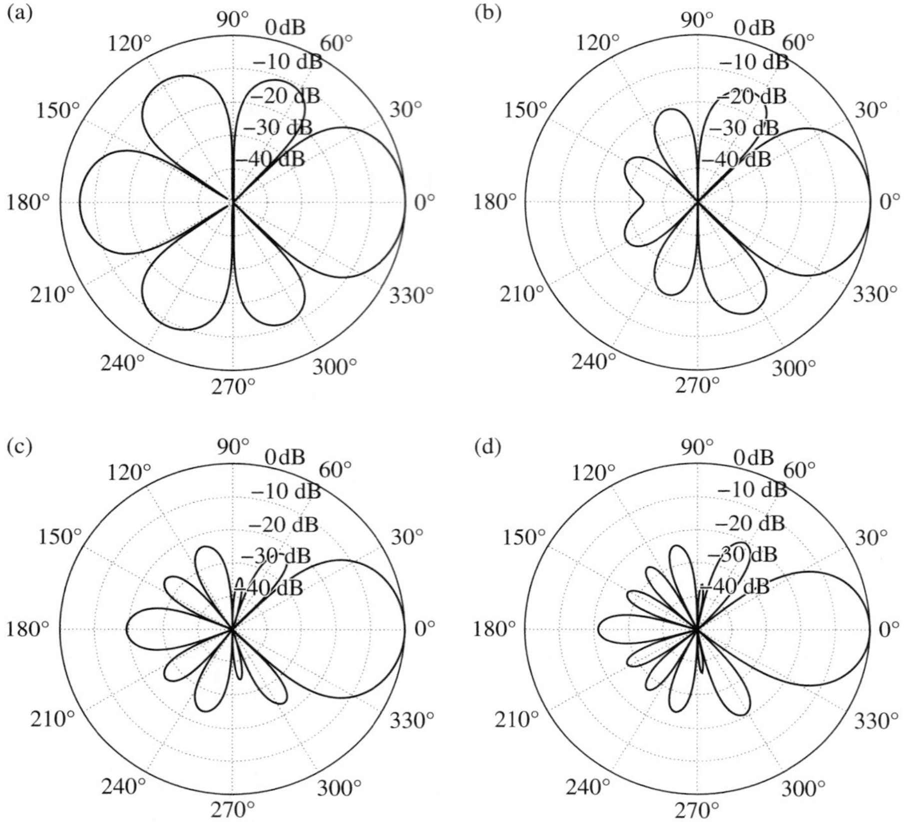

Figure 7.21 Beampatterns of the MN/NS beamformer for several frequencies with $M = 8$ and $\alpha = 1e-5$: (a) $f = 1$ kHz, (b) $f = 2$ kHz, (c) $f = 4$ kHz, and (d) $f = 8$ kHz.

The two Hermitian matrices $\mathbf{\Gamma}_{\mathbf{x}}(f)$ and $\mathbf{\Gamma}_{0,\pi}(f)$ can be jointly diagonalized as follows [15]:

$$\mathbf{T}^H(f)\mathbf{\Gamma}_{\mathbf{x}}(f)\mathbf{T}(f) = \mathbf{\Lambda}(f), \tag{7.100}$$

$$\mathbf{T}^H(f)\mathbf{\Gamma}_{0,\pi}(f)\mathbf{T}(f) = \mathbf{I}_M, \tag{7.101}$$

where

$$\mathbf{T}(f) = \begin{bmatrix} \mathbf{t}_1(f) & \mathbf{t}_2(f) & \cdots & \mathbf{t}_M(f) \end{bmatrix} \tag{7.102}$$

is a full-rank square matrix (of size $M \times M$),

$$\begin{aligned} \mathbf{t}_1(f) &= \frac{\mathbf{\Gamma}_{0,\pi}^{-1}(f)\mathbf{d}\left(f,1\right)}{\sqrt{\mathbf{d}^H\left(f,1\right)\mathbf{\Gamma}_{0,\pi}^{-1}(f)\mathbf{d}\left(f,1\right)}} \\ &= \sqrt{\mathcal{D}_{\max}\left(f,1\right)}\mathbf{h}_{\mathrm{SD}}(f) \end{aligned} \tag{7.103}$$

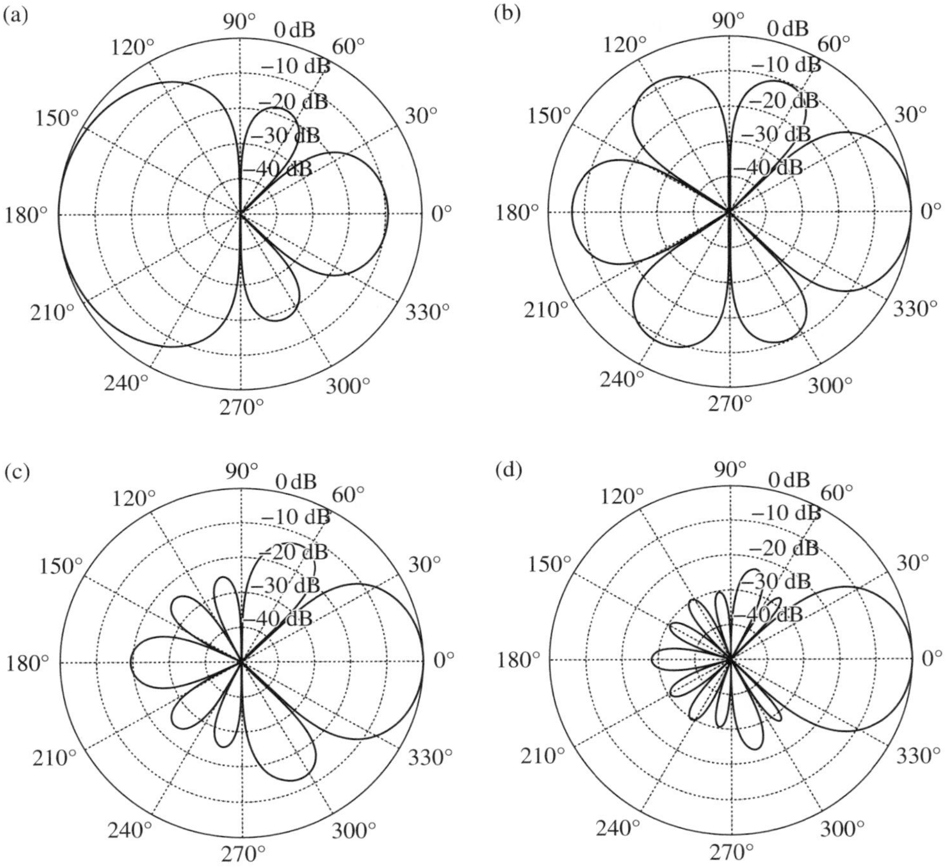

Figure 7.22 Beampatterns of the MN/NS beamformer for several frequencies with $M = 8$ and $\alpha = 1e - 3$: (a) $f = 1$ kHz, (b) $f = 2$ kHz, (c) $f = 4$ kHz, and (d) $f = 8$ kHz.

is the first eigenvector of the matrix $\mathbf{\Gamma}_{0,\pi}^{-1}(f)\mathbf{\Gamma}_{\mathbf{x}}(f)$,

$$\mathbf{\Lambda}(f) = \text{diag}\left[\lambda_1(f), 0, \dots, 0\right] \tag{7.104}$$

is a diagonal matrix (of size $M \times M$), and

$$\begin{aligned}\lambda_1(f) &= \mathbf{d}^H\left(f, 1\right)\mathbf{\Gamma}_{0,\pi}^{-1}(f)\mathbf{d}\left(f, 1\right) \\ &= \mathcal{D}_{\max}\left(f, 1\right)\end{aligned} \tag{7.105}$$

is the only nonnull eigenvalue of $\mathbf{\Gamma}_{0,\pi}^{-1}(f)\mathbf{\Gamma}_{\mathbf{x}}(f)$, whose corresponding eigenvector is $\mathbf{t}_1(f)$. It is important to observe that neither $\mathbf{T}(f)$ nor $\lambda_1(f)$ depend on the statistics of the signals. It can be checked from (7.100) that

$$\mathbf{t}_i^H(f)\mathbf{d}\left(f, 1\right) = 0,\ i = 2, 3, \dots, M. \tag{7.106}$$

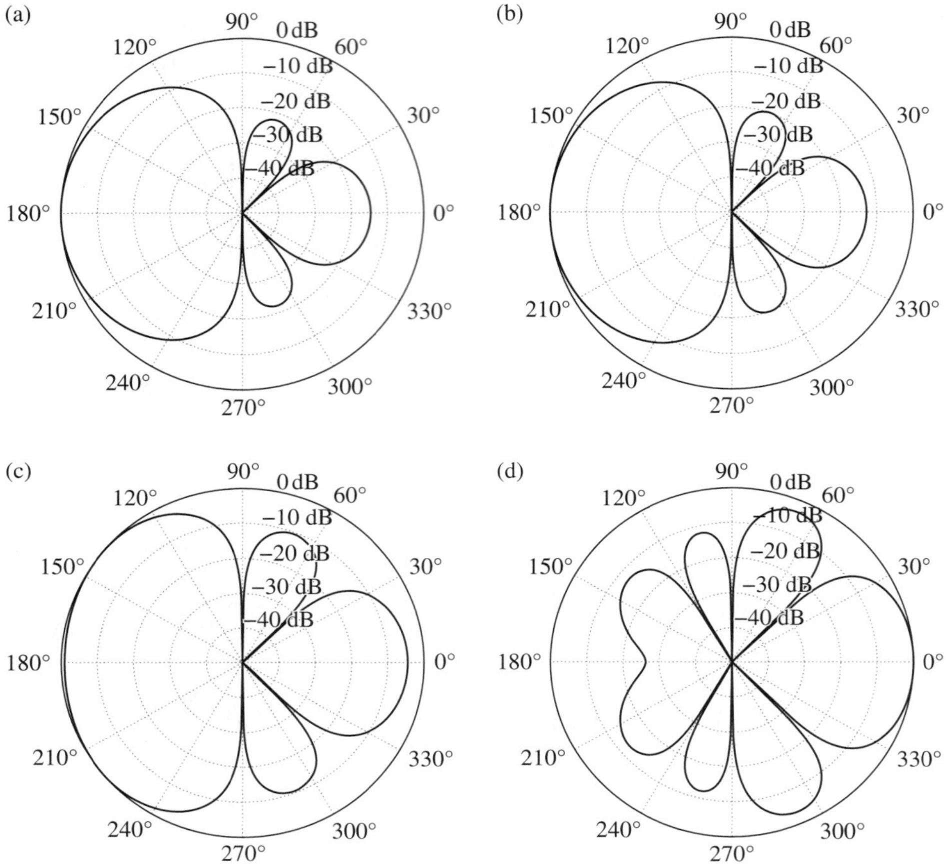

Figure 7.23 Beampatterns of the MN/NS beamformer for several frequencies with $M = 8$ and $\alpha = 1$: (a) $f = 1$ kHz, (b) $f = 2$ kHz, (c) $f = 4$ kHz, and (d) $f = 8$ kHz.

7.6.2 Compromising Between WNG and DF

Let us define the matrix of size $M \times N$:

$$\mathbf{T}_{1:N}(f) = \begin{bmatrix} \mathbf{t}_1(f) & \mathbf{t}_2(f) & \cdots & \mathbf{t}_N(f) \end{bmatrix}, \tag{7.107}$$

with $1 \leq N \leq M$. We consider beamformers that have the form:

$$\mathbf{h}_{1:N}(f) = \mathbf{T}_{1:N}(f)\mathbf{a}(f), \tag{7.108}$$

where

$$\mathbf{a}(f) = \begin{bmatrix} A_1(f) & A_2(f) & \cdots & A_N(f) \end{bmatrix}^T \neq \mathbf{0} \tag{7.109}$$

Table 7.1 Fixed beamformers.

Beamformer	
DS	$\mathbf{h}_{\mathrm{DS}}\left(f,\cos\theta_{\mathrm{d}}\right) = \dfrac{\mathbf{d}\left(f,\cos\theta_{\mathrm{d}}\right)}{M}$
Maximum DF	$\mathbf{h}_{\mathrm{mDF}}\left(f,\cos\theta_{\mathrm{d}}\right) = \dfrac{\mathbf{\Gamma}_{0,\pi}^{-1}(f)\mathbf{d}\left(f,\cos\theta_{\mathrm{d}}\right)}{\mathbf{d}^H\left(f,\cos\theta_{\mathrm{d}}\right)\mathbf{\Gamma}_{0,\pi}^{-1}(f)\mathbf{d}\left(f,\cos\theta_{\mathrm{d}}\right)}$
Superdirective	$\mathbf{h}_{\mathrm{SD}}(f) = \dfrac{\mathbf{\Gamma}_{0,\pi}^{-1}(f)\mathbf{d}\left(f,1\right)}{\mathbf{d}^H\left(f,1\right)\mathbf{\Gamma}_{0,\pi}^{-1}(f)\mathbf{d}\left(f,1\right)}$
Robust SD	$\mathbf{h}_{\mathrm{R},\alpha}(f) = \dfrac{\mathbf{\Gamma}_{\alpha}^{-1}(f)\mathbf{d}\left(f,1\right)}{\mathbf{d}^H\left(f,1\right)\mathbf{\Gamma}_{\alpha}^{-1}(f)\mathbf{d}\left(f,1\right)}$
Minimum norm	$\mathbf{h}_{\mathrm{MN}}\left(f,\cos\theta_{\mathrm{d}}\right) = \mathbf{C}\left(f,\theta_{\mathrm{d}},\theta_{1:N}\right)\left[\mathbf{C}^H\left(f,\theta_{\mathrm{d}},\theta_{1:N}\right)\mathbf{C}\left(f,\theta_{\mathrm{d}},\theta_{1:N}\right)\right]^{-1}\mathbf{i}_{\mathrm{c}}$
Null steering	$\mathbf{h}_{\mathrm{NS}}\left(f,\cos\theta_{\mathrm{d}}\right) = \mathbf{\Gamma}_{0,\pi}^{-1}(f)\mathbf{C}\left(f,\theta_{\mathrm{d}},\theta_{1:N}\right) \times \left[\mathbf{C}^H\left(f,\theta_{\mathrm{d}},\theta_{1:N}\right)\mathbf{\Gamma}_{0,\pi}^{-1}(f)\mathbf{C}\left(f,\theta_{\mathrm{d}},\theta_{1:N}\right)\right]^{-1}\mathbf{i}_{\mathrm{c}}$
MN/NS	$\mathbf{h}_{\alpha}\left(f,\cos\theta_{\mathrm{d}}\right) = \mathbf{\Gamma}_{\alpha}^{-1}(f)\mathbf{C}\left(f,\theta_{\mathrm{d}},\theta_{1:N}\right) \times \left[\mathbf{C}^H\left(f,\theta_{\mathrm{d}},\theta_{1:N}\right)\mathbf{\Gamma}_{\alpha}^{-1}(f)\mathbf{C}\left(f,\theta_{\mathrm{d}},\theta_{1:N}\right)\right]^{-1}\mathbf{i}_{\mathrm{c}}$

is a vector of length N. Substituting (7.108) into (7.4), we find that

$$
\begin{aligned}
Z(f) &= \mathbf{a}^H(f)\mathbf{T}_{1:N}^H(f)\mathbf{d}\left(f,1\right)X(f) + \mathbf{a}^H(f)\mathbf{T}_{1:N}^H(f)\mathbf{v}(f) \\
&= A_1^*(f)\sqrt{\lambda_1(f)}X(f) + \mathbf{a}^H(f)\mathbf{T}_{1:N}^H(f)\mathbf{v}(f).
\end{aligned}
\tag{7.110}
$$

Since the distortionless constraint is desired, it is clear from the previous expression that we always choose

$$
A_1(f) = \frac{1}{\sqrt{\lambda_1(f)}}. \tag{7.111}
$$

Now, we need to determine the other elements of $\mathbf{a}(f)$.

With the beamformer given in (7.108), the WNG and the DF are, respectively,

$$
\begin{aligned}
\mathcal{W}\left[\mathbf{h}_{1:N}(f)\right] &= \frac{\left|\mathbf{h}_{1:N}^H(f)\mathbf{d}\left(f,1\right)\right|^2}{\mathbf{h}_{1:N}^H(f)\mathbf{h}_{1:N}(f)} \\
&= \frac{\left|\mathbf{a}^H(f)\mathbf{T}_{1:N}^H(f)\mathbf{d}\left(f,1\right)\right|^2}{\mathbf{a}^H(f)\mathbf{T}_{1:N}^H(f)\mathbf{T}_{1:N}(f)\mathbf{a}(f)}
\end{aligned}
\tag{7.112}
$$

and

$$\mathcal{D}\left[\mathbf{h}_{1:N}(f)\right] = \frac{\left|\mathbf{h}_{1:N}^H(f)\mathbf{d}\left(f,1\right)\right|^2}{\mathbf{h}_{1:N}^H(f)\mathbf{\Gamma}_{0,\pi}(f)\mathbf{h}_{1:N}(f)} \tag{7.113}$$
$$= \frac{\left|\mathbf{a}^H(f)\mathbf{T}_{1:N}^H(f)\mathbf{d}\left(f,1\right)\right|^2}{\mathbf{a}^H(f)\mathbf{T}_{1:N}^H(f)\mathbf{\Gamma}_{0,\pi}(f)\mathbf{T}_{1:N}(f)\mathbf{a}(f)}$$
$$= \frac{\left|\mathbf{a}^H(f)\mathbf{T}_{1:N}^H(f)\mathbf{d}\left(f,1\right)\right|^2}{\mathbf{a}^H(f)\mathbf{a}(f)}.$$

The maximization of the DF or, equivalently, the minimization of $\mathbf{h}_{1:N}^H(f)\mathbf{\Gamma}_{0,\pi}(f)$ $\mathbf{h}_{1:N}(f)$ subject to $\mathbf{h}_{1:N}^H(f)\mathbf{d}\left(f,1\right) = 1$, leads to the conventional superdirective beamformer:

$$\mathbf{h}_{\text{SD}}(f) = \frac{\mathbf{T}_{1:N}(f)\mathbf{T}_{1:N}^H(f)\mathbf{d}\left(f,1\right)}{\mathbf{d}^H\left(f,1\right)\mathbf{T}_{1:N}(f)\mathbf{T}_{1:N}^H(f)\mathbf{d}\left(f,1\right)} \tag{7.114}$$
$$= \frac{\mathbf{t}_1(f)\mathbf{t}_1^H(f)\mathbf{d}\left(f,1\right)}{\left|\mathbf{t}_1^H(f)\mathbf{d}\left(f,1\right)\right|^2}$$
$$= \frac{\mathbf{\Gamma}_{0,\pi}^{-1}(f)\mathbf{d}\left(f,1\right)}{\mathbf{d}^H\left(f,1\right)\mathbf{\Gamma}_{0,\pi}^{-1}(f)\mathbf{d}\left(f,1\right)}.$$

The most useful subspace beamformer is derived by maximizing the WNG. This is equivalent to minimizing $\mathbf{h}_{1:N}^H(f)\mathbf{h}_{1:N}(f)$ subject to $\mathbf{h}_{1:N}^H(f)\mathbf{d}\left(f,1\right) = 1$. We easily find

$$\mathbf{h}_{1:N}(f) = \frac{\mathbf{P}_{\mathbf{T}_{1:N}}(f)\mathbf{d}\left(f,1\right)}{\mathbf{d}^H\left(f,1\right)\mathbf{P}_{\mathbf{T}_{1:N}}(f)\mathbf{d}\left(f,1\right)}, \tag{7.115}$$

where

$$\mathbf{P}_{\mathbf{T}_{1:N}}(f) = \mathbf{T}_{1:N}(f)\left[\mathbf{T}_{1:N}^H(f)\mathbf{T}_{1:N}f\right]^{-1}\mathbf{T}_{1:N}^H(f). \tag{7.116}$$

For $N = 1$, we get

$$\mathbf{h}_{1:1}(f) = \frac{\mathbf{t}_1(f)}{\mathbf{d}^H\left(f,1\right)\mathbf{t}_1(f)} \tag{7.117}$$
$$= \mathbf{h}_{\text{SD}}(f),$$

which is the conventional superdirective beamformer, and for $N = M$, we obtain

$$\mathbf{h}_{1:M}(f) = \frac{\mathbf{d}\left(f,1\right)}{\mathbf{d}^H\left(f,1\right)\mathbf{d}\left(f,1\right)} \tag{7.118}$$
$$= \mathbf{h}_{\text{DS}}\left(f,1\right),$$

which is the DS beamformer. Therefore, by playing with N, we obtain different beamformers whose performances are in between the performances of $\mathbf{h}_{\mathrm{SD}}(f)$ and $\mathbf{h}_{\mathrm{DS}}(f,1)$. To make matters a bit more complicated, we can also make N frequency dependent. Indeed, at high frequencies, it may be desirable to take values of N lower than at low frequencies.

With the proposed beamformer, the WNG is

$$\begin{aligned}\mathcal{W}\left[\mathbf{h}_{1:N}(f)\right] &= \mathbf{d}^H(f,1)\,\mathbf{P}_{\mathbf{T}_{1:N}}(f)\mathbf{d}(f,1) \\ &= \lambda_1(f)\mathbf{i}^T\left[\mathbf{T}_{1:N}^H(f)\mathbf{T}_{1:N}(f)\right]^{-1}\mathbf{i},\end{aligned} \tag{7.119}$$

where $\mathbf{i}$ is the first column of the $N \times N$ identity matrix, $\mathbf{I}_N$, with

$$\begin{aligned}\mathcal{W}\left[\mathbf{h}_{1:1}(f)\right] &= \frac{\left|\mathbf{t}_1^H(f)\mathbf{d}(f,1)\right|^2}{\mathbf{t}_1^H(f)\mathbf{t}_1(f)} \\ &= \frac{\lambda_1(f)}{\mathbf{t}_1^H(f)\mathbf{t}_1(f)} \\ &= \frac{\left[\mathbf{d}^H(f,1)\,\boldsymbol{\Gamma}_{0,\pi}^{-1}(f)\mathbf{d}(f,1)\right]^2}{\mathbf{d}^H(f,1)\,\boldsymbol{\Gamma}_{0,\pi}^{-2}(f)\mathbf{d}(f,1)} \leq M\end{aligned} \tag{7.120}$$

and

$$\mathcal{W}\left[\mathbf{h}_{1:M}(f)\right] = M. \tag{7.121}$$

The DF is

$$\begin{aligned}\mathcal{D}\left[\mathbf{h}_{1:N}(f)\right] &= \frac{\left[\mathbf{d}^H(f,1)\,\mathbf{P}_{\mathbf{T}_{1:N}}(f)\mathbf{d}(f,1)\right]^2}{\mathbf{d}^H(f,1)\,\mathbf{P}_{\mathbf{T}_{1:N}}(f)\boldsymbol{\Gamma}_{0,\pi}(f)\mathbf{P}_{\mathbf{T}_{1:N}}(f)\mathbf{d}(f,1)} \\ &= \lambda_1(f)\frac{\left\{\mathbf{i}^T\left[\mathbf{T}_{1:N}^H(f)\mathbf{T}_{1:N}(f)\right]^{-1}\mathbf{i}\right\}^2}{\mathbf{i}^T\left[\mathbf{T}_{1:N}^H(f)\mathbf{T}_{1:N}(f)\right]^{-2}\mathbf{i}},\end{aligned} \tag{7.122}$$

with

$$\mathcal{D}\left[\mathbf{h}_{1:1}(f)\right] = \lambda_1(f) \leq M^2 \tag{7.123}$$

and

$$\begin{aligned}\mathcal{D}\left[\mathbf{h}_{1:M}(f)\right] &= \frac{\left[\mathbf{d}^H(f,1)\,\mathbf{d}(f,1)\right]^2}{\mathbf{d}^H(f,1)\,\boldsymbol{\Gamma}_{0,\pi}(f)\mathbf{d}(f,1)} \\ &= \frac{M^2}{\mathbf{d}^H(f,1)\,\boldsymbol{\Gamma}_{0,\pi}(f)\mathbf{d}(f,1)} \geq 1.\end{aligned} \tag{7.124}$$

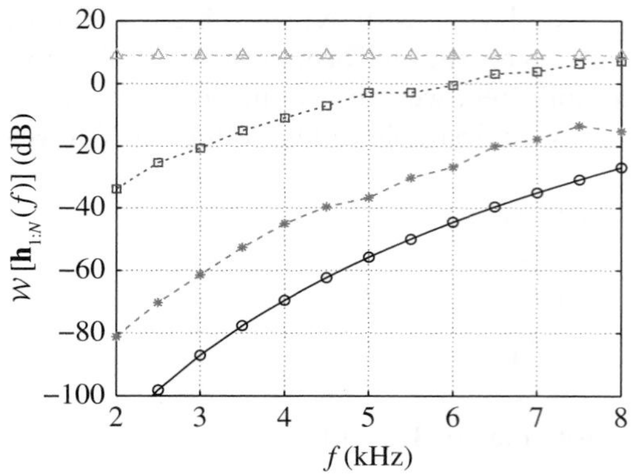

Figure 7.24 WNG of the subspace beamformer as a function of frequency, for several values of N: $N = 1$ (solid line with circles), $N = 2$ (dashed line with asterisks), $N = 4$ (dotted line with squares), and $N = 8$ (dash-dot line with triangles).

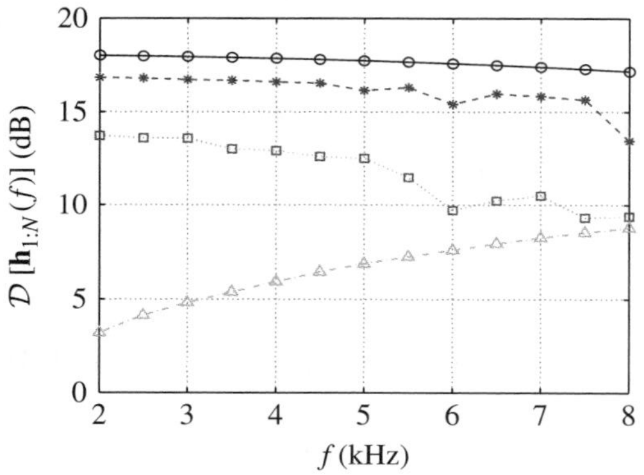

Figure 7.25 DF of the subspace beamformer as a function of frequency, for several values of N: $N = 1$ (solid line with circles), $N = 2$ (dashed line with asterisks), $N = 4$ (dotted line with squares), and $N = 8$ (dash-dot line with triangles).

We also deduce a useful relationship between the WNG and the DF:

$$\frac{\mathcal{D}\left[\mathbf{h}_{1:N}(f)\right]}{\mathcal{W}\left[\mathbf{h}_{1:N}(f)\right]} = \frac{\mathbf{i}^T\left[\mathbf{T}_{1:N}^H(f)\mathbf{T}_{1:N}(f)\right]^{-1}\mathbf{i}}{\mathbf{i}^T\left[\mathbf{T}_{1:N}^H(f)\mathbf{T}_{1:N}(f)\right]^{-2}\mathbf{i}}, \tag{7.125}$$

where

$$\frac{1}{M} \leq \frac{\mathcal{D}\left[\mathbf{h}_{1:N}(f)\right]}{\mathcal{W}\left[\mathbf{h}_{1:N}(f)\right]} < \infty. \tag{7.126}$$

We should always have

$$M^2 \geq \mathcal{D}\left[\mathbf{h}_{1:1}(f)\right] \geq \mathcal{D}\left[\mathbf{h}_{1:2}(f)\right] \geq \cdots \geq \mathcal{D}\left[\mathbf{h}_{1:M}(f)\right] \tag{7.127}$$

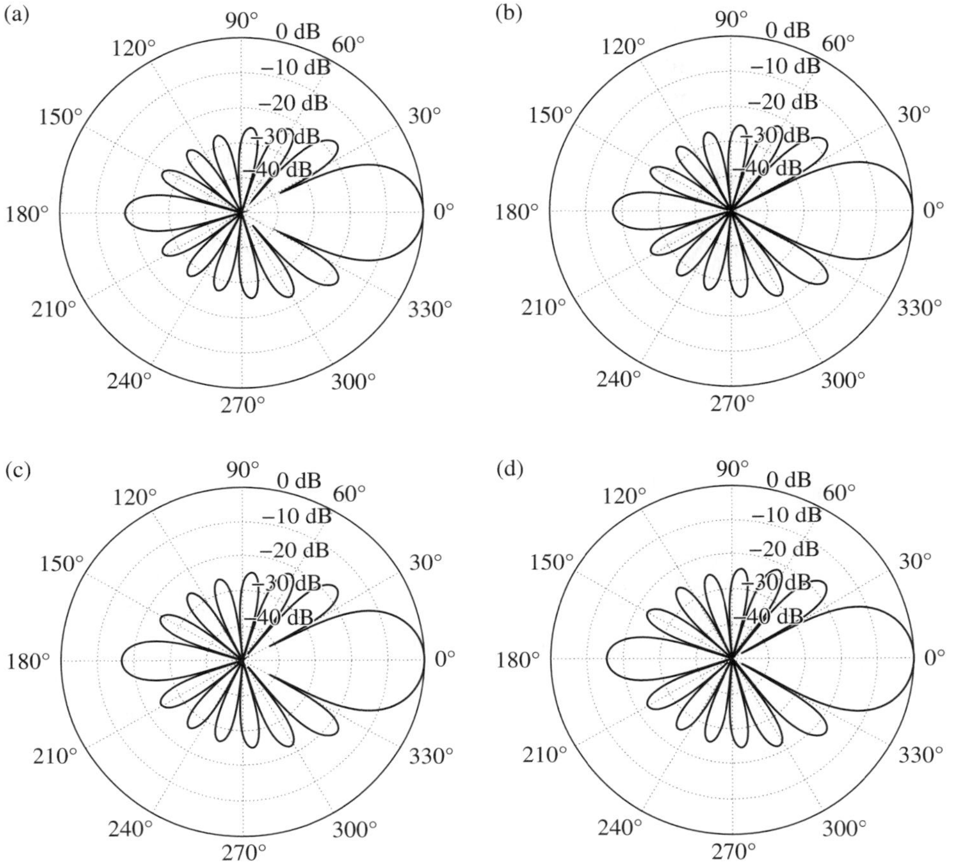

Figure 7.26 Beampatterns of the subspace beamformer for $N = 1$ and several frequencies: (a) $f = 2$ kHz, (b) $f = 4$ kHz, (c) $f = 6$ kHz, and (d) $f = 8$ kHz.

and

$$M = \mathcal{W}\left[\mathbf{h}_{1:M}(f)\right] \geq \mathcal{W}\left[\mathbf{h}_{1:M-1}(f)\right] \geq \cdots \geq \mathcal{W}\left[\mathbf{h}_{1:1}(f)\right]. \tag{7.128}$$

Clearly, the beamformer $\mathbf{h}_{1:N}(f)$ is much more useful and practical than the beamformer $\mathbf{h}_{\mathrm{R},\alpha}(f)$ for control of white noise amplification while giving a reasonably good DF.

Example 7.6.1 Returning to Example 7.5.1, we now employ the subspace beamformer, $\mathbf{h}_{1:N}(f)$, given in (7.115). To demonstrate the performance of the subspace beamformer, we choose $M = 8$ and $\delta = 1$ cm. Figure 7.24 shows plots of the WNG, $\mathcal{W}\left[\mathbf{h}_{1:N}(f)\right]$, as a function of frequency, for several values of N. For a given frequency, as the value of N increases, the WNG of the subspace beamformer increases, and is upper bounded by the WNG of the DS beamformer.

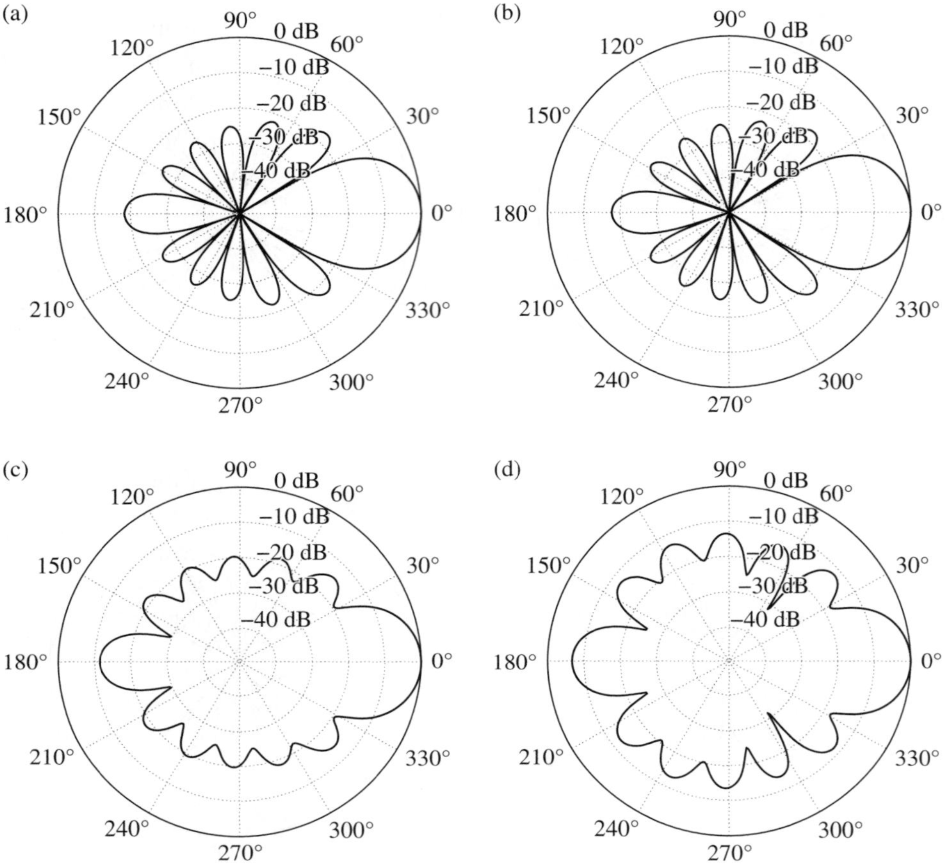

Figure 7.27 Beampatterns of the subspace beamformer for $N = 2$ and several frequencies: (a) $f = 2$ kHz, (b) $f = 4$ kHz, (c) $f = 6$ kHz, and (d) $f = 8$ kHz.

Figure 7.25 shows plots of the DF, $\mathcal{D}\left[\mathbf{h}_{1:N}(f)\right]$, as a function of frequency, for several values of N. For a given frequency, as the value of N decreases, the DF of the subspace beamformer increases, and is upper bounded by the DF of the superdirective beamformer.

Figures 7.26–7.29 show beampatterns, $\left|\mathcal{B}\left[\mathbf{h}_{1:N}(f), \cos\theta\right]\right|$, for several values of N, and several frequencies. The main beam is in the direction of the desired signal; in other words, $\theta_{\mathrm{d}} = 0°$. ■

Problems

7.1 Show that, in the distortionless case, the WNG can be written as

$$\mathcal{W}\left[\mathbf{h}(f)\right] = \mathcal{W}_{\max} \cos^2\left[\mathbf{d}\left(f, \cos\theta_{\mathrm{d}}\right), \mathbf{h}(f)\right].$$

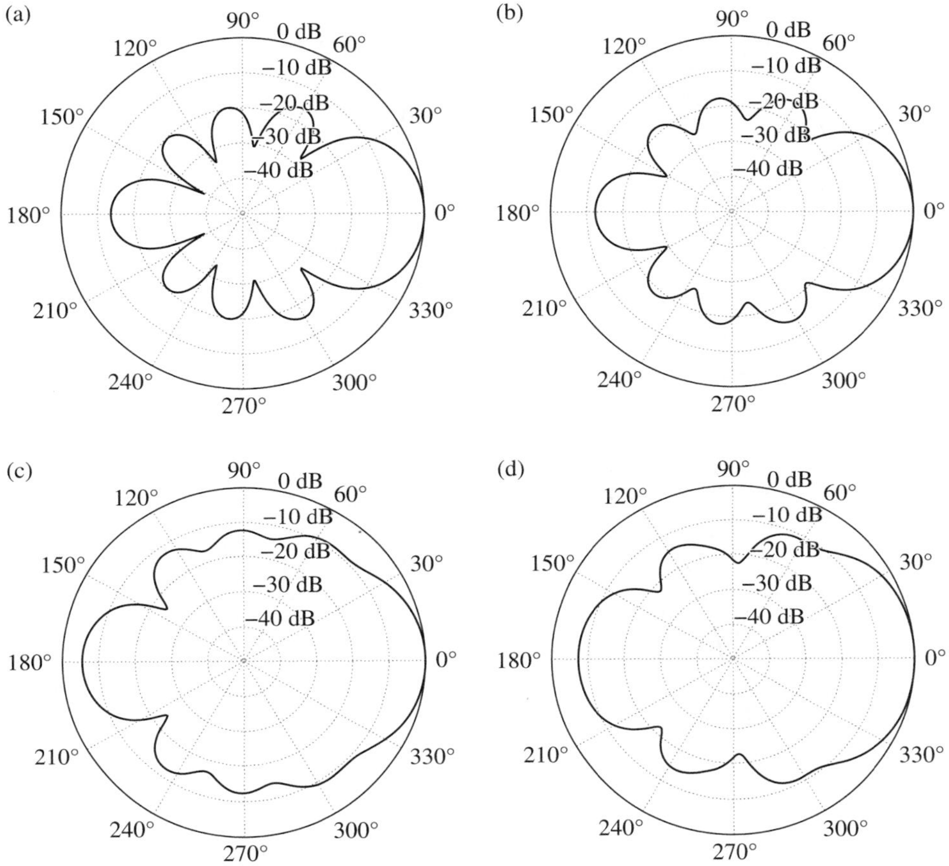

Figure 7.28 Beampatterns of the subspace beamformer for $N = 4$ and several frequencies: (a) $f = 2$ kHz, (b) $f = 4$ kHz, (c) $f = 6$ kHz, and (d) $f = 8$ kHz.

7.2 Using the Cauchy–Schwarz inequality, show that the maximum DF is

$$\begin{aligned}\mathcal{D}_{\max}\left(f, \cos\theta_{\rm d}\right) &= \mathbf{d}^H\left(f, \cos\theta_{\rm d}\right)\mathbf{\Gamma}_{0,\pi}^{-1}(f)\mathbf{d}\left(f, \cos\theta_{\rm d}\right) \\ &= \mathrm{tr}\left[\mathbf{\Gamma}_{0,\pi}^{-1}(f)\mathbf{d}\left(f, \cos\theta_{\rm d}\right)\mathbf{d}^H\left(f, \cos\theta_{\rm d}\right)\right] \\ &\leq M\mathrm{tr}\left[\mathbf{\Gamma}_{0,\pi}^{-1}(f)\right].\end{aligned}$$

7.3 Show that, in the distortionless case, the DF can be written as

$$\mathcal{D}\left[\mathbf{h}(f)\right] = \mathcal{D}_{\max}\left(f, \cos\theta_{\rm d}\right)\cos^2\left[\mathbf{\Gamma}_{0,\pi}^{-1/2}(f)\mathbf{d}\left(f, \cos\theta_{\rm d}\right), \mathbf{\Gamma}_{0,\pi}^{1/2}(f)\mathbf{h}(f)\right].$$

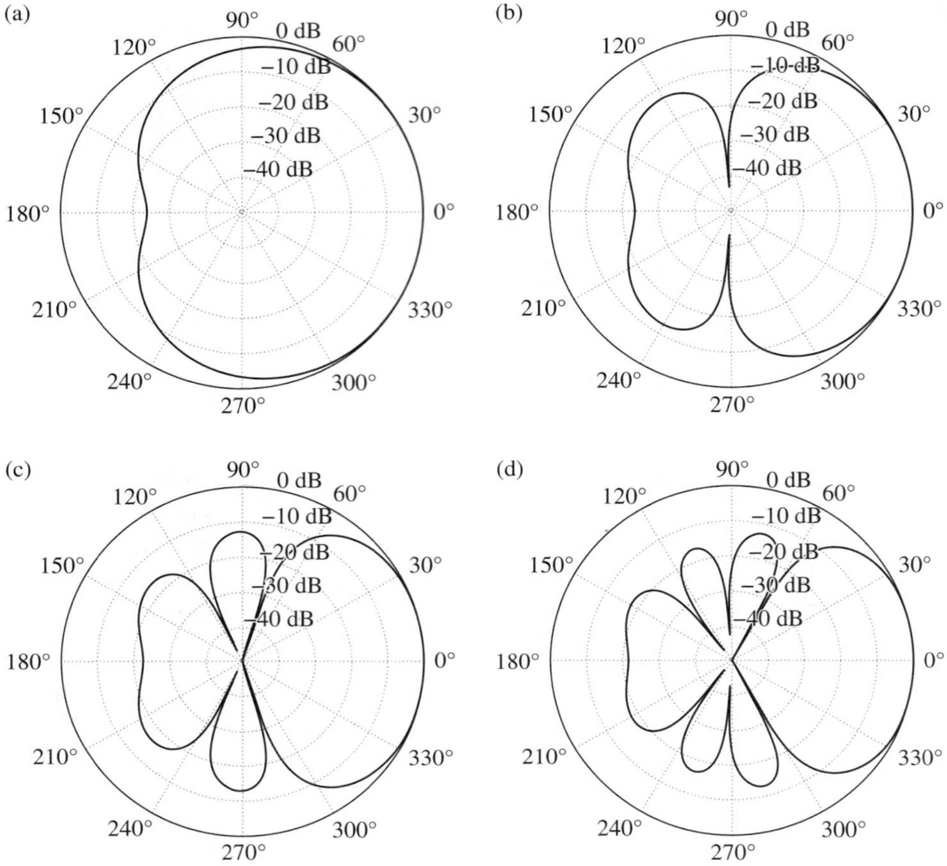

Figure 7.29 Beampatterns of the subspace beamformer for $N = 8$ and several frequencies: (a) $f = 2$ kHz, (b) $f = 4$ kHz, (c) $f = 6$ kHz, and (d) $f = 8$ kHz.

7.4 Show that the condition to prevent spatial aliasing is

$$\frac{\delta}{\lambda} < \frac{1}{2}.$$

7.5 Show that the DS beamformer:

$$\mathbf{h}_{\mathrm{DS}}\left(f, \cos\theta_{\mathrm{d}}\right) = \frac{\mathbf{d}\left(f, \cos\theta_{\mathrm{d}}\right)}{M},$$

maximizes the WNG:

$$\min_{\mathbf{h}(f)} \mathbf{h}^H(f)\mathbf{h}(f) \quad \text{subject to} \quad \mathbf{h}^H(f)\mathbf{d}\left(f, \cos\theta_{\mathrm{d}}\right) = 1.$$

7.6 Show that with the DS beamformer, the DF is given by

$$\mathcal{D}\left[\mathbf{h}_{\mathrm{DS}}\left(f,\cos\theta_{\mathrm{d}}\right)\right] = \frac{M^2}{\mathbf{d}^H\left(f,\cos\theta_{\mathrm{d}}\right)\mathbf{\Gamma}_{0,\pi}(f)\mathbf{d}\left(f,\cos\theta_{\mathrm{d}}\right)}.$$

7.7 Show that the DS beamformer never amplifies the diffuse noise: $\mathcal{D}\left[\mathbf{h}_{\mathrm{DS}}\left(f,\cos\theta_{\mathrm{d}}\right)\right] \geq 1$.

7.8 Show that the beampattern of the DS beamformer is given by

$$\left|\mathcal{B}\left[\mathbf{h}_{\mathrm{DS}}\left(f,\cos\theta_{\mathrm{d}}\right),\cos\theta\right]\right|^2 = \frac{1}{M^2}\left|\frac{1-e^{\jmath M2\pi f\tau_0\left(\cos\theta-\cos\theta_{\mathrm{d}}\right)}}{1-e^{\jmath 2\pi f\tau_0\left(\cos\theta-\cos\theta_{\mathrm{d}}\right)}}\right|^2.$$

7.9 Show that the DF of the DS beamformer can be written as

$$\begin{aligned}\mathcal{D}\left[\mathbf{h}_{\mathrm{DS}}\left(f,\cos\theta_{\mathrm{d}}\right)\right] &= \mathcal{D}_{\max}\left(f,\cos\theta_{\mathrm{d}}\right)\\ &\quad\times\cos^2\left[\mathbf{\Gamma}_{0,\pi}^{1/2}(f)\mathbf{d}\left(f,\cos\theta_{\mathrm{d}}\right),\mathbf{\Gamma}_{0,\pi}^{-1/2}(f)\mathbf{d}\left(f,\cos\theta_{\mathrm{d}}\right)\right].\end{aligned}$$

7.10 Let $\sigma_1(f)$ and $\sigma_M(f)$ be the maximum and minimum eigenvalues of $\mathbf{\Gamma}_{0,\pi}(f)$, respectively. Show that

$$\frac{4\sigma_1(f)\sigma_M(f)}{\left[\sigma_1(f)+\sigma_M(f)\right]^2} \leq \frac{\mathcal{D}\left[\mathbf{h}_{\mathrm{DS}}\left(f,\cos\theta_{\mathrm{d}}\right)\right]}{\mathcal{D}_{\max}\left(f,\cos\theta_{\mathrm{d}}\right)} \leq 1.$$

7.11 Show that the beamformer that maximizes the DF is given by

$$\mathbf{h}_{\mathrm{mDF}}\left(f,\cos\theta_{\mathrm{d}}\right) = \frac{\mathbf{\Gamma}_{0,\pi}^{-1}(f)\mathbf{d}\left(f,\cos\theta_{\mathrm{d}}\right)}{\mathbf{d}^H\left(f,\cos\theta_{\mathrm{d}}\right)\mathbf{\Gamma}_{0,\pi}^{-1}(f)\mathbf{d}\left(f,\cos\theta_{\mathrm{d}}\right)}.$$

7.12 Show that the WNG and the DF of the maximum DF beamformer are given by

$$\begin{aligned}\mathcal{W}\left[\mathbf{h}_{\mathrm{mDF}}\left(f,\cos\theta_{\mathrm{d}}\right)\right] &= \frac{\left[\mathbf{d}^H\left(f,\cos\theta_{\mathrm{d}}\right)\mathbf{\Gamma}_{0,\pi}^{-1}(f)\mathbf{d}\left(f,\cos\theta_{\mathrm{d}}\right)\right]^2}{\mathbf{d}^H\left(f,\cos\theta_{\mathrm{d}}\right)\mathbf{\Gamma}_{0,\pi}^{-2}(f)\mathbf{d}\left(f,\cos\theta_{\mathrm{d}}\right)},\\ \mathcal{D}\left[\mathbf{h}_{\mathrm{mDF}}\left(f,\cos\theta_{\mathrm{d}}\right)\right] &= \mathbf{d}^H\left(f,\cos\theta_{\mathrm{d}}\right)\mathbf{\Gamma}_{0,\pi}^{-1}(f)\mathbf{d}\left(f,\cos\theta_{\mathrm{d}}\right).\end{aligned}$$

7.13 Let $\sigma_1(f)$ and $\sigma_M(f)$ be the maximum and minimum eigenvalues of $\mathbf{\Gamma}_{0,\pi}(f)$, respectively. Show that

$$\frac{4\sigma_1(f)\sigma_M(f)}{\left[\sigma_1(f)+\sigma_M(f)\right]^2} \leq \frac{\mathcal{W}\left[\mathbf{h}_{\mathrm{mDF}}\left(f,\cos\theta_{\mathrm{d}}\right)\right]}{\mathcal{W}_{\max}} \leq 1.$$

7.14 Show that the DS and maximum DF beamformers are related by

$$\mathcal{D}_{\max}\left(f,\cos\theta_{\mathrm{d}}\right)\mathbf{\Gamma}_{0,\pi}(f)\mathbf{h}_{\mathrm{mDF}}\left(f,\cos\theta_{\mathrm{d}}\right) = \mathcal{W}_{\max}\mathbf{h}_{\mathrm{DS}}\left(f,\cos\theta_{\mathrm{d}}\right).$$

7.15 Show that for $M = 2$ sensors and small δ, the maximum DF can be approximated by

$$\mathcal{D}_{\max}\left(f, \cos\theta_{\mathrm{d}}\right) \approx 3\cos^2\theta_{\mathrm{d}} + 1.$$

7.16 Show that minimizing

$$\frac{1}{\mathcal{D}\left[\mathbf{h}(f)\right]} + \epsilon \frac{1}{\mathcal{W}\left[\mathbf{h}(f)\right]}$$

with the distortionless constraint yields the robust superdirective beamformer:

$$\mathbf{h}_{\mathrm{R},\epsilon}(f) = \frac{\left[\mathbf{\Gamma}_{0,\pi}(f) + \epsilon\mathbf{I}_M\right]^{-1}\mathbf{d}\left(f,1\right)}{\mathbf{d}^H\left(f,1\right)\left[\mathbf{\Gamma}_{0,\pi}(f) + \epsilon\mathbf{I}_M\right]^{-1}\mathbf{d}\left(f,1\right)}.$$

7.17 Show that the robust superdirective beamformer can be expressed as

$$\mathbf{h}_{\mathrm{R},\alpha}(f) = \frac{\mathbf{\Gamma}_{\alpha}^{-1}(f)\mathbf{d}\left(f,1\right)}{\mathbf{d}^H\left(f,1\right)\mathbf{\Gamma}_{\alpha}^{-1}(f)\mathbf{d}\left(f,1\right)},$$

where $\mathbf{\Gamma}_{\alpha}(f) = \left[1-\alpha(f)\right]\mathbf{\Gamma}_{0,\pi}(f) + \alpha(f)\mathbf{I}_M$, and $\epsilon(f) = \frac{\alpha(f)}{1-\alpha(f)}$.

7.18 Let $\mathbf{\Gamma}_{\alpha^-}(f) = \left[1-\alpha(f)\right]\mathbf{\Gamma}_{0,\pi}(f) - \alpha(f)\mathbf{I}_M$. Show that for $0 \leq \alpha \leq 0.5$,

$$\mathbf{\Gamma}_{\alpha}(f) \approx (1-\alpha)^{-2}\mathbf{\Gamma}_{0,\pi}^{-1}(f)\mathbf{\Gamma}_{\alpha^-}(f)\mathbf{\Gamma}_{0,\pi}^{-1}(f),$$

and for $0.5 < \alpha \leq 1$,

$$\mathbf{\Gamma}_{\alpha}(f) \approx -\alpha^{-2}\mathbf{\Gamma}_{\alpha^-}(f).$$

7.19 Show that maximizing the WNG with the constraint $\mathbf{C}^H\left(f, \theta_{\mathrm{d}}, \theta_{1:N}\right)\mathbf{h}(f) = \mathbf{i}_{\mathrm{c}}$ yields the MN beamformer:

$$\mathbf{h}_{\mathrm{MN}}\left(f, \cos\theta_{\mathrm{d}}\right) = \mathbf{C}\left(f, \theta_{\mathrm{d}}, \theta_{1:N}\right)\left[\mathbf{C}^H\left(f, \theta_{\mathrm{d}}, \theta_{1:N}\right)\mathbf{C}\left(f, \theta_{\mathrm{d}}, \theta_{1:N}\right)\right]^{-1}\mathbf{i}_{\mathrm{c}}.$$

7.20 Show that maximizing the DF with the constraint $\mathbf{C}^H\left(f, \theta_{\mathrm{d}}, \theta_{1:N}\right)\mathbf{h}(f) = \mathbf{i}_{\mathrm{c}}$ yields the NS beamformer:

$$\mathbf{h}_{\mathrm{NS}}\left(f, \cos\theta_{\mathrm{d}}\right) = \mathbf{\Gamma}_{0,\pi}^{-1}(f)\mathbf{C}\left(f, \theta_{\mathrm{d}}, \theta_{1:N}\right) \times \left[\mathbf{C}^H\left(f, \theta_{\mathrm{d}}, \theta_{1:N}\right)\mathbf{\Gamma}_{0,\pi}^{-1}(f)\mathbf{C}\left(f, \theta_{\mathrm{d}}, \theta_{1:N}\right)\right]^{-1}\mathbf{i}_{\mathrm{c}}.$$

7.21 Show that with a beamformer of the form $\mathbf{h}_{1:N}(f) = \mathbf{T}_{1:N}(f)\mathbf{a}(f)$, the WNG and the DF are

$$\mathcal{W}\left[\mathbf{h}_{1:N}(f)\right] = \frac{\left|\mathbf{a}^H(f)\mathbf{T}_{1:N}^H(f)\mathbf{d}\left(f,1\right)\right|^2}{\mathbf{a}^H(f)\mathbf{T}_{1:N}^H(f)\mathbf{T}_{1:N}(f)\mathbf{a}(f)},$$

$$\mathcal{D}\left[\mathbf{h}_{1:N}(f)\right] = \frac{\left|\mathbf{a}^H(f)\mathbf{T}_{1:N}^H(f)\mathbf{d}\left(f,1\right)\right|^2}{\mathbf{a}^H(f)\mathbf{a}(f)}.$$

7.22 Show the following relationship between the WNG and the DF:

$$\frac{\mathcal{D}\left[\mathbf{h}_{1:N}(f)\right]}{\mathcal{W}\left[\mathbf{h}_{1:N}(f)\right]} = \frac{\mathbf{i}^T\left[\mathbf{T}_{1:N}^H(f)\mathbf{T}_{1:N}(f)\right]^{-1}\mathbf{i}}{\mathbf{i}^T\left[\mathbf{T}_{1:N}^H(f)\mathbf{T}_{1:N}(f)\right]^{-2}\mathbf{i}}.$$

7.23 Show that with the beamformer $\mathbf{h}_{1:N}(f)$, we can control the WNG and the DF by

$$M^2 \geq \mathcal{D}\left[\mathbf{h}_{1:1}(f)\right] \geq \mathcal{D}\left[\mathbf{h}_{1:2}(f)\right] \geq \cdots \geq \mathcal{D}\left[\mathbf{h}_{1:M}(f)\right]$$

and

$$M = \mathcal{W}\left[\mathbf{h}_{1:M}(f)\right] \geq \mathcal{W}\left[\mathbf{h}_{1:M-1}(f)\right] \geq \cdots \geq \mathcal{W}\left[\mathbf{h}_{1:1}(f)\right].$$

References

1 B. D. Van Veen and K. M. Buckley, "Beamforming: a versatile approach to spatial filtering," *IEEE Acoust., Speech, Signal Process. Mag.*, vol. 5, pp. 4–24, Apr. 1988.

2 D. H. Johnson and D. E. Dudgeon, *Array Signal Processing: Concepts and Techniques.* Signal Processing Series. Englewood Cliffs, NJ: Prentice-Hall, 1993.

3 R. A. Monzingo and T. W. Miller, *Introduction to Adaptive Arrays.* Raleigh, NC: SciTech, 2004.

4 J. P. Dmochowski and J. Benesty, "Microphone arrays: fundamental concepts," in *Speech Processing in Modern Communication–Challenges and Perspectives*, I. Cohen, J. Benesty, and S. Gannot, (eds). Berlin, Germany: Springer-Verlag, 2010.

5 H. L. Van Trees, *Optimum Array Processing: Part IV of Detection, Estimation, and Modulation Theory.* New York, NY: John Wiley & Sons, Inc., 2002.

6 R. Berkun, I. Cohen, and J. Benesty, "Combined beamformers for robust broadband regularized superdirective beamforming," *IEEE/ACM Trans. Audio, Speech, Language Process.*, vol. 23, pp. 877–886, May 2015.

7 G. A. F. Seber, *A Matrix Handbook for Statisticians.* Hoboken, NJ: John Wiley & Sons, Inc., 2008.

8 C. Pan, J. Chen, and J. Benesty, "Performance study of the MVDR beamformer as a function of the source incidence angle," *IEEE/ACM Trans. Audio, Speech, Language Process.*, vol. 22, pp. 67–79, Jan. 2014.

9 J. M. Kates and M. R. Weiss, "A comparison of hearing-aid array-processing techniques," *J. Acoust. Soc. Am.*, vol. 99, pp. 3138–3148, May 1996.

10 A. I. Uzkov, "An approach to the problem of optimum directive antenna design," *Comptes Rendus (Doklady) de l'Academie des Sciences de l'URSS*, vol. LIII, no. 1, pp. 35–38, 1946.

11 H. Cox, R. M. Zeskind, and T. Kooij, "Practical supergain," *IEEE Trans. Acoust., Speech, Signal Process.*, vol. ASSP-34, pp. 393–398, June 1986.

12 H. Cox, R. M. Zeskind, and M. M. Owen, "Robust adaptive beamforming," *IEEE Trans. Acoust., Speech, Signal Process.*, vol. ASSP-35, pp. 1365–1376, Oct. 1987.

13 K. B. Petersen and M. S. Pedersen, *The Matrix Cookbook*. http://matrixcookbook.com, 2012.

14 C. Li, J. Benesty, G. Huang, and J. Chen, "Subspace superdirective beamformers based on joint diagonalization," in *Proc. IEEE ICASSP*, 2016, pp. 400–404.

15 J. N. Franklin, *Matrix Theory*. Englewood Cliffs, NJ: Prentice-Hall, 1968.

8

Adaptive Beamforming

In the previous chapter, we developed fixed beamformers, which do not depend on the statistics of the array data. This was possible because a model for the noise field was used. These beamformers are easy to implement and can work pretty well in different scenarios. However, in very challenging environments with multipath propagation, the performance of these algorithms, in terms of noise reduction, may be limited. Therefore, it is necessary to develop optimal linear filters that take into consideration the statistics of the incoming data. The resulting beamformers are referred to as adaptive beamformers. This is what we describe in this chapter. We will see that many useful adaptive beamformers can be derived, and each one of them can be expressed in different, but equivalent ways, depending on what statistics need to be estimated.

8.1 Signal Model, Problem Formulation, and Array Model

We consider a ULA consisting of M omnidirectional sensors and the signal model of the previous chapter:

$$\begin{aligned}\mathbf{y}(f) &= \mathbf{x}(f) + \mathbf{v}(f) \\ &= \mathbf{d}\left(f, \cos\theta_{\mathrm{d}}\right) X(f) + \mathbf{v}(f),\end{aligned} \tag{8.1}$$

where $\mathbf{y}(f)$ is observation signal vector (of length M), $\mathbf{d}\left(f, \cos\theta_{\mathrm{d}}\right)$ is the steering vector associated with the desired signal, $X(f)$, impinging on the array from the direction θ_{d}, and $\mathbf{v}(f)$ is the noise signal vector. The correlation matrix of $\mathbf{y}(f)$ is

$$\begin{aligned}\mathbf{\Phi}_{\mathbf{y}}(f) &= \mathbf{\Phi}_{\mathbf{x}}(f) + \mathbf{\Phi}_{\mathbf{v}}(f) \\ &= \phi_X(f)\mathbf{d}\left(f, \cos\theta_{\mathrm{d}}\right)\mathbf{d}^H\left(f, \cos\theta_{\mathrm{d}}\right) + \mathbf{\Phi}_{\mathbf{v}}(f),\end{aligned} \tag{8.2}$$

where $\mathbf{\Phi}_{\mathbf{x}}(f)$ and $\mathbf{\Phi}_{\mathbf{v}}(f)$ are the correlation matrices of $\mathbf{x}(f)$ and $\mathbf{v}(f)$, respectively, and $\phi_X(f)$ is the variance of $X(f)$.

Fundamentals of Signal Enhancement and Array Signal Processing, First Edition.
Jacob Benesty, Israel Cohen, and Jingdong Chen.

Companion website: www.wiley.com/go/benesty/arraysignalprocessing

Beamforming or linear filtering [1] consists of applying a complex-valued linear filter, $\mathbf{h}(f)$, of length M to $\mathbf{y}(f)$:

$$\begin{aligned} Z(f) &= \mathbf{h}^H(f)\mathbf{y}(f) \\ &= \mathbf{h}^H(f)\left[\mathbf{x}(f) + \mathbf{v}(f)\right] \\ &= X_{\mathrm{fd}}(f) + V_{\mathrm{rn}}(f), \end{aligned} \tag{8.3}$$

where $Z(f)$ is, in general, the estimate of the desired signal, and $X_{\mathrm{fd}}(f)$ and $V_{\mathrm{rn}}(f)$ are the filtered desired signal and residual noise, respectively. Assuming that $X_{\mathrm{fd}}(f)$ and $V_{\mathrm{rn}}(f)$ are uncorrelated, the variance of $Z(f)$ is

$$\begin{aligned} \phi_Z(f) &= \phi_{X_{\mathrm{fd}}}(f) + \phi_{V_{\mathrm{rn}}}(f) \\ &= \phi_X(f)\left|\mathbf{h}^H(f)\mathbf{d}\left(f, \cos\theta_{\mathrm{d}}\right)\right|^2 + \mathbf{h}^H(f)\mathbf{\Phi}_{\mathbf{v}}(f)\mathbf{h}(f). \end{aligned} \tag{8.4}$$

Our objective in this chapter is to design and describe beamformers that depend on the statistics of the signals as well as the knowledge of the direction of the desired signal. These so-called adaptive beamformers can usually adapt fairly quickly to changes in the environments in which they operate. They do not rely on some model of the noise field, as the fixed beamformers described in Chapter 7 do.

8.2 Performance Measures

In this section, we briefly recall both the main narrowband and broadband performance measures. The narrowband and broadband input SNRs are, respectively,

$$\mathrm{iSNR}(f) = \frac{\phi_X(f)}{\phi_{V_1}(f)} \tag{8.5}$$

and

$$\mathrm{iSNR} = \frac{\int_f \phi_X(f)df}{\int_f \phi_{V_1}(f)df}, \tag{8.6}$$

where $\phi_{V_1}(f)$ is the variance of $V_1(f)$, which is the first element of the vector $\mathbf{v}(f)$.

From (8.4), we deduce the narrowband output SNR:

$$\begin{aligned} \mathrm{oSNR}\left[\mathbf{h}(f)\right] &= \frac{\phi_{X_{\mathrm{fd}}}(f)}{\phi_{V_{\mathrm{rn}}}(f)} \\ &= \frac{\phi_X(f)\left|\mathbf{h}^H(f)\mathbf{d}\left(f, \cos\theta_{\mathrm{d}}\right)\right|^2}{\mathbf{h}^H(f)\mathbf{\Phi}_{\mathbf{v}}(f)\mathbf{h}(f)} \end{aligned} \tag{8.7}$$

and the broadband output SNR:

$$\mathrm{oSNR}(\mathbf{h}) = \frac{\int_f \phi_X(f) \left|\mathbf{h}^H(f)\mathbf{d}\left(f, \cos\theta_{\mathrm{d}}\right)\right|^2 df}{\int_f \mathbf{h}^H(f)\mathbf{\Phi}_{\mathbf{v}}(f)\mathbf{h}(f)df}. \tag{8.8}$$

It follows from the definitions of the input and output SNRs that the narrowband and broadband array gains are, respectively,

$$\mathcal{G}\left[\mathbf{h}(f)\right] = \frac{\mathrm{oSNR}\left[\mathbf{h}(f)\right]}{\mathrm{iSNR}(f)}, \tag{8.9}$$

$$\mathcal{G}(\mathbf{h}) = \frac{\mathrm{oSNR}(\mathbf{h})}{\mathrm{iSNR}}. \tag{8.10}$$

Adaptive beamformers should be designed in such a way that $\mathcal{G}\left[\mathbf{h}(f)\right] > 1$ and $\mathcal{G}(\mathbf{h}) > 1$.

Other useful definitions to quantify noise reduction are the narrowband noise reduction factor:

$$\xi_{\mathrm{n}}\left[\mathbf{h}(f)\right] = \frac{\phi_{V_1}(f)}{\mathbf{h}^H(f)\mathbf{\Phi}_{\mathbf{v}}(f)\mathbf{h}(f)} \tag{8.11}$$

and the broadband noise reduction factor:

$$\xi_{\mathrm{n}}(\mathbf{h}) = \frac{\int_f \phi_{V_1}(f)df}{\int_f \mathbf{h}^H(f)\mathbf{\Phi}_{\mathbf{v}}(f)\mathbf{h}(f)df}. \tag{8.12}$$

In the distortionless case; in other words,

$$\mathbf{h}^H(f)\mathbf{d}\left(f, \cos\theta_{\mathrm{d}}\right) = 1, \tag{8.13}$$

the noise reduction factor coincides with the array gain for both the narrowband and broadband measures.

In order to quantify distortion of the desired signal due to the beamforming operation, we define the narrowband desired signal reduction factor:

$$\xi_{\mathrm{d}}\left[\mathbf{h}(f)\right] = \frac{1}{\left|\mathbf{h}^H(f)\mathbf{d}\left(f, \cos\theta_{\mathrm{d}}\right)\right|^2} \tag{8.14}$$

and the broadband desired signal reduction factor:

$$\xi_{\mathrm{d}}(\mathbf{h}) = \frac{\int_f \phi_X(f)df}{\int_f \phi_X(f)\left|\mathbf{h}^H(f)\mathbf{d}\left(f, \cos\theta_{\mathrm{d}}\right)\right|^2 df}. \tag{8.15}$$

In the distortionless case, we have $\xi_{\mathrm{d}} = 1$, but when distortion occurs, we have $\xi_{\mathrm{d}} > 1$.

An alternative measure to the desired signal reduction factor is the desired signal distortion index. We have the following definitions:

- the narrowband desired signal distortion index

$$\upsilon_{\mathrm{d}}\left[\mathbf{h}(f)\right] = \left|\mathbf{h}^H(f)\mathbf{d}\left(f, \cos\theta_{\mathrm{d}}\right) - 1\right|^2 \tag{8.16}$$

- and the broadband desired signal distortion index

$$\upsilon_{\mathrm{d}}\left(\mathbf{h}\right) = \frac{\int_f \phi_X(f)\left|\mathbf{h}^H(f)\mathbf{d}\left(f, \cos\theta_{\mathrm{d}}\right) - 1\right|^2 df}{\int_f \phi_X(f)df}. \tag{8.17}$$

The error signal between the estimated and desired signals at the frequency f is given by

$$\begin{aligned}\mathcal{E}\left(f\right) &= Z(f) - X(f) \\ &= X_{\mathrm{fd}}(f) + V_{\mathrm{rn}}(f) - X(f) \\ &= \mathcal{E}_{\mathrm{d}}\left(f\right) + \mathcal{E}_{\mathrm{n}}\left(f\right),\end{aligned} \tag{8.18}$$

where

$$\mathcal{E}_{\mathrm{d}}\left(f\right) = \left[\mathbf{h}^H(f)\mathbf{d}\left(f, \cos\theta_{\mathrm{d}}\right) - 1\right] X(f) \tag{8.19}$$

is the desired signal distortion due to the beamformer and

$$\mathcal{E}_{\mathrm{n}}\left(f\right) = \mathbf{h}^H(f)\mathbf{v}(f) \tag{8.20}$$

represents the residual noise. Assuming that $\mathcal{E}_{\mathrm{d}}\left(f\right)$ and $\mathcal{E}_{\mathrm{n}}\left(f\right)$ are incoherent, the narrowband MSE can be expressed as

$$\begin{aligned}J\left[\mathbf{h}(f)\right] &= E\left[\left|\mathcal{E}\left(f\right)\right|^2\right] \\ &= E\left[\left|\mathcal{E}_{\mathrm{d}}\left(f\right)\right|^2\right] + E\left[\left|\mathcal{E}_{\mathrm{n}}\left(f\right)\right|^2\right] \\ &= J_{\mathrm{d}}\left[\mathbf{h}(f)\right] + J_{\mathrm{n}}\left[\mathbf{h}(f)\right] \\ &= \phi_X(f) + \mathbf{h}^H(f)\mathbf{\Phi}_{\mathbf{y}}(f)\mathbf{h}(f) - \phi_X(f)\mathbf{h}^H(f)\mathbf{d}\left(f, \cos\theta_{\mathrm{d}}\right) \\ &\quad - \phi_X(f)\mathbf{d}^H\left(f, \cos\theta_{\mathrm{d}}\right)\mathbf{h}(f),\end{aligned} \tag{8.21}$$

where

$$\begin{aligned}J_{\mathrm{d}}\left[\mathbf{h}(f)\right] &= \phi_X(f)\left|\mathbf{h}^H(f)\mathbf{d}\left(f, \cos\theta_{\mathrm{d}}\right) - 1\right|^2 \\ &= \phi_X(f)\upsilon_{\mathrm{d}}\left[\mathbf{h}(f)\right]\end{aligned} \tag{8.22}$$

and

$$J_n\left[\mathbf{h}(f)\right] = \mathbf{h}^H(f)\mathbf{\Phi}_\mathbf{v}(f)\mathbf{h}(f) \tag{8.23}$$
$$= \frac{\phi_{V_1}(f)}{\xi_n\left[\mathbf{h}(f)\right]}.$$

We have the following classical relationships:

$$\frac{J_d\left[\mathbf{h}(f)\right]}{J_n\left[\mathbf{h}(f)\right]} = \text{iSNR}(f) \times \xi_n\left[\mathbf{h}(f)\right] \times \upsilon_d\left[\mathbf{h}(f)\right] \tag{8.24}$$
$$= \text{oSNR}\left[\mathbf{h}(f)\right] \times \xi_d\left[\mathbf{h}(f)\right] \times \upsilon_d\left[\mathbf{h}(f)\right].$$

8.3 Adaptive Beamformers

In this section, we show how to design different kinds of adaptive beamformers. For each one of them, we give several equivalent formulations, depending on what (second-order) statistics we want or need to estimate.

8.3.1 Wiener

The Wiener beamformer is found by minimizing $J\left[\mathbf{h}(f)\right]$, the narrowband MSE from (8.21). We can easily obtain

$$\mathbf{h}_W\left(f, \cos\theta_d\right) = \phi_X(f)\mathbf{\Phi}_\mathbf{y}^{-1}(f)\mathbf{d}\left(f, \cos\theta_d\right). \tag{8.25}$$

In this filter, we need to estimate $\phi_X(f)$ and $\mathbf{\Phi}_\mathbf{y}(f)$. The latter quantity is easy to estimate from the observations, but the former is not. Let

$$\mathbf{\Gamma}_\mathbf{y}(f) = \frac{\mathbf{\Phi}_\mathbf{y}(f)}{\phi_{Y_1}(f)} \tag{8.26}$$

be the pseudo-coherence matrix of the observations, where $\phi_{Y_1}(f)$ is the variance of $Y_1(f)$. We can rewrite (8.25) as

$$\mathbf{h}_W\left(f, \cos\theta_d\right) = \frac{\text{iSNR}(f)}{1 + \text{iSNR}(f)}\mathbf{\Gamma}_\mathbf{y}^{-1}(f)\mathbf{d}\left(f, \cos\theta_d\right) \tag{8.27}$$
$$= H_W(f)\mathbf{\Gamma}_\mathbf{y}^{-1}(f)\mathbf{d}\left(f, \cos\theta_d\right),$$

where

$$H_W(f) = \frac{\text{iSNR}(f)}{1 + \text{iSNR}(f)} \tag{8.28}$$

is the single-channel Wiener gain (see Chapter 3). Now, instead of estimating $\phi_X(f)$ as in (8.25), we need to estimate the narrowband input SNR, iSNR(f) or, equivalently, $H_W(f)$.

The Wiener filter can also be expressed as a function of the statistics of the observation and noise signals:

$$\mathbf{h}_{\mathrm{W}}\left(f, \cos\theta_{\mathrm{d}}\right) = \left[\mathbf{I}_M - \mathbf{\Phi}_{\mathbf{y}}^{-1}(f)\mathbf{\Phi}_{\mathbf{v}}(f)\right]\mathbf{i}_{\mathrm{i}}, \tag{8.29}$$

where $\mathbf{i}_{\mathrm{i}}$ is the first column of $\mathbf{I}_M$.

Determining the inverse of $\mathbf{\Phi}_{\mathbf{y}}(f)$ from (8.2) with the Woodbury identity, we get

$$\mathbf{\Phi}_{\mathbf{y}}^{-1}(f) = \mathbf{\Phi}_{\mathbf{v}}^{-1}(f) - \frac{\mathbf{\Phi}_{\mathbf{v}}^{-1}(f)\mathbf{d}\left(f, \cos\theta_{\mathrm{d}}\right)\mathbf{d}^H\left(f, \cos\theta_{\mathrm{d}}\right)\mathbf{\Phi}_{\mathbf{v}}^{-1}(f)}{\phi_X^{-1}(f) + \mathbf{d}^H\left(f, \cos\theta_{\mathrm{d}}\right)\mathbf{\Phi}_{\mathbf{v}}^{-1}(f)\mathbf{d}\left(f, \cos\theta_{\mathrm{d}}\right)}. \tag{8.30}$$

Substituting (8.30) into (8.25) gives

$$\begin{aligned}\mathbf{h}_{\mathrm{W}}\left(f, \cos\theta_{\mathrm{d}}\right) &= \frac{\phi_X(f)\mathbf{\Phi}_{\mathbf{v}}^{-1}(f)\mathbf{d}\left(f, \cos\theta_{\mathrm{d}}\right)}{1 + \phi_X(f)\mathbf{d}^H\left(f, \cos\theta_{\mathrm{d}}\right)\mathbf{\Phi}_{\mathbf{v}}^{-1}(f)\mathbf{d}\left(f, \cos\theta_{\mathrm{d}}\right)} \\ &= \frac{\mathbf{\Phi}_{\mathbf{v}}^{-1}(f)\mathbf{\Phi}_{\mathbf{y}}(f) - \mathbf{I}_M}{1 - M + \mathrm{tr}\left[\mathbf{\Phi}_{\mathbf{v}}^{-1}(f)\mathbf{\Phi}_{\mathbf{y}}(f)\right]}\mathbf{i}_{\mathrm{i}}.\end{aligned} \tag{8.31}$$

In the second equation of (8.31), $\mathbf{h}_{\mathrm{W}}\left(f, \cos\theta_{\mathrm{d}}\right)$ depends on the statistics of the observation and noise signals and the matrix $\mathbf{\Phi}_{\mathbf{v}}(f)$ is inverted, while in the formulaton given in (8.29), $\mathbf{h}_{\mathrm{W}}\left(f, \cos\theta_{\mathrm{d}}\right)$ depends on the same statistics but the matrix $\mathbf{\Phi}_{\mathbf{y}}(f)$ is inverted.

We already know from the previous chapters that the Wiener beamformer maximizes the narrowband array gain but does not necessarily maximize the broadband array gain, but it definitely makes the latter greater than 1. Distortion is obviously expected and is increased when the input SNR is decreased. However, if we increase the number of sensors, we decrease distortion.

Example 8.3.1 Consider a ULA of M sensors, as shown in Figure 7.1. Suppose that a desired signal impinges on the ULA from the direction θ_{d}, and that two statistically independent interferences impinge on the ULA from directions θ_1 and θ_2. Assume that the desired signal is a harmonic pulse of T samples:

$$x(t) = \begin{cases} A\sin\left(2\pi f_0 t + \phi\right), & 0 \le t \le T-1 \\ 0, & t < 0,\ t \ge T \end{cases},$$

with fixed amplitude A and frequency f_0, and random phase ϕ, uniformly distributed on the interval from 0 to 2π. Assume that the interferences $u_1(t)$ and $u_2(t)$ are IID white Gaussian noise, $u_1(t), u_2(t) \sim \mathcal{N}\left(0, \sigma_u^2\right)$, uncorrelated with $x(t)$. In addition, the sensors contain thermal white Gaussian noise, $w_m(t) \sim \mathcal{N}\left(0, \sigma_w^2\right)$, the signals of which are mutually uncorrelated. The noisy received signals are given by $y_m(t) = x_m(t) + v_m(t)$, $m = 1, \ldots, M$, where $v_m(t) = u_m(t) + w_m(t)$, $m = 1, \ldots, M$ are the interference-plus-noise signals.

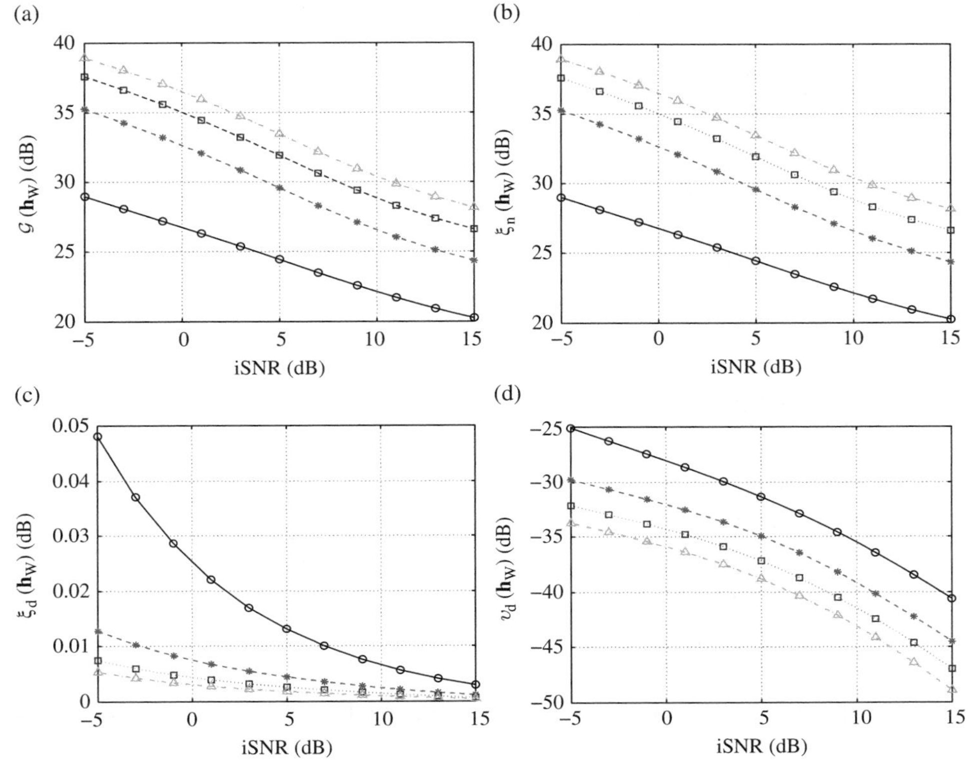

Figure 8.1 (a) The broadband gain in SNR, (b) the broadband noise reduction factor, (c) the broadband desired signal reduction factor, and (d) the broadband desired signal distortion index of the Wiener beamformer as a function of the broadband input SNR, for different numbers of sensors, M: $M = 10$ (solid line with circles), $M = 20$ (dashed line with asterisks), $M = 30$ (dotted line with squares), and $M = 40$ (dash-dot line with triangles).

Following Example 5.4.1, we choose a sampling interval T_s that satisfies $T_s = \delta/c$. The variance of $X(f)$ is given by

$$\phi_X(f) = \frac{A^2}{4} D_T^2 \left[\pi \left(f + f_0\right)\right] + \frac{A^2}{4} D_T^2 \left[\pi \left(f - f_0\right)\right],$$

where

$$D_T(x) = \frac{\sin (Tx)}{\sin (x)}.$$

The correlation matrices of $\mathbf{x}(f)$ and $\mathbf{v}(f)$ are given by

$$\begin{aligned}
\mathbf{\Phi}_\mathbf{x}(f) &= \phi_X(f) \mathbf{d}\left(f, \cos\theta_d\right) \mathbf{d}^H\left(f, \cos\theta_d\right), \\
\mathbf{\Phi}_\mathbf{v}(f) &= T\sigma_u^2 \mathbf{d}\left(f, \cos\theta_1\right) \mathbf{d}^H\left(f, \cos\theta_1\right) \\
&\quad + T\sigma_u^2 \mathbf{d}\left(f, \cos\theta_2\right) \mathbf{d}^H\left(f, \cos\theta_2\right) + T\sigma_w^2 \mathbf{I}_M.
\end{aligned}$$

The narrowband and broadband input SNRs are, respectively,

$$\begin{aligned}\text{iSNR}(f) &= \frac{\phi_X(f)}{\phi_{V_1}(f)} \\ &= \frac{A^2}{4T\left(2\sigma_u^2+\sigma_w^2\right)} D_T^2\left[\pi\left(f+f_0\right)\right] \\ &\quad + \frac{A^2}{4T\left(2\sigma_u^2+\sigma_w^2\right)} D_T^2\left[\pi\left(f-f_0\right)\right]\end{aligned}$$

and

$$\begin{aligned}\text{iSNR} &= \frac{\int_f \phi_X(f)df}{\int_f \phi_{V_1}(f)df} \\ &= \frac{\sum_t E\left[\left|x_1(t)\right|^2\right]}{\sum_t E\left[\left|v_1(t)\right|^2\right]} \\ &= \frac{A^2}{2\left(2\sigma_u^2+\sigma_w^2\right)},\end{aligned}$$

where we have used Parseval's identity. The Wiener beamformer, $\mathbf{h}_\text{W}\left(f,\cos\theta_\text{d}\right)$, is obtained from (8.29).

To demonstrate the performance of the Wiener beamformer, we choose $A = 0.5$, $f_0 = 0.1c/\delta$, $T = 500$, $\theta_\text{d} = 70°$, $\theta_1 = 30°$, $\theta_2 = 50°$, and $\sigma_w^2 = 0.01\sigma_u^2$. Figure 8.1 shows plots of the broadband gain in SNR, $\mathcal{G}\left(\mathbf{h}_\text{W}\right)$, the broadband noise reduction factor, $\xi_\text{n}\left(\mathbf{h}_\text{W}\right)$, the broadband desired signal reduction factor, $\xi_\text{d}\left(\mathbf{h}_\text{W}\right)$, and the broadband desired signal distortion index, $\upsilon_\text{d}\left(\mathbf{h}_\text{W}\right)$, as a function of the broadband input SNR, for different numbers of sensors, M. For a given broadband input SNR, as the number of sensors increases, the broadband gain in SNR and the broadband noise reduction factor increase, while the broadband desired signal reduction factor and the broadband desired signal distortion index decrease.

Figure 8.2 shows beampatterns, $\left|\mathcal{B}\left[\mathbf{h}_\text{W}\left(f,\cos\theta_\text{d}\right),\cos\theta\right]\right|$, for $f = f_0$ and different numbers of sensors, M. The main beam is in the direction of the desired signal, θ_d, and there are nulls in the directions of the interferences, θ_1 and θ_2. As the number of sensors increases, the width of the main beam decreases, and the nulls in the directions of the interferences become deeper. ■

8.3.2 MVDR

The MVDR beamformer proposed by Capon [2, 3] is obtained by minimizing the narrowband MSE of the residual noise, $J_\text{r}\left[\mathbf{h}(f)\right]$, subject to the distortionless constraint:

$$\min_{\mathbf{h}(f)} \mathbf{h}^H(f)\mathbf{\Phi}_\mathbf{v}(f)\mathbf{h}(f) \quad \text{subject to} \quad \mathbf{h}^H(f)\mathbf{d}\left(f,\cos\theta_\text{d}\right) = 1. \tag{8.32}$$

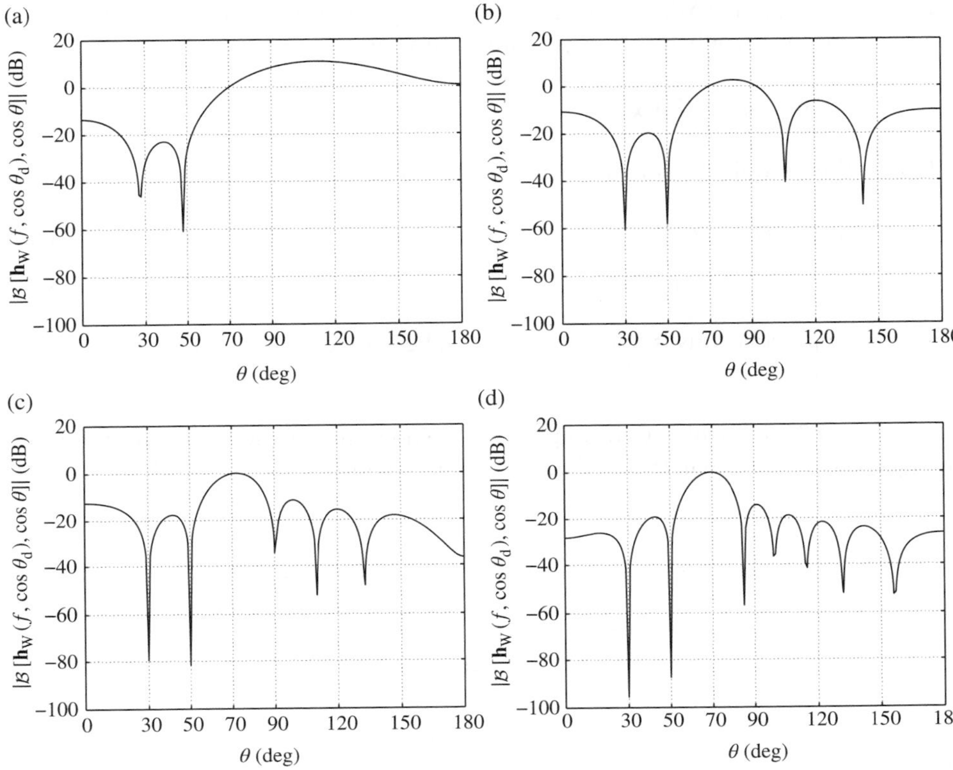

Figure 8.2 Beampatterns of the Wiener beamformer for $f = f_0$ and different numbers of sensors, M: (a) $M = 10$, (b) $M = 20$, (c) $M = 30$, and (d) $M = 40$.

The solution to this optimization problem is

$$\mathbf{h}_{\text{MVDR}}\left(f, \cos\theta_d\right) = \frac{\mathbf{\Phi}_{\mathbf{v}}^{-1}(f)\mathbf{d}\left(f, \cos\theta_d\right)}{\mathbf{d}^H\left(f, \cos\theta_d\right)\mathbf{\Phi}_{\mathbf{v}}^{-1}(f)\mathbf{d}\left(f, \cos\theta_d\right)}, \tag{8.33}$$

which depends on the statistics of the noise only.

Using the Woodbury identity again, it is easy to show that the MVDR beamformer is also

$$\begin{aligned}\mathbf{h}_{\text{MVDR}}\left(f, \cos\theta_d\right) &= \frac{\mathbf{\Phi}_{\mathbf{y}}^{-1}(f)\mathbf{d}\left(f, \cos\theta_d\right)}{\mathbf{d}^H\left(f, \cos\theta_d\right)\mathbf{\Phi}_{\mathbf{y}}^{-1}(f)\mathbf{d}\left(f, \cos\theta_d\right)} \\ &= \frac{\mathbf{\Gamma}_{\mathbf{y}}^{-1}(f)\mathbf{d}\left(f, \cos\theta_d\right)}{\mathbf{d}^H\left(f, \cos\theta_d\right)\mathbf{\Gamma}_{\mathbf{y}}^{-1}(f)\mathbf{d}\left(f, \cos\theta_d\right)}.\end{aligned} \tag{8.34}$$

This formulation is important and practical, since it depends on the statistics of the observations only, and these are easy to estimate in practice.

It is clear that the MVDR beamformer maximizes the narrowband array gain. However, for the broadband array gain, we always have

$$1 \leq \mathcal{G}\left(\mathbf{h}_{\text{MVDR}}\right) \leq \mathcal{G}\left(\mathbf{h}_{\text{W}}\right). \tag{8.35}$$

From a theoretical point of view, it is also clear that we have

$$v_{\text{d}}\left[\mathbf{h}_{\text{MVDR}}\left(f, \cos\theta_{\text{d}}\right)\right] = 0, \tag{8.36}$$

$$v_{\text{d}}\left(\mathbf{h}_{\text{MVDR}}\right) = 0. \tag{8.37}$$

However, in practice, this is not true in general because of reverberation, which is not taken into account in our model.

Example 8.3.2 Returning to Example 8.3.1, we now employ the MVDR beamformer, $\mathbf{h}_{\text{MVDR}}\left(f, \cos\theta_{\text{d}}\right)$, given in (8.34). Figure 8.3 shows plots of the broadband gain in SNR,

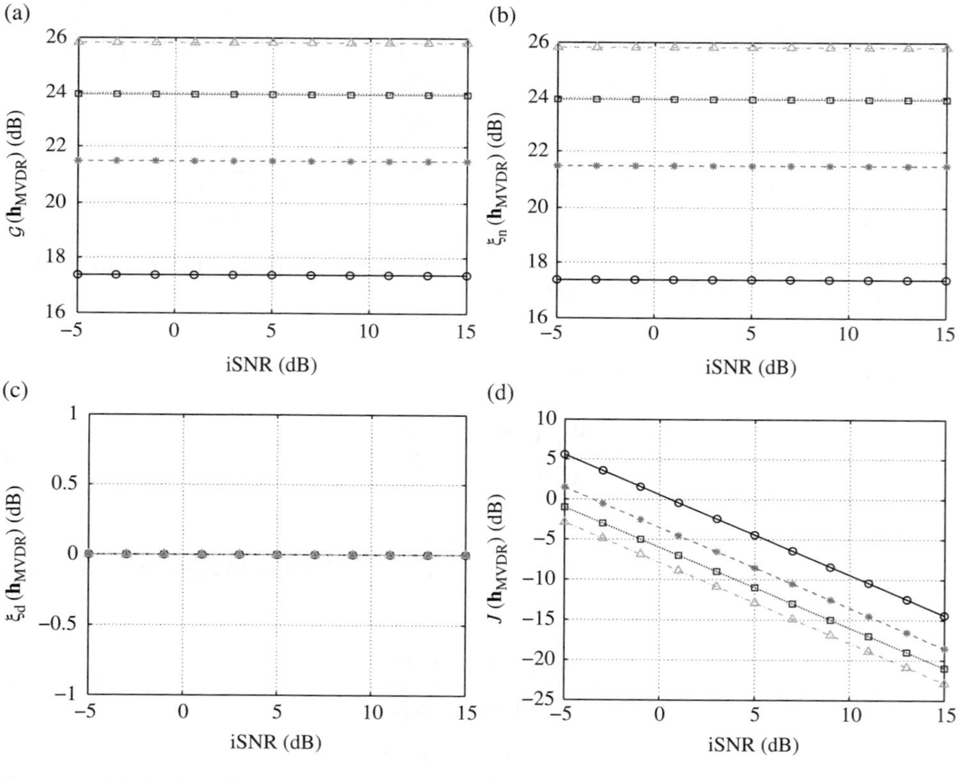

Figure 8.3 (a) The broadband gain in SNR, (b) the broadband noise reduction factor, (c) the broadband desired signal reduction factor, and (d) the broadband MSE of the MVDR beamformer as a function of the broadband input SNR, for different numbers of sensors, M: $M = 10$ (solid line with circles), $M = 20$ (dashed line with asterisks), $M = 30$ (dotted line with squares), and $M = 40$ (dash-dot line with triangles).

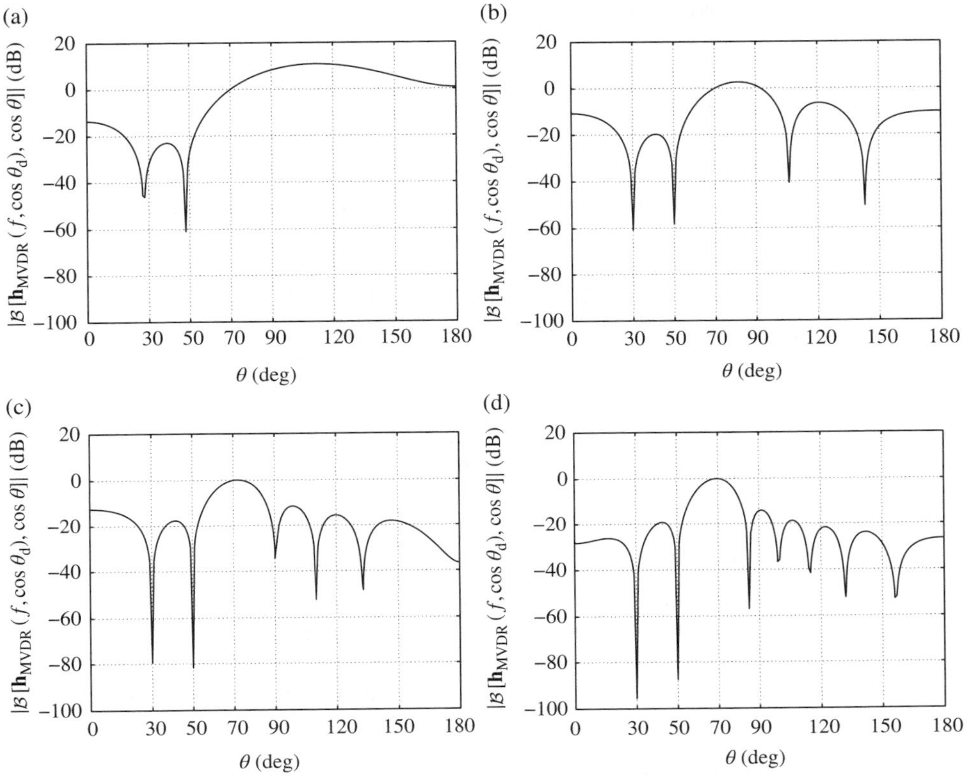

Figure 8.4 Beampatterns of the MVDR beamformer for $f = f_0$ and different numbers of sensors, M: (a) $M = 10$, (b) $M = 20$, (c) $M = 30$, and (d) $M = 40$.

$\mathcal{G}\left(\mathbf{h}_{\mathrm{MVDR}}\right)$, the broadband noise reduction factor, $\xi_{\mathrm{n}}\left(\mathbf{h}_{\mathrm{MVDR}}\right)$, the broadband desired signal reduction factor, $\xi_{\mathrm{d}}\left(\mathbf{h}_{\mathrm{MVDR}}\right)$, and the broadband MSE, $J\left(\mathbf{h}_{\mathrm{MVDR}}\right)$, as a function of the broadband input SNR, for different numbers of sensors, M. For a given broadband input SNR, as the number of sensors increases, the broadband gain in SNR and the broadband noise reduction factor increase, while the broadband MSE decreases.

Figure 8.4 shows beampatterns, $\left|\mathcal{B}\left[\mathbf{h}_{\mathrm{MVDR}}\left(f, \cos\theta_{\mathrm{d}}\right), \cos\theta\right]\right|$, for $f = f_0$ and different numbers of sensors, M. The main beam is in the direction of the desired signal, θ_{d}, and there are nulls in the directions of the interferences, θ_1 and θ_2. In particular, $\left|\mathcal{B}\left[\mathbf{h}_{\mathrm{MVDR}}\left(f, \cos\theta_{\mathrm{d}}\right), \cos\theta\right]\right|$ is 1 for $\theta = \theta_{\mathrm{d}}$. As the number of sensors increases, the width of the main beam decreases, and the nulls in the directions of the interferences become deeper. ■

8.3.3 Tradeoff

In order to make a better compromise between noise reduction and desired signal distortion, we can minimize the narrowband desired signal distortion index with the constraint that the narrowband noise reduction factor is equal to a positive value that is greater than 1:

$$\min_{\mathbf{h}(f)} J_{\mathrm{d}}\left[\mathbf{h}(f)\right] \quad \text{subject to} \quad J_{\mathrm{n}}\left[\mathbf{h}(f)\right] = \aleph \phi_{V_1}(f), \tag{8.38}$$

where $0 < \aleph < 1$ to ensure that we get some noise reduction. By using a Lagrange multiplier, $\mu > 0$, to adjoin the constraint to the cost function, we get the tradeoff beamformer:

$$\begin{aligned}\mathbf{h}_{\mathrm{T},\mu}\left(f, \cos\theta_{\mathrm{d}}\right) &= \phi_X(f)\left[\boldsymbol{\Phi}_{\mathbf{x}}(f) + \mu\boldsymbol{\Phi}_{\mathbf{v}}(f)\right]^{-1}\mathbf{d}\left(f, \cos\theta_{\mathrm{d}}\right) \\ &= \frac{\phi_X(f)\boldsymbol{\Phi}_{\mathbf{v}}^{-1}(f)\mathbf{d}\left(f, \cos\theta_{\mathrm{d}}\right)}{\mu + \phi_X(f)\mathbf{d}^H\left(f, \cos\theta_{\mathrm{d}}\right)\boldsymbol{\Phi}_{\mathbf{v}}^{-1}(f)\mathbf{d}\left(f, \cos\theta_{\mathrm{d}}\right)} \\ &= \frac{\boldsymbol{\Phi}_{\mathbf{v}}^{-1}(f)\boldsymbol{\Phi}_{\mathbf{y}}(f) - \mathbf{I}_M}{\mu - M + \mathrm{tr}\left[\boldsymbol{\Phi}_{\mathbf{v}}^{-1}(f)\boldsymbol{\Phi}_{\mathbf{y}}(f)\right]}\mathbf{i}_{\mathrm{i}}.\end{aligned} \tag{8.39}$$

We can see that for

- $\mu = 1$, $\mathbf{h}_{\mathrm{T},1}\left(f, \cos\theta_{\mathrm{d}}\right) = \mathbf{h}_{\mathrm{W}}\left(f, \cos\theta_{\mathrm{d}}\right)$, which is the Wiener beamformer
- $\mu = 0$, $\mathbf{h}_{\mathrm{T},0}\left(f, \cos\theta_{\mathrm{d}}\right) = \mathbf{h}_{\mathrm{MVDR}}\left(f, \cos\theta_{\mathrm{d}}\right)$, which is the MVDR beamformer
- $\mu > 1$, the result is a beamformer with low residual noise at the expense of high desired signal distortion (as compared to Wiener)
- $\mu < 1$, the result is a beamformer with high residual noise and low desired signal distortion (as compared to Wiener).

A more useful way to express the tradeoff beamformer is

$$\begin{aligned}\mathbf{h}_{\mathrm{T},\mu}\left(f, \cos\theta_{\mathrm{d}}\right) &= \left[(1-\mu)\mathbf{d}\left(f, \cos\theta_{\mathrm{d}}\right)\mathbf{d}^H\left(f, \cos\theta_{\mathrm{d}}\right) + \mu\frac{1+\mathrm{iSNR}(f)}{\mathrm{iSNR}(f)}\boldsymbol{\Gamma}_{\mathbf{y}}(f)\right]^{-1} \\ &\quad \times \mathbf{d}\left(f, \cos\theta_{\mathrm{d}}\right),\end{aligned} \tag{8.40}$$

or, equivalently, with the help of the Woodbury identity:

$$\begin{aligned}\mathbf{h}_{\mathrm{T},\mu}\left(f, \cos\theta_{\mathrm{d}}\right) &= \frac{\mathrm{iSNR}(f)}{1+\mathrm{iSNR}(f)} \\ &\quad \times \frac{\boldsymbol{\Gamma}_{\mathbf{y}}^{-1}(f)\mathbf{d}\left(f, \cos\theta_{\mathrm{d}}\right)}{\mu + (1-\mu)\frac{\mathrm{iSNR}(f)}{1+\mathrm{iSNR}(f)}\mathbf{d}^H\left(f, \cos\theta_{\mathrm{d}}\right)\boldsymbol{\Gamma}_{\mathbf{y}}^{-1}(f)\mathbf{d}\left(f, \cos\theta_{\mathrm{d}}\right)}.\end{aligned} \tag{8.41}$$

The previous expression depends only on the estimation of the statistics of the observations as well as the estimation of the narrowband input SNR. We can simplify (8.41) by writing it as a function of $H_{\mathrm{W}}(f)$:

$$\mathbf{h}_{\mathrm{T},\mu}\left(f, \cos\theta_{\mathrm{d}}\right) = \frac{H_{\mathrm{W}}(f)\boldsymbol{\Gamma}_{\mathbf{y}}^{-1}(f)\mathbf{d}\left(f, \cos\theta_{\mathrm{d}}\right)}{\mu + (1-\mu)H_{\mathrm{W}}(f)\mathbf{d}^H\left(f, \cos\theta_{\mathrm{d}}\right)\boldsymbol{\Gamma}_{\mathbf{y}}^{-1}(f)\mathbf{d}\left(f, \cos\theta_{\mathrm{d}}\right)}. \tag{8.42}$$

Obviously, the tradeoff beamformer also maximizes the narrowband array gain, $\forall \mu \geq 0$. However, for the broadband array gain, we always have for $\mu \geq 1$,

$$1 \leq \mathcal{G}\left(\mathbf{h}_{\mathrm{MVDR}}\right) \leq \mathcal{G}\left(\mathbf{h}_{\mathrm{W}}\right) \leq \mathcal{G}\left(\mathbf{h}_{\mathrm{T},\mu}\right), \tag{8.43}$$

and for $0 \leq \mu \leq 1$,

$$1 \leq \mathcal{G}\left(\mathbf{h}_{\mathrm{MVDR}}\right) \leq \mathcal{G}\left(\mathbf{h}_{\mathrm{T},\mu}\right) \leq \mathcal{G}\left(\mathbf{h}_{\mathrm{W}}\right). \tag{8.44}$$

Distortion of the desired signal, on the other hand, depends quite a lot on the values of μ. However, the closer the value of μ is to 0, the less distorted the desired signal.

Example 8.3.3 Returning to Example 8.3.1, we now employ the tradeoff beamformer, $\mathbf{h}_{\mathrm{T},\mu}\left(f, \cos\theta_{\mathrm{d}}\right)$, given in (8.39). Figure 8.5 shows plots of the broadband gain in SNR, $\mathcal{G}\left(\mathbf{h}_{\mathrm{T},\mu}\right)$, the broadband noise reduction factor, $\xi_{\mathrm{n}}\left(\mathbf{h}_{\mathrm{T},\mu}\right)$, the broadband desired signal

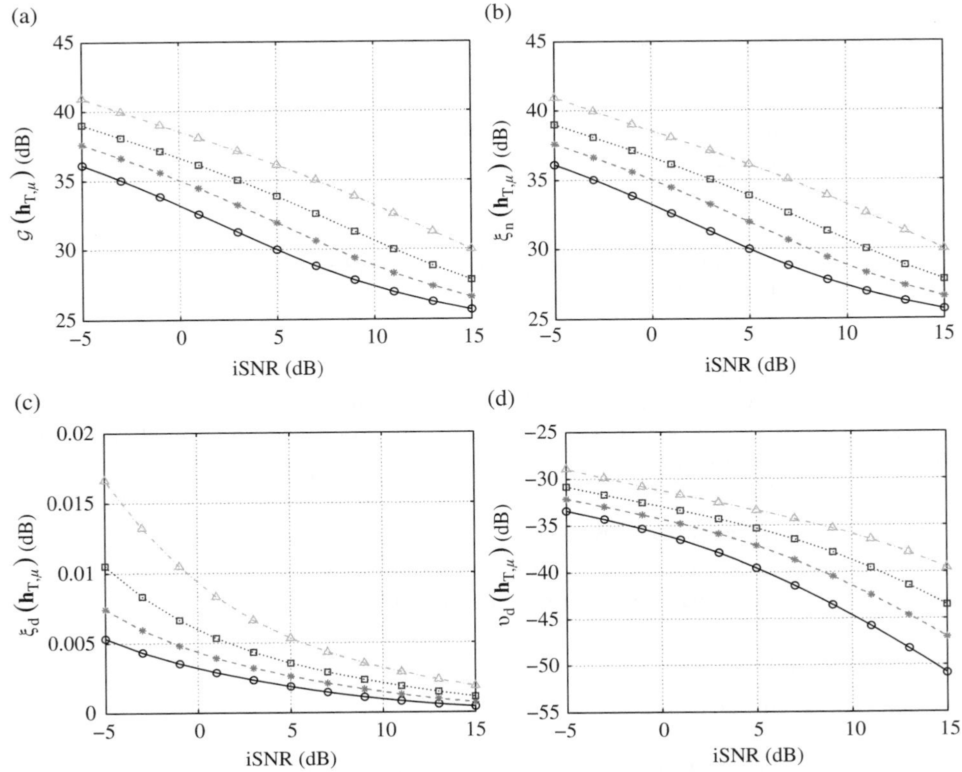

Figure 8.5 (a) The broadband gain in SNR, (b) the broadband noise reduction factor, (c) the broadband desired signal reduction factor, and (d) the broadband desired signal distortion index of the tradeoff beamformer as a function of the broadband input SNR, for $M = 30$ and several values of μ: $\mu = 0.5$ (solid line with circles), $\mu = 1$ (dashed line with asterisks), $\mu = 2$ (dotted line with squares), and $\mu = 5$ (dash-dot line with triangles).

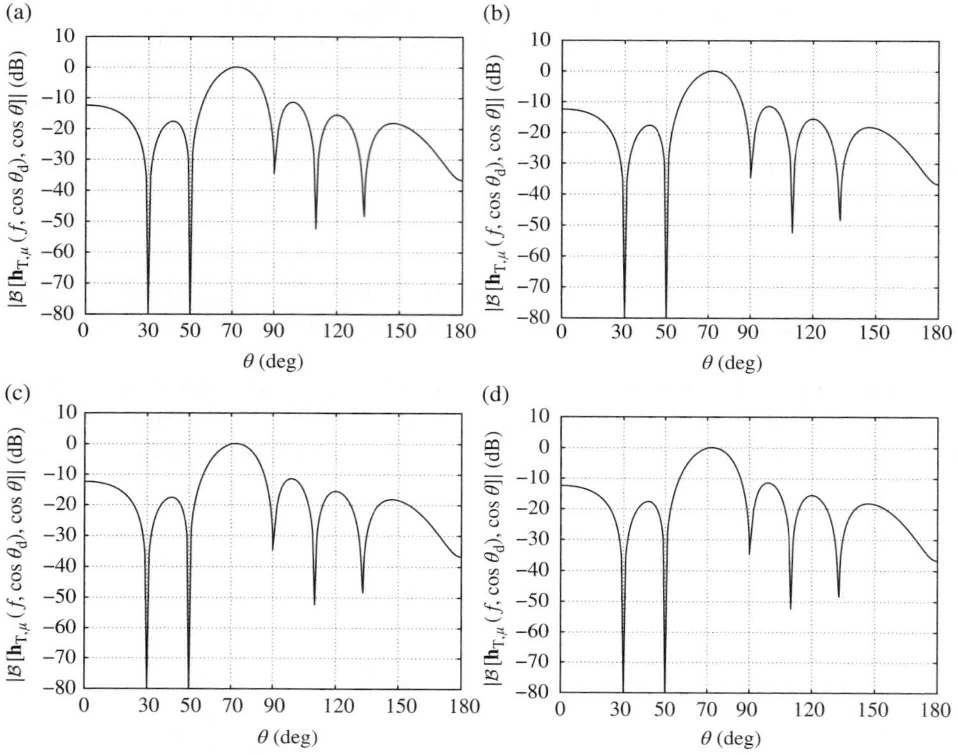

Figure 8.6 Beampatterns of the tradeoff beamformer for $f = f_0$, $M = 30$ and several values of μ: (a) $\mu = 0.5$, (b) $\mu = 1$, (c) $\mu = 2$, and (d) $\mu = 5$.

reduction factor, $\xi_{\mathrm{d}}\left(\mathbf{h}_{\mathrm{T},\mu}\right)$, and the broadband desired signal distortion index, $\upsilon_{\mathrm{d}}\left(\mathbf{h}_{\mathrm{T},\mu}\right)$, as a function of the broadband input SNR, for $M = 30$ and several values of μ. For a given broadband input SNR, the higher the value of μ, the higher are the broadband gain in SNR and the broadband noise reduction factor, but at the expense of a higher broadband desired signal reduction factor and a higher broadband desired signal distortion index.

Figure 8.6 shows beampatterns, $\left|\mathcal{B}\left[\mathbf{h}_{\mathrm{T},\mu}\left(f, \cos\theta_{\mathrm{d}}\right), \cos\theta\right]\right|$, for $f = f_0$, $M = 30$ and several values of μ. The main beam is in the direction of the desired signal, θ_{d}, and there are nulls in the directions of the interferences, θ_1 and θ_2. ■

8.3.4 Maximum Array Gain

We can express the narrowband array gain as

$$\mathcal{G}\left[\mathbf{h}(f)\right] = \frac{\phi_{V_1}(f)\mathbf{h}^H(f)\mathbf{d}\left(f, \cos\theta_{\mathrm{d}}\right)\mathbf{d}^H\left(f, \cos\theta_{\mathrm{d}}\right)\mathbf{h}(f)}{\mathbf{h}^H(f)\mathbf{\Phi}_{\mathbf{v}}(f)\mathbf{h}(f)}. \tag{8.45}$$

The maximum array gain beamformer, $\mathbf{h}_{\max}\left(f, \cos\theta_{\mathrm{d}}\right)$, is obtained by maximizing the array gain as given above. In (8.45), we recognize the generalized Rayleigh quotient [4].

It is well known that this quotient is maximized with the maximum eigenvector of the matrix $\phi_{V_1}(f)\mathbf{\Phi}_{\mathbf{v}}^{-1}(f)\mathbf{d}\left(f, \cos\theta_{\mathrm{d}}\right)\mathbf{d}^H\left(f, \cos\theta_{\mathrm{d}}\right)$. Let us denote by $\lambda_{\max}\left(f, \cos\theta_{\mathrm{d}}\right)$ the maximum eigenvalue corresponding to this maximum eigenvector. Since the rank of the matrix mentioned is equal to 1, we have

$$\begin{aligned}\lambda_{\max}\left(f, \cos\theta_{\mathrm{d}}\right) &= \mathrm{tr}\left[\phi_{V_1}(f)\mathbf{\Phi}_{\mathbf{v}}^{-1}(f)\mathbf{d}\left(f, \cos\theta_{\mathrm{d}}\right)\mathbf{d}^H\left(f, \cos\theta_{\mathrm{d}}\right)\right] \\ &= \phi_{V_1}(f)\mathbf{d}^H\left(f, \cos\theta_{\mathrm{d}}\right)\mathbf{\Phi}_{\mathbf{v}}^{-1}(f)\mathbf{d}\left(f, \cos\theta_{\mathrm{d}}\right).\end{aligned} \tag{8.46}$$

As a result,

$$\begin{aligned}\mathcal{G}\left[\mathbf{h}_{\max}\left(f, \cos\theta_{\mathrm{d}}\right)\right] &= \lambda_{\max}\left(f, \cos\theta_{\mathrm{d}}\right) \\ &= \mathcal{G}_{\max}\left(f, \cos\theta_{\mathrm{d}}\right),\end{aligned} \tag{8.47}$$

which corresponds to the maximum possible narrowband array gain.

Obviously, we also have

$$\mathbf{h}_{\max}\left(f, \cos\theta_{\mathrm{d}}\right) = \varsigma(f)\mathbf{\Gamma}_{\mathbf{y}}^{-1}(f)\mathbf{d}\left(f, \cos\theta_{\mathrm{d}}\right), \tag{8.48}$$

where $\varsigma(f) \neq 0$ is an arbitrary frequency-dependent complex number. We can observe that all beamformers derived so far are equivalent up to a scaling factor.

8.3.5 LCMV

Assume that we have N interferences, with $N < M$, impinging on the array from the directions $\theta_1 \neq \theta_2 \neq \cdots \neq \theta_N \neq \theta_{\mathrm{d}}$. We would like to place nulls in the directions θ_n, $n = 1, 2, \ldots, N$, with a beamformer $\mathbf{h}(f)$, and, meanwhile, recover the desired source coming from the direction θ_{d}. Combining all these constraints together, we get the constraint equation:

$$\mathbf{C}^H\left(f, \theta_{\mathrm{d}}, \theta_{1:N}\right)\mathbf{h}(f) = \mathbf{i}_{\mathrm{c}}, \tag{8.49}$$

where

$$\mathbf{C}\left(f, \theta_{\mathrm{d}}, \theta_{1:N}\right) = \begin{bmatrix} \mathbf{d}\left(f, \theta_{\mathrm{d}}\right) & \mathbf{d}\left(f, \theta_1\right) & \cdots & \mathbf{d}\left(f, \theta_N\right) \end{bmatrix} \tag{8.50}$$

is the constraint matrix of size $M \times (N + 1)$ the $N + 1$ columns of which are linearly independent and

$$\mathbf{i}_{\mathrm{c}} = \begin{bmatrix} 1 & 0 & \cdots & 0 \end{bmatrix}^T \tag{8.51}$$

is a vector of length $N + 1$.

The most convenient way to solve this problem is by minimizing the narrowband MSE of the residual noise, $J_{\mathrm{r}}\left[\mathbf{h}(f)\right]$, subject to (8.49):

$$\min_{\mathbf{h}(f)} \mathbf{h}^H(f)\mathbf{\Phi}_{\mathbf{v}}(f)\mathbf{h}(f) \quad \text{subject to} \quad \mathbf{C}^H\left(f, \theta_{\mathrm{d}}, \theta_{1:N}\right)\mathbf{h}(f) = \mathbf{i}_{\mathrm{c}}. \tag{8.52}$$

The solution to this optimization problem gives the well-known LCMV beamformer [5, 6]:

$$\mathbf{h}_{\mathrm{LCMV}}\left(f, \cos\theta_{\mathrm{d}}\right) = \mathbf{\Phi}_{\mathbf{v}}^{-1}(f)\mathbf{C}\left(f, \theta_{\mathrm{d}}, \theta_{1:N}\right) \times \left[\mathbf{C}^{H}\left(f, \theta_{\mathrm{d}}, \theta_{1:N}\right)\mathbf{\Phi}_{\mathbf{v}}^{-1}(f)\mathbf{C}\left(f, \theta_{\mathrm{d}}, \theta_{1:N}\right)\right]^{-1}\mathbf{i}_{\mathrm{c}}, \tag{8.53}$$

which depends on the statistics of the noise only.

It can be shown that a more useful formulation of the LCMV beamformer is

$$\mathbf{h}_{\mathrm{LCMV}}\left(f, \cos\theta_{\mathrm{d}}\right) = \mathbf{\Gamma}_{\mathbf{y}}^{-1}(f)\mathbf{C}\left(f, \theta_{\mathrm{d}}, \theta_{1:N}\right) \times \left[\mathbf{C}^{H}\left(f, \theta_{\mathrm{d}}, \theta_{1:N}\right)\mathbf{\Gamma}_{\mathbf{y}}^{-1}(f)\mathbf{C}\left(f, \theta_{\mathrm{d}}, \theta_{1:N}\right)\right]^{-1}\mathbf{i}_{\mathrm{c}}. \tag{8.54}$$

The last expression depends on the statistics of the observations only, and these should be easy to estimate.

Example 8.3.4 Returning to Example 8.3.1, we now employ the LCMV beamformer, $\mathbf{h}_{\mathrm{LCMV}}\left(f, \cos\theta_{\mathrm{d}}\right)$, given in (8.54). Figure 8.7 shows plots of the broadband gain in SNR,

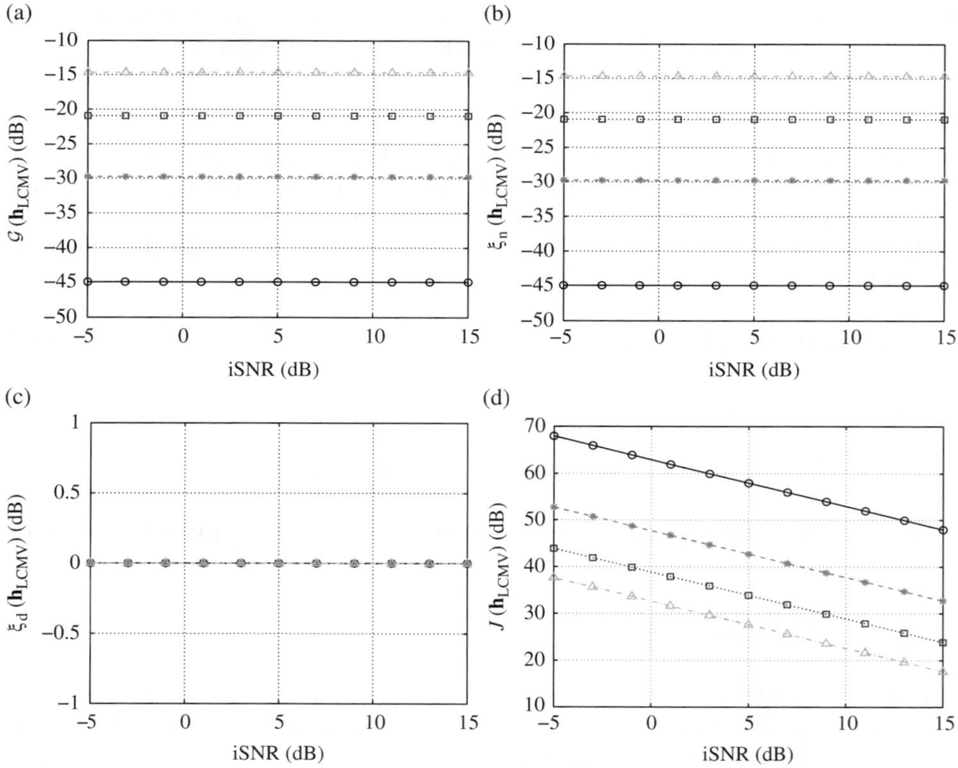

Figure 8.7 (a) The broadband gain in SNR, (b) the broadband noise reduction factor, (c) the broadband desired signal reduction factor, and (d) the broadband MSE of the LCMV beamformer as a function of the broadband input SNR, for different numbers of sensors, M: $M = 10$ (solid line with circles), $M = 20$ (dashed line with asterisks), $M = 30$ (dotted line with squares), and $M = 40$ (dash-dot line with triangles).

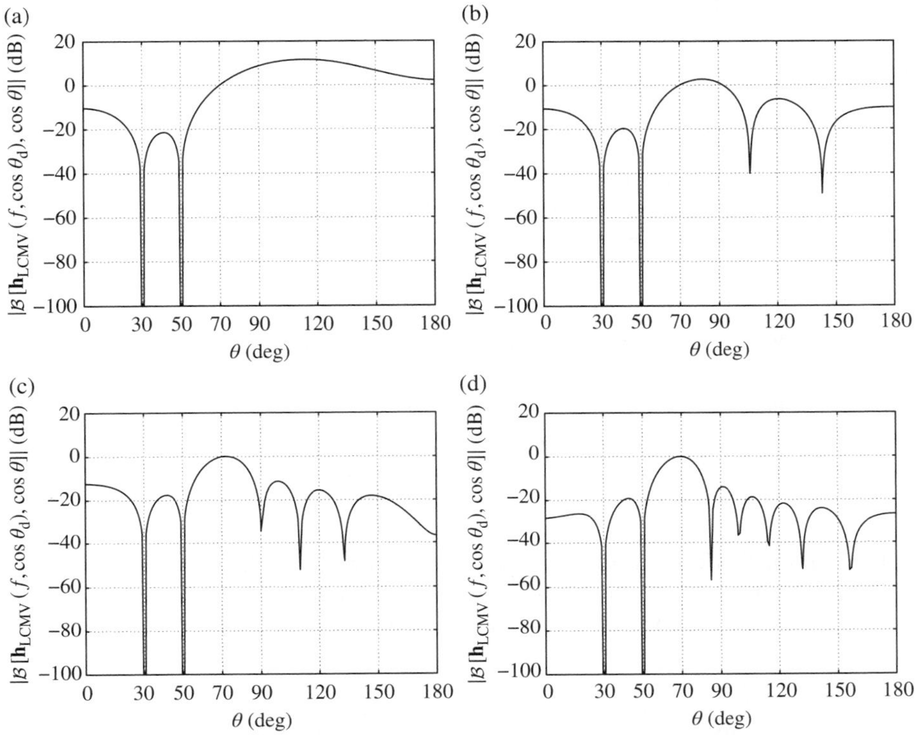

Figure 8.8 Beampatterns of the LCMV beamformer for $f = f_0$ and different numbers of sensors, M: (a) $M = 10$, (b) $M = 20$, (c) $M = 30$, and (d) $M = 40$.

$\mathcal{G}\left(\mathbf{h}_{\text{LCMV}}\right)$, the broadband noise reduction factor, $\xi_{\text{n}}\left(\mathbf{h}_{\text{LCMV}}\right)$, the broadband desired signal reduction factor, $\xi_{\text{d}}\left(\mathbf{h}_{\text{LCMV}}\right)$, and the broadband MSE, $J\left(\mathbf{h}_{\text{LCMV}}\right)$, as a function of the broadband input SNR, for different numbers of sensors, M. For a given broadband input SNR, as the number of sensors increases, the broadband gain in SNR and the broadband noise reduction factor increase, while the broadband MSE decreases.

Figure 8.8 shows beampatterns, $\left|\mathcal{B}\left[\mathbf{h}_{\text{LCMV}}\left(f, \cos\theta_{\text{d}}\right), \cos\theta\right]\right|$, for $f = f_0$ and different numbers of sensors, M. The main beam is in the direction of the desired signal, θ_{d}, and there are nulls in the directions of the interferences, θ_1 and θ_2. In particular, $\left|\mathcal{B}\left[\mathbf{h}_{\text{LCMV}}\left(f, \cos\theta_{\text{d}}\right), \cos\theta\right]\right|$ is 1 for $\theta = \theta_{\text{d}}$, and is identically zero for $\theta = \theta_1$ and $\theta = \theta_2$. As the number of sensors increases, the width of the main beam decreases. ■

In Table 8.1, we summarize all the optimal adaptive beamformers described in this section.

8.4 SNR Estimation

From Table 8.1, we can see that all the beamformers depend on the statistics of the observations, $\mathbf{\Gamma}_{\mathbf{y}}(f)$, while some of them depend also on the narrowband input SNR, iSNR(f), or, equivalently, on $H_{\text{W}}(f)$. In practice, while it is easy to get an estimate for

Table 8.1 Adaptive beamformers.

Beamformer	
Wiener	$\mathbf{h}_{\mathrm{W}}\left(f, \cos\theta_{\mathrm{d}}\right) = H_{\mathrm{W}}(f)\mathbf{\Gamma}_{\mathbf{y}}^{-1}(f)\mathbf{d}\left(f, \cos\theta_{\mathrm{d}}\right)$
MVDR	$\mathbf{h}_{\mathrm{MVDR}}\left(f, \cos\theta_{\mathrm{d}}\right) = \dfrac{\mathbf{\Gamma}_{\mathbf{y}}^{-1}(f)\mathbf{d}\left(f, \cos\theta_{\mathrm{d}}\right)}{\mathbf{d}^H\left(f, \cos\theta_{\mathrm{d}}\right)\mathbf{\Gamma}_{\mathbf{y}}^{-1}(f)\mathbf{d}\left(f, \cos\theta_{\mathrm{d}}\right)}$
Tradeoff	$\mathbf{h}_{\mathrm{T},\mu}\left(f, \cos\theta_{\mathrm{d}}\right) = \dfrac{H_{\mathrm{W}}(f)\mathbf{\Gamma}_{\mathbf{y}}^{-1}(f)\mathbf{d}\left(f, \cos\theta_{\mathrm{d}}\right)}{\mu + (1-\mu)H_{\mathrm{W}}(f)\mathbf{d}^H\left(f, \cos\theta_{\mathrm{d}}\right)\mathbf{\Gamma}_{\mathbf{y}}^{-1}(f)\mathbf{d}\left(f, \cos\theta_{\mathrm{d}}\right)}$
Maximum array gain	$\mathbf{h}_{\max}\left(f, \cos\theta_{\mathrm{d}}\right) = \varsigma(f)\mathbf{\Gamma}_{\mathbf{y}}^{-1}(f)\mathbf{d}\left(f, \cos\theta_{\mathrm{d}}\right), \ \varsigma(f) \neq 0$
LCMV	$\mathbf{h}_{\mathrm{LCMV}}\left(f, \cos\theta_{\mathrm{d}}\right) = \mathbf{\Gamma}_{\mathbf{y}}^{-1}(f)\mathbf{C}\left(f, \theta_{\mathrm{d}}, \theta_{1:N}\right) \times \left[\mathbf{C}^H\left(f, \theta_{\mathrm{d}}, \theta_{1:N}\right)\mathbf{\Gamma}_{\mathbf{y}}^{-1}(f)\mathbf{C}\left(f, \theta_{\mathrm{d}}, \theta_{1:N}\right)\right]^{-1}\mathbf{i}_{\mathrm{c}}$

$\mathbf{\Gamma}_{\mathbf{y}}(f)$, it is not for $\mathrm{iSNR}(f)$. In this section, we show one possible way to estimate this SNR. In fact, it is much more straightforward to estimate $H_{\mathrm{W}}(f)$ as explained below.

We can express the pseudo-coherence matrix of the observations as

$$\begin{aligned}\mathbf{\Gamma}_{\mathbf{y}}(f) &= \frac{\mathrm{iSNR}(f)}{1+\mathrm{iSNR}(f)}\mathbf{d}\left(f, \cos\theta_{\mathrm{d}}\right)\mathbf{d}^H\left(f, \cos\theta_{\mathrm{d}}\right) + \frac{1}{1+\mathrm{iSNR}(f)}\mathbf{\Gamma}_{\mathbf{v}}(f) \\ &= H_{\mathrm{W}}(f)\mathbf{d}\left(f, \cos\theta_{\mathrm{d}}\right)\mathbf{d}^H\left(f, \cos\theta_{\mathrm{d}}\right) + \left[1 - H_{\mathrm{W}}(f)\right]\mathbf{\Gamma}_{\mathbf{v}}(f),\end{aligned} \tag{8.55}$$

where

$$\mathbf{\Gamma}_{\mathbf{v}}(f) = \frac{\mathbf{\Phi}_{\mathbf{v}}(f)}{\phi_{V_1}(f)}. \tag{8.56}$$

Let us assume that we are in the presence of the spherically isotropic noise. In this case, $\mathbf{\Gamma}_{\mathbf{v}}(f)$ coincides with $\mathbf{\Gamma}_{0,\pi}(f)$, which is defined in Chapter 7. Since $\mathbf{y}(f)$ is observable, it is easy to estimate $\mathbf{\Gamma}_{\mathbf{y}}(f)$. We denote this estimate by $\widehat{\mathbf{\Gamma}}_{\mathbf{y}}(f)$. By following the approaches developed in [7–10], we can write the components of the matrix $\widehat{\mathbf{\Gamma}}_{\mathbf{y}}(f)$ as

$$\begin{aligned}\Re\left\{\left[\widehat{\mathbf{\Gamma}}_{\mathbf{y}}(f)\right]_{ij}\right\} &= \widehat{H}_{\mathrm{W}}(f)\Re\left[D_i\left(f, \cos\theta_{\mathrm{d}}\right)D_j^*\left(f, \cos\theta_{\mathrm{d}}\right)\right] \\ &\quad + \left[1 - \widehat{H}_{\mathrm{W}}(f)\right]\left[\mathbf{\Gamma}_{0,\pi}(f)\right]_{ij},\end{aligned} \tag{8.57}$$

$$\Im\left\{\left[\widehat{\mathbf{\Gamma}}_{\mathbf{y}}(f)\right]_{ij}\right\} = \widehat{H}_{\mathrm{W}}(f)\Im\left[D_i\left(f, \cos\theta_{\mathrm{d}}\right)D_j^*\left(f, \cos\theta_{\mathrm{d}}\right)\right], \tag{8.58}$$

for $i \neq j,\ i,j = 1, 2, \dots, M$, where $\Re[\cdot]$ and $\Im[\cdot]$ are the real part and imaginary part operators, respectively, $\widehat{H}_{\mathrm{W}}(f)$ is the estimate of $H_{\mathrm{W}}(f)$, and $D_m\left(f, \cos\theta_{\mathrm{d}}\right)$ is the mth element of $\mathbf{d}\left(f, \cos\theta_{\mathrm{d}}\right)$. We deduce from the previous expressions that

$$\widehat{H}_{\mathrm{W}}(f) = \frac{\Re\left\{\left[\widehat{\mathbf{\Gamma}}_{\mathbf{y}}(f)\right]_{ij}\right\} - \left[\mathbf{\Gamma}_{0,\pi}(f)\right]_{ij}}{\Re\left[D_i\left(f, \cos\theta_{\mathrm{d}}\right) D_j^*\left(f, \cos\theta_{\mathrm{d}}\right)\right] - \left[\mathbf{\Gamma}_{0,\pi}(f)\right]_{ij}},\ i \neq j, \tag{8.59}$$

$$\widehat{H}_{\mathrm{W}}(f) = \frac{\Im\left\{\left[\widehat{\mathbf{\Gamma}}_{\mathbf{y}}(f)\right]_{ij}\right\}}{\Im\left[D_i\left(f, \cos\theta_{\mathrm{d}}\right) D_j^*\left(f, \cos\theta_{\mathrm{d}}\right)\right]},\ i \neq j. \tag{8.60}$$

To get a much more reliable estimate, it is better to average (8.59) and (8.60) over all possible sensor combinations [7–10], resulting in the estimator:

$$\begin{aligned}\widehat{H}_{\mathrm{W}}(f) &\\ &= \frac{1}{M(M-1)} \sum_{i=1}^{M-1} \sum_{j=i+1}^{M} \frac{\Re\left\{\left[\widehat{\mathbf{\Gamma}}_{\mathbf{y}}(f)\right]_{ij}\right\} - \left[\mathbf{\Gamma}_{0,\pi}(f)\right]_{ij}}{\Re\left[D_i\left(f, \cos\theta_{\mathrm{d}}\right) D_j^*\left(f, \cos\theta_{\mathrm{d}}\right)\right] - \left[\mathbf{\Gamma}_{0,\pi}(f)\right]_{ij}} \\ &\quad + \frac{1}{M(M-1)} \sum_{i=1}^{M-1} \sum_{j=i+1}^{M} \frac{\Im\left\{\left[\widehat{\mathbf{\Gamma}}_{\mathbf{y}}(f)\right]_{ij}\right\}}{\Im\left[D_i\left(f, \cos\theta_{\mathrm{d}}\right) D_j^*\left(f, \cos\theta_{\mathrm{d}}\right)\right]}.\end{aligned} \tag{8.61}$$

Obviously, in practice, it is much better to estimate $H_{\mathrm{W}}(f)$ than iSNR(f) since the former is bounded ($0 \leq H_{\mathrm{W}}(f) \leq 1$) while the latter is not. If the estimate of the Wiener gain is greater than 1, then we should force it to 1, and if it is negative, we should put it to 0.

It is possible to estimate the single-channel Wiener gain directly from (8.55), in a much simpler way, by pre- and post-multiplying both sides of (8.55) by $\mathbf{d}^H\left(f, \cos\theta_{\mathrm{d}}\right)$ and $\mathbf{d}\left(f, \cos\theta_{\mathrm{d}}\right)$, respectively, and by replacing $\mathbf{\Gamma}_{\mathbf{y}}(f)$ and $\mathbf{\Gamma}_{\mathbf{v}}(f)$ with $\widehat{\mathbf{\Gamma}}_{\mathbf{y}}(f)$ and $\mathbf{\Gamma}_{0,\pi}(f)$, respectively. We easily obtain

$$\widehat{H}_{\mathrm{W}}(f) = \frac{\mathbf{d}^H\left(f, \cos\theta_{\mathrm{d}}\right)\left[\widehat{\mathbf{\Gamma}}_{\mathbf{y}}(f) - \mathbf{\Gamma}_{0,\pi}(f)\right]\mathbf{d}\left(f, \cos\theta_{\mathrm{d}}\right)}{M^2 - \mathbf{d}^H\left(f, \cos\theta_{\mathrm{d}}\right)\mathbf{\Gamma}_{0,\pi}(f)\mathbf{d}\left(f, \cos\theta_{\mathrm{d}}\right)}. \tag{8.62}$$

Example 8.4.1 Consider a ULA of M sensors, as shown in Figure 7.1. Suppose that a desired signal impinges on the ULA from the direction $\theta_\mathrm{d} = 70°$. Assume that the desired signal is a harmonic pulse of T samples:

$$x(t) = \begin{cases} A \sin\left(2\pi f_0 t + \phi\right), & 0 \leq t \leq T-1 \\ 0, & t < 0,\ t \geq T \end{cases},$$

with fixed amplitude A and frequency f_0, and random phase ϕ, uniformly distributed on the interval from 0 to 2π. Assume that the interference $u_m(t)$ is a diffuse noise uncorrelated with $x(t)$. In addition, the sensors contain thermal white Gaussian noise, $w_m(t) \sim \mathcal{N}\left(0, \sigma_w^2\right)$, the signals of which are mutually uncorrelated. The noisy received signals are given by $y_m(t) = x_m(t) + v_m(t)$, $m = 1, \ldots, M$, where $v_m(t) = u_m(t) + w_m(t)$, $m = 1, \ldots, M$ are the interference-plus-noise signals.

The pseudo-coherence matrix of the noise can be written as

$$\mathbf{\Gamma}_\mathbf{v}(f) = (1-\alpha)\mathbf{\Gamma}_{0,\pi}(f) + \alpha \mathbf{I}_M, \tag{8.63}$$

where α $(0 \leq \alpha \leq 1)$ is related to the ratio between the powers of the thermal and diffuse noises. Figures 8.9 and 8.10 show plots of the estimators $\widehat{H}_\mathrm{W}(f)$, given by (8.61) and (8.62), respectively, as a function of the narrowband input SNR for several values of α and different numbers of sensors, M. The theoretical plot is indicated by thick solid line, and the pseudo-coherence matrix of the observations $\mathbf{\Gamma}_\mathbf{y}(f)$ is obtained by

$$\begin{aligned}\mathbf{\Gamma}_\mathbf{y}(f) = & \frac{\mathrm{iSNR}(f)}{1+\mathrm{iSNR}(f)} \mathbf{d}\left(f, \cos\theta_\mathrm{d}\right) \mathbf{d}^H\left(f, \cos\theta_\mathrm{d}\right) \\ & + \frac{(1-\alpha)\mathbf{\Gamma}_{0,\pi}(f) + \alpha\mathbf{I}_M}{1+\mathrm{iSNR}(f)}.\end{aligned} \tag{8.64}$$

As the number of sensors is larger and as the value of α is smaller, the estimators (8.61) and (8.62) are closer to the theoretical values. Generally, the estimator (8.62) produces better results than (8.61).

Now we set α to 0.001, and estimate the pseudo-coherence matrix of the observations using K random snapshots:

$$\widehat{\mathbf{\Phi}}_\mathbf{y}(f) = \frac{1}{K}\sum_{k=1}^{K} \mathbf{y}_k(f)\mathbf{y}_k^H(f), \tag{8.65}$$

$$\widehat{\mathbf{\Gamma}}_\mathbf{y}(f) = \frac{\widehat{\mathbf{\Phi}}_\mathbf{y}(f)}{\widehat{\phi}_{Y_1}(f)}, \tag{8.66}$$

where $\mathbf{y}_k(f)$ is a random snapshot of $\mathbf{y}(f)$. Figures 8.11 and 8.12 show plots of the estimators $\widehat{H}_\mathrm{W}(f)$, given by (8.61) and (8.62), respectively, as a function of the narrowband input SNR for different numbers of snapshots, K, and different numbers of sensors, M. The theoretical plot is indicated by thick solid line. As the number of sensors is larger or as the number of snapshots is larger, the estimators (8.61) and (8.62) are closer to the theoretical values. Generally, the estimator (8.62) produces better results than (8.61). ■

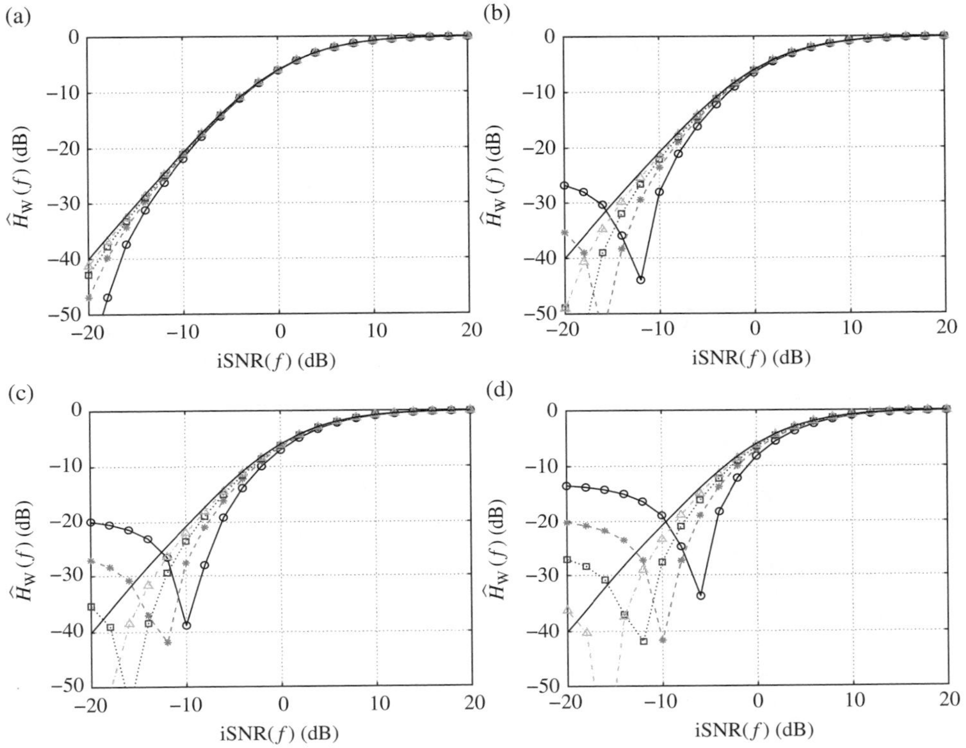

Figure 8.9 The estimator $\widehat{H}_W(f)$, given by (8.61), as a function of the narrowband input SNR for several values of α and different numbers of sensors, M: (a) $\alpha = 0.001$, (b) $\alpha = 0.005$, (c) $\alpha = 0.01$, and (d) $\alpha = 0.02$. Each figure shows the theoretical plot (thick solid line), and estimates for $M = 2$ (solid line with circles), $M = 5$ (dashed line with asterisks), $M = 10$ (dotted line with squares), and $M = 20$ (dash-dot line with triangles).

8.5 DOA Estimation

In practice, the direction-of-arrival (DOA) of the desired signal, θ_d, may not always be known. Therefore, it is of great interest to be able to estimate this angle. Obviously, the literature on this subject is extremely rich [11] and many different approaches can be derived, depending on several factors. In this section, we propose a method that naturally flows from the perspective developed throughout this text.

Let us assume that an estimate of the pseudo-coherence matrix of the observations, $\mathbf{\Gamma}_\mathbf{y}(f)$, is

$$\begin{aligned}\widehat{\mathbf{\Gamma}}_\mathbf{y}(f) &= \frac{\widehat{\mathbf{\Phi}}_\mathbf{y}(f)}{\widehat{\phi}_{Y_1}(f)} \\ &= \frac{\text{iSNR}(f)}{1+\text{iSNR}(f)}\mathbf{d}\left(f, \cos\theta_d\right)\mathbf{d}^H\left(f, \cos\theta_d\right) + \frac{1}{1+\text{iSNR}(f)}\mathbf{\Gamma}_{0,\pi}(f),\end{aligned} \quad (8.67)$$

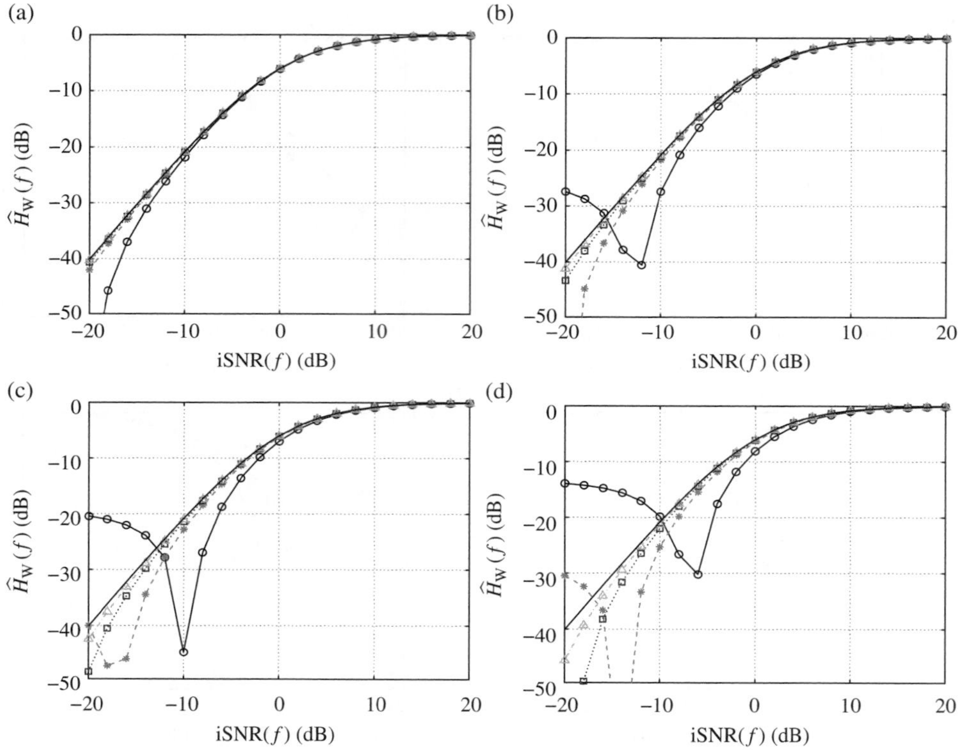

Figure 8.10 The estimator $\widehat{H}_W(f)$, given by (8.62), as a function of the narrowband input SNR for several values of α and different numbers of sensors, M: (a) $\alpha = 0.001$, (b) $\alpha = 0.005$, (c) $\alpha = 0.01$, and (d) $\alpha = 0.02$. Each figure shows the theoretical plot (thick solid line), and estimates for $M = 2$ (solid line with circles), $M = 5$ (dashed line with asterisks), $M = 10$ (dotted line with squares), and $M = 20$ (dash-dot line with triangles).

where $\widehat{\mathbf{\Phi}}_\mathbf{y}(f)$ and $\widehat{\phi}_{Y_1}(f)$ are estimates of $\mathbf{\Phi}_\mathbf{y}(f)$ and $\phi_{Y_1}(f)$, respectively. In (8.67), we explicitly assume that $\mathbf{\Gamma}_{0,\pi}(f)$ is an estimate of $\mathbf{\Gamma}_\mathbf{v}(f)$.

The pseudo-coherence matrix corresponding to a source signal coming from the direction θ may be written as

$$\mathbf{\Gamma}_\mathbf{x}\left(f, \cos\theta\right) = \mathbf{d}\left(f, \cos\theta\right) \mathbf{d}^H\left(f, \cos\theta\right), \tag{8.68}$$

which is a rank-1 matrix.

Using the joint diagonalization technique [4], the two matrices $\widehat{\mathbf{\Gamma}}_\mathbf{y}(f)$ and $\mathbf{\Gamma}_{0,\pi}(f)$ can be decomposed as

$$\mathbf{T}^H(f)\widehat{\mathbf{\Gamma}}_\mathbf{y}(f)\mathbf{T}(f) = \mathbf{\Lambda}(f), \tag{8.69}$$

$$\mathbf{T}^H(f)\mathbf{\Gamma}_{0,\pi}(f)\mathbf{T}(f) = \mathbf{I}_M, \tag{8.70}$$

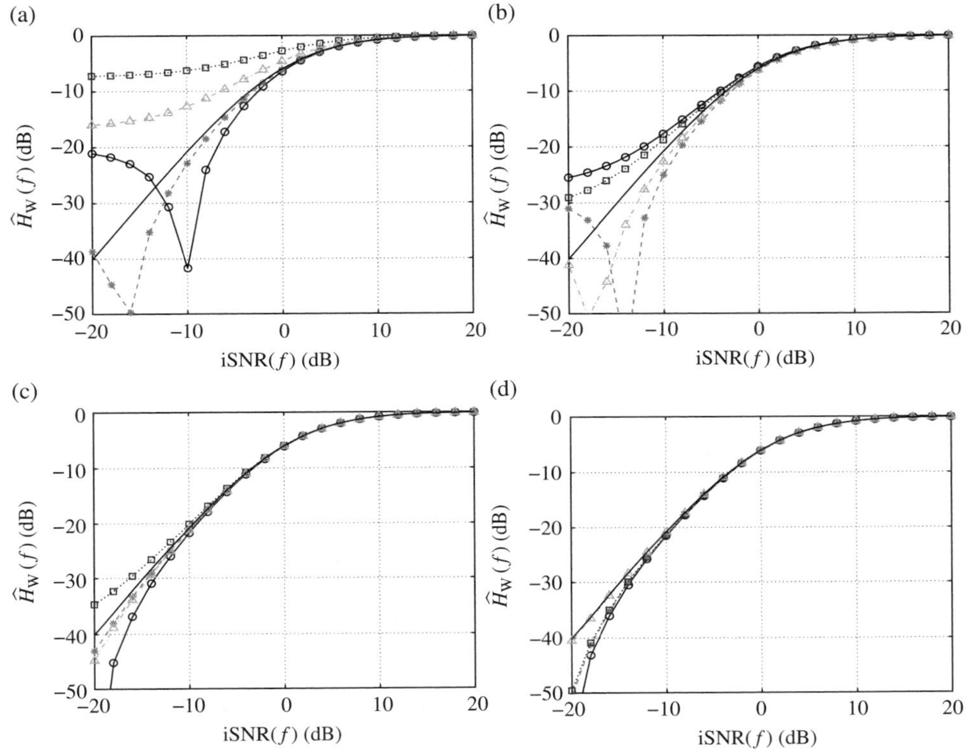

Figure 8.11 The estimator $\widehat{H}_W(f)$, given by (8.61), as a function of the narrowband input SNR for different numbers of snapshots, K, and different numbers of sensors, M: (a) $K = 10^3$, (b) $K = 10^4$, (c) $K = 10^5$, and (d) $K = 10^6$. Each figure shows the theoretical plot (thick solid line), and estimates for $M = 2$ (solid line with circles), $M = 5$ (dashed line with asterisks), $M = 10$ (dotted line with squares), and $M = 20$ (dash-dot line with triangles).

where

$$\mathbf{T}(f) = \begin{bmatrix} \mathbf{t}_1(f) & \mathbf{t}_2(f) & \cdots & \mathbf{t}_M(f) \end{bmatrix} \tag{8.71}$$

is a full-rank square matrix and

$$\mathbf{\Lambda}(f) = \text{diag}\left[\lambda_1(f), \lambda_2(f), \dots, \lambda_M(f)\right] \tag{8.72}$$

is a diagonal matrix with $\lambda_1(f) \geq \lambda_2(f) \geq \cdots \geq \lambda_M(f) > 0$. For $\theta = \theta_\mathrm{d}$, we have

$$\begin{aligned}\mathbf{T}^H(f)\widehat{\mathbf{\Gamma}}_\mathbf{y}(f)\mathbf{T}(f) = {} & \frac{\text{iSNR}(f)}{1+\text{iSNR}(f)}\mathbf{T}^H(f)\mathbf{\Gamma}_\mathbf{x}\left(f, \cos\theta_\mathrm{d}\right)\mathbf{T}(f) \\ & + \frac{1}{1+\text{iSNR}(f)}\mathbf{I}_M,\end{aligned} \tag{8.73}$$

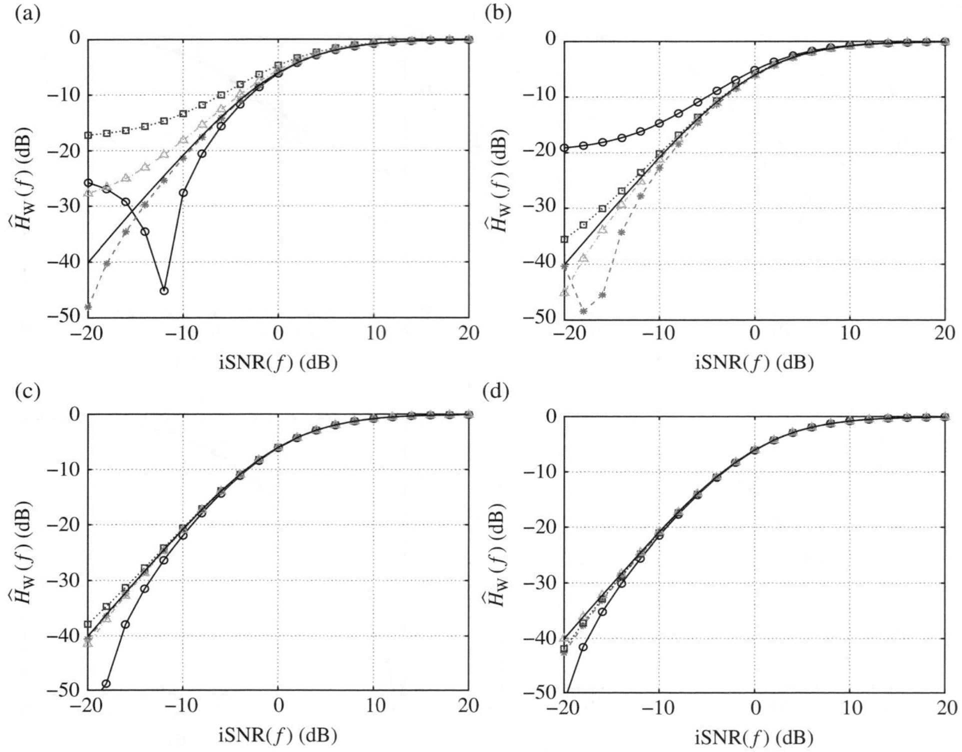

Figure 8.12 The estimator $\widehat{H}_W(f)$, given by (8.62), as a function of the narrowband input SNR for different numbers of snapshots, K, and different numbers of sensors, M: (a) $K = 10^3$, (b) $K = 10^4$, (c) $K = 10^5$, and (d) $K = 10^6$. Each figure shows the theoretical plot (thick solid line), and estimates for $M = 2$ (solid line with circles), $M = 5$ (dashed line with asterisks), $M = 10$ (dotted line with squares), and $M = 20$ (dash-dot line with triangles).

where $\mathbf{T}^H(f)\mathbf{\Gamma}_\mathbf{x}\left(f, \cos\theta_\mathrm{d}\right)\mathbf{T}(f)$ is a diagonal matrix whose first diagonal element is

$$\lambda_{\mathbf{x},1}(f) = \frac{\left[1 + \mathrm{iSNR}(f)\right]\lambda_1(f) - 1}{\mathrm{iSNR}(f)} > 0 \tag{8.74}$$

and the other diagonal elements are zero. However, for $\theta \neq \theta_\mathrm{d}$, the matrix $\mathbf{T}^H(f)\mathbf{\Gamma}_\mathbf{x}\left(f, \cos\theta\right)\mathbf{T}(f)$ is no longer diagonal but its rank is still equal to 1. Consequently, we can take advantage of this property to estimate the DOA. Indeed, it is easy to observe that

$$\left|\mathbf{t}_i^H(f)\mathbf{d}\left(f, \cos\theta_\mathrm{d}\right)\right|^2 = 0,\ i = 2, 3, \ldots, M \tag{8.75}$$

but

$$\left|\mathbf{t}_i^H(f)\mathbf{d}\left(f, \cos\theta\right)\right|^2 > 0,\ i = 2, 3, \ldots, M,\ \theta \neq \theta_\mathrm{d}. \tag{8.76}$$

As a result, the previous equations may be combined and used as a good criterion for the estimation of $\theta_{\rm d}$:

$$\widehat{\theta}_{\rm d} = \arg\min_{\theta} \sum_{i=2}^{M} \left| \mathbf{t}_i^H(f)\mathbf{d}\left(f, \cos\theta\right) \right|^2. \tag{8.77}$$

The last expression corresponds to a narrowband estimation of the desired angle. A more reliable estimator can be obtained by integrating the criterion over a range of frequencies:

$$\widehat{\theta}_{\rm d} = \arg\min_{\theta} \int_{f_1}^{f_2} \sum_{i=2}^{M} \left| \mathbf{t}_i^H(f)\mathbf{d}\left(f, \cos\theta\right) \right|^2 df, \tag{8.78}$$

which corresponds to a broadband estimation of $\theta_{\rm d}$.

This approach is a generalization of the well-known MUSIC (multiple signal classification) algorithm [12, 13], which was originally developed for spatially white noise, to the spherically isotropic noise field. Obviously, this approach works for more than one desired angle. But the number of desired angles must be smaller than M.

A byproduct of this method is that the input SNR can be easily estimated. From (8.73), we deduce that

$$\mathbf{t}_i^H(f)\widehat{\boldsymbol{\Gamma}}_{\rm y}(f)\mathbf{t}_i(f) = \frac{1}{1 + \text{iSNR}(f)}, \ i = 2, 3, \ldots, M. \tag{8.79}$$

As a result, an estimate of the input SNR is

$$\widehat{\text{iSNR}}(f) = \frac{1}{M-1} \sum_{i=2}^{M} \frac{1}{\mathbf{t}_i^H(f)\widehat{\boldsymbol{\Gamma}}_{\rm y}(f)\mathbf{t}_i(f)} - 1. \tag{8.80}$$

Example 8.5.1 Returning to Example 8.4.1, we set the narrowband input SNR to $\text{iSNR}(f) = -10$ dB, compute the pseudo-coherence matrix of the observations, $\boldsymbol{\Gamma}_{\rm y}(f)$, using (8.64), obtain $\mathbf{T}(f)$ using (8.69) and (8.70), and calculate the following function:

$$R(f, \theta) = \sum_{i=2}^{M} \left| \mathbf{t}_i^H(f)\mathbf{d}\left(f, \cos\theta\right) \right|^2. \tag{8.81}$$

According to (8.77), the minimum of $R(f, \theta)$ is obtained for $\theta = \widehat{\theta}_{\rm d}$. Figure 8.13 shows plots of $R(f, \theta)$ as a function of θ for several values of α and different numbers of sensors, M. The value of $\theta_{\rm d}$ is better estimated for smaller values of α and larger number of sensors. As the value of α is smaller, $\widehat{\theta}_{\rm d}$ can be obtained using a larger number of sensors with a better accuracy than that obtained with fewer sensors.

Now we set α to 10^{-20}, and estimate the pseudo-coherence matrix of the observations using (8.65) and (8.66) with K random snapshots. Figure 8.14 shows plots of $R(f, \theta)$ as a function of θ for different numbers of snapshots, K, and different numbers of sensors, M. For a small number of snapshots, a good estimate of $\theta_{\rm d}$ requires a larger number of

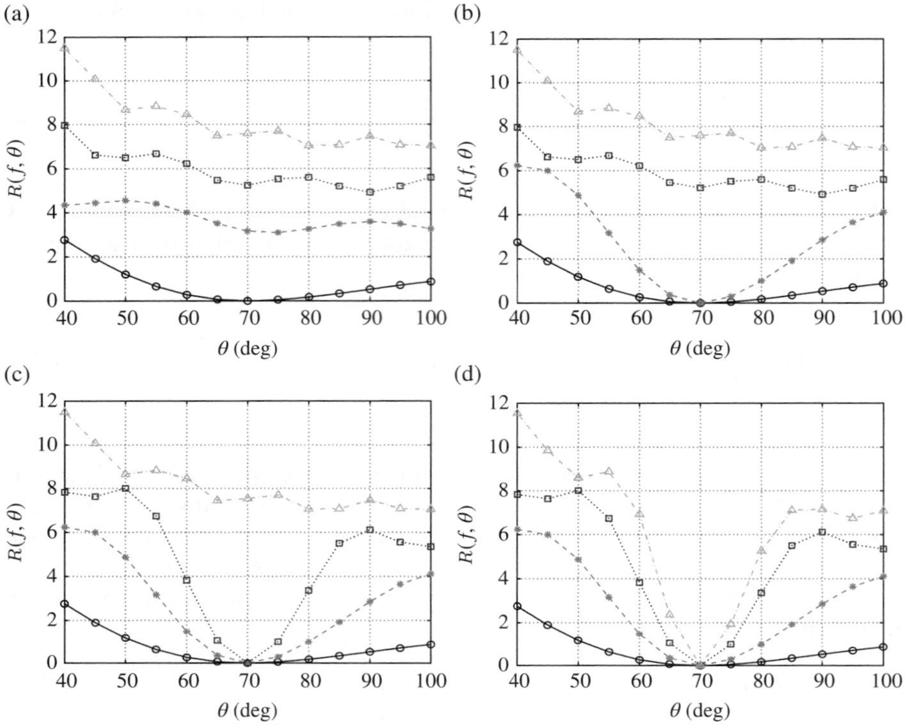

Figure 8.13 The function $R(f, \theta)$, given by (8.81), as a function of θ for several values of α and different numbers of sensors, M: (a) $\alpha = 10^{-5}$, (b) $\alpha = 10^{-10}$, (c) $\alpha = 10^{-15}$, and (d) $\alpha = 10^{-20}$. Each figure shows plots for $M = 3$ (solid line with circles), $M = 6$ (dashed line with asterisks), $M = 9$ (dotted line with squares), and $M = 12$ (dash-dot line with triangles).

sensors. As the number of snapshots is larger, we can estimate θ_d using a smaller number of sensors, but still a better estimate is obtained using a larger number of sensors. ■

8.6 A Spectral Coherence Perspective

The coherence function, which is a bounded function, is a fundamental measure in linear systems. It describes how two complex signals are linearly related. In this section, we show how the coherence can be used as an alternative to the MSE criterion to derive all kinds of adaptive beamformers, since all beamforming techniques discussed in this text are fundamentally linear filtering. What we propose in the following is far from exhaustive; much more can be done.

8.6.1 Definitions

It is of great importance to know how much of $X(f)$ or $V_1(f)$ is contained in the estimator $Z(f)$. The best second-order statistics based measure to evaluate this is via the magnitude squared coherence (MSC). We define the MSC between $Z(f)$ and $X(f)$ as

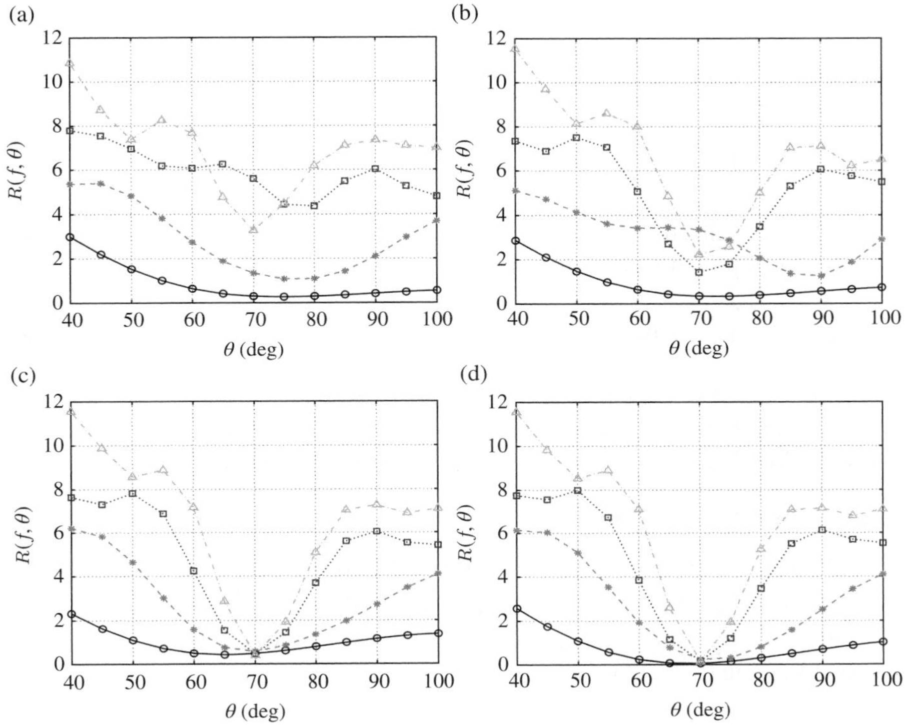

Figure 8.14 The function $R(f,\theta)$, given by (8.81), as a function of θ for different numbers of snapshots, K, and different numbers of sensors, M: (a) $K = 30$, (b) $K = 50$, (c) $K = 300$, and (d) $K = 1000$. Each figure shows plots for $M = 3$ (solid line with circles), $M = 6$ (dashed line with asterisks), $M = 9$ (dotted line with squares), and $M = 12$ (dash-dot line with triangles).

$$\begin{aligned}\left|\gamma_{ZX}\left[\mathbf{h}(f)\right]\right|^2 &= \frac{\left|E\left[Z(f)X^*(f)\right]\right|^2}{E\left[|Z(f)|^2\right]E\left[|X(f)|^2\right]} \\ &= \frac{\phi_X(f)\left|\mathbf{h}^H(f)\mathbf{d}\left(f,\cos\theta_{\mathrm{d}}\right)\right|^2}{\mathbf{h}^H(f)\mathbf{\Phi}_{\mathbf{y}}(f)\mathbf{h}(f)} \\ &= \frac{\mathbf{h}^H(f)\mathbf{\Phi}_{\mathbf{x}}(f)\mathbf{h}(f)}{\mathbf{h}^H(f)\mathbf{\Phi}_{\mathbf{y}}(f)\mathbf{h}(f)}.\end{aligned} \tag{8.82}$$

In the same manner, we define the MSC between $Z(f)$ and $V_1(f)$ as

$$\begin{aligned}\left|\gamma_{ZV_1}\left[\mathbf{h}(f)\right]\right|^2 &= \frac{\left|E\left[Z(f)V_1^*(f)\right]\right|^2}{E\left[|Z(f)|^2\right]E\left[\left|V_1(f)\right|^2\right]} \\ &= \frac{\phi_{V_1}(f)\left|\mathbf{h}^H(f)\boldsymbol{\rho}_{\mathbf{v}V_1}(f)\right|^2}{\mathbf{h}^H(f)\mathbf{\Phi}_{\mathbf{y}}(f)\mathbf{h}(f)},\end{aligned} \tag{8.83}$$

where

$$\boldsymbol{\rho}_{\mathbf{v}V_1}(f) = \begin{bmatrix} 1 & \rho_{V_2V_1}(f) & \cdots & \rho_{V_MV_1}(f) \end{bmatrix}^T = \frac{E\left[\mathbf{v}(f)V_1^*(f)\right]}{E\left[|V_1(f)|^2\right]} \tag{8.84}$$

is the partially normalized – with respect to $V_1(f)$ – coherence vector (of length M) between $\mathbf{v}(f)$ and $V_1(f)$. It can be shown that

$$\left|\gamma_{ZX}\left[\mathbf{h}(f)\right]\right|^2 + \left|\gamma_{ZV_1}\left[\mathbf{h}(f)\right]\right|^2 \leq 1. \tag{8.85}$$

We see how the MSCs defined above, which resemble the generalized Rayleigh quotient [4], depend explicitly on the beamformer $\mathbf{h}(f)$. This observation suggests that we can use the MSC as a criterion to derive optimal adaptive beamformers.

8.6.2 Derivation of Optimal Beamformers

Intuitively, it makes sense to maximize or minimize the MSC in order to find an estimate of $X(f)$ or $V_1(f)$. For example, it is clear that the maximization of $\left|\gamma_{ZX}\left[\mathbf{h}(f)\right]\right|^2$ will give a good estimate of $X(f)$ since, in this case, the coherence between $Z(f)$ and $X(f)$ will be maximal, implying that $Z(f)$ is close to $X(f)$.

8.6.2.1 Coherence Between Beamformer Output and Desired Signal

Here, we consider the MSC between $Z(f)$ and $X(f)$, which is defined in (8.82). A maximal (resp. minimal) MSC implies that $Z(f)$ is an estimate of $X(f)$ [resp. $V_1(f)$].

It is obvious that the maximization of (8.82) leads to an estimate of the desired signal. In (8.82), we recognize the generalized Rayleigh quotient [4], which is maximized with the maximum eigenvector, $\mathbf{t}_1\left(f, \cos\theta_{\mathrm{d}}\right)$, of the matrix $\boldsymbol{\Phi}_{\mathbf{y}}^{-1}(f)\boldsymbol{\Phi}_{\mathbf{x}}(f)$. Let us denote by $\lambda_1\left(f, \cos\theta_{\mathrm{d}}\right)$ the maximum eigenvalue corresponding to $\mathbf{t}_1\left(f, \cos\theta_{\mathrm{d}}\right)$. Since the rank of the mentioned matrix is equal to 1, we have

$$\mathbf{t}_1\left(f, \cos\theta_{\mathrm{d}}\right) = \frac{\boldsymbol{\Phi}_{\mathbf{y}}^{-1}(f)\mathbf{d}\left(f, \cos\theta_{\mathrm{d}}\right)}{\sqrt{\mathbf{d}^H\left(f, \cos\theta_{\mathrm{d}}\right)\boldsymbol{\Phi}_{\mathbf{y}}^{-1}(f)\mathbf{d}\left(f, \cos\theta_{\mathrm{d}}\right)}}, \tag{8.86}$$

$$\lambda_1\left(f, \cos\theta_{\mathrm{d}}\right) = \mathrm{tr}\left[\boldsymbol{\Phi}_{\mathbf{y}}^{-1}(f)\boldsymbol{\Phi}_{\mathbf{x}}(f)\right] = \phi_X(f)\mathbf{d}^H\left(f, \cos\theta_{\mathrm{d}}\right)\boldsymbol{\Phi}_{\mathbf{y}}^{-1}(f)\mathbf{d}\left(f, \cos\theta_{\mathrm{d}}\right), \tag{8.87}$$

and the maximum coherence is

$$\left|\gamma_{ZX,\max}\left(f, \cos\theta_{\mathrm{d}}\right)\right|^2 = \lambda_1\left(f, \cos\theta_{\mathrm{d}}\right). \tag{8.88}$$

As a result, the optimal filter is

$$\mathbf{h}_\alpha\left(f, \cos\theta_{\mathrm{d}}\right) = \alpha(f)\boldsymbol{\Phi}_{\mathbf{y}}^{-1}(f)\mathbf{d}\left(f, \cos\theta_{\mathrm{d}}\right), \tag{8.89}$$

where $\alpha(f) \neq 0$ is an arbitrary complex number. Hence, the estimate of $X(f)$ is

$$\widehat{X}_\alpha(f) = \mathbf{h}_\alpha^H\left(f, \cos\theta_\mathrm{d}\right)\mathbf{y}(f). \tag{8.90}$$

The narrowband output SNR is then

$$\mathrm{oSNR}\left[\mathbf{h}_\alpha\left(f, \cos\theta_\mathrm{d}\right)\right] = \frac{\mathbf{h}_\alpha^H\left(f, \cos\theta_\mathrm{d}\right)\mathbf{\Phi}_\mathbf{x}(f)\mathbf{h}_\alpha\left(f, \cos\theta_\mathrm{d}\right)}{\mathbf{h}_\alpha^H\left(f, \cos\theta_\mathrm{d}\right)\mathbf{\Phi}_\mathbf{v}(f)\mathbf{h}_\alpha\left(f, \cos\theta_\mathrm{d}\right)} \tag{8.91}$$

and it is not hard to find how it is related to the MSC:

$$\lambda_1\left(f, \cos\theta_\mathrm{d}\right) = \frac{\mathrm{oSNR}\left[\mathbf{h}_\alpha\left(f, \cos\theta_\mathrm{d}\right)\right]}{1 + \mathrm{oSNR}\left[\mathbf{h}_\alpha\left(f, \cos\theta_\mathrm{d}\right)\right]}. \tag{8.92}$$

Since the MSC is maximized, so is the narrowband output SNR. We deduce that

$$\mathrm{oSNR}\left[\mathbf{h}_\alpha\left(f, \cos\theta_\mathrm{d}\right)\right] = \frac{\lambda_1\left(f, \cos\theta_\mathrm{d}\right)}{1 - \lambda_1\left(f, \cos\theta_\mathrm{d}\right)} \geq \mathrm{iSNR}(f). \tag{8.93}$$

Now, we need to determine $\alpha(f)$. There are at least two ways to find this parameter. The first one is from the MSE between $X(f)$ and $\widehat{X}_\alpha(f)$:

$$J\left[\mathbf{h}_\alpha\left(f, \cos\theta_\mathrm{d}\right)\right] = E\left[\left|X(f) - \mathbf{h}_\alpha^H\left(f, \cos\theta_\mathrm{d}\right)\mathbf{y}(f)\right|^2\right]. \tag{8.94}$$

The second possibility is to use the distortion-based MSE:

$$J_\mathrm{d}\left[\mathbf{h}_\alpha\left(f, \cos\theta_\mathrm{d}\right)\right] = E\left[\left|X(f) - \mathbf{h}_\alpha^H\left(f, \cos\theta_\mathrm{d}\right)\mathbf{x}(f)\right|^2\right]. \tag{8.95}$$

The minimization of $J\left[\mathbf{h}_\alpha\left(f, \cos\theta_\mathrm{d}\right)\right]$ with respect to $\alpha(f)$ leads to

$$\alpha(f) = \phi_X(f). \tag{8.96}$$

Substituting this value into (8.89), we get the conventional Wiener beamformer:

$$\mathbf{h}_\mathrm{W}\left(f, \cos\theta_\mathrm{d}\right) = \phi_X(f)\mathbf{\Phi}_\mathbf{y}^{-1}(f)\mathbf{d}\left(f, \cos\theta_\mathrm{d}\right). \tag{8.97}$$

By minimizing $J_\mathrm{d}\left[\mathbf{h}_\alpha\left(f, \cos\theta_\mathrm{d}\right)\right]$ with respect to $\alpha(f)$, we obtain

$$\alpha(f) = \frac{1}{\mathbf{d}^H\left(f, \cos\theta_\mathrm{d}\right)\mathbf{\Phi}_\mathbf{y}^{-1}(f)\mathbf{d}\left(f, \cos\theta_\mathrm{d}\right)} \tag{8.98}$$

and substituting the previous result into (8.89) gives the classical MVDR beamformer:

$$\mathbf{h}_\mathrm{MVDR}\left(f, \cos\theta_\mathrm{d}\right) = \frac{\mathbf{\Phi}_\mathbf{y}^{-1}(f)\mathbf{d}\left(f, \cos\theta_\mathrm{d}\right)}{\mathbf{d}^H\left(f, \cos\theta_\mathrm{d}\right)\mathbf{\Phi}_\mathbf{y}^{-1}(f)\mathbf{d}\left(f, \cos\theta_\mathrm{d}\right)}. \tag{8.99}$$

Another approach is to find the beamformer that minimizes (8.82). Then, the beamformer output will be an estimate of $V_1(f)$. The matrix $\mathbf{\Phi}_{\mathbf{y}}^{-1}(f)\mathbf{\Phi}_{\mathbf{x}}(f)$ has $M-1$ eigenvalues equal to 0, since its rank is equal to 1. Let $\mathbf{t}_2(f), \mathbf{t}_3(f), \ldots, \mathbf{t}_M(f)$ be the corresponding eigenvectors. The beamformer:

$$\mathbf{h}_{\boldsymbol{\alpha},V_1}(f) = \mathbf{T}_{2:M}(f)\boldsymbol{\alpha}(f), \tag{8.100}$$

where

$$\mathbf{T}_{2:M}(f) = \begin{bmatrix} \mathbf{t}_2(f) & \mathbf{t}_3(f) & \cdots & \mathbf{t}_M(f) \end{bmatrix} \tag{8.101}$$

is a matrix of size $M \times (M-1)$ and

$$\boldsymbol{\alpha}(f) = \begin{bmatrix} \alpha_2(f) & \alpha_3(f) & \cdots & \alpha_M(f) \end{bmatrix}^T \neq \mathbf{0} \tag{8.102}$$

is a vector of length $M-1$, which minimizes (8.82), since

$$\left|\gamma_{ZX}\left[\mathbf{h}_{\boldsymbol{\alpha},V_1}(f)\right]\right|^2 = \left|\gamma_{ZX,\min}(f)\right|^2 = 0. \tag{8.103}$$

Therefore, the estimates of $V_1(f)$ and $X(f)$ are, respectively,

$$\begin{aligned} \widehat{V}_{1,\boldsymbol{\alpha}}(f) &= \mathbf{h}_{\boldsymbol{\alpha},V_1}^H(f)\mathbf{y}(f) \\ &= \mathbf{h}_{\boldsymbol{\alpha},V_1}^H(f)\mathbf{v}(f) \end{aligned} \tag{8.104}$$

and

$$\begin{aligned} \widehat{X}_{\boldsymbol{\alpha}}(f) &= Y_1(f) - \widehat{V}_{1,\boldsymbol{\alpha}}(f) \\ &= \mathbf{h}_{\boldsymbol{\alpha}}^H(f)\mathbf{y}(f), \end{aligned} \tag{8.105}$$

where

$$\mathbf{h}_{\boldsymbol{\alpha}}(f) = \mathbf{i}_{\mathrm{i}} - \mathbf{h}_{\boldsymbol{\alpha},V_1}(f) \tag{8.106}$$

is the equivalent filter for the estimation of $X(f)$. We can express (8.105) as

$$\widehat{X}_{\boldsymbol{\alpha}}(f) = X(f) + V_1(f) - \mathbf{h}_{\boldsymbol{\alpha},V_1}^H(f)\mathbf{v}(f). \tag{8.107}$$

We see that the previous estimate is distortionless since the desired signal is not filtered at all.

There is at least one interesting way to find $\boldsymbol{\alpha}(f)$. It is obtained from the power of the residual noise:

$$J_{\mathrm{r}}\left[\mathbf{h}_{\boldsymbol{\alpha},V_1}(f)\right] = E\left[\left|V_1(f) - \boldsymbol{\alpha}^H(f)\mathbf{T}_{2:M}^H(f)\mathbf{v}(f)\right|^2\right]. \tag{8.108}$$

The minimization of the previous expression with respect to $\boldsymbol{\alpha}(f)$ gives

$$\boldsymbol{\alpha}(f) = \mathbf{T}_{2:M}^H(f)\mathbf{\Phi}_{\mathbf{v}}(f)\mathbf{i}_{\mathrm{i}}. \tag{8.109}$$

As a result,

$$\mathbf{h}_{\boldsymbol{\alpha},V_1}(f) = \mathbf{T}_{2:M}(f)\mathbf{T}_{2:M}^H(f)\mathbf{\Phi}_{\mathbf{v}}(f)\mathbf{i}_{\mathrm{i}} \tag{8.110}$$

and

$$\mathbf{h}_{\boldsymbol{\alpha}}(f) = \left[\mathbf{I}_M - \mathbf{T}_{2:M}(f)\mathbf{T}_{2:M}^H(f)\mathbf{\Phi}_{\mathbf{v}}(f)\right]\mathbf{i}_{\mathrm{i}}. \tag{8.111}$$

By using the properties of the joint diagonalization, it can easily be shown that

$$\mathbf{h}_{\boldsymbol{\alpha}}(f) = \mathbf{h}_{\mathrm{MVDR}}\left(f, \cos\theta_{\mathrm{d}}\right). \tag{8.112}$$

This is another interesting way to write the MVDR beamformer.

8.6.2.2 Coherence Between Beamformer Output and Noise Signal

Now, we consider the MSC between $Z(f)$ and $V_1(f)$. A maximal (resp. minimal) MSC implies that $Z(f)$ is an estimate of $V_1(f)$ [resp. $X(f)$].

The rank of the matrix $\phi_{V_1}(f)\mathbf{\Phi}_{\mathbf{y}}^{-1}(f)\boldsymbol{\rho}_{\mathbf{v}V_1}(f)\boldsymbol{\rho}_{\mathbf{v}V_1}^H(f) = \mathbf{\Phi}_{\mathbf{y}}^{-1}(f)\mathbf{\Phi}_{\mathbf{v}_{\mathrm{C}}}(f)$, where $\mathbf{\Phi}_{\mathbf{v}_{\mathrm{C}}}(f) = \phi_{V_1}(f)\boldsymbol{\rho}_{\mathbf{v}V_1}(f)\boldsymbol{\rho}_{\mathbf{v}V_1}^H(f)$, is equal to 1, so its only nonnull and positive eigenvalue is

$$\lambda_{1,V_1}(f) = \phi_{V_1}(f)\boldsymbol{\rho}_{\mathbf{v}V_1}^H(f)\mathbf{\Phi}_{\mathbf{y}}^{-1}(f)\boldsymbol{\rho}_{\mathbf{v}V_1}(f), \tag{8.113}$$

the corresponding eigenvector of which is

$$\mathbf{t}_{1,V_1}(f) = \frac{\mathbf{\Phi}_{\mathbf{y}}^{-1}(f)\boldsymbol{\rho}_{\mathbf{v}V_1}(f)}{\sqrt{\boldsymbol{\rho}_{\mathbf{v}V_1}^H(f)\mathbf{\Phi}_{\mathbf{y}}^{-1}(f)\boldsymbol{\rho}_{\mathbf{v}V_1}(f)}}. \tag{8.114}$$

As a result, the beamformer that maximizes (8.83) is

$$\mathbf{h}_{\beta,V_1}(f) = \beta(f)\mathbf{\Phi}_{\mathbf{y}}^{-1}(f)\boldsymbol{\rho}_{\mathbf{v}V_1}(f), \tag{8.115}$$

where $\beta(f) \neq 0$ is an arbitrary complex number and the maximum coherence is

$$\left|\gamma_{ZV_1,\max}(f)\right|^2 = \lambda_{1,V_1}(f). \tag{8.116}$$

This beamformer output gives an estimate of $V_1(f)$:

$$\widehat{V}_{1,\beta}(f) = \mathbf{h}_{\beta,V_1}^H(f)\mathbf{y}(f). \tag{8.117}$$

We deduce that the estimate of the desired signal is

$$\begin{aligned}\widehat{X}_{\beta}(f) &= Y_1(f) - \widehat{V}_{1,\beta}(f) \\ &= \mathbf{h}_{\beta}^H(f)\mathbf{y}(f),\end{aligned} \tag{8.118}$$

where

$$\mathbf{h}_{\beta}(f) = \mathbf{i}_{\mathrm{i}} - \mathbf{h}_{\beta,V_1}(f) \tag{8.119}$$

is the equivalent beamformer for the estimation of $X(f)$. This beamformer will always distort the desired signal since $\mathbf{h}_{\beta}^{H}(f)\mathbf{x}(f) \neq 0$.

One way to find $\beta(f)$ is from the MSE between $X(f)$ and $\widehat{X}_{\beta}(f)$, or, equivalently, $V_1(f)$ and $\widehat{V}_{1,\beta}(f)$:

$$J\left[\mathbf{h}_{\beta,V_1}(f)\right] = E\left[\left|V_1(f) - \mathbf{h}_{\beta,V_1}^{H}(f)\mathbf{y}(f)\right|^2\right]. \tag{8.120}$$

Indeed, the optimization of the previous expression leads to

$$\beta(f) = \phi_{V_1}(f). \tag{8.121}$$

Therefore, we have

$$\mathbf{h}_{\beta,V_1}(f) = \phi_{V_1}(f)\mathbf{\Phi}_{\mathbf{y}}^{-1}(f)\boldsymbol{\rho}_{\mathbf{v}V_1}(f) \tag{8.122}$$

and the equivalent beamformer for the estimation of $X(f)$ is

$$\begin{aligned}\mathbf{h}_{\beta}(f) &= \left[\mathbf{I}_M - \mathbf{\Phi}_{\mathbf{y}}^{-1}(f)\mathbf{\Phi}_{\mathbf{v}_{\mathrm{C}}}(f)\right]\mathbf{i}_{\mathrm{i}} \\ &= \left[\mathbf{I}_M - \mathbf{\Phi}_{\mathbf{y}}^{-1}(f)\mathbf{\Phi}_{\mathbf{v}}(f)\right]\mathbf{i}_{\mathrm{i}} \\ &= \mathbf{h}_{\mathrm{W}}\left(f, \cos\theta_{\mathrm{d}}\right),\end{aligned} \tag{8.123}$$

which is the classical Wiener beamformer.

The other way to find $\beta(f)$ is from the power of the residual noise:

$$J_{\mathrm{r}}\left[\mathbf{h}_{\beta,V_1}(f)\right] = E\left[\left|V_1(f) - \mathbf{h}_{\beta,V_1}^{H}(f)\mathbf{v}(f)\right|^2\right]. \tag{8.124}$$

In this case, we easily find the minimum noise (minN) beamformer for the estimation of $X(f)$:

$$\mathbf{h}_{\mathrm{minN}}(f) = \left\{\mathbf{I}_M - \frac{\mathbf{\Phi}_{\mathbf{y}}^{-1}(f)\mathbf{\Phi}_{\mathbf{v}_{\mathrm{C}}}(f)\mathbf{\Phi}_{\mathbf{y}}^{-1}(f)\mathbf{\Phi}_{\mathbf{v}}(f)}{\mathrm{tr}\left[\mathbf{\Phi}_{\mathbf{y}}^{-1}(f)\mathbf{\Phi}_{\mathbf{v}}(f)\mathbf{\Phi}_{\mathbf{y}}^{-1}(f)\mathbf{\Phi}_{\mathbf{v}_{\mathrm{C}}}(f)\right]}\right\}\mathbf{i}_{\mathrm{i}}. \tag{8.125}$$

This beamformer will reduce more noise than $\mathbf{h}_{\mathrm{W}}\left(f, \cos\theta_{\mathrm{d}}\right)$ but it will introduce much more distortion.

Let $\mathbf{t}_{2,V_1}(f), \mathbf{t}_{3,V_1}(f), \ldots, \mathbf{t}_{M,V_1}(f)$ be the eigenvectors corresponding to the $M-1$ null eigenvalues of the matrix $\mathbf{\Phi}_{\mathbf{y}}^{-1}(f)\mathbf{\Phi}_{\mathbf{v}_{\mathrm{C}}}(f)$. Let us form the beamformer:

$$\mathbf{h}_{\boldsymbol{\beta}}(f) = \mathbf{T}_{2:M,V_1}(f)\boldsymbol{\beta}(f), \tag{8.126}$$

where

$$\mathbf{T}_{2:M,V_1}(f) = \left[\begin{array}{cccc} \mathbf{t}_{2,V_1}(f) & \mathbf{t}_{23,V_1}(f) & \cdots & \mathbf{t}_{M,V_1}(f) \end{array} \right] \quad (8.127)$$

is a matrix of size $M \times (M-1)$ and

$$\boldsymbol{\beta}(f) = \left[\begin{array}{cccc} \beta_2(f) & \beta_3(f) & \cdots & \beta_M(f) \end{array} \right]^T \neq \mathbf{0} \quad (8.128)$$

is a vector of length $M-1$. It can be verified that $\mathbf{h}_{\boldsymbol{\beta}}(f)$ minimizes (8.83), since

$$\left| \gamma_{ZV_1} \left[\mathbf{h}_{\boldsymbol{\beta}}(f) \right] \right|^2 = \left| \gamma_{ZV_1,\min}(f) \right|^2 = 0. \quad (8.129)$$

Therefore, the beamformer output can be considered as an estimate of the desired signal:

$$\widehat{X}_{\boldsymbol{\beta}}(f) = \mathbf{h}_{\boldsymbol{\beta}}^H(f)\mathbf{y}(f). \quad (8.130)$$

The MSE between $X(f)$ and $\widehat{X}_{\boldsymbol{\beta}}(f)$ is then

$$J\left[\mathbf{h}_{\boldsymbol{\beta}}(f) \right] = E\left[\left| X(f) - \boldsymbol{\beta}^H(f)\mathbf{T}_{2:M,V_1}^H(f)\mathbf{y}(f) \right|^2 \right]. \quad (8.131)$$

The minimization of the previous expression gives

$$\boldsymbol{\beta}(f) = \mathbf{T}_{2:M,V_1}^H(f)\boldsymbol{\Phi}_{\mathbf{x}}(f)\mathbf{i}_{\mathrm{i}}. \quad (8.132)$$

We deduce a reduced-rank Wiener beamformer:

$$\begin{aligned} \mathbf{h}_{\mathrm{RRW}}\left(f, \cos\theta_{\mathrm{d}}\right) &= \mathbf{T}_{2:M,V_1}(f)\mathbf{T}_{2:M,V_1}^H(f)\boldsymbol{\Phi}_{\mathbf{x}}(f)\mathbf{i}_{\mathrm{i}} \\ &= \phi_X(f)\mathbf{T}_{2:M,V_1}(f)\mathbf{T}_{2:M,V_1}^H(f)\mathbf{d}\left(f, \cos\theta_{\mathrm{d}}\right). \end{aligned} \quad (8.133)$$

We can also minimize the residual noise subject to the distortionless constraint:

$$\begin{aligned} &\min_{\boldsymbol{\beta}(f)} \boldsymbol{\beta}^H(f)\mathbf{T}_{2:M,V_1}^H(f)\boldsymbol{\Phi}_{\mathbf{v}}(f)\mathbf{T}_{2:M,V_1}(f)\boldsymbol{\beta}(f) \\ &\quad \text{subject to} \quad \boldsymbol{\beta}^H(f)\mathbf{T}_{2:M,V_1}^H(f)\mathbf{d}\left(f, \cos\theta_{\mathrm{d}}\right) = 1. \end{aligned} \quad (8.134)$$

We find that the optimal solution is

$$\mathbf{h}_{\mathrm{LCMV},2}\left(f, \cos\theta_{\mathrm{d}}\right) = \frac{\mathbf{P}_{\mathbf{v}}(f)\mathbf{d}\left(f, \cos\theta_{\mathrm{d}}\right)}{\mathbf{d}^H\left(f, \cos\theta_{\mathrm{d}}\right)\mathbf{P}_{\mathbf{v}}(f)\mathbf{d}\left(f, \cos\theta_{\mathrm{d}}\right)}, \quad (8.135)$$

where

$$\mathbf{P}_{\mathbf{v}}(f) = \mathbf{T}_{2:M,V_1}(f)\left[\mathbf{T}_{2:M,V_1}^H(f)\boldsymbol{\Phi}_{\mathbf{v}}(f)\mathbf{T}_{2:M,V_1}(f) \right]^{-1}\mathbf{T}_{2:M,V_1}^H(f). \quad (8.136)$$

The filter $\mathbf{h}_{\mathrm{LCMV},2}\left(f, \cos\theta_{\mathrm{d}}\right)$ is a particular form of the LCMV beamformer since it places a null at the coherent component of the noise signal.

Problems

8.1 Show that the narrowband MSE can be expressed as

$$J\left[\mathbf{h}(f)\right] = \phi_X(f) + \mathbf{h}^H(f)\mathbf{\Phi}_{\mathbf{y}}(f)\mathbf{h}(f) - \phi_X(f)\mathbf{h}^H(f)\mathbf{d}\left(f, \cos\theta_{\mathrm{d}}\right) \\ - \phi_X(f)\mathbf{d}^H\left(f, \cos\theta_{\mathrm{d}}\right)\mathbf{h}(f).$$

8.2 Show that the MSEs are related to the different performance measures by

$$\frac{J_{\mathrm{d}}\left[\mathbf{h}(f)\right]}{J_{\mathrm{n}}\left[\mathbf{h}(f)\right]} = \mathrm{iSNR}(f) \times \xi_{\mathrm{n}}\left[\mathbf{h}(f)\right] \times \upsilon_{\mathrm{d}}\left[\mathbf{h}(f)\right] \\ = \mathrm{oSNR}\left[\mathbf{h}(f)\right] \times \xi_{\mathrm{d}}\left[\mathbf{h}(f)\right] \times \upsilon_{\mathrm{d}}\left[\mathbf{h}(f)\right].$$

8.3 Show that by minimizing the narrowband MSE, $J\left[\mathbf{h}(f)\right]$, we obtain the Wiener beamformer:

$$\mathbf{h}_{\mathrm{W}}\left(f, \cos\theta_{\mathrm{d}}\right) = \phi_X(f)\mathbf{\Phi}_{\mathbf{y}}^{-1}(f)\mathbf{d}\left(f, \cos\theta_{\mathrm{d}}\right).$$

8.4 Show that the Wiener beamformer can be written as

$$\mathbf{h}_{\mathrm{W}}\left(f, \cos\theta_{\mathrm{d}}\right) = \frac{\mathrm{iSNR}(f)}{1 + \mathrm{iSNR}(f)}\mathbf{\Gamma}_{\mathbf{y}}^{-1}(f)\mathbf{d}\left(f, \cos\theta_{\mathrm{d}}\right).$$

8.5 Show that the Wiener beamformer can be expressed as a function of the statistics of the observation and noise signals by

$$\mathbf{h}_{\mathrm{W}}\left(f, \cos\theta_{\mathrm{d}}\right) = \left[\mathbf{I}_M - \mathbf{\Phi}_{\mathbf{y}}^{-1}(f)\mathbf{\Phi}_{\mathbf{v}}(f)\right]\mathbf{i}_{\mathrm{i}}.$$

8.6 Using the Woodbury identity, show that the inverse of $\mathbf{\Phi}_{\mathbf{y}}(f)$ is given by

$$\mathbf{\Phi}_{\mathbf{y}}^{-1}(f) = \mathbf{\Phi}_{\mathbf{v}}^{-1}(f) - \frac{\mathbf{\Phi}_{\mathbf{v}}^{-1}(f)\mathbf{d}\left(f, \cos\theta_{\mathrm{d}}\right)\mathbf{d}^H\left(f, \cos\theta_{\mathrm{d}}\right)\mathbf{\Phi}_{\mathbf{v}}^{-1}(f)}{\phi_X^{-1}(f) + \mathbf{d}^H\left(f, \cos\theta_{\mathrm{d}}\right)\mathbf{\Phi}_{\mathbf{v}}^{-1}(f)\mathbf{d}\left(f, \cos\theta_{\mathrm{d}}\right)}.$$

8.7 Show that the Wiener beamformer can be written as

$$\mathbf{h}_{\mathrm{W}}\left(f, \cos\theta_{\mathrm{d}}\right) = \frac{\mathbf{\Phi}_{\mathbf{v}}^{-1}(f)\mathbf{\Phi}_{\mathbf{y}}(f) - \mathbf{I}_M}{1 - M + \mathrm{tr}\left[\mathbf{\Phi}_{\mathbf{v}}^{-1}(f)\mathbf{\Phi}_{\mathbf{y}}(f)\right]}\mathbf{i}_{\mathrm{i}}.$$

8.8 Show that by minimizing the narrowband MSE of the residual noise, $J_{\mathrm{r}}\left[\mathbf{h}(f)\right]$, subject to the distortionless constraint, we obtain the MVDR beamformer:

$$\mathbf{h}_{\mathrm{MVDR}}\left(f, \cos\theta_{\mathrm{d}}\right) = \frac{\boldsymbol{\Phi}_{\mathbf{v}}^{-1}(f)\mathbf{d}\left(f, \cos\theta_{\mathrm{d}}\right)}{\mathbf{d}^{H}\left(f, \cos\theta_{\mathrm{d}}\right)\boldsymbol{\Phi}_{\mathbf{v}}^{-1}(f)\mathbf{d}\left(f, \cos\theta_{\mathrm{d}}\right)}.$$

8.9 Using the Woodbury identity, show that the MVDR beamformer can be expressed as a function of the statistics of the observation by

$$\mathbf{h}_{\mathrm{MVDR}}\left(f, \cos\theta_{\mathrm{d}}\right) = \frac{\boldsymbol{\Gamma}_{\mathbf{y}}^{-1}(f)\mathbf{d}\left(f, \cos\theta_{\mathrm{d}}\right)}{\mathbf{d}^{H}\left(f, \cos\theta_{\mathrm{d}}\right)\boldsymbol{\Gamma}_{\mathbf{y}}^{-1}(f)\mathbf{d}\left(f, \cos\theta_{\mathrm{d}}\right)}.$$

8.10 Show that by minimizing the narrowband desired signal distortion index with the constraint that the narrowband noise reduction factor is equal to a positive value, we obtain the tradeoff beamformer:

$$\mathbf{h}_{\mathrm{T},\mu}\left(f, \cos\theta_{\mathrm{d}}\right) = \phi_{X}(f)\left[\boldsymbol{\Phi}_{\mathbf{x}}(f) + \mu\boldsymbol{\Phi}_{\mathbf{v}}(f)\right]^{-1}\mathbf{d}\left(f, \cos\theta_{\mathrm{d}}\right),$$

where $\mu > 0$ is a Lagrange multiplier.

8.11 Show that the tradeoff beamformer can be written as

$$\begin{aligned}\mathbf{h}_{\mathrm{T},\mu}\left(f, \cos\theta_{\mathrm{d}}\right) &= \frac{\phi_{X}(f)\boldsymbol{\Phi}_{\mathbf{v}}^{-1}(f)\mathbf{d}\left(f, \cos\theta_{\mathrm{d}}\right)}{\mu + \phi_{X}(f)\mathbf{d}^{H}\left(f, \cos\theta_{\mathrm{d}}\right)\boldsymbol{\Phi}_{\mathbf{v}}^{-1}(f)\mathbf{d}\left(f, \cos\theta_{\mathrm{d}}\right)} \\ &= \frac{\boldsymbol{\Phi}_{\mathbf{v}}^{-1}(f)\boldsymbol{\Phi}_{\mathbf{y}}(f) - \mathbf{I}_{M}}{\mu - M + \mathrm{tr}\left[\boldsymbol{\Phi}_{\mathbf{v}}^{-1}(f)\boldsymbol{\Phi}_{\mathbf{y}}(f)\right]}\mathbf{i}_{\mathrm{i}}.\end{aligned}$$

8.12 Show that for

a) $\mu = 1$, the tradeoff beamformer is identical to the Wiener beamformer: $\mathbf{h}_{\mathrm{T},1}\left(f, \cos\theta_{\mathrm{d}}\right) = \mathbf{h}_{\mathrm{W}}\left(f, \cos\theta_{\mathrm{d}}\right)$

b) $\mu = 0$, the tradeoff beamformer is identical to the MVDR beamformer: $\mathbf{h}_{\mathrm{T},0}\left(f, \cos\theta_{\mathrm{d}}\right) = \mathbf{h}_{\mathrm{MVDR}}\left(f, \cos\theta_{\mathrm{d}}\right)$

c) $\mu > 1$, the tradeoff beamformer, compared to the Wiener beamformer, gives lower residual noise at the expense of higher desired signal distortion.

8.13 Show that the tradeoff beamformer can be written as

$$\mathbf{h}_{\mathrm{T},\mu}\left(f,\cos\theta_{\mathrm{d}}\right) = \left[(1-\mu)\mathbf{d}\left(f,\cos\theta_{\mathrm{d}}\right)\mathbf{d}^{H}\left(f,\cos\theta_{\mathrm{d}}\right) + \mu\frac{1+\mathrm{iSNR}(f)}{\mathrm{iSNR}(f)}\mathbf{\Gamma}_{\mathbf{y}}(f)\right]^{-1} \times \mathbf{d}\left(f,\cos\theta_{\mathrm{d}}\right).$$

8.14 Show that the tradeoff beamformer can be written as

$$\mathbf{h}_{\mathrm{T},\mu}\left(f,\cos\theta_{\mathrm{d}}\right) = \frac{H_{\mathrm{W}}(f)\mathbf{\Gamma}_{\mathbf{y}}^{-1}(f)\mathbf{d}\left(f,\cos\theta_{\mathrm{d}}\right)}{\mu+(1-\mu)H_{\mathrm{W}}(f)\mathbf{d}^{H}\left(f,\cos\theta_{\mathrm{d}}\right)\mathbf{\Gamma}_{\mathbf{y}}^{-1}(f)\mathbf{d}\left(f,\cos\theta_{\mathrm{d}}\right)}.$$

8.15 Show that the broadband array gains of the tradeoff, MVDR, and Wiener beamformers are related by

$$1 \leq \mathcal{G}\left(\mathbf{h}_{\mathrm{MVDR}}\right) \leq \mathcal{G}\left(\mathbf{h}_{\mathrm{W}}\right) \leq \mathcal{G}\left(\mathbf{h}_{\mathrm{T},\mu}\right)$$

for $\mu \geq 1$, and

$$1 \leq \mathcal{G}\left(\mathbf{h}_{\mathrm{MVDR}}\right) \leq \mathcal{G}\left(\mathbf{h}_{\mathrm{T},\mu}\right) \leq \mathcal{G}\left(\mathbf{h}_{\mathrm{W}}\right)$$

for $0 \leq \mu \leq 1$.

8.16 Show that the maximum array gain is given by

$$\mathcal{G}_{\max}\left(f,\cos\theta_{\mathrm{d}}\right) = \phi_{V_1}(f)\mathbf{d}^{H}\left(f,\cos\theta_{\mathrm{d}}\right)\mathbf{\Phi}_{\mathbf{v}}^{-1}(f)\mathbf{d}\left(f,\cos\theta_{\mathrm{d}}\right).$$

8.17 Show that the maximum array gain beamformer, $\mathbf{h}_{\max}\left(f,\cos\theta_{\mathrm{d}}\right)$, which maximizes the array gain, is given by

$$\mathbf{h}_{\max}\left(f,\cos\theta_{\mathrm{d}}\right) = \varsigma(f)\mathbf{\Gamma}_{\mathbf{y}}^{-1}(f)\mathbf{d}\left(f,\cos\theta_{\mathrm{d}}\right),$$

where $\varsigma(f) \neq 0$ is an arbitrary frequency-dependent complex number.

8.18 Show that by minimizing the narrowband MSE of the residual noise, $J_{\mathrm{r}}\left[\mathbf{h}(f)\right]$, subject to the constraint $\mathbf{C}^{H}\left(f,\theta_{\mathrm{d}},\theta_{1:N}\right)\mathbf{h}(f) = \mathbf{i}_{\mathrm{c}}$, we obtain the LCMV beamformer:

$$\mathbf{h}_{\mathrm{LCMV}}\left(f,\cos\theta_{\mathrm{d}}\right) = \mathbf{\Phi}_{\mathbf{v}}^{-1}(f)\mathbf{C}\left(f,\theta_{\mathrm{d}},\theta_{1:N}\right) \times \left[\mathbf{C}^{H}\left(f,\theta_{\mathrm{d}},\theta_{1:N}\right)\mathbf{\Phi}_{\mathbf{v}}^{-1}(f)\mathbf{C}\left(f,\theta_{\mathrm{d}},\theta_{1:N}\right)\right]^{-1}\mathbf{i}_{\mathrm{c}}.$$

8.19 Show that the LCMV beamformer can be written as

$$\mathbf{h}_{\mathrm{LCMV}}\left(f,\cos\theta_{\mathrm{d}}\right) = \mathbf{\Gamma}_{\mathbf{y}}^{-1}(f)\mathbf{C}\left(f,\theta_{\mathrm{d}},\theta_{1:N}\right) \times \left[\mathbf{C}^{H}\left(f,\theta_{\mathrm{d}},\theta_{1:N}\right)\mathbf{\Gamma}_{\mathbf{y}}^{-1}(f)\mathbf{C}\left(f,\theta_{\mathrm{d}},\theta_{1:N}\right)\right]^{-1}\mathbf{i}_{\mathrm{c}}.$$

8.20 Show that the single-channel Wiener gain $H_W(f)$ is related to the pseudo-coherence matrix, $\mathbf{\Gamma}_\mathbf{y}(f)$, of the observations by

$$H_W(f) = \frac{\mathbf{d}^H\left(f, \cos\theta_\mathrm{d}\right)\left[\mathbf{\Gamma}_\mathbf{y}(f) - \mathbf{\Gamma}_{0,\pi}(f)\right]\mathbf{d}\left(f, \cos\theta_\mathrm{d}\right)}{M^2 - \mathbf{d}^H\left(f, \cos\theta_\mathrm{d}\right)\mathbf{\Gamma}_{0,\pi}(f)\mathbf{d}\left(f, \cos\theta_\mathrm{d}\right)}.$$

8.21 Show that the MSC between $Z(f)$ and $X(f)$ plus the MSC between $Z(f)$ and $V_1(f)$ is equal to 1:

$$\left|\gamma_{ZX}\left[\mathbf{h}(f)\right]\right|^2 + \left|\gamma_{ZV_1}\left[\mathbf{h}(f)\right]\right|^2 \leq 1.$$

8.22 Show that maximization of the MSC between $Z(f)$ and $X(f)$ yields the filter:

$$\mathbf{h}_\alpha\left(f, \cos\theta_\mathrm{d}\right) = \alpha(f)\mathbf{\Phi}_\mathbf{y}^{-1}(f)\mathbf{d}\left(f, \cos\theta_\mathrm{d}\right),$$

where $\alpha(f) \neq 0$ is an arbitrary complex number.

8.23 Show that with the filter $\mathbf{h}_\alpha\left(f, \cos\theta_\mathrm{d}\right)$, the narrowband output SNR is related to the MSC between $Z(f)$ and $X(f)$ by

$$\left|\gamma_{ZX,\max}\left(f, \cos\theta_\mathrm{d}\right)\right|^2 = \frac{\mathrm{oSNR}\left[\mathbf{h}_\alpha\left(f, \cos\theta_\mathrm{d}\right)\right]}{1 + \mathrm{oSNR}\left[\mathbf{h}_\alpha\left(f, \cos\theta_\mathrm{d}\right)\right]}.$$

8.24 Show that the filter $\mathbf{h}_\alpha\left(f, \cos\theta_\mathrm{d}\right)$ that minimizes the distortion-based MSE,$J_\mathrm{d}\left[\mathbf{h}_\alpha\left(f, \cos\theta_\mathrm{d}\right)\right]$, is identical to the MVDR beamformer:

$$\mathbf{h}_{\mathrm{MVDR}}\left(f, \cos\theta_\mathrm{d}\right) = \frac{\mathbf{\Phi}_\mathbf{y}^{-1}(f)\mathbf{d}\left(f, \cos\theta_\mathrm{d}\right)}{\mathbf{d}^H\left(f, \cos\theta_\mathrm{d}\right)\mathbf{\Phi}_\mathbf{y}^{-1}(f)\mathbf{d}\left(f, \cos\theta_\mathrm{d}\right)}.$$

References

1 J. Benesty, J. Chen, and Y. Huang, *Microphone Array Signal Processing*. Berlin, Germany: Springer-Verlag, 2008.

2 J. Capon, "High resolution frequency-wavenumber spectrum analysis," *Proc. IEEE*, vol. 57, pp. 1408–1418, Aug. 1969.

3 R. T. Lacoss, "Data adaptive spectral analysis methods," *Geophysics*, vol. 36, pp. 661–675, Aug. 1971.

4 J. N. Franklin, *Matrix Theory*. Englewood Cliffs, NJ: Prentice-Hall, 1968.

5 A. Booker and C. Y. Ong, "Multiple constraint adaptive filtering," *Geophysics*, vol. 36, pp. 498–509, June 1971.

6 O. Frost, "An algorithm for linearly constrained adaptive array processing," *Proc. IEEE*, vol. 60, pp. 926–935, Jan. 1972.

7 R. Zelinski, "A microphone array with adaptive post-filtering for noise reduction in reverberant rooms," in *Proc. IEEE ICASSP*, vol. 5, 1988, pp. 2578–2581.

8 J. Meyer and K. Uwe Simmer, "Multi-channel speech enhancement in a car environment using Wiener filtering and spectral subtraction," in *Proc. IEEE ICASSP*, vol. 2, 1997, pp. 1167–1170.

9 I. A. McCowan and H. Bourlard, "Microphone array post-filter based on noise field coherence," *IEEE Trans. Speech, Audio Process.*, vol. 11, pp. 709–716, Nov. 2003.

10 S. Lefkimmiatis and P. Maragos, "A generalized estimation approach for linear and nonlinear microphone array post-filters," *Speech Communication*, vol. 49, pp. 657–666, 2007.

11 H. L. van Trees, *Optimum Array Processing: Part IV of Detection, Estimation, and Modulation Theory*. New York, NY: John Wiley & Sons, Inc., 2002.

12 G. Bienvenu and L. Kopp, "Adaptivity to background noise spatial coherence for high resolution passive methods," in *Proc. IEEE ICASSP*, 1980, pp. 307–310.

13 R. O. Schmidt, "Multiple emitter location and signal parameter estimation," *IEEE Trans. Antennas Propag.*, vol. AP-34, pp. 276–280, Mar. 1986.

9

Differential Beamforming

Differential beamforming is a subcategory of classical fixed beamforming. Differential beamformers have two great advantages. The first is that the corresponding beampatterns are almost frequency invariant, which is extremely important when we deal with broadband signals. The second is that they give the highest gains in diffuse noise. These two characteristics make differential beamforming very useful and practical in many applications. However, the main drawback of this approach is white noise amplification. In this chapter, we derive and describe differential beamformers of different orders. We explain the advantages as well as the main problems of this method. We give many design examples. Finally, we show how to deal with the white noise amplification problem.

9.1 Signal Model, Problem Formulation, and Array Model

Again, we consider the signal model of Chapters 7 and 8, which consists of a unique desired source impinging on a ULA of M omnidirectional sensors. The observation vector is then [1]

$$\begin{aligned}\mathbf{y}(f) &= \begin{bmatrix} Y_1(f) & Y_2(f) & \cdots & Y_M(f) \end{bmatrix}^T \\ &= \mathbf{x}(f) + \mathbf{v}(f) \\ &= \mathbf{d}\left(f, \cos\theta_{\mathrm{d}}\right) X(f) + \mathbf{v}(f),\end{aligned} \tag{9.1}$$

where $Y_m(f)$ is the mth sensor signal, $\mathbf{x}(f) = \mathbf{d}\left(f, \cos\theta_{\mathrm{d}}\right) X(f)$,

$$\mathbf{d}\left(f, \cos\theta_{\mathrm{d}}\right) = \begin{bmatrix} 1 & e^{-\jmath 2\pi f \delta \cos\theta_{\mathrm{d}}/c} & \cdots & e^{-\jmath (M-1) 2\pi f \delta \cos\theta_{\mathrm{d}}/c} \end{bmatrix}^T \tag{9.2}$$

is the steering vector, $X(f)$ is the desired source signal, and $\mathbf{v}(f)$ is the additive noise signal vector of length M.

To ensure that differential beamforming takes place, the following two assumptions are made [2–5]:

i) The sensor spacing, δ, is much smaller than the wavelength, $\lambda = c/f$; that is, $\delta \ll \lambda$ (this implies that $f\delta \ll c$). This assumption is required so that the true acoustic

Fundamentals of Signal Enhancement and Array Signal Processing, First Edition.
Jacob Benesty, Israel Cohen, and Jingdong Chen.

Companion website: www.wiley.com/go/benesty/arraysignalprocessing

pressure differentials can be approximated by finite differences of the sensors' outputs.

ii) The desired source signal propagates from the angle $\theta_{\mathrm{d}} = 0$ (endfire direction). Therefore, (9.1) becomes

$$\mathbf{y}(f) = \mathbf{d}\left(f, 1\right) X(f) + \mathbf{v}(f), \tag{9.3}$$

and, at the endfire, the value of the beamformer beampattern should always be equal to 1 (or maximal).

Assumption (i) also implies that we can make a good approximation of the exponential function that appears in the steering vector with the first few elements of its series expansion, so frequency-invariant beamforming may be possible. Because of the symmetry of the steering vector, the only directions in which we can design any desired beampatterns are at the endfires (0 and π); in other directions, the beampattern design is very limited; that is why assumption (ii) is of great importance.

With the conventional linear approach, the beamformer output is simply [1]:

$$\begin{aligned} Z(f) &= \sum_{m=1}^{M} H_m^*(f) Y_m(f) \\ &= \mathbf{h}^H(f)\mathbf{y}(f) \\ &= \mathbf{h}^H(f)\mathbf{d}\left(f, 1\right) X(f) + \mathbf{h}^H(f)\mathbf{v}(f), \end{aligned} \tag{9.4}$$

where $Z(f)$ is, in general, the estimate of the desired signal, $X(f)$, and $\mathbf{h}(f)$ is the beamformer of length M. In our context, the distortionless constraint is wanted:

$$\mathbf{h}^H(f)\mathbf{d}\left(f, 1\right) = 1. \tag{9.5}$$

This means that the value of the beamformer beampattern is equal to 1 at $\theta = 0$ and smaller than 1 at $\theta \neq 0$.

In this chapter, we examine differential sensor arrays (DSAs) of different orders.

9.2 Beampatterns

We recall that the beampattern or directivity pattern, which describes the sensitivity of the beamformer to a plane wave impinging on the array from the direction θ, is defined as

$$\begin{aligned} \mathcal{B}\left[\mathbf{h}(f), \cos\theta\right] &= \mathbf{d}^H\left(f, \cos\theta\right)\mathbf{h}(f) \\ &= \sum_{m=1}^{M} H_m(f) e^{\jmath(m-1)2\pi f\delta\cos\theta/c}. \end{aligned} \tag{9.6}$$

The frequency-independent beampattern of a theoretical Nth-order DSA is well known. It is defined as [3]:

$$\begin{aligned}\mathcal{B}\left(\mathbf{a}_N, \cos\theta\right) &= \sum_{n=0}^{N} a_{N,n} \cos^n \theta \\ &= \mathbf{a}_N^T \mathbf{p}\left(\cos\theta\right),\end{aligned} \tag{9.7}$$

where $a_{N,n}$, $n = 0, 1, \ldots, N$ are real coefficients and

$$\begin{aligned}\mathbf{a}_N &= \left[\begin{array}{cccc} a_{N,0} & a_{N,1} & \cdots & a_{N,N} \end{array}\right]^T, \\ \mathbf{p}\left(\cos\theta\right) &= \left[\begin{array}{cccc} 1 & \cos\theta & \cdots & \cos^N\theta \end{array}\right]^T,\end{aligned}$$

are vectors of length $N+1$. The different values of the coefficients $a_{N,n}$, $n = 0, 1, \ldots, N$ determine the different directivity patterns of the Nth-order DSA. It may be convenient to use a normalization convention for the coefficients. For that, in the direction of the desired signal – that is, for $\theta = 0$ – we would like the beampattern to be equal to 1; that is, $\mathcal{B}\left(\mathbf{a}_N, 1\right) = 1$. Therefore, we have

$$\sum_{n=0}^{N} a_{N,n} = 1. \tag{9.8}$$

As a result, we may choose the first coefficient as

$$a_{N,0} = 1 - \sum_{n=1}^{N} a_{N,n}. \tag{9.9}$$

Since $\cos\theta$ is an even function, so is $\mathcal{B}\left(\mathbf{a}_N, \cos\theta\right)$. Therefore, on a polar plot, $\mathcal{B}\left(\mathbf{a}_N, \cos\theta\right)$ is symmetric about the axis $0 - \pi$ and any DSA beampattern design can be restricted to this range. All useful beampatterns have at least one null in some direction. It follows from (9.7) that an Nth-order directivity pattern has at most N (distinct) nulls in this range.

9.3 Front-to-back Ratios

The front-to-back ratio (FBR) is defined as the ratio of the power of the output of the array for signals propagating from the front-half plane to the output power for signals arriving from the rear-half plane [6]. This ratio, for the spherically isotropic (diffuse) noise field, is mathematically defined as [6]:

$$\mathcal{F}\left[\mathbf{h}(f)\right] = \frac{\displaystyle\int_0^{\pi/2} \left|\mathcal{B}\left[\mathbf{h}(f), \cos\theta\right]\right|^2 \sin\theta d\theta}{\displaystyle\int_{\pi/2}^{\pi} \left|\mathcal{B}\left[\mathbf{h}(f), \cos\theta\right]\right|^2 \sin\theta d\theta} \tag{9.10}$$

$$= \frac{\mathbf{h}^H(f)\mathbf{\Gamma}_{0,\pi/2}(f)\mathbf{h}(f)}{\mathbf{h}^H(f)\mathbf{\Gamma}_{\pi/2,\pi}(f)\mathbf{h}(f)},$$

where

$$\mathbf{\Gamma}_{0,\pi/2}(f) = \int_0^{\pi/2} \mathbf{d}\left(f, \cos\theta\right) \mathbf{d}^H\left(f, \cos\theta\right) \sin\theta d\theta, \tag{9.11}$$

$$\mathbf{\Gamma}_{\pi/2,\pi}(f) = \int_{\pi/2}^{\pi} \mathbf{d}\left(f, \cos\theta\right) \mathbf{d}^H\left(f, \cos\theta\right) \sin\theta d\theta. \tag{9.12}$$

Now, let us compute the entries of the matrix:

$$\mathbf{\Gamma}_{\psi_1,\psi_2}(f) = \mathcal{N}_{\psi_1,\psi_2} \int_{\psi_1}^{\psi_2} \mathbf{d}\left(f, \cos\theta\right) \mathbf{d}^H\left(f, \cos\theta\right) \sin\theta d\theta, \tag{9.13}$$

where $0 \leq \psi_1 \leq \psi_2 \leq \pi$ and

$$\begin{aligned} \mathcal{N}_{\psi_1,\psi_2} &= \frac{1}{\int_{\psi_1}^{\psi_2} \sin\theta d\theta} \\ &= \frac{1}{\cos\psi_1 - \cos\psi_2} \end{aligned} \tag{9.14}$$

is a normalization term. The (i,j)th element (with $i,j = 1,2,\ldots,M$) of $\mathbf{\Gamma}_{\psi_1,\psi_2}(f)$ can be written as

$$\begin{aligned} \left[\mathbf{\Gamma}_{\psi_1,\psi_2}(f)\right]_{ij} &= \mathcal{N}_{\psi_1,\psi_2} \int_{\psi_1}^{\psi_2} e^{-\jmath 2\pi f(i-1)\tau_0 \cos\theta} e^{\jmath 2\pi f(j-1)\tau_0 \cos\theta} \sin\theta d\theta \\ &= \mathcal{N}_{\psi_1,\psi_2} \int_{\psi_1}^{\psi_2} e^{\jmath 2\pi f(j-i)\tau_0 \cos\theta} \sin\theta d\theta \\ &= -\mathcal{N}_{\psi_1,\psi_2} \int_{\cos\psi_1}^{\cos\psi_2} e^{\jmath 2\pi f(j-i)\tau_0 u} du \\ &= \mathcal{N}_{\psi_1,\psi_2} \int_{\cos\psi_2}^{\cos\psi_1} e^{\jmath 2\pi f(j-i)\tau_0 u} du, \end{aligned} \tag{9.15}$$

where $\tau_0 = \delta/c$. Therefore, we deduce that

$$\left[\mathbf{\Gamma}_{\psi_1,\psi_2}(f)\right]_{ij} = \mathcal{N}_{\psi_1,\psi_2} \frac{e^{\jmath 2\pi f(j-i)\tau_0 \cos\psi_1} - e^{\jmath 2\pi f(j-i)\tau_0 \cos\psi_2}}{\jmath 2\pi f(j-i)\tau_0}, \tag{9.16}$$

with

$$\left[\mathbf{\Gamma}_{\psi_1,\psi_2}(f)\right]_{mm} = 1,\ m = 1,2,\ldots,M. \tag{9.17}$$

As a result, the elements of the $M \times M$ matrices $\boldsymbol{\Gamma}_{0,\pi/2}(f)$ and $\boldsymbol{\Gamma}_{\pi/2,\pi}(f)$ are, respectively,

$$\left[\boldsymbol{\Gamma}_{0,\pi/2}(f)\right]_{ij} = \frac{e^{\jmath 2\pi f(j-i)\tau_0} - 1}{\jmath 2\pi f(j-i)\tau_0} \tag{9.18}$$

and

$$\left[\boldsymbol{\Gamma}_{\pi/2,\pi}(f)\right]_{ij} = \frac{1 - e^{-\jmath 2\pi f(j-i)\tau_0}}{\jmath 2\pi f(j-i)\tau_0}, \tag{9.19}$$

with $\left[\boldsymbol{\Gamma}_{0,\pi/2}(f)\right]_{mm} = \left[\boldsymbol{\Gamma}_{\pi/2,\pi}(f)\right]_{mm} = 1,\ m = 1, 2, \ldots, M$.

For the spherically isotropic noise field, the frequency-independent FBR of a theoretical Nth-order DSA is defined as [3]:

$$\mathcal{F}\left(\mathbf{a}_N\right) = \frac{\int_0^{\pi/2} \mathcal{B}^2\left(\mathbf{a}_N, \cos\theta\right) \sin\theta d\theta}{\int_{\pi/2}^{\pi} \mathcal{B}^2\left(\mathbf{a}_N, \cos\theta\right) \sin\theta d\theta}. \tag{9.20}$$

9.4 Array Gains

From Chapter 7, we know that the array gain is given by

$$\mathcal{G}\left[\mathbf{h}(f)\right] = \frac{\left|\mathbf{h}^H(f)\mathbf{d}\left(f,1\right)\right|^2}{\mathbf{h}^H(f)\boldsymbol{\Gamma}_{\mathbf{v}}(f)\mathbf{h}(f)}, \tag{9.21}$$

where $\boldsymbol{\Gamma}_{\mathbf{v}}(f)$ is the pseudo-coherence matrix of $\mathbf{v}(f)$.

The WNG is directly deduced from (9.21) by taking $\boldsymbol{\Gamma}_{\mathbf{v}}(f) = \mathbf{I}_M$. We obtain

$$\mathcal{W}\left[\mathbf{h}(f)\right] = \frac{\left|\mathbf{h}^H(f)\mathbf{d}\left(f,1\right)\right|^2}{\mathbf{h}^H(f)\mathbf{h}(f)} \tag{9.22}$$

and we can easily show that the maximum WNG is

$$\mathcal{W}_{\max} = M. \tag{9.23}$$

The DF, which is the array gain in the diffuse (spherically isotropic) noise field, is given by

$$\mathcal{D}\left[\mathbf{h}(f)\right] = \frac{\left|\mathbf{h}^H(f)\mathbf{d}\left(f,1\right)\right|^2}{\mathbf{h}^H(f)\boldsymbol{\Gamma}_{0,\pi}(f)\mathbf{h}(f)} \tag{9.24}$$

and the maximum DF is

$$\mathcal{D}_{\max}(f) = \mathbf{d}^H\left(f, 1\right) \mathbf{\Gamma}_{0,\pi}^{-1}(f) \mathbf{d}\left(f, 1\right). \tag{9.25}$$

We also have [7]:

$$\lim_{\delta \to 0} \mathcal{D}_{\max}(f) = M^2. \tag{9.26}$$

For the spherically isotropic noise field, the frequency-independent DF of a theoretical Nth-order DSA is defined as [3]:

$$\mathcal{D}\left(\mathbf{a}_N\right) = \frac{\mathcal{B}^2\left(\mathbf{a}_N, 1\right)}{\frac{1}{2}\int_0^{\pi} \mathcal{B}^2\left(\mathbf{a}_N, \cos\theta\right) \sin\theta d\theta}. \tag{9.27}$$

9.5 Examples of Theoretical Differential Beamformers

The most well-known and frequently studied Nth-order DSA beampatterns are the dipole, the cardioid, the hypercardioid, and the supercardioid. In the following, we show how they are obtained.

The Nth-order dipole has a unique null, with multiplicity N in the direction $\pi/2$. Its beampattern is then given by

$$\mathcal{B}_{N,\mathrm{Dp}}\left(\cos\theta\right) = \cos^N\theta, \tag{9.28}$$

implying that $a_{N,N} = 1$ and $a_{N,N-1} = a_{N,N-2} = \cdots = a_{N,0} = 0$.

The Nth-order cardioid has a unique null with multiplicity N in the direction π. Its beampattern is then given by

$$\begin{aligned}\mathcal{B}_{N,\mathrm{Cd}}\left(\cos\theta\right) &= \frac{1}{2^N}\left(1 + \cos\theta\right)^N \\ &= \sum_{n=0}^{N} \frac{N!}{2^N n!(N-n)!} \cos^n\theta,\end{aligned} \tag{9.29}$$

implying that

$$a_{N,n} = \frac{N!}{2^N n!(N-n)!},\ n = 0, 1, \ldots, N. \tag{9.30}$$

The coefficients of the Nth-order hypercardioid can be obtained by maximizing the DF, $\mathcal{D}\left(\mathbf{a}_N\right)$, given in (9.27). It can be shown that [3]:

$$\mathcal{D}\left(\mathbf{a}_N\right) = \frac{\mathbf{a}_N^T \mathbf{1}\mathbf{1}^T \mathbf{a}_N}{\mathbf{a}_N^T \mathbf{H}_N \mathbf{a}_N}, \tag{9.31}$$

where

$$\mathbf{1} = \begin{bmatrix} 1 & 1 & \cdots & 1 \end{bmatrix}^T$$

is a vector of length $N+1$ and $\mathbf{H}_N$ is a Hankel matrix – of size $(N+1) \times (N+1)$ – the elements of which are given by

$$\left[\mathbf{H}_N\right]_{ij} = \begin{cases} \dfrac{1}{1+i+j}, & \text{if } i+j \text{ even} \\ 0, & \text{otherwise} \end{cases}, \tag{9.32}$$

with $i, j = 0, 1, \ldots, N$. In (9.31), we notice the generalized Rayleigh quotient. Therefore, the vector $\mathbf{a}_N$ that maximizes $\mathcal{D}\left(\mathbf{a}_N\right)$ is the eigenvector corresponding to the maximum eigenvalue of the matrix $\mathbf{H}_N^{-1}\mathbf{1}\mathbf{1}^T$:

$$\mathbf{a}_{N,\max} = \frac{\mathbf{H}_N^{-1}\mathbf{1}}{\mathbf{1}^T\mathbf{H}_N^{-1}\mathbf{1}}. \tag{9.33}$$

As a result, the beampattern of the Nth-order hypercardioid is

$$\mathcal{B}_{N,\mathrm{Hd}}\left(\cos\theta\right) = \frac{\mathbf{1}^T\mathbf{H}_N^{-1}\mathbf{p}\left(\cos\theta\right)}{\mathbf{1}^T\mathbf{H}_N^{-1}\mathbf{1}}. \tag{9.34}$$

The coefficients of the Nth-order supercardioid can be obtained by maximizing the FBR, $\mathcal{F}\left(\mathbf{a}_N\right)$, defined in (9.20). It can be shown that [3]:

$$\mathcal{F}\left(\mathbf{a}_N\right) = \frac{\mathbf{a}_N^T\mathbf{H}_N''\mathbf{a}_N}{\mathbf{a}_N^T\mathbf{H}_N'\mathbf{a}_N}, \tag{9.35}$$

where $\mathbf{H}_N'$ and $\mathbf{H}_N''$ are two Hankel matrices – of size $(N+1) \times (N+1)$ – the elements of which are given by, respectively,

$$\left[\mathbf{H}_N'\right]_{ij} = \frac{(-1)^{i+j}}{1+i+j} \tag{9.36}$$

and

$$\left[\mathbf{H}_N''\right]_{ij} = \frac{1}{1+i+j}, \tag{9.37}$$

with $i, j = 0, 1, \ldots, N$. Let us denote by $\mathbf{a}_{N,\max}'$ the eigenvector corresponding to the maximum eigenvalue of $\mathbf{H}_N'^{-1}\mathbf{H}_N''$. Then, $\mathbf{a}_{N,\max}'$ maximizes the FBR and the beampattern of the Nth-order supercardioid is

$$\mathcal{B}_{N,\mathrm{Sd}}\left(\cos\theta\right) = \frac{\mathbf{a}_{N,\max}'^T\mathbf{p}\left(\cos\theta\right)}{\mathbf{a}_{N,\max}'^T\mathbf{p}\left(1\right)}. \tag{9.38}$$

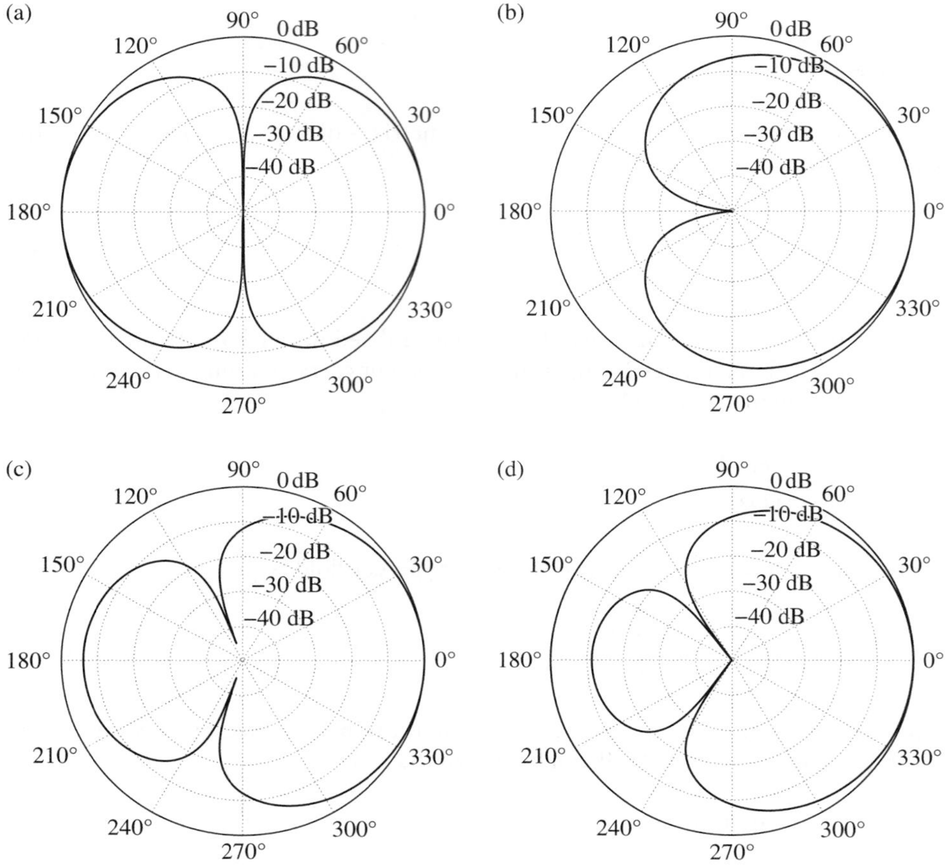

Figure 9.1 First-order directivity patterns: (a) dipole, (b) cardioid, (c) hypercardioid, and (d) supercardioid.

The best-known first-order directivity patterns are expressed as

$$\mathcal{B}_{1,\mathrm{Dp}}(\cos\theta) = \cos\theta, \tag{9.39}$$

$$\mathcal{B}_{1,\mathrm{Cd}}(\cos\theta) = \frac{1}{2} + \frac{1}{2}\cos\theta, \tag{9.40}$$

$$\mathcal{B}_{1,\mathrm{Hd}}(\cos\theta) = \frac{1}{4} + \frac{3}{4}\cos\theta, \tag{9.41}$$

$$\mathcal{B}_{1,\mathrm{Sd}}(\cos\theta) = \frac{\sqrt{3}-1}{2} + \frac{3-\sqrt{3}}{2}\cos\theta. \tag{9.42}$$

Figure 9.1 shows these different polar beampatterns. What exactly is shown are the values of the magnitude squared beampattern in decibels; that is, $10\log_{10}\mathcal{B}^2\left(\mathbf{a}_N, \cos\theta\right)$.

The most useful second-order directivity patterns are given by

$$\mathcal{B}_{2,\mathrm{Dp}}(\cos\theta) = \cos^2\theta, \tag{9.43}$$

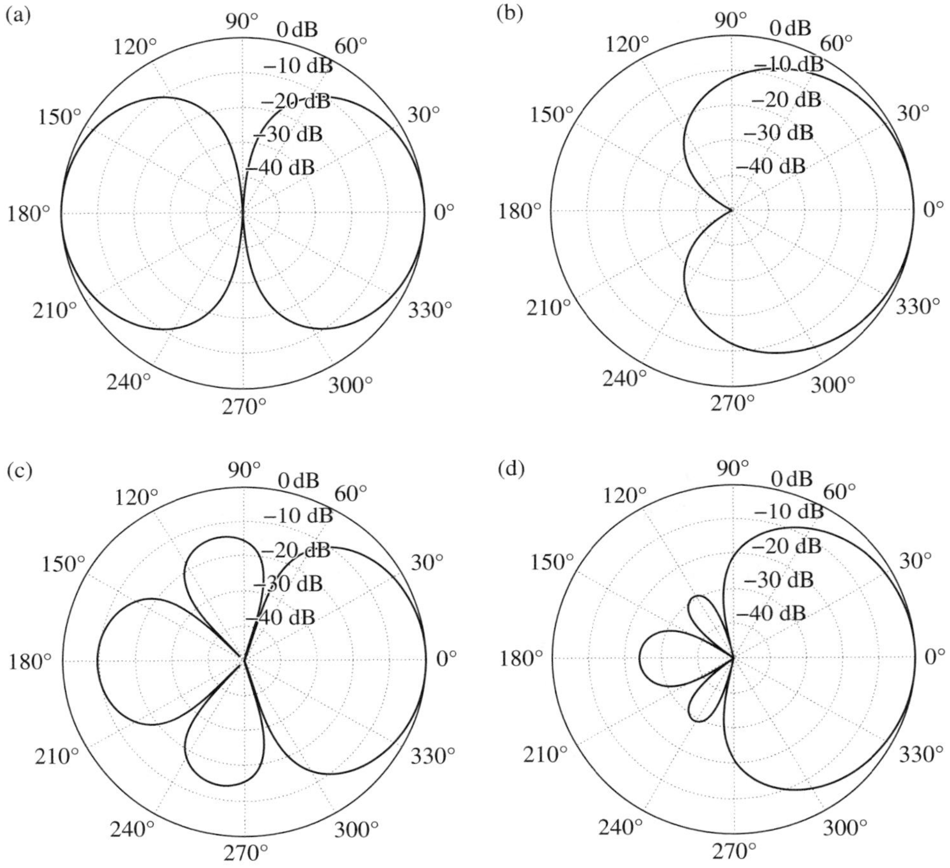

Figure 9.2 Second-order directivity patterns: (a) dipole, (b) cardioid, (c) hypercardioid, and (d) supercardioid.

$$\mathcal{B}_{2,\mathrm{Cd}}(\cos\theta) = \frac{1}{4} + \frac{1}{2}\cos\theta + \frac{1}{4}\cos^2\theta, \tag{9.44}$$

$$\mathcal{B}_{2,\mathrm{Hd}}(\cos\theta) = -\frac{1}{6} + \frac{1}{3}\cos\theta + \frac{5}{6}\cos^2\theta, \tag{9.45}$$

$$\mathcal{B}_{2,\mathrm{Sd}}(\cos\theta) = \frac{1}{2\left(3+\sqrt{7}\right)} + \frac{\sqrt{7}}{3+\sqrt{7}}\cos\theta + \frac{5}{2\left(3+\sqrt{7}\right)}\cos^2\theta. \tag{9.46}$$

Figure 9.2 depicts the different second-order directivity patterns given above.

The most important third-order directivity patterns are expressed as

$$\mathcal{B}_{3,\mathrm{Dp}}(\cos\theta) = \cos^3\theta, \tag{9.47}$$

$$\mathcal{B}_{3,\mathrm{Cd}}(\cos\theta) = \frac{1}{8} + \frac{3}{8}\cos\theta + \frac{3}{8}\cos^2\theta + \frac{1}{8}\cos^3\theta, \tag{9.48}$$

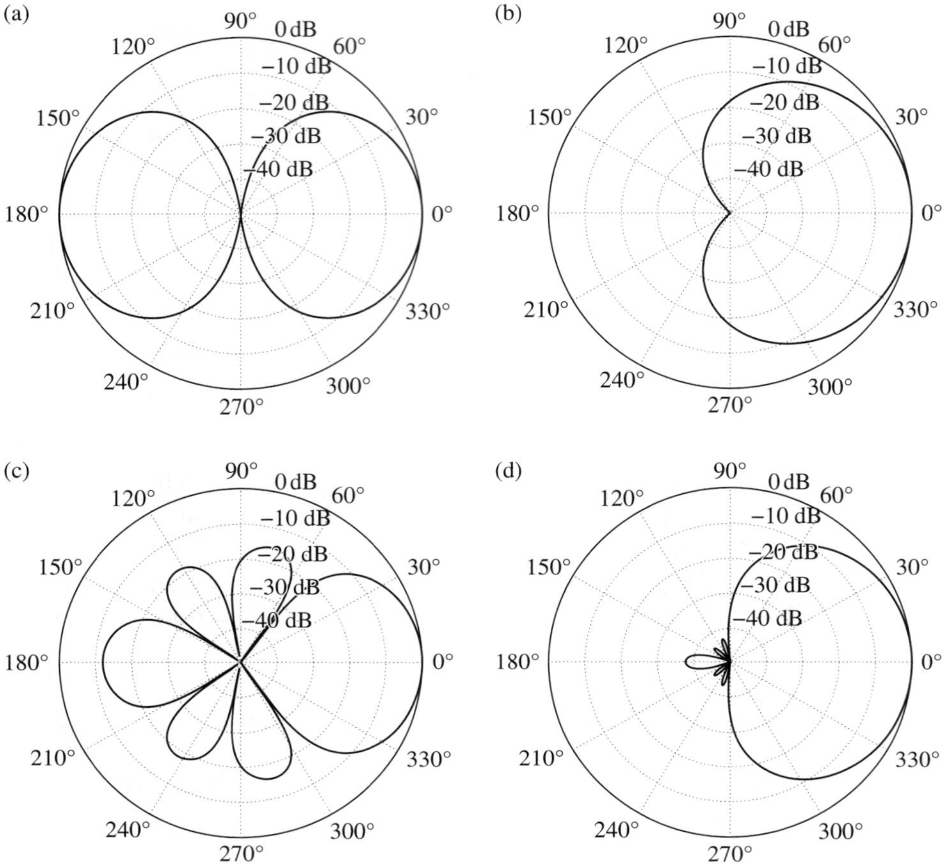

Figure 9.3 Third-order directivity patterns: (a) dipole, (b) cardioid, (c) hypercardioid, and (d) supercardioid.

$$\mathcal{B}_{3,\mathrm{Hd}}(\cos\theta) = -\frac{3}{32} - \frac{15}{32}\cos\theta + \frac{15}{32}\cos^2\theta + \frac{35}{32}\cos^3\theta, \tag{9.49}$$

$$\mathcal{B}_{3,\mathrm{Sd}}(\cos\theta) \approx 0.0184 + 0.2004\cos\theta + 0.4750\cos^2\theta + 0.3061\cos^3\theta. \tag{9.50}$$

Figure 9.3 depicts the different third-order directivity patterns given above.

The approach to designing DSAs described in this chapter is based, mostly, on the obvious observation that any useful theoretical frequency-independent DSA beampattern has a one at the angle $\theta = 0$ and a number of nulls in some specific directions (with $\theta \gg 0$). In the most obvious design, which is also the conventional way to do differential beamforming, the number of sensors is equal to the order plus one that is, $M = N + 1$.

9.6 First-order Design

9.6.1 Principle

First-order DSAs are designed with two sensors. In this case, we have exactly two constraints to fulfill. The first constraint is the distortionless response (a one at the angle $\theta = 0$) and the second constraint is a null in the interval $0 < \theta \leq \pi$. Thus, these two constraints can be written as

$$\mathbf{d}^H(f,1)\,\mathbf{h}(f) = 1, \tag{9.51}$$

$$\mathbf{d}^H(f,\alpha_{1,1})\,\mathbf{h}(f) = 0, \tag{9.52}$$

where $\alpha_{1,1} = \cos\theta_{1,1}$ is given by design (a null at the angle $\theta_{1,1}$) with $-1 \leq \alpha_{1,1} < 1$. We can express (9.51)–(9.52) as

$$\begin{aligned}\begin{bmatrix} \mathbf{d}^H(f,1) \\ \mathbf{d}^H(f,\alpha_{1,1}) \end{bmatrix}\mathbf{h}(f) &= \begin{bmatrix} 1 & e^{j2\pi f\tau_0} \\ 1 & e^{j2\pi f\tau_0\alpha_{1,1}} \end{bmatrix}\mathbf{h}(f) \\ &= \begin{bmatrix} 1 \\ 0 \end{bmatrix}.\end{aligned} \tag{9.53}$$

The last expression is a linear system of two equations and two unknowns, for which the solution is

$$\mathbf{h}_1(f) = \frac{1}{1 - e^{j2\pi f\tau_0(1-\alpha_{1,1})}}\begin{bmatrix} 1 \\ -e^{-j2\pi f\tau_0\alpha_{1,1}} \end{bmatrix}. \tag{9.54}$$

Substituting (9.54) into (9.6), we find that the beampattern is

$$\mathcal{B}\left[\mathbf{h}_1(f),\cos\theta\right] = \frac{1 - e^{j2\pi f\tau_0(\cos\theta-\alpha_{1,1})}}{1 - e^{j2\pi f\tau_0(1-\alpha_{1,1})}}. \tag{9.55}$$

Using assumption (i) and the approximation:

$$e^x \approx 1 + x, \tag{9.56}$$

we can approximate (9.55) as

$$\mathcal{B}\left[\mathbf{h}_1(f),\cos\theta\right] \approx \frac{1}{1-\alpha_{1,1}}\cos\theta - \frac{\alpha_{1,1}}{1-\alpha_{1,1}}, \tag{9.57}$$

which resembles the theoretical first-order DSA. Most importantly, the beampattern is frequency invariant. This is a useful feature since, as we can see, differential beamforming tends to lead to broadband beamformers, which are important in applications dealing with broadband signals such as speech.

It is not hard to find that the DF is

$$\mathcal{D}\left[\mathbf{h}_1(f)\right] = \frac{1 - \cos\left[2\pi f\tau_0\left(1-\alpha_{1,1}\right)\right]}{1 - \text{sinc}\left(2\pi f\tau_0\right)\cos\left(2\pi f\tau_0\alpha_{1,1}\right)}. \tag{9.58}$$

Using the approximations:

$$\cos x \approx 1 - \frac{x^2}{2}, \tag{9.59}$$

$$\text{sinc}\, x \approx 1 - \frac{x^2}{6}, \tag{9.60}$$

the DF becomes

$$\mathcal{D}\left[\mathbf{h}_1(f)\right] \approx \frac{\left(1 - \alpha_{1,1}\right)^2}{\alpha_{1,1}^2 + \frac{1}{3}}. \tag{9.61}$$

We observe that the DF is almost frequency independent as long as δ is small. Also, the value of $\alpha_{1,1}$ that maximizes (9.61) is equal to $-\frac{1}{3}$, which corresponds to the hypercardioid and leads to a DF of 4; this is the maximum possible DF for $M = 2$.

The WNG is

$$\begin{aligned} \mathcal{W}\left[\mathbf{h}_1(f)\right] &= \frac{1}{2}\left|1 - e^{\jmath 2\pi f \tau_0 (1-\alpha_{1,1})}\right|^2 \\ &= 1 - \cos\left[2\pi f \tau_0 \left(1 - \alpha_{1,1}\right)\right], \end{aligned} \tag{9.62}$$

which we can approximate as

$$\mathcal{W}\left[\mathbf{h}_1(f)\right] \approx \frac{1}{2}\left[2\pi f \tau_0 \left(1 - \alpha_{1,1}\right)\right]^2. \tag{9.63}$$

Some observations are in order. First, the WNG is very much frequency dependent. Second, the WNG is much larger at high than at low frequencies. Third, the WNG can be smaller than 1, especially at low frequencies, implying white noise amplification. Finally, it is obvious that the WNG is maximized for $\alpha_{1,1} = -1$, which corresponds to the cardioid.

9.6.2 Design Examples

In this section, important particular cases of first-order DSAs with two sensors are numerically described. Depending on the value of $\alpha_{1,1}$ we find four useful first-order DSAs:

- dipole: $\alpha_{1,1} = 0$
- cardioid: $\alpha_{1,1} = -1$
- hypercardioid: $\alpha_{1,1} = -\frac{1}{3}$
- supercardioid: $\alpha_{1,1} = \frac{1-\sqrt{3}}{3-\sqrt{3}}$.

Figure 9.4 displays the patterns – with $\mathbf{h}_1(f)$ defined as in (9.54) – of the first-order dipole, cardioid, hypercardioid, and supercardioid for a low frequency ($f = 0.5$ kHz) and a small value of δ ($\delta = 1$ cm). Figure 9.5 shows the patterns for a high frequency ($f = 7$ kHz) and a small value of δ ($\delta = 1$ cm). As long as the sensor spacing is small, the beampatterns of the first-order DSAs are frequency independent.

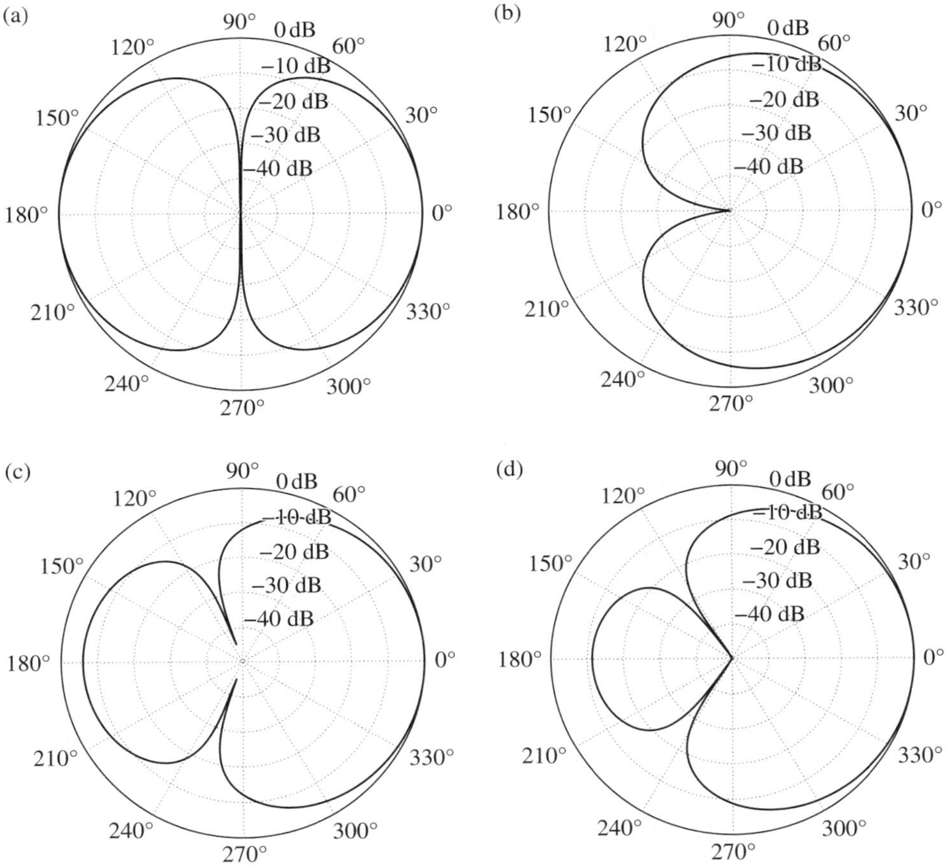

Figure 9.4 Beampatterns of the first-order DSAs for $f = 0.5$ kHz and $\delta = 1$ cm: (a) dipole, (b) cardioid, (c) hypercardioid, and (d) supercardioid.

Figures 9.6 and 9.7 display the patterns of the first-order dipole, cardioid, hypercardioid, and supercardioid for a value of δ equal to 4 cm. In this case, the sensor spacing is too large, which causes deterioration of the beampatterns at high frequencies.

Figure 9.8 shows plots of the DF, $\mathcal{D}\left[\mathbf{h}_1(f)\right]$, as a function of frequency, for the dipole, cardioid, hypercardioid, and supercardioid and several values of δ. Corresponding plots of the WNG, $\mathcal{W}\left[\mathbf{h}_1(f)\right]$, as a function of frequency are depicted in Figure 9.9. We observe that increasing the sensor spacing enables us to increase the WNG, especially at low frequencies. Accordingly, if we do not want to amplify the white (or sensor) noise, the sensor spacing must be large. However, a large value of δ is in conflict with the DSA assumption, which states that δ should be small. Therefore, there is always a tradeoff between white noise amplification, especially at low frequencies, and a frequency-independent directivity pattern at high frequencies. The sensor spacing should be selected according to this compromise.

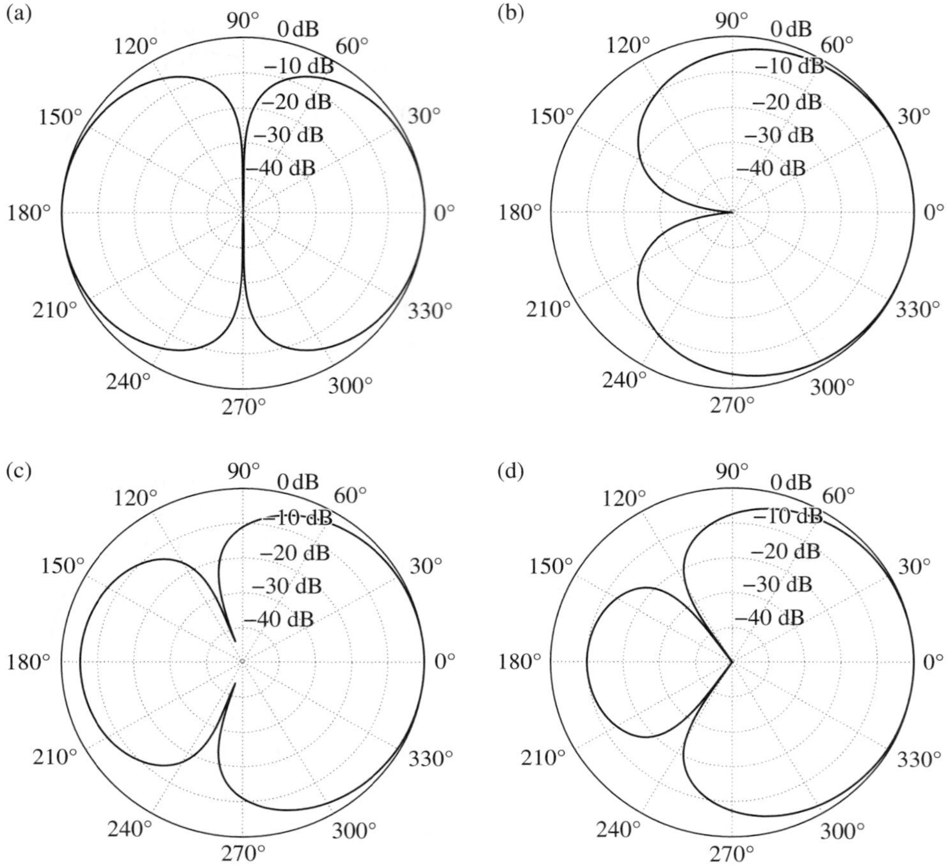

Figure 9.5 Beampatterns of the first-order DSAs for $f = 7$ kHz and $\delta = 1$ cm: (a) dipole, (b) cardioid, (c) hypercardioid, and (d) supercardioid.

9.7 Second-order Design

9.7.1 Principle

Any second-order DSA can be realized with three sensors. Therefore, we assume that we have exactly three sensors. As a result, we have three constraints to fulfill, with the first one being, as usual, at the angle $\theta = 0$. We deduce that the general linear system of equations to design any second-order differential array is

$$\begin{bmatrix} \mathbf{d}^H(f,1) \\ \mathbf{d}^H(f,\alpha_{2,1}) \\ \mathbf{d}^H(f,\alpha_{2,2}) \end{bmatrix} \mathbf{h}(f) = \begin{bmatrix} 1 \\ \beta_{2,1} \\ \beta_{2,2} \end{bmatrix}, \tag{9.64}$$

where $-1 \leq \alpha_{2,1} = \cos\theta_{2,1} < 1$, $-1 \leq \alpha_{2,2} = \cos\theta_{2,2} < 1$, $\alpha_{2,1} \neq \alpha_{2,2}$, $-1 \leq \beta_{2,1} \leq 1$, and $-1 \leq \beta_{2,2} \leq 1$. The parameter $\alpha_{2,i}$ is a chosen direction and $\beta_{2,i}$ is its corresponding

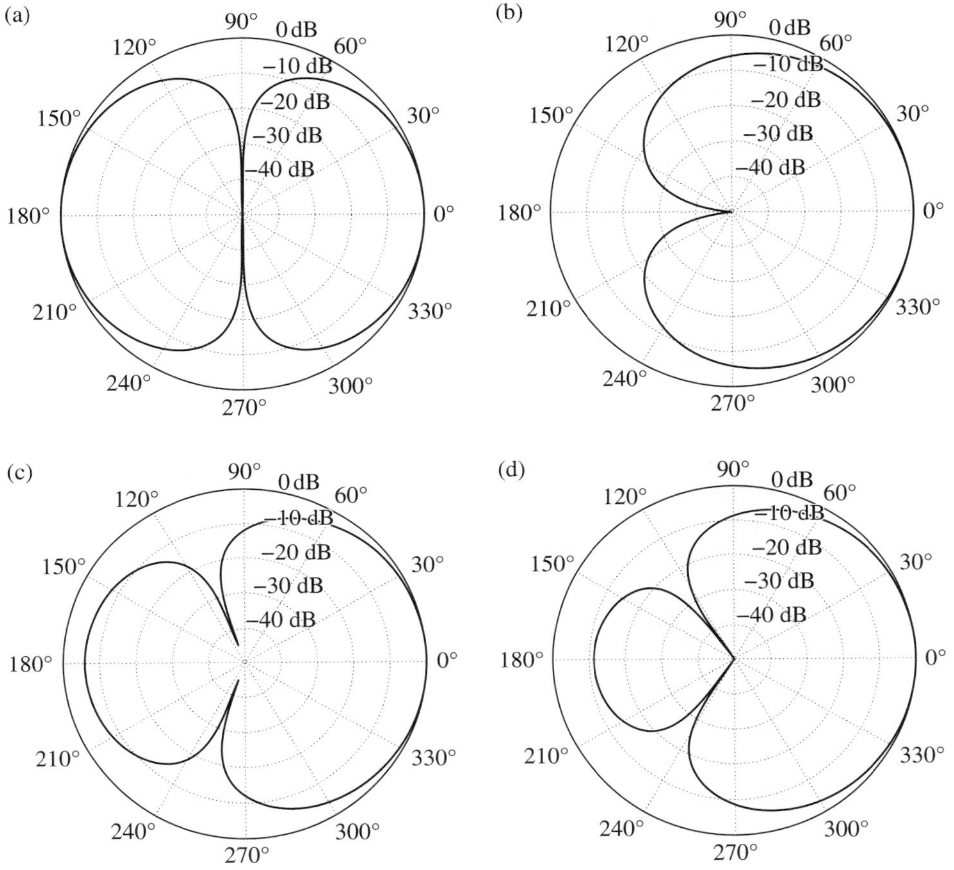

Figure 9.6 Beampatterns of the first-order DSAs for $f = 0.5$ kHz and $\delta = 4$ cm: (a) dipole, (b) cardioid, (c) hypercardioid, and (d) supercardioid.

value on the given desired beampattern. We should always privilege the zeroes of the beampattern.

Let us denote by

$$\mathbf{V}(f) = \begin{bmatrix} \mathbf{d}^H\left(f, 1\right) \\ \mathbf{d}^H\left(f, \alpha_{2,1}\right) \\ \mathbf{d}^H\left(f, \alpha_{2,2}\right) \end{bmatrix} = \begin{bmatrix} 1 & v_1(f) & v_1^2(f) \\ 1 & v_2(f) & v_2^2(f) \\ 1 & v_3(f) & v_3^2(f) \end{bmatrix} \tag{9.65}$$

the 3×3 Vandermonde matrix that appears in (9.64), where $v_1(f) = e^{\jmath 2\pi f \tau_0}$, $v_2(f) = e^{\jmath 2\pi f \tau_0 \alpha_{2,1}}$, and $v_3(f) = e^{\jmath 2\pi f \tau_0 \alpha_{2,2}}$. From the decomposition $\mathbf{V}^{-1}(f) = \mathbf{U}(f)\mathbf{L}(f)$ [4], where

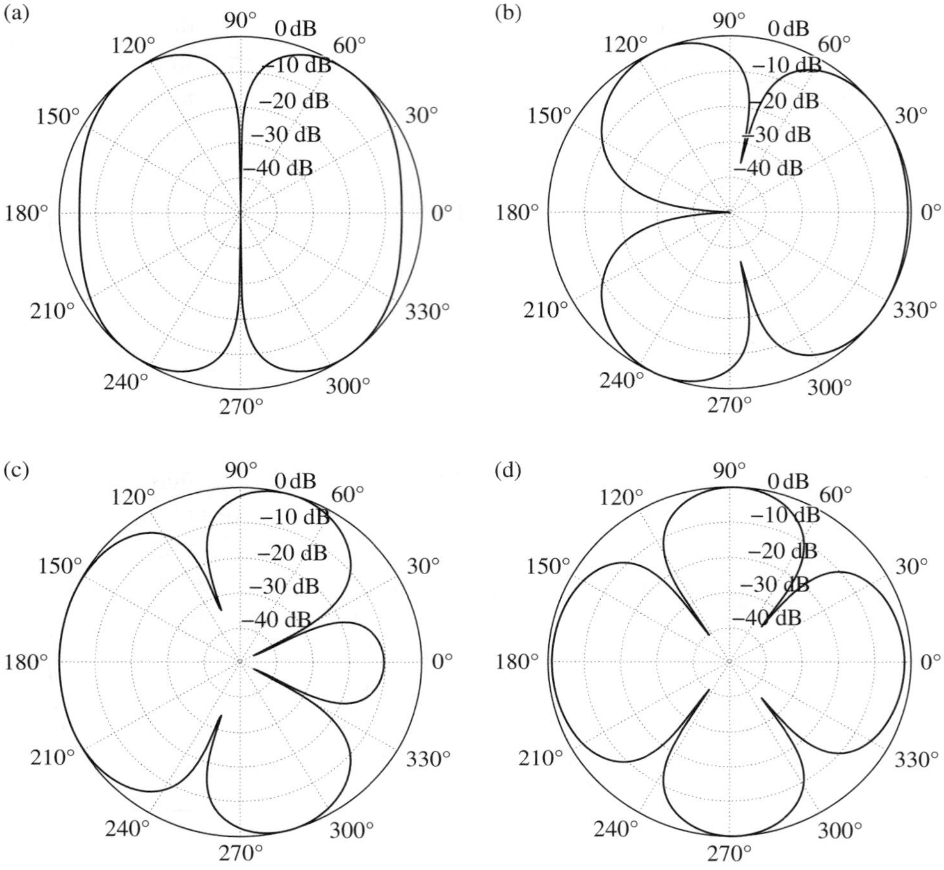

Figure 9.7 Beampatterns of the first-order DSAs for $f = 7$ kHz and $\delta = 4$ cm: (a) dipole, (b) cardioid, (c) hypercardioid, and (d) supercardioid.

$$\mathbf{U}(f) = \begin{bmatrix} 1 & -\nu_1(f) & \nu_1(f)\nu_2(f) \\ 0 & 1 & -\left[\nu_1(f) + \nu_2(f)\right] \\ 0 & 0 & 1 \end{bmatrix} \tag{9.66}$$

and[1]

$$\mathbf{L}(f) = \begin{bmatrix} 1 & 0 & 0 \\ \dfrac{1}{\nu_1 - \nu_2} & \dfrac{1}{\nu_2 - \nu_1} & 0 \\ \dfrac{1}{(\nu_1 - \nu_2)(\nu_1 - \nu_3)} & \dfrac{1}{(\nu_2 - \nu_1)(\nu_2 - \nu_3)} & \dfrac{1}{(\nu_3 - \nu_1)(\nu_3 - \nu_2)} \end{bmatrix}, \tag{9.67}$$

1 In some matrices, we drop the dependency on f to simplify the presentation.

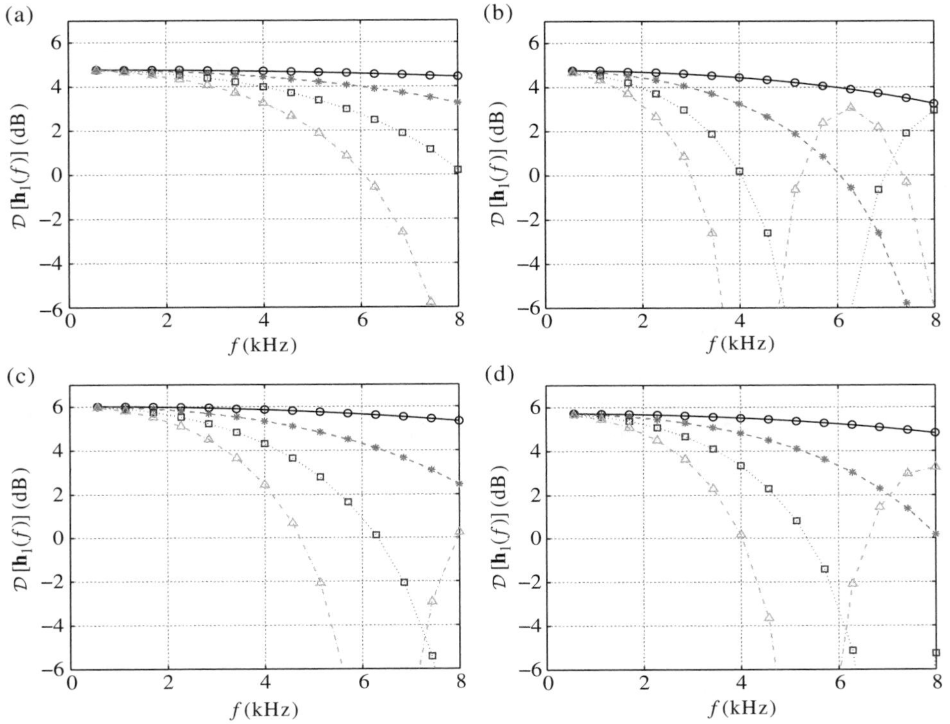

Figure 9.8 DF of the first-order DSAs as a function of frequency, for several values of δ: $\delta = 1$ cm (solid line with circles), $\delta = 2$ cm (dashed line with asterisks), $\delta = 3$ cm (dotted line with squares), and $\delta = 4$ cm (dash-dot line with triangles). (a) Dipole, (b) cardioid, (c) hypercardioid, and (d) supercardioid.

we find that the inverse of $\mathbf{V}(f)$ is

$$\mathbf{V}^{-1}(f) = \begin{bmatrix} \frac{\nu_2\nu_3}{(\nu_2-\nu_1)(\nu_3-\nu_1)} & -\frac{\nu_1\nu_3}{(\nu_2-\nu_1)(\nu_3-\nu_2)} & \frac{\nu_1\nu_2}{(\nu_3-\nu_1)(\nu_3-\nu_2)} \\ -\frac{\nu_2+\nu_3}{(\nu_2-\nu_1)(\nu_3-\nu_1)} & \frac{\nu_1+\nu_3}{(\nu_2-\nu_1)(\nu_3-\nu_2)} & -\frac{\nu_1+\nu_2}{(\nu_3-\nu_1)(\nu_3-\nu_2)} \\ \frac{1}{(\nu_2-\nu_1)(\nu_3-\nu_1)} & -\frac{1}{(\nu_2-\nu_1)(\nu_3-\nu_2)} & \frac{1}{(\nu_3-\nu_1)(\nu_3-\nu_2)} \end{bmatrix}. \tag{9.68}$$

This inverse can be of great help in designing second-order DSAs. We deduce that the beamformer is

$$\mathbf{h}_2(f) = \mathbf{U}(f)\mathbf{L}(f)\begin{bmatrix} 1 \\ \beta_{2,1} \\ \beta_{2,2} \end{bmatrix}. \tag{9.69}$$

While this approach is very general, it is not applicable to beampatterns that have a zero with multiplicity greater than 1. Let us show how to design a beampattern that has a zero, $\alpha_{2,1}$, with multiplicity 2. The theoretical DSA beampattern of such a case is

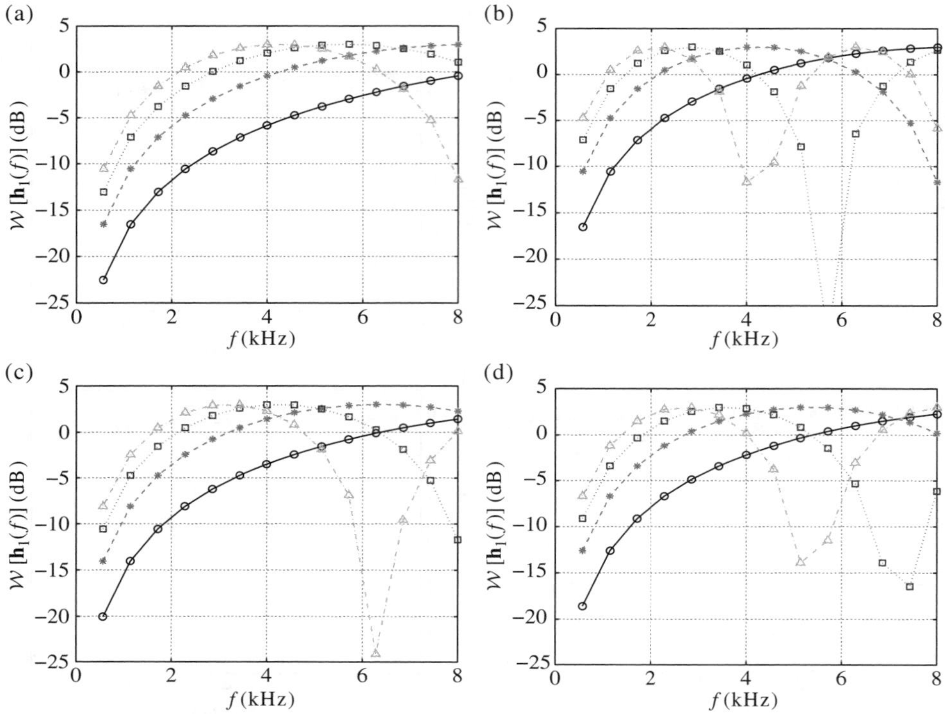

Figure 9.9 WNG of the first-order DSAs as a function of frequency, for several values of δ: $\delta = 1$ cm (solid line with circles), $\delta = 2$ cm (dashed line with asterisks), $\delta = 3$ cm (dotted line with squares), and $\delta = 4$ cm (dash-dot line with triangles). (a) Dipole, (b) cardioid, (c) hypercardioid, and (d) supercardioid.

$$\mathcal{B}\left(\alpha_{2,1}, \alpha\right) = \frac{1}{\left(1 - \alpha_{2,1}\right)^2} \left(\alpha - \alpha_{2,1}\right)^2, \tag{9.70}$$

where $\alpha = \cos\theta$. It is clear that the derivative of $\mathcal{B}\left(\alpha_{2,1}, \alpha\right)$ with respect to α at $\alpha_{2,1}$ is

$$\left. \frac{d\mathcal{B}\left(\alpha_{2,1}, \alpha\right)}{d\alpha} \right|_{\alpha=\alpha_{2,1}} = 0. \tag{9.71}$$

Applying this property to the beamformer beampattern, we get

$$\left. \frac{d\mathcal{B}\left[\mathbf{h}(f), \alpha\right]}{d\alpha} \right|_{\alpha=\alpha_{2,1}} = \jmath 2\pi f \tau_0 \left[\mathbf{\Sigma d}\left(f, \alpha_{2,1}\right)\right]^H \mathbf{h}(f) = 0, \tag{9.72}$$

where

$$\mathbf{\Sigma} = \text{diag}\left(0, 1, 2\right) \tag{9.73}$$

is a diagonal matrix. From (9.72), we deduce the constraint equation:

$$\left[\boldsymbol{\Sigma}\mathbf{d}\left(f, \alpha_{2,1}\right)\right]^H \mathbf{h}(f) = 0. \tag{9.74}$$

Combining the distortionless constraint, the null constraint in the direction $\alpha_{2,1}$; that is,

$$\mathbf{d}\left(f, \alpha_{2,1}\right) \mathbf{h}(f) = 0, \tag{9.75}$$

and (9.74), we obtain

$$\begin{bmatrix} \mathbf{d}^H\left(f, 1\right) \\ \mathbf{d}^H\left(f, \alpha_{2,1}\right) \\ \left[\boldsymbol{\Sigma}\mathbf{d}\left(f, \alpha_{2,1}\right)\right]^H \end{bmatrix} \mathbf{h}(f) = \begin{bmatrix} 1 \\ 0 \\ 0 \end{bmatrix}. \tag{9.76}$$

It is straightforward to see that the solution is

$$\mathbf{h}_{2,0}(f) = \frac{1}{\left[1 - e^{\jmath 2\pi f \tau_0\left(1-\alpha_{2,1}\right)}\right]^2} \begin{bmatrix} 1 \\ -2e^{-\jmath 2\pi f \tau_0 \alpha_{2,1}} \\ e^{-\jmath 4\pi f \tau_0 \alpha_{2,1}} \end{bmatrix}. \tag{9.77}$$

Because of the different particular constraints, it is obvious that the beampattern has the form:

$$\mathcal{B}\left[\mathbf{h}_{2,0}(f), \cos\theta\right] = \frac{\left[1 - e^{\jmath 2\pi f \tau_0\left(\cos\theta-\alpha_{2,1}\right)}\right]^2}{\left[1 - e^{\jmath 2\pi f \tau_0\left(1-\alpha_{2,1}\right)}\right]^2}. \tag{9.78}$$

With assumption (i) and the approximation in (9.56), we can rewrite this beampattern as

$$\mathcal{B}\left[\mathbf{h}_{2,0}(f), \cos\theta\right] \approx \frac{1}{\left(1 - \alpha_{2,1}\right)^2} \left(\cos\theta - \alpha_{2,1}\right)^2, \tag{9.79}$$

which is the expected result.

We find that the WNG is

$$\begin{aligned} \mathcal{W}\left[\mathbf{h}_{2,0}(f)\right] &= \frac{1}{6} \left|1 - e^{\jmath 2\pi f \tau_0\left(1-\alpha_{2,1}\right)}\right|^4 \\ &= \frac{2}{3} \left\{1 - \cos\left[2\pi f \tau_0 \left(1 - \alpha_{2,1}\right)\right]\right\}^2, \end{aligned} \tag{9.80}$$

which can be approximated as

$$\mathcal{W}\left[\mathbf{h}_{2,0}(f)\right] \approx \frac{1}{6} \left[2\pi f \tau_0 \left(1 - \alpha_{2,1}\right)\right]^4. \tag{9.81}$$

The WNG of the beamformer with second-order design is much worse than the WNG of the beamformer with first-order design.

9.7.2 Design Examples

In this section, we design and compare two second-order DSAs. The first is a second-order cardioid with $\mathbf{h}_{2,0}(f)$ and $\alpha_{2,1} = -1$, which has a unique multiple null at $\theta = \pi$. The second DSA is a second-order cardioid with $\mathbf{h}_2(f)$, $\alpha_{2,1} = -1,\ \beta_{2,1} = 0, \alpha_{2,2} = 0,\ \beta_{2,2} = 0$, which has two distinct nulls at $\theta = \frac{\pi}{2}$ and π.

Figures 9.10 and 9.11 display the patterns of the two second-order cardioids for low and high frequencies and two values of δ. As long as the sensor spacing is small, the beampatterns of the second-order DSAs are frequency independent. When the sensor spacing is too large, the beampatterns at high frequencies deteriorate.

Figure 9.12 shows plots of the DFs of the two second-order cardioids, $\mathcal{D}\left[\mathbf{h}_{2,0}(f)\right]$ and $\mathcal{D}\left[\mathbf{h}_2(f)\right]$, as a function of frequency for several values of δ. Corresponding plots of the WNG, $\mathcal{W}\left[\mathbf{h}_{2,0}(f)\right]$ and $\mathcal{W}\left[\mathbf{h}_2(f)\right]$, as a function of frequency are depicted in Figure 9.13. We observe that the DF of the second-order cardioid with two distinct nulls is higher

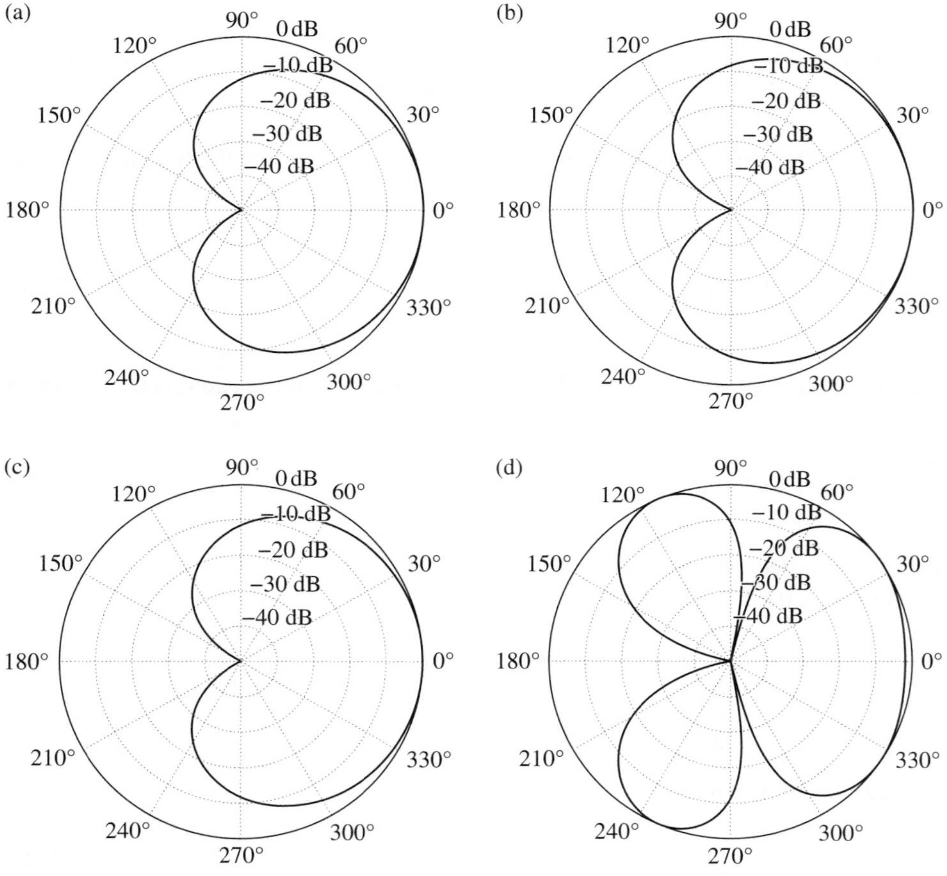

Figure 9.10 Beampatterns of the second-order cardioid, $\mathbf{h}_{2,0}(f)$, with a unique multiple null at $\theta = \pi$, for low and high frequencies, and two values of δ: (a) $f = 0.5$ kHz, $\delta = 1$ cm, (b) $f = 7$ kHz, $\delta = 1$ cm, (c) $f = 0.5$ kHz, $\delta = 4$ cm, and (d) $f = 7$ kHz, $\delta = 4$ cm.

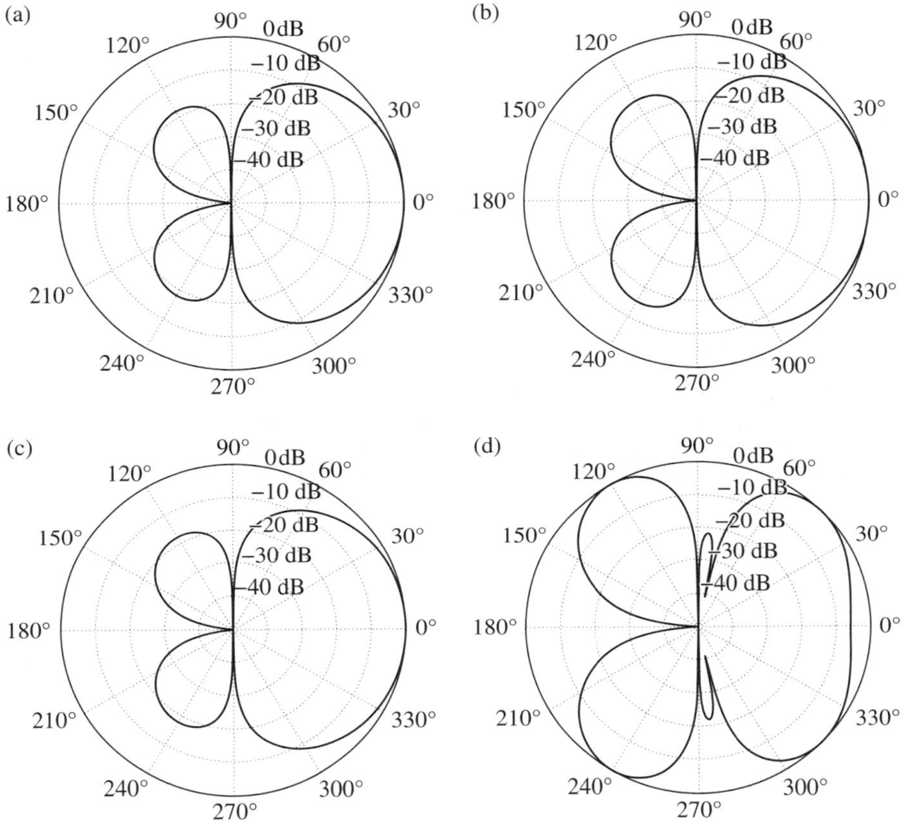

Figure 9.11 Beampatterns of the second-order cardioid, $\mathbf{h}_2(f)$, with two distinct nulls at $\theta = \frac{\pi}{2}$ and π, for low and high frequencies, and two values of δ: (a) $f = 0.5$ kHz, $\delta = 1$ cm, (b) $f = 7$ kHz, $\delta = 1$ cm, (c) $f = 0.5$ kHz, $\delta = 4$ cm, and (d) $f = 7$ kHz, $\delta = 4$ cm.

than that of the second-order cardioid with a unique multiple null, but at the expense of lower WNG. Furthermore, similar to the first-order DSAs, increasing the sensor spacing enables the WNG to be increased, especially at low frequencies. However, a large value of δ contradicts the DSA assumption, which results in deterioration of the beampatterns at high frequencies. Therefore, the value of δ should be a compromise between white noise amplification at low frequencies and a frequency-independent directivity pattern at high frequencies.

9.8 Third-order Design

9.8.1 Principle

We start this section by deriving an important family of third-order differential beamformers, the beampatterns of which have three distinct nulls. This can be done with

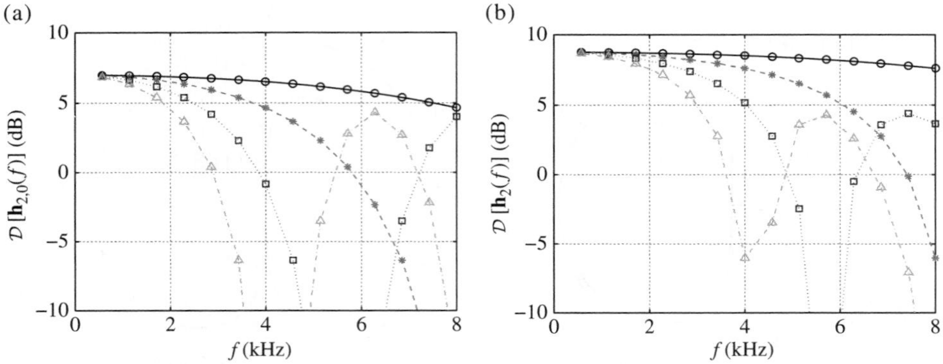

Figure 9.12 DF of second-order DSAs as a function of frequency, for several values of δ: $\delta = 1$ cm (solid line with circles), $\delta = 2$ cm (dashed line with asterisks), $\delta = 3$ cm (dotted line with squares), and $\delta = 4$ cm (dash-dot line with triangles). (a) Second-order cardioid, $\mathbf{h}_{2,0}(f)$, with a unique multiple null at $\theta = \pi$, and (b) second-order cardioid, $\mathbf{h}_2(f)$, with two distinct nulls at $\theta = \frac{\pi}{2}$ and π.

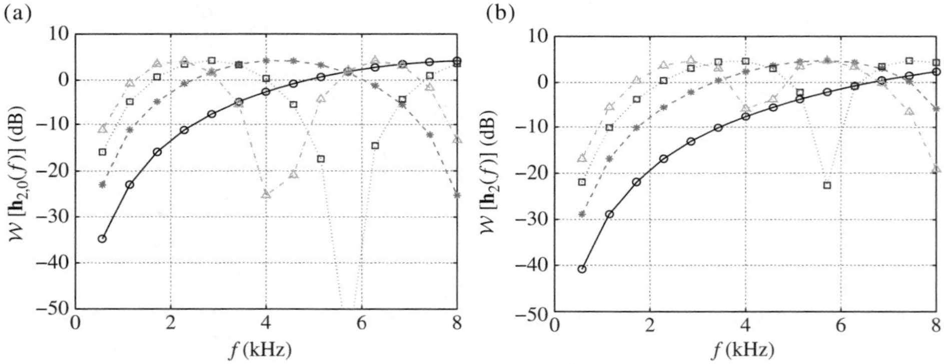

Figure 9.13 WNG of second-order DSAs as a function of frequency, for several values of δ: $\delta = 1$ cm (solid line with circles), $\delta = 2$ cm (dashed line with asterisks), $\delta = 3$ cm (dotted line with squares), and $\delta = 4$ cm (dash-dot line with triangles). (a) Second-order cardioid, $\mathbf{h}_{2,0}(f)$, with a unique multiple null at $\theta = \pi$, and (b) second-order cardioid, $\mathbf{h}_2(f)$, with two distinct nulls at $\theta = \frac{\pi}{2}$ and π.

exactly four omnidirectional sensors. It is clear that the linear system of four equations tailored for the derivation of such beamformers is

$$\begin{bmatrix} \mathbf{d}^H(f, 1) \\ \mathbf{d}^H(f, \alpha_{3,1}) \\ \mathbf{d}^H(f, \alpha_{3,2}) \\ \mathbf{d}^H(f, \alpha_{3,3}) \end{bmatrix} \mathbf{h}(f) = \begin{bmatrix} 1 \\ 0 \\ 0 \\ 0 \end{bmatrix}, \tag{9.82}$$

where $-1 \leq \alpha_{3,1} = \cos\theta_{3,1} < 1$, $-1 \leq \alpha_{3,2} = \cos\theta_{3,2} < 1$, $-1 \leq \alpha_{3,3} = \cos\theta_{3,3} < 1$, and $\alpha_{3,1} \neq \alpha_{3,2} \neq \alpha_{3,3}$. We denote by

$$\mathbf{V}(f) = \begin{bmatrix} \mathbf{d}^H(f,1) \\ \mathbf{d}^H(f,\alpha_{3,1}) \\ \mathbf{d}^H(f,\alpha_{3,2}) \\ \mathbf{d}^H(f,\alpha_{3,3}) \end{bmatrix} = \begin{bmatrix} 1 & v_1(f) & v_1^2(f) & v_1^3(f) \\ 1 & v_2(f) & v_2^2(f) & v_2^3(f) \\ 1 & v_3(f) & v_3^2(f) & v_3^3(f) \\ 1 & v_4(f) & v_4^2(f) & v_4^3(f) \end{bmatrix} \quad (9.83)$$

the 4×4 Vandermonde matrix that appears in (9.82), where $v_1(f) = e^{\jmath 2\pi f \tau_0}$, $v_2(f) = e^{\jmath 2\pi f \tau_0 \alpha_{3,1}}$, $v_3(f) = e^{\jmath 2\pi f \tau_0 \alpha_{3,2}}$, and $v_4(f) = e^{\jmath 2\pi f \tau_0 \alpha_{3,3}}$. Because of the structure of the vector on the right-hand side of (9.82), we only need to compute the first column of $\mathbf{V}^{-1}(f)$ to find $\mathbf{h}(f)$. Using the decomposition $\mathbf{V}^{-1}(f) = \mathbf{U}(f)\mathbf{L}(f)$ [4], the matrix $\mathbf{U}(f)$, and the first column of $\mathbf{L}(f)$, we find that the first column of $\mathbf{V}^{-1}(f)$ is

$$\mathbf{V}^{-1}(f; :, 1) = \begin{bmatrix} \dfrac{v_2 v_3 v_4}{(v_2 - v_1)(v_3 - v_1)(v_4 - v_1)} \\ -\dfrac{v_2 v_3 + v_3 v_4 + v_2 v_4}{(v_2 - v_1)(v_3 - v_1)(v_4 - v_1)} \\ \dfrac{v_2 + v_3 + v_4}{(v_2 - v_1)(v_3 - v_1)(v_4 - v_1)} \\ -\dfrac{1}{(v_2 - v_1)(v_3 - v_1)(v_4 - v_1)} \end{bmatrix}. \quad (9.84)$$

From the previous expression, we easily find that the solution is

$$\mathbf{h}_3(f) = \frac{1}{\left[1 - e^{\jmath 2\pi f \tau_0 (1-\alpha_{3,1})}\right]\left[1 - e^{\jmath 2\pi f \tau_0 (1-\alpha_{3,2})}\right]\left[1 - e^{\jmath 2\pi f \tau_0 (1-\alpha_{3,3})}\right]} \times \begin{bmatrix} 1 \\ -e^{-\jmath 2\pi f \tau_0 \alpha_{3,1}} - e^{-\jmath 2\pi f \tau_0 \alpha_{3,2}} - e^{-\jmath 2\pi f \tau_0 \alpha_{3,3}} \\ e^{-\jmath 2\pi f \tau_0 (\alpha_{3,1}+\alpha_{3,2})} + e^{-\jmath 2\pi f \tau_0 (\alpha_{3,2}+\alpha_{3,3})} + e^{-\jmath 2\pi f \tau_0 (\alpha_{3,1}+\alpha_{3,3})} \\ -e^{-\jmath 2\pi f \tau_0 (\alpha_{3,1}+\alpha_{3,2}+\alpha_{3,3})} \end{bmatrix}. \quad (9.85)$$

Now, let us derive differential beamformers with beampatterns having a unique null in the direction $\alpha_{3,1}$ with multiplicity 3. Using the facts that

$$\left. \frac{d\mathcal{B}\left[\mathbf{h}(f), \alpha\right]}{d\alpha} \right|_{\alpha=\alpha_{3,1}} = \jmath 2\pi f \tau_0 \left[\mathbf{\Sigma d}\left(f, \alpha_{3,1}\right)\right]^H \mathbf{h}(f) = 0 \quad (9.86)$$

and

$$\left. \frac{d^2\mathcal{B}\left[\mathbf{h}(f), \alpha\right]}{d\alpha^2} \right|_{\alpha=\alpha_{3,1}} = \left(\jmath 2\pi f \tau_0\right)^2 \left[\mathbf{\Sigma}^2\mathbf{d}\left(f, \alpha_{3,1}\right)\right]^H \mathbf{h}(f) = 0, \quad (9.87)$$

where

$$\mathbf{\Sigma} = \text{diag}\,(0, 1, 2, 3) \tag{9.88}$$

is a diagonal matrix, we easily find that the linear system to solve is

$$\begin{bmatrix} \mathbf{d}^H\left(f, 1\right) \\ \mathbf{d}^H\left(f, \alpha_{3,1}\right) \\ \left[\mathbf{\Sigma}\mathbf{d}\left(f, \alpha_{3,1}\right)\right]^H \\ \left[\mathbf{\Sigma}^2\mathbf{d}\left(f, \alpha_{3,1}\right)\right]^H \end{bmatrix} \mathbf{h}(f) = \begin{bmatrix} 1 \\ 0 \\ 0 \\ 0 \end{bmatrix}. \tag{9.89}$$

We deduce that the solution is

$$\mathbf{h}_{3,0}(f) = \frac{1}{\left[1 - e^{\jmath 2\pi f \tau_0 \left(1-\alpha_{3,1}\right)}\right]^3} \begin{bmatrix} 1 \\ -3e^{-\jmath 2\pi f \tau_0 \alpha_{3,1}} \\ 3e^{-\jmath 4\pi f \tau_0 \alpha_{2,1}} \\ -e^{-\jmath 6\pi f \tau_0 \alpha_{2,1}} \end{bmatrix}. \tag{9.90}$$

The beampattern corresponding to the beamformer $\mathbf{h}_{3,0}(f)$ is

$$\mathcal{B}\left[\mathbf{h}_{3,0}(f), \cos\theta\right] = \frac{\left[1 - e^{\jmath 2\pi f \tau_0 \left(\cos\theta - \alpha_{3,1}\right)}\right]^3}{\left[1 - e^{\jmath 2\pi f \tau_0 \left(1-\alpha_{3,1}\right)}\right]^3} \tag{9.91}$$

and can be approximated as

$$\mathcal{B}\left[\mathbf{h}_{3,0}(f), \cos\theta\right] \approx \frac{1}{\left(1 - \alpha_{3,1}\right)^3} \left(\cos\theta - \alpha_{3,1}\right)^3, \tag{9.92}$$

which is identical to the theoretical third-order DSA beampattern with a unique null with multiplicity 3.

The WNG is

$$\begin{aligned} \mathcal{W}\left[\mathbf{h}_{3,0}(f)\right] &= \frac{1}{20} \left|1 - e^{\jmath 2\pi f \tau_0 \left(1-\alpha_{3,1}\right)}\right|^6 \\ &= \frac{2}{5} \left\{1 - \cos\left[2\pi f \tau_0 \left(1 - \alpha_{3,1}\right)\right]\right\}^3, \end{aligned} \tag{9.93}$$

which we can approximate as

$$\mathcal{W}\left[\mathbf{h}_{3,0}(f)\right] \approx \frac{1}{20} \left[2\pi f \tau_0 \left(1 - \alpha_{3,1}\right)\right]^6. \tag{9.94}$$

The generalization of the beamformers $\mathbf{h}_3(f)$ and $\mathbf{h}_{3,0}(f)$ to any order is straightforward.

It is also possible to derive differential beamformers directly from some of the performance measures. There are two possibilities.

The first beamformer is obtained by maximizing the DF as defined in (9.24). Considering the distortionless constraint, we easily get the hypercardioid of order $M-1$:

$$\mathbf{h}_{\mathrm{Hd}}(f) = \frac{\mathbf{\Gamma}_{0,\pi}^{-1}(f)\mathbf{d}\left(f,1\right)}{\mathbf{d}^H\left(f,1\right)\mathbf{\Gamma}_{0,\pi}^{-1}(f)\mathbf{d}\left(f,1\right)}, \tag{9.95}$$

which is the superdirective beamformer described in Chapter 7.

The second differential beamformer is obtained by maximizing the FBR as defined in (9.10). If we denote by $\mathbf{t}_1(f)$ the eigenvector corresponding to the maximum eigenvalue of the matrix $\mathbf{\Gamma}_{\pi/2,\pi}^{-1}(f)\mathbf{\Gamma}_{0,\pi/2}(f)$ and taking into account the distortionless constraint, we get the supercardioid of order $M-1$:

$$\mathbf{h}_{\mathrm{Sd}}(f) = \frac{\mathbf{t}_1(f)}{\mathbf{d}^H\left(f,1\right)\mathbf{t}_1(f)}. \tag{9.96}$$

9.8.2 Design Examples

In this section, we design and compare four third-order DSAs. The first is a third-order cardioid with $\mathbf{h}_{3,0}(f)$ and $\alpha_{3,1} = -1$, and has a unique multiple null at $\theta = \pi$. The second DSA is a third-order DSA with $\mathbf{h}_3(f)$, $\alpha_{3,1} = 0$, $\alpha_{3,2} = -\frac{1}{2}$, and $\alpha_{3,3} = -1$, and has three distinct nulls at $\theta = \frac{\pi}{2}, \frac{2\pi}{3}$ and π. The third and fourth DSAs are, respectively, the third-order hypercardioid with $\mathbf{h}_{\mathrm{Hd}}(f)$ and the third-order supercardioid with $\mathbf{h}_{\mathrm{Sd}}(f)$.

Figures 9.14–9.17 display the patterns of the four third-order DSAs for low and high frequencies and two values of δ. As long as the sensor spacing is small, the beampatterns of the third-order DSAs are frequency independent. When the sensor spacing is too large, the beampatterns at high frequencies deteriorate.

Figure 9.18 shows plots of the DFs of the four third-order DSAs, $\mathcal{D}\left[\mathbf{h}_{3,0}(f)\right], \mathcal{D}\left[\mathbf{h}_3(f)\right]$, $\mathcal{D}\left[\mathbf{h}_{\mathrm{Hd}}(f)\right]$, and $\mathcal{D}\left[\mathbf{h}_{\mathrm{Sd}}(f)\right]$, as a function of frequency for several values of δ. Corresponding plots of the WNG, $\mathcal{W}\left[\mathbf{h}_{3,0}(f)\right]$, $\mathcal{W}\left[\mathbf{h}_3(f)\right]$, $\mathcal{W}\left[\mathbf{h}_{\mathrm{Hd}}(f)\right]$, and $\mathcal{W}\left[\mathbf{h}_{\mathrm{Sd}}(f)\right]$, as a function of frequency are depicted in Figure 9.19. We observe that the highest DF is obtained with the third-order hypercardioid, but at the cost of the lowest WNG. The highest WNG is obtained with the third-order cardioid that has a unique multiple null, but at the cost of the lowest DF. The DF of the third-order cardioid with three distinct nulls is higher than that of the third-order cardioid with a unique multiple null, but at the expense of lower WNG. Furthermore, similar to the first- and second-order DSAs, increasing the sensor spacing enables the WNG to be increased, especially at low frequencies. However, a large value of δ contradicts the DSA assumption, which results in deterioration of the beampatterns at high frequencies. Therefore, the value of δ should be a compromise between white noise amplification at low frequencies and a frequency-independent directivity pattern at high frequencies.

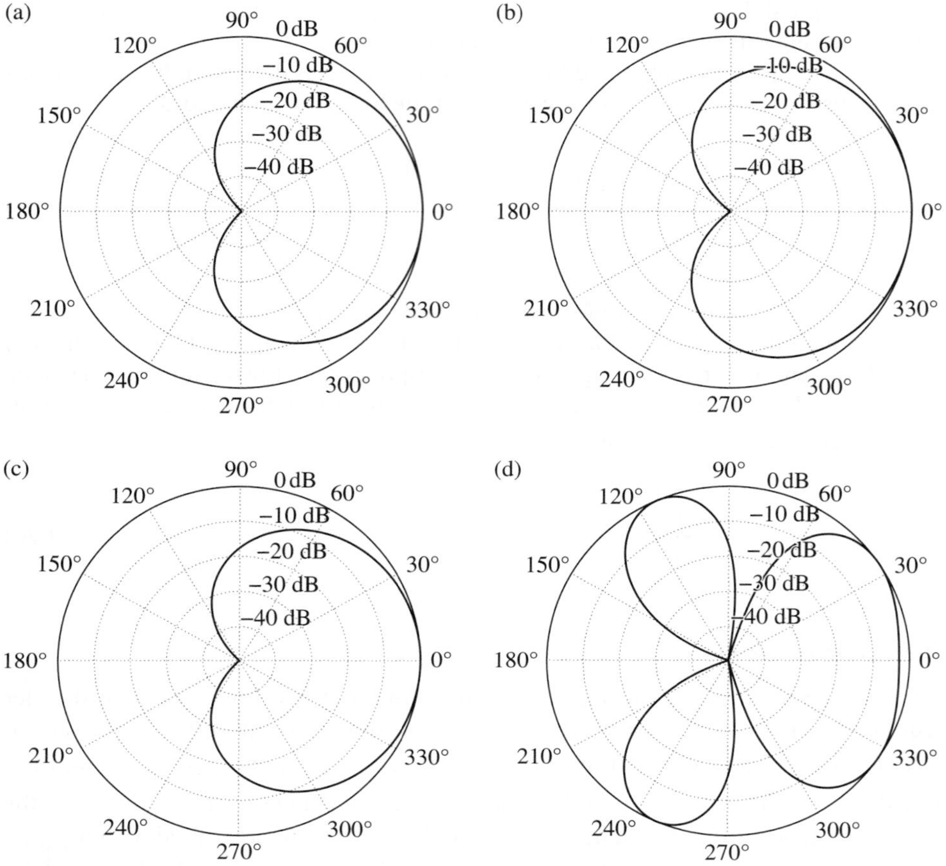

Figure 9.14 Beampatterns of the third-order cardioid, $\mathbf{h}_{3,0}(f)$, with a unique multiple null at $\theta = \pi$, for low and high frequencies, and two values of δ: (a) $f = 0.5$ kHz, $\delta = 1$ cm, (b) $f = 7$ kHz, $\delta = 1$ cm, (c) $f = 0.5$ kHz, $\delta = 4$ cm, and (d) $f = 7$ kHz, $\delta = 4$ cm.

9.9 Minimum-norm Beamformers

9.9.1 Principle

In the three previous sections, we could see that the major drawback of DSAs is white noise amplification. As the order increases, the amplification of white noise worsens. The best way to deal with this fundamental problem is to disconnect the order of the DSAs from the number of sensors and increase the latter for a fixed order. Consequently, we can use this degree of freedom to maximize the WNG [4].

We know from the previous sections that any DSA of order N can be designed by solving the linear system of $N + 1$ equations:

$$\mathbf{D}\left(f, \boldsymbol{\alpha}\right)\mathbf{h}(f) = \boldsymbol{\beta}, \tag{9.97}$$

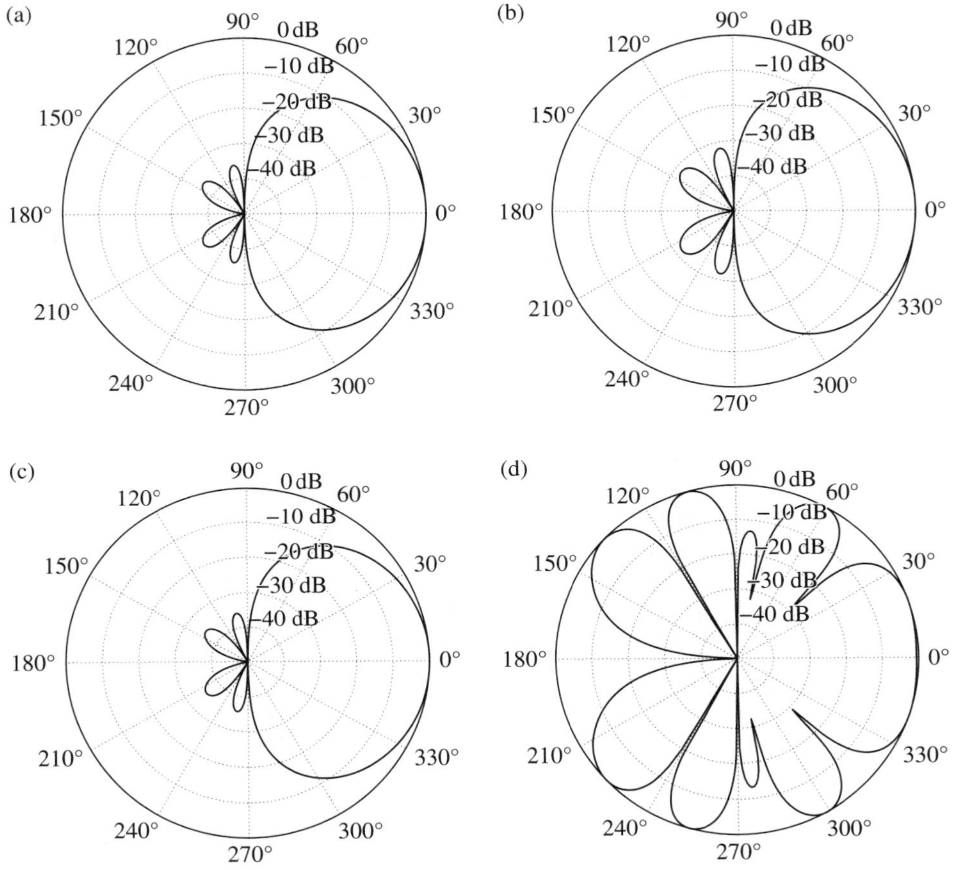

Figure 9.15 Beampatterns of the third-order DSA, $\mathbf{h}_3(f)$, with three distinct nulls at $\theta = \frac{\pi}{2}, \frac{2\pi}{3}$, and π, for low and high frequencies, and two values of δ: (a) $f = 0.5$ kHz, $\delta = 1$ cm, (b) $f = 7$ kHz, $\delta = 1$ cm, (c) $f = 0.5$ kHz, $\delta = 4$ cm, and (d) $f = 7$ kHz, $\delta = 4$ cm.

where

$$\mathbf{D}\left(f, \boldsymbol{\alpha}\right) = \begin{bmatrix} \mathbf{d}^H\left(f, 1\right) \\ \mathbf{d}^H\left(f, \alpha_{N,1}\right) \\ \vdots \\ \mathbf{d}^H\left(f, \alpha_{N,N}\right) \end{bmatrix} \tag{9.98}$$

is the constraint matrix of size $(N+1) \times M$, M is the number of sensors,

$$\mathbf{d}\left(f, \alpha_{N,n}\right) = \begin{bmatrix} 1 & e^{-\jmath 2\pi f \tau_0 \alpha_{N,n}} & \cdots & e^{-\jmath (M-1) 2\pi f \tau_0 \alpha_{N,n}} \end{bmatrix}^T, \quad n = 1, 2, \ldots, N \tag{9.99}$$

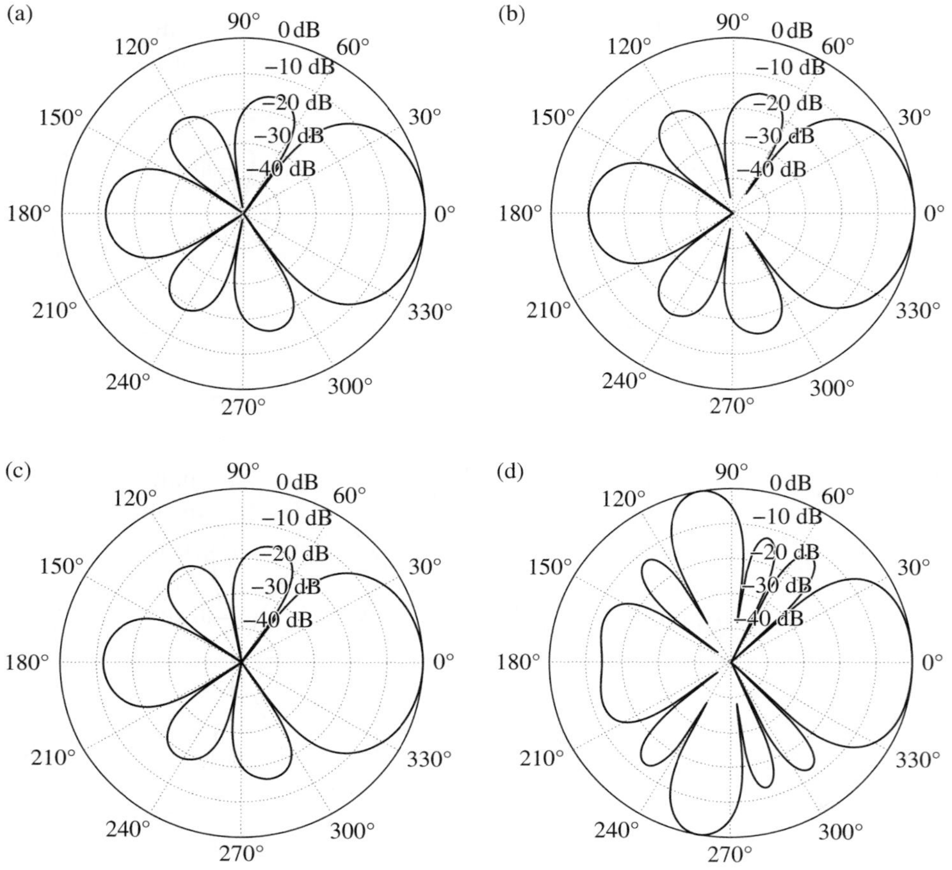

Figure 9.16 Beampatterns of the third-order hypercardioid, $\mathbf{h}_{\text{Hd}}(f)$, for low and high frequencies, and two values of δ: (a) $f = 0.5$ kHz, $\delta = 1$ cm, (b) $f = 7$ kHz, $\delta = 1$ cm, (c) $f = 0.5$ kHz, $\delta = 4$ cm, and (d) $f = 7$ kHz, $\delta = 4$ cm.

is a steering vector of length M,

$$\mathbf{h}(f) = \begin{bmatrix} H_1(f) & H_2(f) & \cdots & H_M(f) \end{bmatrix}^T \tag{9.100}$$

is a filter of length M, and

$$\boldsymbol{\alpha} = \begin{bmatrix} 1 & \alpha_{N,1} & \cdots & \alpha_{N,N} \end{bmatrix}^T, \tag{9.101}$$

$$\boldsymbol{\beta} = \begin{bmatrix} 1 & \beta_{N,1} & \cdots & \beta_{N,N} \end{bmatrix}^T, \tag{9.102}$$

are vectors of length $N+1$ containing the design coefficients of the directivity pattern. In previous sections, only the case $M = N+1$ was considered. This is also the case in all known approaches in the literature [3]. But, obviously from (9.97), nothing prevents us from taking $M > N+1$.

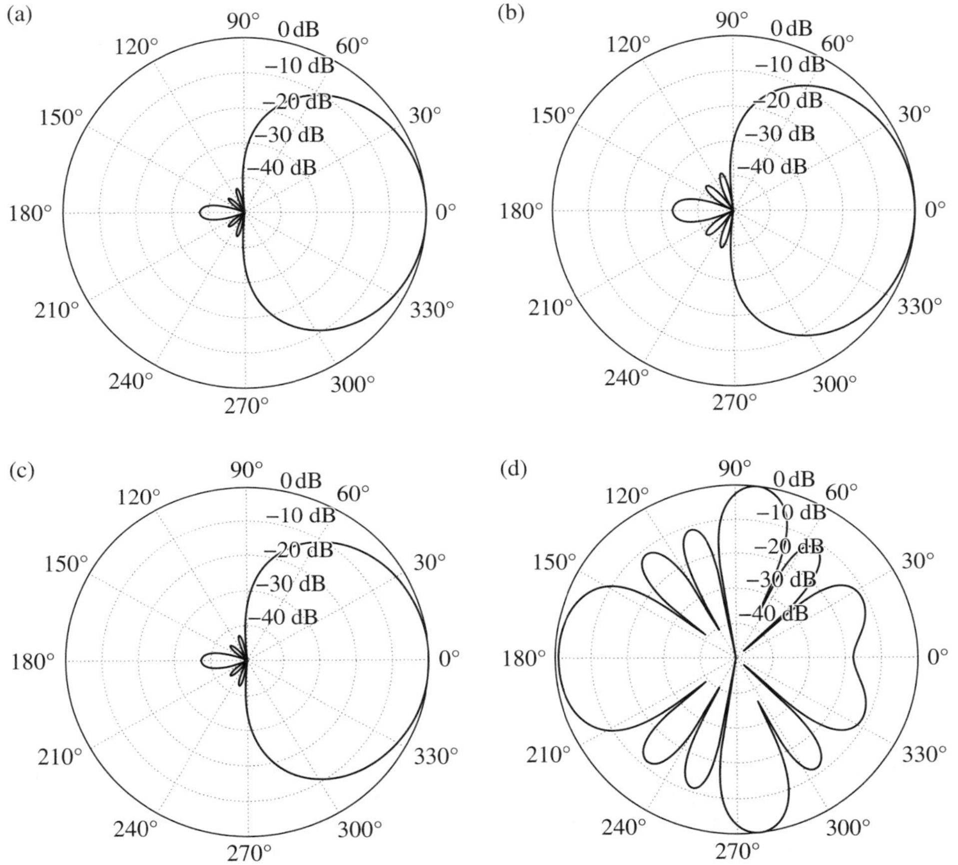

Figure 9.17 Beampatterns of the third-order supercardioid, $\mathbf{h}_{\mathrm{Sd}}(f)$, for low and high frequencies, and two values of δ: (a) $f = 0.5$ kHz, $\delta = 1$ cm, (b) $f = 7$ kHz, $\delta = 1$ cm, (c) $f = 0.5$ kHz, $\delta = 4$ cm, and (d) $f = 7$ kHz, $\delta = 4$ cm.

Now, assume that $M \geq N + 1$, then we can maximize the WNG subject to (9.97):

$$\min_{\mathbf{h}(f)} \mathbf{h}^H(f)\mathbf{h}(f) \quad \text{subject to} \quad \mathbf{D}\left(f, \boldsymbol{\alpha}\right)\mathbf{h}(f) = \boldsymbol{\beta}. \tag{9.103}$$

Obviously, the solution of the above problem is

$$\mathbf{h}_{\mathrm{MN}}\left(f, \boldsymbol{\alpha}, \boldsymbol{\beta}\right) = \mathbf{D}^H\left(f, \boldsymbol{\alpha}\right)\left[\mathbf{D}\left(f, \boldsymbol{\alpha}\right)\mathbf{D}^H\left(f, \boldsymbol{\alpha}\right)\right]^{-1}\boldsymbol{\beta}, \tag{9.104}$$

which is the minimum-norm solution of (9.97). The vectors $\boldsymbol{\alpha}$ and $\boldsymbol{\beta}$ of length $N + 1$ determine the beampattern and the order of the DSA. Basically, the lengths of these vectors determine (roughly) the order of the DSA while their values determine the beampattern. Meanwhile, the length, M, of the minimum-norm beamformer, $\mathbf{h}_{\mathrm{MN}}\left(f, \boldsymbol{\alpha}, \boldsymbol{\beta}\right)$,

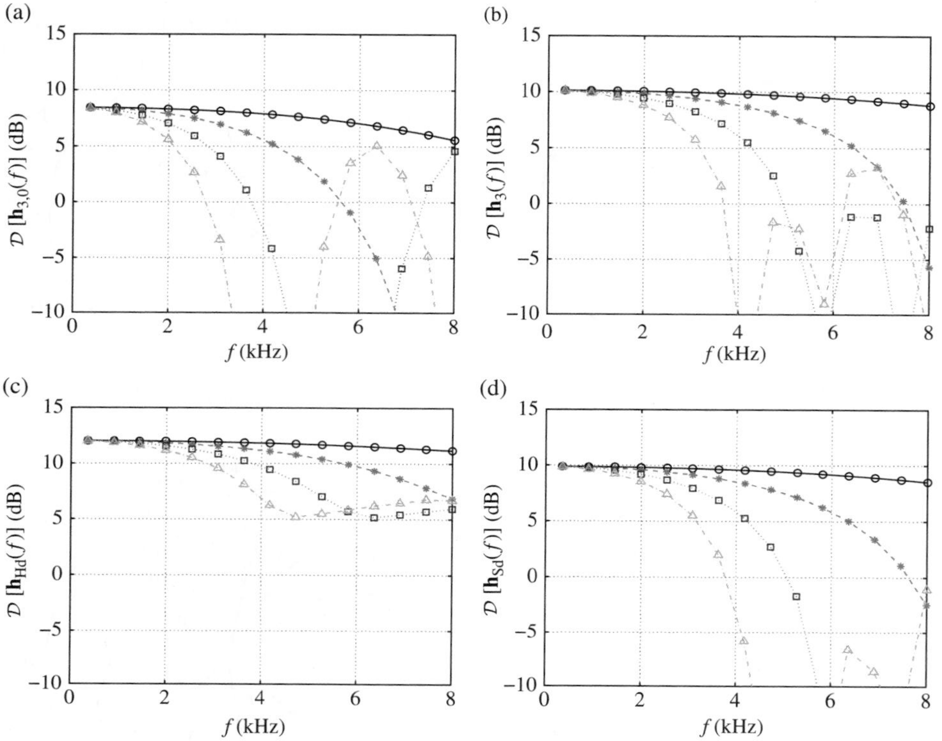

Figure 9.18 DF of third-order DSAs as a function of frequency, for several values of δ: $\delta = 1$ cm (solid line with circles), $\delta = 2$ cm (dashed line with asterisks), $\delta = 3$ cm (dotted line with squares), and $\delta = 4$ cm (dash-dot line with triangles). (a) Third-order cardioid, $\mathbf{h}_{3,0}(f)$, with a unique multiple null at $\theta = \pi$, (b) third-order DSA, $\mathbf{h}_3(f)$, with three distinct nulls at $\theta = \frac{\pi}{2}, \frac{2\pi}{3}$, and π, (c) third-order hypercardioid with $\mathbf{h}_{\mathrm{Hd}}(f)$, and (d) third-order supercardioid with $\mathbf{h}_{\mathrm{Sd}}(f)$.

can be much larger than $N + 1$, which will help make it robust against white noise amplification. In this case, the WNG should approach M and the order of the DSA may not be equal to N anymore, but the Nth-order DSA fundamental constraints will always be fulfilled. Because of this, the resulting shape of the directivity pattern may be slightly different than that obtained with $M = N + 1$.

It is easy to see that the beampattern, the WNG, and the DF of the minimum-norm beamformer are, respectively,

$$\begin{aligned} \mathcal{B}\left[\mathbf{h}_{\mathrm{MN}}\left(f, \boldsymbol{\alpha}, \boldsymbol{\beta}\right), \cos\theta\right] &= \mathbf{d}^H\left(f, \cos\theta\right) \mathbf{h}_{\mathrm{MN}}\left(f, \boldsymbol{\alpha}, \boldsymbol{\beta}\right) \\ &= \mathbf{d}^H\left(f, \cos\theta\right) \mathbf{D}^H\left(f, \boldsymbol{\alpha}\right) \left[\mathbf{D}\left(f, \boldsymbol{\alpha}\right) \mathbf{D}^H\left(f, \boldsymbol{\alpha}\right)\right]^{-1} \boldsymbol{\beta}, \end{aligned} \tag{9.105}$$

$$\mathcal{W}\left[\mathbf{h}_{\mathrm{MN}}\left(f, \boldsymbol{\alpha}, \boldsymbol{\beta}\right)\right] = \frac{1}{\boldsymbol{\beta}^T \left[\mathbf{D}\left(f, \boldsymbol{\alpha}\right) \mathbf{D}^H\left(f, \boldsymbol{\alpha}\right)\right]^{-1} \boldsymbol{\beta}}, \tag{9.106}$$

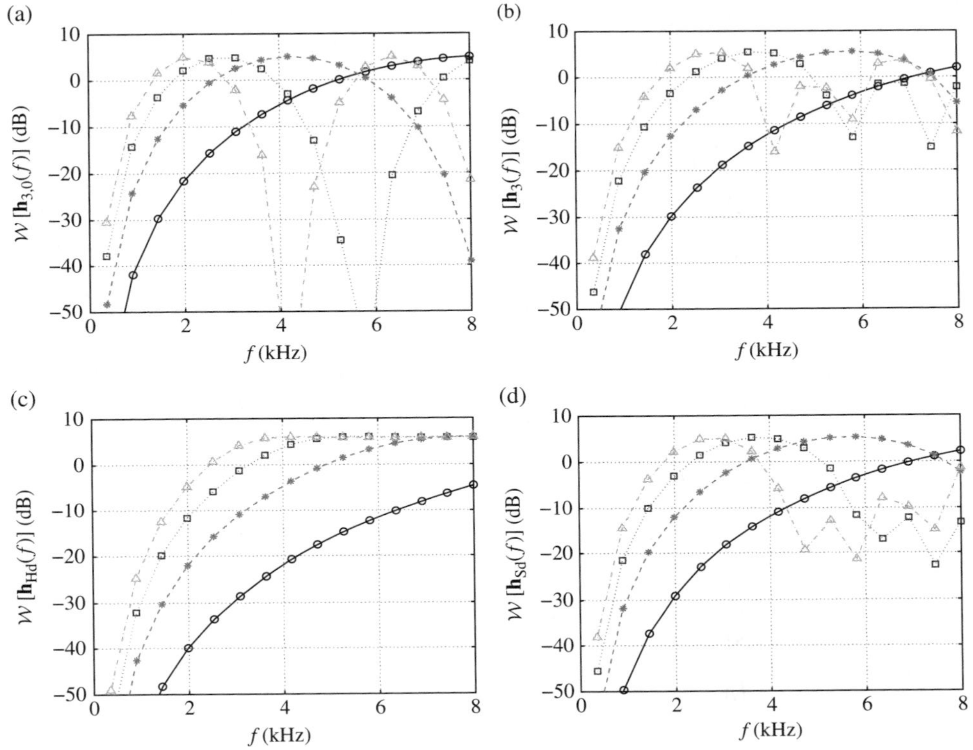

Figure 9.19 WNG of third-order DSAs as a function of frequency, for several values of δ: $\delta = 1$ cm (solid line with circles), $\delta = 2$ cm (dashed line with asterisks), $\delta = 3$ cm (dotted line with squares), and $\delta = 4$ cm (dash-dot line with triangles). (a) Third-order cardioid, $\mathbf{h}_{3,0}(f)$, with a unique multiple null at $\theta = \pi$, (b) third-order DSA, $\mathbf{h}_3(f)$, with three distinct nulls at $\theta = \frac{\pi}{2}, \frac{2\pi}{3}$, and π, (c) third-order hypercardioid with $\mathbf{h}_{\mathrm{Hd}}(f)$, and (d) third-order supercardioid with $\mathbf{h}_{\mathrm{Sd}}(f)$.

and

$$\mathcal{D}\left[\mathbf{h}_{\mathrm{MN}}\left(f,\boldsymbol{\alpha},\boldsymbol{\beta}\right)\right] = \frac{1}{\mathbf{h}_{\mathrm{MN}}^H\left(f,\boldsymbol{\alpha},\boldsymbol{\beta}\right)\boldsymbol{\Gamma}_{0,\pi}(f)\mathbf{h}_{\mathrm{MN}}\left(f,\boldsymbol{\alpha},\boldsymbol{\beta}\right)}. \tag{9.107}$$

In the same way, we can design a robust DSA with a beampattern having a null in the direction $\alpha_{N,1}$ with multiplicity N. The constraint equation is

$$\mathbf{D}_0\left(f,\alpha_{N,1}\right)\mathbf{h}(f) = \mathbf{i}_1, \tag{9.108}$$

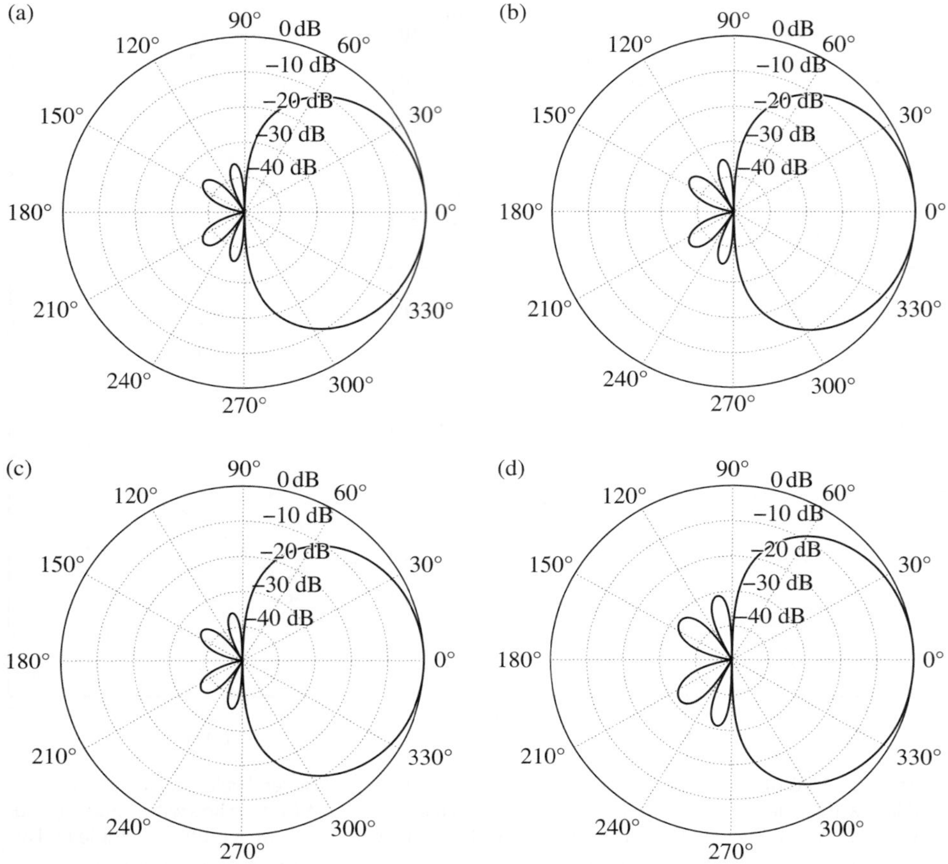

Figure 9.20 Beampatterns of a third-order DSA with three distinct nulls, $\mathbf{h}_{\rm MN}(f, \boldsymbol{\alpha}, \boldsymbol{\beta})$, for low and high frequencies, and two values of M: (a) $f = 0.5$ kHz, $M = 4$, (b) $f = 7$ kHz, $M = 4$, (c) $f = 0.5$ kHz, $M = 8$, and (d) $f = 7$ kHz, $M = 8$.

where

$$\mathbf{D}_0\left(f, \alpha_{N,1}\right) = \begin{bmatrix} \mathbf{d}^H\left(f, 1\right) \\ \mathbf{d}^H\left(f, \alpha_{N,1}\right) \\ \left[\mathbf{\Sigma d}\left(f, \alpha_{N,1}\right)\right]^H \\ \vdots \\ \left[\mathbf{\Sigma}^{N-1}\mathbf{d}\left(f, \alpha_{N,1}\right)\right]^H \end{bmatrix} \tag{9.109}$$

is a matrix of size $(N+1) \times M$, $\mathbf{i}_1$ is the first column of the $(N+1) \times (N+1)$ identity matrix, $\mathbf{I}_{N+1}$, and

$$\mathbf{\Sigma} = \text{diag}\left(0, 1, \ldots, M-1\right) \tag{9.110}$$

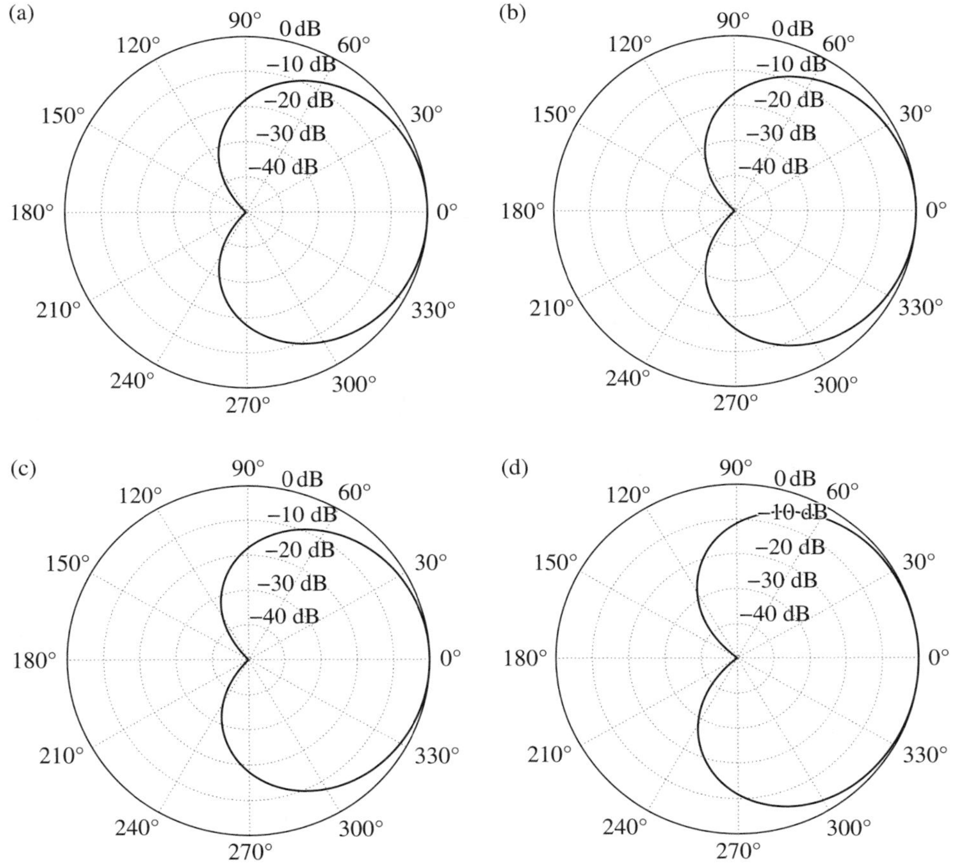

Figure 9.21 Beampatterns of a third-order cardioid, $\mathbf{h}_{\mathrm{MN},0}\left(f, \alpha_{3,1}\right)$, with $\alpha_{3,1} = -1$, for low and high frequencies, and two values of M: (a) $f = 0.5$ kHz, $M = 4$, (b) $f = 7$ kHz, $M = 4$, (c) $f = 0.5$ kHz, $M = 8$, and (d) $f = 7$ kHz, $M = 8$.

is a diagonal matrix. Assuming that $M \geq N + 1$, the maximization of the WNG subject to (9.108) leads to the minimum-norm beamformer:

$$\mathbf{h}_{\mathrm{MN},0}\left(f, \alpha_{N,1}\right) = \mathbf{D}_0^H\left(f, \alpha_{N,1}\right)\left[\mathbf{D}_0\left(f, \alpha_{N,1}\right)\mathbf{D}_0^H\left(f, \alpha_{N,1}\right)\right]^{-1}\mathbf{i}_1. \tag{9.111}$$

9.9.2 Design Examples

In this section, we demonstrate the effectiveness of the minimum-norm filter in the design of robust DSAs. Fundamentally, we exploit the fact that we have many more sensors than the order of the DSA. We design and compare two third-order DSAs with different numbers of sensors. The first is a a third-order DSA with three distinct nulls with $\mathbf{h}_{\mathrm{MN}}\left(f, \boldsymbol{\alpha}, \boldsymbol{\beta}\right)$. In this scenario, we have

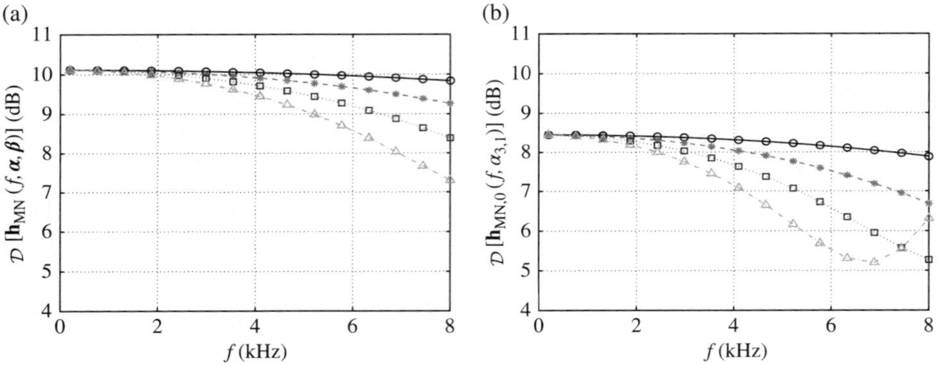

Figure 9.22 DF of third-order DSAs with minimum-norm filters as a function of frequency, for different values of M: $M = 4$ (solid line with circles), $M = 6$ (dashed line with asterisks), $M = 8$ (dotted line with squares), and $M = 10$ (dash-dot line with triangles). (a) Third-order DSA with three distinct nulls, $\mathbf{h}_{\mathrm{MN}}(f, \boldsymbol{\alpha}, \boldsymbol{\beta})$, and (b) third-order cardioid, $\mathbf{h}_{\mathrm{MN},0}\left(f, \alpha_{3,1}\right)$, with $\alpha_{3,1} = -1$.

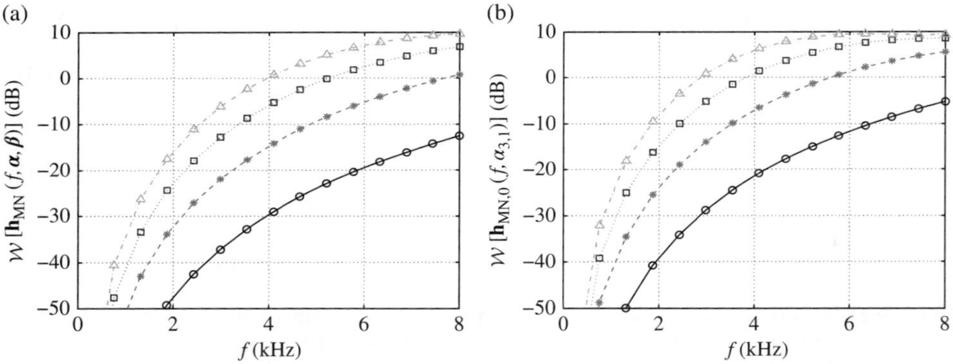

Figure 9.23 WNG of third-order DSAs with minimum-norm filters as a function of frequency, for different values of M: $M = 4$ (solid line with circles), $M = 6$ (dashed line with asterisks), $M = 8$ (dotted line with squares), and $M = 10$ (dash-dot line with triangles). (a) Third-order DSA with three distinct nulls, $\mathbf{h}_{\mathrm{MN}}(f, \boldsymbol{\alpha}, \boldsymbol{\beta})$, and (b) third-order cardioid, $\mathbf{h}_{\mathrm{MN},0}\left(f, \alpha_{3,1}\right)$, with $\alpha_{3,1} = -1$.

$$\boldsymbol{\alpha} = \begin{bmatrix} 1 & 0 & -\frac{1}{2} & -1 \end{bmatrix}^T, \tag{9.112}$$

$$\boldsymbol{\beta} = \begin{bmatrix} 1 & 0 & 0 & 0 \end{bmatrix}^T. \tag{9.113}$$

The second DSA is a third-order cardioid with $\mathbf{h}_{\mathrm{MN},0}\left(f, \alpha_{3,1}\right)$ and $\alpha_{3,1} = -1$. In both cases, the interelement spacing is $\delta = 5$ mm.

Figures 9.20 and 9.21 display the patterns of the two minimum-norm third-order DSAs for low and high frequencies, and two values of M. At low frequencies, the

Table 9.1 Differential beamformers.

Beamformer	
First-order	$\mathbf{h}_1(f) = \frac{1}{1 - e^{\jmath 2\pi f \tau_0 (1-\alpha_{1,1})}} \begin{bmatrix} 1 \\ -e^{-\jmath 2\pi f \tau_0 \alpha_{1,1}} \end{bmatrix}$
Second-order	$\mathbf{h}_2(f) = \mathbf{U}(f)\mathbf{L}(f) \begin{bmatrix} 1 \\ \beta_{2,1} \\ \beta_{2,2} \end{bmatrix}$
	$\mathbf{h}_{2,0}(f) = \frac{1}{\left[1 - e^{\jmath 2\pi f \tau_0 (1-\alpha_{2,1})}\right]^2} \begin{bmatrix} 1 \\ -2e^{-\jmath 2\pi f \tau_0 \alpha_{2,1}} \\ e^{-\jmath 4\pi f \tau_0 \alpha_{2,1}} \end{bmatrix}$
Third-order	Equation 9.85
	$\mathbf{h}_{3,0}(f) = \frac{1}{\left[1 - e^{\jmath 2\pi f \tau_0 (1-\alpha_{3,1})}\right]^3} \begin{bmatrix} 1 \\ -3e^{-\jmath 2\pi f \tau_0 \alpha_{3,1}} \\ 3e^{-\jmath 4\pi f \tau_0 \alpha_{2,1}} \\ -e^{-\jmath 6\pi f \tau_0 \alpha_{2,1}} \end{bmatrix}$
Hypercardioid	$\mathbf{h}_{\mathrm{Hd}}(f) = \frac{\mathbf{\Gamma}_{0,\pi}^{-1}(f)\mathbf{d}(f,1)}{\mathbf{d}^H(f,1)\mathbf{\Gamma}_{0,\pi}^{-1}(f)\mathbf{d}(f,1)}$
Supercardioid	$\mathbf{h}_{\mathrm{Sd}}(f) = \frac{\mathbf{t}_1(f)}{\mathbf{d}^H(f,1)\mathbf{t}_1(f)}$
Minimum-norm	$\mathbf{h}_{\mathrm{MN}}(f, \boldsymbol{\alpha}, \boldsymbol{\beta}) = \mathbf{D}^H(f, \boldsymbol{\alpha}) \left[\mathbf{D}(f, \boldsymbol{\alpha}) \mathbf{D}^H(f, \boldsymbol{\alpha})\right]^{-1} \boldsymbol{\beta}$
	$\mathbf{h}_{\mathrm{MN},0}(f, \alpha_{N,1}) =$ $\mathbf{D}_0^H(f, \alpha_{N,1}) \left[\mathbf{D}_0(f, \alpha_{N,1}) \mathbf{D}_0^H(f, \alpha_{N,1})\right]^{-1} \mathbf{i}_1$

patterns for $M = 8$ look similar to the patterns for $M = 4$. At high frequencies, the patterns for $M = 8$ look less directional than the patterns for $M = 4$.

Figure 9.22 shows plots of the DFs of the two third-order DSAs, $\mathcal{D}\left[\mathbf{h}_{\mathrm{MN}}(f, \boldsymbol{\alpha}, \boldsymbol{\beta})\right]$ and $\mathcal{D}\left[\mathbf{h}_{\mathrm{MN},0}(f, \alpha_{3,1})\right]$, as a function of frequency for several values of M. Corresponding plots of the WNG, $\mathcal{W}\left[\mathbf{h}_{\mathrm{MN}}(f, \boldsymbol{\alpha}, \boldsymbol{\beta})\right]$ and $\mathcal{W}\left[\mathbf{h}_{\mathrm{MN},0}(f, \alpha_{3,1})\right]$, as a function of frequency are depicted in Figure 9.23.

For $M = 4$, the DF is almost constant up to 8 kHz. As M increases, the frequency range for which the DF is constant decreases, but at high frequencies we can get much higher WNG than at low frequencies. Increasing the number of sensors enables the WNG to be increased. However, as the number of sensors increases, the DF at high frequencies decreases. Furthermore, for a given number of senors, the DF of the third-order DSA with three distinct nulls is higher than that of the third-order cardioid with a unique multiple null, but at the expense of lower WNG.

Table 9.1 summarizes all DSAs described in this chapter.

Problems

9.1 Using the definition of the frequency-independent DF of a theoretical Nth-order DSA (9.27), show that

$$\mathcal{D}\left(\mathbf{a}_N\right) = \frac{\mathbf{a}_N^T \mathbf{1}\mathbf{1}^T \mathbf{a}_N}{\mathbf{a}_N^T \mathbf{H}_N \mathbf{a}_N},$$

where $\mathbf{1}$ is a vector of ones, and $\mathbf{H}_N$ is a Hankel matrix.

9.2 Show that the coefficients of the Nth-order hypercardioid are given by

$$\mathbf{a}_{N,\max} = \frac{\mathbf{H}_N^{-1}\mathbf{1}}{\mathbf{1}^T \mathbf{H}_N^{-1}\mathbf{1}}.$$

9.3 Using the definition of the frequency-independent FBR of a theoretical Nth-order DSA (9.20), show that

$$\mathcal{F}\left(\mathbf{a}_N\right) = \frac{\mathbf{a}_N^T \mathbf{H}_N'' \mathbf{a}_N}{\mathbf{a}_N^T \mathbf{H}_N' \mathbf{a}_N},$$

where $\mathbf{H}_N'$ and $\mathbf{H}_N''$ are Hankel matrices.

9.4 Show that the beampattern of the Nth-order supercardioid is

$$\mathcal{B}_{N,\text{Sd}}\left(\cos\theta\right) = \frac{\mathbf{a}_{N,\max}'^T \mathbf{p}\left(\cos\theta\right)}{\mathbf{a}_{N,\max}'^T \mathbf{p}\left(1\right)},$$

where $\mathbf{a}_{N,\max}'$ is the eigenvector corresponding to the maximum eigenvalue of $\mathbf{H}_N'^{-1}\mathbf{H}_N''$.

9.5 Show that the directivity pattern of the first-order hypercardioid can be expressed as

$$\mathcal{B}_{1,\text{Hd}}\left(\cos\theta\right) = \frac{1}{4} + \frac{3}{4}\cos\theta.$$

9.6 Show that the directivity pattern of the first-order supercardioid can be expressed as

$$\mathcal{B}_{1,\text{Sd}}\left(\cos\theta\right) = \frac{\sqrt{3}-1}{2} + \frac{3-\sqrt{3}}{2}\cos\theta.$$

9.7 Show that the directivity pattern of the second-order hypercardioid can be expressed as

$$\mathcal{B}_{2,\text{Hd}}\left(\cos\theta\right) = -\frac{1}{6} + \frac{1}{3}\cos\theta + \frac{5}{6}\cos^2\theta.$$

9.8 Show that the directivity pattern of the second-order supercardioid can be expressed as

$$\mathcal{B}_{2,\mathrm{Sd}}(\cos\theta) = \frac{1}{2\left(3+\sqrt{7}\right)} + \frac{\sqrt{7}}{3+\sqrt{7}}\cos\theta + \frac{5}{2\left(3+\sqrt{7}\right)}\cos^2\theta.$$

9.9 Show that the directivity pattern of the third-order hypercardioid can be expressed as

$$\mathcal{B}_{3,\mathrm{Hd}}(\cos\theta) = -\frac{3}{32} - \frac{15}{32}\cos\theta + \frac{15}{32}\cos^2\theta + \frac{35}{32}\cos^3\theta.$$

9.10 Show that the directivity pattern of the third-order supercardioid can be expressed as

$$\mathcal{B}_{3,\mathrm{Sd}}(\cos\theta) \approx 0.0184 + 0.2004\cos\theta + 0.4750\cos^2\theta + 0.3061\cos^3\theta.$$

9.11 Show that the beampattern, the DF, and the WNG of the first-order DSA can be approximated as

$$\mathcal{B}\left[\mathbf{h}_1(f), \cos\theta\right] \approx \frac{1}{1-\alpha_{1,1}}\cos\theta - \frac{\alpha_{1,1}}{1-\alpha_{1,1}},$$

$$\mathcal{D}\left[\mathbf{h}_1(f)\right] \approx \frac{\left(1-\alpha_{1,1}\right)^2}{\alpha_{1,1}^2 + \frac{1}{3}},$$

$$\mathcal{W}\left[\mathbf{h}_1(f)\right] \approx \frac{1}{2}\left[2\pi f\tau_0\left(1-\alpha_{1,1}\right)\right]^2.$$

9.12 Show that the inverse of the Vandermonde matrix $\mathbf{V}(f)$ that appears in (9.64) is given by

$$\mathbf{V}^{-1}(f) =$$
$$\begin{bmatrix} \frac{\nu_2\nu_3}{\left(\nu_2-\nu_1\right)\left(\nu_3-\nu_1\right)} & -\frac{\nu_1\nu_3}{\left(\nu_2-\nu_1\right)\left(\nu_3-\nu_2\right)} & \frac{\nu_1\nu_2}{\left(\nu_3-\nu_1\right)\left(\nu_3-\nu_2\right)} \\ -\frac{\nu_2+\nu_3}{\left(\nu_2-\nu_1\right)\left(\nu_3-\nu_1\right)} & \frac{\nu_1+\nu_3}{\left(\nu_2-\nu_1\right)\left(\nu_3-\nu_2\right)} & -\frac{\nu_1+\nu_2}{\left(\nu_3-\nu_1\right)\left(\nu_3-\nu_2\right)} \\ \frac{1}{\left(\nu_2-\nu_1\right)\left(\nu_3-\nu_1\right)} & -\frac{1}{\left(\nu_2-\nu_1\right)\left(\nu_3-\nu_2\right)} & \frac{1}{\left(\nu_3-\nu_1\right)\left(\nu_3-\nu_2\right)} \end{bmatrix}.$$

9.13 Show that in the case of a second-order DSA with a zero of multiplicity 2 in the beampattern:

a) the beamformer is given by

$$\mathbf{h}_{2,0}(f) = \frac{1}{\left[1-e^{\jmath 2\pi f\tau_0\left(1-\alpha_{2,1}\right)}\right]^2}\begin{bmatrix} 1 \\ -2e^{-\jmath 2\pi f\tau_0\alpha_{2,1}} \\ e^{-\jmath 4\pi f\tau_0\alpha_{2,1}} \end{bmatrix},$$

b) the beampattern can be written as

$$\mathcal{B}\left[\mathbf{h}_{2,0}(f), \cos\theta\right] \approx \frac{1}{\left(1-\alpha_{2,1}\right)^2}\left(\cos\theta - \alpha_{2,1}\right)^2,$$

c) the WNG can be approximated as

$$\mathcal{W}\left[\mathbf{h}_{2,0}(f)\right] \approx \frac{1}{6}\left[2\pi f \tau_0 \left(1-\alpha_{2,1}\right)\right]^4.$$

9.14 Show that the first column of the inverse of the Vandermonde matrix $\mathbf{V}(f)$ that appears in (9.82) is given by

$$\mathbf{V}^{-1}(f; :, 1) = \begin{bmatrix} \dfrac{v_2 v_3 v_4}{\left(v_2 - v_1\right)\left(v_3 - v_1\right)\left(v_4 - v_1\right)} \\ -\dfrac{v_2 v_3 + v_3 v_4 + v_2 v_4}{\left(v_2 - v_1\right)\left(v_3 - v_1\right)\left(v_4 - v_1\right)} \\ \dfrac{v_2 + v_3 + v_4}{\left(v_2 - v_1\right)\left(v_3 - v_1\right)\left(v_4 - v_1\right)} \\ -\dfrac{1}{\left(v_2 - v_1\right)\left(v_3 - v_1\right)\left(v_4 - v_1\right)} \end{bmatrix}.$$

9.15 Show that the third-order DSA beamformer is given by

$$\mathbf{h}_3(f) = \frac{1}{\left[1 - e^{\jmath 2\pi f \tau_0 (1-\alpha_{3,1})}\right]\left[1 - e^{\jmath 2\pi f \tau_0 (1-\alpha_{3,2})}\right]\left[1 - e^{\jmath 2\pi f \tau_0 (1-\alpha_{3,3})}\right]}$$
$$\times \begin{bmatrix} 1 \\ -e^{-\jmath 2\pi f \tau_0 \alpha_{3,1}} - e^{-\jmath 2\pi f \tau_0 \alpha_{3,2}} - e^{-\jmath 2\pi f \tau_0 \alpha_{3,3}} \\ e^{-\jmath 2\pi f \tau_0 \left(\alpha_{3,1}+\alpha_{3,2}\right)} + e^{-\jmath 2\pi f \tau_0 \left(\alpha_{3,2}+\alpha_{3,3}\right)} + e^{-\jmath 2\pi f \tau_0 \left(\alpha_{3,1}+\alpha_{3,3}\right)} \\ -e^{-\jmath 2\pi f \tau_0 \left(\alpha_{3,1}+\alpha_{3,2}+\alpha_{3,3}\right)} \end{bmatrix}.$$

9.16 Show that in the case of a third-order DSA with a zero of multiplicity 3 in the beampattern:

a) the beamformer is given by

$$\mathbf{h}_{3,0}(f) = \frac{1}{\left[1 - e^{\jmath 2\pi f \tau_0 (1-\alpha_{3,1})}\right]^3} \begin{bmatrix} 1 \\ -3e^{-\jmath 2\pi f \tau_0 \alpha_{3,1}} \\ 3e^{-\jmath 4\pi f \tau_0 \alpha_{2,1}} \\ -e^{-\jmath 6\pi f \tau_0 \alpha_{2,1}} \end{bmatrix},$$

b) the beampattern can be approximated as

$$\mathcal{B}\left[\mathbf{h}_{3,0}(f), \cos\theta\right] \approx \frac{1}{\left(1-\alpha_{3,1}\right)^3}\left(\cos\theta - \alpha_{3,1}\right)^3,$$

c) the WNG can be approximated as

$$\mathcal{W}\left[\mathbf{h}_{3,0}(f)\right] \approx \frac{1}{20}\left[2\pi f \tau_0 \left(1 - \alpha_{3,1}\right)\right]^6 .$$

9.17 Show that for $M \geq N + 1$, the minimum-norm beamformer maximizes the WNG subject to $\mathbf{D}\left(f, \boldsymbol{\alpha}\right)\mathbf{h}(f) = \boldsymbol{\beta}$, and is given by

$$\mathbf{h}_{\mathrm{MN}}\left(f, \boldsymbol{\alpha}, \boldsymbol{\beta}\right) = \mathbf{D}^H\left(f, \boldsymbol{\alpha}\right)\left[\mathbf{D}\left(f, \boldsymbol{\alpha}\right)\mathbf{D}^H\left(f, \boldsymbol{\alpha}\right)\right]^{-1}\boldsymbol{\beta}.$$

9.18 Show that the beampattern, the WNG, and the DF of the minimum-norm beamformer are given by

$$\begin{aligned}
\mathcal{B}\left[\mathbf{h}_{\mathrm{MN}}\left(f, \boldsymbol{\alpha}, \boldsymbol{\beta}\right), \cos\theta\right] &= \mathbf{d}^H\left(f, \cos\theta\right)\mathbf{D}^H\left(f, \boldsymbol{\alpha}\right) \times \\
&\quad \left[\mathbf{D}\left(f, \boldsymbol{\alpha}\right)\mathbf{D}^H\left(f, \boldsymbol{\alpha}\right)\right]^{-1}\boldsymbol{\beta}, \\
\mathcal{W}\left[\mathbf{h}_{\mathrm{MN}}\left(f, \boldsymbol{\alpha}, \boldsymbol{\beta}\right)\right] &= \frac{1}{\boldsymbol{\beta}^T\left[\mathbf{D}\left(f, \boldsymbol{\alpha}\right)\mathbf{D}^H\left(f, \boldsymbol{\alpha}\right)\right]^{-1}\boldsymbol{\beta}}, \\
\mathcal{D}\left[\mathbf{h}_{\mathrm{MN}}\left(f, \boldsymbol{\alpha}, \boldsymbol{\beta}\right)\right] &= \frac{1}{\mathbf{h}_{\mathrm{MN}}^H\left(f, \boldsymbol{\alpha}, \boldsymbol{\beta}\right)\boldsymbol{\Gamma}_{0,\pi}(f)\mathbf{h}_{\mathrm{MN}}\left(f, \boldsymbol{\alpha}, \boldsymbol{\beta}\right)}.
\end{aligned}$$

9.19 Show that the minimum-norm beamformer whose beampattern has a null in the direction $\alpha_{N,1}$ with multiplicity N is given by

$$\mathbf{h}_{\mathrm{MN},0}\left(f, \alpha_{N,1}\right) = \mathbf{D}_0^H\left(f, \alpha_{N,1}\right)\left[\mathbf{D}_0\left(f, \alpha_{N,1}\right)\mathbf{D}_0^H\left(f, \alpha_{N,1}\right)\right]^{-1}\mathbf{i}_1.$$

References

1 J. Benesty, J. Chen, and Y. Huang, *Microphone Array Signal Processing*. Berlin, Germany: Springer-Verlag, 2008.
2 G. W. Elko and J. Meyer, "Microphone arrays," in *Springer Handbook of Speech Processing*, J. Benesty, M. M. Sondhi, and Y. Huang (eds). Berlin, Germany: Springer-Verlag, 2008, Chapter 50, pp. 1021–1041.
3 G. W. Elko, "Superdirectional microphone arrays," in *Acoustic Signal Processing for Telecommunication*, S. L. Gay and J. Benesty (eds). Boston, MA: Kluwer Academic Publishers, 2000, Chapter 10, pp. 181–237.
4 J. Benesty and J. Chen, *Study and Design of Differential Microphone Arrays*. Berlin, Germany: Springer-Verlag, 2012.
5 J. Chen, J. Benesty, and C. Pan "On the design and implementation of linear differential microphone arrays," *J. Acoust. Soc. Am.*, vol. 136, pp. 3097–3113, Dec. 2014.
6 R. N. Marshall and W. R. Harry, "A new microphone providing uniform directivity over an extended frequency range," *J. Acoust. Soc. Am.*, vol. 12, pp. 481–497, 1941.
7 A. I. Uzkov, "An approach to the problem of optimum directive antenna design," *Comptes Rendus (Doklady) de l'Academie des Sciences de l'URSS*, vol. LIII, no. 1, pp. 35–38, 1946.

10

Beampattern Design

We again assume that we have a uniform linear array (ULA). Because of the symmetry of the steering vector associated with a ULA, the only directions where we can design a symmetric beampattern are at the endfires (i.e., 0 and π). Since we are interested in frequency-invariant beampatterns, the distance between two successive sensors must be small, as explained in Chapter 9. This makes sense since, contrary to many approaches proposed in the literature, we can now design any desired frequency-invariant symmetric beampattern without any specific constraints. Therefore, the beampatterns that we outline in this chapter are similar to those obtained with differential sensor arrays (DSAs). After revisiting the definitions of the beampatterns and showing some relationships between them, we explain the different techniques for beampattern design.

10.1 Beampatterns Revisited

From Chapter 7, we know that the beampattern corresponding to a filter $\mathbf{h}(f)$, of length M, applied to a ULA is

$$\begin{aligned} \mathcal{B}\left[\mathbf{h}(f), \cos\theta\right] &= \mathbf{d}^H\left(f, \cos\theta\right)\mathbf{h}(f) \\ &= \sum_{m=1}^{M} H_m(f) e^{\jmath \bar{f}_m \cos\theta}, \end{aligned} \tag{10.1}$$

where we define

$$\begin{aligned} \bar{f}_m &= \frac{2\pi\delta}{c}(m-1)f \\ &= 2\pi\tau_0(m-1)f \end{aligned} \tag{10.2}$$

to simplify the notation. We recall that $\mathbf{h}(f)$ is designed so that the array looks in the direction $\theta = 0$ (or $\theta = \pi$). For a fixed $\mathbf{h}(f)$, it is obvious that $\mathcal{B}\left[\mathbf{h}(f), \cos\theta\right]$ is even and periodic with respect to the variable θ:

$$\mathcal{B}\left[\mathbf{h}(f), \cos(-\theta)\right] = \mathcal{B}\left[\mathbf{h}(f), \cos\theta\right] \tag{10.3}$$

Fundamentals of Signal Enhancement and Array Signal Processing, First Edition.
Jacob Benesty, Israel Cohen, and Jingdong Chen.

Companion website: www.wiley.com/go/benesty/arraysignalprocessing

and

$$\mathcal{B}\left[\mathbf{h}(f), \cos(\theta + 2\pi)\right] = \mathcal{B}\left[\mathbf{h}(f), \cos\theta\right]. \tag{10.4}$$

As a result, the analysis and design of a desired beampattern is limited to $\theta \in [0, \pi]$.

Let $\mathcal{B}(\theta)$ be a real even periodic function with period 2π and such that $\int_0^{\pi} |\mathcal{B}(\theta)|\, d\theta$ exists. In this case, it is well known that $\mathcal{B}(\theta)$ can be written in terms of its Fourier cosine series [1]:

$$\mathcal{B}(\theta) = \sum_{n=0}^{\infty} b_n \cos(n\theta), \tag{10.5}$$

where

$$\begin{cases} b_0 = \dfrac{1}{\pi}\displaystyle\int_0^{\pi} \mathcal{B}(\theta)\, d\theta \\ b_i = \dfrac{2}{\pi}\displaystyle\int_0^{\pi} \mathcal{B}(\theta)\cos(i\theta)\, d\theta, \quad i \geq 1 \end{cases}.$$

Now, if we limit this series to order N, $\mathcal{B}(\theta)$ can be approximated by [2, 3]:

$$\begin{aligned} \mathcal{B}\left(\mathbf{b}_N, \cos\theta\right) &= \sum_{n=0}^{N} b_{N,n} \cos(n\theta) \\ &= \mathbf{b}_N^T \mathbf{p}_C(\cos\theta), \end{aligned} \tag{10.6}$$

where $b_{N,n}$, $n = 0, 1, \ldots, N$ are real coefficients and

$$\begin{aligned} \mathbf{b}_N &= \begin{bmatrix} b_{N,0} & b_{N,1} & \cdots & b_{N,N} \end{bmatrix}^T, \\ \mathbf{p}_C(\cos\theta) &= \begin{bmatrix} 1 & \cos\theta & \cdots & \cos(N\theta) \end{bmatrix}^T, \end{aligned}$$

are vectors of length $N+1$. The function $\mathcal{B}\left(\mathbf{b}_N, \cos\theta\right)$ is, in fact, a very general definition of a frequency-independent directivity pattern of order N. It is very much related to the directivity pattern of the Nth-order DSA defined in Chapter 9:

$$\begin{aligned} \mathcal{B}\left(\mathbf{a}_N, \cos\theta\right) &= \sum_{n=0}^{N} a_{N,n} \cos^n\theta \\ &= \mathbf{a}_N^T \mathbf{p}(\cos\theta), \end{aligned} \tag{10.7}$$

and any DSA beampattern can be designed with $\mathcal{B}\left(\mathbf{b}_N, \cos\theta\right)$. Indeed, we know from the usual trigonometric identities that

$$\cos^n\theta = \sum_i b(n,i)\cos\left[(n-2i)\theta\right], \tag{10.8}$$

where $b(n, i)$ are binomial coefficients. Substituting (10.8) into (10.7), we deduce that any DSA beampattern can be written as a general beampattern, $\mathcal{B}\left(\mathbf{b}_N, \cos\theta\right)$. It is well known that

$$\cos(n\theta) = T_n(\cos\theta), \tag{10.9}$$

where $T_n(\cdot)$ is the nth Chebyshev polynomial of the first kind [4], which has the recurrence relation:

$$T_{n+1}(\cos\theta) = 2\cos\theta \times T_n(\cos\theta) - T_{n-1}(\cos\theta), \tag{10.10}$$

with

$$\begin{cases} T_0(\cos\theta) = 1 \\ T_1(\cos\theta) = \cos\theta \end{cases}.$$

Thus, $\cos(n\theta)$ can be expressed as a sum of powers of $\cos\theta$. Consequently, any general beampattern can be written as a DSA beampattern. We can then conclude that $\mathcal{B}\left(\mathbf{b}_N, \cos\theta\right)$ and $\mathcal{B}\left(\mathbf{a}_N, \cos\theta\right)$ are strictly equivalent. An even more general definition of a frequency-independent beampattern with orthogonal polynomials can be found in the paper by Pan et al. [5]. Basically, this shows that any even periodic function (here a desired beampattern) can be designed or approximated by its Fourier cosine series, which also corresponds to the theoretical Nth-order DSA beampattern.

For convenience, we give the relations between the coefficients $b_{N,n}$, $n = 0, 1, \ldots, N$ of $\mathcal{B}\left(\mathbf{b}_N, \cos\theta\right)$ and the coefficients $a_{N,n}$, $n = 0, 1, \ldots, N$ of $\mathcal{B}\left(\mathbf{a}_N, \cos\theta\right)$ for the first three orders:

- $N = 1$: $b_{1,0} = a_{1,0}$, $b_{1,1} = a_{1,1}$;
- $N = 2$: $b_{2,0} = a_{2,0} + \frac{a_{2,2}}{2}$, $b_{2,1} = a_{2,1}$, $b_{2,2} = \frac{a_{2,2}}{2}$; and
- $N = 3$: $b_{3,0} = a_{3,0} + \frac{a_{3,2}}{2}$, $b_{3,1} = a_{3,1} + \frac{3a_{3,3}}{4}$, $b_{3,2} = \frac{a_{3,2}}{2}$, $b_{3,3} = \frac{a_{3,3}}{4}$.

Now, in order to be able to design any desired beampattern, $\mathcal{B}\left(\mathbf{b}_N, \cos\theta\right)$, with $\mathcal{B}\left[\mathbf{h}(f), \cos\theta\right]$, where $\mathbf{h}(f)$ needs to be found accordingly, we have to approximate the exponential function that appears in (10.1) in terms of Chebyshev polynomials, as will become clearer soon. Since the complex-valued exponential function is infinitely differentiable and even with respect to the variable θ, we can find the complex-valued coefficients c_n, $n = 0, 1, 2, \ldots$ such that

$$e^{\jmath \bar{f}_m \cos\theta} = \lim_{N\to\infty} \sum_{n=0}^{N} c_n \cos(n\theta). \tag{10.11}$$

By limiting the above series to a fixed N, we propose to find the coefficients c_n, $n = 0, 1, \ldots, N$, in the best possible way in a least-squares error (LSE) sense, by minimizing the criterion:

$$\mathrm{LSE}\left(\mathbf{c}_N\right) = \frac{1}{\pi}\int_0^{\pi} \left| e^{-\jmath \bar{f}_m \cos\theta} - \sum_{n=0}^{N} c_n^* \cos(n\theta) \right|^2 d\theta \tag{10.12}$$

$$
\begin{aligned}
&= \frac{1}{\pi} \int_0^{\pi} \left| e^{-\jmath \overline{f}_m \cos\theta} - \mathbf{c}_N^H \mathbf{p}_{\mathrm{C}} (\cos\theta) \right|^2 d\theta \\
&= 1 - \mathbf{v}_{\mathrm{C}}^H \left(\jmath \overline{f}_m \right) \mathbf{c}_N - \mathbf{c}_N^H \mathbf{v}_{\mathrm{C}} \left(\jmath \overline{f}_m \right) + \mathbf{c}_N^H \mathbf{M}_{\mathrm{C}} \mathbf{c}_N,
\end{aligned}
$$

where

$$
\begin{aligned}
\mathbf{c}_N &= \begin{bmatrix} c_0 & c_1 & \cdots & c_N \end{bmatrix}^T, \\
\mathbf{v}_{\mathrm{C}} \left(\jmath \overline{f}_m \right) &= \frac{1}{\pi} \int_0^{\pi} e^{\jmath \overline{f}_m \cos\theta} \mathbf{p}_{\mathrm{C}} (\cos\theta)\, d\theta, \\
\mathbf{M}_{\mathrm{C}} &= \frac{1}{\pi} \int_0^{\pi} \mathbf{p}_{\mathrm{C}} (\cos\theta)\, \mathbf{p}_{\mathrm{C}}^T (\cos\theta)\, d\theta.
\end{aligned}
$$

The minimization of the LSE criterion gives the optimal solution:

$$
\mathbf{c}_N = \mathbf{M}_{\mathrm{C}}^{-1} \mathbf{v}_{\mathrm{C}} \left(\jmath \overline{f}_m \right). \tag{10.13}
$$

Let us have a closer look at $\mathbf{v}_{\mathrm{C}} \left(\jmath \overline{f}_m \right)$ and $\mathbf{M}_{\mathrm{C}}$. The elements of the vector $\mathbf{v}_{\mathrm{C}} \left(\jmath \overline{f}_m \right)$ are

$$
\begin{aligned}
\left[\mathbf{v}_{\mathrm{C}} \left(\jmath \overline{f}_m \right) \right]_{n+1} &= \frac{1}{\pi} \int_0^{\pi} e^{\jmath \overline{f}_m \cos\theta} \cos (n\theta)\, d\theta \\
&= I_n \left(\jmath \overline{f}_m \right) \\
&= \jmath^n J_n \left(\overline{f}_m \right),
\end{aligned} \tag{10.14}
$$

with $n = 0, 1, \ldots, N$, where

$$
I_n (z) = \frac{1}{\pi} \int_0^{\pi} e^{z \cos\theta} \cos (n\theta)\, d\theta \tag{10.15}
$$

is the integral representation of the modified Bessel function of the first kind [4] and

$$
\begin{aligned}
J_n (z) &= \frac{\jmath^{-n}}{\pi} \int_0^{\pi} e^{\jmath z \cos\theta} \cos (n\theta)\, d\theta \\
&= \jmath^{-n} I_n (\jmath z)
\end{aligned} \tag{10.16}
$$

is the integral representation of the Bessel function of the first kind [4]. The elements of the matrix $\mathbf{M}_{\mathrm{C}}$ are

$$
\left[\mathbf{M}_{\mathrm{C}} \right]_{i+1,j+1} = \frac{1}{\pi} \int_0^{\pi} \cos (i\theta) \cos \left(j\theta \right) d\theta, \tag{10.17}
$$

with $i, j = 0, 1, \ldots, N$. It can be checked that

$$\begin{cases} \left[\mathbf{M}_\mathrm{C}\right]_{1,1} = 1, & \\ \left[\mathbf{M}_\mathrm{C}\right]_{i+1,i+1} = \frac{1}{2}, & i \geq 1 \\ \left[\mathbf{M}_\mathrm{C}\right]_{i+1,j+1} = 0, & i \neq j \end{cases}.$$

This is a consequence of the fact that Chebyshev polynomials are orthogonal. Therefore, the matrix $\mathbf{M}_\mathrm{C}$ is diagonal:

$$\mathbf{M}_\mathrm{C} = \mathrm{diag}\left(1, \frac{1}{2}, \ldots, \frac{1}{2}\right). \tag{10.18}$$

We deduce that the exponential function given in (10.11) can be expressed as [3]:

$$\begin{aligned} e^{\jmath \bar{f}_m \cos\theta} &= J_0\left(\bar{f}_m\right) + 2\sum_{n=1}^{\infty} \jmath^n J_n\left(\bar{f}_m\right)\cos(n\theta) \\ &= \sum_{n=0}^{\infty} \mathcal{J}_n J_n\left(\bar{f}_m\right)\cos(n\theta), \end{aligned} \tag{10.19}$$

where

$$\mathcal{J}_n = \begin{cases} 1, & n = 0 \\ 2\jmath^n, & n = 1, 2, \ldots, N \end{cases}.$$

Equation 10.19 is actually the well-known Jacobi–Anger expansion [6, 7], which represents an expansion of plane waves into a series of cylindrical waves. Using (10.19) in the definition of the beampattern corresponding to $\mathbf{h}(f)$, we obtain

$$\begin{aligned} \mathcal{B}\left[\mathbf{h}(f), \cos\theta\right] &= \sum_{m=1}^{M} H_m(f) e^{\jmath \bar{f}_m \cos\theta} \\ &= \sum_{m=1}^{M} H_m(f) \sum_{n=0}^{\infty} \mathcal{J}_n J_n\left(\bar{f}_m\right)\cos(n\theta) \\ &= \sum_{n=0}^{\infty} \cos(n\theta)\left[\sum_{m=1}^{M} \mathcal{J}_n J_n\left(\bar{f}_m\right) H_m(f)\right]. \end{aligned} \tag{10.20}$$

If we limit the expansion to the order N, $\mathcal{B}\left[\mathbf{h}(f), \cos\theta\right]$ can be approximated by

$$\mathcal{B}_N\left[\mathbf{h}(f), \cos\theta\right] = \sum_{n=0}^{N} \cos(n\theta)\left[\sum_{m=1}^{M} \mathcal{J}_n J_n\left(\bar{f}_m\right) H_m(f)\right]. \tag{10.21}$$

For $m = 1, \bar{f}_1 = 0$, so that $J_0\left(\bar{f}_1\right) = 1$ and $J_n\left(\bar{f}_1\right) = 0,\ n = 1, 2, \ldots, N$. We will see how to use (10.21) in order to design any desired symmetric beampattern or, equivalently, any desired DSA beampattern of any order. Next, we explain the different approaches used.

10.2 Nonrobust Approach

In the nonrobust approach, it is always assumed that the number of sensors is equal to the order plus 1: $M = N + 1$. This is how all DSA beampatterns have been traditionally designed [8, 9]. Because of this relation between the number of sensors and the DSA order, the white noise amplification problem gets much worse quickly as the order increases; in this sense, this technique is a nonrobust one.

The beampattern in (10.21) can be rewritten as

$$\mathcal{B}_{M-1}\left[\mathbf{h}(f), \cos\theta\right] = \sum_{i=0}^{M-1} \cos(i\theta)\, \overline{\mathbf{b}}_i^T(f)\mathbf{h}(f), \tag{10.22}$$

where $M \geq 2$ and

$$\overline{\mathbf{b}}_i(f) = \jmath^i \left[\begin{array}{cccc} J_i\left(\overline{f}_1\right) & J_i\left(\overline{f}_2\right) & \cdots & J_i\left(\overline{f}_M\right) \end{array}\right]^T. \tag{10.23}$$

In the proposed beampattern design, we would like to find the filter $\mathbf{h}(f)$ in such a way that $\mathcal{B}_{M-1}\left[\mathbf{h}(f), \cos\theta\right]$ is an $(M-1)$th-order frequency-invariant DSA beampattern; that is,

$$\mathcal{B}_{M-1}\left[\mathbf{h}(f), \cos\theta\right] = \mathcal{B}\left(\mathbf{b}_{M-1}, \cos\theta\right), \tag{10.24}$$

where $\mathcal{B}\left(\mathbf{b}_{M-1}, \cos\theta\right)$ is defined in (10.6). By simple identification, we easily find that

$$\overline{\mathbf{B}}_{M-1}(f)\mathbf{h}(f) = \mathbf{b}_{M-1}, \tag{10.25}$$

where

$$\overline{\mathbf{B}}_{M-1}(f) = \begin{bmatrix} \overline{\mathbf{b}}_0^T(f) \\ \overline{\mathbf{b}}_1^T(f) \\ \vdots \\ \overline{\mathbf{b}}_{M-1}^T(f) \end{bmatrix} \tag{10.26}$$

is an $M \times M$ matrix. Assuming that $\overline{\mathbf{B}}_{M-1}(f)$ is a full-rank matrix, we find that the nonrobust filter for beampattern design is

$$\mathbf{h}_{\mathrm{NR}}(f) = \overline{\mathbf{B}}_{M-1}^{-1}(f)\mathbf{b}_{M-1}. \tag{10.27}$$

Let us take the example of $M = 2$. It is easy to check that

$$\mathcal{B}_1\left[\mathbf{h}(f), \cos\theta\right] = H_1(f) + J_0\left(\overline{f}_2\right)H_2(f) + 2\jmath J_1\left(\overline{f}_2\right)H_2(f)\cos\theta, \tag{10.28}$$

$$\mathcal{B}\left(\mathbf{b}_1, \cos\theta\right) = b_{1,0} + b_{1,1}\cos\theta. \tag{10.29}$$

Identifying the two previous expressions, we get

$$H_{2,\mathrm{NR}}(f) = \frac{b_{1,1}}{2\jmath J_1\left(\overline{f}_2\right)} \tag{10.30}$$

and

$$H_{1,\mathrm{NR}}(f) = -J_0\left(\overline{f}_2\right)H_{2,\mathrm{NR}}(f) + b_{1,0}. \tag{10.31}$$

Therefore, with this approach, we can design any first-order DSA beampattern.

Depending on the values of $b_{1,0}$ and $b_{1,1}$ we find four useful first-order DSAs:

- dipole: $b_{1,0} = 0$ and $b_{1,1} = 1$
- cardioid: $b_{1,0} = \frac{1}{2}$ and $b_{1,1} = \frac{1}{2}$
- hypercardioid: $b_{1,0} = \frac{1}{4}$ and $b_{1,1} = \frac{3}{4}$
- supercardioid: $b_{1,0} = \frac{\sqrt{3}-1}{2}$ and $b_{1,1} = \frac{3-\sqrt{3}}{2}$.

Figure 10.1 displays the patterns – with $\mathbf{h}_{\mathrm{NR}}(f)$ defined in (10.27) – of the first-order dipole, cardioid, hypercardioid, and supercardioid for $f = 1$ kHz and $\delta = 0.5$ cm. Comparing the patterns of Figures 10.1 and 9.1, we observe that the designed patterns have less explicit nulls than the corresponding first-order directivity patterns. This is due to the Jacobi–Anger series approximation. Figure 10.2 shows plots of the DF, $\mathcal{D}\left[\mathbf{h}_{\mathrm{NR}}(f)\right]$, as a function of frequency, for the dipole, cardioid, hypercardioid, and supercardioid, and several values of δ. Corresponding plots of the WNG, $\mathcal{W}\left[\mathbf{h}_{\mathrm{NR}}(f)\right]$, as a function of frequency are depicted in Figure 10.3. We observe that for a small sensor spacing, the first-order DSAs give an approximately constant DF while the WNG is negative. The white noise amplification is especially high at low frequencies. Increasing the sensor spacing enables the WNG to be increased, but reduces the DF, especially at high frequencies. A large value of δ contradicts the DSA assumption, which results in deterioration of the beampatterns at high frequencies.

10.3 Robust Approach

In the robust scenario, the number of sensors is greater than the DSA order plus 1: $M > N + 1$. By taking advantage of the fact that we have many more sensors than the order, we can control white noise amplification; in this sense, this technique is a robust one. Again, we would like to find the filter $\mathbf{h}(f)$ in such a way that $\mathcal{B}_N\left[\mathbf{h}(f), \cos\theta\right]$ is an Nth-order frequency-invariant DSA beampattern; that is,

$$\mathcal{B}_N\left[\mathbf{h}(f), \cos\theta\right] = \mathcal{B}\left(\mathbf{b}_N, \cos\theta\right). \tag{10.32}$$

By simple identification, we easily find that

$$\overline{\mathbf{B}}_N(f)\mathbf{h}(f) = \mathbf{b}_N, \tag{10.33}$$

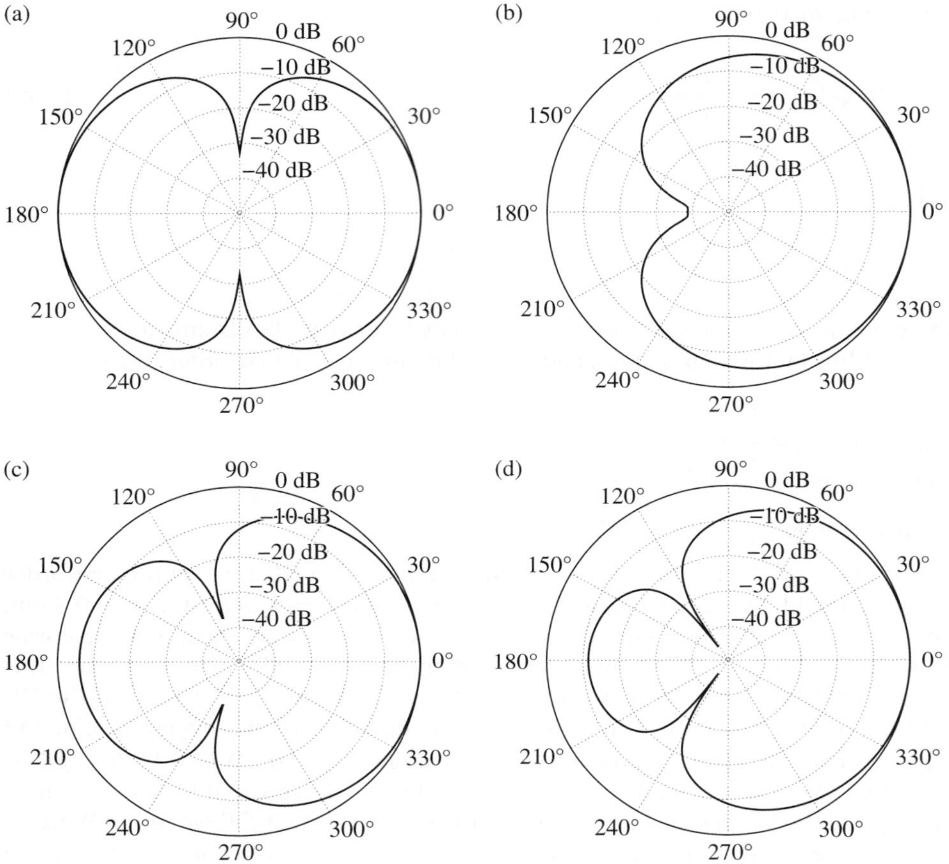

Figure 10.1 Beampatterns of the nonrobust first-order DSAs: (a) dipole, (b) cardioid, (c) hypercardioid, and (d) supercardioid. $M = 2$, $\delta = 0.5$ cm, and $f = 1$ kHz.

where

$$\overline{\mathbf{B}}_N(f) = \begin{bmatrix} \overline{\mathbf{b}}_0^T(f) \\ \overline{\mathbf{b}}_1^T(f) \\ \vdots \\ \overline{\mathbf{b}}_N^T(f) \end{bmatrix} \tag{10.34}$$

is now an $(N + 1) \times M$ matrix. Assuming that $\overline{\mathbf{B}}_N^H(f)$ is a full-column rank matrix and taking the minimum-norm solution of (10.33), we find that the robust filter for beampattern design is

$$\mathbf{h}_{\mathrm{R}}(f) = \overline{\mathbf{B}}_N^H(f) \left[\overline{\mathbf{B}}_N(f) \overline{\mathbf{B}}_N^H(f) \right]^{-1} \mathbf{b}_N. \tag{10.35}$$

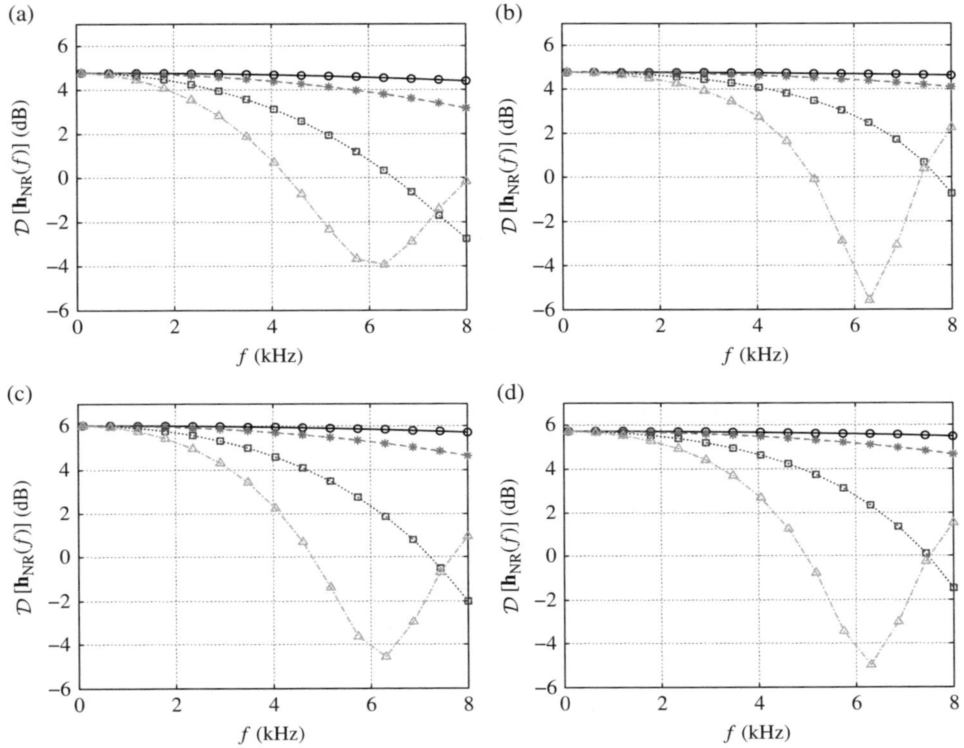

Figure 10.2 DF of the nonrobust first-order DSAs as a function of frequency, for $M = 2$ and several values of δ: $\delta = 0.1$ cm (solid line with circles), $\delta = 1$ cm (dashed line with asterisks), $\delta = 2$ cm (dotted line with squares), and $\delta = 3$ cm (dash-dot line with triangles). (a) Dipole, (b) cardioid, (c) hypercardioid, and (d) supercardioid.

Let us take the example of the first-order case ($N = 1$) with $M > 2$. We still want to find the coefficients $H_m(f)$, $m = 1, 2, \ldots, M$ in such a way that $\mathcal{B}_1\left[\mathbf{h}(f), \cos\theta\right] = \mathcal{B}_1\left(\mathbf{b}_1, \cos\theta\right)$. It is not hard to get

$$\begin{bmatrix} J_1\left(\bar{f}_2\right) & J_1\left(\bar{f}_3\right) & \cdots & J_1\left(\bar{f}_M\right) \end{bmatrix} \begin{bmatrix} H_2(f) \\ H_3(f) \\ \vdots \\ H_M(f) \end{bmatrix} = \frac{b_{1,1}}{2\jmath} \tag{10.36}$$

and

$$H_1(f) + \sum_{i=2}^{M} J_0\left(\bar{f}_i\right) H_i(f) = b_{1,0}. \tag{10.37}$$

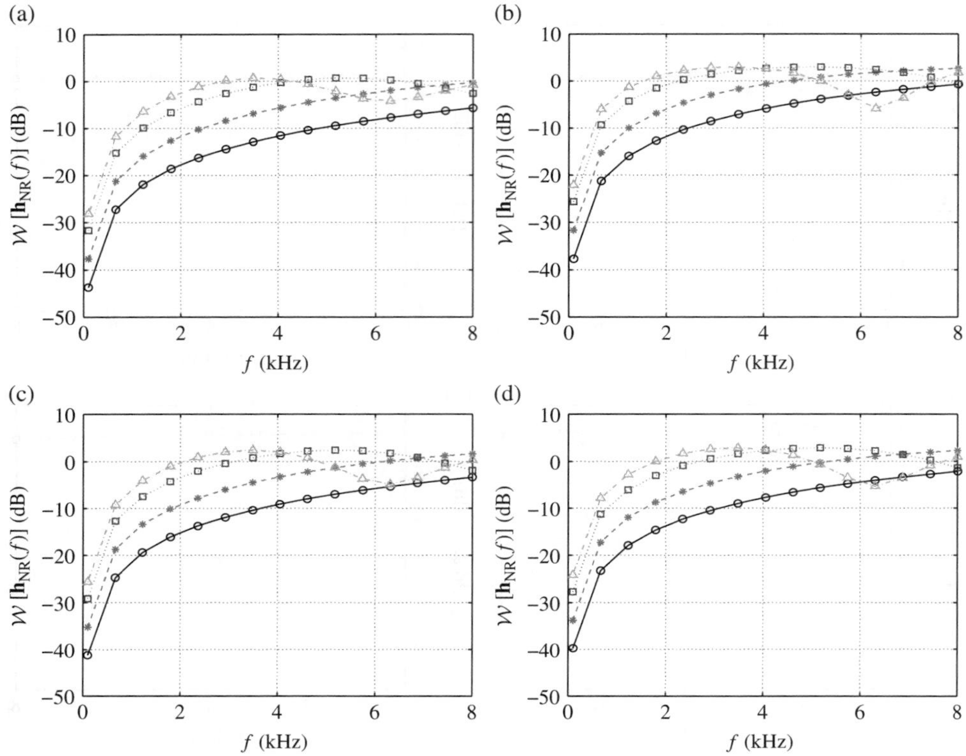

Figure 10.3 WNG of the nonrobust first-order DSAs as a function of frequency, for $M = 2$ and several values of δ: $\delta = 0.1$ cm (solid line with circles), $\delta = 1$ cm (dashed line with asterisks), $\delta = 2$ cm (dotted line with squares), and $\delta = 3$ cm (dash-dot line with triangles). (a) Dipole, (b) cardioid, (c) hypercardioid, and (d) supercardioid.

Taking the minimum-norm solution of (10.36), it is clear that the filter coefficients are as follows:

$$H_{i,\mathrm{R}}(f) = \frac{J_1\left(\bar{f}_i\right) b_{1,1}}{2\jmath \sum_{j=2}^{M} J_1^2\left(\bar{f}_j\right)}, \; i = 2, 3, \ldots, M \tag{10.38}$$

and

$$H_{1,\mathrm{R}}(f) = -\sum_{i=2}^{M} J_0\left(\bar{f}_i\right) H_{i,\mathrm{R}}(f) + b_{1,0}. \tag{10.39}$$

The robust filter, $\mathbf{h}_{\mathrm{R}}(f)$, the components of which are given in (10.39) and (10.38), is the minimum-norm filter for the design of first-order DSA beampatterns. The WNG with $\mathbf{h}_{\mathrm{R}}(f)$ should be much better than that with $\mathbf{h}_{\mathrm{NR}}(f)$.

Figures 10.4–10.7 display the patterns – with $\mathbf{h}_{\mathrm{R}}(f)$ defined in (10.35) – of the first-order dipole, cardioid, hypercardioid, and supercardioid for $f = 1$ kHz, $\delta = 0.5$ cm,

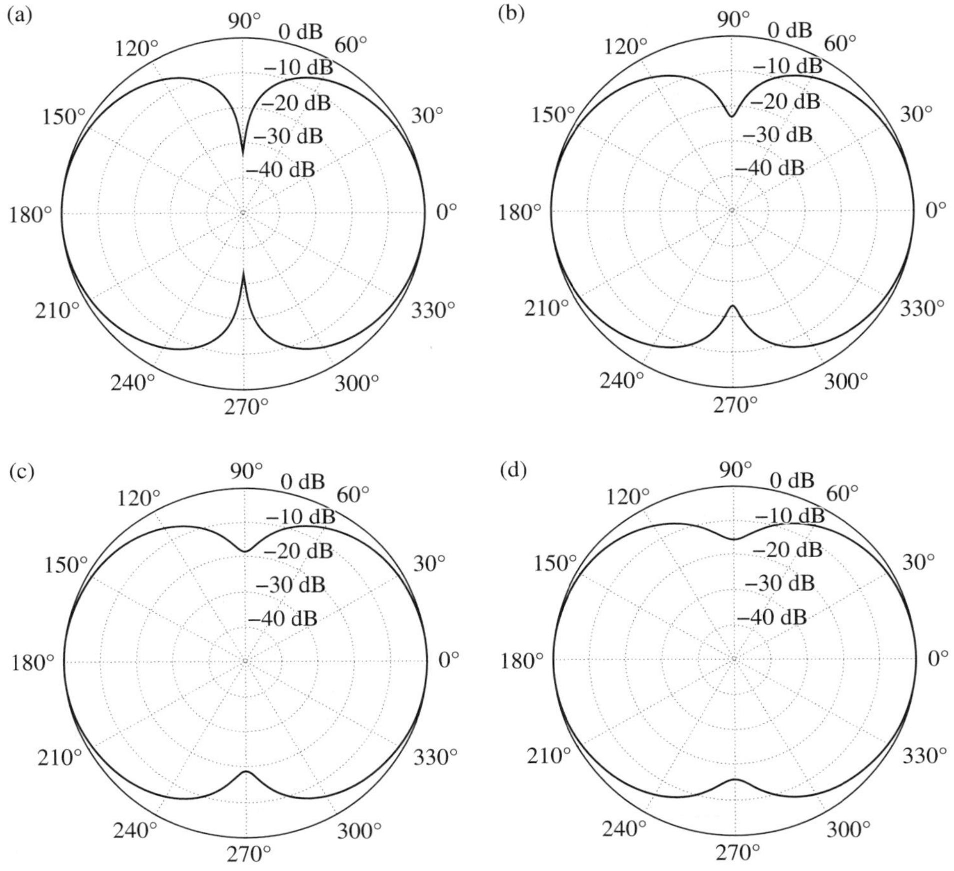

Figure 10.4 Beampatterns of the robust first-order dipole for $f = 1$ kHz, $\delta = 0.5$ cm, and several values of M: (a) $M = 2$, (b) $M = 4$, (c) $M = 6$, and (d) $M = 8$.

and several values of M. Figure 10.8 shows plots of the DF, $\mathcal{D}\left[\mathbf{h}_{\mathrm{R}}(f)\right]$, as a function of frequency, for the dipole, cardioid, hypercardioid, and supercardioid, and several values of M. Corresponding plots of the WNG, $\mathcal{W}\left[\mathbf{h}_{\mathrm{R}}(f)\right]$, as a function of frequency are depicted in Figure 10.9. We can see that the WNG is considerably improved as M increases, while the beampatterns and the DFs do not change much. It is clear that the WNG with $\mathbf{h}_{\mathrm{R}}(f)$ is much better than that with $\mathbf{h}_{\mathrm{NR}}(f)$. The larger the number of sensors, the more robust is the first-order DSA against white noise amplification.

10.4 Frequency-invariant Beampattern Design

Let us define the criterion:

$$\begin{aligned} J_{\mathrm{FI}}\left[\mathbf{h}(f)\right] &= \frac{1}{\pi}\int_0^{\pi}\left|\mathcal{B}\left[\mathbf{h}(f), \cos\theta\right]\right|^2 d\theta \\ &= \mathbf{h}^H(f)\boldsymbol{\Gamma}_{\mathrm{C}}(f)\mathbf{h}(f), \end{aligned} \tag{10.40}$$

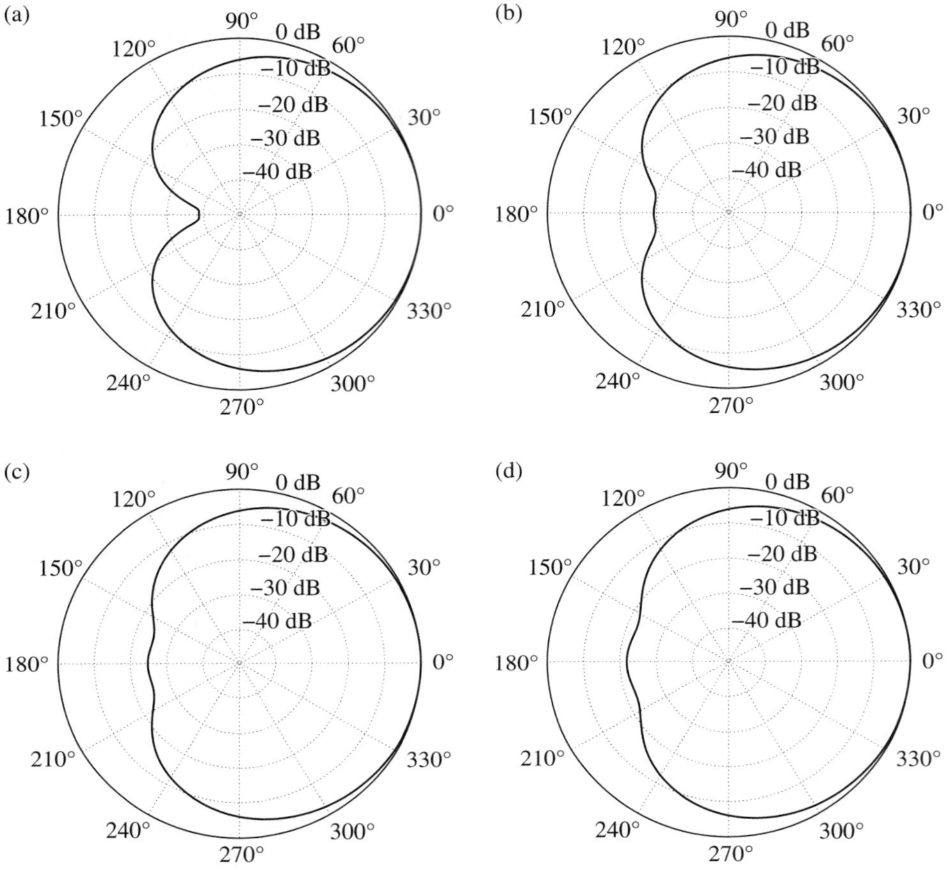

Figure 10.5 Beampatterns of the robust first-order cardioid for $f = 1$ kHz, $\delta = 0.5$ cm, and several values of M: (a) $M = 2$, (b) $M = 4$, (c) $M = 6$, and (d) $M = 8$.

where

$$\mathbf{\Gamma}_{\mathrm{C}}(f) = \frac{1}{\pi} \int_0^{\pi} \mathbf{d}\left(f, \cos\theta\right) \mathbf{d}^H\left(f, \cos\theta\right) d\theta. \tag{10.41}$$

The (i, j)th (with $i, j = 1, 2, \ldots, M$) element of the $M \times M$ matrix $\mathbf{\Gamma}_{\mathrm{C}}(f)$ can be computed as

$$\begin{aligned} \left[\mathbf{\Gamma}_{\mathrm{C}}(f)\right]_{i,j} &= \frac{1}{\pi} \int_0^{\pi} e^{\jmath 2\pi f (j-i)\tau_0 \cos\theta} d\theta \\ &= I_0 \left[\jmath 2\pi f (j-i)\tau_0\right]. \end{aligned} \tag{10.42}$$

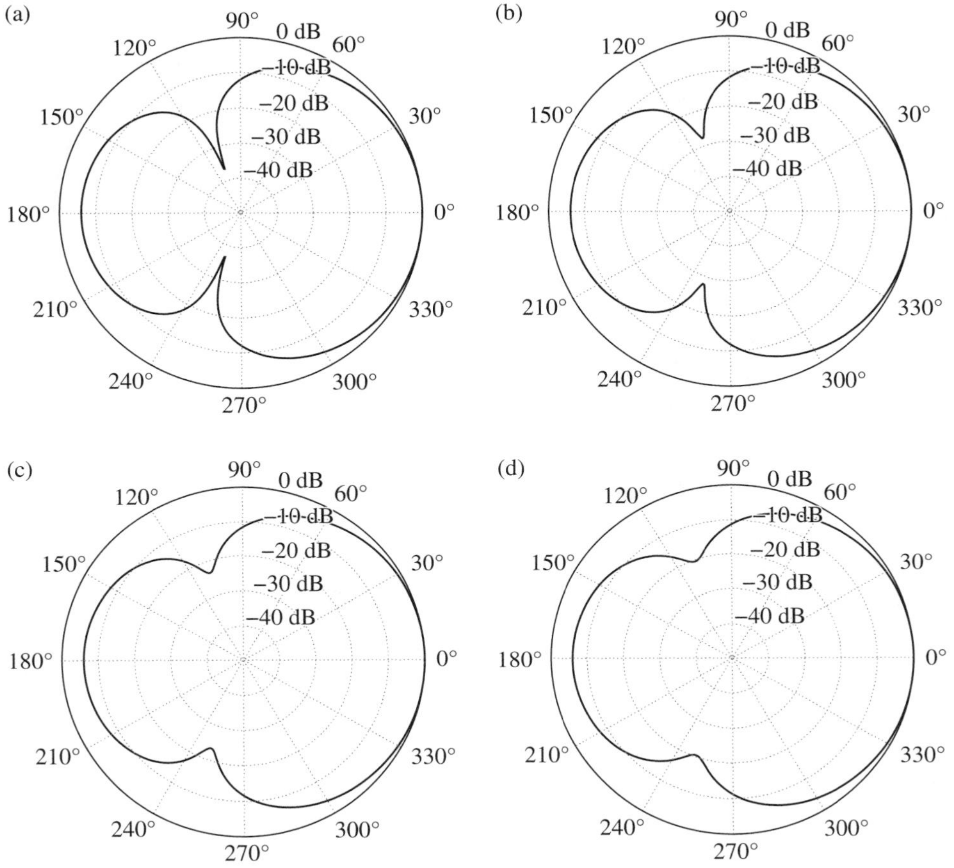

Figure 10.6 Beampatterns of the robust first-order hypercardioid for $f = 1$ kHz, $\delta = 0.5$ cm, and several values of M: (a) $M = 2$, (b) $M = 4$, (c) $M = 6$, and (d) $M = 8$.

In order to design a frequency-invariant beampattern, we can minimize $J_{\mathrm{FI}}\left[\mathbf{h}(f)\right]$ subject to (10.33):

$$\min_{\mathbf{h}(f)} \mathbf{h}^H(f)\mathbf{\Gamma}_{\mathrm{C}}(f)\mathbf{h}(f) \quad \text{subject to} \quad \overline{\mathbf{B}}_N(f)\mathbf{h}(f) = \mathbf{b}_N. \tag{10.43}$$

We easily find that the corresponding filter is

$$\mathbf{h}_{\mathrm{FI}}(f) = \mathbf{\Gamma}_{\mathrm{C}}^{-1}(f)\overline{\mathbf{B}}_N^H(f)\left[\overline{\mathbf{B}}_N(f)\mathbf{\Gamma}_{\mathrm{C}}^{-1}(f)\overline{\mathbf{B}}_N^H(f)\right]^{-1}\mathbf{b}_N. \tag{10.44}$$

When it comes to white noise amplification, the filter $\mathbf{h}_{\mathrm{FI}}(f)$ is usually much worse than the previous two derived filters $\mathbf{h}_{\mathrm{NR}}(f)$ and $\mathbf{h}_{\mathrm{R}}(f)$.

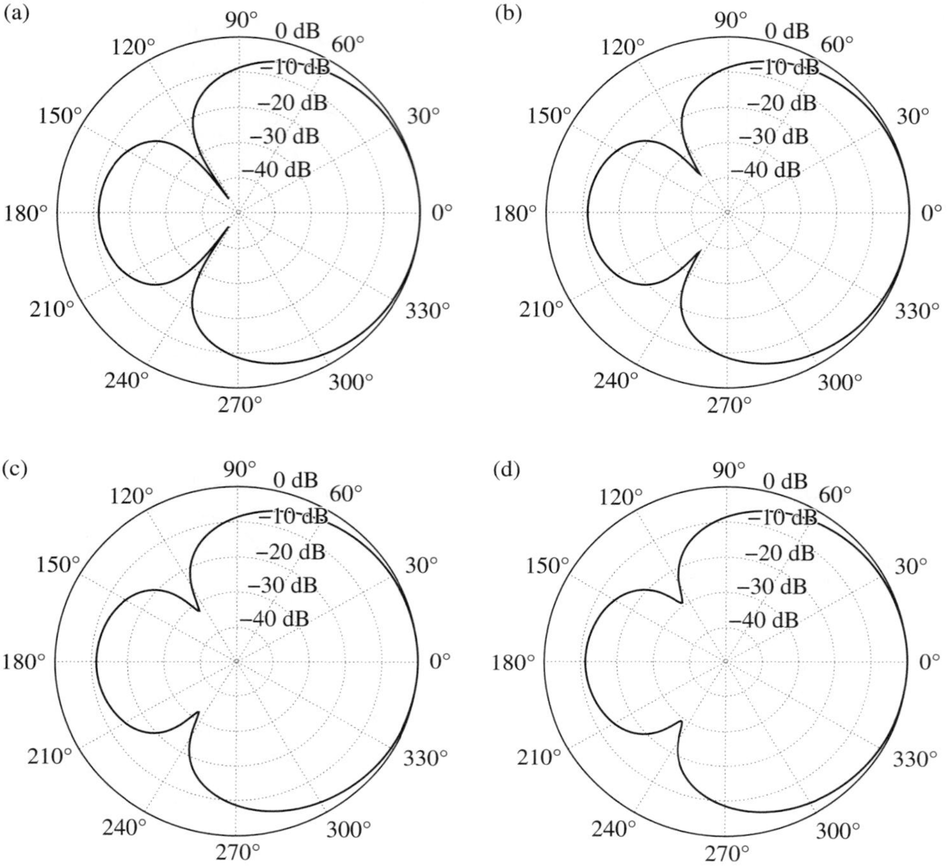

Figure 10.7 Beampatterns of the robust first-order supercardioid for $f = 1$ kHz, $\delta = 0.5$ cm, and several values of M: (a) $M = 2$, (b) $M = 4$, (c) $M = 6$, and (d) $M = 8$.

To give a better compromise for the level of white noise amplification, we can use the following filter:

$$\mathbf{h}_{\mathrm{FI},\epsilon}(f) = \mathbf{\Gamma}_{\mathrm{C},\epsilon}^{-1}(f)\overline{\mathbf{B}}_N^H(f)\left[\overline{\mathbf{B}}_N(f)\mathbf{\Gamma}_{\mathrm{C},\epsilon}^{-1}(f)\overline{\mathbf{B}}_N^H(f)\right]^{-1}\mathbf{b}_N, \tag{10.45}$$

where

$$\mathbf{\Gamma}_{\mathrm{C},\epsilon}(f) = \mathbf{\Gamma}_{\mathrm{C}}(f) + \epsilon\mathbf{I}_M, \tag{10.46}$$

with $\epsilon \geq 0$ being the regularization parameter. It is clear that $\mathbf{h}_{\mathrm{FI},0}(f) = \mathbf{h}_{\mathrm{FI}}(f)$ and $\mathbf{h}_{\mathrm{FI},\infty}(f) = \mathbf{h}_{\mathrm{R}}(f)$.

Figures 10.10–10.13 display the patterns – with $\mathbf{h}_{\mathrm{FI},\epsilon}(f)$ defined in (10.45) – of the first-order dipole, cardioid, hypercardioid, and supercardioid for $f = 1$ kHz, $\delta = 0.5$ cm, $M = 6$, and several values of ϵ. Figure 10.14 shows plots of the DF, $\mathcal{D}\left[\mathbf{h}_{\mathrm{FI},\epsilon}(f)\right]$, as

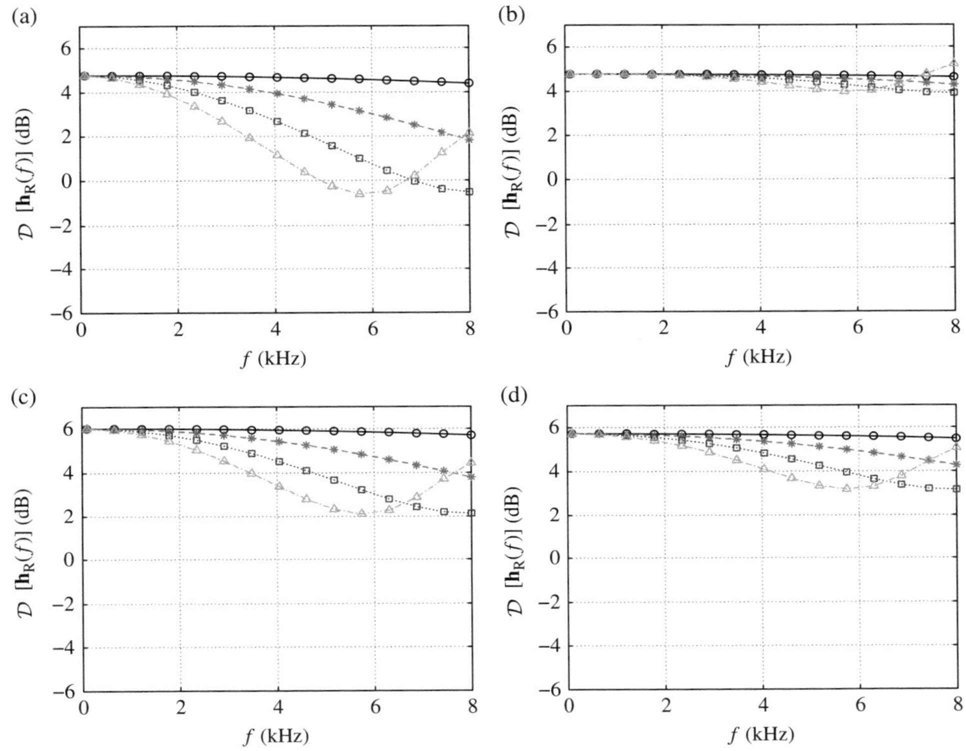

Figure 10.8 DF of the robust first-order DSAs as a function of frequency, for $\delta = 0.5$ cm and several values of M: $M = 2$ (solid line with circles), $M = 4$ (dashed line with asterisks), $M = 6$ (dotted line with squares), and $M = 8$ (dash-dot line with triangles). (a) Dipole, (b) cardioid, (c) hypercardioid, and (d) supercardioid.

a function of frequency, for the dipole, cardioid, hypercardioid, and supercardioid, and several values of ϵ. Corresponding plots of the WNG, $\mathcal{W}\left[\mathbf{h}_{\mathrm{FI},\epsilon}(f)\right]$, as a function of frequency are depicted in Figure 10.15. We can see that the WNG is considerably improved as ϵ increases, while the beampatterns and the DFs do not change much. The larger the value of ϵ, the more robust is the frequency-invariant first-order DSA against white noise amplification, but at the expense of less explicit nulls.

10.5 Least-squares Method

Let us define the error signal between the array beampattern and the desired directivity pattern:

$$\begin{aligned}\mathcal{E}\left[\mathbf{h}(f), \cos\theta\right] &= \mathcal{B}\left[\mathbf{h}(f), \cos\theta\right] - \mathcal{B}\left(\mathbf{b}_N, \cos\theta\right) \\ &= \mathbf{d}^H\left(f, \cos\theta\right)\mathbf{h}(f) - \mathbf{p}_{\mathrm{C}}^T\left(\cos\theta\right)\mathbf{b}_N.\end{aligned} \quad (10.47)$$

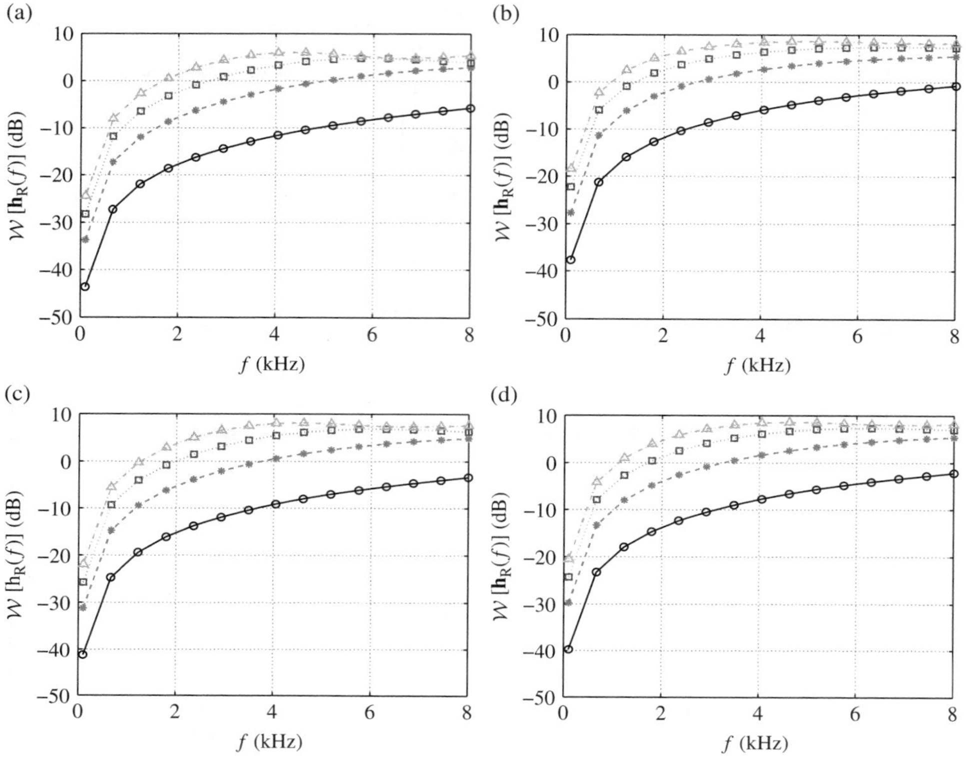

Figure 10.9 WNG of the robust first-order DSAs as a function of frequency, for $\delta = 0.5$ cm and several values of M: $M = 2$ (solid line with circles), $M = 4$ (dashed line with asterisks), $M = 6$ (dotted line with squares), and $M = 8$ (dash-dot line with triangles). (a) Dipole, (b) cardioid, (c) hypercardioid, and (d) supercardioid.

Then, the least-squares (LS) method consists of minimizing the LSE criterion:

$$\begin{aligned} \mathrm{LSE}\left[\mathbf{h}(f)\right] &= \frac{1}{\pi}\int_0^{\pi}\left|\mathcal{E}\left[\mathbf{h}(f), \cos\theta\right]\right|^2 d\theta \\ &= \mathbf{h}^H(f)\mathbf{\Gamma}_{\mathrm{C}}(f)\mathbf{h}(f) - \mathbf{h}^H(f)\mathbf{\Gamma}_{\mathbf{dp}_{\mathrm{C}}}(f)\mathbf{b}_N \\ &\quad - \mathbf{b}_N^T\mathbf{\Gamma}_{\mathbf{dp}_{\mathrm{C}}}^H(f)\mathbf{h}(f) + \mathbf{b}_N^T\mathbf{M}_{\mathrm{C}}\mathbf{b}_N, \end{aligned} \tag{10.48}$$

where

$$\mathbf{\Gamma}_{\mathbf{dp}_{\mathrm{C}}}(f) = \frac{1}{\pi}\int_0^{\pi}\mathbf{d}\left(f, \cos\theta\right)\mathbf{p}_{\mathrm{C}}^T\left(\cos\theta\right)d\theta, \tag{10.49}$$

and $\mathbf{\Gamma}_{\mathrm{C}}(f)$ and $\mathbf{M}_{\mathrm{C}}$ are defined in (10.41) and (10.18), respectively. The minimization of (10.48) gives the LS filter:

$$\mathbf{h}_{\mathrm{LS}}(f) = \mathbf{\Gamma}_{\mathrm{C}}^{-1}(f)\mathbf{\Gamma}_{\mathbf{dp}_{\mathrm{C}}}(f)\mathbf{b}_N. \tag{10.50}$$

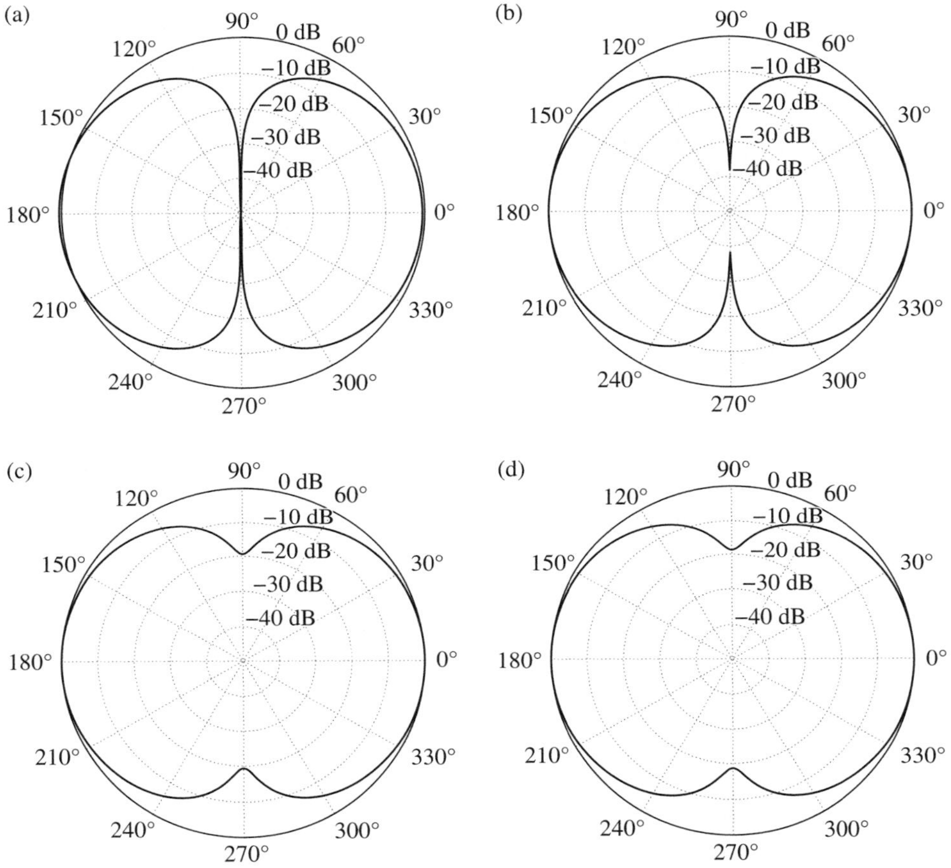

Figure 10.10 Beampatterns of the frequency-invariant first-order dipole for $f = 1$ kHz, $\delta = 0.5$ cm, $M = 6$, and several values of ϵ: (a) $\epsilon = 0$, (b) $\epsilon = 10^{-5}$, (c) $\epsilon = 10^{-3}$, and (d) $\epsilon = 0.1$.

It is also easy to find the regularized LS filter:

$$\mathbf{h}_{\mathrm{LS},\epsilon}(f) = \mathbf{\Gamma}_{\mathrm{C},\epsilon}^{-1}(f)\mathbf{\Gamma}_{\mathbf{dp}_{\mathrm{C}}}(f)\mathbf{b}_N. \tag{10.51}$$

Another more useful idea is to minimize the LSE criterion subject to the distortionless constraint [2]:

$$\min_{\mathbf{h}(f)} \mathrm{LSE}\left[\mathbf{h}(f)\right] \quad \text{subject to} \quad \mathbf{h}^H(f)\mathbf{d}\left(f,1\right) = 1. \tag{10.52}$$

We easily obtain the constrained LS (CLS) filter:

$$\mathbf{h}_{\mathrm{CLS}}(f) = \mathbf{h}_{\mathrm{LS}}(f) - \frac{1 - \mathbf{d}^H\left(f,1\right)\mathbf{h}_{\mathrm{LS}}(f)}{\mathbf{d}^H\left(f,1\right)\mathbf{\Gamma}_{\mathrm{C}}^{-1}(f)\mathbf{d}\left(f,1\right)}\mathbf{\Gamma}_{\mathrm{C}}^{-1}(f)\mathbf{d}\left(f,1\right). \tag{10.53}$$

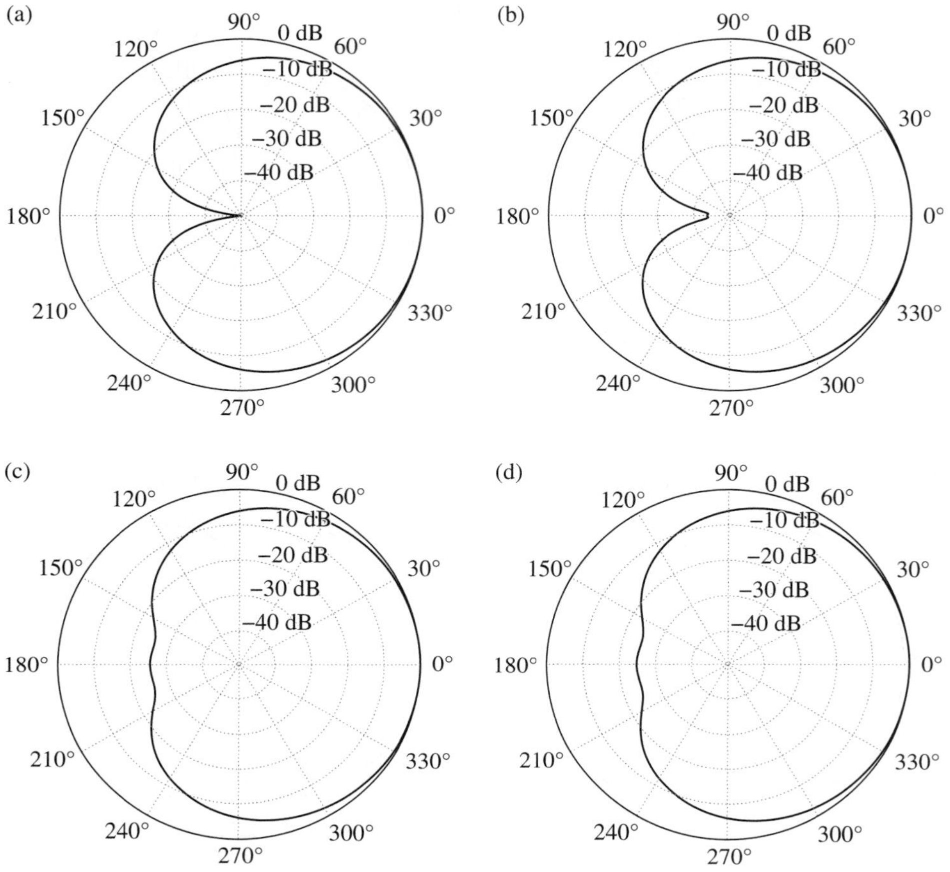

Figure 10.11 Beampatterns of the frequency-invariant first-order cardioid for $f = 1$ kHz, $\delta = 0.5$ cm, $M = 6$, and several values of ϵ: (a) $\epsilon = 0$, (b) $\epsilon = 10^{-5}$, (c) $\epsilon = 10^{-3}$, and (d) $\epsilon = 0.1$.

A more robust version is the regularized CLS filter is:

$$\mathbf{h}_{\mathrm{CLS},\epsilon}(f) = \mathbf{h}_{\mathrm{LS},\epsilon}(f) - \frac{1 - \mathbf{d}^H(f,1)\,\mathbf{h}_{\mathrm{LS},\epsilon}(f)}{\mathbf{d}^H(f,1)\,\mathbf{\Gamma}_{\mathrm{C},\epsilon}^{-1}(f)\mathbf{d}(f,1)}\mathbf{\Gamma}_{\mathrm{C},\epsilon}^{-1}(f)\mathbf{d}(f,1). \tag{10.54}$$

The error signal defined in (10.47) can also be expressed as

$$\begin{aligned} \mathcal{E}\left[\mathbf{h}(f), \cos\theta\right] &= \sum_{i=0}^{\infty} \cos(i\theta)\,\overline{\mathbf{b}}_i^T(f)\mathbf{h}(f) - \sum_{i=0}^{N} b_{N,i}\cos(i\theta) \\ &= \sum_{i=0}^{N} \cos(i\theta)\,\overline{\mathbf{b}}_i^T(f)\mathbf{h}(f) - \sum_{i=0}^{N} b_{N,i}\cos(i\theta) \\ &\quad + \sum_{i=N+1}^{\infty} \cos(i\theta)\,\overline{\mathbf{b}}_i^T(f)\mathbf{h}(f). \end{aligned} \tag{10.55}$$

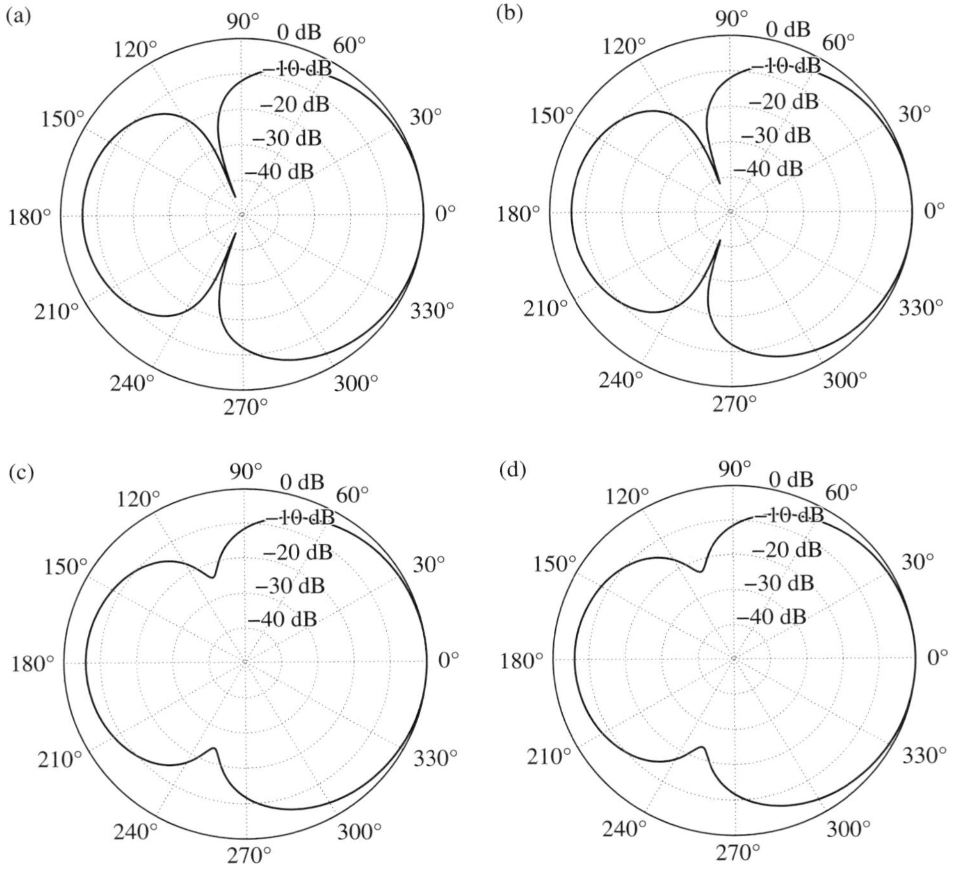

Figure 10.12 Beampatterns of the frequency-invariant first-order hypercardioid for $f = 1$ kHz, $\delta = 0.5$ cm, $M = 6$, and several values of ϵ: (a) $\epsilon = 0$, (b) $\epsilon = 10^{-5}$, (c) $\epsilon = 10^{-3}$, and (d) $\epsilon = 0.1$.

Using the constraint $\overline{\mathbf{B}}_N(f)\mathbf{h}(f) = \mathbf{b}_N$ [or, equivalently, $\overline{\mathbf{b}}_i^T(f)\mathbf{h}(f) = b_{N,i}$, $i = 0, 1, \ldots, N$] in the first element on the right-hand side of the previous expression, the error simplifies to

$$\mathcal{E}\left[\mathbf{h}(f), \cos\theta\right] = \sum_{i=N+1}^{\infty} \cos(i\theta)\, \overline{\mathbf{b}}_i^T(f)\mathbf{h}(f). \tag{10.56}$$

Therefore, the (constrained) LSE criterion is also

$$\text{LSE}\left[\mathbf{h}(f)\right] = \frac{1}{\pi}\int_0^{\pi} \left| \sum_{i=N+1}^{\infty} \cos(i\theta)\, \overline{\mathbf{b}}_i^T(f)\mathbf{h}(f) \right|^2 d\theta. \tag{10.57}$$

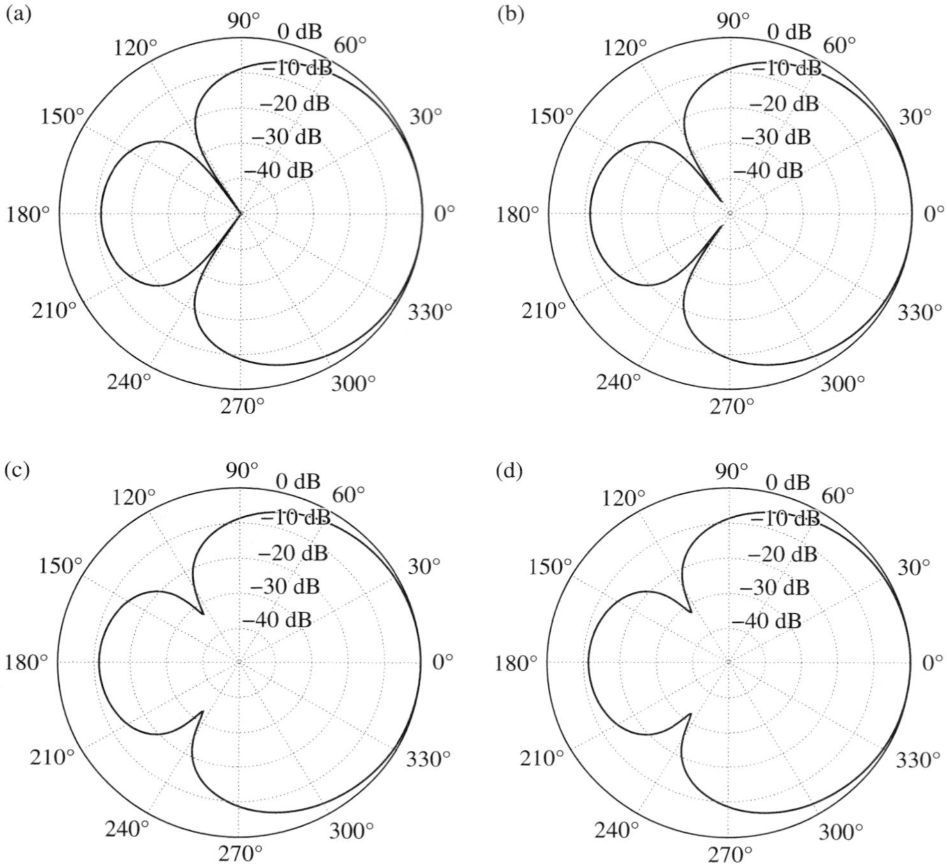

Figure 10.13 Beampatterns of the frequency-invariant first-order supercardioid for $f = 1$ kHz, $\delta = 0.5$ cm, $M = 6$, and several values of ϵ: (a) $\epsilon = 0$, (b) $\epsilon = 10^{-5}$, (c) $\epsilon = 10^{-3}$, and (d) $\epsilon = 0.1$.

Using (10.56), the criterion defined in (10.40) can be expressed as

$$\begin{aligned} J_{\mathrm{FI}}\left[\mathbf{h}(f)\right] &= \frac{1}{\pi}\int_0^{\pi}\left|\mathcal{E}\left[\mathbf{h}(f),\cos\theta\right] + \mathcal{B}\left(\mathbf{b}_N,\cos\theta\right)\right|^2 d\theta \\ &= \frac{1}{\pi}\int_0^{\pi}\left|\sum_{i=N+1}^{\infty}\cos(i\theta)\,\overline{\mathbf{b}}_i^T(f)\mathbf{h}(f) + \sum_{i=0}^{N} b_{N,i}\cos(i\theta)\right|^2 d\theta. \end{aligned} \tag{10.58}$$

Using the orthogonality property of the Chebyshev polynomials, the previous expression simplifies to

$$J_{\mathrm{FI}}\left[\mathbf{h}(f)\right] = \mathrm{LSE}\left[\mathbf{h}(f)\right] + \frac{1}{\pi}\int_0^{\pi}\left|\mathcal{B}\left(\mathbf{b}_N,\cos\theta\right)\right|^2 d\theta, \tag{10.59}$$

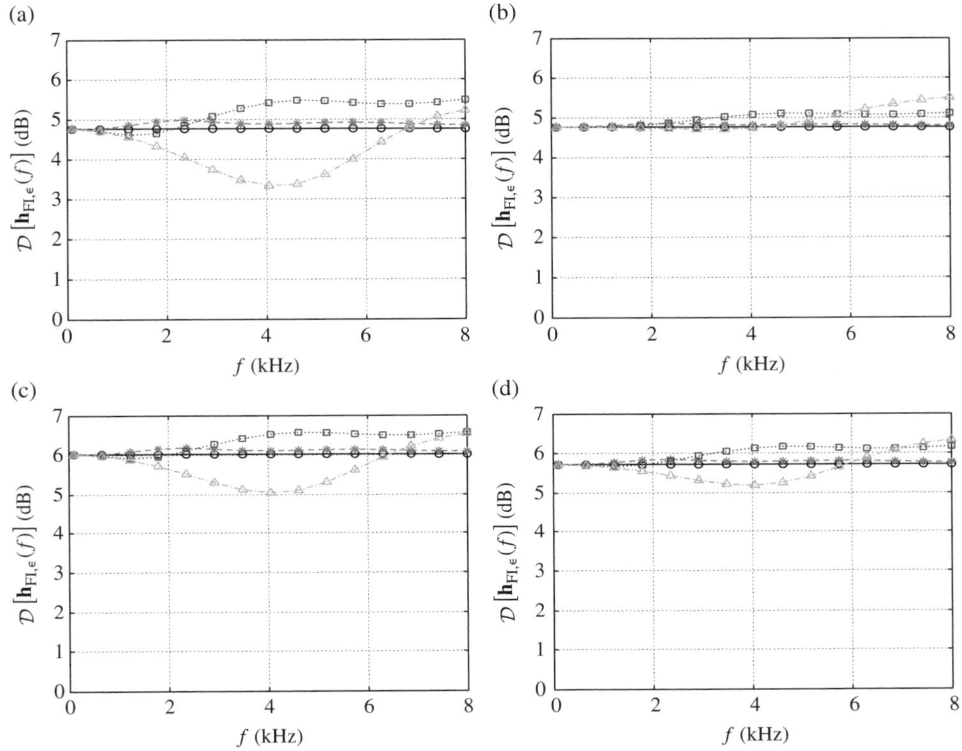

Figure 10.14 DF of the frequency-invariant first-order DSAs as a function of frequency, for $\delta = 0.5$ cm, $M = 6$, and several values of ϵ: $\epsilon = 0$ (solid line with circles), $\epsilon = 10^{-5}$ (dashed line with asterisks), $\epsilon = 10^{-3}$ (dotted line with squares), and $\epsilon = 0.1$ (dash-dot line with triangles). (a) Dipole, (b) cardioid, (c) hypercardioid, and (d) supercardioid.

where the second term on the right-hand side of (10.59) does not depend on $\mathbf{h}(f)$. This shows that minimizing $J_{\mathrm{FI}}\left[\mathbf{h}(f)\right]$ subject to the constraint $\overline{\mathbf{B}}_N(f)\mathbf{h}(f) = \mathbf{b}_N$ is equivalent to minimizing LSE $\left[\mathbf{h}(f)\right]$ subject to the same constraint.

Figures 10.16 displays the patterns – with $\mathbf{h}_{\mathrm{LS},\epsilon}(f)$ defined in (10.51) – of the first-order supercardioid for $f = 1$ kHz, $\delta = 0.5$ cm, $M = 6$, and several values of ϵ. Corresponding plots of the first-order supercardioid, obtained with $\mathbf{h}_{\mathrm{CLS},\epsilon}(f)$ – as defined in (10.54) – are depicted in Figure 10.17. Figure 10.18 shows plots of the DFs, $\mathcal{D}\left[\mathbf{h}_{\mathrm{LS},\epsilon}(f)\right]$ and $\mathcal{D}\left[\mathbf{h}_{\mathrm{CLS},\epsilon}(f)\right]$, as a function of frequency, for the first-order supercardioid and several values of ϵ. Corresponding plots of the WNGs, $\mathcal{W}\left[\mathbf{h}_{\mathrm{LS},\epsilon}(f)\right]$ and $\mathcal{W}\left[\mathbf{h}_{\mathrm{CLS},\epsilon}(f)\right]$, as a function of frequency are depicted in Figure 10.19. We observe that the WNG is considerably improved as ϵ increases, while the beampatterns and the DFs do not change much as long as ϵ is not too large. The larger is ϵ, the more robust are the regularized LS and CLS first-order DSAs against white noise amplification, but at the expense of less explicit nulls.

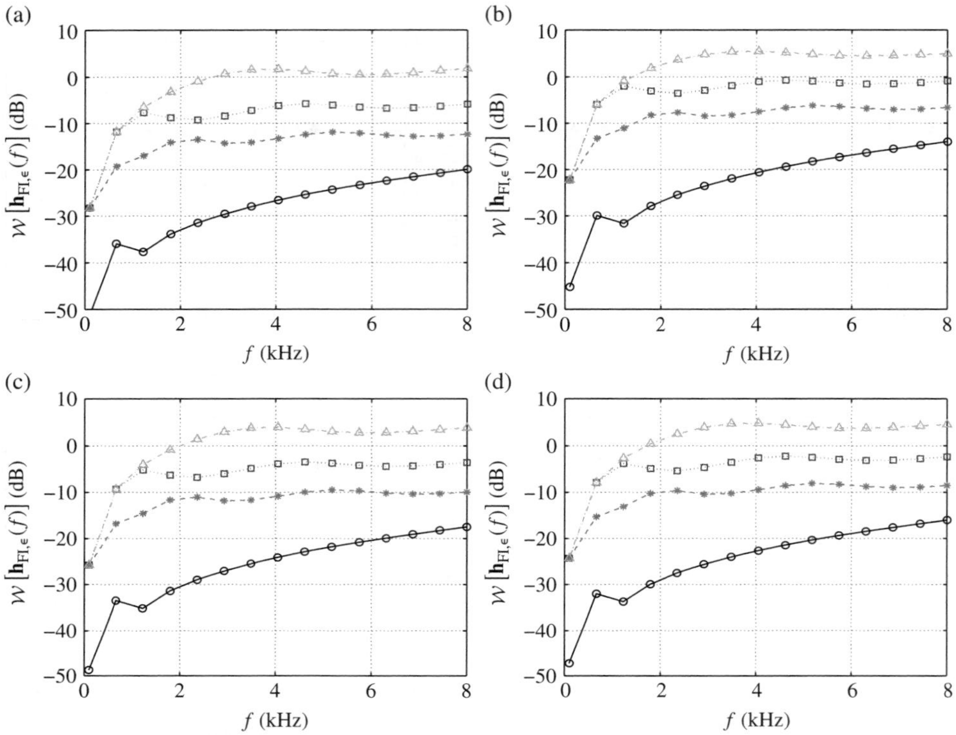

Figure 10.15 WNG of the frequency-invariant first-order DSAs as a function of frequency, for $\delta = 0.5$ cm, $M = 6$, and several values of ϵ: $\epsilon = 0$ (solid line with circles), $\epsilon = 10^{-5}$ (dashed line with asterisks), $\epsilon = 10^{-3}$ (dotted line with squares), and $\epsilon = 0.1$ (dash-dot line with triangles). (a) Dipole, (b) cardioid, (c) hypercardioid, and (d) supercardioid.

10.6 Joint Optimization

Here, of course, we assume that $M > N+1$. This gives us much more flexibility to design beampatterns with different compromises thanks to the array redundancy.

We denote by $\mathbf{h}'(f)$, the filter of length $N + 1$, which is equal to the filter $\mathbf{h}_{\mathrm{NR}}(f)$ derived in Section 10.2 with $N+1 = M$. We are interested in the class of filters of length $M(> N + 1)$, whose form is

$$\mathbf{h}(f) = \mathbf{H}'(f)\mathbf{g}(f), \tag{10.60}$$

where $\mathbf{H}'(f)$ is a matrix of size $M \times (M - N)$, with

$$\mathbf{H}'^{H}(f) = \begin{bmatrix} \mathbf{h}'^{H}(f) & & \mathbf{0}_{1\times(M-N-1)} \\ 0 & \mathbf{h}'^{H}(f) & \mathbf{0}_{1\times(M-N-2)} \\ \vdots & \ddots & \\ \mathbf{0}_{1\times(M-N-1)} & & \mathbf{h}'^{H}(f) \end{bmatrix}, \tag{10.61}$$

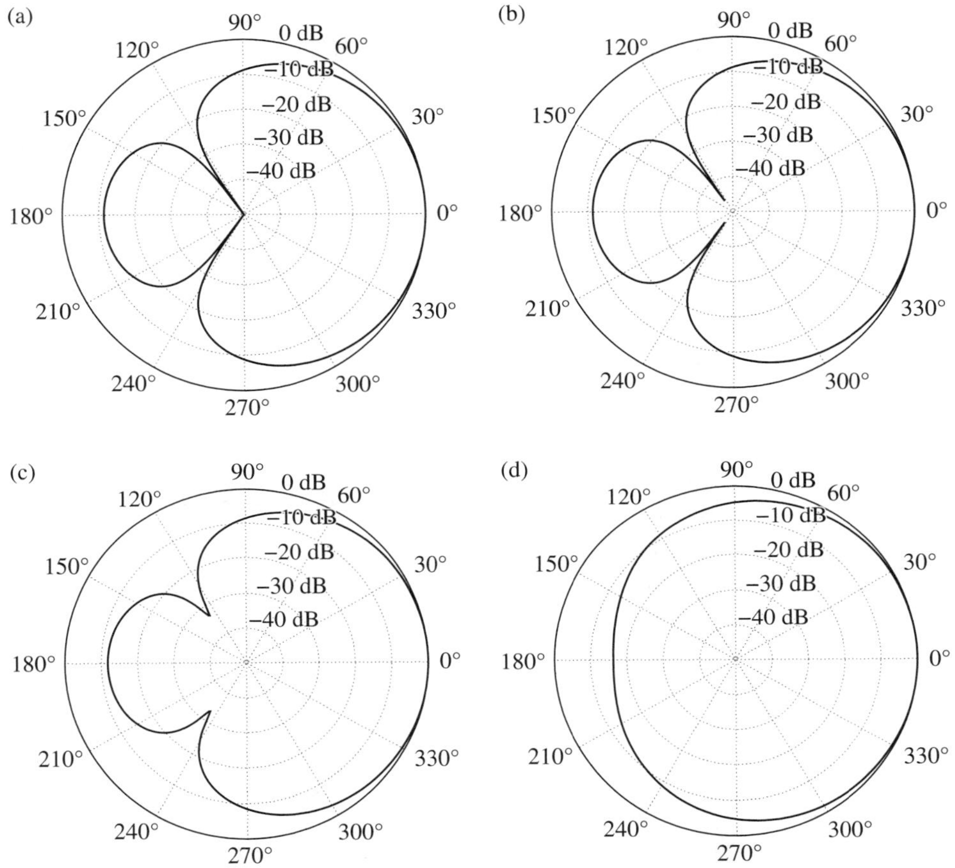

Figure 10.16 Beampatterns of the regularized LS first-order supercardioid for $f = 1$ kHz, $\delta = 0.5$ cm, $M = 6$, and several values of ϵ: (a) $\epsilon = 0$, (b) $\epsilon = 10^{-5}$, (c) $\epsilon = 10^{-3}$, and (d) $\epsilon = 0.1$.

$\mathbf{h}'(f) = \mathbf{h}_{\text{NR}}(f)$, and $\mathbf{g}(f) \neq \mathbf{0}$ is a filter of length $M - N$. The fundamental property of the class of filters defined in (10.60) is that they preserve the nulls of $\mathbf{h}'(f) = \mathbf{h}_{\text{NR}}(f)$. Indeed, if θ_0 is a null of $\mathbf{h}'(f)$, it can be verified that, thanks to the structure of the steering vector, we have

$$\mathbf{h}^H(f)\mathbf{d}\left(f, \cos\theta_0\right) = \mathbf{g}^H(f)\widetilde{\mathbf{d}}\left(f, \cos\theta_0\right) \times 0 = 0, \tag{10.62}$$

where

$$\widetilde{\mathbf{d}}\left(f, \cos\theta_0\right) = \begin{bmatrix} 1 & e^{-\jmath 2\pi f \tau_0 \cos\theta_0} & \cdots & e^{-\jmath (M-N-1)2\pi f \tau_0 \cos\theta_0} \end{bmatrix}^T. \tag{10.63}$$

At this point, it is important to mention that what characterizes the different array beampatterns is the different directions of their nulls; so when the nulls are preserved, the shape of the beampatterns is also mostly preserved. Now, we can adjust the filter

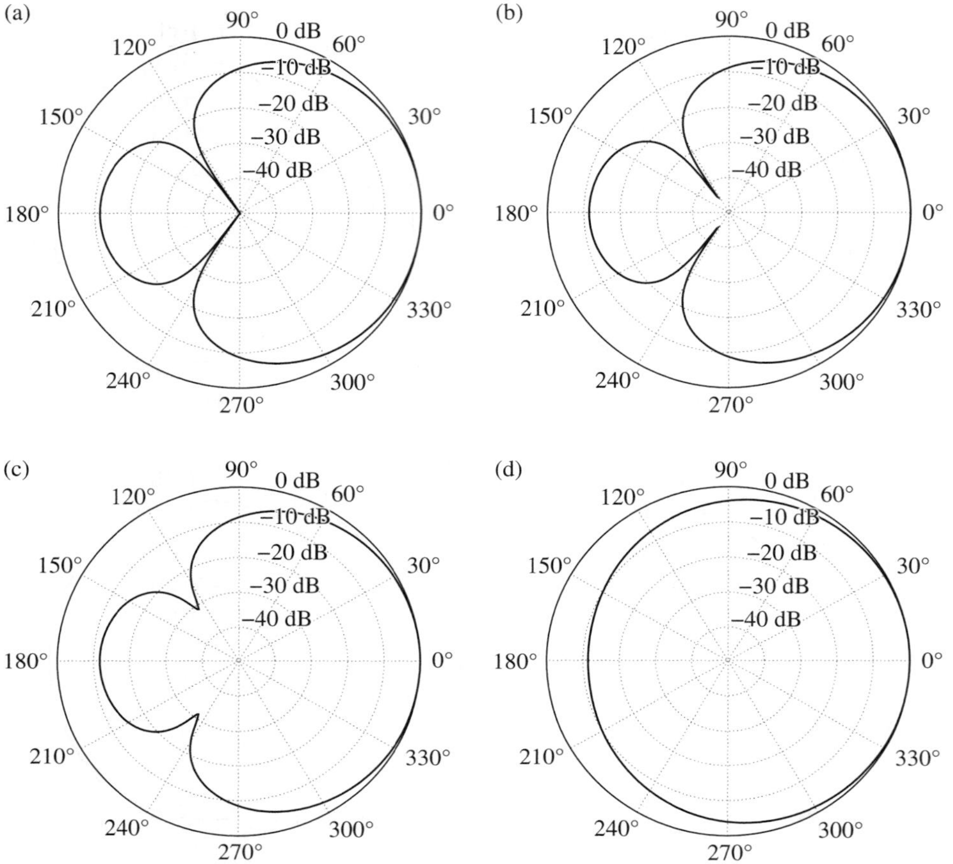

Figure 10.17 Beampatterns of the regularized CLS first-order supercardioid for $f = 1$ kHz, $\delta = 0.5$ cm, $M = 6$, and several values of ϵ: (a) $\epsilon = 0$, (b) $\epsilon = 10^{-5}$, (c) $\epsilon = 10^{-3}$, and (d) $\epsilon = 0.1$.

$\mathbf{g}(f)$ and its dimension to improve the WNG and/or the frequency invariance of the beampatterns.

At $\theta = 0$, we have

$$\begin{aligned} \mathbf{H}'^H(f)\mathbf{d}\left(f,1\right) &= \left[\begin{array}{cccc} 1 & e^{-\jmath 2\pi f\tau_0} & \cdots & e^{-\jmath(M-N-1)2\pi f\tau_0} \end{array}\right]^T \\ &= \widetilde{\mathbf{d}}\left(f,1\right). \end{aligned} \tag{10.64}$$

As a result, the distortionless constraint for the filter $\mathbf{h}(f)$ or, equivalently, the filter $\mathbf{g}(f)$ is

$$\mathbf{h}^H(f)\mathbf{d}\left(f,1\right) = \mathbf{g}^H(f)\widetilde{\mathbf{d}}\left(f,1\right) = 1. \tag{10.65}$$

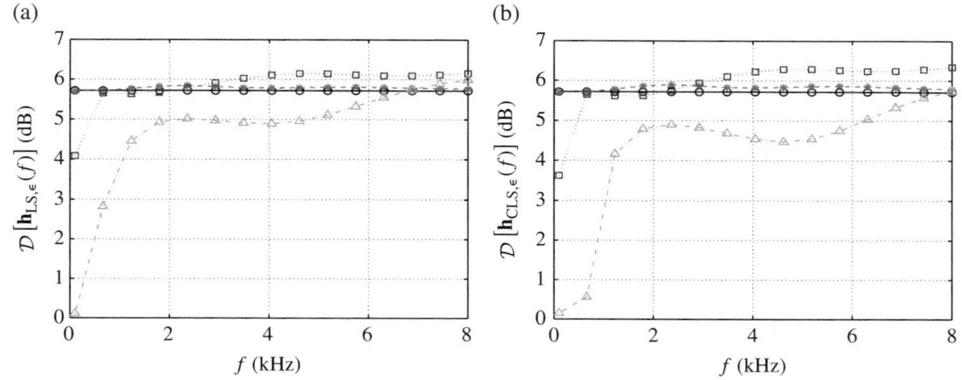

Figure 10.18 DF of first-order supercardioids as a function of frequency, for $\delta = 0.5$ cm, $M = 6$, and several values of ϵ: $\epsilon = 0$ (solid line with circles), $\epsilon = 10^{-5}$ (dashed line with asterisks), $\epsilon = 10^{-3}$ (dotted line with squares), and $\epsilon = 0.1$ (dash-dot line with triangles). (a) Regularized LS and (b) regularized CLS.

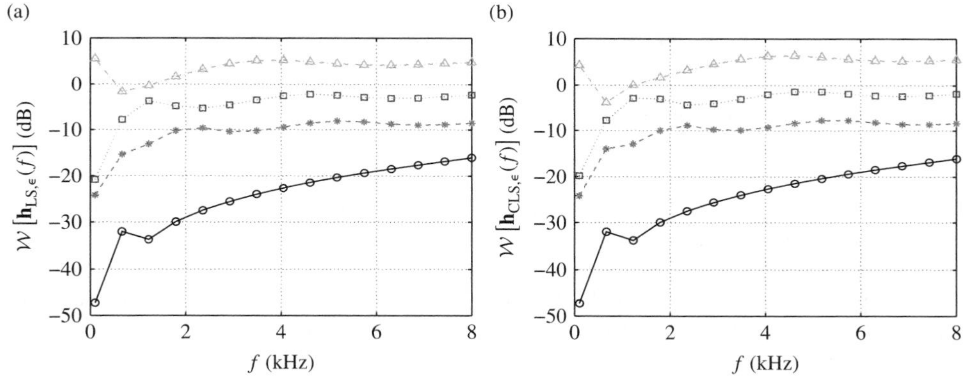

Figure 10.19 WNG of first-order supercardioids as a function of frequency, for $\delta = 0.5$ cm, $M = 6$, and several values of ϵ: $\epsilon = 0$ (solid line with circles), $\epsilon = 10^{-5}$ (dashed line with asterisks), $\epsilon = 10^{-3}$ (dotted line with squares), and $\epsilon = 0.1$ (dash-dot line with triangles). (a) Regularized LS and (b) regularized CLS.

Using (10.60), we can express the WNG and the beampattern as, respectively,

$$\begin{aligned} \mathcal{W}\left[\mathbf{h}(f)\right] &= \frac{\left|\mathbf{h}^H(f)\mathbf{d}\left(f,1\right)\right|^2}{\mathbf{h}^H(f)\mathbf{h}(f)} \\ &= \frac{\left|\mathbf{g}^H(f)\widetilde{\mathbf{d}}\left(f,1\right)\right|^2}{\mathbf{g}^H(f)\mathbf{H}'^H(f)\mathbf{H}'(f)\mathbf{g}(f)} \\ &= \mathcal{W}\left[\mathbf{g}(f)\right] \end{aligned} \tag{10.66}$$

and

$$\begin{aligned}\mathcal{B}\left[\mathbf{h}(f), \cos\theta\right] &= \mathbf{d}^H\left(f, \cos\theta\right)\mathbf{h}(f) \\ &= \mathbf{d}^H\left(f, \cos\theta\right)\mathbf{H}'(f)\mathbf{g}(f) \\ &= \mathcal{B}\left[\mathbf{g}(f), \cos\theta\right].\end{aligned} \tag{10.67}$$

With the proposed approach, the best way to improve the robustness of the filter with respect to white noise amplification is to maximize the WNG as given in (10.66):

$$\min_{\mathbf{g}(f)} \mathbf{g}^H(f)\mathbf{H}'^H(f)\mathbf{H}'(f)\mathbf{g}(f) \quad \text{subject to} \quad \mathbf{g}^H(f)\widetilde{\mathbf{d}}\left(f, 1\right) = 1. \tag{10.68}$$

We obtain the maximum WNG (MWNG) filter:

$$\mathbf{g}_{\text{MWNG}}(f) = \frac{\left[\mathbf{H}'^H(f)\mathbf{H}'(f)\right]^{-1}\widetilde{\mathbf{d}}\left(f, 1\right)}{\widetilde{\mathbf{d}}^H\left(f, 1\right)\left[\mathbf{H}'^H(f)\mathbf{H}'(f)\right]^{-1}\widetilde{\mathbf{d}}\left(f, 1\right)}. \tag{10.69}$$

As a result, the global MWNG filter is

$$\mathbf{h}_{\text{MWNG}}(f) = \mathbf{H}'(f)\mathbf{g}_{\text{MWNG}}(f). \tag{10.70}$$

This filter is equivalent to the robust filter, $\mathbf{h}_{\text{R}}(f)$, derived in Section 10.3. While $\mathbf{h}_{\text{MWNG}}(f)$ greatly improves the WNG, the designed beampattern diverges from the desired one as the frequency increases.

Figure 10.20 displays the patterns – with $\mathbf{h}_{\text{MWNG}}(f)$ as defined in (10.70) – of the first-order supercardioid for $f = 1$ kHz, $\delta = 0.5$ cm, and several values of M. Figure 10.21 shows plots of the DF and WNG, $\mathcal{D}\left[\mathbf{h}_{\text{MWNG}}(f)\right]$ and $\mathcal{W}\left[\mathbf{h}_{\text{MWNG}}(f)\right]$, as a function of frequency, for the first-order supercardioid and several values of M. We observe that the WNG is considerably improved as M increases, while the beampatterns and the DFs do not change much.

Let us define the error signal between the array beampattern and the desired directivity pattern:

$$\begin{aligned}\mathcal{E}\left[\mathbf{h}(f), \cos\theta\right] &= \mathcal{B}\left[\mathbf{h}(f), \cos\theta\right] - \mathcal{B}\left(\mathbf{b}_N, \cos\theta\right) \\ &= \mathbf{d}^H\left(f, \cos\theta\right)\mathbf{H}'(f)\mathbf{g}(f) - \mathbf{p}_{\text{C}}^T\left(\cos\theta\right)\mathbf{b}_N \\ &= \mathcal{E}\left[\mathbf{g}(f), \cos\theta\right].\end{aligned} \tag{10.71}$$

The LSE criterion can be expressed as

$$\begin{aligned}\text{LSE}\left[\mathbf{g}(f)\right] &= \frac{1}{\pi}\int_0^{\pi}\left|\mathcal{E}\left[\mathbf{g}(f), \cos\theta\right]\right|^2 d\theta \\ &= \mathbf{g}^H(f)\mathbf{H}'^H(f)\mathbf{\Gamma}_{\text{C}}(f)\mathbf{H}'(f)\mathbf{g}(f) - \mathbf{g}^H(f)\mathbf{H}'^H(f)\mathbf{\Gamma}_{\mathbf{dp}_{\text{C}}}(f)\mathbf{b}_N \\ &\quad - \mathbf{b}_N^T\mathbf{\Gamma}_{\mathbf{dp}_{\text{C}}}^H(f)\mathbf{H}'(f)\mathbf{g}(f) + \mathbf{b}_N^T\mathbf{M}_{\text{C}}\mathbf{b}_N.\end{aligned} \tag{10.72}$$

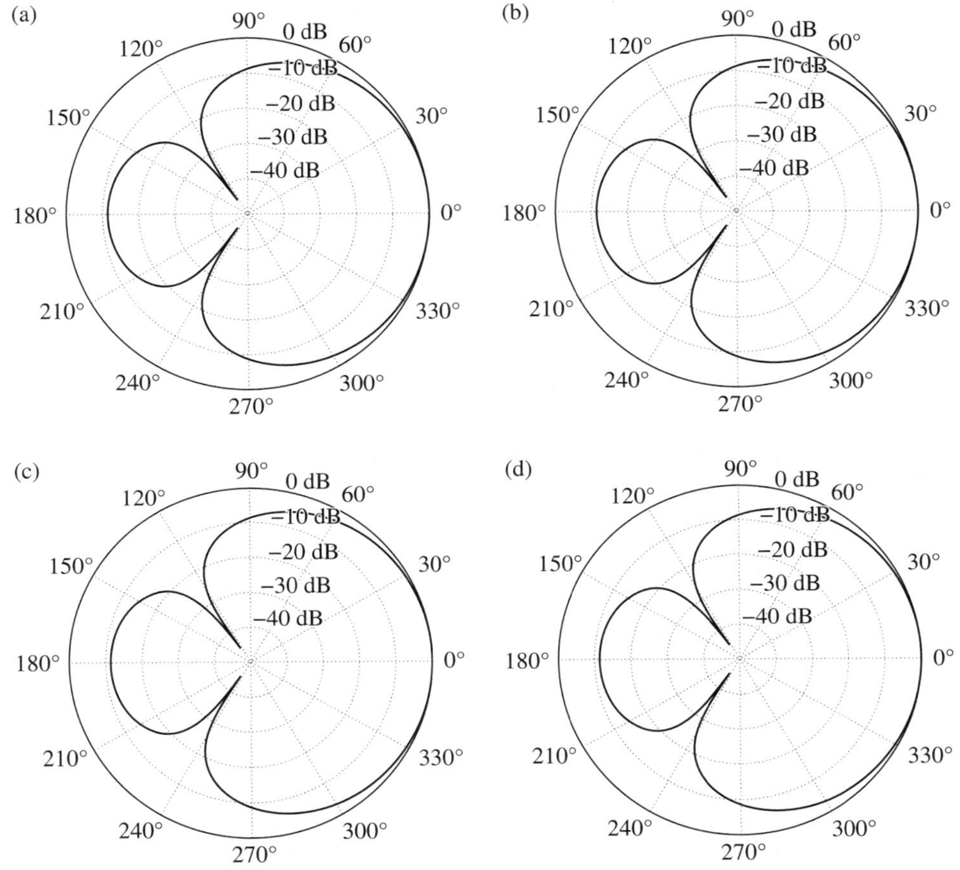

Figure 10.20 Beampatterns of the MWNG first-order supercardioid for $f = 1$ kHz, $\delta = 0.5$ cm, and several values of M: (a) $M = 3$, (b) $M = 4$, (c) $M = 6$, and (d) $M = 8$.

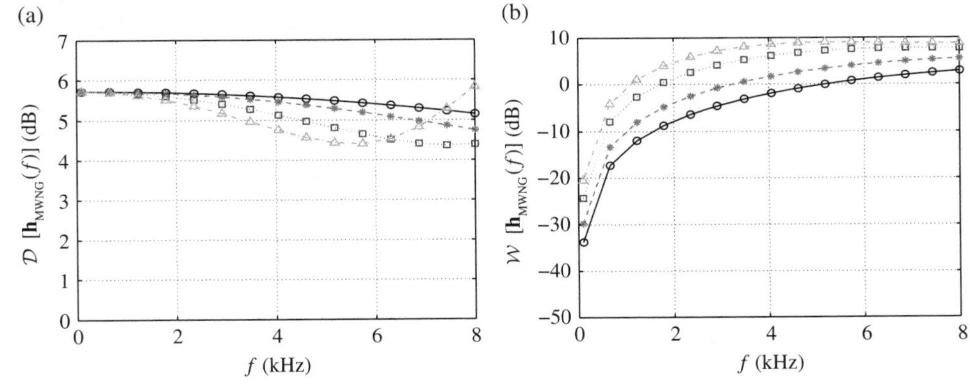

Figure 10.21 (a) DF and (b) WNG of the MWNG first-order supercardioid as a function of frequency, for $\delta = 0.5$ cm and several values of M: $M = 3$ (solid line with circles), $M = 4$ (dashed line with asterisks), $M = 6$ (dotted line with squares), and $M = 8$ (dash-dot line with triangles).

In order to get frequency-invariant beampatterns, we can minimize the LSE criterion subject to the distortionless constraint:

$$\min_{\mathbf{g}(f)} \text{LSE}\left[\mathbf{g}(f)\right] \quad \text{subject to} \quad \mathbf{g}^H(f)\widetilde{\mathbf{d}}\left(f,1\right) = 1, \tag{10.73}$$

from which we deduce the constrained LS (CLS) filter:

$$\mathbf{g}_{\text{CLS}}(f) = \mathbf{g}_{\text{LS}}(f) + \frac{1 - \widetilde{\mathbf{d}}^H\left(f,1\right)\mathbf{g}_{\text{LS}}(f)}{\widetilde{\mathbf{d}}^H\left(f,1\right)\mathbf{R}^{-1}(f)\widetilde{\mathbf{d}}\left(f,1\right)}\mathbf{R}^{-1}(f)\widetilde{\mathbf{d}}\left(f,1\right), \tag{10.74}$$

where

$$\mathbf{g}_{\text{LS}}(f) = \mathbf{R}^{-1}(f)\mathbf{H}'^H(f)\boldsymbol{\Gamma}_{\mathbf{dp}_\text{C}}(f)\mathbf{b}_N \tag{10.75}$$

is the LS filter obtained by minimizing LSE $\left[\mathbf{g}(f)\right]$ and

$$\mathbf{R}(f) = \mathbf{H}'^H(f)\boldsymbol{\Gamma}_\text{C}(f)\mathbf{H}'(f). \tag{10.76}$$

As a result, the global CLS filter is

$$\mathbf{h}_{\text{CLS},2}(f) = \mathbf{H}'(f)\mathbf{g}_{\text{CLS}}(f). \tag{10.77}$$

This filter is mostly equivalent to $\mathbf{h}_{\text{CLS}}(f)$, as derived in Section 10.5. While $\mathbf{h}_{\text{CLS},2}(f)$ leads to very nice frequency-invariant responses, it suffers severely from white noise amplification.

Figures 10.22 displays the patterns – with $\mathbf{h}_{\text{CLS},2}(f)$ as defined in (10.77) – of the first-order supercardioid for $f = 1$ kHz, $\delta = 0.5$ cm, and several values of M. Figure 10.23 shows plots of the DF and WNG, $\mathcal{D}\left[\mathbf{h}_{\text{CLS},2}(f)\right]$ and $\mathcal{W}\left[\mathbf{h}_{\text{CLS},2}(f)\right]$, as a function of frequency, for the first-order supercardioid and several values of M. We observe that the beampatterns and the DFs are approximately frequency invariant, but the WNG is very low, and becomes even worse as M increases.

In order to give a compromise between the WNG and frequency-invariant beampatterns, we should jointly optimize the two previous approaches. Let us define the criterion:

$$J_\aleph\left[\mathbf{g}(f)\right] = \aleph\text{LSE}\left[\mathbf{g}(f)\right] + (1-\aleph)\mathbf{g}^H(f)\mathbf{H}'^H(f)\mathbf{H}'(f)\mathbf{g}(f), \tag{10.78}$$

where $\aleph \in [0,1]$ controls the tradeoff between the WNG and the error beampattern. Taking into account the distortionless constraint, the optimization problem is

$$\min_{\mathbf{g}(f)} J_\aleph\left[\mathbf{g}(f)\right] \quad \text{subject to} \quad \mathbf{g}^H(f)\widetilde{\mathbf{d}}\left(f,1\right) = 1. \tag{10.79}$$

We find that the tradeoff filter is

$$\mathbf{g}_{\text{T},\aleph}(f) = \mathbf{g}_{\text{U},\aleph}(f) + \frac{1 - \widetilde{\mathbf{d}}^H\left(f,1\right)\mathbf{g}_{\text{U},\aleph}(f)}{\widetilde{\mathbf{d}}^H\left(f,1\right)\mathbf{R}_\aleph^{-1}(f)\widetilde{\mathbf{d}}\left(f,1\right)}\mathbf{R}_\aleph^{-1}(f)\widetilde{\mathbf{d}}\left(f,1\right), \tag{10.80}$$

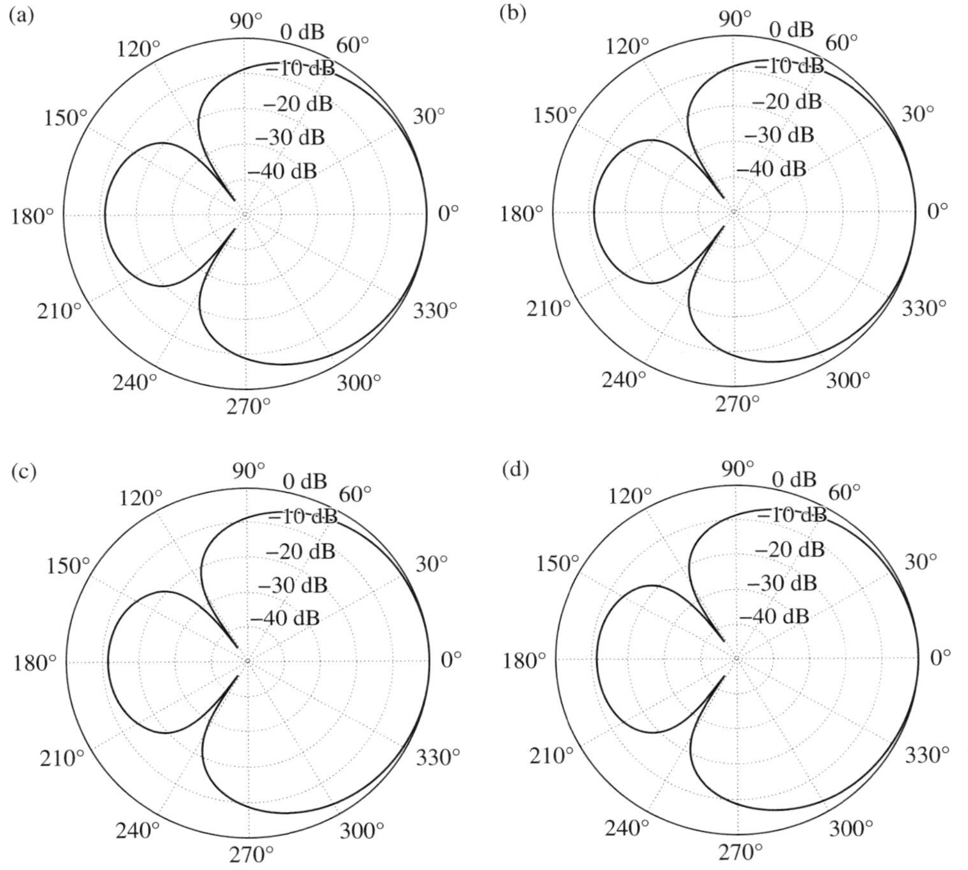

Figure 10.22 Beampatterns of the CLS first-order supercardioid for $f = 1$ kHz, $\delta = 0.5$ cm, and several values of M: (a) $M = 3$, (b) $M = 4$, (c) $M = 6$, and (d) $M = 8$.

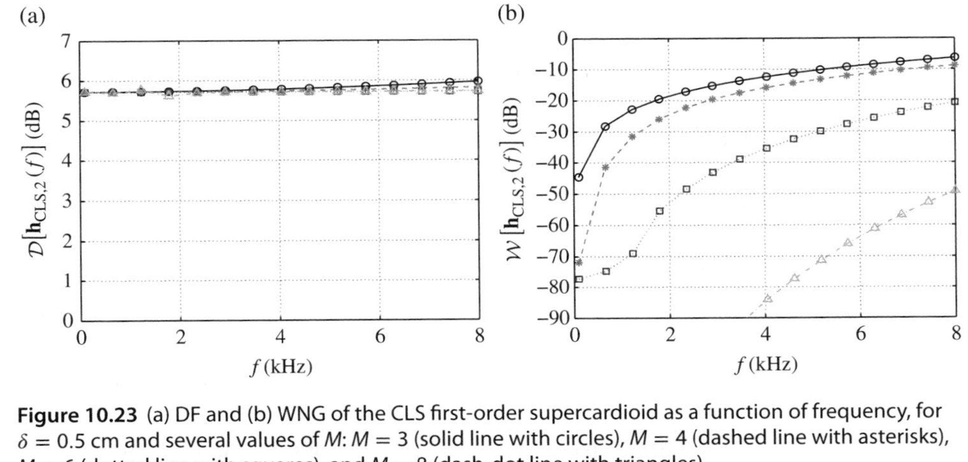

Figure 10.23 (a) DF and (b) WNG of the CLS first-order supercardioid as a function of frequency, for $\delta = 0.5$ cm and several values of M: $M = 3$ (solid line with circles), $M = 4$ (dashed line with asterisks), $M = 6$ (dotted line with squares), and $M = 8$ (dash-dot line with triangles).

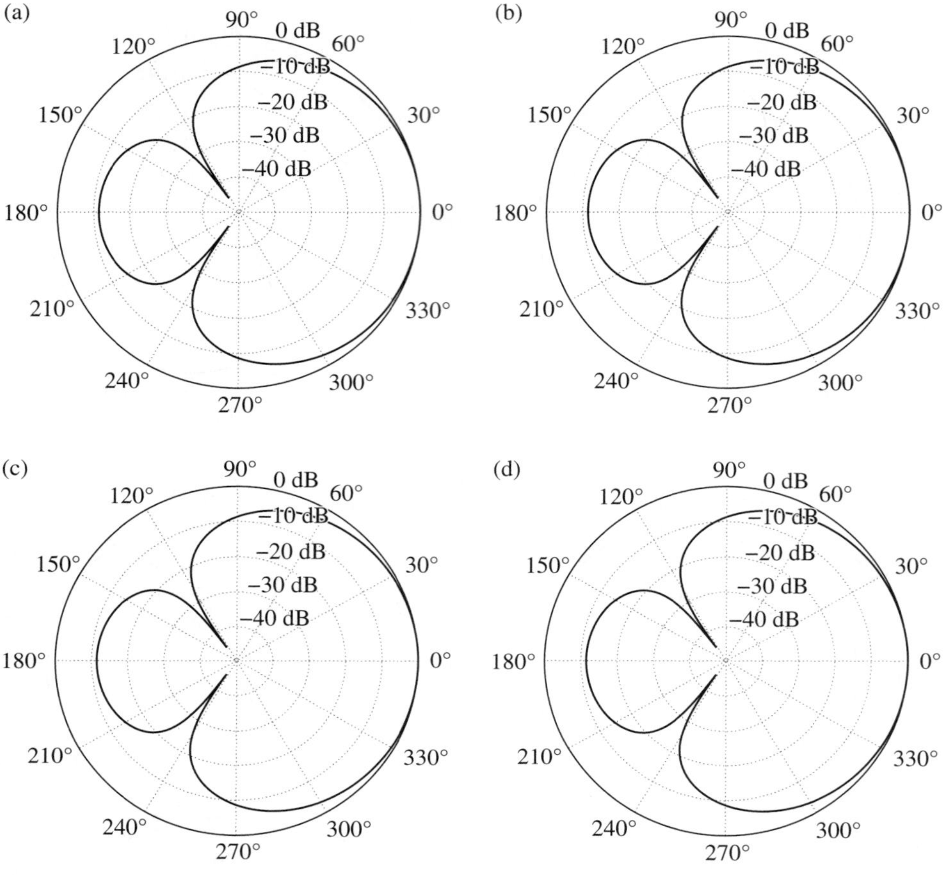

Figure 10.24 Beampatterns of the jointly optimized first-order supercardioid for $f = 1$ kHz, $\delta = 0.5$ cm, $M = 6$, and several values of $\aleph$: (a) $\aleph = 0$, (b) $\aleph = 0.5$, (c) $\aleph = 0.9$, and (d) $\aleph = 0.99$.

where

$$\mathbf{g}_{\mathrm{U},\aleph}(f) = \aleph \mathbf{R}_{\aleph}^{-1}(f)\mathbf{H}'^{H}(f)\mathbf{\Gamma}_{\mathbf{dp}_{\mathrm{C}}}(f)\mathbf{b}_N \tag{10.81}$$

is the unconstrained filter obtained by minimizing $J_{\aleph}\left[\mathbf{g}(f)\right]$ and

$$\mathbf{R}_{\aleph}(f) = \aleph \mathbf{R}(f) + (1 - \aleph)\mathbf{H}'^{H}(f)\mathbf{H}'(f). \tag{10.82}$$

Consequently, the global tradeoff filter from the proposed joint optimization is

$$\mathbf{h}_{\mathrm{T},\aleph}(f) = \mathbf{H}'(f)\mathbf{g}_{\mathrm{T},\aleph}(f). \tag{10.83}$$

Obviously, in the two extreme cases, we have $\mathbf{h}_{\mathrm{T},0}(f) = \mathbf{h}_{\mathrm{MWNG}}(f)$ and $\mathbf{h}_{\mathrm{T},1}(f) = \mathbf{h}_{\mathrm{CLS},2}(f)$.

Figure 10.24 displays the patterns – with $\mathbf{h}_{\mathrm{T},\aleph}(f)$ as defined in (10.83) – of the first-order supercardioid for $f = 1$ kHz, $\delta = 0.5$ cm, $M = 6$, and several values of $\aleph$.

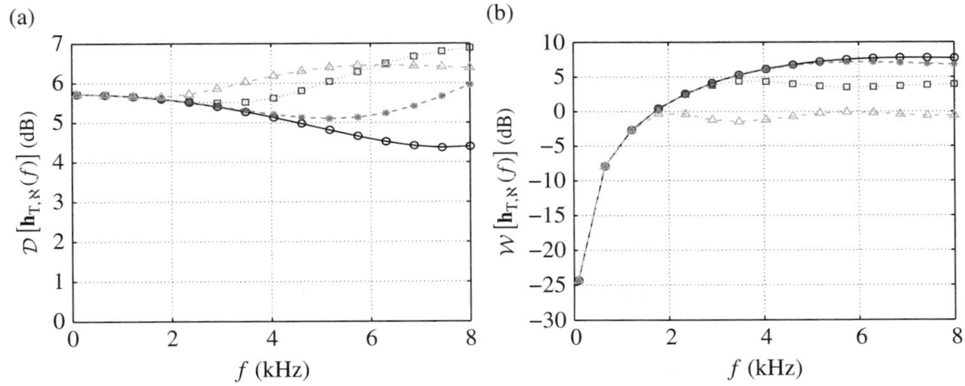

Figure 10.25 (a) DF and (b) WNG of the jointly optimized first-order supercardioid as a function of frequency, for $\delta = 0.5$ cm, $M = 6$, and several values of $\aleph$: $\aleph = 0$ (solid line with circles), $\aleph = 0.5$ (dashed line with asterisks), $\aleph = 0.9$ (dotted line with squares), and $\aleph = 0.99$ (dash-dot line with triangles).

Table 10.1 Filters for beampattern design.

Filter	
Nonrobust	$\mathbf{h}_{\mathrm{NR}}(f) = \overline{\mathbf{B}}_{M-1}^{-1}(f)\mathbf{b}_{M-1}$
Robust	$\mathbf{h}_{\mathrm{R}}(f) = \overline{\mathbf{B}}_{N}^{H}(f)\left[\overline{\mathbf{B}}_{N}(f)\overline{\mathbf{B}}_{N}^{H}(f)\right]^{-1}\mathbf{b}_N$
Frequency invariant	$\mathbf{h}_{\mathrm{FI}}(f) = \mathbf{\Gamma}_{\mathrm{C}}^{-1}(f)\overline{\mathbf{B}}_{N}^{H}(f)\left[\overline{\mathbf{B}}_{N}(f)\mathbf{\Gamma}_{\mathrm{C}}^{-1}(f)\overline{\mathbf{B}}_{N}^{H}(f)\right]^{-1}\mathbf{b}_N$
	$\mathbf{h}_{\mathrm{FI},\epsilon}(f) = \mathbf{\Gamma}_{\mathrm{C},\epsilon}^{-1}(f)\overline{\mathbf{B}}_{N}^{H}(f)\left[\overline{\mathbf{B}}_{N}(f)\mathbf{\Gamma}_{\mathrm{C},\epsilon}^{-1}(f)\overline{\mathbf{B}}_{N}^{H}(f)\right]^{-1}\mathbf{b}_N$
Least squares	$\mathbf{h}_{\mathrm{LS}}(f) = \mathbf{\Gamma}_{\mathrm{C}}^{-1}(f)\mathbf{\Gamma}_{\mathbf{dp}_{\mathrm{C}}}(f)\mathbf{b}_N$
	$\mathbf{h}_{\mathrm{LS},\epsilon}(f) = \mathbf{\Gamma}_{\mathrm{C},\epsilon}^{-1}(f)\mathbf{\Gamma}_{\mathbf{dp}_{\mathrm{C}}}(f)\mathbf{b}_N$
	$\mathbf{h}_{\mathrm{CLS}}(f) = \mathbf{h}_{\mathrm{LS}}(f) - \dfrac{1-\mathbf{d}^H(f,1)\,\mathbf{h}_{\mathrm{LS}}(f)}{\mathbf{d}^H(f,1)\,\mathbf{\Gamma}_{\mathrm{C}}^{-1}(f)\mathbf{d}(f,1)}\mathbf{\Gamma}_{\mathrm{C}}^{-1}(f)\mathbf{d}(f,1)$
	$\mathbf{h}_{\mathrm{CLS},\epsilon}(f) = \mathbf{h}_{\mathrm{LS},\epsilon}(f) - \dfrac{1-\mathbf{d}^H(f,1)\,\mathbf{h}_{\mathrm{LS},\epsilon}(f)}{\mathbf{d}^H(f,1)\,\mathbf{\Gamma}_{\mathrm{C},\epsilon}^{-1}(f)\mathbf{d}(f,1)}\mathbf{\Gamma}_{\mathrm{C},\epsilon}^{-1}(f)\mathbf{d}(f,1)$
	$\mathbf{h}_{\mathrm{CLS},2}(f) = \mathbf{H}'(f)\mathbf{g}_{\mathrm{CLS}}(f)$
Maximum WNG	$\mathbf{h}_{\mathrm{MWNG}}(f) = \dfrac{\mathbf{H}'(f)\left[\mathbf{H}'^H(f)\mathbf{H}'(f)\right]^{-1}\widetilde{\mathbf{d}}(f,1)}{\widetilde{\mathbf{d}}^H(f,1)\left[\mathbf{H}'^H(f)\mathbf{H}'(f)\right]^{-1}\widetilde{\mathbf{d}}(f,1)}$
Joint optimization	$\mathbf{h}_{\mathrm{T},\aleph}(f) = \mathbf{H}'(f)\mathbf{g}_{\mathrm{T},\aleph}(f)$

Figure 10.25 shows plots of the DF and WNG, $\mathcal{D}\left[\mathbf{h}_{\mathrm{T},\aleph}(f)\right]$ and $\mathcal{W}\left[\mathbf{h}_{\mathrm{T},\aleph}(f)\right]$, as a function of frequency, for the first-order supercardioid and several values of $\aleph$. We observe that $\aleph$ gives a compromise between the WNG and frequency-invariant beampatterns. The DF at high frequencies is improved as $\aleph$ increases, while the WNG is significantly

higher than that with $\mathbf{h}_{\mathrm{CLS},2}(f)$. Compared to the CLS filter, the jointly optimized filter is considerably more robust against white noise amplification, but leads to slightly less frequency-invariant responses.

To conclude this chapter, we present in Table 10.1 most of the filters outlined for beampattern design.

Problems

10.1 Show that the minimization of the LSE criterion yields

$$\mathbf{c}_N = \mathbf{M}_{\mathrm{C}}^{-1}\mathbf{v}_{\mathrm{C}}\left(\jmath\bar{f}_m\right).$$

10.2 Show that the elements of the vector $\mathbf{v}_{\mathrm{C}}\left(\jmath\bar{f}_m\right)$ are

$$\left[\mathbf{v}_{\mathrm{C}}\left(\jmath\bar{f}_m\right)\right]_{n+1} = \jmath^n J_n\left(\bar{f}_m\right),$$

where $J_n(z)$ is the Bessel function of the first kind.

10.3 Show that the elements of the matrix $\mathbf{M}_{\mathrm{C}}$ are

$$\left[\mathbf{M}_{\mathrm{C}}\right]_{i+1,j+1} = \frac{1}{\pi}\int_0^{\pi}\cos(i\theta)\cos(j\theta)\,d\theta.$$

10.4 Prove the Jacobi–Anger expansion:

$$e^{\jmath\bar{f}_m\cos\theta} = \sum_{n=0}^{\infty}\mathcal{J}_n J_n\left(\bar{f}_m\right)\cos(n\theta),$$

where

$$\mathcal{J}_n = \begin{cases} 1, & n = 0 \\ 2\jmath^n, & n = 1, 2, \ldots, N \end{cases}.$$

10.5 Show that the beampattern can be approximated by

$$\mathcal{B}_N\left[\mathbf{h}(f), \cos\theta\right] = \sum_{n=0}^{N}\cos(n\theta)\left[\sum_{m=1}^{M}\mathcal{J}_n J_n\left(\bar{f}_m\right)H_m(f)\right].$$

10.6 Show that with the nonrobust filter, $\mathbf{h}_{\mathrm{NR}}(f)$, the first-order beampattern is given by

$$\mathcal{B}_1\left[\mathbf{h}(f), \cos\theta\right] = H_1(f) + J_0\left(\bar{f}_2\right)H_2(f) + 2\jmath J_1\left(\bar{f}_2\right)H_2(f)\cos\theta.$$

10.7 Show that the minimum-norm filter for the design of first-order DSA beampatterns is given by

$$H_{i,\mathrm{R}}(f) = \frac{J_1\left(\overline{f}_i\right) b_{1,1}}{2\jmath \sum_{j=2}^{M} J_1^2\left(\overline{f}_j\right)}, \ i = 2, 3, \ldots, M,$$

$$H_{1,\mathrm{R}}(f) = -\sum_{i=2}^{M} J_0\left(\overline{f}_i\right) H_{i,\mathrm{R}}(f) + b_{1,0}.$$

10.8 Show that by minimizing $J_{\mathrm{FI}}\left[\mathbf{h}(f)\right]$ subject to $\overline{\mathbf{B}}_N(f)\mathbf{h}(f) = \mathbf{b}_N$ and $\mathbf{h}^H(f)\mathbf{h}(f) = \delta_\epsilon$, we obtain the filter:

$$\mathbf{h}_{\mathrm{FI},\epsilon}(f) = \mathbf{\Gamma}_{\mathrm{C},\epsilon}^{-1}(f)\overline{\mathbf{B}}_N^H(f)\left[\overline{\mathbf{B}}_N(f)\mathbf{\Gamma}_{\mathrm{C},\epsilon}^{-1}(f)\overline{\mathbf{B}}_N^H(f)\right]^{-1}\mathbf{b}_N,$$

where $\mathbf{\Gamma}_{\mathrm{C},\epsilon}(f) = \mathbf{\Gamma}_{\mathrm{C}}(f) + \epsilon\mathbf{I}_M$.

10.9 Show that the LSE between the array beampattern and the desired directivity pattern can be written as

$$\begin{aligned}\mathrm{LSE}\left[\mathbf{h}(f)\right] &= \mathbf{h}^H(f)\mathbf{\Gamma}_{\mathrm{C}}(f)\mathbf{h}(f) - \mathbf{h}^H(f)\mathbf{\Gamma}_{\mathbf{dp}_\mathrm{C}}(f)\mathbf{b}_N \\ &\quad - \mathbf{b}_N^T\mathbf{\Gamma}_{\mathbf{dp}_\mathrm{C}}^H(f)\mathbf{h}(f) + \mathbf{b}_N^T\mathbf{M}_\mathrm{C}\mathbf{b}_N.\end{aligned}$$

10.10 Show that by minimizing the LSE with a constraint on the coefficients, we obtain the regularized LS filter:

$$\mathbf{h}_{\mathrm{LS},\epsilon}(f) = \mathbf{\Gamma}_{\mathrm{C},\epsilon}^{-1}(f)\mathbf{\Gamma}_{\mathbf{dp}_\mathrm{C}}(f)\mathbf{b}_N.$$

10.11 Show that by minimizing the LSE subject to the distortionless constraint and a constraint on the coefficients, we obtain the regularized CLS filter:

$$\mathbf{h}_{\mathrm{CLS},\epsilon}(f) = \mathbf{h}_{\mathrm{LS},\epsilon}(f) - \frac{1 - \mathbf{d}^H\left(f,1\right)\mathbf{h}_{\mathrm{LS},\epsilon}(f)}{\mathbf{d}^H\left(f,1\right)\mathbf{\Gamma}_{\mathrm{C},\epsilon}^{-1}(f)\mathbf{d}\left(f,1\right)}\mathbf{\Gamma}_{\mathrm{C},\epsilon}^{-1}(f)\mathbf{d}\left(f,1\right).$$

10.12 Show that with the constraint $\overline{\mathbf{B}}_N(f)\mathbf{h}(f) = \mathbf{b}_N$, the error signal between the array beampattern and the desired directivity pattern can be expressed as

$$\mathcal{E}\left[\mathbf{h}(f), \cos\theta\right] = \sum_{i=N+1}^{\infty} \cos\left(i\theta\right)\overline{\mathbf{b}}_i^T(f)\mathbf{h}(f).$$

10.13 Using the orthogonality property of the Chebyshev polynomials, show that the criterion $J_{\mathrm{FI}}\left[\mathbf{h}(f)\right]$ can be expressed as

$$J_{\mathrm{FI}}\left[\mathbf{h}(f)\right] = \mathrm{LSE}\left[\mathbf{h}(f)\right] + \frac{1}{\pi}\int_0^\pi \left|\mathcal{B}\left(\mathbf{b}_N, \cos\theta\right)\right|^2 d\theta,$$

where

$$\mathrm{LSE}\left[\mathbf{h}(f)\right] = \frac{1}{\pi}\int_0^{\pi}\left|\sum_{i=N+1}^{\infty}\cos(i\theta)\,\overline{\mathbf{b}}_i^T(f)\mathbf{h}(f)\right|^2 d\theta.$$

10.14 Show that the filters defined in (10.60) preserve the nulls of $\mathbf{h}'(f) = \mathbf{h}_{\mathrm{NR}}(f)$; that is, if θ_0 is a null of $\mathbf{h}'(f)$, then

$$\mathbf{h}^H(f)\mathbf{d}\left(f,\cos\theta_0\right) = \mathbf{g}^H(f)\widetilde{\mathbf{d}}\left(f,\cos\theta_0\right)\times 0 = 0,$$

where

$$\widetilde{\mathbf{d}}\left(f,\cos\theta_0\right) = \left[\begin{array}{cccc} 1 & e^{-j2\pi f\tau_0\cos\theta_0} & \cdots & e^{-j(M-N-1)2\pi f\tau_0\cos\theta_0}\end{array}\right]^T.$$

10.15 Show that by maximizing the WNG subject to the distortionless constraint, we obtain the MWNG filter:

$$\mathbf{h}_{\mathrm{MWNG}}(f) = \frac{\mathbf{H}'(f)\left[\mathbf{H}'^H(f)\mathbf{H}'(f)\right]^{-1}\widetilde{\mathbf{d}}\left(f,1\right)}{\widetilde{\mathbf{d}}^H\left(f,1\right)\left[\mathbf{H}'^H(f)\mathbf{H}'(f)\right]^{-1}\widetilde{\mathbf{d}}\left(f,1\right)}.$$

10.16 Show that by minimizing $J_{\aleph}\left[\mathbf{g}(f)\right]$ subject to the distortionless constraint, we obtain the tradeoff filter:

$$\mathbf{g}_{\mathrm{T},\aleph}(f) = \mathbf{g}_{\mathrm{U},\aleph}(f) + \frac{1-\widetilde{\mathbf{d}}^H\left(f,1\right)\mathbf{g}_{\mathrm{U},\aleph}(f)}{\widetilde{\mathbf{d}}^H\left(f,1\right)\mathbf{R}_{\aleph}^{-1}(f)\widetilde{\mathbf{d}}\left(f,1\right)}\mathbf{R}_{\aleph}^{-1}(f)\widetilde{\mathbf{d}}\left(f,1\right),$$

where

$$\mathbf{g}_{\mathrm{U},\aleph}(f) = \aleph\mathbf{R}_{\aleph}^{-1}(f)\mathbf{H}'^H(f)\boldsymbol{\Gamma}_{\mathbf{dp}_{\mathrm{C}}}(f)\mathbf{b}_N$$

is the unconstrained filter obtained by minimizing $J_{\aleph}\left[\mathbf{g}(f)\right]$ and

$$\mathbf{R}_{\aleph}(f) = \aleph\mathbf{R}(f) + (1-\aleph)\mathbf{H}'^H(f)\mathbf{H}'(f).$$

References

1 H. F. Davis, *Fourier Series and Orthogonal Functions*. New York: Dover, 1989.
2 L. Zhao, J. Benesty, and J. Chen, "Optimal design of directivity patterns for endfire linear microphone arrays," in *Proc. IEEE ICASSP*, 2015, pp. 295–299.
3 L. Zhao, J. Benesty, and J. Chen, "Design of robust differential microphone arrays with the Jacobi–Anger expansion," *Applied Acoustics*, vol. 110, pp. 194–206, Sep. 2016.
4 M. Abramowitz and I. A. Stegun (eds), *Handbook of Mathematical Functions with Formulas, Graphs, and Mathematical Tables*. New York, NY: Dover, 1970.
5 C. Pan, J. Benesty, and J. Chen, "Design of robust differential microphone arrays with orthogonal polynomials," *J. Acoust. Soc. Am.*, vol. 138, pp. 1079–1089, Aug. 2015.

6 D. Colton and R. Krees, *Inverse Acoustics and Electromagnetic Scattering Theory*, 2nd edn. Berlin, Germany: Springer-Verlag, 1998.

7 A. Cuyt, V. B. Petersen, B. Verdonk, H. Waadeland, and W. B. Jones, *Handbook of Continued Fractions for Special Functions*. Berlin, Germany: Springer-Verlag, 2008.

8 G. W. Elko, "Superdirectional microphone arrays," in *Acoustic Signal Processing for Telecommunication*, S. L. Gay and J. Benesty (eds). Boston, MA: Kluwer Academic Publishers, 2000, Chapter 10, pp. 181–237.

9 J. Benesty and J. Chen, *Study and Design of Differential Microphone Arrays*. Berlin, Germany: Springer-Verlag, 2012.

11

Beamforming in the Time Domain

This chapter is concerned with beamforming in the time domain, which has the advantage of being more intuitive than beamforming in the frequency domain. Furthermore, the approach depicted here is broadband in nature. We describe the time-domain signal model that we adopt and explain how broadband beamforming works. Then we define many performance measures that are essential for the derivation and analysis of broadband beamformers. Some measures are only relevant for fixed beamforming while others are more relevant for adaptive beamforming. Finally, we show in great detail how to derive three classes of time-domain beamformers: fixed, adaptive, and differential.

11.1 Signal Model and Problem Formulation

We consider a desired broadband source signal, $x(t)$, in the far-field that propagates in an anechoic acoustic environment, and impinges on a ULA consisting of M omnidirectional sensors, where the distance between two successive sensors is equal to δ. Sensor 1 is chosen as the reference. In this scenario, the signal measured at the mth sensor is given by [1]:

$$\begin{aligned} y_m(t) &= x\left[t - \Delta - f_s\tau_m\left(\cos\theta_d\right)\right] + v_m(t) \\ &= x_m(t) + v_m(t),\ m = 1, 2, \ldots, M, \end{aligned} \tag{11.1}$$

where Δ is the propagation time from the position of the source (desired signal), $x(t)$, to sensor 1, f_s is the sampling frequency,

$$\tau_m\left(\cos\theta_d\right) = (m-1)\frac{\delta\cos\theta_d}{c} \tag{11.2}$$

is the delay between the first and mth sensors, θ_d is the direction, c is the speed of the waves in the medium, and $v_m(t)$ is the noise picked up by the mth sensor. For the sake of simplicity, we assume that

$$0 \leq \frac{f_s\delta\cos\theta_d}{c} \in \mathbb{Z}. \tag{11.3}$$

Fundamentals of Signal Enhancement and Array Signal Processing, First Edition.
Jacob Benesty, Israel Cohen, and Jingdong Chen.

Companion website: www.wiley.com/go/benesty/arraysignalprocessing

This clearly restricts θ_d, but simplifies the signal model. In what follows, we generalize the signal model when assumption (11.3) is not satisfied. Under this assumption, we can express (11.1) as

$$y_m(t) = \mathbf{g}_m^T\left(\cos\theta_d\right)\mathbf{x}'\left(t-\Delta\right) + v_m(t), \tag{11.4}$$

where

$$\mathbf{g}_m\left(\cos\theta_d\right) = \begin{bmatrix} 0 & \cdots & 0 & 1 & 0 & \cdots & 0 \end{bmatrix}^T \tag{11.5}$$

is a vector of length $L_g \geq f_s\tau_m\left(\cos\theta_d\right)+1$, the $\left[f_s\tau_m\left(\cos\theta_d\right)+1\right]$th component of which is equal to 1 and

$$\begin{aligned}\mathbf{x}'\left(t-\Delta\right) = \big[\; & x\left(t-\Delta\right) \quad x\left(t-\Delta-1\right) \quad \cdots \\ & x\left[t-\Delta-f_s\tau_m\left(\cos\theta_d\right)\right] \quad \cdots \quad x\left(t-\Delta-L_g+1\right)\;\big]^T.\end{aligned} \tag{11.6}$$

The vector $\mathbf{g}_m\left(\cos\theta_d\right)$ is a 1-sparse vector and the position of the 1 depends on both θ_d and m, with

$$\mathbf{g}_1\left(\cos\theta_d\right) = \begin{bmatrix} 1 & 0 & \cdots & 0 \end{bmatrix}^T. \tag{11.7}$$

By considering L_h successive time samples of the mth sensor signal, (11.4) becomes a vector of length L_h:

$$\mathbf{y}_m(t) = \mathbf{G}_m\left(\cos\theta_d\right)\mathbf{x}\left(t-\Delta\right) + \mathbf{v}_m(t), \tag{11.8}$$

where

$$\mathbf{G}_m\left(\cos\theta_d\right) = \begin{bmatrix} \mathbf{g}_m^T\left(\cos\theta_d\right) & 0 & 0 & \cdots & 0 \\ 0 & \mathbf{g}_m^T\left(\cos\theta_d\right) & 0 & \cdots & 0 \\ \vdots & \vdots & \ddots & & \vdots \\ 0 & 0 & 0 & \cdots & \mathbf{g}_m^T\left(\cos\theta_d\right) \end{bmatrix} \tag{11.9}$$

is a Sylvester matrix of size $L_h \times L$, with $L = L_g + L_h - 1$,

$$\mathbf{x}\left(t-\Delta\right) = \begin{bmatrix} x\left(t-\Delta\right) & x\left(t-\Delta-1\right) & \cdots & x\left(t-\Delta-L+1\right) \end{bmatrix}^T \tag{11.10}$$

is a vector of length L, and

$$\mathbf{v}_m(t) = \begin{bmatrix} v_m(t) & v_m(t-1) & \cdots & v_m(t-L_h+1) \end{bmatrix}^T \tag{11.11}$$

is a vector of length L_h. Figure 11.1 illustrates the multichannel signal model in the time domain. By concatenating the observations from the M sensors, we get the vector of length ML_h:

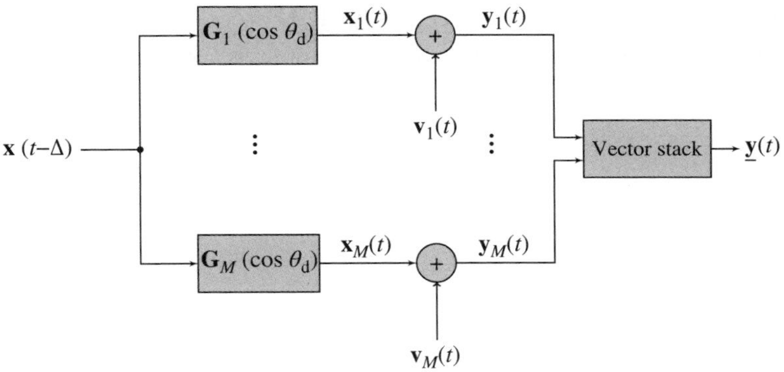

Figure 11.1 Multichannel signal model in the time domain.

$$\begin{aligned}\underline{\mathbf{y}}(t) &= \begin{bmatrix} \mathbf{y}_1^T(t) & \mathbf{y}_2^T(t) & \cdots & \mathbf{y}_M^T(t) \end{bmatrix}^T \\ &= \underline{\mathbf{G}}\left(\cos\theta_{\mathrm{d}}\right)\mathbf{x}(t-\Delta) + \underline{\mathbf{v}}(t) \\ &= \underline{\mathbf{x}}(t) + \underline{\mathbf{v}}(t),\end{aligned} \tag{11.12}$$

where

$$\underline{\mathbf{G}}\left(\cos\theta_{\mathrm{d}}\right) = \begin{bmatrix} \mathbf{G}_1\left(\cos\theta_{\mathrm{d}}\right) \\ \mathbf{G}_2\left(\cos\theta_{\mathrm{d}}\right) \\ \vdots \\ \mathbf{G}_M\left(\cos\theta_{\mathrm{d}}\right) \end{bmatrix} \tag{11.13}$$

is a matrix of size $ML_h \times L$,

$$\underline{\mathbf{v}}(t) = \begin{bmatrix} \mathbf{v}_1^T(t) & \mathbf{v}_2^T(t) & \cdots & \mathbf{v}_M^T(t) \end{bmatrix}^T \tag{11.14}$$

is a vector of length ML_h, and

$$\begin{aligned}\underline{\mathbf{x}}(t) &= \begin{bmatrix} \mathbf{x}_1^T(t) & \mathbf{x}_2^T(t) & \cdots & \mathbf{x}_M^T(t) \end{bmatrix}^T \\ &= \underline{\mathbf{G}}\left(\cos\theta_{\mathrm{d}}\right)\mathbf{x}(t-\Delta),\end{aligned} \tag{11.15}$$

with $\mathbf{x}_m(t) = \mathbf{G}_m\left(\cos\theta_{\mathrm{d}}\right)\mathbf{x}(t-\Delta)$.

From (11.12), we deduce that the correlation matrix (of size $ML_h \times ML_h$) of $\underline{\mathbf{y}}(t)$ is

$$\begin{aligned}\mathbf{R}_{\underline{\mathbf{y}}} &= E\left[\underline{\mathbf{y}}(t)\underline{\mathbf{y}}^T(t)\right] \\ &= \mathbf{R}_{\underline{\mathbf{x}}} + \mathbf{R}_{\underline{\mathbf{v}}} \\ &= \underline{\mathbf{G}}\left(\cos\theta_{\mathrm{d}}\right)\mathbf{R}_{\mathbf{x}}\underline{\mathbf{G}}^T\left(\cos\theta_{\mathrm{d}}\right) + \mathbf{R}_{\underline{\mathbf{v}}},\end{aligned} \tag{11.16}$$

where $\mathbf{R}_{\underline{\mathbf{x}}}$, $\mathbf{R}_{\underline{\mathbf{v}}}$, and $\mathbf{R}_{\mathbf{x}}$ are the correlation matrices of $\underline{\mathbf{x}}(t)$, $\underline{\mathbf{v}}(t)$, and $\mathbf{x}(t-\Delta)$, respectively. We always assume that $\mathbf{R}_{\underline{\mathbf{v}}}$ has full rank. However, to fully exploit the spatial information, as in the frequency domain, the matrix $\mathbf{R}_{\underline{\mathbf{x}}} = \underline{\mathbf{G}}\left(\cos\theta_{\mathrm{d}}\right)\mathbf{R}_{\mathbf{x}}\underline{\mathbf{G}}^T\left(\cos\theta_{\mathrm{d}}\right)$ must be rank deficient. Since the size of $\underline{\mathbf{G}}\left(\cos\theta_{\mathrm{d}}\right)$ is $ML_h \times L$ and the size of $\mathbf{R}_{\mathbf{x}}$ is $L \times L$, the condition for that is

$$ML_h > L \tag{11.17}$$

or, equivalently,

$$L_h > \frac{L_g - 1}{M - 1}. \tag{11.18}$$

We see that as M increases, the minimal value of L_h decreases.

Our objective is to design all kinds of time-domain or broadband beamformers with a real-valued spatiotemporal filter of length ML_h:

$$\underline{\mathbf{h}} = \left[\begin{array}{cccc} \mathbf{h}_1^T & \mathbf{h}_2^T & \cdots & \mathbf{h}_M^T \end{array}\right]^T, \tag{11.19}$$

where $\mathbf{h}_m,\ m = 1, 2, \ldots, M$ are temporal filters of length L_h.

To generalize the signal model when assumption (11.3) is not satisfied, we resort to Shannon's sampling theorem [2, 3], which implies that

$$\begin{aligned} x_m(t) &= x\left[t - \Delta - f_{\mathrm{s}}\tau_m\left(\cos\theta_{\mathrm{d}}\right)\right] \\ &= \sum_{n=-\infty}^{\infty} x\left(t - \Delta - n\right)\operatorname{sinc}\left[n - f_{\mathrm{s}}\tau_m\left(\cos\theta_{\mathrm{d}}\right)\right] \\ &\approx \sum_{n=-P-L_h+1}^{P} x\left(t - \Delta - n\right)\operatorname{sinc}\left[n - f_{\mathrm{s}}\tau_m\left(\cos\theta_{\mathrm{d}}\right)\right] \end{aligned} \tag{11.20}$$

for $P \gg f_{\mathrm{s}}\tau_m\left(\cos\theta_{\mathrm{d}}\right)$. Hence, we can simply redefine $\mathbf{x}\left(t - \Delta\right)$ as a vector of length $L = 2P + L_h$ with

$$\begin{aligned} \mathbf{x}\left(t - \Delta\right) = \Big[\ & x\left(t - \Delta + P + L_h - 1\right) \quad x\left(t - \Delta + P + L_h - 2\right) \\ & \cdots \quad x\left(t - \Delta - P\right)\ \Big]^T \end{aligned} \tag{11.21}$$

and redefine $\mathbf{G}_m\left(\cos\theta_{\mathrm{d}}\right)$ as a Toeplitz matrix of size $L_h \times L$ with the elements:

$$\left[\mathbf{G}_m\left(\cos\theta_{\mathrm{d}}\right)\right]_{i,j} = \operatorname{sinc}\left[-P - L_h + 1 - i + j - f_{\mathrm{s}}\tau_m\left(\cos\theta_{\mathrm{d}}\right)\right], \tag{11.22}$$

where $i = 1, \ldots, L_h, j = 1, \ldots, L$.

11.2 Broadband Beamforming

By applying the spatiotemporal filter, $\underline{\mathbf{h}}$, to the observation signal vector, $\underline{\mathbf{y}}(t)$, we obtain the output of the broadband beamformer, as illustrated in Figure 11.2:

$$\begin{aligned} z(t) &= \sum_{m=1}^{M} \mathbf{h}_m^T \mathbf{y}_m(t) \\ &= \underline{\mathbf{h}}^T \underline{\mathbf{y}}(t) \\ &= x_{\text{fd}}(t) + v_{\text{rn}}(t), \end{aligned} \tag{11.23}$$

where

$$\begin{aligned} x_{\text{fd}}(t) &= \sum_{m=1}^{M} \mathbf{h}_m^T \mathbf{G}_m \left(\cos\theta_{\text{d}}\right) \mathbf{x}\left(t - \Delta\right) \\ &= \underline{\mathbf{h}}^T \underline{\mathbf{G}} \left(\cos\theta_{\text{d}}\right) \mathbf{x}\left(t - \Delta\right) \end{aligned} \tag{11.24}$$

is the filtered desired signal and

$$\begin{aligned} v_{\text{rn}}(t) &= \sum_{m=1}^{M} \mathbf{h}_m^T \mathbf{v}_m(t) \\ &= \underline{\mathbf{h}}^T \underline{\mathbf{v}}(t) \end{aligned} \tag{11.25}$$

is the residual noise. We deduce that the variance of $z(t)$ is

$$\begin{aligned} \sigma_z^2 &= \underline{\mathbf{h}}^T \mathbf{R}_{\underline{\mathbf{y}}} \underline{\mathbf{h}} \\ &= \sigma_{x_{\text{fd}}}^2 + \sigma_{v_{\text{rn}}}^2, \end{aligned} \tag{11.26}$$

where

$$\begin{aligned} \sigma_{x_{\text{fd}}}^2 &= \underline{\mathbf{h}}^T \underline{\mathbf{G}} \left(\cos\theta_{\text{d}}\right) \mathbf{R}_{\mathbf{x}} \underline{\mathbf{G}}^T \left(\cos\theta_{\text{d}}\right) \underline{\mathbf{h}} \\ &= \underline{\mathbf{h}}^T \mathbf{R}_{\underline{\mathbf{x}}} \underline{\mathbf{h}} \end{aligned} \tag{11.27}$$

is the variance of $x_{\text{fd}}(t)$ and

$$\sigma_{v_{\text{rn}}}^2 = \underline{\mathbf{h}}^T \mathbf{R}_{\underline{\mathbf{v}}} \underline{\mathbf{h}} \tag{11.28}$$

is the variance of $v_{\text{rn}}(t)$.

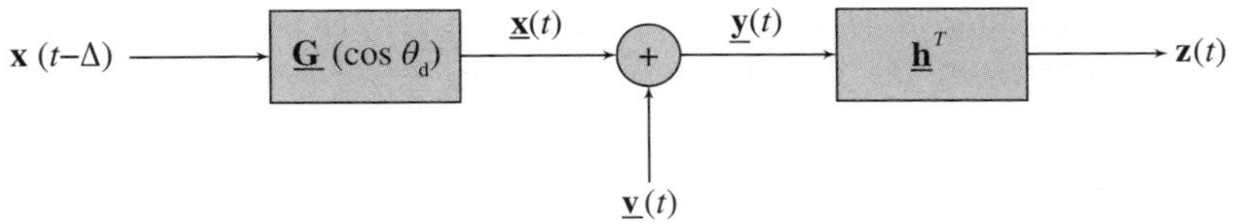

Figure 11.2 Block diagram of broadband beamforming in the time domain.

In principle, any element of the vector $\mathbf{x}(t-\Delta)$ can be considered as the desired signal. Therefore, from (11.24), we see that the distortionless constraint is

$$\underline{\mathbf{h}}^T \underline{\mathbf{G}}\left(\cos\theta_{\mathrm{d}}\right) = \mathbf{i}_l^T, \tag{11.29}$$

where $\mathbf{i}_l$ is the lth column of the $L \times L$ identity matrix, $\mathbf{I}_L$.

11.3 Performance Measures

In this section, we define all relevant performance measures for the derivation and analysis of fixed and adaptive beamformers in the time domain. For fixed beamforming only, since we are concerned with broadband signals, we assume for convenience that the source signal, $x(t)$, is white; this way, the whole spectrum is taken into account.

Since sensor 1 is the reference, the input SNR is computed from the first L_h components of $\underline{\mathbf{y}}(t)$ as defined in (11.12): $\mathbf{y}_1(t) = \mathbf{x}_1(t) + \mathbf{v}_1(t)$. We easily find that

$$\begin{aligned} \mathrm{iSNR} &= \frac{\mathrm{tr}\left(\mathbf{R}_{\mathbf{x}_1}\right)}{\mathrm{tr}\left(\mathbf{R}_{\mathbf{v}_1}\right)} \\ &= \frac{\sigma_x^2}{\sigma_{v_1}^2}, \end{aligned} \tag{11.30}$$

where $\mathbf{R}_{\mathbf{x}_1}$ and $\mathbf{R}_{\mathbf{v}_1}$ are the correlation matrices of $\mathbf{x}_1(t)$ and $\mathbf{v}_1(t)$, respectively, and σ_x^2 and $\sigma_{v_1}^2$ are the variances of $x(t)$ and $v_1(t)$, respectively.

The output SNR is obtained from (11.26). It is given by

$$\begin{aligned} \mathrm{oSNR}\left(\underline{\mathbf{h}}\right) &= \frac{\sigma_{x_{\mathrm{fd}}}^2}{\sigma_{v_{\mathrm{rn}}}^2} \\ &= \frac{\underline{\mathbf{h}}^T \underline{\mathbf{G}}\left(\cos\theta_{\mathrm{d}}\right) \mathbf{R}_{\mathbf{x}} \underline{\mathbf{G}}^T\left(\cos\theta_{\mathrm{d}}\right) \underline{\mathbf{h}}}{\underline{\mathbf{h}}^T \mathbf{R}_{\underline{\mathbf{v}}} \underline{\mathbf{h}}} \\ &= \frac{\sigma_x^2}{\sigma_{v_1}^2} \times \frac{\underline{\mathbf{h}}^T \underline{\mathbf{G}}\left(\cos\theta_{\mathrm{d}}\right) \underline{\mathbf{G}}^T\left(\cos\theta_{\mathrm{d}}\right) \underline{\mathbf{h}}}{\underline{\mathbf{h}}^T \boldsymbol{\Gamma}_{\underline{\mathbf{v}}} \underline{\mathbf{h}}}, \end{aligned} \tag{11.31}$$

where

$$\boldsymbol{\Gamma}_{\underline{\mathbf{v}}} = \frac{\mathbf{R}_{\underline{\mathbf{v}}}}{\sigma_{v_1}^2} \tag{11.32}$$

is the pseudo-correlation matrix of $\underline{\mathbf{v}}(t)$. The third line of (11.31) is valid for fixed beamforming only. We see from (11.31) that the array gain is

$$\begin{aligned} \mathcal{G}\left(\underline{\mathbf{h}}\right) &= \frac{\mathrm{oSNR}\left(\underline{\mathbf{h}}\right)}{\mathrm{iSNR}} \\ &= \frac{\underline{\mathbf{h}}^T \underline{\mathbf{G}}\left(\cos\theta_{\mathrm{d}}\right) \underline{\mathbf{G}}^T\left(\cos\theta_{\mathrm{d}}\right) \underline{\mathbf{h}}}{\underline{\mathbf{h}}^T \boldsymbol{\Gamma}_{\underline{\mathbf{v}}} \underline{\mathbf{h}}}. \end{aligned} \tag{11.33}$$

The white noise gain (WNG) is obtained by taking $\mathbf{\Gamma}_{\underline{\mathbf{v}}} = \mathbf{I}_{ML_h}$ in (11.33), where $\mathbf{I}_{ML_h}$ is the $ML_h \times ML_h$ identity matrix:

$$\mathcal{W}\left(\underline{\mathbf{h}}\right) = \frac{\underline{\mathbf{h}}^T \underline{\mathbf{G}}\left(\cos\theta_{\mathrm{d}}\right) \underline{\mathbf{G}}^T\left(\cos\theta_{\mathrm{d}}\right) \underline{\mathbf{h}}}{\underline{\mathbf{h}}^T \underline{\mathbf{h}}}. \tag{11.34}$$

We define the broadband beampattern or broadband directivity pattern as

$$\left|\mathcal{B}\left(\underline{\mathbf{h}}, \cos\theta\right)\right|^2 = \underline{\mathbf{h}}^T \underline{\mathbf{G}}\left(\cos\theta\right) \underline{\mathbf{G}}^T\left(\cos\theta\right) \underline{\mathbf{h}}. \tag{11.35}$$

In the time domain, the definition of the directivity factor (DF) is

$$\begin{aligned} \mathcal{D}\left(\underline{\mathbf{h}}\right) &= \frac{\left|\mathcal{B}\left(\underline{\mathbf{h}}, \cos\theta_{\mathrm{d}}\right)\right|^2}{\frac{1}{2}\int_0^{\pi} \left|\mathcal{B}\left(\underline{\mathbf{h}}, \cos\theta\right)\right|^2 \sin\theta d\theta} \\ &= \frac{\underline{\mathbf{h}}^T \underline{\mathbf{G}}\left(\cos\theta_{\mathrm{d}}\right) \underline{\mathbf{G}}^T\left(\cos\theta_{\mathrm{d}}\right) \underline{\mathbf{h}}}{\underline{\mathbf{h}}^T \mathbf{\Gamma}_{\mathrm{T},0,\pi} \underline{\mathbf{h}}}, \end{aligned} \tag{11.36}$$

where

$$\mathbf{\Gamma}_{\mathrm{T},0,\pi} = \frac{1}{2}\int_0^{\pi} \underline{\mathbf{G}}\left(\cos\theta\right) \underline{\mathbf{G}}^T\left(\cos\theta\right) \sin\theta d\theta \tag{11.37}$$

is a matrix of size $ML_h \times ML_h$, which is the equivalent form of $\mathbf{\Gamma}_{0,\pi}(f)$ in the time domain. Note that an explicit expression for $\mathbf{\Gamma}_{\mathrm{T},0,\pi}$ is not available. In practice, we compute the DF in the time domain directly from the first line of (11.36) with numerical integration.

In the same manner, we define the broadband front-to-back ratio (FBR) as

$$\begin{aligned} \mathcal{F}\left(\underline{\mathbf{h}}\right) &= \frac{\frac{1}{2}\int_0^{\pi/2} \left|\mathcal{B}\left(\underline{\mathbf{h}}, \cos\theta\right)\right|^2 \sin\theta d\theta}{\frac{1}{2}\int_{\pi/2}^{\pi} \left|\mathcal{B}\left(\underline{\mathbf{h}}, \cos\theta\right)\right|^2 \sin\theta d\theta} \\ &= \frac{\underline{\mathbf{h}}^T \mathbf{\Gamma}_{\mathrm{T},0,\pi/2} \underline{\mathbf{h}}}{\underline{\mathbf{h}}^T \mathbf{\Gamma}_{\mathrm{T},\pi/2,\pi} \underline{\mathbf{h}}}, \end{aligned} \tag{11.38}$$

where

$$\mathbf{\Gamma}_{\mathrm{T},0,\pi/2} = \frac{1}{2}\int_0^{\pi/2} \underline{\mathbf{G}}\left(\cos\theta\right) \underline{\mathbf{G}}^T\left(\cos\theta\right) \sin\theta d\theta, \tag{11.39}$$

$$\mathbf{\Gamma}_{\mathrm{T},\pi/2,\pi} = \frac{1}{2}\int_{\pi/2}^{\pi} \underline{\mathbf{G}}\left(\cos\theta\right) \underline{\mathbf{G}}^T\left(\cos\theta\right) \sin\theta d\theta. \tag{11.40}$$

Now, let us define the error signal between the estimated and desired signals:

$$\begin{aligned} e(t) &= z(t) - \mathbf{i}_l^T \mathbf{x}(t-\Delta) \\ &= x_{\mathrm{fd}}(t) + v_{\mathrm{rn}}(t) - \mathbf{i}_l^T \mathbf{x}(t-\Delta). \end{aligned} \tag{11.41}$$

This error can be rewritten as

$$e(t) = e_{\mathrm{d}}(t) + e_{\mathrm{n}}(t), \tag{11.42}$$

where

$$\begin{aligned} e_{\mathrm{d}}(t) &= x_{\mathrm{fd}}(t) - \mathbf{i}_l^T \mathbf{x}(t-\Delta) \\ &= \left[\underline{\mathbf{G}}^T \left(\cos\theta_{\mathrm{d}}\right) \underline{\mathbf{h}} - \mathbf{i}_l\right]^T \mathbf{x}(t-\Delta) \end{aligned} \tag{11.43}$$

and

$$\begin{aligned} e_{\mathrm{n}}(t) &= v_{\mathrm{rn}}(t) \\ &= \underline{\mathbf{h}}^T \underline{\mathbf{v}}(t) \end{aligned} \tag{11.44}$$

are, respectively, the desired signal distortion due to the beamformer and the residual noise. Therefore, the MSE criterion is

$$\begin{aligned} J\left(\underline{\mathbf{h}}\right) &= E\left[e^2(t)\right] \\ &= \sigma_x^2 - 2\underline{\mathbf{h}}^T \underline{\mathbf{G}}\left(\cos\theta_{\mathrm{d}}\right) \mathbf{R}_{\mathbf{x}} \mathbf{i}_l + \underline{\mathbf{h}}^T \mathbf{R}_{\underline{\mathbf{y}}} \underline{\mathbf{h}} \\ &= J_{\mathrm{d}}\left(\underline{\mathbf{h}}\right) + J_{\mathrm{n}}\left(\underline{\mathbf{h}}\right), \end{aligned} \tag{11.45}$$

where

$$\begin{aligned} J_{\mathrm{d}}\left(\underline{\mathbf{h}}\right) &= E\left[e_{\mathrm{d}}^2(t)\right] \\ &= \left[\underline{\mathbf{G}}^T \left(\cos_{\mathrm{d}}\theta\right) \underline{\mathbf{h}} - \mathbf{i}_l\right]^T \mathbf{R}_{\mathbf{x}} \left[\underline{\mathbf{G}}^T \left(\cos_{\mathrm{d}}\theta\right) \underline{\mathbf{h}} - \mathbf{i}_l\right] \\ &= \sigma_x^2 \upsilon_{\mathrm{d}}\left(\underline{\mathbf{h}}\right) \end{aligned} \tag{11.46}$$

and

$$\begin{aligned} J_{\mathrm{n}}\left(\underline{\mathbf{h}}\right) &= E\left[e_{\mathrm{n}}^2(t)\right] \\ &= \underline{\mathbf{h}}^T \mathbf{R}_{\underline{\mathbf{v}}} \underline{\mathbf{h}} \\ &= \frac{\sigma_{v_1}^2}{\xi_{\mathrm{n}}\left(\underline{\mathbf{h}}\right)}, \end{aligned} \tag{11.47}$$

with

$$\begin{aligned} \upsilon_{\mathrm{d}}\left(\underline{\mathbf{h}}\right) &= \frac{E\left\{\left[x_{\mathrm{fd}}(t) - \mathbf{i}_l^T \mathbf{x}(t-\Delta)\right]^2\right\}}{\sigma_x^2} \\ &= \frac{\left[\underline{\mathbf{G}}^T \left(\cos_{\mathrm{d}}\theta\right) \underline{\mathbf{h}} - \mathbf{i}_l\right]^T \mathbf{R}_{\mathbf{x}} \left[\underline{\mathbf{G}}^T \left(\cos_{\mathrm{d}}\theta\right) \underline{\mathbf{h}} - \mathbf{i}_l\right]}{\sigma_x^2} \end{aligned} \tag{11.48}$$

being the desired signal distortion index and

$$\xi_{\mathrm{n}}\left(\underline{\mathbf{h}}\right) = \frac{\sigma_{v_1}^2}{\underline{\mathbf{h}}^T \mathbf{R}_{\underline{\mathbf{v}}} \underline{\mathbf{h}}} \tag{11.49}$$

being the noise reduction factor. We deduce that

$$\begin{aligned} \frac{J_{\mathrm{d}}\left(\underline{\mathbf{h}}\right)}{J_{\mathrm{n}}\left(\underline{\mathbf{h}}\right)} &= \mathrm{iSNR} \times \xi_{\mathrm{n}}\left(\underline{\mathbf{h}}\right) \times \upsilon_{\mathrm{d}}\left(\underline{\mathbf{h}}\right) \\ &= \mathrm{oSNR}\left(\underline{\mathbf{h}}\right) \times \xi_{\mathrm{d}}\left(\underline{\mathbf{h}}\right) \times \upsilon_{\mathrm{d}}\left(\underline{\mathbf{h}}\right), \end{aligned} \tag{11.50}$$

where

$$\xi_{\mathrm{d}}\left(\underline{\mathbf{h}}\right) = \frac{\sigma_x^2}{\underline{\mathbf{h}}^T \mathbf{R}_{\underline{\mathbf{x}}} \underline{\mathbf{h}}} \tag{11.51}$$

is the desired signal reduction factor.

11.4 Fixed Beamformers

In this section, we show how to derive the most conventional time-domain fixed beamformers from the WNG and the DF.

11.4.1 Delay and Sum

The classical delay-and-sum (DS) beamformer in the time domain is derived by maximizing the WNG subject to the distortionless constraint. This is equivalent to

$$\min_{\underline{\mathbf{h}}} \underline{\mathbf{h}}^T \underline{\mathbf{h}} \quad \text{subject to} \quad \underline{\mathbf{h}}^T \underline{\mathbf{G}}\left(\cos\theta_{\mathrm{d}}\right) = \mathbf{i}_l^T. \tag{11.52}$$

We easily obtain

$$\underline{\mathbf{h}}_{\mathrm{DS}}\left(\cos\theta_{\mathrm{d}}\right) = \underline{\mathbf{G}}\left(\cos\theta_{\mathrm{d}}\right)\left[\underline{\mathbf{G}}^T\left(\cos\theta_{\mathrm{d}}\right)\underline{\mathbf{G}}\left(\cos\theta_{\mathrm{d}}\right)\right]^{-1}\mathbf{i}_l. \tag{11.53}$$

Therefore, the WNG is

$$\mathcal{W}\left[\underline{\mathbf{h}}_{\mathrm{DS}}\left(\cos\theta_{\mathrm{d}}\right)\right] = \frac{1}{\mathbf{i}_l^T\left[\underline{\mathbf{G}}^T\left(\cos\theta_{\mathrm{d}}\right)\underline{\mathbf{G}}\left(\cos\theta_{\mathrm{d}}\right)\right]^{-1}\mathbf{i}_l}. \tag{11.54}$$

It can be checked that the matrix product $\mathbf{G}_m^T\left(\cos\theta_{\mathrm{d}}\right)\mathbf{G}_m\left(\cos\theta_{\mathrm{d}}\right)$ is a diagonal matrix, the elements of which are 0 or 1. As a result, the matrix $\underline{\mathbf{G}}^T\left(\cos\theta_{\mathrm{d}}\right)\underline{\mathbf{G}}\left(\cos\theta_{\mathrm{d}}\right) = \sum_{m=1}^{M}\mathbf{G}_m^T\left(\cos\theta_{\mathrm{d}}\right)\mathbf{G}_m\left(\cos\theta_{\mathrm{d}}\right)$ is also a diagonal matrix the elements of which are between 0 and M. We conclude that the position of the 1 in $\mathbf{i}_l$ must coincide with the position of the maximum element of the diagonal of $\underline{\mathbf{G}}^T(\cos\theta_\mathrm{d})\,\underline{\mathbf{G}}(\cos\theta_\mathrm{d})$. In this case, we have

$$\mathcal{W}\left[\underline{\mathbf{h}}_\mathrm{DS}(\cos\theta_\mathrm{d})\right] = M \tag{11.55}$$

and

$$\underline{\mathbf{h}}_\mathrm{DS}(\cos\theta_\mathrm{d}) = \underline{\mathbf{G}}(\cos\theta_\mathrm{d})\,\frac{\mathbf{i}_l}{M}. \tag{11.56}$$

In the rest, it is always assumed that the position of the 1 in $\mathbf{i}_l$ is chosen such that $\mathbf{i}_l^T\left[\underline{\mathbf{G}}^T(\cos\theta_\mathrm{d})\,\underline{\mathbf{G}}(\cos\theta_\mathrm{d})\right]^{-1}\mathbf{i}_l = 1/M$.

Example 11.4.1 Consider a ULA of M sensors. Suppose that a desired signal impinges on the ULA from the direction θ_d. Assume that $f_\mathrm{s} = 8$ kHz, $P = 25$, and $L_h = 30$. Figures 11.3–11.5 show broadband beampatterns, $\left|\mathcal{B}\left[\underline{\mathbf{h}}_\mathrm{DS}(\cos\theta_\mathrm{d}), \cos\theta\right]\right|$, for different source directions θ_d and several values of M and δ. The main beam is in the direction of the desired signal, θ_d. As the number of sensors, M, increases, or as the intersensor

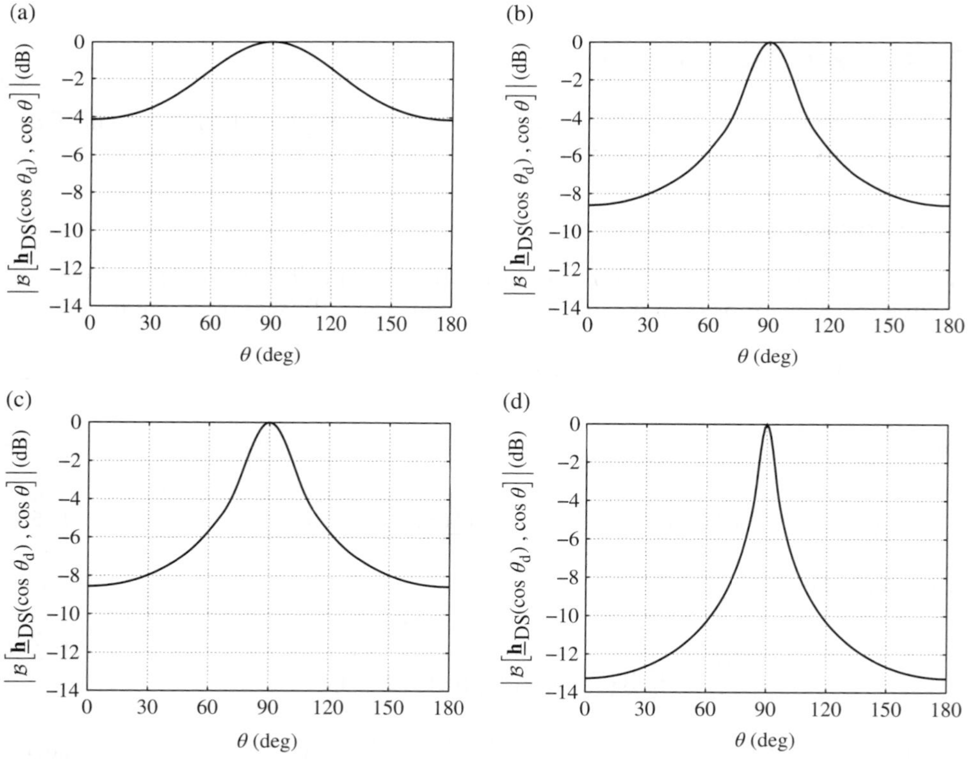

Figure 11.3 Broadband beampatterns of the DS beamformer for $\theta_\mathrm{d} = 90°$, and several values of M and δ: (a) $M = 10$, $\delta = 1$ cm, (b) $M = 30$, $\delta = 1$ cm, (c) $M = 10$, $\delta = 3$ cm, and (d) $M = 30$, $\delta = 3$ cm.

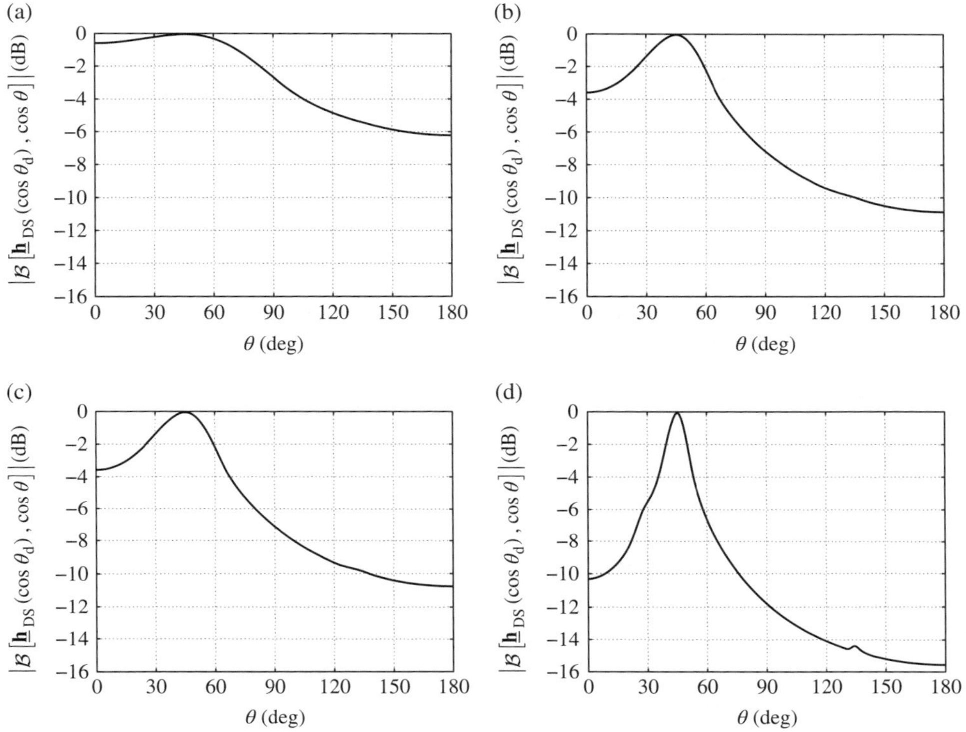

Figure 11.4 Broadband beampatterns of the DS beamformer for $\theta_{\mathrm{d}} = 45°$, and several values of M and δ: (a) $M = 10$, $\delta = 1$ cm, (b) $M = 30$, $\delta = 1$ cm, (c) $M = 10$, $\delta = 3$ cm, and (d) $M = 30$, $\delta = 3$ cm.

spacing, δ, increases, the width of the main beam decreases, and the values obtained for $\theta \neq \theta_{\mathrm{d}}$ become lower. Figure 11.6 shows plots of the DF, $\mathcal{D}\left[\underline{\mathbf{h}}_{\mathrm{DS}}\left(\cos\theta_{\mathrm{d}}\right)\right]$, and the WNG, $\mathcal{W}\left[\underline{\mathbf{h}}_{\mathrm{DS}}\left(\cos\theta_{\mathrm{d}}\right)\right]$, as a function of δ for $\theta_{\mathrm{d}} = 90°$ and several values of M. As the number of sensors increases, both the DF and the WNG of the DS beamformer increase. For a given M, the DF of the DS beamformer increases as a function of δ. ■

11.4.2 Maximum DF

Let $\underline{\mathbf{t}}_1\left(\cos\theta_{\mathrm{d}}\right)$ be the eigenvector corresponding to the maximum eigenvalue, $\underline{\lambda}_1\left(\cos\theta_{\mathrm{d}}\right)$, of the matrix $\mathbf{\Gamma}_{\mathrm{T},0,\pi}^{-1}\underline{\mathbf{G}}\left(\cos\theta_{\mathrm{d}}\right)\underline{\mathbf{G}}^T\left(\cos\theta_{\mathrm{d}}\right)$. It is obvious that the maximum DF beamformer is

$$\underline{\mathbf{h}}_{\mathrm{max}}\left(\cos\theta_{\mathrm{d}}\right) = \varsigma\underline{\mathbf{t}}_1\left(\cos\theta_{\mathrm{d}}\right), \tag{11.57}$$

where $\varsigma \neq 0$ is an arbitrary real number, and the maximum DF is

$$\mathcal{D}_{\mathrm{max}}\left(\cos\theta_{\mathrm{d}}\right) = \underline{\lambda}_1\left(\cos\theta_{\mathrm{d}}\right). \tag{11.58}$$

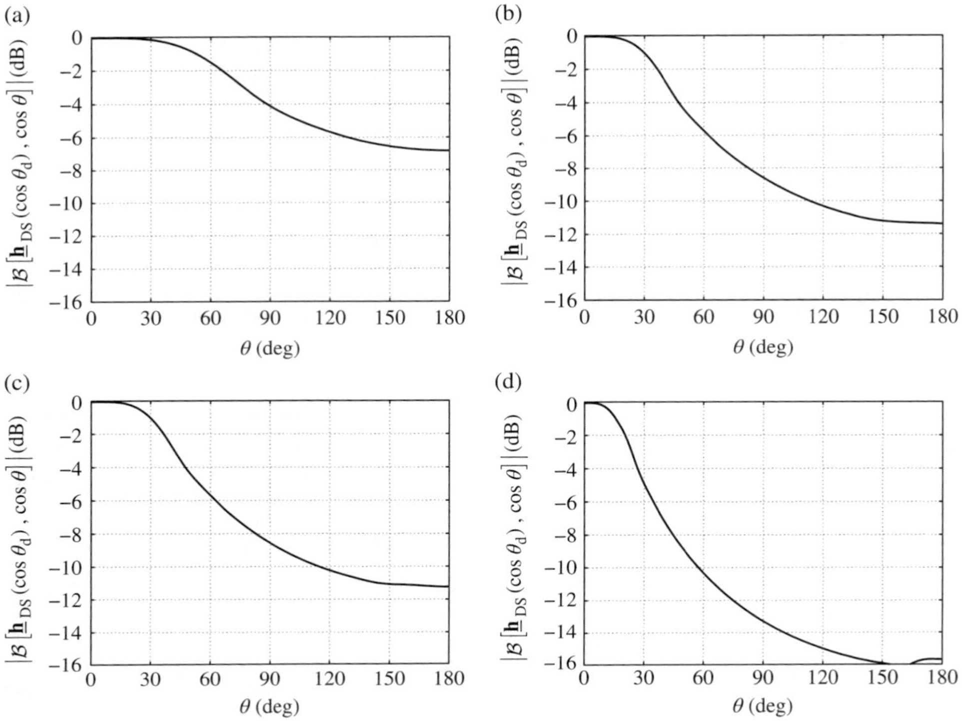

Figure 11.5 Broadband beampatterns of the DS beamformer for $\theta_d = 0°$, and several values of M and δ: (a) $M = 10$, $\delta = 1$ cm, (b) $M = 30$, $\delta = 1$ cm, (c) $M = 10$, $\delta = 3$ cm, and (d) $M = 30$, $\delta = 3$ cm.

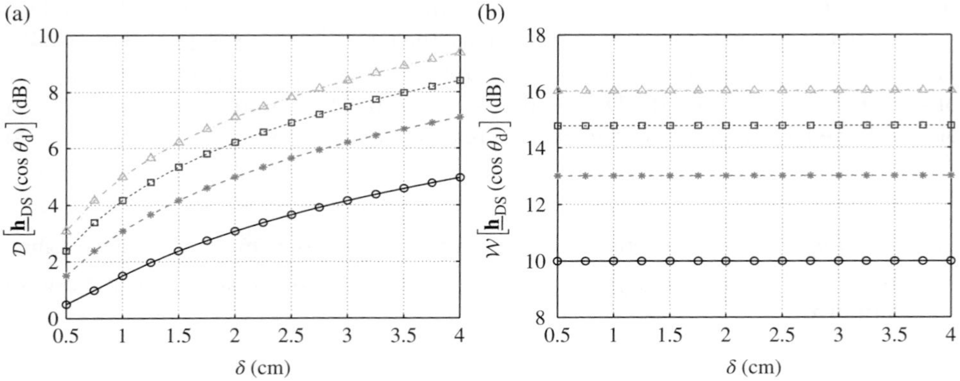

Figure 11.6 (a) DF and (b) WNG of the DS beamformer as a function of δ for $\theta_d = 90°$ and several values of M: $M = 10$ (solid line with circles), $M = 20$ (dashed line with asterisks), $M = 30$ (dotted line with squares), and $M = 40$ (dash-dot line with triangles).

Therefore,

$$\mathcal{D}_{\max}\left(\cos\theta_{\mathrm{d}}\right) \geq \mathcal{D}\left(\underline{\mathbf{h}}\right), \ \forall \underline{\mathbf{h}}. \tag{11.59}$$

While $\underline{\mathbf{h}}_{\max}\left(\cos\theta_{\mathrm{d}}\right)$ maximizes the DF, it cannot be distortionless.

11.4.3 Distortionless Maximum DF

To find the distortionless maximum DF beamformer, we need to minimize the denominator of the DF subject to the distortionless constraint in the numerator of the DF:

$$\min_{\underline{\mathbf{h}}} \underline{\mathbf{h}}^T \mathbf{\Gamma}_{\mathrm{T},0,\pi} \underline{\mathbf{h}} \quad \text{subject to} \quad \underline{\mathbf{h}}^T \underline{\mathbf{G}}\left(\cos\theta_{\mathrm{d}}\right) = \mathbf{i}_l^T. \tag{11.60}$$

Then, it is clear that the distortionless maximum DF beamformer is

$$\underline{\mathbf{h}}_{\mathrm{mDF}}\left(\cos\theta_{\mathrm{d}}\right) = \mathbf{\Gamma}_{\mathrm{T},0,\pi}^{-1} \underline{\mathbf{G}}\left(\cos\theta_{\mathrm{d}}\right)\left[\underline{\mathbf{G}}^T\left(\cos\theta_{\mathrm{d}}\right) \mathbf{\Gamma}_{\mathrm{T},0,\pi}^{-1} \underline{\mathbf{G}}\left(\cos\theta_{\mathrm{d}}\right)\right]^{-1} \mathbf{i}_l. \tag{11.61}$$

We deduce that the corresponding DF is

$$\mathcal{D}\left[\underline{\mathbf{h}}_{\mathrm{mDF}}\left(\cos\theta_{\mathrm{d}}\right)\right] = \frac{1}{\mathbf{i}_l^T\left[\underline{\mathbf{G}}^T\left(\cos\theta_{\mathrm{d}}\right) \mathbf{\Gamma}_{\mathrm{T},0,\pi}^{-1} \underline{\mathbf{G}}\left(\cos\theta_{\mathrm{d}}\right)\right]^{-1} \mathbf{i}_l}. \tag{11.62}$$

Example 11.4.2 Returning to Example 11.4.1, we now employ the distortionless maximum DF beamformer, $\underline{\mathbf{h}}_{\mathrm{mDF}}\left(\cos\theta_{\mathrm{d}}\right)$, given in (11.61). Figures 11.7–11.9 show broadband beampatterns, $\left|\mathcal{B}\left[\underline{\mathbf{h}}_{\mathrm{mDF}}\left(\cos\theta_{\mathrm{d}}\right), \cos\theta\right]\right|$, for different source directions θ_{d} and several values of M and δ. The main beam is in the direction of the desired signal, θ_{d}. As the number of sensors, M, increases, or as the intersensor spacing, δ, increases, the width of the main beam decreases, and the values obtained for $\theta \neq \theta_{\mathrm{d}}$ generally become lower. Figure 11.10 shows plots of the DF, $\mathcal{D}\left[\underline{\mathbf{h}}_{\mathrm{mDF}}\left(\cos\theta_{\mathrm{d}}\right)\right]$, and the WNG, $\mathcal{W}\left[\underline{\mathbf{h}}_{\mathrm{mDF}}\left(\cos\theta_{\mathrm{d}}\right)\right]$, as a function of δ for $\theta_{\mathrm{d}} = 0°$ and several values of M. Compared to the DS beamformer, the distortionless maximum DF beamformer gives higher DF, but lower WNG (cf. Figures 11.6 and 11.10). For a sufficiently small δ, as the number of sensors increases, both the DF and the WNG of the DS beamformer increase. For a given M and a sufficiently small δ, the DF of the distortionless maximum DF beamformer increases as a function of δ. The WNG of the distortionless maximum DF beamformer is significantly lower than 0 dB, which implies that the distortionless maximum DF beamformer amplifies the white noise. ■

11.4.4 Superdirective

The time-domain superdirective beamformer is simply a particular case of the distortionless maximum DF beamformer, where $\theta_{\mathrm{d}} = 0$ and δ is small. We get

$$\underline{\mathbf{h}}_{\mathrm{SD}} = \mathbf{\Gamma}_{\mathrm{T},0,\pi}^{-1} \underline{\mathbf{G}}\left(\underline{\mathbf{G}}^T \mathbf{\Gamma}_{\mathrm{T},0,\pi}^{-1} \underline{\mathbf{G}}\right)^{-1} \mathbf{i}_l, \tag{11.63}$$

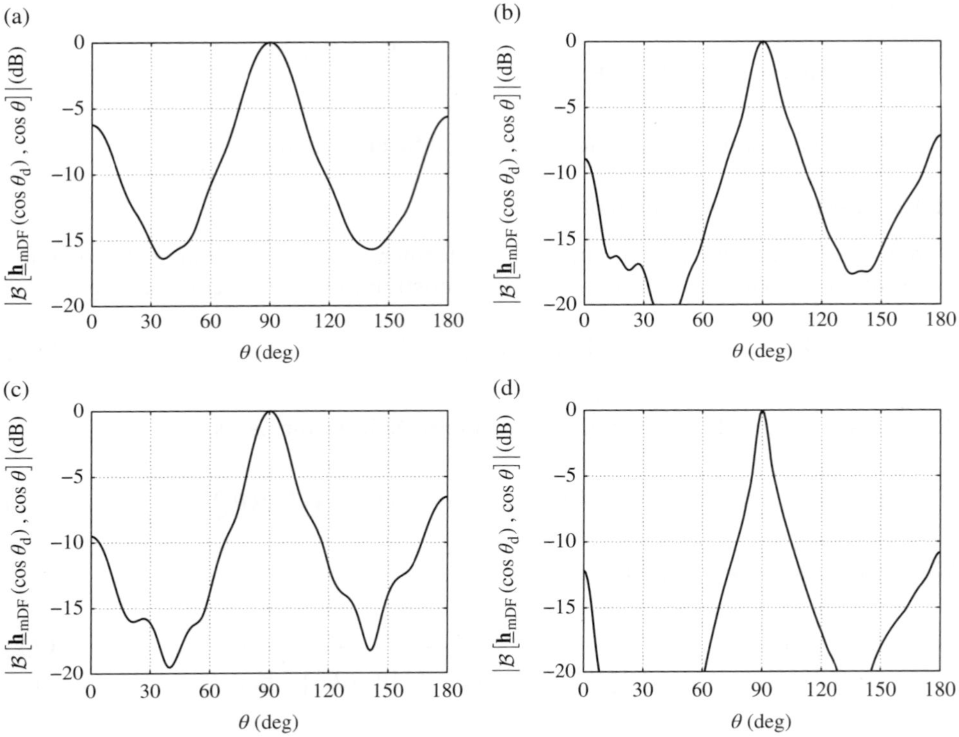

Figure 11.7 Broadband beampatterns of the distortionless maximum DF beamformer for $\theta_{\mathrm{d}} = 90°$, and several values of M and δ: (a) $M = 10$, $\delta = 1$ cm, (b) $M = 30$, $\delta = 1$ cm, (c) $M = 10$, $\delta = 3$ cm, and (d) $M = 30$, $\delta = 3$ cm.

where $\underline{\mathbf{G}} = \underline{\mathbf{G}}(\cos 0)$. The corresponding DF is

$$\mathcal{D}\left(\underline{\mathbf{h}}_{\mathrm{SD}}\right) = \frac{1}{\mathbf{i}_l^T\left(\underline{\mathbf{G}}^T\mathbf{\Gamma}_{\mathrm{T},0,\pi}^{-1}\underline{\mathbf{G}}\right)^{-1}\mathbf{i}_l}. \tag{11.64}$$

This gain should approach M^2 for a small value of δ.

Following the ideas of Cox et al. [4, 5], we can easily derive the time-domain robust superdirective beamformer:

$$\underline{\mathbf{h}}_{\mathrm{R},\epsilon} = \mathbf{\Gamma}_{\mathrm{T},0,\pi,\epsilon}^{-1}\underline{\mathbf{G}}\left(\underline{\mathbf{G}}^T\mathbf{\Gamma}_{\mathrm{T},0,\pi,\epsilon}^{-1}\underline{\mathbf{G}}\right)^{-1}\mathbf{i}_l, \tag{11.65}$$

where

$$\mathbf{\Gamma}_{\mathrm{T},0,\pi,\epsilon} = \mathbf{\Gamma}_{\mathrm{T},0,\pi} + \epsilon\mathbf{I}_{ML_h}, \tag{11.66}$$

with $\epsilon \geq 0$. We see that $\underline{\mathbf{h}}_{\mathrm{R},0} = \underline{\mathbf{h}}_{\mathrm{SD}}$ and $\underline{\mathbf{h}}_{\mathrm{R},\infty} = \underline{\mathbf{h}}_{\mathrm{DS}}(1)$.

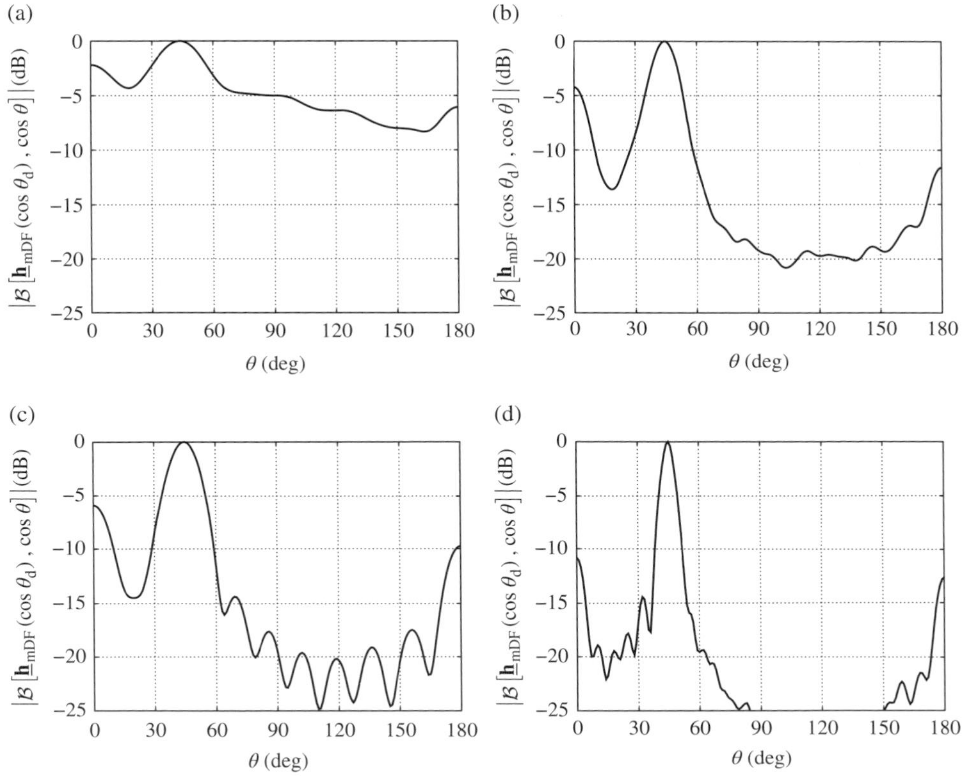

Figure 11.8 Broadband beampatterns of the distortionless maximum DF beamformer for $\theta_d = 45°$, and several values of M and δ: (a) $M = 10, \delta = 1$ cm, (b) $M = 30, \delta = 1$ cm, (c) $M = 10, \delta = 3$ cm, and (d) $M = 30, \delta = 3$ cm.

Example 11.4.3 Returning to Example 11.4.1, we now employ the robust superdirective beamformer, $\underline{\mathbf{h}}_{R,\epsilon}$, given in (11.65). Figure 11.11 shows broadband beampatterns, $\left|\mathcal{B}\left(\underline{\mathbf{h}}_{R,\epsilon}, \cos\theta\right)\right|$, for $M = 10$, $\delta = 1$ cm, and several values of ϵ. The main beam is in the direction of the desired signal, $\theta_d = 0$. As the value of ϵ increases, the width of the main beam increases, and the sidelobe level also increases (lower DF). Figure 11.12 shows plots of the DF, $\mathcal{D}\left(\underline{\mathbf{h}}_{R,\epsilon}\right)$, and the WNG, $\mathcal{W}\left(\underline{\mathbf{h}}_{R,\epsilon}\right)$, as a function of δ for several values of ϵ. For a given δ, as the value of ϵ increases, the WNG of the robust superdirective beamformer increases at the expense of a lower DF. For a given ϵ and a sufficiently small δ, both the DF and the WNG of the robust superdirective beamformer increase as a function of δ. ■

11.4.5 Null Steering

We assume that we have an undesired source impinging on the array from the direction $\theta_n \neq \theta_d$. The objective is to completely cancel this source while recovering the desired

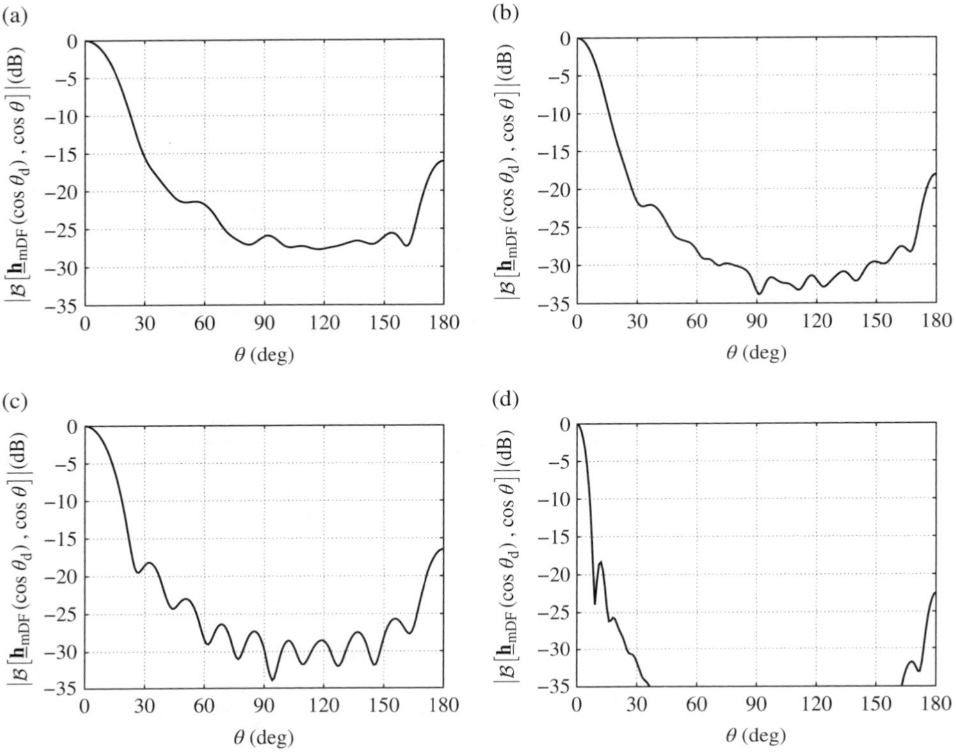

Figure 11.9 Broadband beampatterns of the distortionless maximum DF beamformer for $\theta_{\mathrm{d}} = 0°$, and several values of M and δ: (a) $M = 10$, $\delta = 1$ cm, (b) $M = 30$, $\delta = 1$ cm, (c) $M = 10$, $\delta = 3$ cm, and (d) $M = 30$, $\delta = 3$ cm.

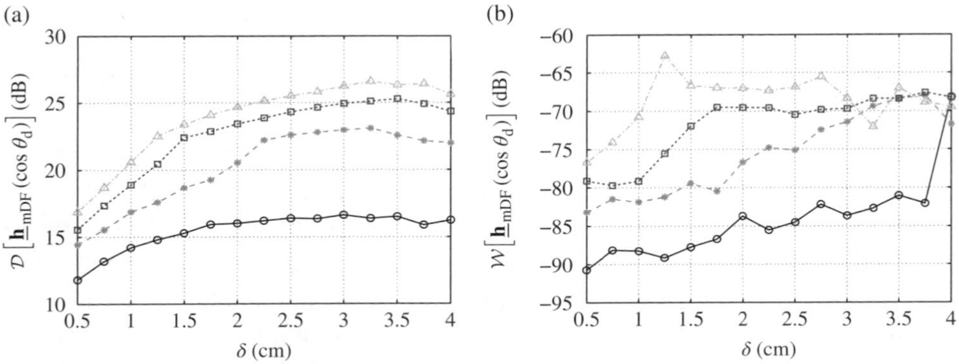

Figure 11.10 (a) DF and (b) WNG of the distortionless maximum DF beamformer as a function of δ for $\theta_{\mathrm{d}} = 0°$ and several values of M: $M = 10$ (solid line with circles), $M = 20$ (dashed line with asterisks), $M = 30$ (dotted line with squares), and $M = 40$ (dash-dot line with triangles).

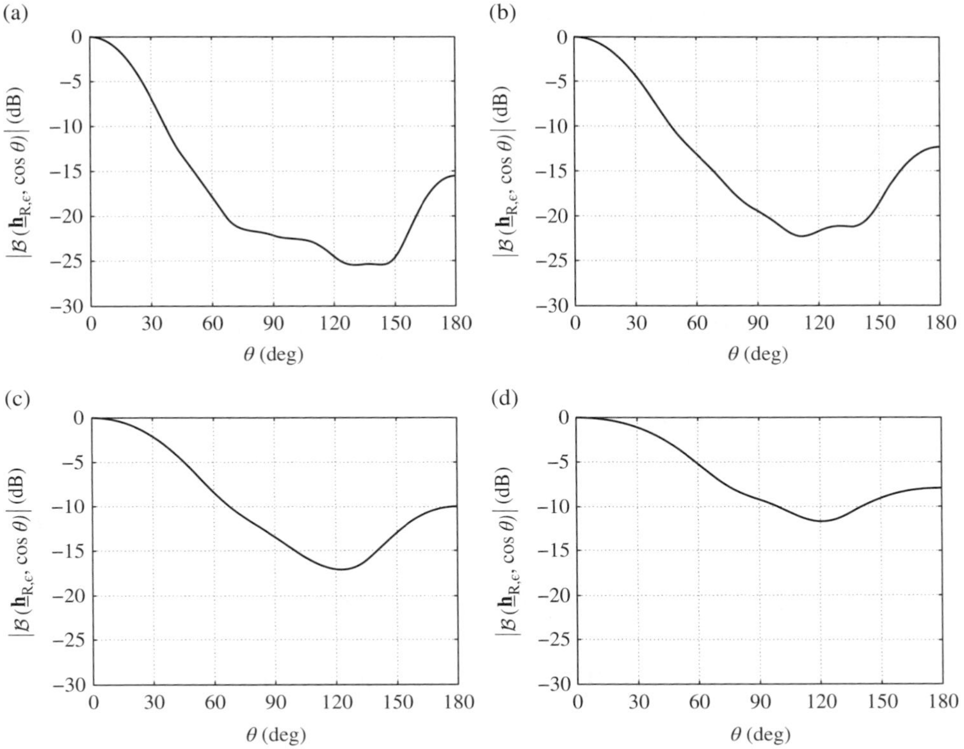

Figure 11.11 Broadband beampatterns of the robust superdirective beamformer for $M = 10$, $\delta = 1$ cm, and several values of ϵ: (a) $\epsilon = 10^{-5}$, (b) $\epsilon = 10^{-3}$, (c) $\epsilon = 0.1$, and (d) $\epsilon = 1$.

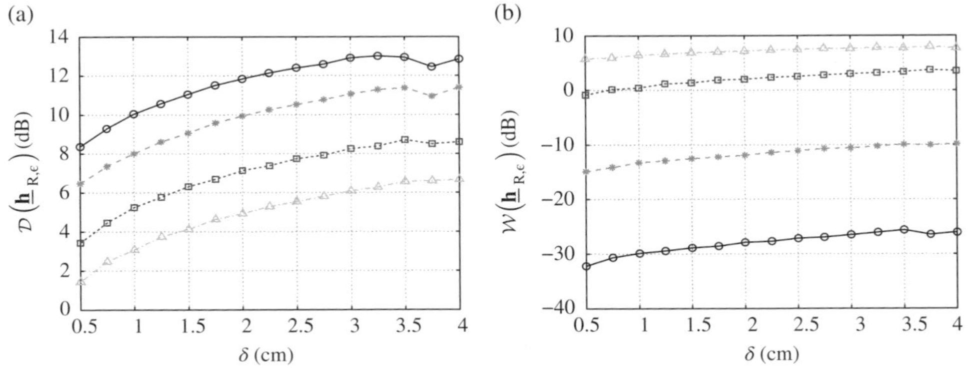

Figure 11.12 (a) DF and (b) WNG of the robust superdirective beamformer as a function of δ for $M = 10$ and several values of ϵ: $\epsilon = 10^{-5}$ (solid line with circles), $\epsilon = 10^{-3}$ (dashed line with asterisks), $\epsilon = 0.1$ (dotted line with squares), and $\epsilon = 1$ (dash-dot line with triangles).

source impinging on the array from the direction θ_d. Then, it is obvious that the constraint equation is

$$\underline{\mathbf{C}}^T(\theta_\mathrm{d}, \theta_\mathrm{n})\,\underline{\mathbf{h}} = \begin{bmatrix} \mathbf{i}_l \\ \mathbf{0} \end{bmatrix}, \tag{11.67}$$

where

$$\underline{\mathbf{C}}(\theta_\mathrm{d}, \theta_\mathrm{n}) = \begin{bmatrix} \underline{\mathbf{G}}(\cos\theta_\mathrm{d}) & \underline{\mathbf{G}}(\cos\theta_\mathrm{n}) \end{bmatrix} \tag{11.68}$$

is the constraint matrix of size $ML_h \times 2L$ and $\mathbf{0}$ is the zero vector of length L.

Depending on what we desire, there are different ways to achieve the goal explained above. We present two of these methods.

The first obvious beamformer is obtained by maximizing the WNG and by taking (11.67) into account:

$$\min_{\underline{\mathbf{h}}} \underline{\mathbf{h}}^T\underline{\mathbf{h}} \quad \text{subject to} \quad \underline{\mathbf{C}}^T(\theta_\mathrm{d}, \theta_\mathrm{n})\,\underline{\mathbf{h}} = \begin{bmatrix} \mathbf{i}_l \\ \mathbf{0} \end{bmatrix}. \tag{11.69}$$

From this criterion, we find the minimum-norm (MN) beamformer:

$$\underline{\mathbf{h}}_\mathrm{MN}(\cos\theta_\mathrm{d}) = \underline{\mathbf{C}}(\theta_\mathrm{d}, \theta_\mathrm{n})\left[\underline{\mathbf{C}}^T(\theta_\mathrm{d}, \theta_\mathrm{n})\,\underline{\mathbf{C}}(\theta_\mathrm{d}, \theta_\mathrm{n})\right]^{-1}\begin{bmatrix} \mathbf{i}_l \\ \mathbf{0} \end{bmatrix}, \tag{11.70}$$

which is also the minimum-norm solution of (11.67).

The other beamformer is obtained by maximizing the DF and by taking (11.67) into account:

$$\min_{\underline{\mathbf{h}}} \underline{\mathbf{h}}^T\mathbf{\Gamma}_{\mathrm{T},0,\pi}\underline{\mathbf{h}} \quad \text{subject to} \quad \underline{\mathbf{C}}^T(\theta_\mathrm{d}, \theta_\mathrm{n})\,\underline{\mathbf{h}} = \begin{bmatrix} \mathbf{i}_l \\ \mathbf{0} \end{bmatrix}. \tag{11.71}$$

We easily find the null steering (NS) beamformer:

$$\begin{aligned} \underline{\mathbf{h}}_\mathrm{NS}(\cos\theta_\mathrm{d}) &= \mathbf{\Gamma}_{\mathrm{T},0,\pi}^{-1}\underline{\mathbf{C}}(\theta_\mathrm{d}, \theta_\mathrm{n}) \\ &\quad \times \left[\underline{\mathbf{C}}^T(\theta_\mathrm{d}, \theta_\mathrm{n})\,\mathbf{\Gamma}_{\mathrm{T},0,\pi}^{-1}\underline{\mathbf{C}}(\theta_\mathrm{d}, \theta_\mathrm{n})\right]^{-1}\begin{bmatrix} \mathbf{i}_l \\ \mathbf{0} \end{bmatrix}. \end{aligned} \tag{11.72}$$

Example 11.4.4 Consider a ULA of M sensors. Suppose that a desired signal impinges on the ULA from the direction $\theta_\mathrm{d} = 0°$, and an undesired interference impinges on the ULA from the direction $\theta_\mathrm{n} = 90°$. Assume that $f_s = 8$ kHz, $P = 25$, and $L_h = 30$.

Figure 11.13 shows broadband beampatterns, $\left|\mathcal{B}\left[\underline{\mathbf{h}}_\mathrm{MN}(\cos\theta_\mathrm{d}), \cos\theta\right]\right|$, for several values of M and δ. Clearly, the beam is in the direction of the desired signal, θ_d, and the null is in the direction of the interfering signal, θ_n. As the number of sensors, M, increases, or as the intersensor spacing, δ, increases, the width of the main beam and the level of the sidelobe decrease. Figure 11.14 shows plots of the DF, $\mathcal{D}\left[\underline{\mathbf{h}}_\mathrm{MN}(\cos\theta_\mathrm{d})\right]$, and the WNG, $\mathcal{W}\left[\underline{\mathbf{h}}_\mathrm{MN}(\cos\theta_\mathrm{d})\right]$, as a function of δ for several values of M. For a small δ,

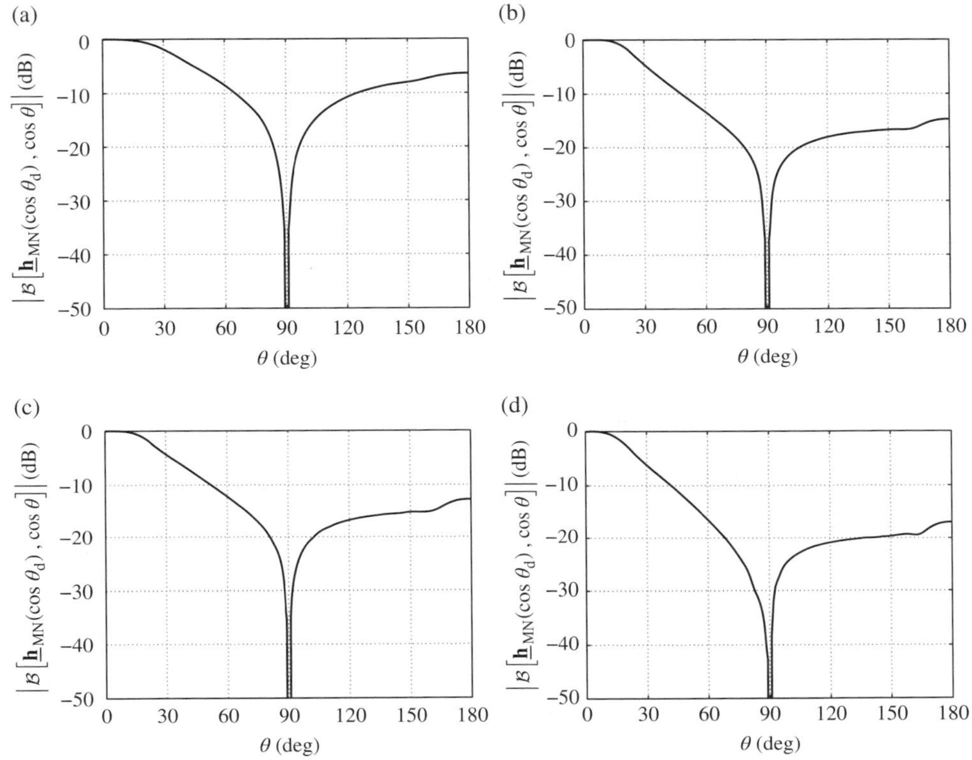

Figure 11.13 Broadband beampatterns of the MN beamformer for $\theta_{\mathrm{d}} = 0°$, $\theta_{\mathrm{n}} = 90°$, and several values of M and δ: (a) $M = 20$, $\delta = 2$ cm, (b) $M = 40$, $\delta = 2$ cm, (c) $M = 20$, $\delta = 4$ cm, and (d) $M = 40$, $\delta = 4$ cm.

both the DF and the WNG increase as M increases. However, for a large δ, the DF and the WNG of the MN beamformer are less sensitive to M, if M is sufficiently large. For a given M and small δ, both the DF and the WNG of the MN beamformer increase as a function of δ.

Figure 11.15 shows broadband beampatterns, $\left|\mathcal{B}\left[\underline{\mathbf{h}}_{\mathrm{NS}}\left(\cos\theta_{\mathrm{d}}\right),\cos\theta\right]\right|$, for several values of M and δ. Here again, the beam is in the direction of the desired signal, and the null is in the direction of the interfering signal. As the number of sensors, M, increases, or as the intersensor spacing, δ, increases, the width of the main beam and the level of the sidelobe decrease. Figure 11.16 shows plots of the DF, $\mathcal{D}\left[\underline{\mathbf{h}}_{\mathrm{NS}}\left(\cos\theta_{\mathrm{d}}\right)\right]$, and the WNG, $\mathcal{W}\left[\underline{\mathbf{h}}_{\mathrm{NS}}\left(\cos\theta_{\mathrm{d}}\right)\right]$, as a function of δ for several values of M. For a small δ, both the DF and the WNG of the NS beamformer increase as M increases. For a given M and small δ, the DF of the NS beamformer increases as a function of δ. Compared with the MN beamformer, the NS beamformer gives higher DF, but lower WNG. The WNG of the NS beamformer is significantly lower than 0 dB, which implies that the NS beamformer amplifies the white noise. ■

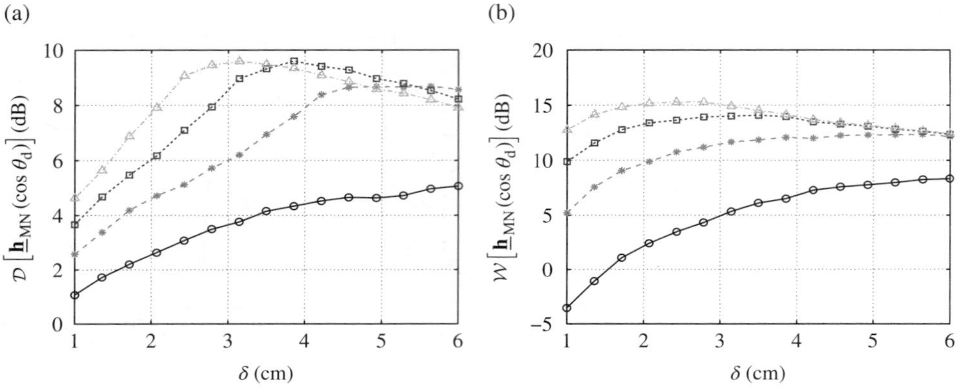

Figure 11.14 (a) DF and (b) WNG of the MN beamformer as a function of δ, for $\theta_d = 0°$, $\theta_n = 90°$, and several values of M: $M = 10$ (solid line with circles), $M = 20$ (dashed line with asterisks), $M = 30$ (dotted line with squares), and $M = 40$ (dash-dot line with triangles).

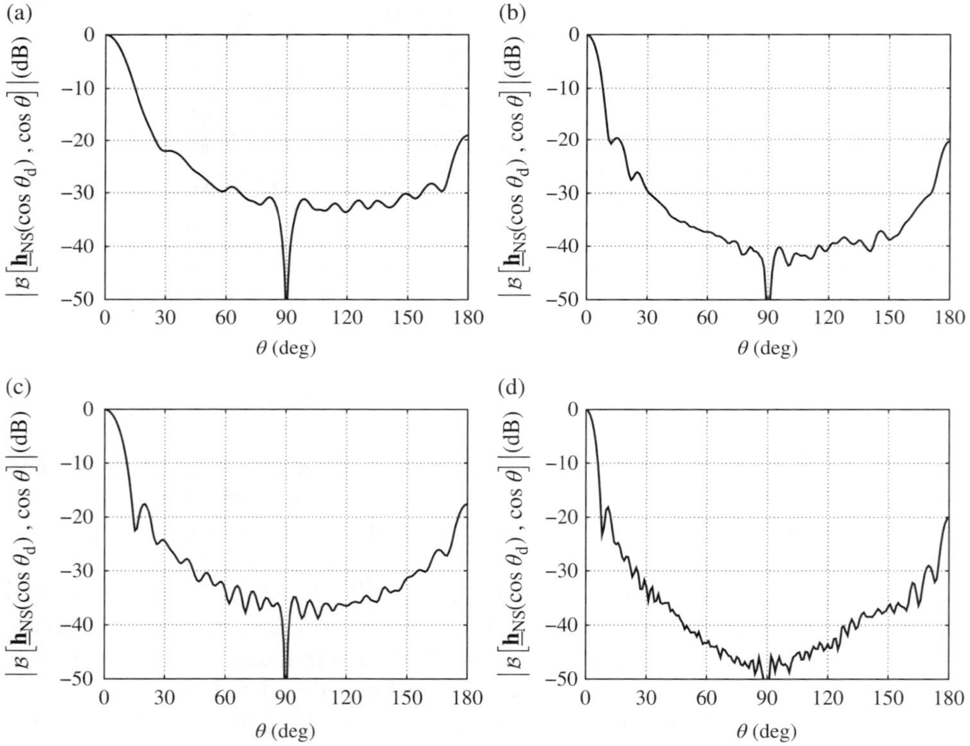

Figure 11.15 Broadband beampatterns of the NS beamformer for $\theta_d = 0°$, $\theta_n = 90°$, and several values of M and δ: (a) $M = 20$, $\delta = 2$ cm, (b) $M = 40$, $\delta = 2$ cm, (c) $M = 20$, $\delta = 4$ cm, and (d) $M = 40$, $\delta = 4$ cm.

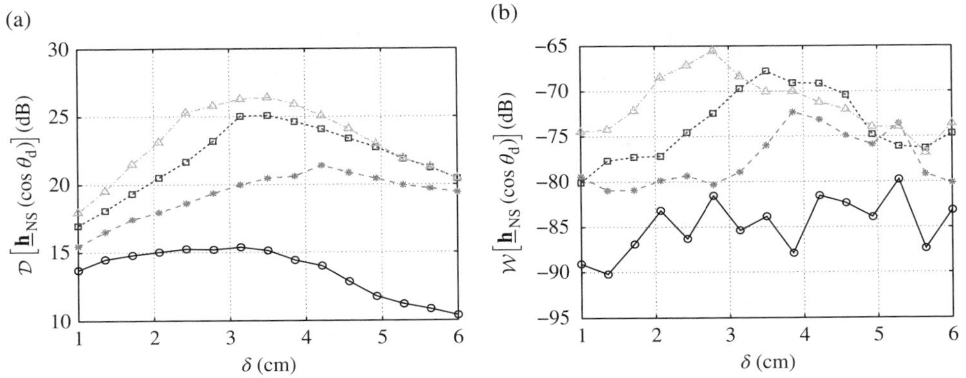

Figure 11.16 (a) DF and (b) WNG of the NS beamformer as a function of δ, for $\theta_{\mathrm{d}} = 0°$, $\theta_{\mathrm{n}} = 90°$, and several values of M: $M = 10$ (solid line with circles), $M = 20$ (dashed line with asterisks), $M = 30$ (dotted line with squares), and $M = 40$ (dash-dot line with triangles).

Table 11.1 Fixed beamformers in the time domain.

Beamformer	
Delay-and-sum	$\underline{\mathbf{h}}_{\mathrm{DS}}\left(\cos\theta_{\mathrm{d}}\right) = \underline{\mathbf{G}}\left(\cos\theta_{\mathrm{d}}\right)\dfrac{\mathbf{i}_l}{M}$
Maximum DF	$\underline{\mathbf{h}}_{\max}\left(\cos\theta_{\mathrm{d}}\right) = \varsigma\underline{\mathbf{t}}_1\left(\cos\theta_{\mathrm{d}}\right),\ \varsigma \neq 0$
Distortionless max. DF	$\underline{\mathbf{h}}_{\mathrm{mDF}}\left(\cos\theta_{\mathrm{d}}\right) = \mathbf{\Gamma}_{\mathrm{T},0,\pi}^{-1}\underline{\mathbf{G}}\left(\cos\theta_{\mathrm{d}}\right)\times \left[\underline{\mathbf{G}}^T\left(\cos\theta_{\mathrm{d}}\right)\mathbf{\Gamma}_{\mathrm{T},0,\pi}^{-1}\underline{\mathbf{G}}\left(\cos\theta_{\mathrm{d}}\right)\right]^{-1}\mathbf{i}_l$
Superdirective	$\underline{\mathbf{h}}_{\mathrm{SD}} = \mathbf{\Gamma}_{\mathrm{T},0,\pi}^{-1}\underline{\mathbf{G}}\left(\underline{\mathbf{G}}^T\mathbf{\Gamma}_{\mathrm{T},0,\pi}^{-1}\underline{\mathbf{G}}\right)^{-1}\mathbf{i}_l$
Robust SD	$\underline{\mathbf{h}}_{\mathrm{R},\epsilon} = \mathbf{\Gamma}_{\mathrm{T},0,\pi,\epsilon}^{-1}\underline{\mathbf{G}}\left(\underline{\mathbf{G}}^T\mathbf{\Gamma}_{\mathrm{T},0,\pi,\epsilon}^{-1}\underline{\mathbf{G}}\right)^{-1}\mathbf{i}_l$
Minimum norm	$\underline{\mathbf{h}}_{\mathrm{MN}}\left(\cos\theta_{\mathrm{d}}\right) = \underline{\mathbf{C}}\left(\theta_{\mathrm{d}},\theta_{\mathrm{n}}\right)\left[\underline{\mathbf{C}}^T\left(\theta_{\mathrm{d}},\theta_{\mathrm{n}}\right)\underline{\mathbf{C}}\left(\theta_{\mathrm{d}},\theta_{\mathrm{n}}\right)\right]^{-1}\begin{bmatrix}\mathbf{i}_l\\ \mathbf{0}\end{bmatrix}$
Null steering	$\underline{\mathbf{h}}_{\mathrm{NS}}\left(\cos\theta_{\mathrm{d}}\right) = \mathbf{\Gamma}_{\mathrm{T},0,\pi}^{-1}\underline{\mathbf{C}}\left(\theta_{\mathrm{d}},\theta_{\mathrm{n}}\right)\times \left[\underline{\mathbf{C}}^T\left(\theta_{\mathrm{d}},\theta_{\mathrm{n}}\right)\mathbf{\Gamma}_{\mathrm{T},0,\pi}^{-1}\underline{\mathbf{C}}\left(\theta_{\mathrm{d}},\theta_{\mathrm{n}}\right)\right]^{-1}\begin{bmatrix}\mathbf{i}_l\\ \mathbf{0}\end{bmatrix}$

In Table 11.1, we summarize all the time-domain fixed beamformers derived in this section.

11.5 Adaptive Beamformers

Most of the adaptive beamformers are easily derived from the time-domain MSE criterion defined in (11.45). Below, we give some important examples.

11.5.1 Wiener

From the minimization of the MSE criterion, $J\left(\underline{\mathbf{h}}\right)$, we find the Wiener beamformer:

$$\begin{aligned}\underline{\mathbf{h}}_{\mathrm{W}}\left(\cos\theta_{\mathrm{d}}\right) &= \mathbf{R}_{\underline{\mathbf{y}}}^{-1}\underline{\mathbf{G}}\left(\cos\theta_{\mathrm{d}}\right)\mathbf{R}_{\mathbf{x}}\mathbf{i}_{l} \\ &= \mathbf{R}_{\underline{\mathbf{y}}}^{-1}\underline{\mathbf{G}}\left(\cos\theta_{\mathrm{d}}\right)\mathbf{R}_{\mathbf{x}}\underline{\mathbf{G}}^{T}\left(\cos\theta_{\mathrm{d}}\right)\underline{\mathbf{i}} \\ &= \mathbf{R}_{\underline{\mathbf{y}}}^{-1}\mathbf{R}_{\underline{\mathbf{x}}}\underline{\mathbf{i}},\end{aligned} \tag{11.73}$$

where $\underline{\mathbf{i}}$ is a vector of length ML_h whose all elements are 0 except for one entry, which is equal to 1 in the appropriate position. This Wiener beamformer can be rewritten as

$$\underline{\mathbf{h}}_{\mathrm{W}}\left(\cos\theta_{\mathrm{d}}\right) = \left(\mathbf{I}_{ML_h} - \mathbf{R}_{\underline{\mathbf{y}}}^{-1}\mathbf{R}_{\underline{\mathbf{v}}}\right)\underline{\mathbf{i}}. \tag{11.74}$$

This expression depends on the statistics of the observations and noise.

Determining the inverse of $\mathbf{R}_{\underline{\mathbf{y}}}$ from (11.16) with the Woodbury identity, we get

$$\begin{aligned}\mathbf{R}_{\underline{\mathbf{y}}}^{-1} &= \mathbf{R}_{\underline{\mathbf{v}}}^{-1} - \mathbf{R}_{\underline{\mathbf{v}}}^{-1}\underline{\mathbf{G}}\left(\cos\theta_{\mathrm{d}}\right) \\ &\quad\times\left[\mathbf{R}_{\mathbf{x}}^{-1} + \underline{\mathbf{G}}^{T}\left(\cos\theta_{\mathrm{d}}\right)\mathbf{R}_{\underline{\mathbf{v}}}^{-1}\underline{\mathbf{G}}\left(\cos\theta_{\mathrm{d}}\right)\right]^{-1}\underline{\mathbf{G}}^{T}\left(\cos\theta_{\mathrm{d}}\right)\mathbf{R}_{\underline{\mathbf{v}}}^{-1}.\end{aligned} \tag{11.75}$$

Substituting (11.75) into (11.73) leads to another useful formulation of the Wiener beamformer:

$$\begin{aligned}\underline{\mathbf{h}}_{\mathrm{W}}\left(\cos\theta_{\mathrm{d}}\right) &= \mathbf{R}_{\underline{\mathbf{v}}}^{-1}\underline{\mathbf{G}}\left(\cos\theta_{\mathrm{d}}\right) \\ &\quad\times\left[\mathbf{R}_{\mathbf{x}}^{-1} + \underline{\mathbf{G}}^{T}\left(\cos\theta_{\mathrm{d}}\right)\mathbf{R}_{\underline{\mathbf{v}}}^{-1}\underline{\mathbf{G}}\left(\cos\theta_{\mathrm{d}}\right)\right]^{-1}\mathbf{i}_{l}.\end{aligned} \tag{11.76}$$

The output SNR with the Wiener beamformer is greater than the input SNR but the estimated desired signal is distorted. This distortion is supposed to decrease when the number of sensors increases.

Example 11.5.1 Consider a ULA of M sensors. Suppose that a desired signal, $x(t)$, with the autocorrelation sequence:

$$E\left[x(t)x(t')\right] = \alpha^{|t-t'|},\ -1 < \alpha < 1$$

impinges on the ULA from the direction $\theta_{\mathrm{d}} = 0°$. Assume that an undesired white Gaussian noise interference, $u(t)$, impinges on the ULA from the direction $\theta_{\mathrm{n}} = 90°$, $u(t) \sim \mathcal{N}\left(0, \sigma_u^2\right)$, which is uncorrelated with $x(t)$. In addition, the sensors contain thermal white Gaussian noise, $w_m(t) \sim \mathcal{N}\left(0, \sigma_w^2\right)$, the signals of which are mutually uncorrelated. The noisy received signals are given by $y_m(t) = x_m(t) + v_m(t)$, $m = 1, \ldots, M$, where $v_m(t) = u_m(t) + w_m(t)$, $m = 1, \ldots, M$ are the interference-plus-noise signals.

The elements of the $L \times L$ matrix $\mathbf{R}_\mathbf{x}$ are

$$\left[\mathbf{R}_\mathbf{x}\right]_{i,j} = \alpha^{|i-j|},\ i,j = 1,\dots,L.$$

The $ML_h \times ML_h$ correlation matrix of $\underline{\mathbf{x}}(t)$ is

$$\mathbf{R}_{\underline{\mathbf{x}}} = \underline{\mathbf{G}}\left(\cos\theta_\mathrm{d}\right)\mathbf{R}_\mathbf{x}\underline{\mathbf{G}}^T\left(\cos\theta_\mathrm{d}\right).$$

Since the interference is in the broadside direction, the $ML_h \times ML_h$ correlation matrix of $\underline{\mathbf{v}}(t)$ is

$$\mathbf{R}_{\underline{\mathbf{v}}} = \mathbf{1}_M \otimes \sigma_u^2\mathbf{I}_{L_h} + \sigma_w^2\mathbf{I}_{ML_h},$$

where $\otimes$ is the Kronecker product and $\mathbf{1}_M$ is an $M \times M$ matrix of all ones.

To demonstrate the performance of the Wiener beamformer, we choose $f_s = 8$ kHz, $\delta = 3$ cm, $\alpha = 0.8$, $\sigma_w^2 = 0.1\sigma_u^2$, $P = 20$, and $L_h = 30$. Figure 11.17 shows plots of the array gain, $\mathcal{G}\left[\underline{\mathbf{h}}_\mathrm{W}\left(\cos\theta_\mathrm{d}\right)\right]$, the noise reduction factor, $\xi_\mathrm{n}\left[\underline{\mathbf{h}}_\mathrm{W}\left(\cos\theta_\mathrm{d}\right)\right]$, the desired signal reduction factor, $\xi_\mathrm{d}\left[\underline{\mathbf{h}}_\mathrm{W}\left(\cos\theta_\mathrm{d}\right)\right]$, and the desired signal distortion index, $\upsilon_\mathrm{d}\left[\underline{\mathbf{h}}_\mathrm{W}\left(\cos\theta_\mathrm{d}\right)\right]$, as a function of the input SNR, for different numbers of sensors, M. For a given input SNR, as the number of sensors increases, the array gain and the noise reduction factor increase, while the desired signal reduction factor and the desired signal distortion index decrease.

Figure 11.18 shows broadband beampatterns, $\left|\mathcal{B}\left[\underline{\mathbf{h}}_\mathrm{W}\left(\cos\theta_\mathrm{d}\right), \cos\theta\right]\right|$, for different numbers of sensors, M. The main beam is in the direction of the desired signal, θ_d, and there is a null in the direction of the interference, θ_n. As the number of sensors increases, the width of the main beam decreases, and the null in the direction of the interference becomes deeper. ■

11.5.2 MVDR

From the optimization of the criterion:

$$\min_{\underline{\mathbf{h}}} \underline{\mathbf{h}}^T\mathbf{R}_{\underline{\mathbf{v}}}\underline{\mathbf{h}} \quad \text{subject to} \quad \underline{\mathbf{h}}^T\underline{\mathbf{G}}\left(\cos\theta_\mathrm{d}\right) = \mathbf{i}_l^T, \tag{11.77}$$

we find the MVDR beamformer:

$$\underline{\mathbf{h}}_\mathrm{MVDR}\left(\cos\theta_\mathrm{d}\right) = \mathbf{R}_{\underline{\mathbf{v}}}^{-1}\underline{\mathbf{G}}\left(\cos\theta_\mathrm{d}\right)\left[\underline{\mathbf{G}}^T\left(\cos\theta_\mathrm{d}\right)\mathbf{R}_{\underline{\mathbf{v}}}^{-1}\underline{\mathbf{G}}\left(\cos\theta_\mathrm{d}\right)\right]^{-1}\mathbf{i}_l. \tag{11.78}$$

It can be shown that the MVDR beamformer is also

$$\underline{\mathbf{h}}_\mathrm{MVDR}\left(\cos\theta_\mathrm{d}\right) = \mathbf{R}_{\underline{\mathbf{y}}}^{-1}\underline{\mathbf{G}}\left(\cos\theta_\mathrm{d}\right)\left[\underline{\mathbf{G}}^T\left(\cos\theta_\mathrm{d}\right)\mathbf{R}_{\underline{\mathbf{y}}}^{-1}\underline{\mathbf{G}}\left(\cos\theta_\mathrm{d}\right)\right]^{-1}\mathbf{i}_l. \tag{11.79}$$

This formulation is more useful in practice as it depends on the statistics of the observations only.

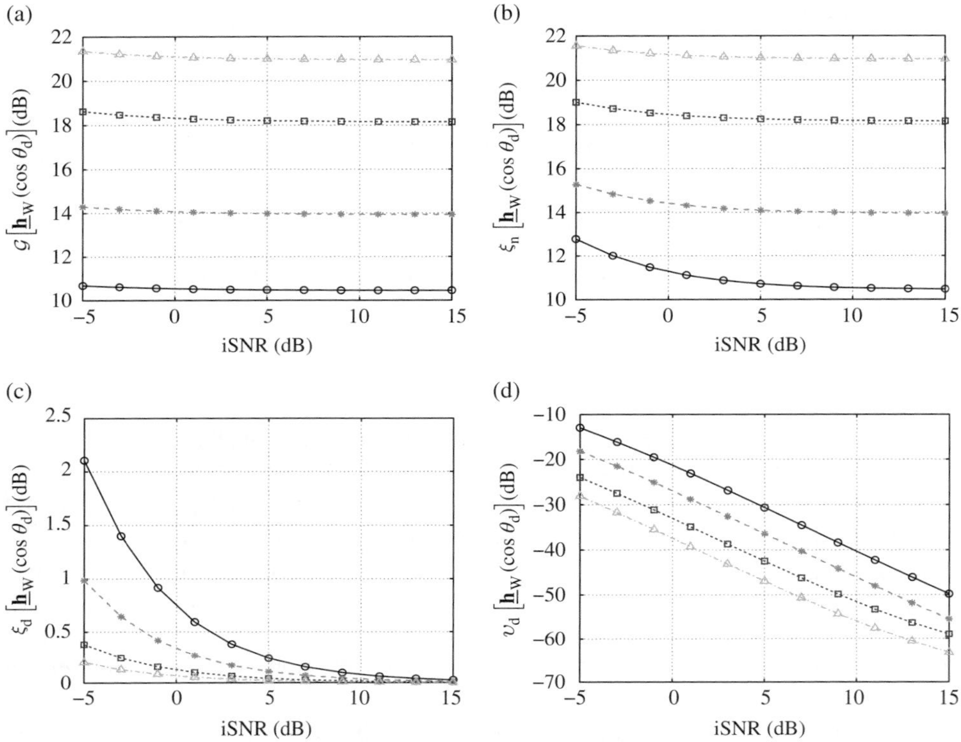

Figure 11.17 (a) The array gain, (b) the noise reduction factor, (c) the desired signal reduction factor, and (d) the desired signal distortion index of the Wiener beamformer as a function of the input SNR, for different numbers of sensors, M: $M = 4$ (solid line with circles), $M = 6$ (dashed line with asterisks), $M = 10$ (dotted line with squares), and $M = 15$ (dash-dot line with triangles).

We always have

$$\mathrm{oSNR}\left[\underline{\mathbf{h}}_{\mathrm{MVDR}}\left(\cos\theta_{\mathrm{d}}\right)\right] \leq \mathrm{oSNR}\left[\underline{\mathbf{h}}_{\mathrm{W}}\left(\cos\theta_{\mathrm{d}}\right)\right]. \tag{11.80}$$

Also, with the signal model given in (11.1), the MVDR beamformer does not distort the desired signal. However, in practice, since this model does not include reverberation, $\underline{\mathbf{h}}_{\mathrm{MVDR}}\left(\cos\theta_{\mathrm{d}}\right)$ may no longer be distortionless.

Example 11.5.2 Returning to Example 11.5.1, we now employ the MVDR beamformer, $\underline{\mathbf{h}}_{\mathrm{MVDR}}\left(\cos\theta_{\mathrm{d}}\right)$, given in (11.79). Figure 11.19 shows plots of the array gain, $\mathcal{G}\left[\underline{\mathbf{h}}_{\mathrm{MVDR}}\left(\cos\theta_{\mathrm{d}}\right)\right]$, the noise reduction factor, $\xi_{\mathrm{n}}\left[\underline{\mathbf{h}}_{\mathrm{MVDR}}\left(\cos\theta_{\mathrm{d}}\right)\right]$, the desired signal reduction factor, $\xi_{\mathrm{d}}\left[\underline{\mathbf{h}}_{\mathrm{MVDR}}\left(\cos\theta_{\mathrm{d}}\right)\right]$, and the MSE, $J\left[\underline{\mathbf{h}}_{\mathrm{MVDR}}\left(\cos\theta_{\mathrm{d}}\right)\right]$, as a function of the input SNR, for different numbers of sensors, M. For a given input SNR, as the number of sensors increases, the array gain and the noise reduction factor increase, while the MSE decreases.

Figure 11.20 shows broadband beampatterns, $\left|\mathcal{B}\left[\underline{\mathbf{h}}_{\mathrm{MVDR}}\left(\cos\theta_{\mathrm{d}}\right), \cos\theta\right]\right|$, for different numbers of sensors, M. The main beam is in the direction of the desired signal, θ_{d}, and

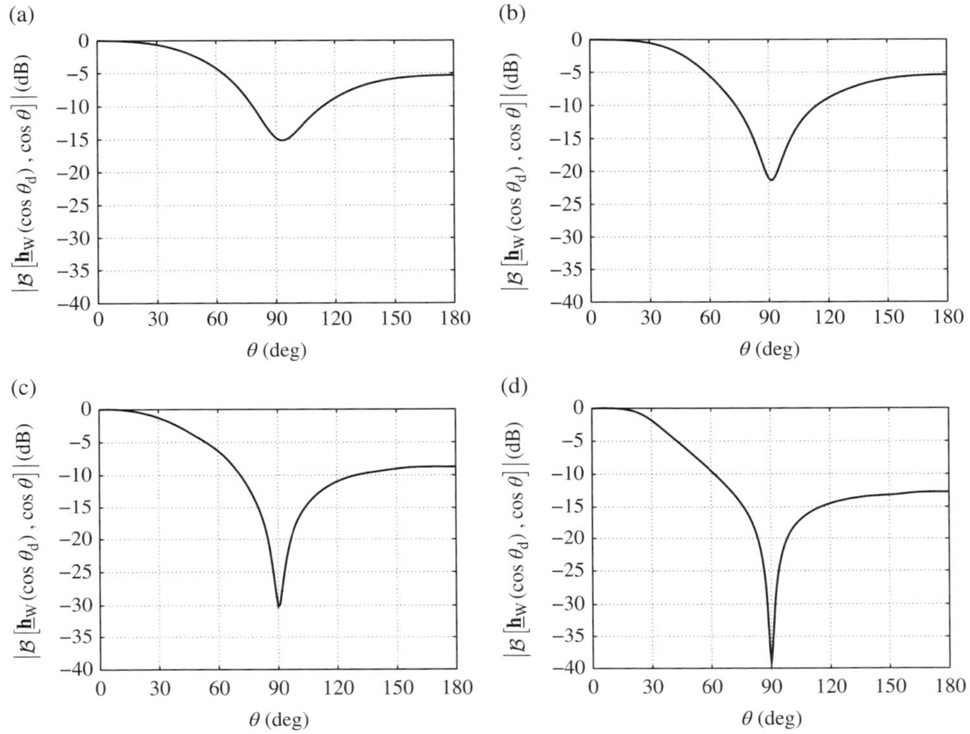

Figure 11.18 Broadband beampatterns of the Wiener beamformer for different numbers of sensors, M: (a) $M = 4$, (b) $M = 6$, (c) $M = 10$, and (d) $M = 15$.

there is a null in the direction of the interference, θ_n. As the number of sensors increases, the width of the main beam decreases, the null in the direction of the interference becomes deeper, and the level of the sidelobe decreases. ■

11.5.3 Tradeoff

The easiest way to compromise between desired signal distortion and noise reduction is to optimize the criterion:

$$\min_{\underline{\mathbf{h}}} J_d\left(\underline{\mathbf{h}}\right) \quad \text{subject to} \quad J_n\left(\underline{\mathbf{h}}\right) = \aleph\sigma_{v_1}^2, \tag{11.81}$$

where $0 < \aleph < 1$ to ensure that we get some noise reduction. By using a Lagrange multiplier, $\mu > 0$, to adjoin the constraint to the cost function, we get the tradeoff beamformer:

$$\begin{aligned}\underline{\mathbf{h}}_{T,\mu}\left(\cos\theta_d\right) &= \mathbf{R}_{\underline{\mathbf{v}}}^{-1}\underline{\mathbf{G}}\left(\cos\theta_d\right)\\ &\times\left[\mu\mathbf{R}_{\mathbf{x}}^{-1} + \underline{\mathbf{G}}^T\left(\cos\theta_d\right)\mathbf{R}_{\underline{\mathbf{v}}}^{-1}\underline{\mathbf{G}}\left(\cos\theta_d\right)\right]^{-1}\mathbf{i}_l.\end{aligned} \tag{11.82}$$

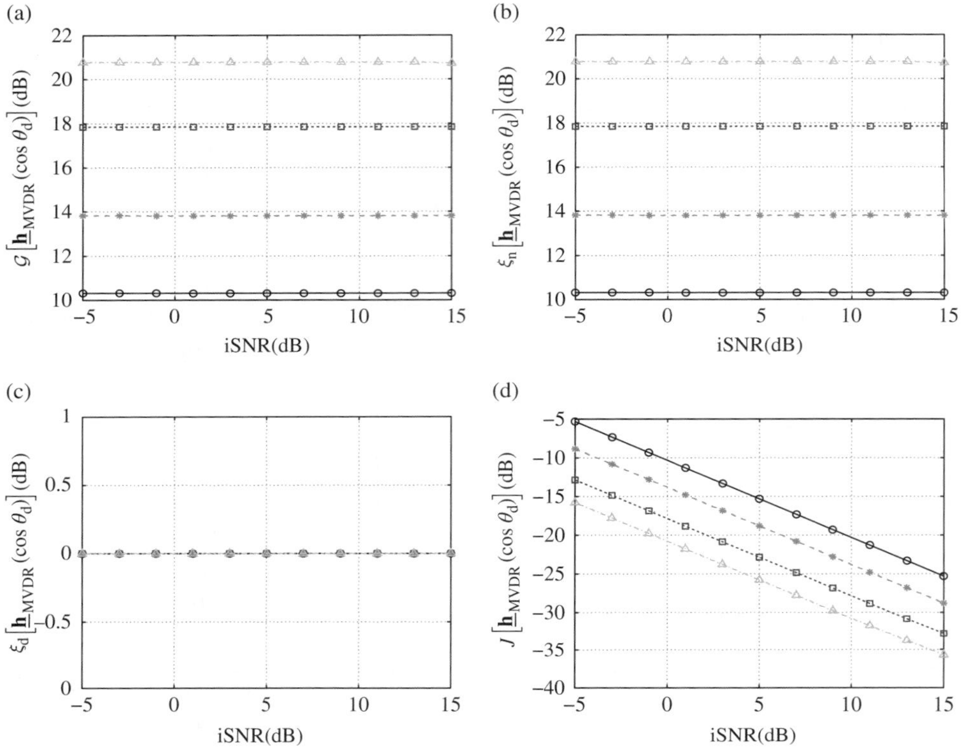

Figure 11.19 (a) The array gain, (b) the noise reduction factor, (c) the desired signal reduction factor, and (d) the MSE of the MVDR beamformer as a function of the input SNR, for different numbers of sensors, M: $M = 4$ (solid line with circles), $M = 6$ (dashed line with asterisks), $M = 10$ (dotted line with squares), and $M = 15$ (dash-dot line with triangles).

We can see that for:

- $\mu = 1$, $\underline{\mathbf{h}}_{\mathrm{T},1}\left(\cos\theta_{\mathrm{d}}\right) = \underline{\mathbf{h}}_{\mathrm{W}}\left(\cos\theta_{\mathrm{d}}\right)$, which is the Wiener beamformer
- $\mu = 0$, $\underline{\mathbf{h}}_{\mathrm{T},0}\left(\cos\theta_{\mathrm{d}}\right) = \underline{\mathbf{h}}_{\mathrm{MVDR}}\left(\cos\theta_{\mathrm{d}}\right)$, which is the MVDR beamformer
- $\mu > 1$, the result is a beamformer with low residual noise at the expense of high desired signal distortion (as compared to Wiener)
- $\mu < 1$, the result is a beamformer with high residual noise and low desired signal distortion (as compared to Wiener).

Example 11.5.3 Returning to Example 11.5.1, we now employ the tradeoff beamformer, $\underline{\mathbf{h}}_{\mathrm{T},\mu}\left(\cos\theta_{\mathrm{d}}\right)$, given in (11.82). Figure 11.21 shows plots of the array gain, $\mathcal{G}\left[\underline{\mathbf{h}}_{\mathrm{T},\mu}\left(\cos\theta_{\mathrm{d}}\right)\right]$, the noise reduction factor, $\xi_{\mathrm{n}}\left[\underline{\mathbf{h}}_{\mathrm{T},\mu}\left(\cos\theta_{\mathrm{d}}\right)\right]$, the desired signal reduction factor, $\xi_{\mathrm{d}}\left[\underline{\mathbf{h}}_{\mathrm{T},\mu}\left(\cos\theta_{\mathrm{d}}\right)\right]$, and the desired signal distortion index, $\upsilon_{\mathrm{d}}\left[\underline{\mathbf{h}}_{\mathrm{T},\mu}\left(\cos\theta_{\mathrm{d}}\right)\right]$, as a function of the input SNR, for $M = 10$ and several values of μ. For a given input SNR, the higher the value of μ, the higher are the array gain and the noise reduction

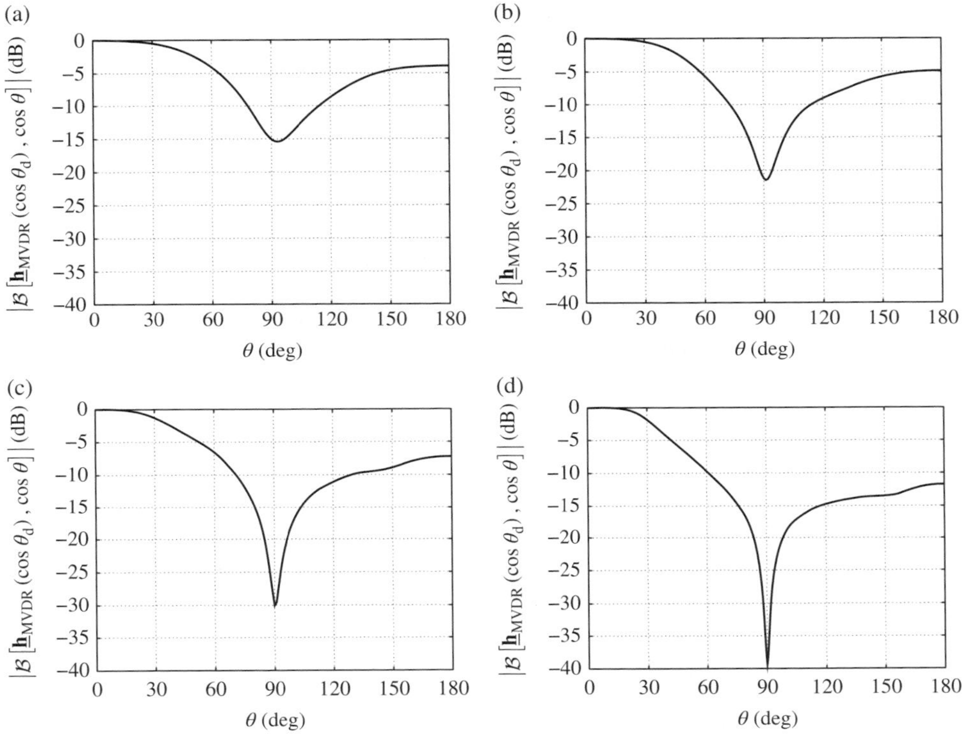

Figure 11.20 Broadband beampatterns of the MVDR beamformer for different numbers of sensors, M: (a) $M = 4$, (b) $M = 6$, (c) $M = 10$, and (d) $M = 15$.

factor, but at the expense of higher desired signal reduction factor and higher desired signal distortion index.

Figure 11.22 shows broadband beampatterns, $\left|\mathcal{B}\left[\underline{\mathbf{h}}_{\mathrm{T},\mu}\left(\cos\theta_{\mathrm{d}}\right), \cos\theta\right]\right|$, for $M = 10$ and several values of μ. The main beam is in the direction, θ_{d}, of the desired signal, and there is a null in the direction of the interference, θ_{n}. ■

11.5.4 Maximum SNR

Let us denote by $\underline{\mathbf{t}}_1'\left(\cos\theta_{\mathrm{d}}\right)$ the eigenvector corresponding to the maximum eigenvalue, $\underline{\lambda}_1'\left(\cos\theta_{\mathrm{d}}\right)$, of the matrix $\mathbf{R}_{\underline{\mathbf{v}}}^{-1}\underline{\mathbf{G}}\left(\cos\theta_{\mathrm{d}}\right)\mathbf{R}_{\mathbf{x}}\underline{\mathbf{G}}^T\left(\cos\theta_{\mathrm{d}}\right)$. It is clear that the beamformer:

$$\underline{\mathbf{h}}_{\max}\left(\cos\theta_{\mathrm{d}}\right) = \varsigma\underline{\mathbf{t}}_1'\left(\cos\theta_{\mathrm{d}}\right), \tag{11.83}$$

where $\varsigma \neq 0$ is an arbitrary real number, maximizes the output SNR, as defined in (11.31). With the maximum SNR beamformer, $\underline{\mathbf{h}}_{\max}\left(\cos\theta_{\mathrm{d}}\right)$, the output SNR is

$$\mathrm{oSNR}\left[\underline{\mathbf{h}}_{\max}\left(\cos\theta_{\mathrm{d}}\right)\right] = \underline{\lambda}_1'\left(\cos\theta_{\mathrm{d}}\right) \tag{11.84}$$

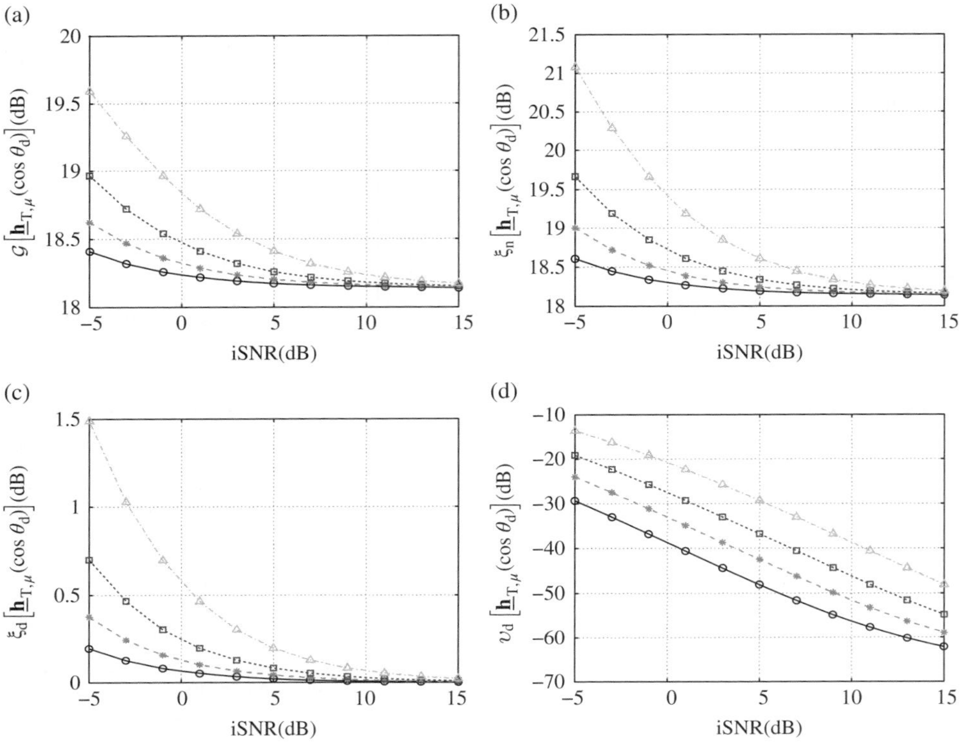

Figure 11.21 (a) The array gain, (b) the noise reduction factor, (c) the desired signal reduction factor, and (d) the desired signal distortion index of the tradeoff beamformer as a function of the input SNR, for $M = 10$ and several values of μ: $\mu = 0.5$ (solid line with circles), $\mu = 1$ (dashed line with asterisks), $\mu = 2$ (dotted line with squares), and $\mu = 5$ (dash-dot line with triangles).

and

$$\mathrm{oSNR}\left[\underline{\mathbf{h}}_{\max}\left(\cos\theta_{\mathrm{d}}\right)\right] \geq \mathrm{oSNR}\left(\underline{\mathbf{h}}\right), \ \forall \underline{\mathbf{h}}. \tag{11.85}$$

The parameter ς can be found by minimizing distortion or the MSE. Substituting (11.83) into (11.45) we obtain

$$J\left(\underline{\mathbf{h}}\right) = \sigma_x^2 - 2\varsigma \underline{\mathbf{t}}_{1}^{\prime T}\left(\cos\theta_{\mathrm{d}}\right) \underline{\mathbf{G}}\left(\cos\theta_{\mathrm{d}}\right) \mathbf{R}_{\mathbf{x}} \mathbf{i}_l + \varsigma^2 \underline{\mathbf{t}}_{1}^{\prime T}\left(\cos\theta_{\mathrm{d}}\right) \mathbf{R}_{\underline{\mathbf{y}}} \underline{\mathbf{t}}_{1}^{\prime}\left(\cos\theta_{\mathrm{d}}\right). \tag{11.86}$$

Therefore, the value of ς that minimizes the MSE is given by

$$\varsigma = \frac{\underline{\mathbf{t}}_{1}^{\prime T}\left(\cos\theta_{\mathrm{d}}\right) \underline{\mathbf{G}}\left(\cos\theta_{\mathrm{d}}\right) \mathbf{R}_{\mathbf{x}} \mathbf{i}_l}{\underline{\mathbf{t}}_{1}^{\prime T}\left(\cos\theta_{\mathrm{d}}\right) \mathbf{R}_{\underline{\mathbf{y}}} \underline{\mathbf{t}}_{1}^{\prime}\left(\cos\theta_{\mathrm{d}}\right)}. \tag{11.87}$$

Example 11.5.4 Returning to Example 11.5.1, we now employ the maximum SNR beamformer, $\underline{\mathbf{h}}_{\max}\left(\cos\theta_{\mathrm{d}}\right)$, given in (11.83) with the value of ς that minimizes the MSE.

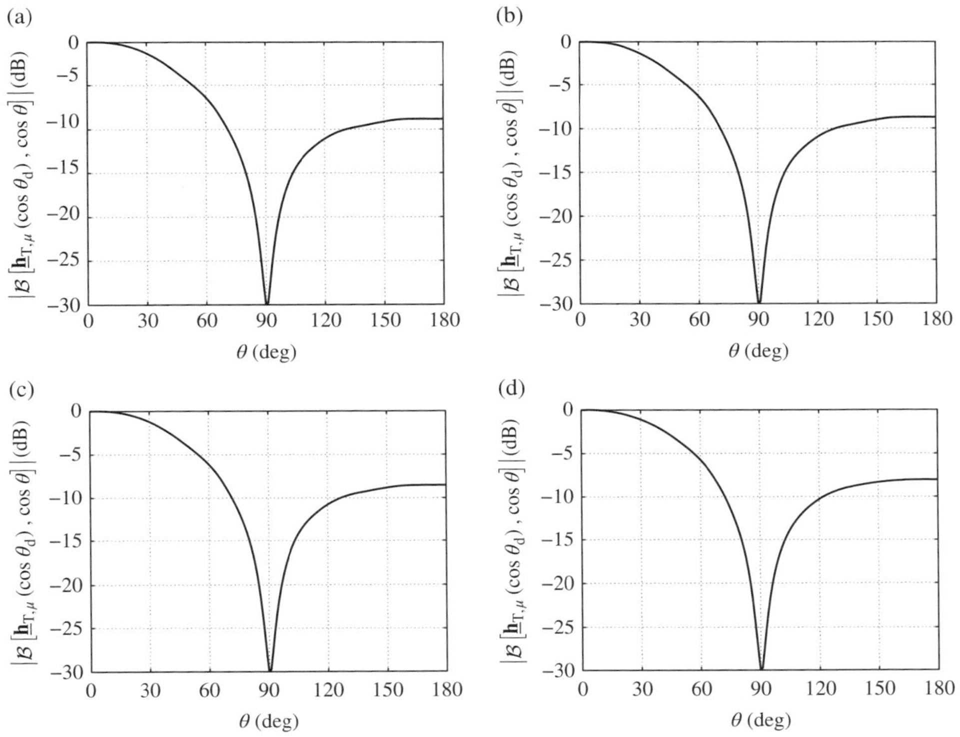

Figure 11.22 Broadband beampatterns of the tradeoff beamformer for $M = 10$ and several values of μ: (a) $\mu = 0.5$, (b) $\mu = 1$, (c) $\mu = 2$, and (d) $\mu = 5$.

Figure 11.23 shows plots of the array gain, $\mathcal{G}\left[\underline{\mathbf{h}}_{\max}\left(\cos\theta_{\mathrm{d}}\right)\right]$, the noise reduction factor, $\xi_{\mathrm{n}}\left[\underline{\mathbf{h}}_{\max}\left(\cos\theta_{\mathrm{d}}\right)\right]$, the desired signal reduction factor, $\xi_{\mathrm{d}}\left[\underline{\mathbf{h}}_{\max}\left(\cos\theta_{\mathrm{d}}\right)\right]$, and the desired signal distortion index, $\upsilon_{\mathrm{d}}\left[\underline{\mathbf{h}}_{\max}\left(\cos\theta_{\mathrm{d}}\right)\right]$, as a function of the input SNR, for different numbers of sensors, M. For a given input SNR, as the number of sensors increases, the array gain and noise reduction factor increase, while the desired signal reduction factor and desired signal distortion index decrease.

Figure 11.24 shows broadband beampatterns, $\left|\mathcal{B}\left[\underline{\mathbf{h}}_{\max}\left(\cos\theta_{\mathrm{d}}\right), \cos\theta\right]\right|$, for different numbers of sensors, M. The main beam is in the direction of the desired signal, θ_{d}, and there is a null in the direction of the interference, θ_{n}. As the number of sensors increases, the null in the direction of the interference becomes deeper. ■

11.5.5 LCMV

We assume that we have an undesired source impinging on the array from the direction $\theta_{\mathrm{n}} \neq \theta_{\mathrm{d}}$. The objective is to completely cancel this source while recovering the desired source impinging on the array from the direction θ_{d}. Then, it is obvious that the constraint equation is identical to the one given in (11.67).

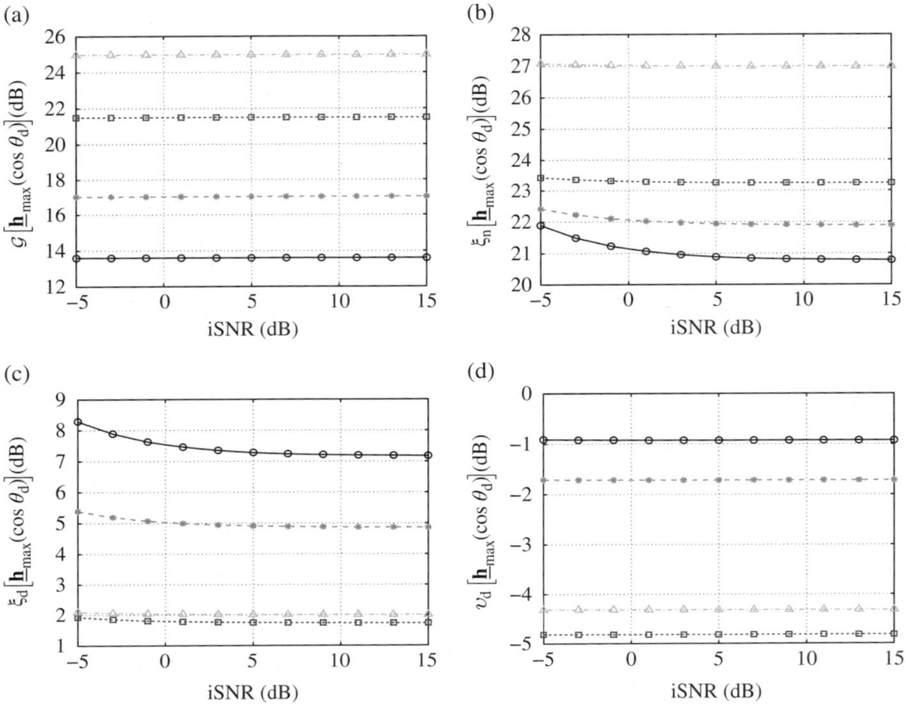

Figure 11.23 (a) The array gain, (b) the noise reduction factor, (c) the desired signal reduction factor, and (d) the desired signal distortion index of the maximum SNR beamformer as a function of the input SNR, for different numbers of sensors, M: $M = 4$ (solid line with circles), $M = 6$ (dashed line with asterisks), $M = 10$ (dotted line with squares), and $M = 15$ (dash-dot line with triangles).

The above problem is solved by minimizing the MSE of the residual noise, $J_{\mathrm{r}}\left(\underline{\mathbf{h}}\right)$, subject (11.67):

$$\min_{\underline{\mathbf{h}}} \underline{\mathbf{h}}^T \mathbf{R}_{\underline{\mathbf{v}}} \underline{\mathbf{h}} \quad \text{subject to} \quad \underline{\mathbf{C}}^T\left(\theta_{\mathrm{d}}, \theta_{\mathrm{n}}\right) \underline{\mathbf{h}} = \left[\begin{array}{c} \mathbf{i}_l \\ \mathbf{0} \end{array} \right] \tag{11.88}$$

The solution to this optimization problem gives the well-known LCMV beamformer [6, 7]:

$$\begin{aligned} \underline{\mathbf{h}}_{\mathrm{LCMV}}\left(\cos\theta_{\mathrm{d}}\right) &= \mathbf{R}_{\underline{\mathbf{v}}}^{-1} \underline{\mathbf{C}}\left(\theta_{\mathrm{d}}, \theta_{\mathrm{n}}\right) \\ &\quad \times \left[\underline{\mathbf{C}}^T\left(\theta_{\mathrm{d}}, \theta_{\mathrm{n}}\right) \mathbf{R}_{\underline{\mathbf{v}}}^{-1} \underline{\mathbf{C}}\left(\theta_{\mathrm{d}}, \theta_{\mathrm{n}}\right)\right]^{-1} \left[\begin{array}{c} \mathbf{i}_l \\ \mathbf{0} \end{array} \right], \end{aligned} \tag{11.89}$$

which depends on the statistics of the noise only.

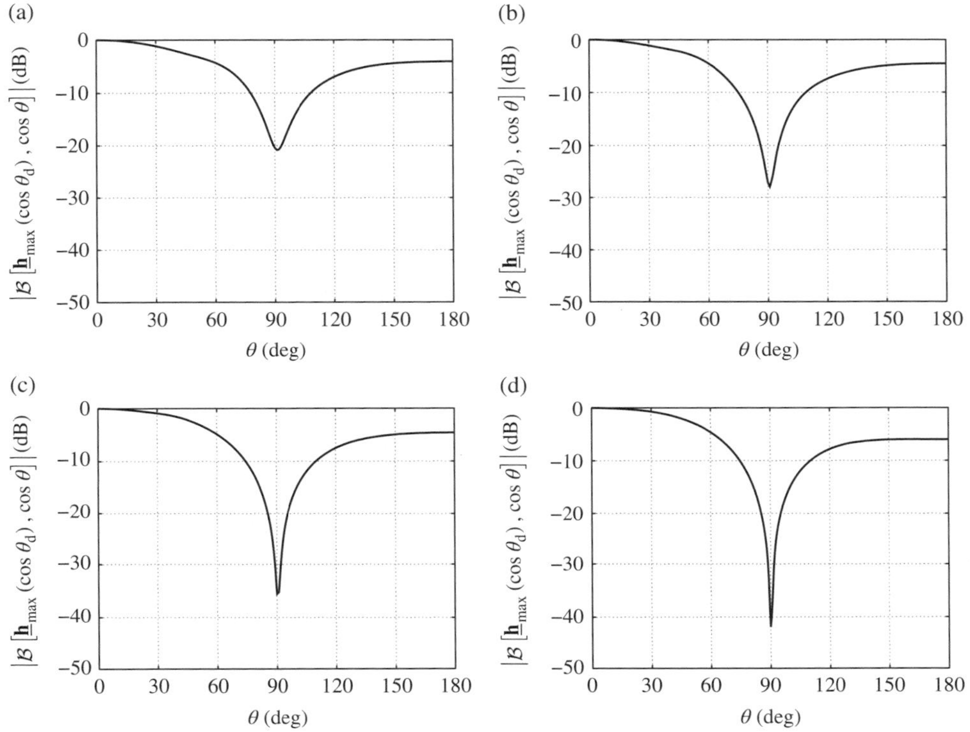

Figure 11.24 Broadband beampatterns of the maximum SNR beamformer for different numbers of sensors, M: (a) $M = 4$, (b) $M = 6$, (c) $M = 10$, and (d) $M = 15$.

It can be shown that a more useful formulation of the LCMV beamformer is

$$\underline{\mathbf{h}}_{\mathrm{LCMV}}\left(\cos\theta_{\mathrm{d}}\right) = \mathbf{R}_{\underline{\mathbf{y}}}^{-1}\underline{\mathbf{C}}\left(\theta_{\mathrm{d}},\theta_{\mathrm{n}}\right) \times \left[\underline{\mathbf{C}}^{T}\left(\theta_{\mathrm{d}},\theta_{\mathrm{n}}\right)\mathbf{R}_{\underline{\mathbf{y}}}^{-1}\underline{\mathbf{C}}\left(\theta_{\mathrm{d}},\theta_{\mathrm{n}}\right)\right]^{-1}\left[\begin{array}{c}\mathbf{i}_{l}\\ \mathbf{0}\end{array}\right]. \quad (11.90)$$

This depends on the statistics of the observations only, which should be easy to estimate.

Example 11.5.5 Returning to Example 11.5.1, we now employ the LCMV beamformer, $\underline{\mathbf{h}}_{\mathrm{LCMV}}\left(\cos\theta_{\mathrm{d}}\right)$, given in (11.90). Figure 11.25 shows plots of the array gain, $\mathcal{G}\left[\underline{\mathbf{h}}_{\mathrm{LCMV}}\left(\cos\theta_{\mathrm{d}}\right)\right]$, the noise reduction factor, $\xi_{\mathrm{n}}\left[\underline{\mathbf{h}}_{\mathrm{LCMV}}\left(\cos\theta_{\mathrm{d}}\right)\right]$, the desired signal reduction factor, $\xi_{\mathrm{d}}\left[\underline{\mathbf{h}}_{\mathrm{LCMV}}\left(\cos\theta_{\mathrm{d}}\right)\right]$, and the MSE, $J\left[\underline{\mathbf{h}}_{\mathrm{LCMV}}\left(\cos\theta_{\mathrm{d}}\right)\right]$, as a function of the input SNR, for different numbers of sensors, M. For a given input SNR, as the number of sensors increases, the array gain and the noise reduction factor slightly increase.

Figure 11.26 shows broadband beampatterns, $\left|\mathcal{B}\left[\underline{\mathbf{h}}_{\mathrm{LCMV}}\left(\cos\theta_{\mathrm{d}}\right),\cos\theta\right]\right|$, for different numbers of sensors, M. The main beam is in the direction of the desired

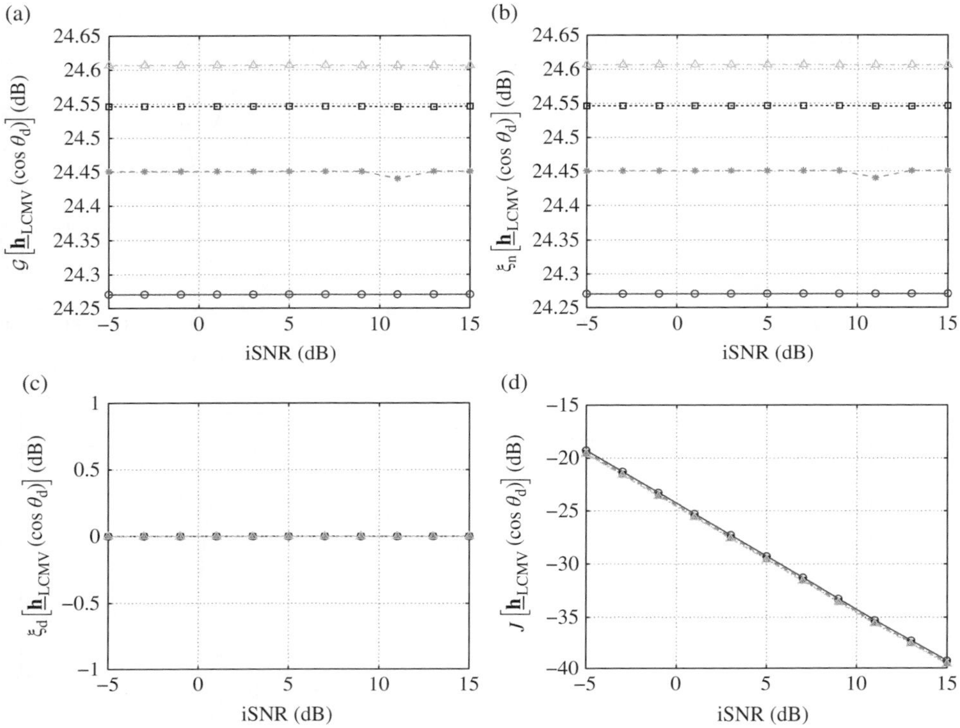

Figure 11.25 (a) The array gain, (b) the noise reduction factor, (c) the desired signal reduction factor, and (d) the MSE of the LCMV beamformer as a function of the input SNR, for different numbers of sensors, M: $M = 30$ (solid line with circles), $M = 35$ (dashed line with asterisks), $M = 40$ (dotted line with squares), and $M = 45$ (dash-dot line with triangles).

signal, $\theta_{\rm d}$, and there is a null in the direction of the interference, $\theta_{\rm n}$. In particular, $\left|\mathcal{B}\left[\underline{\mathbf{h}}_{\rm LCMV}\left(\cos\theta_{\rm d}\right), \cos\theta\right]\right|$ is 1 for $\theta = \theta_{\rm d}$, and is identically zero for $\theta_{\rm n}$. ■

In Table 11.2, we summarize all the time-domain adaptive beamformers derived in this section.

11.6 Differential Beamformers

As we usually do in differential beamforming, we assume in this section that the interelement spacing, δ, is small and the desired source signal propagates from the endfire; that is, $\theta_{\rm d} = 0$. To simplify the presentation, we write $\mathbf{G}_m(\cos 0) = \mathbf{G}_m$.

11.6.1 First Order

It is well known that the design of a first-order differential beamformer requires at least two sensors [8, 9]. First, we assume that we have exactly two sensors ($M = 2$). In this

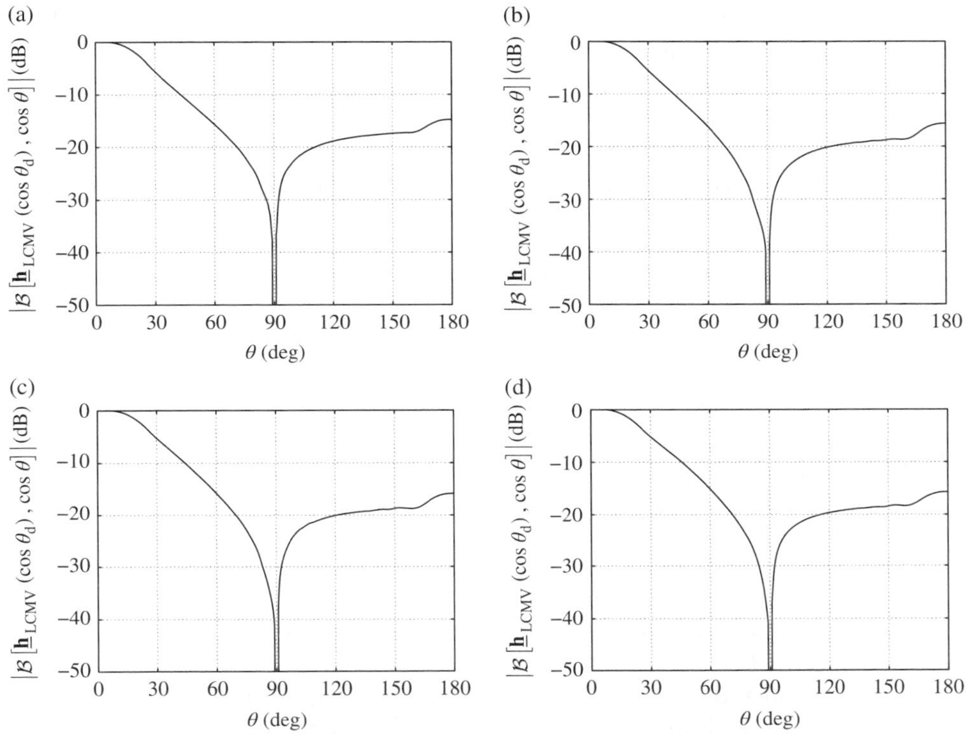

Figure 11.26 Broadband beampatterns of the LCMV beamformer for different numbers of sensors, M: (a) $M = 30$, (b) $M = 35$, (c) $M = 40$, and (d) $M = 45$.

Table 11.2 Adaptive beamformers in the time domain.

Beamformer	
Wiener	$\underline{\mathbf{h}}_{\text{W}}(\cos\theta_{\text{d}}) = \mathbf{R}_{\underline{\mathbf{v}}}^{-1}\underline{\mathbf{G}}(\cos\theta_{\text{d}}) \times \left[\mathbf{R}_{\mathbf{x}}^{-1} + \underline{\mathbf{G}}^T(\cos\theta_{\text{d}})\mathbf{R}_{\underline{\mathbf{v}}}^{-1}\underline{\mathbf{G}}(\cos\theta_{\text{d}})\right]^{-1}\mathbf{i}_l$
MVDR	$\underline{\mathbf{h}}_{\text{MVDR}}(\cos\theta_{\text{d}}) = \mathbf{R}_{\underline{\mathbf{v}}}^{-1}\underline{\mathbf{G}}(\cos\theta_{\text{d}}) \times \left[\underline{\mathbf{G}}^T(\cos\theta_{\text{d}})\mathbf{R}_{\underline{\mathbf{v}}}^{-1}\underline{\mathbf{G}}(\cos\theta_{\text{d}})\right]^{-1}\mathbf{i}_l$
Tradeoff	$\underline{\mathbf{h}}_{\text{T},\mu}(\cos\theta_{\text{d}}) = \mathbf{R}_{\underline{\mathbf{v}}}^{-1}\underline{\mathbf{G}}(\cos\theta_{\text{d}}) \times \left[\mu\mathbf{R}_{\mathbf{x}}^{-1} + \underline{\mathbf{G}}^T(\cos\theta_{\text{d}})\mathbf{R}_{\underline{\mathbf{v}}}^{-1}\underline{\mathbf{G}}(\cos\theta_{\text{d}})\right]^{-1}\mathbf{i}_l,\ \mu \geq 0$
Maximum SNR	$\underline{\mathbf{h}}_{\max}(\cos\theta_{\text{d}}) = \varsigma\underline{\mathbf{t}}_1'(\cos\theta_{\text{d}}),\ \varsigma \neq 0$
LCMV	$\underline{\mathbf{h}}_{\text{LCMV}}(\cos\theta_{\text{d}}) = \mathbf{R}_{\underline{\mathbf{v}}}^{-1}\underline{\mathbf{C}}(\theta_{\text{d}}, \theta_{\text{n}}) \times \left[\underline{\mathbf{C}}^T(\theta_{\text{d}}, \theta_{\text{n}})\mathbf{R}_{\underline{\mathbf{v}}}^{-1}\underline{\mathbf{C}}(\theta_{\text{d}}, \theta_{\text{n}})\right]^{-1}\begin{bmatrix}\mathbf{i}_l \\ \mathbf{0}\end{bmatrix}$

case, we have two constraints to fulfill; the distortionless one given in (11.29) and a constraint with a null in the direction $\theta_1 \in \left[\frac{\pi}{2}, \pi\right]$:

$$\underline{\mathbf{h}}^T \underline{\mathbf{G}}\left(\cos\theta_1\right) = \mathbf{0}^T. \tag{11.91}$$

Combining these two constraints together, we get the following linear system to solve

$$\begin{bmatrix} \mathbf{G}_1^T & \mathbf{G}_2^T \\ \mathbf{G}_1^T\left(\cos\theta_1\right) & \mathbf{G}_2^T\left(\cos\theta_1\right) \end{bmatrix} \begin{bmatrix} \mathbf{h}_1 \\ \mathbf{h}_2 \end{bmatrix} = \begin{bmatrix} \mathbf{i}_l \\ \mathbf{0} \end{bmatrix} \tag{11.92}$$

or, equivalently,

$$\underline{\mathbf{C}}_{1,2}^T\left(\theta_1\right) \underline{\mathbf{h}}_{1,2} = \underline{\mathbf{i}}_1, \tag{11.93}$$

where $\underline{\mathbf{C}}_{1,2}^T\left(\theta_1\right)$ is a matrix[1] of size $2L \times 2L_h$. Since $L = L_g + L_h - 1 \geq L_h$, we deduce from (11.93) the least-squares (LS) filter:

$$\underline{\mathbf{h}}_{1,2;\mathrm{LS}} = \left[\underline{\mathbf{C}}_{1,2}\left(\theta_1\right) \underline{\mathbf{C}}_{1,2}^T\left(\theta_1\right)\right]^{-1} \underline{\mathbf{C}}_{1,2}\left(\theta_1\right) \underline{\mathbf{i}}_1. \tag{11.94}$$

The performance of this beamformer may not be satisfactory in practice as far as the WNG is concerned.

Let us assume that the number of sensors is given and equal to $M > 2$. We still have two constraints to fulfill and the linear system to solve is now

$$\begin{bmatrix} \mathbf{G}_1^T & \mathbf{G}_2^T & \cdots & \mathbf{G}_M^T \\ \mathbf{G}_1^T\left(\cos\theta_1\right) & \mathbf{G}_2^T\left(\cos\theta_1\right) & \cdots & \mathbf{G}_M^T\left(\cos\theta_1\right) \end{bmatrix} \begin{bmatrix} \mathbf{h}_1 \\ \mathbf{h}_2 \\ \vdots \\ \mathbf{h}_M \end{bmatrix} = \underline{\mathbf{i}}_1 \tag{11.95}$$

or, equivalently,

$$\underline{\mathbf{C}}_{1,M}^T\left(\theta_1\right) \underline{\mathbf{h}}_{1,M} = \underline{\mathbf{i}}_1, \tag{11.96}$$

where $\underline{\mathbf{C}}_{1,M}^T\left(\theta_1\right)$ is a matrix of size $2L \times ML_h$. We can always find the length L_h in such a way that $ML_h = 2L$:

$$L_h = 2\frac{L_g - 1}{M - 2}. \tag{11.97}$$

In this case, $\underline{\mathbf{C}}_{1,M}^T\left(\theta_1\right)$ is a square matrix and the solution to (11.96) is exact:

$$\underline{\mathbf{h}}_{1,M;\mathrm{E}} = \underline{\mathbf{C}}_{1,M}^{-T}\left(\theta_1\right) \underline{\mathbf{i}}_1. \tag{11.98}$$

---

1 The subscript 1 corresponds to the differential beamformer order and the subscript 2 corresponds to the number of sensors.

If the value of L_h given in (11.97) is not an integer, we can always take it such as $ML_h > 2L$. As a consequence, we get the minimum-norm (MN) filter:

$$\underline{\mathbf{h}}_{1,M;\mathrm{MN}} = \underline{\mathbf{C}}_{1,M}\left(\theta_1\right)\left[\underline{\mathbf{C}}_{1,M}^T\left(\theta_1\right)\underline{\mathbf{C}}_{1,M}\left(\theta_1\right) + \epsilon \mathbf{I}_{2L}\right]^{-1}\underline{\mathbf{i}}_1, \tag{11.99}$$

where $\epsilon \geq 0$ is the regularization parameter.

Example 11.6.1 In this example, we demonstrate the MN filter in the design of the robust dipole ($\theta_1 = \pi/2$) and cardioid ($\theta_1 = \pi$). We choose $f_s = 8$ kHz, $\delta = 1$ cm, $P = 10$, and $\epsilon = 10^{-4}$.

Figure 11.27 shows plots of the DF, $\mathcal{D}\left(\underline{\mathbf{h}}_{1,M;\mathrm{MN}}\right)$, and the WNG, $\mathcal{W}\left(\underline{\mathbf{h}}_{1,M;\mathrm{MN}}\right)$, of the first-order dipole as a function of L_h for several values of M. From this figure, we can choose an appropriate length L_h that is sufficiently large to maintain high DF and WNG. Figure 11.28 shows broadband beampatterns, $\left|\mathcal{B}\left(\underline{\mathbf{h}}_{1,M;\mathrm{MN}}, \cos\theta\right)\right|$, of the first-order dipole for $L_h = 30$ and several values of M.

Figure 11.29 shows plots of the DF, $\mathcal{D}\left(\underline{\mathbf{h}}_{1,M;\mathrm{MN}}\right)$, and the WNG, $\mathcal{W}\left(\underline{\mathbf{h}}_{1,M;\mathrm{MN}}\right)$, of the first-order cardioid as a function of L_h for several values of M. Again, from this figure we can choose an appropriate length L_h. Figure 11.30 shows broadband beampatterns, $\left|\mathcal{B}\left(\underline{\mathbf{h}}_{1,M;\mathrm{MN}}, \cos\theta\right)\right|$, of the first-order cardioid for $L_h = 30$ and several values of M. ■

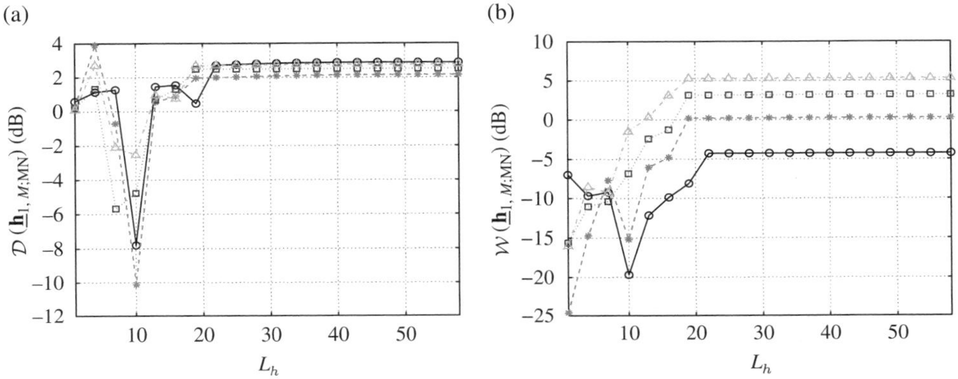

Figure 11.27 (a) DF and (b) WNG of the first-order dipole as a function of L_h for several values of M: $M = 4$ (solid line with circles), $M = 6$ (dashed line with asterisks), $M = 8$ (dotted line with squares), and $M = 10$ (dash-dot line with triangles).

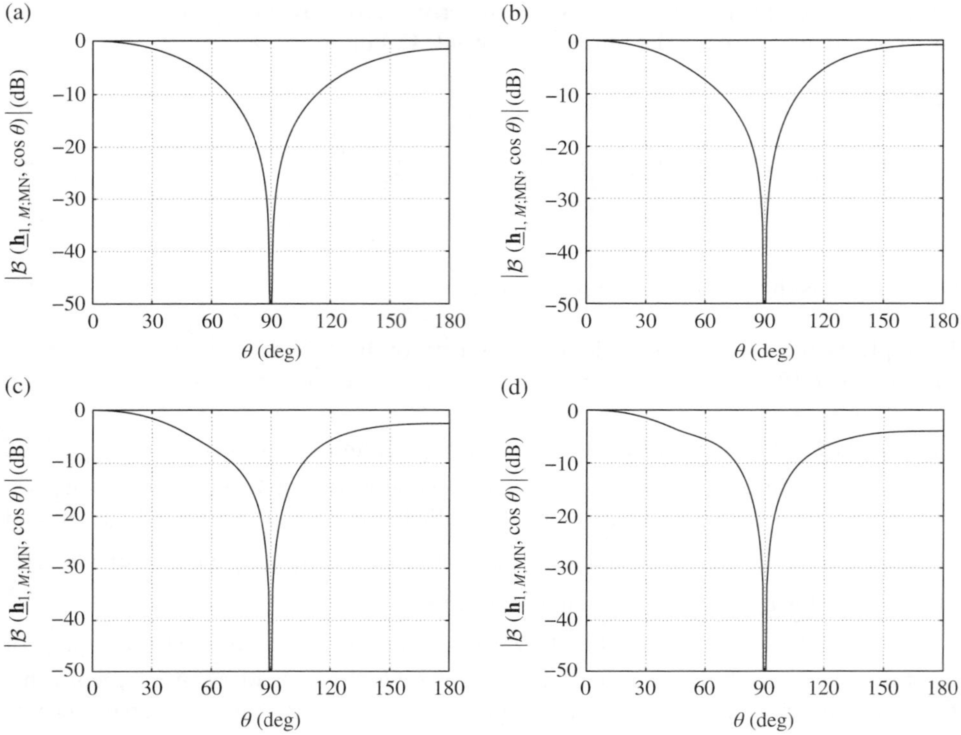

Figure 11.28 Broadband beampatterns of the first-order dipole for $L_h = 25$ and several values of M: (a) $M = 4$, (b) $M = 6$, (c) $M = 8$, and (d) $M = 10$.

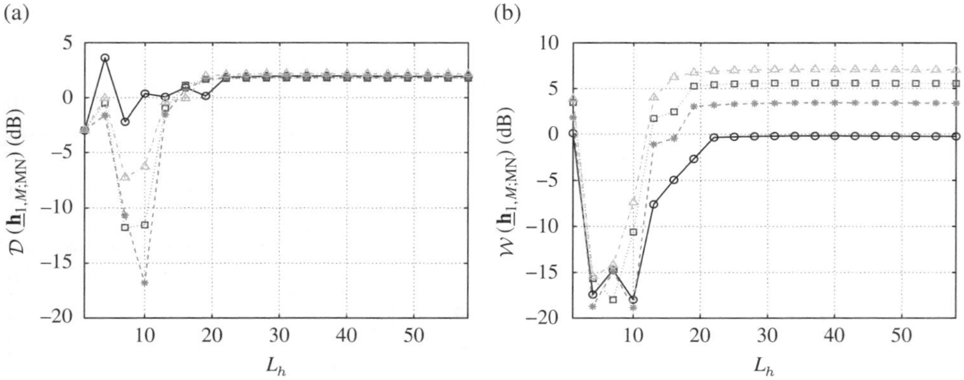

Figure 11.29 (a) DF and (b) WNG of the first-order cardioid as a function of L_h for several values of M: $M = 4$ (solid line with circles), $M = 6$ (dashed line with asterisks), $M = 8$ (dotted line with squares), and $M = 10$ (dash-dot line with triangles).

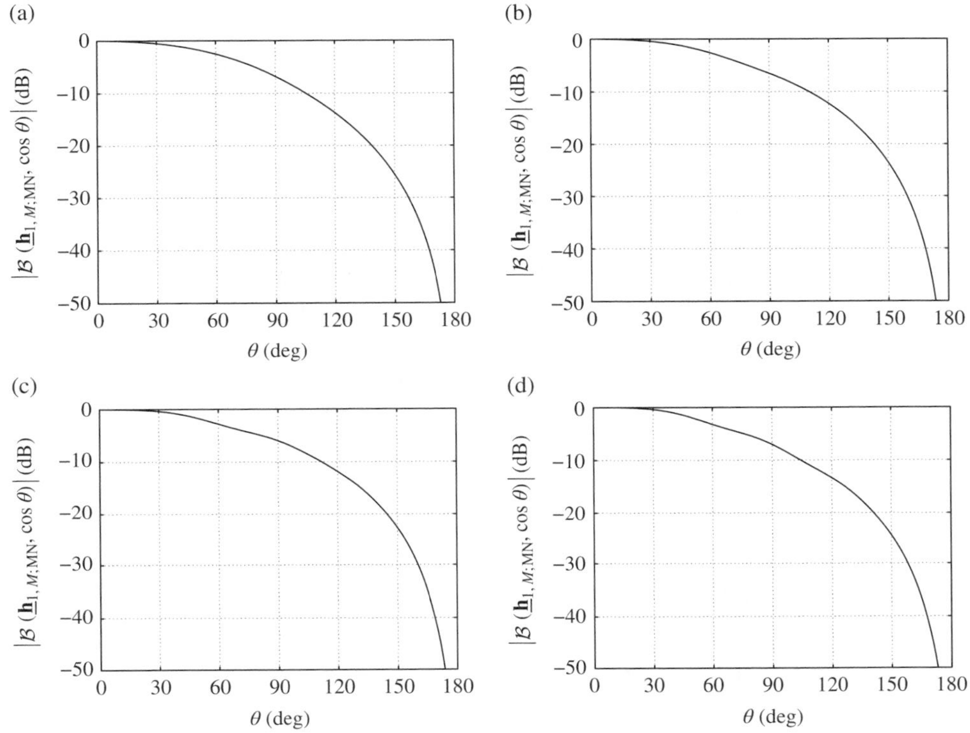

Figure 11.30 Broadband beampatterns of the first-order cardioid for $L_h = 25$ and several values of M: (a) $M = 4$, (b) $M = 6$, (c) $M = 8$, and (d) $M = 10$.

11.6.2 Second Order

The design of a second-order differential beamformer requires at least three sensors and exactly three constraints. For $M = 3$, we need to solve the linear system:

$$\begin{bmatrix} \mathbf{G}_1^T & \mathbf{G}_2^T & \mathbf{G}_3^T \\ \mathbf{G}_1^T\left(\cos\theta_1\right) & \mathbf{G}_2^T\left(\cos\theta_1\right) & \mathbf{G}_3^T\left(\cos\theta_1\right) \\ \mathbf{G}_1^T\left(\cos\theta_2\right) & \mathbf{G}_2^T\left(\cos\theta_2\right) & \mathbf{G}_3^T\left(\cos\theta_2\right) \end{bmatrix} \begin{bmatrix} \mathbf{h}_1 \\ \mathbf{h}_2 \\ \mathbf{h}_3 \end{bmatrix} = \begin{bmatrix} \mathbf{i}_l \\ \alpha_1 \mathbf{i}_l \\ \alpha_2 \mathbf{i}_l \end{bmatrix}, \tag{11.100}$$

where $\theta_1, \theta_2 \in \left[\frac{\pi}{2}, \pi\right]$, with $\theta_1 \neq \theta_2$, are the directions in which attenuations are desired, and α_1, α_2, with $0 \leq \alpha_1, \alpha_2 \leq 1$, are the attenuation parameters. Equivalently, we can express (11.100) as

$$\underline{\mathbf{C}}_{2,3}^T\left(\theta_{1:2}\right)\underline{\mathbf{h}}_{2,3} = \underline{\mathbf{i}}_2\left(\alpha_{1:2}\right), \tag{11.101}$$

where $\underline{\mathbf{C}}_{2,3}^T\left(\theta_{1:2}\right)$ is a matrix of size $3L \times 3L_h$. We deduce the LS solution:

$$\underline{\mathbf{h}}_{2,3;\mathrm{LS}} = \left[\underline{\mathbf{C}}_{2,3}\left(\theta_{1:2}\right)\underline{\mathbf{C}}_{2,3}^T\left(\theta_{1:2}\right)\right]^{-1}\underline{\mathbf{C}}_{2,3}\left(\theta_{1:2}\right)\underline{\mathbf{i}}_2\left(\alpha_{1:2}\right). \tag{11.102}$$

For $M > 3$, the linear system in (11.101) becomes

$$\underline{\mathbf{C}}_{2,M}^{T}\left(\theta_{1:2}\right)\underline{\mathbf{h}}_{2,M}=\underline{\mathbf{i}}_{2}\left(\alpha_{1:2}\right), \tag{11.103}$$

where

$$\underline{\mathbf{C}}_{2,M}^{T}\left(\theta_{1:2}\right)=\begin{bmatrix}\mathbf{G}_{1}^{T} & \mathbf{G}_{2}^{T} & \cdots & \mathbf{G}_{M}^{T}\\ \mathbf{G}_{1}^{T}\left(\cos\theta_{1}\right) & \mathbf{G}_{2}^{T}\left(\cos\theta_{1}\right) & \cdots & \mathbf{G}_{M}^{T}\left(\cos\theta_{1}\right)\\ \mathbf{G}_{1}^{T}\left(\cos\theta_{2}\right) & \mathbf{G}_{2}^{T}\left(\cos\theta_{2}\right) & \cdots & \mathbf{G}_{M}^{T}\left(\cos\theta_{2}\right)\end{bmatrix} \tag{11.104}$$

is a matrix of size $3L \times ML_h$. We can always find the length L_h in such a way that $ML_h = 3L$:

$$L_h = 3\frac{L_g - 1}{M - 3}. \tag{11.105}$$

In this scenario, $\underline{\mathbf{C}}_{2,M}^{T}\left(\theta_{1:2}\right)$ is a square matrix and the solution to (11.103) is exact:

$$\underline{\mathbf{h}}_{2,M;\mathrm{E}}=\underline{\mathbf{C}}_{2,M}^{-T}\left(\theta_{1:2}\right)\underline{\mathbf{i}}_{2}\left(\alpha_{1:2}\right). \tag{11.106}$$

If the value of L_h given in (11.105) is not an integer, we can always take it such as $ML_h > 3L$. As a consequence, we get the MN filter:

$$\underline{\mathbf{h}}_{2,M;\mathrm{MN}}=\underline{\mathbf{C}}_{2,M}\left(\theta_{1:2}\right)\left[\underline{\mathbf{C}}_{2,M}^{T}\left(\theta_{1:2}\right)\underline{\mathbf{C}}_{2,M}\left(\theta_{1:2}\right)+\epsilon\mathbf{I}_{3L}\right]^{-1}\underline{\mathbf{i}}_{2}\left(\alpha_{1:2}\right), \tag{11.107}$$

where $\epsilon \geq 0$ is the regularization parameter.

Example 11.6.2 In this example, we demonstrate the MN filter in the design of the robust second-order cardioid ($\theta_1 = \pi/2, \theta_2 = \pi, \alpha_1 = \alpha_2 = 0$). We choose $f_s = 8$ kHz, $\delta = 1$ cm, $P = 10$, and $\epsilon = 10^{-5}$.

Figure 11.31 shows plots of the DF, $\mathcal{D}\left(\underline{\mathbf{h}}_{2,M;\mathrm{MN}}\right)$, and the WNG, $\mathcal{W}\left(\underline{\mathbf{h}}_{2,M;\mathrm{MN}}\right)$, of the second-order cardioid as a function of L_h for several values of M. From this figure, we can choose an appropriate length L_h that is sufficiently large to maintain high DF and WNG. Figure 11.32 shows broadband beampatterns, $\left|\mathcal{B}\left(\underline{\mathbf{h}}_{2,M;\mathrm{MN}}, \cos\theta\right)\right|$, of the second-order cardioid for $L_h = 30$ and several values of M. ■

11.6.3 General Order

From what we have shown in the two previous subsections, it is straightforward to design any differential beamformer of order $N \geq 1$. We may consider two cases: $N = M - 1$ and $N < M - 1$.

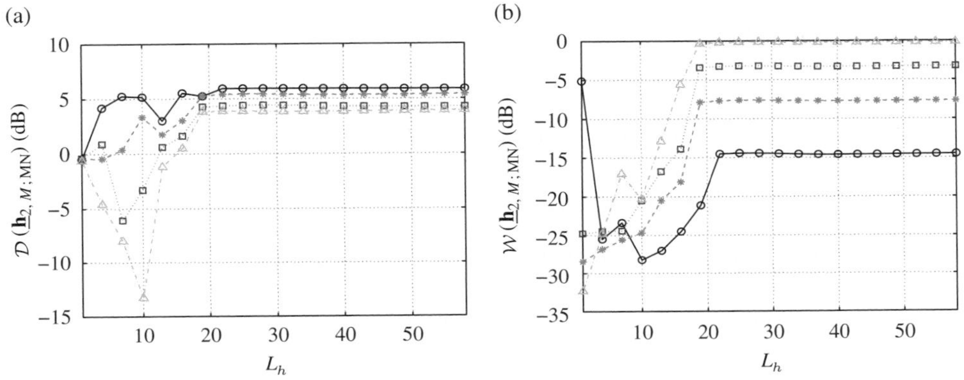

Figure 11.31 (a) DF and (b) WNG of the second-order cardioid as a function of L_h for several values of M: $M = 4$ (solid line with circles), $M = 6$ (dashed line with asterisks), $M = 8$ (dotted line with squares), and $M = 10$ (dash-dot line with triangles).

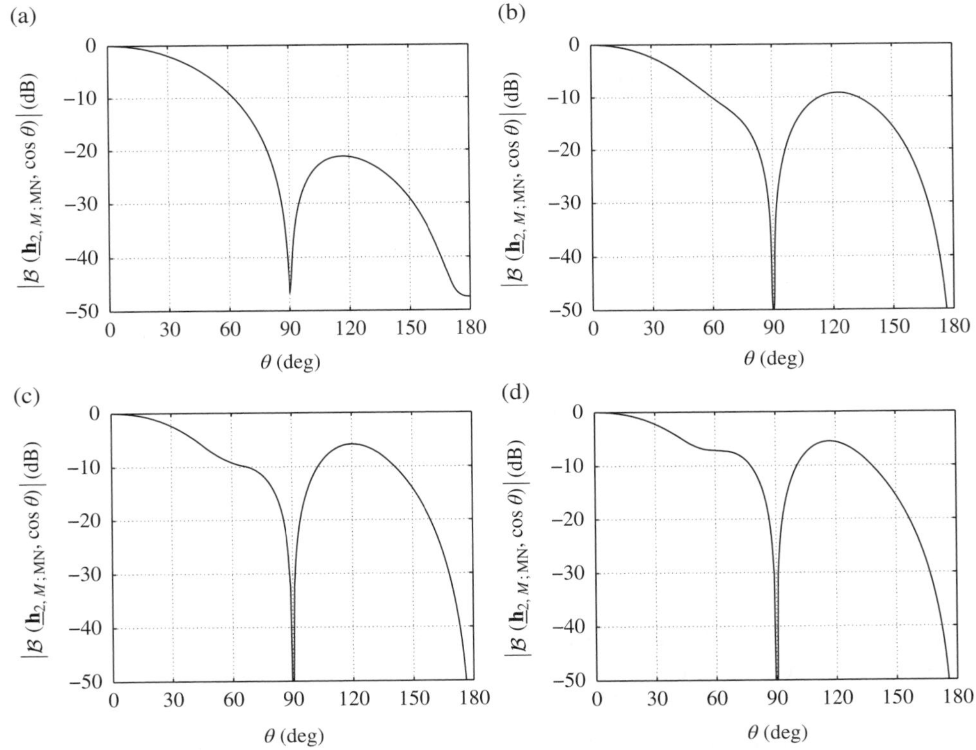

Figure 11.32 Broadband beampatterns of the second-order cardioid for $L_h = 25$ and several values of M: (a) $M = 4$, (b) $M = 6$, (c) $M = 8$, and (d) $M = 10$.

The number of constraints is exactly equal to $N+1$, so the constraint matrix of size $(N+1)L \times ML_h$ is defined as

$$\underline{\mathbf{C}}_{N,M}^T\left(\theta_{1:N}\right) = \begin{bmatrix} \mathbf{G}_1^T & \mathbf{G}_2^T & \cdots & \mathbf{G}_M^T \\ \mathbf{G}_1^T\left(\cos\theta_1\right) & \mathbf{G}_2^T\left(\cos\theta_1\right) & \cdots & \mathbf{G}_M^T\left(\cos\theta_1\right) \\ \vdots & \vdots & \ddots & \vdots \\ \mathbf{G}_1^T\left(\cos\theta_N\right) & \mathbf{G}_2^T\left(\cos\theta_N\right) & \cdots & \mathbf{G}_M^T\left(\cos\theta_N\right) \end{bmatrix}, \quad (11.108)$$

where $\theta_n \in (0, \pi]$, $n = 1, 2, \dots, N$, with $\theta_1 \neq \theta_2 \neq \cdots \neq \theta_N$, are the directions in which attenuations are desired.

For $N = M - 1$, we can derive the LS beamformer:

$$\begin{aligned}\underline{\mathbf{h}}_{M-1,M;\mathrm{LS}} &= \left[\underline{\mathbf{C}}_{M-1,M}\left(\theta_{1:N}\right)\underline{\mathbf{C}}_{M-1,M}^T\left(\theta_{1:N}\right)\right]^{-1} \\ &\quad \times \mathbf{C}_{M-1,M}\left(\theta_{1:N}\right)\underline{\mathbf{i}}_N\left(\alpha_{1:N}\right),\end{aligned} \quad (11.109)$$

where

$$\underline{\mathbf{i}}_N\left(\alpha_{1:N}\right) = \begin{bmatrix} \mathbf{i}_l^T & \alpha_1\mathbf{i}_l^T & \cdots & \alpha_N\mathbf{i}_l^T \end{bmatrix}^T \quad (11.110)$$

is a vector of length NL and α_n, $n = 1, 2, \dots, N$, are the attenuation parameters, with $0 \leq \alpha_n \leq 1$.

For $N < M-1$, it is always possible to find the length L_h in such a way that $ML_h = NL$:

$$L_h = N\frac{L_g - 1}{M - N}. \quad (11.111)$$

As a result, we find the exact filter:

$$\underline{\mathbf{h}}_{N,M;\mathrm{E}} = \underline{\mathbf{C}}_{N,M}^{-T}\left(\theta_{1:N}\right)\underline{\mathbf{i}}_N\left(\alpha_{1:N}\right). \quad (11.112)$$

By taking $ML_h > NL$, we can obtain the MN filter:

$$\begin{aligned}\underline{\mathbf{h}}_{N,M;\mathrm{MN}} &= \underline{\mathbf{C}}_{N,M}\left(\theta_{1:N}\right)\left[\underline{\mathbf{C}}_{N,M}^T\left(\theta_{1:N}\right)\underline{\mathbf{C}}_{N,M}\left(\theta_{1:N}\right) + \epsilon\mathbf{I}_{(N+1)L}\right]^{-1} \\ &\quad \times \underline{\mathbf{i}}_N\left(\alpha_{1:N}\right),\end{aligned} \quad (11.113)$$

where $\epsilon \geq 0$ is the regularization parameter.

Example 11.6.3 In this example, we demonstrate the MN filter in the design of a third-order differential beamformer with three distinct nulls ($\theta_1 = \pi/2, \theta_2 = 3\pi/4, \theta_3 = \pi, \alpha_1 = \alpha_2 = \alpha_3 = 0$). We choose $f_s = 8$ kHz, $\delta = 1$ cm, $P = 10$, and $\epsilon = 10^{-5}$.

Figure 11.33 shows plots of the DF, $\mathcal{D}\left(\underline{\mathbf{h}}_{3,M;\mathrm{MN}}\right)$, and the WNG, $\mathcal{W}\left(\underline{\mathbf{h}}_{3,M;\mathrm{MN}}\right)$, of a third-order differential beamformer as a function of L_h for several values of M. From this figure, we can choose an appropriate length L_h that is sufficiently large to maintain high DF and WNG. Figure 11.34 shows broadband beampatterns, $\left|\mathcal{B}\left(\underline{\mathbf{h}}_{3,M;\mathrm{MN}}, \cos\theta\right)\right|$, for $L_h = 30$ and several values of M. ■

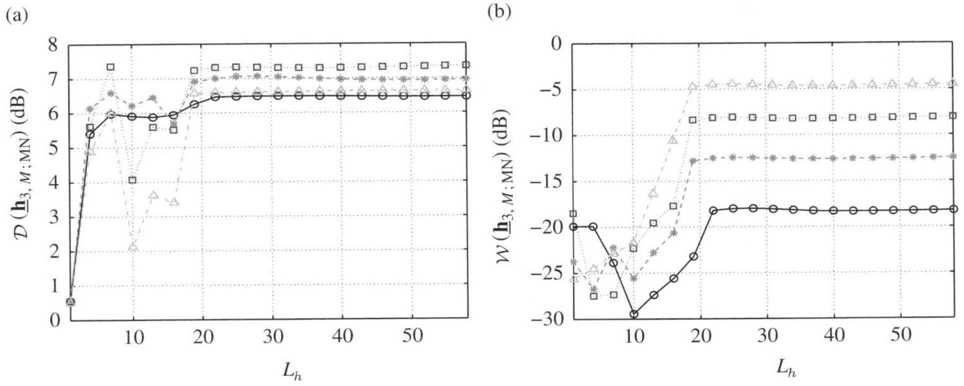

Figure 11.33 (a) DF and (b) WNG of a third-order differential beamformer with three distinct nulls ($\theta_1 = \pi/2, \theta_2 = 3\pi/4, \theta_3 = \pi$) as a function of L_h for several values of M: $M = 4$ (solid line with circles), $M = 6$ (dashed line with asterisks), $M = 8$ (dotted line with squares), and $M = 10$ (dash-dot line with triangles).

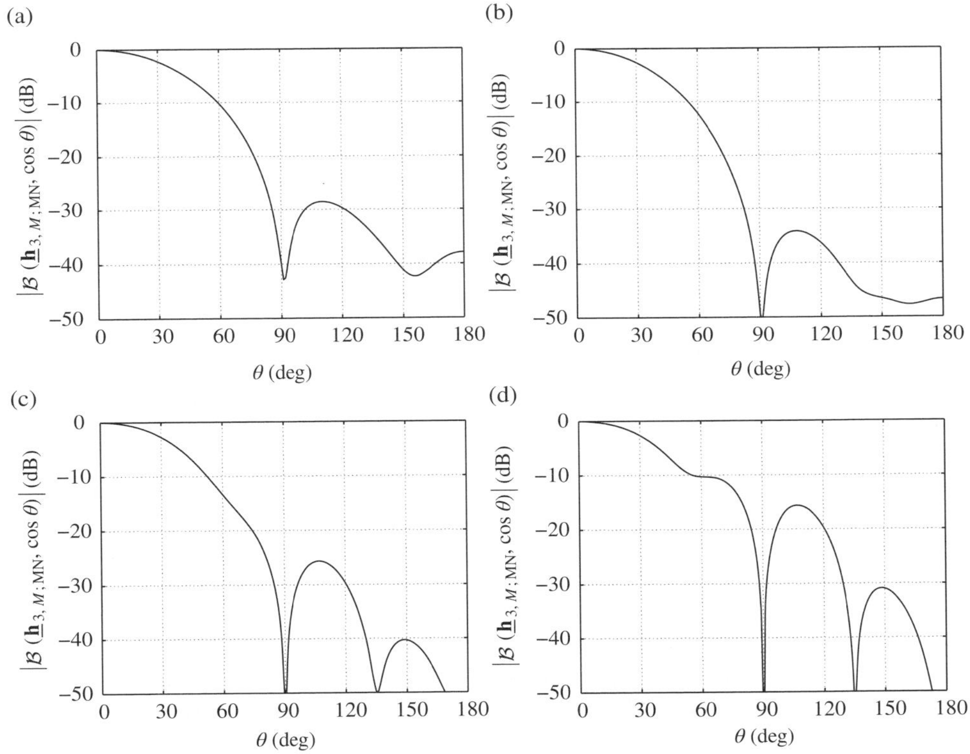

Figure 11.34 Broadband beampatterns of a third-order differential beamformer with three distinct nulls ($\theta_1 = \pi/2, \theta_2 = 3\pi/4, \theta_3 = \pi$) for $L_h = 30$ and several values of M: (a) $M = 4$, (b) $M = 6$, (c) $M = 8$, and (d) $M = 10$.

11.6.4 Hypercardioid

Traditionally, the hypercardioid is derived from the DF definition. Here, the hypercardioid of order $N = M - 1$ is obtained by maximizing the DF as defined in (11.36) and taking into account the distortionless constraint. We get the superdirective beamformer:

$$\underline{\mathbf{h}}_{\mathrm{Hd}} = \mathbf{\Gamma}_{\mathrm{T},0,\pi}^{-1}\underline{\mathbf{G}}\left(\underline{\mathbf{G}}^T\mathbf{\Gamma}_{\mathrm{T},0,\pi}^{-1}\underline{\mathbf{G}}\right)^{-1}\mathbf{i}_l. \tag{11.114}$$

Therefore, the hypercardioid of order $N = M - 1$ and the superdirective beamformer are identical.

11.6.5 Supercardioid

Traditionally, the supercardioid is derived from the FBR definition [10]. Next, we show how to derive it in our context.

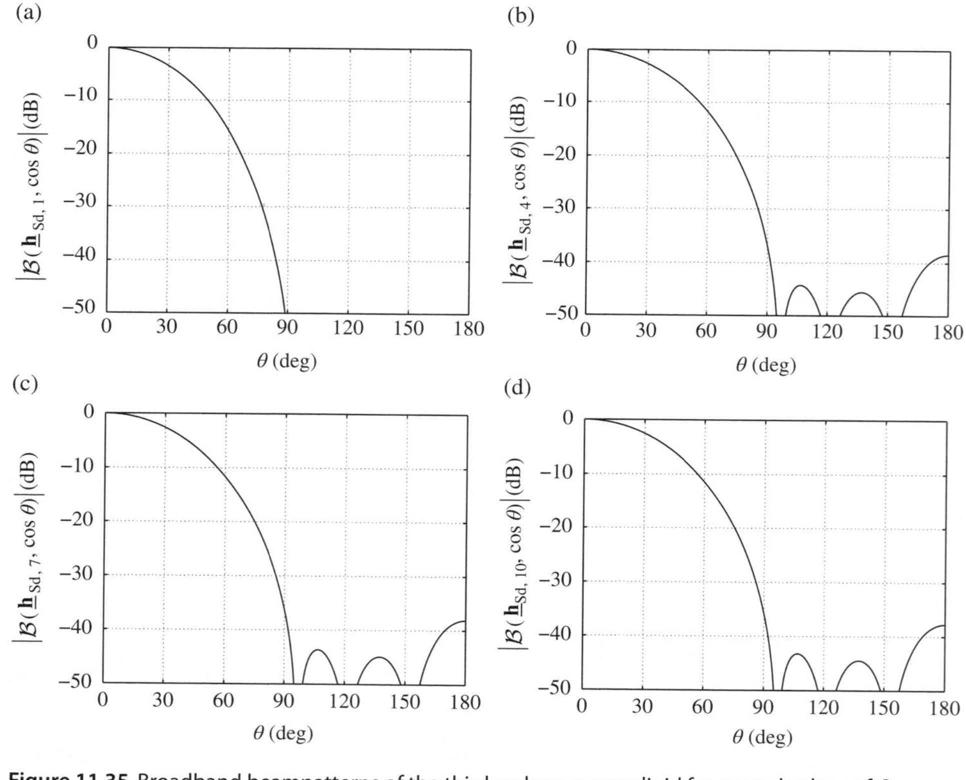

Figure 11.35 Broadband beampatterns of the third-order supercardioid for several values of Q: (a) $Q = 1$, $\mathcal{F}\left(\underline{\mathbf{h}}_{\mathrm{Sd},1}\right) = 285,980$, (b) $Q = 4$, $\mathcal{F}\left(\underline{\mathbf{h}}_{\mathrm{Sd},4}\right) = 7,194$, (c) $Q = 7$, $\mathcal{F}\left(\underline{\mathbf{h}}_{\mathrm{Sd},7}\right) = 6,457$, and (d) $Q = 10$, $\mathcal{F}\left(\underline{\mathbf{h}}_{\mathrm{Sd},10}\right) = 5,994$.

Table 11.3 Differential beamformers in the time domain.

Beamformer	
First order	$\underline{\mathbf{h}}_{1,2;\mathrm{LS}} = \left[\underline{\mathbf{C}}_{1,2}(\theta_1)\underline{\mathbf{C}}_{1,2}^T(\theta_1)\right]^{-1}\underline{\mathbf{C}}_{1,2}(\theta_1)\underline{\mathbf{i}}_1$
	$\underline{\mathbf{h}}_{1,M;\mathrm{E}} = \underline{\mathbf{C}}_{1,M}^{-T}(\theta_1)\underline{\mathbf{i}}_1$
	$\underline{\mathbf{h}}_{1,M;\mathrm{MN}} = \underline{\mathbf{C}}_{1,M}(\theta_1)\left[\underline{\mathbf{C}}_{1,M}^T(\theta_1)\underline{\mathbf{C}}_{1,M}(\theta_1)\right]^{-1}\underline{\mathbf{i}}_1$
Second order	$\underline{\mathbf{h}}_{2,3;\mathrm{LS}} = \left[\underline{\mathbf{C}}_{2,3}(\theta_{1:2})\underline{\mathbf{C}}_{2,3}^T(\theta_{1:2})\right]^{-1}\underline{\mathbf{C}}_{2,3}(\theta_{1:2})\underline{\mathbf{i}}_2(\alpha_{1:2})$
	$\underline{\mathbf{h}}_{2,M;\mathrm{E}} = \underline{\mathbf{C}}_{2,M}^{-T}(\theta_{1:2})\underline{\mathbf{i}}_2(\alpha_{1:2})$
	$\underline{\mathbf{h}}_{2,M;\mathrm{MN}} = \underline{\mathbf{C}}_{2,M}(\theta_{1:2})\left[\underline{\mathbf{C}}_{2,M}^T(\theta_{1:2})\underline{\mathbf{C}}_{2,M}(\theta_{1:2})\right]^{-1}\underline{\mathbf{i}}_2(\alpha_{1:2})$
General order	$\underline{\mathbf{h}}_{M-1,M;\mathrm{LS}} = \left[\underline{\mathbf{C}}_{M-1,M}(\theta_{1:N})\underline{\mathbf{C}}_{M-1,M}^T(\theta_{1:N})\right]^{-1} \times \underline{\mathbf{C}}_{M-1,M}(\theta_{1:N})\underline{\mathbf{i}}_N(\alpha_{1:N})$
	$\underline{\mathbf{h}}_{N,M;\mathrm{E}} = \underline{\mathbf{C}}_{N,M}^{-T}(\theta_{1:N})\underline{\mathbf{i}}_N(\alpha_{1:N})$
	$\underline{\mathbf{h}}_{N,M;\mathrm{MN}} = \underline{\mathbf{C}}_{N,M}(\theta_{1:N}) \times \left[\underline{\mathbf{C}}_{N,M}^T(\theta_{1:N})\underline{\mathbf{C}}_{N,M}(\theta_{1:N})\right]^{-1}\underline{\mathbf{i}}_N(\alpha_{1:N})$
Hypercardioid	$\underline{\mathbf{h}}_{\mathrm{Hd}} = \mathbf{\Gamma}_{\mathrm{T},0,\pi}^{-1}\underline{\mathbf{G}}\left(\underline{\mathbf{G}}^T\mathbf{\Gamma}_{\mathrm{T},0,\pi}^{-1}\underline{\mathbf{G}}\right)^{-1}\mathbf{i}_l$
Supercardioid	$\underline{\mathbf{h}}_{\mathrm{Sd},Q} = \underline{\mathbf{T}}_Q\left(\underline{\mathbf{T}}_Q^T\underline{\mathbf{G}}\underline{\mathbf{G}}^T\underline{\mathbf{T}}_Q\right)^{-1}\underline{\mathbf{T}}_Q^T\underline{\mathbf{G}}\mathbf{i}_l$

Let $\underline{\mathbf{t}}_1, \underline{\mathbf{t}}_2, \ldots, \underline{\mathbf{t}}_Q$ be the eigenvectors corresponding to the Q largest eigenvalues, $\underline{\lambda}_1, \underline{\lambda}_2, \ldots, \underline{\lambda}_Q$, of the matrix $\mathbf{\Gamma}_{\mathrm{T},\pi/2,\pi}^{-1}\mathbf{\Gamma}_{\mathrm{T},0,\pi/2}$, with $\underline{\lambda}_1 \geq \underline{\lambda}_2 \geq \cdots \geq \underline{\lambda}_Q \geq 0$. We consider beamformers of the form:

$$\underline{\mathbf{h}} = \underline{\mathbf{T}}_Q\mathbf{a}, \tag{11.115}$$

where

$$\underline{\mathbf{T}}_Q = \left[\begin{array}{cccc} \underline{\mathbf{t}}_1 & \underline{\mathbf{t}}_2 & \cdots & \underline{\mathbf{t}}_Q \end{array}\right] \tag{11.116}$$

is a matrix of size $ML_h \times Q$ and $\mathbf{a} \neq \mathbf{0}$ is a vector of length Q. Now, we find $\mathbf{a}$ in such a way that $\underline{\mathbf{h}}$ from (11.115) is distortionless. Substituting (11.115) into (11.29), we find

$$\mathbf{a} = \left(\underline{\mathbf{T}}_Q^T\underline{\mathbf{G}}\underline{\mathbf{G}}^T\underline{\mathbf{T}}_Q\right)^{-1}\underline{\mathbf{T}}_Q^T\underline{\mathbf{G}}\mathbf{i}_l. \tag{11.117}$$

As a result, the supercardioid of order $M-1$ is

$$\underline{\mathbf{h}}_{\mathrm{Sd},Q} = \underline{\mathbf{T}}_Q\left(\underline{\mathbf{T}}_Q^T\underline{\mathbf{G}}\underline{\mathbf{G}}^T\underline{\mathbf{T}}_Q\right)^{-1}\underline{\mathbf{T}}_Q^T\underline{\mathbf{G}}\mathbf{i}_l. \tag{11.118}$$

Note that the FBR as defined in (11.38) decreases as Q increases:

$$\mathcal{F}\left(\underline{\mathbf{h}}_{\mathrm{Sd},Q}\right) \geq \mathcal{F}\left(\underline{\mathbf{h}}_{\mathrm{Sd},Q+1}\right). \tag{11.119}$$

Example 11.6.4 In this example, we demonstrate the third-order supercardioid with a ULA of $M = 4$ sensors. We choose $f_s = 8$ kHz, $\delta = 1$ cm, $P = 10$, and $L_h = 15$. Figure 11.35 shows broadband beampatterns, $\left|\mathcal{B}\left(\underline{\mathbf{h}}_{\mathrm{Sd},Q}, \cos\theta\right)\right|$, for several values of Q. It can be observed that as Q increases, the FBR decreases. ■

In Table 11.3, we summarize all the time-domain differential beamformers derived in this section.

Problems

11.1 Show that the MSE can be expressed as

$$J\left(\underline{\mathbf{h}}\right) = \sigma_x^2 - 2\underline{\mathbf{h}}^T \underline{\mathbf{G}}\left(\cos\theta_{\mathrm{d}}\right) \mathbf{R}_{\mathbf{x}} \mathbf{i}_l + \underline{\mathbf{h}}^T \mathbf{R}_{\underline{\mathbf{y}}} \underline{\mathbf{h}}.$$

11.2 Show that the desired signal distortion index can be expressed as

$$v_{\mathrm{d}}\left(\underline{\mathbf{h}}\right) = \frac{\left[\underline{\mathbf{G}}^T\left(\cos_{\mathrm{d}}\theta\right)\underline{\mathbf{h}} - \mathbf{i}_l\right]^T \mathbf{R}_{\mathbf{x}} \left[\underline{\mathbf{G}}^T\left(\cos_{\mathrm{d}}\theta\right)\underline{\mathbf{h}} - \mathbf{i}_l\right]}{\sigma_x^2}.$$

11.3 Show that the MSEs are related to the different performance measures by

$$\begin{aligned}\frac{J_{\mathrm{d}}\left(\underline{\mathbf{h}}\right)}{J_{\mathrm{n}}\left(\underline{\mathbf{h}}\right)} &= \mathrm{iSNR} \times \xi_{\mathrm{n}}\left(\underline{\mathbf{h}}\right) \times v_{\mathrm{d}}\left(\underline{\mathbf{h}}\right) \\ &= \mathrm{oSNR}\left(\underline{\mathbf{h}}\right) \times \xi_{\mathrm{d}}\left(\underline{\mathbf{h}}\right) \times v_{\mathrm{d}}\left(\underline{\mathbf{h}}\right).\end{aligned}$$

11.4 Show that by maximizing the WNG subject to the distortionless constraint, we obtain the DS beamformer:

$$\underline{\mathbf{h}}_{\mathrm{DS}}\left(\cos\theta_{\mathrm{d}}\right) = \underline{\mathbf{G}}\left(\cos\theta_{\mathrm{d}}\right)\frac{\mathbf{i}_l}{M}.$$

11.5 Show that the WNG of the DS beamformer, $\mathcal{W}\left[\underline{\mathbf{h}}_{\mathrm{DS}}\left(\cos\theta_{\mathrm{d}}\right)\right]$, is equal to M.

11.6 Show that the maximum DF beamformer is given by

$$\underline{\mathbf{h}}_{\max}\left(\cos\theta_{\mathrm{d}}\right) = \varsigma \underline{\mathbf{t}}_1\left(\cos\theta_{\mathrm{d}}\right),$$

where $\underline{\mathbf{t}}_1\left(\cos\theta_{\mathrm{d}}\right)$ is the eigenvector corresponding to the maximum eigenvalue of the matrix $\boldsymbol{\Gamma}_{\mathrm{T},0,\pi}^{-1}\underline{\mathbf{G}}\left(\cos\theta_{\mathrm{d}}\right)\underline{\mathbf{G}}^T\left(\cos\theta_{\mathrm{d}}\right)$ and $\varsigma \neq 0$ is an arbitrary real number.

11.7 Show that the maximum DF is given by

$$\mathcal{D}_{\max}\left(\cos\theta_{\mathrm{d}}\right) = \underline{\lambda}_1\left(\cos\theta_{\mathrm{d}}\right),$$

where $\underline{\lambda}_1\left(\cos\theta_{\mathrm{d}}\right)$ is the maximum eigenvalue of the matrix $\boldsymbol{\Gamma}_{\mathrm{T},0,\pi}^{-1}\underline{\mathbf{G}}\left(\cos\theta_{\mathrm{d}}\right)\underline{\mathbf{G}}^T\left(\cos\theta_{\mathrm{d}}\right)$.

11.8 Show that by maximizing the DF subject to the distortionless constraint, we obtain the distortionless maximum DF beamformer:

$$\underline{\mathbf{h}}_{\mathrm{mDF}}\left(\cos\theta_{\mathrm{d}}\right) = \boldsymbol{\Gamma}_{\mathrm{T},0,\pi}^{-1}\underline{\mathbf{G}}\left(\cos\theta_{\mathrm{d}}\right)\left[\underline{\mathbf{G}}^T\left(\cos\theta_{\mathrm{d}}\right)\boldsymbol{\Gamma}_{\mathrm{T},0,\pi}^{-1}\underline{\mathbf{G}}\left(\cos\theta_{\mathrm{d}}\right)\right]^{-1}\mathbf{i}_l.$$

11.9 Show that the DF of the distortionless maximum DF beamformer is given by

$$\mathcal{D}\left[\underline{\mathbf{h}}_{\mathrm{mDF}}\left(\cos\theta_{\mathrm{d}}\right)\right] = \frac{1}{\mathbf{i}_l^T\left[\underline{\mathbf{G}}^T\left(\cos\theta_{\mathrm{d}}\right)\boldsymbol{\Gamma}_{\mathrm{T},0,\pi}^{-1}\underline{\mathbf{G}}\left(\cos\theta_{\mathrm{d}}\right)\right]^{-1}\mathbf{i}_l}.$$

11.10 Show that by maximizing the DF subject to the constraint:

$$\underline{\mathbf{C}}^T\left(\theta_{\mathrm{d}},\theta_{\mathrm{n}}\right)\underline{\mathbf{h}} = \begin{bmatrix}\mathbf{i}_l\\ \mathbf{0}\end{bmatrix},$$

we obtain the NS beamformer:

$$\begin{aligned}\underline{\mathbf{h}}_{\mathrm{NS}}\left(\cos\theta_{\mathrm{d}}\right) &= \boldsymbol{\Gamma}_{\mathrm{T},0,\pi}^{-1}\underline{\mathbf{C}}\left(\theta_{\mathrm{d}},\theta_{\mathrm{n}}\right)\\ &\quad\times\left[\underline{\mathbf{C}}^T\left(\theta_{\mathrm{d}},\theta_{\mathrm{n}}\right)\boldsymbol{\Gamma}_{\mathrm{T},0,\pi}^{-1}\underline{\mathbf{C}}\left(\theta_{\mathrm{d}},\theta_{\mathrm{n}}\right)\right]^{-1}\begin{bmatrix}\mathbf{i}_l\\ \mathbf{0}\end{bmatrix}.\end{aligned}$$

11.11 Show that by minimizing the MSE, $J\left(\underline{\mathbf{h}}\right)$, we obtain the Wiener beamformer:

$$\underline{\mathbf{h}}_{\mathrm{W}}\left(\cos\theta_{\mathrm{d}}\right) = \mathbf{R}_{\underline{\mathbf{y}}}^{-1}\underline{\mathbf{G}}\left(\cos\theta_{\mathrm{d}}\right)\mathbf{R}_{\mathbf{x}}\underline{\mathbf{G}}^T\left(\cos\theta_{\mathrm{d}}\right)\underline{\mathbf{i}}.$$

11.12 Show that the Wiener beamformer can be written as

$$\begin{aligned}\underline{\mathbf{h}}_{\mathrm{W}}\left(\cos\theta_{\mathrm{d}}\right) &= \mathbf{R}_{\underline{\mathbf{v}}}^{-1}\underline{\mathbf{G}}\left(\cos\theta_{\mathrm{d}}\right)\\ &\quad\times\left[\mathbf{R}_{\mathbf{x}}^{-1} + \underline{\mathbf{G}}^T\left(\cos\theta_{\mathrm{d}}\right)\mathbf{R}_{\underline{\mathbf{v}}}^{-1}\underline{\mathbf{G}}\left(\cos\theta_{\mathrm{d}}\right)\right]^{-1}\mathbf{i}_l.\end{aligned}$$

11.13 Show that the MVDR beamformer is given by

$$\underline{\mathbf{h}}_{\mathrm{MVDR}}\left(\cos\theta_{\mathrm{d}}\right) = \mathbf{R}_{\underline{\mathbf{v}}}^{-1}\underline{\mathbf{G}}\left(\cos\theta_{\mathrm{d}}\right)\left[\underline{\mathbf{G}}^T\left(\cos\theta_{\mathrm{d}}\right)\mathbf{R}_{\underline{\mathbf{v}}}^{-1}\underline{\mathbf{G}}\left(\cos\theta_{\mathrm{d}}\right)\right]^{-1}\mathbf{i}_l.$$

11.14 Show that the MVDR beamformer can be rewritten as

$$\underline{\mathbf{h}}_{\mathrm{MVDR}}\left(\cos\theta_{\mathrm{d}}\right) = \mathbf{R}_{\underline{\mathbf{y}}}^{-1}\underline{\mathbf{G}}\left(\cos\theta_{\mathrm{d}}\right)\left[\underline{\mathbf{G}}^T\left(\cos\theta_{\mathrm{d}}\right)\mathbf{R}_{\underline{\mathbf{y}}}^{-1}\underline{\mathbf{G}}\left(\cos\theta_{\mathrm{d}}\right)\right]^{-1}\mathbf{i}_l.$$

11.15 Show that the tradeoff beamformer is given by

$$\underline{\mathbf{h}}_{\mathrm{T},\mu}\left(\cos\theta_{\mathrm{d}}\right) = \mathbf{R}_{\underline{\mathbf{v}}}^{-1}\underline{\mathbf{G}}\left(\cos\theta_{\mathrm{d}}\right) \times \left[\mu\mathbf{R}_{\mathbf{x}}^{-1} + \underline{\mathbf{G}}^{T}\left(\cos\theta_{\mathrm{d}}\right)\mathbf{R}_{\underline{\mathbf{v}}}^{-1}\underline{\mathbf{G}}\left(\cos\theta_{\mathrm{d}}\right)\right]^{-1}\mathbf{i}_{l},$$

where $\mu > 0$ is a Lagrange multiplier.

11.16 Show that for $\mu > 1$, the tradeoff beamformer $\underline{\mathbf{h}}_{\mathrm{T},\mu}\left(\cos\theta_{\mathrm{d}}\right)$, compared to Wiener beamformer, obtains low residual noise at the expense of high desired signal distortion.

11.17 Show that the beamformer that maximizes the output SNR is given by

$$\underline{\mathbf{h}}_{\max}\left(\cos\theta_{\mathrm{d}}\right) = \varsigma\underline{\mathbf{t}}_{1}'\left(\cos\theta_{\mathrm{d}}\right),$$

where $\underline{\mathbf{t}}_{1}'\left(\cos\theta_{\mathrm{d}}\right)$ is the eigenvector corresponding to the maximum eigenvalue of the matrix $\mathbf{R}_{\underline{\mathbf{v}}}^{-1}\underline{\mathbf{G}}\left(\cos\theta_{\mathrm{d}}\right)\mathbf{R}_{\mathbf{x}}\underline{\mathbf{G}}^{T}\left(\cos\theta_{\mathrm{d}}\right)$ and $\varsigma \neq 0$ is an arbitrary real number.

11.18 Show that the maximum SNR beamformer that minimizes the MSE is given by

$$\underline{\mathbf{h}}_{\max}\left(\cos\theta_{\mathrm{d}}\right) = \frac{\underline{\mathbf{t}}_{1}'\left(\cos\theta_{\mathrm{d}}\right)\underline{\mathbf{t}}_{1}'^{T}\left(\cos\theta_{\mathrm{d}}\right)\underline{\mathbf{G}}\left(\cos\theta_{\mathrm{d}}\right)\mathbf{R}_{\mathbf{x}}\mathbf{i}_{l}}{\underline{\mathbf{t}}_{1}'^{T}\left(\cos\theta_{\mathrm{d}}\right)\mathbf{R}_{\underline{\mathbf{y}}}\underline{\mathbf{t}}_{1}'\left(\cos\theta_{\mathrm{d}}\right)}.$$

11.19 Show that by minimizing the MSE of the residual noise subject to the constraint:

$$\underline{\mathbf{C}}^{T}\left(\theta_{\mathrm{d}},\theta_{\mathrm{n}}\right)\underline{\mathbf{h}} = \begin{bmatrix}\mathbf{i}_{l}\\ \mathbf{0}\end{bmatrix},$$

we obtain the LCMV beamformer:

$$\underline{\mathbf{h}}_{\mathrm{LCMV}}\left(\cos\theta_{\mathrm{d}}\right) = \mathbf{R}_{\underline{\mathbf{v}}}^{-1}\underline{\mathbf{C}}\left(\theta_{\mathrm{d}},\theta_{\mathrm{n}}\right) \times \left[\underline{\mathbf{C}}^{T}\left(\theta_{\mathrm{d}},\theta_{\mathrm{n}}\right)\mathbf{R}_{\underline{\mathbf{v}}}^{-1}\underline{\mathbf{C}}\left(\theta_{\mathrm{d}},\theta_{\mathrm{n}}\right)\right]^{-1}\begin{bmatrix}\mathbf{i}_{l}\\ \mathbf{0}\end{bmatrix}.$$

11.20 Show that the LCMV beamformer can be written as

$$\underline{\mathbf{h}}_{\mathrm{LCMV}}\left(\cos\theta_{\mathrm{d}}\right) = \mathbf{R}_{\underline{\mathbf{y}}}^{-1}\underline{\mathbf{C}}\left(\theta_{\mathrm{d}},\theta_{\mathrm{n}}\right) \times \left[\underline{\mathbf{C}}^{T}\left(\theta_{\mathrm{d}},\theta_{\mathrm{n}}\right)\mathbf{R}_{\underline{\mathbf{y}}}^{-1}\underline{\mathbf{C}}\left(\theta_{\mathrm{d}},\theta_{\mathrm{n}}\right)\right]^{-1}\begin{bmatrix}\mathbf{i}_{l}\\ \mathbf{0}\end{bmatrix}.$$

11.21 Show that the supercardioid of order $M-1$ is given by

$$\underline{\mathbf{h}}_{\mathrm{Sd},Q} = \underline{\mathbf{T}}_{Q}\left(\underline{\mathbf{T}}_{Q}^{T}\underline{\mathbf{G}}\underline{\mathbf{G}}^{T}\underline{\mathbf{T}}_{Q}\right)^{-1}\underline{\mathbf{T}}_{Q}^{T}\underline{\mathbf{G}}\mathbf{i}_{l},$$

where $\underline{\mathbf{T}}_Q = \begin{bmatrix} \underline{\mathbf{t}}_1 & \underline{\mathbf{t}}_2 & \cdots & \underline{\mathbf{t}}_Q \end{bmatrix}$, and $\underline{\mathbf{t}}_1, \underline{\mathbf{t}}_2, \ldots, \underline{\mathbf{t}}_Q$ are the eigenvectors corresponding to the Q largest eigenvalues of the matrix $\mathbf{\Gamma}_{\mathrm{T},\pi/2,\pi}^{-1} \mathbf{\Gamma}_{\mathrm{T},0,\pi/2}$.

11.22 Show that with the supercardioid, $\underline{\mathbf{h}}_{\mathrm{Sd},Q}$, the FBR decreases as Q increases, i.e.,

$$\mathcal{F}\left(\underline{\mathbf{h}}_{\mathrm{Sd},Q}\right) \geq \mathcal{F}\left(\underline{\mathbf{h}}_{\mathrm{Sd},Q+1}\right).$$

References

1 J. Benesty, J. Chen, and Y. Huang, *Microphone Array Signal Processing*. Berlin, Germany: Springer-Verlag, 2008.

2 C. E. Shannon, "Communications in the presence of noise," *Proc. IRE*, vol. 37, pp. 10–21, Jan. 1949.

3 A. J. Jerri, "The Shannon sampling theorem. Its various extensions and applications: a tutorial review," *Proc. IEEE*, vol. 65, pp. 1565–1596, Nov. 1977

4 H. Cox, R. M. Zeskind, and T. Kooij, "Practical supergain," *IEEE Trans. Acoust., Speech, Signal Process.*, vol. ASSP-34, pp. 393–398, Jun. 1986.

5 H. Cox, R. M. Zeskind, and M. M. Owen, "Robust adaptive beamforming," *IEEE Trans. Acoust., Speech, Signal Process.*, vol. ASSP-35, pp. 1365–1376, Oct. 1987.

6 A. Booker and C. Y. Ong, "Multiple constraint adaptive filtering," *Geophysics*, vol. 36, pp. 498–509, Jun. 1971.

7 O. Frost, "An algorithm for linearly constrained adaptive array processing," *Proc. IEEE*, vol. 60, pp. 926–935, Jan. 1972.

8 J. Benesty and J. Chen, *Study and Design of Differential Microphone Arrays*. Berlin, Germany: Springer-Verlag, 2012.

9 G. W. Elko, "Superdirectional microphone arrays," in *Acoustic Signal Processing for Telecommunication*, S. L. Gay and J. Benesty (eds). Boston, MA: Kluwer Academic Publishers, 2000.

10 R. N. Marshall and W. R. Harry, "A new microphone providing uniform directivity over an extended frequency range," *J. Acoust. Soc. Am.*, vol. 12, pp. 481–497, 1941.

Index

Fundamentals of Signal Enhancement and Array Signal Processing, First Edition.
Jacob Benesty, Israel Cohen, and Jingdong Chen.

Companion website: www.wiley.com/go/benesty/arraysignalprocessing

d

e

f

g

h

i